Carbohydrate Mimics

Concepts and Methods

Edited by Yves Chapleur

WILEY-VCH

Carbohydrate Mimics

Concepts and Methods

Edited by Yves Chapleur
with collaboration of Hans-Josef Altenbach, Françoise Chrétien,
José Marco Contelles, Francesco Nicotra, Amélia P. Rauter and
Stanley M. Roberts

Weinheim · Berlin · New York · Chichester · Toronto · Brisbane · Singapore

Editor-in-Chief

Yves Chapleur
UMR CNRS 7565
Université Henri Poincaré-Nancy I
BP 239
F-54506 Vandœuvre
France

Associated Editors

Hans-Josef Altenbach
Bergische Universität Wuppertal
FB9 Organische Chemie
Gausstrasse
D-42097 Wuppertal
Germany

Françoise Chrétien
UMR CNRS 7565
Université Henri Poincaré-Nancy I
BP 239
F-54506 Vandœuvre
France

José Marco Contelles
Instituto de Quimica Organica
General
C.S.I.C.
Juan de la Cervia, 3
E-28006 Madrid
Spain

Francesco Nicotra
Università degli Studi di Milano
Dipartimento di Chimica Organica
e Industriale
Via Venezian 21
I-20133 Milano
Italy

Amélia P. Rauter
Dept of Chemistry and
Biochemistry
Faculdade de Cieñcas
Universidade de Lisboa
P-1700 Lisboa
Portugal

Stanley M. Roberts
Dept of Chemistry
The University of Liverpool
The Robert Robinson Laboratories
GB-Liverpool L69 3BX
UK

This book was carefully produced. Nevertheless, authors, editor and publisher do not warrant the information contained therein to be free of errors. Readers are advised to keep in mind that statements, data, illustrations, procedural details or other items may inadvertently be inaccurate.

Editorial Director: Dr. Gudrun Walter
Production Manager: Peter J. Biel

Library of Congress Card No. applied for
A CIP catalogue record for this book is available from the British Library

Die Deutsche Bibliothek – CIP-Einheitsaufnahme
Carbohydrate mimics : concepts and methods / ed. by Yves Chapleur
with collab. of Hans-Josef Altenbach ... - Weinheim ; Berlin ; New York ; Chichester ; Toronto ; Brisbane ;
Singapore : Wiley-VCH, 1998
  ISBN 3-527-29526-7

© WILEY-VCH Verlag GmbH, D-69469 Weinheim (Federal Republic of Germany), 1998

Printed on acid-free and chlorine-free paper.

All rights reserved (including those of translation into other languages). No part of this book may be reproduced in any form – by photoprinting, microfilm, or any other means – nor transmitted or translated into a machine language without written permission from the publishers. Registered names, trademarks, etc. used in this book, even when not specifically marked as such, are not to be considered unprotected by law.

Printing: Strauss Offsetdruck GmbH, D-69509 Mörlenbach

Printed in the Federal Republic of Germany

Foreword

Nature probably provided us with the first representatives of sugar mimics like C-glycosides as O-glycosides mimics and with some highly hydroxylated alkaloids such as nojirimycin and others related compounds which are potent inhibitors of glycoprotein processing enzymes. These naturally occurring compounds have provided models for the first mimics, closely related to the parent carbohydrate. Except scarce early reports on the synthesis of carba-sugars, the development of the concept of molecular mimicry in the carbohydrate field stayed far behind as compared to the peptides field. The high complexity of carbohydrate moieties and their even more complex assemblies such as glycans, glycosaminoglycans, glycolipids etc, is probably responsible for the slower development of carbohydrate mimics. However in the late eighteen's a growing interest to the analogues of simple sugars or more complex biologically active ones has appeared.

Since that time, a considerable body of work has been devoted to the synthesis of the so-called "azasugars" in which the ring oxygen has been replaced by a nitrogen atom. This contributed largely to the renewal of carbohydrate chemistry in relation with glycobiology. C-glycosides, which have been the subject of intense interest over the years in relation with the preparation of nucleoside analogues for their intrinsic biological properties, are now considered as a starting point of oligosaccharide mimics based on the C-disaccharide concept. A large set of methods dealing with structural modifications around the anomeric centre of carbohydrates has been available in the last ten years and it appeared interesting to try to summarise on the different aspects of the carbohydrate mimics from chemical and biological point of view. It became clear that a meeting would be an ideal starting point. The idea was enthusiastically supported by colleagues through Europe. Thanks to the « Human Capital and Mobility » programme of the European Communities, it was possible to set up a Euroconferences programme allowing the organisation of two meetings on Carbohydrate Mimics in France (1995) and Italy (1996) with the participation of young researchers. The success of these conferences, of limited attendance, leads us to publish most of the contributions of the two meetings in this book.

We have tried to organise this book in two main areas but we are conscious that our choice is clearly subjective. The well established fields of C-glycosides azasugars and carbasugars have now reached a high level of results and are treated as such under the general area of structural modifications around the anomeric centre. A number of methods are now available but imaginative related methods can probably be find in the field of total syntheses of highly complex polyether molecules such as toxins. If the synthesis of mono- or di- saccharides mimics should be considered as routine very soon, the mimicry of complex oligosaccharides and of nucleotides is still a challenge. Outstanding achievements are reviewed in the second main area of this book.

The synthesis of future carbohydrate mimics will undoubtedly take advantage of the existing concepts and methods, some of which are reviewed here. Simpler ligands for carbohydrate protein recognition control are emerging nowadays and the future of carbohydrate mimics is possibly far away from sugar structures and should stay in judiciously designed scaffolds presenting the structural requirements for recognition in a proper way. This could not be achieved economically by a complete rational design and the required molecular diversity will probably come from a combinatorial approach. I am very confident that chemists involved in the carbohydrate field and others will succeed in this approach and provide some clues for a number of biological problems of the moment. It is our hope that this should be a central topic for new meetings and books.

Nancy, March 1997

Y. Chapleur

Editor-in-Chief

Yves Chapleur
URA CNRS 486
Université Henri Poincaré-Nancy I
BP 239
F-54506 Vandoeuvre
France

Associated Editors

Hans-Josef Altenbach
Bergische Universität Wuppertal
FB9 Organische Chemie
Gausstrasse
42097 Wuppertal
Germany

Françoise Chrétien
URA CNRS 486
Université Henri Poincaré-Nancy I
BP 239
F-54506 Vandoeuvre
France

José Marco Contelles
Instituto de Quimica Organica
General
C.S.I.C.
Juan de la Cervia, 3
E-28006, Madrid
Spain

Francesco Nicotra
Università degli Studi di Milano
Dipartimento di Chimica Organica e
Industriale
Via Venezian 21
I-20133 Milano
Italy

Amélia P. Rauter
Dept of Chemistry and Biochemistry
Faculdade de Ciencas
Universidade de Lisboa
1700 Lisboa
Portugal

Stanley M. Roberts
Dept of Chemistry
The University of Liverpool
The Robert Robinson Laboratories
Liverpool L69 3BX
UK

List of Contributors

Claudine Augé
Laboratoire de Chimie Organique
Multifonctionnelle, ICMO
Université de Paris-Sud
91405 Orsay
France

Alain Baudat
Section de Chimie
Université de Lausanne
BCH
CH 1015 Lausanne-Dorigny
Switzerland

Judith Baumgartner
Institute of Organic Chemistry
Technical University Graz
Stremayrgasse 16
A-8010 Graz
Austria

Sophie Beaubras
Department of Medicinal Chemistry
Institut de Recherches Servier
11 Rue des Moulineaux
F-92150 Suresnes
France

David C. Billington
Department of Medicinal Chemistry
Institut de Recherches Servier
11 Rue des Moulineaux
F-92150 Suresnes
France

Guido Börner
Fach. Chemie und Chemietechnik
Univ.-GH Paderborn,
Warburger Str. 100
D-4790 Paderborn
Germany

Giovanni Casiraghi
Dipt Farmaceutico dell' Università
Università di Parma
Viale delle Scienze
I-43100 Parma
Italy

Sylvie Challal
Department of Metabolic Diseases
Institut de Recherches Servier
11 Rue des Moulineaux
F-92150 Suresnes
France

Yves Chapleur
URA CNRS 486
Université Henri Poincaré-Nancy I
BP 239
F-54506 Vandoeuvre
France

Jose-Luis Chiara
Grupo de Carbohidratos, Instituto
de Quimica Organica, C.S.I.C.
Juan de la Cervia, 3
E-28006, Madrid
Spain

Françoise Chrétien
URA CNRS 486
Université Henri Poincaré-Nancy I
BP 239
F-54506 Vandoeuvre
France

Lieve Dumortier
University of Gent
Department of Organic Chemistry
Krijgslaan, 281 (S.4),
B-9000 GENT
Belgium

Alexander Chucholowski
Pharma Division, Preclinical
Research, Dept. PRPV, Bldg. 15/30
F.Hoffmann-La Roche Ltd,
CH-4070 Basel
Switzerland

Joseph Espinal
Department of Metabolic Diseases
Institut de Recherches Servier
11 Rue des Moulineaux
F-92150 Suresnes
France

Frank Dagron
Laboratoire de Chimie Organique
Multifonctionnelle, ICMO
Université de Paris-Sud
91405 Orsay
France

Alfonso Fernández-Mayoralas
Grupo de Carbohidratos
Instituto de Química Orgánica
General
Juan de la Cierva 3
28006 Madrid
Spain

Jean-Claude Depezay
Université René Descartes Paris V
Lab.Chimie et Biochimie
Pharmacolique.et Toxicologiques
45 Rue des Saints-Pères
75270 Paris Cedex 06 France

Rafael Ferritto
Section de Chimie
Université de Lausanne
BCH
CH 1015 Lausanne-Dorigny
Switzerland

Jacques Duhault
Department of Metabolic Diseases
Institut de Recherches Servier
11 Rue des Moulineaux
F-92150 Suresnes
France

Jürgen Fingerle
Pharma Division, Preclinical
Research, Dept. PRPV, Bldg. 15/30
F.Hoffmann-La Roche Ltd,
CH-4070 Basel
Switzerland

Bruce Ganem
Department of Chemistry
Baker Laboratory
Cornell University
Ithaca New York 14853-1301
USA

Herfried Griengl
Institute of Organic Chemistry
Technical University Graz
Stremayrgasse 16
A-8010 Graz
Austria

Stephan Gringard
Fachbereich Chemie und
Chemietechnik
Univ.-GH Paderborn,
Warburger Str. 100
D-4790 Paderborn
Germany

Piet Herdewijn
Rega Institute
Katholieke Universiteit Leuven
Minderbroedersstraat 10
B-3000 Leuven
Belgium

Barbra M. Heskamp
Leiden Institute of Chemistry
Gorlaeus Laboratory
Leiden University
P.O. Box 9502 2300 RA Leiden
The Netherlands

Niggi Iberg
Pharma Division, Preclinical
Research, Dept. PRPV, Bldg. 15/30
F.Hoffmann-La Roche Ltd,
CH-4070 Basel
Switzerland

Anne Imberty
Laboratoire de Synthèse Organique-
CNRS
Faculté des Sciences et Techniques
2 rue de la Houssinière
F-44072 Nantes
France

David J. Jenkins
Dept of Medicinal Chemistry
School of Pharm. and Pharmacol.
University of Bath
Claverton Down
BATH, BA 7AY
UK

Mustapha Khaldi
URA CNRS 486
Université Henri Poincaré-Nancy I
BP 239
F-54506 Vandoeuvre
France

Noureddine Khiar
Instituto de Investigaciones
Químicas C.S.I.C.
Américo Vespucio Isla de la Cartuja
41092 Seville
Spain.

Karin Kraehenbuehl
Section de Chimie
Université de Lausanne
BCH
CH 1015 Lausanne-Dorigny
Switzerland

Karsten Krohn
Fachbereich Chemie und
Chemietechnik
Univ.-GH Paderborn,
Warburger Str. 100
D-4790 Paderborn
Germany

Günter Legler
Institut für Biochemie
Universität Köln
Otto-Fischer-Str. 12-14
D-50674 Köln
Germany

Rémy Lemoine
Laboratoire de Chimie Organique
Multifonctionnelle, ICMO
Université de Paris-Sud
91405 Orsay
France

Christine Le Narvor
Laboratoire de Chimie Organique
Multifonctionnelle, ICMO
Université de Paris-Sud
91405 Orsay
France

Pingli Liu
University of Gent
Department of Organic Chemistry
Krijgslaan, 281 (S.4),
B-9000 GENT
Belgium

André Lubineau
Laboratoire de Chimie Organique
Multifonctionnelle, ICMO
Université de Paris-Sud
91405 Orsay
France

José Marco-Contelles
Instituto de Quimica Organica
General C.S.I.C.
Juan de la Cervia, 3
E-28006, Madrid
Spain

Hans Peter Märki
Pharma Division, Preclinical
Research, Dept. PRPV, Bldg. 15/30
F.Hoffmann-La Roche Ltd,
CH-4070 Basel
Switzerland

Olivier Martin
Dept of Chemistry
State University of New York
PO Box 6016
Binghamton NY 13902-6016
USA

Manuel Martin-Lomas
Instituto de Investigaciones
Químicas C.S.I.C.
Américo Vespucio Isla de la Cartuja
41092 Seville
Spain.

Neil D. Miller
Department of Chemistry and
The Skaggs Inst. of Chemical
Biology
The Scripps Research Institute
10666 North Torrey Pines Road
La Jolla CA 92037 and
Dept of Chemistry and
Biochemistry
University of California San Diego
9500 Gilman Drive
La Jolla CA 92093
USA

Rita Müller
Pharma Division, Preclinical
Research, Dept. PRPV, Bldg. 15/30
F.Hoffmann-La Roche Ltd,
CH-4070 Basel
Switzerland

Kyriacos C. Nicolaou
Department of Chemistry and
The Skaggs Inst. of Chemical
Biology
The Scripps Research Institute
10666 North Torrey Pines Road
La Jolla CA 92037 and
Dept of Chemistry and
Biochemistry
University of California San Diego
9500 Gilman Drive
La Jolla CA   92093
USA

Francesco. Nicotra
Università degli Studi di Milano
Dipartimento di Chimica Organica e
Industriale
Via Venezian 21
I-20133 Milano
Italy

Seichiro Ogawa
Department of Applied Chemistry
Faculty of Science and Technology
Keio University
Hiyoshi, Kohoku-ku
Yokohama, 223
Japan

Lajos Ötvös
Central Research Institute for
Chemistry
Hungarian Academy of Sciences
H-1525. Budapest
P.O. Box 17. Hungary

Michael Paton
Department of Chemistry
The University of Edinburgh
West Mains Road
Edinburgh EH9 3JJ
Scotland UK

Michael Pech
Pharma Division, Preclinical
Research, Dept. PRPV, Bldg. 15/30
F.Hoffmann-La Roche Ltd,
CH-4070 Basel
Switzerland

Serge Perez
Ingénierie Moléculaire
Institut National de la Recherche
Agronomique
BP 1627
F-44316 Nantes
France

Françoise Perron-Sierra
Department of Medicinal Chemistry
Institut de Recherches Servier
11 Rue des Moulineaux
F-92150 Suresnes
France

Magdalena Pfister-Downar
Pharma Division, Preclinical
Research, Dept. PRPV, Bldg. 15/30
F.Hoffmann-La Roche Ltd,
CH-4070 Basel
Switzerland

Isabelle Picard
Department of Medicinal Chemistry
Institut de Recherches Servier
11 Rue des Moulineaux
F-92150 Suresnes
France

Maarten H. D. Postema
Department of Chemistry and
The Skaggs Inst. of Chemical
Biology
The Scripps Research Institute
10666 North Torrey Pines Road
La Jolla CA 92037 and
Dept of Chemistry and
Biochemistry
University of California San Diego
9500 Gilman Drive
La Jolla CA 92093
USA

Barry V. L. Potter
Dept of Medicinal Chemistry
School of Pharm. and Pharmacol.
University of Bath
Claverton Down
Bath, BA 7AY
UK

Gloria Rassu
Istituto per l' Applicazione
delle Tecniche Chimiche Avanzate
del CNR
Via Vienna, 2
I-07100 Sassari
Italy

Andrew M. Riley
Dept of Medicinal Chemistry
School of Pharm. and Pharmacol.
University of Bath
Claverton Down
Bath, BA 7AY
UK

Marianne Rouge
Pharma Division, Preclinical
Research, Dept. PRPV, Bldg. 15/30
F.Hoffmann-La Roche Ltd,
CH-4070 Basel
Switzerland

René Roy
Department of Chemistry
University of Ottawa
Ottawa, Ontario K1N 6N5
Canada

Yogesh S. Sanghvi
Medicinal Chemistry Department
Isis Pharmaceuticals
2292 Faraday Avenue
Carlsbad, California 92008
USA

Gérard Schmid
Pharma Division, Preclinical
Research, Dept. PRPV, Bldg. 15/30
F.Hoffmann-La Roche Ltd,
CH-4070 Basel
Switzerland

Kuniaki Tatsuta
Graduate School Science and Eng.
Advanced Res. Inst.Science Engin.
Waseda University
3-4-1 Ohkubo, Shinjuku
Tokyo, 169
Japan

Thomas Tschopp
Pharma Division
Preclinical Research
F.Hoffmann-La Roche Ltd
CH-4002 Basel
Switzerland

Constant A. A. Van Boeckel,
N.V. Organon
P.O. Box 20
5340 BH Oss
The Netherlands

Jacques H. Van Boom
Leiden Institute of Chemistry
Gorlaeus Laboratory
Leiden University
P.O. Box 9502
2300 RA Leiden
The Netherlands

Maurits Vandewalle
University of Gent
Department of Organic Chemistry
Krijgslaan, 281 (S.4),
B-9000 GENT
Belgium

Johan Van der Eycken
University of Gent
Department of Organic Chemistry
Krijgslaan, 281 (S.4),
B-9000 GENT
Belgium

Gijs A. Van der Marel
Leiden Institute of Chemistry
Gorlaeus Laboratory
Leiden University
P.O. Box 9502
2300 RA Leiden
The Netherlands

Gerrit H. Veeneman
N.V. Organon
P.O. Box 20
5340 BH Oss
The Netherlands

Pierre Vogel
Section de Chimie
Université de Lausanne
BCH
CH 1015 Lausanne-Dorigny
Switzerland

Hans Peter Wessel
Pharma Division, Preclinical
Research, Dept. PRPV, Bldg. 15/30
F.Hoffmann-La Roche Ltd,
CH-4070 Basel
Switzerland

Franca Zanardi
Dipartimento Farmaceutico dell'
Università
Viale delle Scienze
I-43100 Parma
Italy

Diana Zanini
Department of Chemistry
University of Ottawa
Ottawa, Ontario K1N 6N5
Canada

Contents

Foreword V

List of Contributors XIX

1	**Recent Advances in the Synthesis of Cyclic Polyether Marine Natural Products**	1
	K. C. Nicolaou, M. H. D. Postema and N. D. Miller	
1.1	Introduction	1
1.2	Synthetic Technology for the Construction of Cyclic Ethers	3
1.3	Total Synthesis of Brevetoxin B	3
1.4	Recent Developments in Polyether Synthesis	4
1.4.1	Radical-Based Reduction of Thionolactones to Cyclic Ethers	4
1.4.2	Synthesis of 1,1'-Bistetrahydropyran Systems via Copper(I)-Promoted Stille Coupling	7
1.4.3	The Olefin Metathesis Approach to Cyclic Enol Ether Construction	11
1.5	Conclusion	14
1.6	References	15
2	**Stereoselective Synthesis of *C*-Disaccharides, Aza-*C*-Disaccharides and *C*-Glycosides of Carbapyranoses Using the "Naked Sugars"**	19
	P. Vogel, R. Ferritto, K. Kraehenbuehl and A. Baudat	
2.1	Introduction	19
2.2	Synthetic Methods	20
2.3	The "Naked Sugar" Approach	29
2.3	*C*-Glycosides of Carbapyranoses and Analogues	35
2.4	Branched-Chain Carbohydrates through Cross-Aldolisations	38
2.5	Synthesis of an Aza-*C*(1→3)-Disaccharide	40
2.6	Conclusion	42
2.7	References	42

3	**The Nitrile Oxide/Isoxazoline Route to *C*-Disaccharides**	49
	R. M. Paton	
3.1	Introduction	49
3.2	Generation of Nitrile Oxides	51
3.3	(1→2)- and (1→3)-Carbonyl-linked Disaccharides	52
3.4	(1→6)-Hydroxymethylene-linked Disaccharides	56
3.5	Higher-Carbon Dialdoses	60
3.6	Conclusion	63
3.7	References	64

4	**Synthesis of Glycosyl Phosphate Mimics**	67
	F. Nicotra	
4.1	Introduction	67
4.2	Roles of Glycosyl Phosphates	67
4.3	Mimics of Natural Phosphates	68
4.4	Mimics of Glycosyl Phosphates	69
4.4.1	Direct Methods of Glycosyl Methylphosphonate Synthesis	70
4.4.2	Indirect Methods of Glycosyl Methylphosphonate Synthesis	72
4.4.2.1	Synthesis of C-Glycosyl Halides	73
4.4.2.2	Conversion of C-Glycosyl Halides into Phosphonates	75
4.5	Mimics of *N*-Acetyl-α-D-Glucosamine 1-Phosphate and *N*-Acetyl-α-D-Mannosamine 1-Phosphate	77
4.5.1	The Biological Roles	77
4.5.2	Synthesis	79
4.6	Conclusion	83
4.7	References	83

5	**Synthetic Studies on Glycosidase Inhibitors Composed of 5a-Carba-Sugars**	87
	S. Ogawa	
5.1	Introduction	87
5.2	Synthesis of Biologically Active Natural Compounds Containing 5a-Carba-Sugars	89
5.2.1	Synthesis of Validamycins	89
5.2.2	Synthesis of Acarbose	90
5.2.3	Synthesis of Salbostatin	91
5.2.4	Structure-Activity Relationship of Naturally Occurring Glycosidase Inhibitors Composed of 5a-Carba-Sugars	92
5.3	Synthesis of Carba-Oligosaccharides of Biological Interests	94

5.3.1	5a'-Carba-maltoses Linked by Imino, Ether, and Sulfide Linkages	94
5.3.2	Several Carba-Oligosaccharides: Core-Structures of Cell Surface Glycans	95
5.4	Synthesis of Glycolipid Analogues Containing 5a-Carba-sugars	97
5.4.1	Glycosylamide Analogues	97
5.4.2	Glycosylceramide Analogues	98
5.4.3	Glycocerebrosidase Inhibitors	99
5.5	Development of New-Types of Glycosidase Inhibitors	100
5.5.1	Chemical Modification of the Trehalase Inhibitor Trehazolin	100
5.5.2	Development of 5-Amino-1,2,3,4-cyclopentanetetraol Derivatives as New Potent Glycosidase Inhibitors	102
5.5.3	N-Alkyl and N-Phenyl Cyclic Isourea Derivatives of 5-Aminocyclopentanepolyol Derivatives: New Types of Potent α-Glucosidase Inhibitors	103
5.6	References	104
6	**The Dithiane Route from Carbohydrates to Carbocycles**	107
	K. Krohn, G. Börner and S. Gringard	
6.1	Introduction	107
6.2	Three- to Five-membered Carbocyclic Rings from 2-Deoxy ribose	109
6.2.1	Five-membered Ring	109
6.2.2	Three-membered rings	109
6.2.3	Four-membered Ring	111
6.3	Three- to Seven-membered Carbocyclic Rings from Mannose	112
6.3.1	Cyclopropane Derivatives	112
6.3.2	Cyclobutane Derivatives	113
6.3.3	Cyclohexanes and Cycloheptanes	114
6.4	Application to Natural Product Synthesis	115
6.4.1	Validatol	115
6.4.2	Calcitriol	117
6.5	References	120
7	**New Reductive Carbocyclisations of Carbohydrate Derivatives Promoted by Samarium Diiodide**	123
	J. L. Chiara	
7.1	Introduction	123

7.2	Results and Discussion	125
7.2.1	Intramolecular C=O/C=O Reductive Coupling	125
7.2.2	Intramolecular C=O/C=N-OR Reductive Cross-Coupling	131
7.2.3	Ring Contraction of 6-Deoxy-6-Iodo-Hexopyranosides	135
7.3	Conclusions	138
7.4	References	139
8	**A Concise Route to Carba-Hexopyranoses and Carba-Pentofuranoses from Sugar Lactones**	**143**
	F. Chrétien, M. Khaldi and Y. Chapleur	
8.1	Introduction	143
8.2	Results	144
8.2.1	Carbahexopyranoses	144
8.2.2	Carbapentofuranoses	152
8.2.3	Epilogue	153
8.3	Conclusion	154
8.4	References	154
9	**Asymmetric Synthesis of Aminocyclopentitols *via* Free Radical Cycloisomerization of Enantiomerically Pure Alkyne-Tethered Oxime Ethers Derived from Carbohydrates**	**157**
	J. Marco-Contelles	
9.1	Introduction	157
9.2	Results and Discussion	159
9.2.1	Synthesis and Radical Cyclization of D-*manno* Derivatives 6-8	161
9.2.2	Synthesis and Radical Cyclization of D-*ribo* Derivatives **16** and **17**.	163
9.2.3	Synthetic Elaboration of the Carbocyclic Vinyltin Derivatives. Synthesis of Aminocyclopentitols.	165
9.3	Conclusion	167
9.4	References	168
10	**Carbohydrates as Sources of Chiral Inositol Polyphosphates and their Mimics**	**171**
	D. J. Jenkins, A. M. Riley and B. V. L. Potter*	
10.1	Introduction	171

10.2	Synthesis of D-2-deoxy-*myo*-inositol 1,3,4,5-tetrakis-phosphate	173
10.2.1	Improved Preparation of Methyl 4,6-*O*-Benzylidene-α-D-Glucopyranoside and Methyl 4,6-*O*-Benzylidene-α-D-Mannopyranoside	176
10.2.2	Benzylation of Methyl 4,6-*O*-Benzylidene-α-D-Glucopyranoside and Related Reactions	178
10.2.3	Preparation of Methyl 3,4-Di-*O*-benzoyl-2-*O*-benzyl-6-deoxy-α-D-*xylo*-hex-5-enopyranoside **30**	180
10.2.4	Ferrier Carbocyclisation of **30**	180
10.2.5	Attempted Inversion of Stereochemistry at Position 5 of **32**	181
10.2.6	Synthesis of D-2-Deoxy-*myo*-inositol 1,3,4,5 Tetrakisphosphate **5**	183
10.3	Synthesis of Ring-contracted Analogues of Ins(1,4,5)P$_3$	184
10.3.1	Preparation of Methyl 2-*O*-Benzyl-3,4-di-*O*-(*p*-methoxybenzyl)-α-D-Glucopyranoside **58**	185
10.3.2	Preparation of (1*R*, 2*R*, 3*S*, 4*R*, 5*S*)-3-Hydroxy-1,2,4-trisphospho-5-vinyl Cyclopentane **53b**	187
10.3.3	Preparation of (1*R*, 2*R*, 3*S*, 4*R*, 5*S*)-3-Hydroxy-5-hydroxymethyl-1,2,4 trisphospho Cyclopentane **75**	189
10.4	Synthesis of a Carbohydrate-based Inositol Polyphos-phate Mimic Based on Adenophostin A	191
10.4.1	Synthesis of 2,6-Di-*O*-benzyl-3,4-di-*O*-(p-methoxybenzyl) -D-Glucopyranose **93**	193
10.4.2	Synthesis of (2-Hydroxyethyl)-2',3,4-Trisphosphate-α-D-Glucopyranoside **88**	197
10.4.3	Preparation of a Fluorescent Label Based upon (2-Hydroxy- ethyl)-2',3,4-trisphosphate-α-D-Glucopyranoside	199
10.5	Conclusion	201
10.6	References	201
11	**Chemo-enzymatic Total Synthesis of Some Conduritols, Carbasugars and (+)-Fortamine**	**209**
	L. Dumortier, P. Liu, J. Van der Eycken and M. Vandewalle*	
11.1	Introduction	209
11.2	Results	209
11.2.1	Substrates and Enantioselective Enzymatic Hydrolysis	209
11.2.2	The Synthesis of Some Advanced Intermediates	212

11.2.3	Synthesis of Conduritols	213
11.2.4	Synthesis of Carba-sugars	214
11.2.5	Synthesis of (+)-Fortamine (78)	218
11.3	References	220

12	**Chemo-enzymatic Approaches to Enantiopure Carbasugars and Carbanucleosides**	**223**
	J. Baumgartner and H. Griengl	
12.1	Introduction	223
12.2	Enantiopure Carbahexo- and Pentofuranoses	225
12.3	Introduction of the Base	229
12.4	Carbauloses	231
12.5	References	235

13	**Glycomimetics that Inhibit Carbohydrate Metabolism**	**239**
	Bruce Ganem	
13.1	Introduction	239
13.2	Studies on Mannostatin	241
13.3	Electrophilic Additions to Substituted Cyclopentenes	247
13.4	Studies on Allosamizoline: an Aminocyclopentitol from Allosamidin	249
13.5	Studies on Trehazolin and Kerrufaride	252
13.6	References	255

14	**Toward Azaglycoside Mimics: Aza-*C*-glycosyl Compounds and Homoazaglycosides**	**259**
	O. R. Martin	
14.1	Introduction	259
14.2	Aza-*C*-glycosyl Compounds	262
14.3	Homoazaglycosides	270
14.4	1-Deoxyazaseptanoses	276
14.5	Conclusion	279
14.6	References	279

15	**Total Synthesis and Chemical Design of Useful Glycosidase Inhibitors**	**283**
	K. Tatsuta	
15.1	Introduction	283
15.2	Results	284
15.2.1	Cyclophellitols	284

15.2.2	Nagstatins	290
15.2.3	Gualamycin	298
15.3	Conclusion	304
15.4	References	304

16	**Synthesis of Enantiopure Azasugars from D-Mannitol: Carbohydrate and Peptide Mimics**	307
	J. -C. Depezay	
16.1	Introduction	307
16.2	Results	308
16.2.1	To Azasugars *via* Bis-Aziridines	308
16.2.1.1	Nucleophilic Opening of Bis-Aziridines	308
16.2.1.2	Bis-Aziridines Precursors of Azadisaccharides	310
16.2.2	To Azasugars *via* Bis-Epoxides	314
16.2.2.1	Nucleophilic Opening of Bis-Epoxides	314
16.2.2.2	Isomerisation of Polyhydroxylated Piperidines	317
16.2.3	Isomerisation of Polyhydroxylated Azepanes	320
16.2.4	Inhibition Studies	322
16.2.5	Synthesis of Non-Peptide Mimics of Somatostatin	322
16.3	Conclusion	325
16.4	References	325

17	**Furan-, Pyrrole-, and Thiophene-Based Siloxydienes *en route* to Carbohydrate Mimics and Alkaloids**	327
	G. Casiraghi, G. Rassu and F. Zanardi	
17.1	Introduction	327
17.2	Outline of the Strategy	328
17.3	Synthetic Applications	330
17.3.1	Aminosugar Derivatives	330
17.3.2	Hydroxylated α-Amino Acids	332
17.3.3	Bicyclic Hydroxylated Alkaloids	336
17.3.4	Nucleosides and Nucleoside Mimetics	341
17.4	Conclusion	346
17.5	References	346

18	**Travelling through the Potential Energy Surface of Sialyl Lewis^x**	349
	A. Imberty and S. Perez	
18.1	Introduction	349

18.2	Computational Methods	351
18.2.1	Nomenclature	351
18.2.2	Relaxed Maps of Disaccharides	351
18.2.3	Calculation of Potential Energy Surface Using CICADA	351
18.2.4	Determination of Family of Low Energy Conformers	352
18.2.5	Flexibility Indexes	352
18.2.6	Calculating Molecular Properties	352
18.3	Results and Discussion	352
18.3.1	Analysis of the Linkages Conformation	353
18.3.2	Families of Conformers and Energy Minima	356
18.3.3	Flexibility	359
18.3.4	Molecular Properties	360
18.4	Conclusion	362
18.5	References	362
19	**Syntheses of Sulfated Derivatives as Sialyl Lewis^a and Sialyl Lewis^x Analogues**	**365**
	C. Augé, F. Dagron, R. Lemoine, C. Le Narvor and A. Lubineau	
19.1	Introduction	365
19.2	Synthesis of 3'-Sulfated Lewis^a Trisaccharide and Pentasaccharide	367
19.2.1	3'-Sulfated Lewis^a Trisaccharide	367
19.2.2	3'-Sulfated Lewis^a Pentasaccharide	368
19.3	Regioselective Sulfation	369
19.3.1	Stannylene Methodology	369
19.3.2	Cyclic Sulfates	371
19.4	Polysulfated Oligosaccharides in the Lewis^x Series	372
19.4.1	Introduction	372
19.4.2	6'-Sulfated, 3'-Sulfated, 3',6'-Disulfated Lewis^x Pentasaccharides	373
19.4.3	3',6-Disulfated Lewis^x Pentasaccharide	376
19.5	Immunological studies	379
19.5.1	Specificity of the E-Selectin	379
19.5.2	Specificity of the L-Selectin	381
19.6	References	382
20	**Architectonic Neoglycoconjugates: Effects of Shapes and Valencies in Multiple Carbohydrate-Protein Interactions**	**385**

D. Zanini and R. Roy
20.1	Introduction	385
20.2	Neoglycoproteins and Glycopolymers	387
20.3	Glycotelomers	390
20.4	Glycopeptoids	391
20.5	Glycoclusters	393
20.6	Cyclic Glycopeptides	395
20.7	Glycodendrimers	397
20.8	Spherical Glycodendrimers	402
20.9	Cyclodextrins and Glycocalixarenes	407
20.10	Biological Testing	409
20.11	Conclusion	410
20.12	References	410

21 From Glycosaminoglycans to Heparinoid Mimetics with Antiproliferative Activity 417
H. P. Wessel, A. Chucholowski, J. Fingerle, N. Iberg, H. P. Märki, R. Müller, M. Pech, M. Pfister-Downar, M. Rouge, G. Schmid and T. Tschopp

21.1	Introduction	417
21.2	Sulfated Carbohydrates	417
21.3	Spaced Sugars	424
21.4	A New Approach to Oligosaccharide Mimetics	428
21.5	References	429

22 Conduritols and Analogues as Insulin Modulators 433
D.C. Billington, F. Perron-Sierra, I. Picard, S. Beaubras, J. Duhault, J. Espinal and S. Challal

22.1	Introduction	433
22.2	Results	434
22.3	References and Notes	440

23 Strategies for the Synthesis of Inositol Phospho-glycan Second Messengers 443
N. Khiar and M. Martin-Lomas

23.1	Introduction	443
23.2	The Structure of the Inositol Phosphoglycan which Mediates the Intracellular Post-Receptor Action of Insulin	446

23.3	The Synthesis of Building Blocks for the Preparation of Inositolphosphoglycan Second Messengers	447
23.3.1	Synthesis of Conveniently Functionalised Enantiomerically Pure D-*myo* or D-*chiro* Inositol Derivatives	447
23.3.2	Synthesis of Conveniently Functionalised Glucosamine Unit.	450
23.3.3	Synthesis of 6-O-Glycosyl-D-*chiro* and D-*myo*-Inositol Derivatives	450
23.4	Evaluation of the Insulin-Like Activity of 4-*O*-Gluco-samine-*myo*-Inositol-1-Phosphate, 6-*O*-Glucosamine-*myo*-Inositol-1-Phosphate, 6-*O*-Glucosamine-*myo*-Inositol-1,2-Cyclic Phosphate	452
23.5	Toward the Total Synthesis of the Pseudo octa-saccharide V, Putative Second Messenger of the Hormone Insulin.	454
23.5.1	Synthesis of the Tetrasaccharide VII [31]	454
23.5.2	Synthesis of the Tetrasaccharide VII	457
23.6	Conclusion	459
23.7	References	460
24	**The Catalytic Efficiency of Glycoside Hydrolases and Models of the Transition State Based on Substrate Related Inhibitors**	**463**
	G. Legler	
24.1	Introduction	463
24.2	Catalytic Efficiency and the Transition State	464
24.2	Evidence for a Glycosyl Cation-Like Transition State	468
24.3	Evidence for Active Site Carboxylate Groups	472
24.4	Interactions with the Aglycon Site	481
24.5	N^1-Alkyl Gluconaminidines as "Perfect" Transition State Mimics?	483
24.6	Conclusion	485
24.7	References	487
25	**Design and Synthesis of Potential Fucosyl Transferase Inhibitors**	**491**
	G. A. van der Marel, B. M. Heskamp, G. H. Veeneman, C. A. A. van Boeckel and J. H. van Boom*	
25.1	Introduction	491
25.1.1	Role of Selectins in Leucocyte Extravasation	491
25.2	Receptor-based Adhesion Inhibition	494

25.2.1	Ligand-based Adhesion Inhibition	494
25.2.2	Fucosyltransferase inhibitors	498
25.3	Conclusion	505
25.4	References	506

26	**Enzymatic Synthesis of Lactose Analogues Using Glycosidases**	**511**
	A. Fernández-Mayoralas	
26.1	Introduction	511
26.2	Results and Discussion	514
26.4	Conclusion	521
26.5	References	521

27	**Synthesis of Nitrogen Containing Linkers for Antisense Oligonucleotides**	**523**
	Yogesh S. Sanghvi	
27.1	Introduction	523
27.2	Results	524
27.2.1	3´-C-N-O Linked Backbones	524
27.2.1.1	Synthesis of Oxime Dimer **5**	525
27.2.1.2	Synthesis of MI Dimer **6**	526
27.2.1.3	Synthesis of MMI dimer **7**	527
27.2.2	3´-C-N-N Linked Backbone	527
27.2.2.1	Synthesis of MDH dimer **8**	527
27.2.3	3´-C-O-N Linked Backbone	528
27.2.3.1	Synthesis of MOMI dimer **9**	528
27.2.4	3'-O-N-C Linked Backbone	529
27.2.4.1	Synthesis of HMIM dimer **10**	529
27.2.5	3´-C-N-C and C-C-N Linked Backbone	531
27.2.5.1	Synthesis of Amine **1** and **2** Dimers (**11** and **12**)	531
27.2.6	2´-OMe 3´- C-N-O Linked Backbone	532
27.2.6.1	Synthesis of bis 2´-OMe MMI Linked dimer **13**	532
27.3	Conclusion	534
27.4	References	535

28	**Nucleosides and Nucleotides Containing 5-Alkyl Pyrimidines. Chemistry and Molecular Pharmacology**	**537**
	L. Ötvös	
28.1	Chemistry	537

28.1.1	Synthetic Methods	538
28.1.2	Deacylation of Derivatives in the Sugar Moiety	540
28.2	Bioorganic Chemistry	542
28.2.1	Substrate Specificity in Polymerase Catalysed Reactions	542
28.2.1.1	5-n-Alkyl-, 5-(n-1-alkenyl)- and 5-(n-1-alkynyl)dUTPs	543
28.2.1.2	Deoxyuridines Containing Branched Substituents in Position 5	547
28.2.2	Structural Properties and Stability of Polynucleotides Containing 5-Alkyl-2'-deoxyuridines	548
28.3	Molecular Pharmacology	549
28.3.1	Molecular Pharmacology of the Antiherpetic Activity of 5-Isopropyl-2'-deoxyuridine	549
28.3.2	Antisense Drug Properties of Oligonucleotides Containing 5-Alkyl-2'-deoxyuridines	550
28.4	References	551
29	**Anhydrohexitols as Conformationally Constrained Furanose Mimics, Design of an RNA-Receptor**	553
	P. Herdewijn	
29.1	Introduction	553
29.2	Double Stranded RNA Structure	556
29.3	Design of Oligonucleotides with a Six Membered Carbohydrate Moiety	558
29.4	Synthesis of 2',3'-Dideoxy-3'-C-Hydroxymethyl-α-L-*threo*-Pentopyranosyl Nucleoside and 1,5-Anhydro-2,3-Dideoxy-D-*arabino*-Hexitol Nucleoside.	562
29.5	Oligonucleotide Built up from Nucleosides with a Six-membered Carbohydrate-like Moiety	569
29.6	1,5-Anhydrohexitol Nucleoside as a Mimic of a Furanose Nucleoside in its 2'-endo/3'-exo Conformation	573
29.7	References	579
Index		581

1 Recent Advances in the Synthesis of Cyclic Polyether Marine Natural Products

K. C. Nicolaou, M. H. D. Postema and N. D. Miller

1.1 Introduction

The polyether marine natural products are a growing class of naturally occurring substances with impressive structural features and biological properties [1]. These molecules usually contain arrays of fused cyclic ether rings of sizes ranging from five to ten with the six-membered framework being the most common. Figure 1-1 displays a number of representative examples from this class including brevetoxin B (**1**) [2], the first member to be characterized, and maitotoxin (**7**) [3], the most complex compound of the series.

The biological activities of the polyether marine natural products range from neurotoxicity [4] to antifungal action [5] and they have been attributed to interference with, or formation of, ion channels by these compounds. Brevetoxin B, for example, has been shown to open [6] or form [7] sodium channels causing rapid sodium ion influx and thus cell damage [8], whereas maitotoxin's potent toxicity has been attributed to its ability to interfere with calcium channels [9]. It is interesting to note that the shorter hemibrevetoxin B (**3**) [10] and a synthetic, truncated version of brevetoxin B [11], as well as brevetoxin B_3 (**4**) [12] in which the rigid polyether backbone has been disrupted by the cleavage of a C-O bond, no longer possess the potent neurotoxicity of the parent compound, brevetoxin B. Furthermore, it is tempting to postulate that special folding of the larger polyethers, such as maitotoxin, within the membrane may lead to the formation of ion channels.

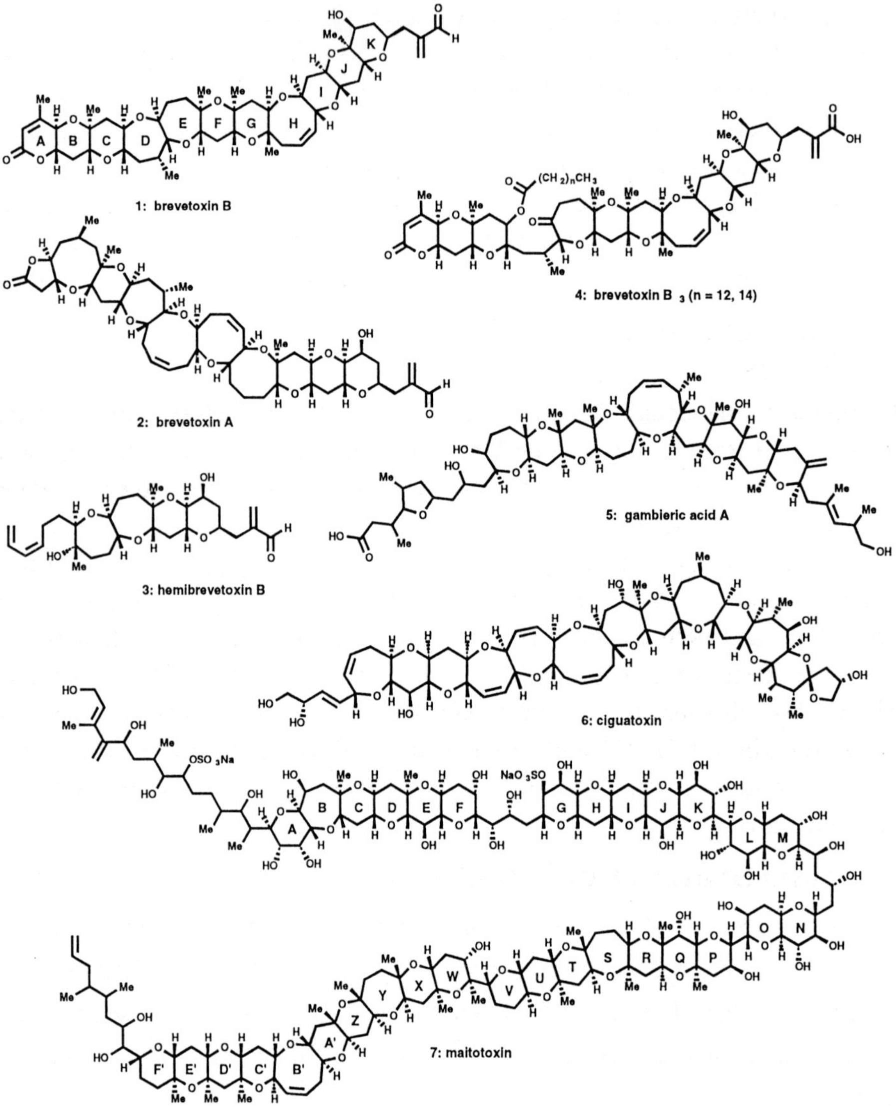

Figure 1-1. Structures of selected neurotoxic marine cyclic polyether natural products.

The intriguing structures of these molecules coupled with their fascinating mode of action led to intense efforts directed towards their total synthesis. The desire to develop analytical methods for their detection in seafood and the marine environment as well as the need to treat victims of the "red tide"[13] and other related catastrophic phenomena added considerably to the rationale for chemical investigations in the field. Finally, the opportunities that these molecules present for the invention and development of new synthetic reactions[14] and for molecular design should not be missed, for they are sure to produce useful science and technology if properly exploited. For these reasons we embarked a number of years ago on a program directed at the total synthesis of a selected number of these target molecules combining the development of novel synthetic strategies and synthetic technology, molecular design and biological studies. While much of our earlier work in this field has already been reviewed [15], in this article we summarize the latest advances in the field from our laboratories.

1.2 Synthetic Technology for the Construction of Cyclic Ethers

The imposing molecular structures of the complex polyether marine natural products gave impetus for the development of numerous new synthetic methods for the construction of cyclic ethers. From our laboratories, these methods included strategies based on intramolecular hydroxy epoxide openings [16] and Michael additions [17], hydroxy dithioketal cyclizations [18], bridging reactions of dithionolactones [19], reductive cyclizations of hydroxy ketones [20], nucleophilic additions to thionolactones [21], intramolecular esterphosphonate-ketone condensations [22] and lactone formation [22]. These enabling technologies allowed us to successfully complete the total syntheses of hemibrevetoxin B (**3**) [23] and brevetoxin B (**1**) [24]. It will suffice here to summarize briefly the application of some of these methods to the total synthesis of brevetoxin B (**1**).

1.3 Total Synthesis of Brevetoxin B

Since a full account of the brevetoxin B project has already appeared [24, 25] only the highlights will be presented here. Figure 1-2 outlines the strategic bond disconnections based on which the synthetic strategy towards the target molecule was developed, whereas Scheme 1-1 summarizes the final stages of the total synthesis.

Figure 1-2. Strategic bond disconnections of the brevetoxin B molecule.

Starting with 2-deoxy-D-ribose (**8**) and D-mannose (**9**), the advanced intermediates, phosphonium salt **10** and aldehyde **11**, were synthesized respectively. These enantiomerically pure fragments were then coupled through the Wittig reaction, yielding, stereoselectively and in high yield, olefinic product **12**. Following deprotection of the appropriate hydroxy group, the resulting hydroxy dithioketal was cyclized[18] under the influence of Ag$^{\oplus}$ to afford the desired oxocene system and thus providing the intact brevetoxin B skeleton. Reductive removal of the remaining ethylthio group under radical conditions, with retention of stereochemistry, furnished compound **13**. The remaining steps in the synthesis were conventional and proceeded in high overall yield as detailed in Scheme 1-1. Despite the linearity of this route, the average yield for each step is an impressive 91%.

1.4 Recent Developments in Polyether Synthesis

1.4.1 Radical-Based Reduction of Thionolactones to Cyclic Ethers

The richness of thiocarbonyl group chemistry [26] has been appropriately exploited for the development of useful methods for the synthesis of cyclic ethers. Our most recent excursion in this area [27] took advantage of the propensity of the thiocarbonyl moiety to undergo attack by tin radicals, a tendency that serves as the basis for the Barton-McCombie deoxygenation reaction [28]. After careful experimentation we found that selection of the appropriate tin hydride reagent leads to an alternative pathway, by-passing the normal Barton-McCombie deoxygenation event (Scheme 1-2) [29].

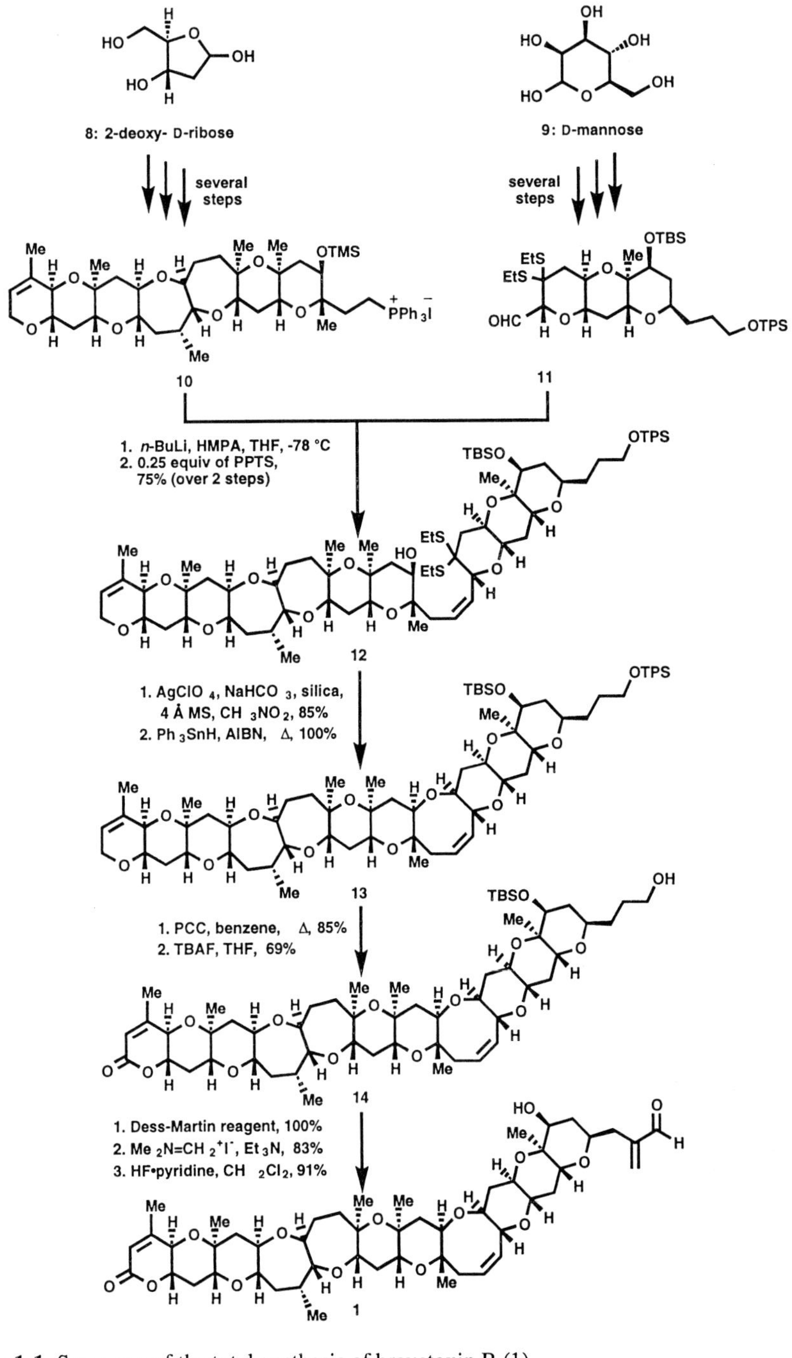

Scheme 1-1. Summary of the total synthesis of brevetoxin B (1).

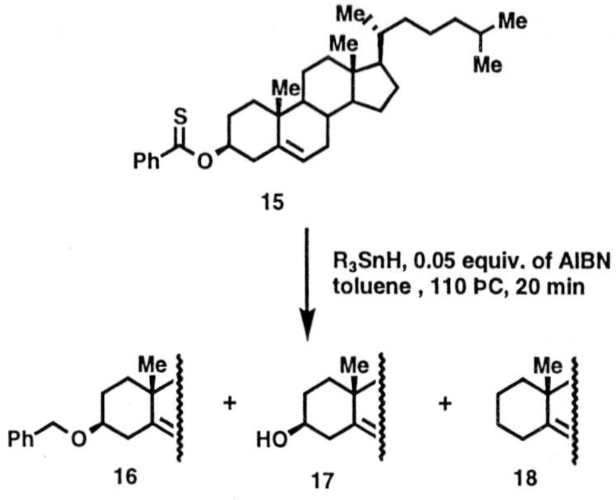

Scheme 1-2. Tin hydride-mediated desulfurization of thionoesters.

Thus, reaction of the cholesterol derived secondary thiobenzoate **15** with 1.5 equiv. of n-Bu$_3$SnH-AIBN (cat.) in refluxing toluene afforded cholesterol (**17**) and the deoxygenated product **18** in 42 and 46% yield respectively, in accord with earlier observations [28]. When the reaction was repeated with 5.0 equiv. of n-Bu$_3$SnH, cholesterol (**17**) and the benzyl ether **16** were isolated in 38 and 49% yield respectively. But when triphenyltin hydride was utilized under otherwise identical conditions, the benzyl ether **16** was formed exclusively and in 97% isolated yield (Table 1-1).

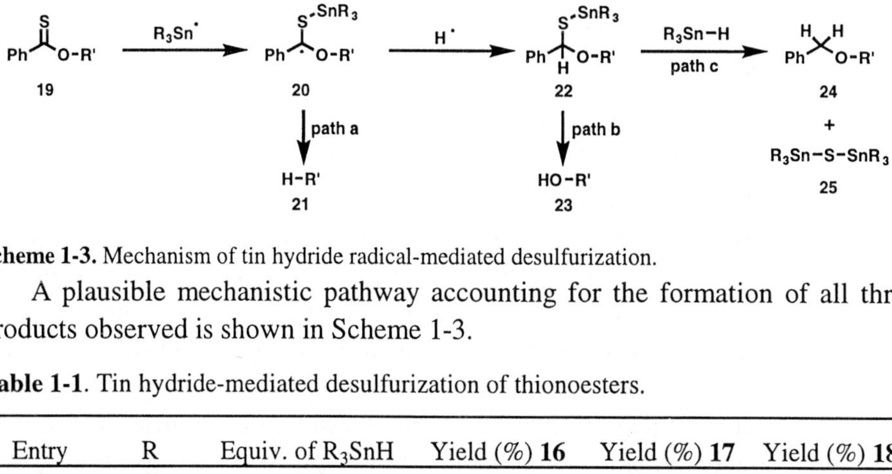

Scheme 1-3. Mechanism of tin hydride radical-mediated desulfurization.

A plausible mechanistic pathway accounting for the formation of all three products observed is shown in Scheme 1-3.

Table 1-1. Tin hydride-mediated desulfurization of thionoesters.

Entry	R	Equiv. of R$_3$SnH	Yield (%) **16**	Yield (%) **17**	Yield (%) **18**
1	n-Bu	1.5	0	42	46
2	n-Bu	5.0	49	38	trace
3	Ph	5.0	97	trace	trace

Table 1-2. Radical desulfurization of thionolactones to cyclic ethers.[a]

Entry	Substrate	Product	Yield (%)
1	(structure)	(structure)	79
2	(structure)	(structure)	77
3	(structure)	(structure)	90
4	(structure)	(structure)	99
5	(structure)	(structure)	72

[a]*Reagents and conditions:* 5.0 equiv of Ph$_3$SnH, 0.05 equiv. of AIBN, toluene (0.01 M based on substrate), 110 °C.

The generality and scope of this procedure was extended to a variety of lactones as illustrated in Table 1-2. Thus, cyclic ethers with medium- and large-ring sizes can easily be derived from lactones via the corresponding thionolactones [30].

1.4.2 Synthesis of 1,1′-Bistetrahydropyran Systems via Copper(I)-Promoted Stille Coupling

With certain structural features of maitotoxin in mind (e.g. K-L, O-P and V-W ring systems, see Fig. 1-1 and Scheme 1-4), we recently initiated a program [31] directed at the development of suitable methodology for the construction of cyclic ether frameworks of the 1,1′-bistetrahydropyran type. The power of palladium(0) [32] as a catalyst to bring together activated vinyl derivatives has been amply demonstrated, as for example in the Stille reaction [33], and served as the basis for our investigations in this area. Indeed, the Stille reaction has been utilized to couple 1-stannyl glycals [34] with a range of simple vinyl or aryl derivatives. However, a common side-reaction in this process is the homo-coupling [35] of the starting components. Indeed, Ley and coworkers [36] exploited this opportunity in the synthesis of symmetrical bisvinyl systems. Since we were interested in the preparation of non-symmetrical bisglycal systems, we focused our attention on a method that would allow for the hetero-coupling of cyclic stannyl enol ethers with cyclic enol triflates to afford the desired products (Scheme 1-5).

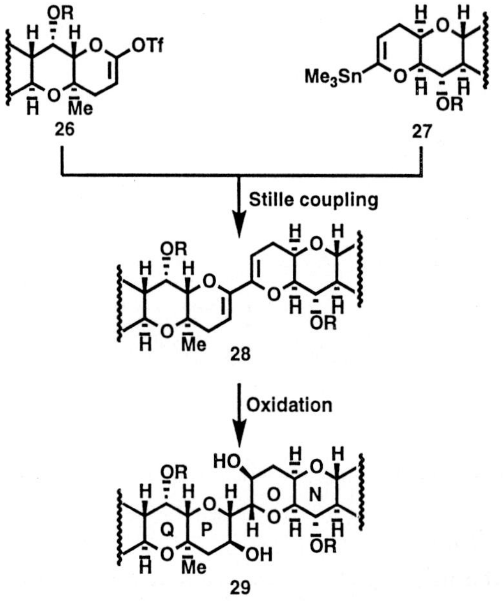

Scheme 1-4. Strategy for the construction of the NO-PQ system present in maitotoxin.

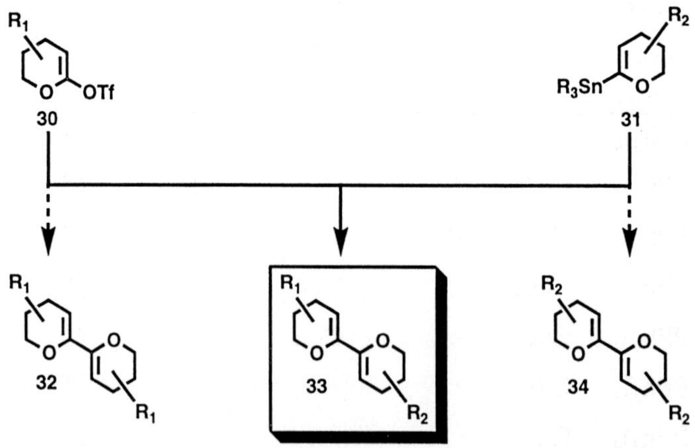

Scheme 1-5. Cross-coupling of cyclic enol triflates with cyclic stannyl enol ethers.

As a simple model system, we initially chose to examine the coupling of enol triflate **36** with trimethylstannane **37**, both easily obtained from tri-*O*-acetyl-D-glucal (**35**), Scheme 1-6.

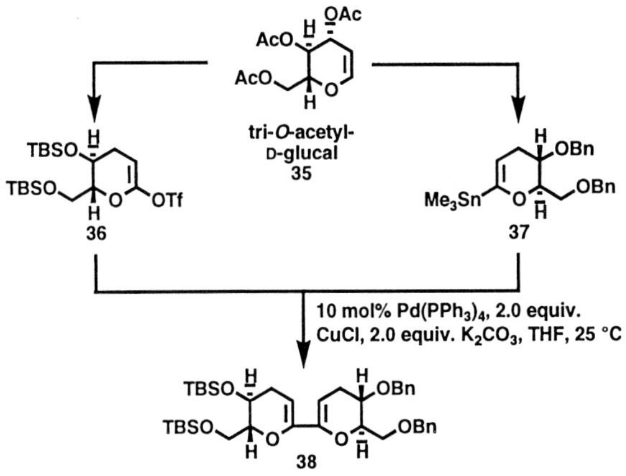

Scheme 1-6. Stille cross-coupling of stannyl enol ether **37** with enol triflate **36**.

As seen from Table 1-3, the optimum conditions for the desired hetero-coupling required the utilization of CuCl (2.0 equiv.) and K_2CO_3 (2.0 equiv.) as additives to the usual $Pd(PPh_3)_4$ catalyst (entry 4). Under these conditions, an 80% yield of hetero-coupled product **38** along with *ca.* 5% combined yield of the undesired homo-dimers derived from each starting material was obtained.

Table 1-3. Stille coupling of stannyl enol ether **37** with enol triflate **36**.

Entry	Catalyst (10 mol%)	Additive (equiv.)	Base (2.0 equiv.)	Temp. (°C)	Time (hrs.)	Yield (%)
1	$Pd(PPh_3)_4$	–	–	25	78	43
2	$Pd(PPh_3)_4$	LiCl (3.0)	–	Δ	12	53
3	$Pd(PPh_3)_4$	CuCl (2.0)	–	Δ	12	49
4	$Pd(PPh_3)_4$	CuCl (2.0)	K_2CO_3	25	1	80
5	$Pd(PPh_3)_4$	CuCN (2.0)	K_2CO_3	25	24	52
6	$Pd(PPh_3)_4$	CuBr (2.0)	K_2CO_3	25	3	68
7	$Pd(PPh_3)_4$	CuI (2.0)	K_2CO_3	25	36	53
8	$Pd(PPh_3)_4$	$CuCl_2$ (2.0)	K_2CO_3	25	12	0
9	–	CuCl (2.0)	K_2CO_3	25	96	0

While the precise role of the copper(I) species [37] in this coupling reaction is a matter of conjecture at the present time, the generality and scope of the reaction is impressive, both in terms of complexity of substrates and effectiveness. Table 1-4 lists but a few examples of such coupling reactions demonstrating the power and versatility of the method.

Table 1-4. Synthesis of complex cyclic polyethers by copper(I)-promoted Stille cross-coupling.[a]

Entry	Enol triflate	Stannyl enol ether	Coupling product	Yield (%)
1	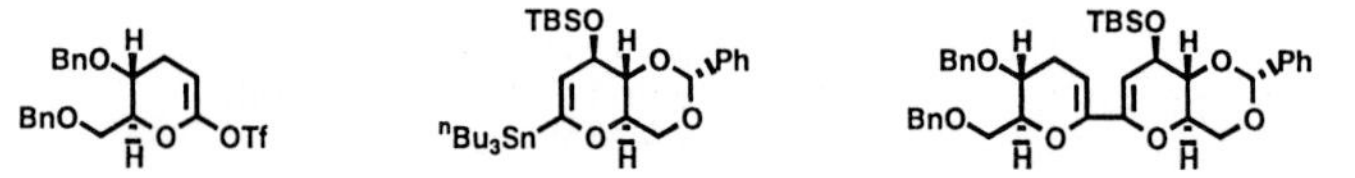			70
2	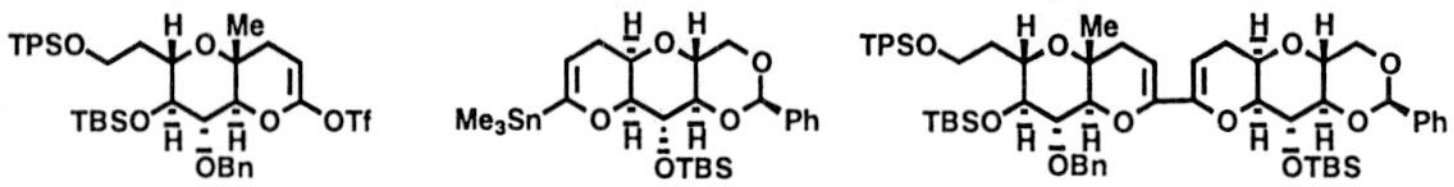			79
3	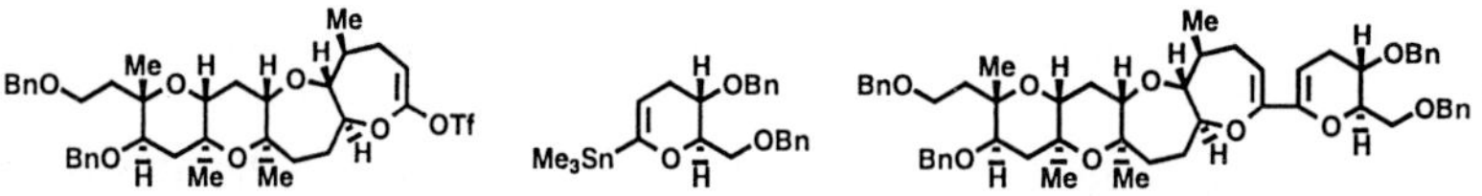			77
4	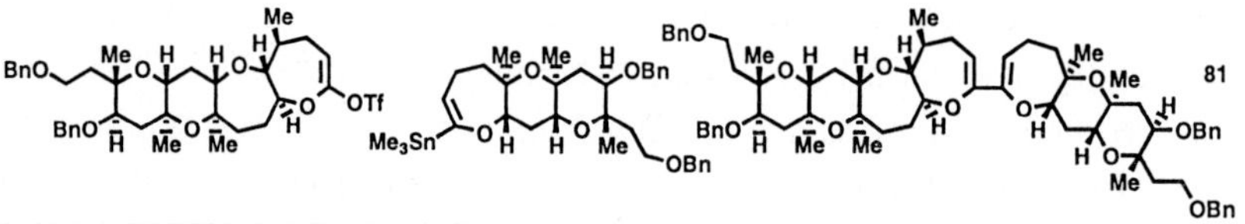			81

[a] Reagents and conditions: 1.0 equiv. of stannyl enol ether, 1.2 equiv. of enol triflate, 10 mol% of Pd(PPh$_3$)$_4$, 2.0 equiv. of CuCl, 2.0 equiv. of K$_2$CO$_3$, THF, 25 °C.

The application of this Pd(0)/Cu(I)-catalyzed reaction to the construction of a maitotoxin related fragment is shown in Scheme 1-7.

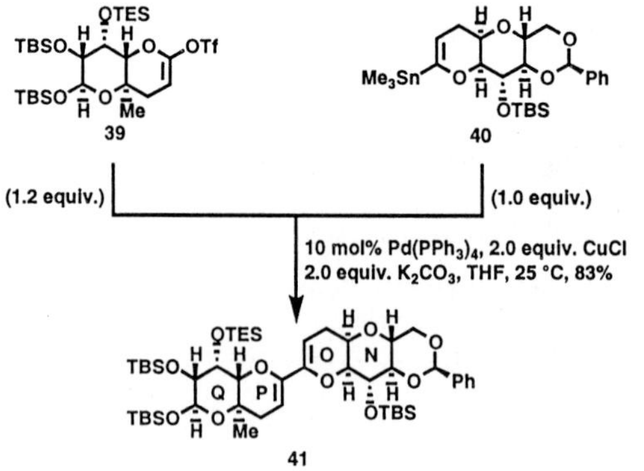

Scheme 1-7. Synthesis of a fragment related to the NO-PQ ring system of maitotoxin.

1.4.3 The Olefin Metathesis Approach to Cyclic Enol Ether Construction

Olefin-olefin metathesis has been a subject of considerable research in recent years [38]. Following the pioneering efforts of Grubbs [39] and Katz [40], this once « industrial only » process has now become a favored method in molecular assembly. In particular, the utilization of stable catalysts [41] that have a wide range of functional group compatibility as demonstrated by Grubbs and coworkers has aided greatly in elevating this reaction to the level of popularity that it enjoys today [42]. If two pendant olefinic partners are metathesized, a new cyclic olefin is formed with the extrusion of the remnant carbons as a second olefin. This process has been termed ring closing metathesis (RCM) [38].

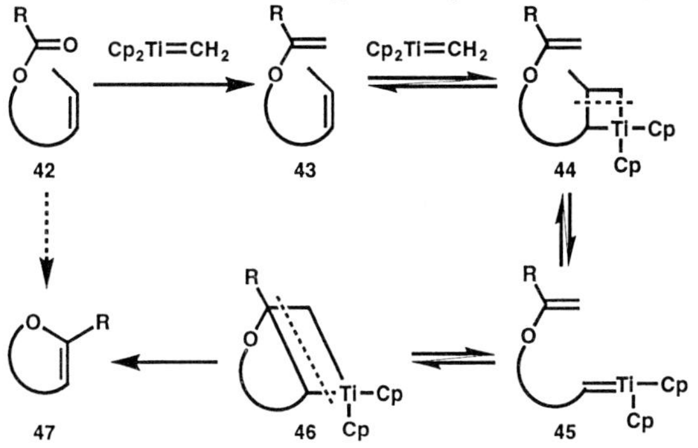

Scheme 1-8. Titanium-mediated metathesis strategy for the conversion of olefinic esters to cyclic enol ethers.

If one of the olefinic partners is an enol ether, the result of metathesis is the generation of a cyclic enol ether. In work related towards cyclic enol ether synthesis, we have recently found [43] that treatment of olefinic esters of the general type **42** (Scheme 1-8) with the Tebbe reagent [44] under appropriate conditions leads to the cyclic enol ether **47** directly.

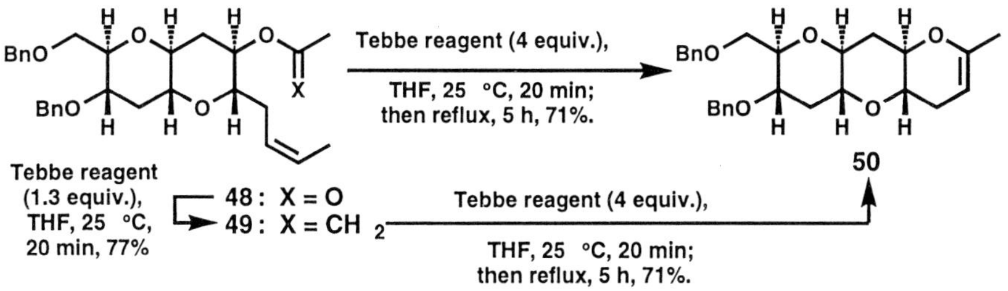

Scheme 1-9. Synthesis of cyclic enol ethers from olefinic esters.

We have proposed that the reaction proceeds via initial formation of the enol ether olefin **43**, which then proceeds to react with a second molecule of titanium methylidene to give the titanacyclobutane **44**. Compound **44** can then undergo fragmentation to **45**, a species that is poised to generate a second titanacyclobutane (**46**), which upon elimination of titanium methylidene gives the product enol ether (**47**).

Table 1-5. Synthesis of cyclic enol ethers from olefinic esters.

Entry	Substrate	Reagent	Product, yield
1	R = Me	TR [a]	R = Me, 71%
2	R = Ph	TR [a]	R = Ph, 70%
3		TR [a]	64%
4		DT [b]	60%
5	R = Me	TR [a]	R = Me, 45%
6	R = Ph	TR [a]	R = Ph, 32%
7		TR [a]	30%

[a] TR = $Cp_2TiCH_2ClAlMe_2$. [b] DT = Cp_2TiMe_2 (see Ref. 47).

Scheme 1-9 shows a specific example of this reaction (**48** → **50**). Furthermore, it was demonstrated that isolation and characterization of the intermediate enol ether-olefin **49** and subsequent re-exposure to a further amount of titanium methylidene gives the same product **50** as the one-pot procedure. This lends support for the proposed mechanism shown in Scheme 1-8, although a mechanism involving initial engagement at the olefin, followed by attack on the carbonyl cannot be ruled out at present.

This reaction is synthetically equivalent to an intramolecular Wittig olefination on an ester function. Previous work by Grubbs *et al* [45]. has shown that these types of structures can also be accessed by metathesis via a two-step protocol involving initial enol ether formation followed by isolation and subjection to a molybdenum-mediated metathesis to give the product. A related closure on a norbornene skeleton involving initial titanacyclobutane formation followed by carbonyl attack has also been published [46]. The generality and scope of this newly developed olefin-metathesis reaction for the construction of cyclic enol ethers are demonstrated by the examples shown in Table 1-5.

To demonstrate the applicability of the reaction in construction of complex molecules, the polycyclic systems **52** (Scheme 1-10) and **58** (Scheme 1-11) have been synthesized. Thus, compound **52** was formed in 61% yield by reaction of olefinic ester **51** with an excess of Tebbe reagent in refluxing THF (Scheme 1-10).

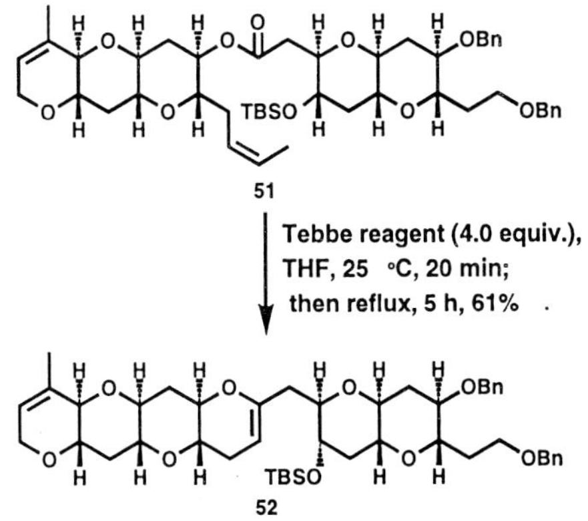

Scheme 1-10. Preparation of hexacycle **52**.

The hexacyclic polyether **58** was synthesized according to Scheme 1-11. Coupling of alcohol **53** with carboxylic acid **54** under the Mukaiyama conditions [48] gave ester **55** (79%) which was subjected to the standard metathesis conditions described above to afford a 64-71% yield of cyclic enol ether **56**. Regio- and stereoselective hydroboration, followed by oxidative work-up gave the corresponding alcohol which was smoothly oxidized to ketone **57** by the action of Dess-Martin reagent. Desilylation provided the expected hydroxy-ketone, that existed exclusively as the lactol form. Reduction with Et$_3$SiH-BF$_3$•OEt$_2$ provided the target polyether **58** with the all *trans*-fused stereochemistry. The power of this strategy lies not only in its convergency, but also in the generality and efficiency of the coupling reaction used to join the two fragments.

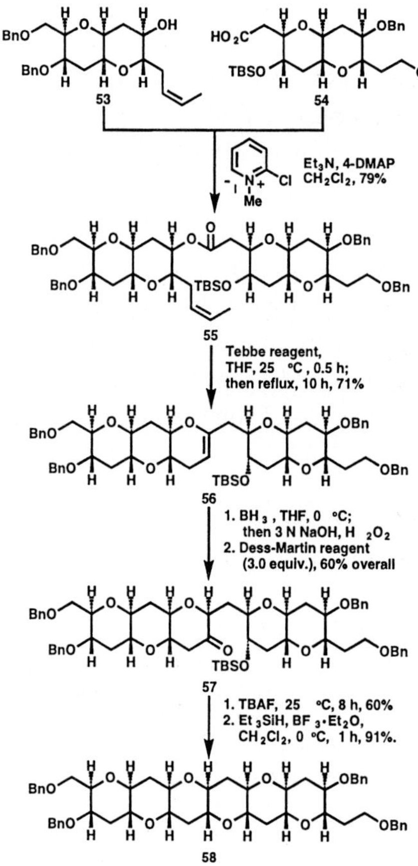

Scheme 1-11. Synthesis of hexacyclic polyether **58**.

1.5 Conclusion

The marine environment has been amply generous to synthetic chemists by providing us with myriads of structural variations of molecular assemblies. These structures excite and stimulate the strategists and the designers of the science as evidenced by the plethora of new synthetic methods developed to address the problems posed by them and by the molecular designs aimed at mimicking their biological action. A particularly fruitful domain within this area has been that of the polyether natural products such as the brevetoxins, ciguatoxin and maitotoxin. As seen from this article, the quest for new strategies and methods for the construction of cyclic ethers continues unabated. It is interesting and important to note that carbohydrates have been ideal starting materials [49] for the construction of many intermediates and target molecules of this class. It is also true, however, that new synthetic technologies developed with these marine natural products in mind served to enrich the arsenal of synthetic chemists in general, and carbohydrate chemists in particular.

Acknowledgements

We would like to extend our gratitude and appreciation to all our co-workers, whose names appear in the references, for their contributions to the research described in this article. Our programs have been financially supported by the National Institutes of Health, Merck, Sharp and Dhome, Glaxo, Inc., Schering-Plough, Pfizer, Hoffman-LaRoche, Amgen, Unitika, the NSERC, and the ALSAM Foundation.

1.6 References

1. T. Yasumoto, M. Murata, *Chem. Rev.* **1993**, *93*, 1897; Faulkner, D. J. *J. Nat. Prod. Chem.* **1994**, *11*, 355 and references cited therein.

2. Y. Lin, M. Risk, M. S. Ray, J. Clardy, J. Golik, J. C. James, K. Nakanishi, *J. Am. Chem. Soc.* **1981**, *103*, 6773.

3. M. Murata, H. Naoki, S. Matsunaga, S. Masayuki, T. Yasumoto, *J. Am. Chem. Soc.* **1994**, *116*, 7098; M. Sasaki, N. Matsumori, M. Murata, K. Tachibana, T. Yasumoto, *Tetrahedron Lett.* **1995**, *56*, 9011.

4. G. Strichartz, N. Castle in *Marine Toxins: Origin, Structure and Molecular Pharmacology* (Eds.: H. Sherwood, G. Strichartz), ACS Symposium Series 418, American Chemical Society, Washington, DC, USA, **1990**, p. 3.

5. H. Nagai, K. Torigoe, M. Satake, M. Murata, T. Yasumoto, H. Hirota, *J. Am. Chem. Soc.* **1992**, *114*, 1102.

6. G. Jeglitsch, K. Rein, D. G. Baden, D. J. Adams, *Biophys. J.* **1994**, *66*, A-323.

7. S. Matile, K. Nakanishi, *Angew. Chem. Int. Ed. Engl.* in press.
8. K. S. Rein, B. Lynn, R. E. Gawley, D. G. Baden, *J. Org. Chem.* **1994**, *59*, 2107.
9. M. Takahashi, Y. Ohizumi, T. Yasumoto, *J. Biol. Chem.* **1982**, *257*, 7287.
10. A. V. K. Prasad, Y. Shimizu, *J. Am. Chem. Soc.* **1989**, *111*, 6476.
11. K. C. Nicolaou, J. Tiebes, E. A. Theodorakis, F. P. J. T. Rutjes, K. Koide, M. Sato, E. Untersteller, *J. Am. Chem. Soc.* **1994**, *116*, 9371; R. E. Gawley, K. S. Rein, G. Jeglitsch, D. G. Adams, E. A. Theodorakis, J. Tiebes, K. C. Nicolaou, D. G. Baden, *Chem. & Biol.* **1995**, *2*, 533.
12. A. Morahashi, S. Masayuki, M. Kazuya, H. Naoki, H. F. Kaspar, T. Yasumoto, *Tetrahedron Lett.* **1995**, *36*, 8995.
13. D. M. Anderson, *Sci. Am.* **1994**, 62.
14. K. C. Nicolaou, *Aldrichim. Acta.* **1993**, *26*, 62; E. Alvarez, M.-L. Candenas, R. Perez, J. L. Ravelo, J. D. Martin, *Chem. Rev.* **1995**, *95*, 1953.
15. K. R. Reddy, G. Skokotas, K. C. Nicolaou, *Gazz. Chim. Ital.* **1993**, *123*, 337.
16. K. C. Nicolaou, C. V. C. Prasad, P. K. Somers, C.-K. Hwang, *J. Am. Chem. Soc.* **1989**, *111*, 5330; K. C. Nicolaou, C. V. C. Prasad, P. K. Somers, C.-K. Hwang, *J. Am. Chem. Soc.* **1989**, *111*, 5335.
17. K. C. Nicolaou, E. A. Theodorakis, F. P. J. T. Rutjes, J. Tiebes, M. Sato, E. Untersteller, X.-Y. Xiao, *J. Am. Chem. Soc.* **1995**, *117*, 1171.
18. K. C. Nicolaou, C. V. C. Prasad, C.-K. Hwang, M. E. Duggan, C. A. Veale, *J. Am. Chem. Soc.* **1989**, *111*, 5321.
19. K. C. Nicolaou, C.-K. Hwang, B. E. Marron, S. A. DeFrees, E. A. Couladouros, Y. Abe, P. J. Caroll, J. P. Snyder, *J. Am. Chem. Soc.* **1990**, *112*, 3040; K. C. Nicolaou, C.-K. Hwang, D. A. Nugiel, *Angew. Chem. Int. Ed. Engl.* **1988**, *27*, 1362.
20. K. C. Nicolaou, C.-K. Hwang, D. A. Nugiel, *J. Am. Chem. Soc.* **1989**, *111*, 4136.
21. K. C. Nicolaou, D. G. McGarry, P. K. Somers, B. H. Kim, W. W. Ogilvie, G. Yiannikouros, C. V. C. Prasad, C. A. Veale, R. R. Hark, *J. Am. Chem. Soc.* **1990**, *112*, 6263.
22. K. C. Nicolaou, F. P. J. T. Rutjes, E. A. Theodorakis, J. Tiebes, M. Sato, E. Untersteller, *J. Am. Chem. Soc.* **1995**, *117*, 1173.

23. K. C. Nicolaou, K. R. Reddy, G. Skokotas, F. Sato, X.-Y. Xiao, C.-K. Hwang, *J. Am. Chem. Soc.* **1993**, *115*, 3558.

24. K. C. Nicolaou, C.-K. Hwang, M. E. Duggan, D. A. Nugiel, Y. Abe, K. Bal Reddy, S. A. DeFrees, D. R. Reddy, R. A. Awartani, S. R. Conley, F. P. J. T. Rutjes, E. A. Theodorakis, *J. Am. Chem. Soc.* **1995**, *117*, 10227; K. C. Nicolaou, E. A. Theodorakis, F. P. J. T. Rutjes, M. Sato, J. Tiebes, X.-Y. Xiao, C.-K. Hwang, M. E. Duggan, Z. Yang, E. A. Couladouros, F. Sato, J. Shin, H.-M. He, T. Bleckman, *J. Am. Chem. Soc.* **1995**, *117*, 10239; K. C. Nicolaou, F. P. J. T. Rutjes, E. A. Theodorakis, J. Tiebes, M. Sato, E. Untersteller, *J. Am. Chem. Soc.* **1995**, *117*, 10252.

25. K. C. Nicolaou, *Angew. Chem. Int. Ed. Engl.* **1996**, *35*, 589.

26. D. H. R. Barton, *Aldrichim. Acta* **1990**, *23*, 3; D. Crich, L. Quintero, *Chem. Rev.* **1989**, *89*, 1413.

27. K. C. Nicolaou, M. Sato, E. A. Theodorakis, N. D. Miller, *J. Chem. Soc., Chem. Commun.* **1995**, 1583.

28. D. H. R. Barton, S. W. McCombie, *J. Chem. Soc., Perkin Trans. 1.* **1975**, 1574.

29. For related tin hydride mediated thiocarbonyl reductions, see: E. M. Acton, R. N. Goerner, H. S. Uh, K. J. Ryan, D. W. Henry, C. E. Cass, G. A. LePage, *J. Med. Chem.* **1975**, *22*, 518; M. J. Robins, J. S. Wilson, F. Hannske, *J. Am. Chem. Soc.* **1983**, *105*, 4059; D. R. Williams, J. L. Moore, *Tetrahedron Lett.* **1983**, *24*, 339.

30. M. P. Cava, M. I. Levinson, *Tetrahedron*, **1985**, *41*, 5061.

31. K. C. Nicolaou, M. Sato, N. D. Miller, J. L. Gunzer, J. Renaud, E. Untersteller, *Angew. Chem. Int. Ed. Engl.* **1996**, *35*, .

32. T. Mitchell, *Synthesis* **1992**, 803; V. Farina in *Comprehensive Organometallic Chemistry II* (Eds.: E. W. Abel, F. G. A. Stone, G. Wilkinson), Pergamon, Oxford, UK, **1995**, Vol. 12, Chapter 3.

33. J. K. Stille, *Angew. Chem. Int. Ed. Engl.* **1986**, *25*, 508.

34. R. W. Friesen, R. W. Loo, *J. Org. Chem.* **1991**, *56*, 4821; E. Dubois, J.-M. Beau, *Carbohydrate Res.* **1992**, *223*, 157; for a recent review of *C*-glycoside synthesis, including palladium mediated approaches, see: M. H. D. Postema, *C-Glycoside Synthesis*, CRC Press, Boca Raton, Florida, **1995**, Chapter 5.

35. For a mechanistic discussion of homo-dimer formation, see: R. W. Friesen, R. W. Loo, C. F. Sturino, *Can. J. Chem.* **1994**, *72*, 1262.

36. R. Downham, P. J. Edwards, D. A. Entwhistle, A. B. Hughes, K. S. Kim, S. V. Ley, *Tetrahedron: Asymmetry* **1995**, *6*, 2403.

37. For leading references on the copper(I)-facilitated Stille coupling, see: V. Farina, S. Kapadia, B. Krishnan, C. Wang, L. S. Liebeskind, *J. Org. Chem.* **1994**, *59*, 5905; J. R. Falck, R. K. Bhatt, Y. Jianhua, *J. Am. Chem. Soc.* **1995**, *117*, 5973; R. A. Gibbs, U. Krishnan, J. M. Dolence, C. D. Poulter, *J. Org. Chem.* **1995**, *60*, 7821 and references cited therein.

38. R. H. Grubbs, S. J. Miller, G. C. Fu, *Acc. Chem. Res.* **1995**, *28*, 446.

39. T. R. Howard, J. B. Lee, R. H. Grubbs, *J. Am. Chem. Soc.* **1980**, *102*, 6876.

40. T. J. Katz, S. J. Lee, M. A. Shippey, *J. Mol. Catal.* **1980**, *8*, 219.

41. S. T. Nguyen, R. H. Grubbs, J. W. Ziller, *J. Am. Chem. Soc.* **1993**, *115*, 9858.

42. H.-G. Schmalz, *Angew. Chem. Int. Ed. Engl.* **1995**, *34,* 1833.

43. K. C. Nicolaou, M. H. D. Postema, C. F. Claiborne, *J. Am. Chem. Soc.* **1996**, *118*, 1565.

44. F. N. Tebbe, G. W. Parshall, G. S. Reddy, *J. Am. Chem. Soc.* **1978**, *100*, 3611; S. H. Pine, R. Zahler, D. A. Evans, R. H. Grubbs, *J. Am. Chem. Soc.* **1980**, *102*, 3270.

45. O. Fujimura, G. C. Fu, R. H. Grubbs, *J. Org. Chem.* **1994**, *59*, 4029; for related work, see: G. C. Fu, R. H. Grubbs, *J. Am. Chem. Soc.* **1993**, *115*, 3800.

46. J. R. Stille, B. D. Santarsiero, R. H. Grubbs, *J. Org. Chem.* **1990**, *55*, 843.

47. N. A. Petasis, E. I. Bzowej, *J. Am. Chem. Soc.* **1990**, *112*, 6392.

48. T. Shibanuma, M. Shiono, T. Mukaiyama, *Chem. Lett.* **1977**, 575.

49. S. Hanessian, *Total Synthesis of Natural Products: The Chiron Approach*, Pergamon Press, Oxford, **1983**; P. Collins, R. Ferrier, *Monosaccharides: Their Chemistry and Their Roles in Natural Products*, John Wiley and Sons, Chichester, England, UK, **1995**; M. Bols, *Carbohydrate Building Blocks*, Wiley-Interscience, New York, NY, USA, **1996**.

2 Stereoselective Synthesis of *C*-Disaccharides, Aza-*C*-Disaccharides and *C*-Glycosides of Carbapyranoses Using the "Naked Sugars".

P. Vogel, R. Ferritto, K. Kraehenbuehl and A. Baudat

2.1 Introduction

Evidence that oligosaccharides play vital roles in intercellular communication and cell-mediated processes has been accompanied by a surge of interest in carbohydrates and analogues [1].The binding of carbohydrate ligands by antibodies and lectins is a well recognized phenomenon for which data concerning structure-binding constant relationships are being collected [2]. The replacement of the interglycosidic oxygen atom in disaccharides by a methylene group generates a class of interesting analogues, namely the *C*-disaccharides, which constitute potential non-hydrolysable epitopes. For some cases, *C*-disaccharides [3] and *C,C*-trisaccharides [4] have been shown to imitate the physical and biological properties of the corresponding *O*-glycosides. These oligosaccharide mimics are also potential inhibitors of the biosynthetic pathway of glycoproteins. Compounds that inhibit glycosidases which are key enzymes in the biosynthesis and processing of glycoproteins [5] may find applications as antibacterial [6], antiviral [7-9], antitumoral [10] or fertility control agents [11]. Inhibitors of α-amylases and other mammalian intestinal carbohydrate-splitting enzymes have aroused medical interest in the treatment of metabolic diseases such as diabetes [12]. Inhibitors of sucrase as well as maltase may bring about a reduction in food consumption and weight gain [13].

Since the first synthesis of a β-(1→6)-*C*-disaccharide (β-D-Glc*p*-CH$_2$(1→6)-α-D-Glc*p*-OMe) by Rouzaud and Sinaÿ [14] several approaches to *C*-disaccharides and analogues have been proposed. The syntheses of *C,C*-trisaccharides have also been reported by Kishi and co-workers [4, 15] (see Table 2-1). For *C*-disaccharides involving two hexose units, one can separate them in two main classes that correspond to linear, long-chain carbohydrates on one hand, and to branched, long-chain carbohydrates on the other hand. In the first class one finds the 8,12-anhydro-7-deoxytredecose derivatives (**L16**) and the 2,6:8,12-

bis(anhydro)-7-deoxytredecitols C-disaccharide 2,6:8,12-bis(anhydro)-7-deoxytredecitols (**L11**). The second class includes 2-deoxy-2-(hexopyranosyl)methyl- (**B12**), 3-deoxy-3-(hexopyranosyl)methyl- 3-deoxy-3-(hexopyranosyl)methyl (**B13**), 4-deoxy-4-(hexopyranosyl)methyl- 4-deoxy-4-(hexopyranosyl)methyl (**B14**) and 5-deoxy-5-(hexofuranosyl)methyl-hexoses 5-deoxy-5-(hexofuranosyl)methyl (**B15**).

2.2 Synthetic Methods

The discovery of natural, bio-active long-chain carbohydrates has stirred a great deal of efforts to synthesize these compounds and analogues [16]. Among the earlier attempts to prepare long-chain sugars through ionic condensation of two hexose sub-units, the addition of the dilithium salt of 3,4-di-*O*-benzyl-5-*O*-trityl-D-arabinose trimethylenedithioacetal (**1**) to 1,2:3,4-di-*O*-isopropylidene-α-D-*galacto*-hexodialdo-1,5-pyranose (**2**) by Paulsen and co-workers [17], and the base-catalyzed addition of 2,3:5,6-di-*O*-isopropylidene-1-deoxy-1-nitro-D-*manno*-furanose (**3**) to **2** by Vasella and co-workers [18] must be mentionned. Because the generation of a non-stabilized carbanion β to an alkoxy group generally leads to the quick β-elimination of the corresponding alkoxide anion, reactions **1 + 2** and **3 + 2** represented first solutions to circumvent this problem for anomeric carbanionic sugar systems. In this context it is useful to mention the recent reports of Kessler and co-workers [18, 19] demonstrating the possibility to generate *C*-glycosides through electrophilic quenching of lithio 2-*O*-lithio-α- and -β-D-glucopyranosyl derivatives (**4, 5**). Another elegant solution to this problem is to use glycosyl samarium(III) diiodides [20, 21] as we shall see later. Alternatively sugar-derived aldehydes can be condensed with highly stabilized anomeric carbanionic species obtained by reductive cleavage (Zn/Cu or CeCl$_3$/NaI) of α-bromo ketones such as glycosyl-2-ulosyl bromide **6** and the 3,6-anhydroanalogue **7** as proposed recently by Gallagher, Lichtenthaler and co-workers [16b].

We cannot give here an exhaustive survey of the various methods proposed to generate *C*-disaccharides, however, Table 1 will help the reader to get an idea of the variety of proposals already made. A selection of *C*-disaccharide syntheses are summarized here-below that should illustrate the general principles used thus far, and will put our own work in a critical perspective.

The first synthesis of a *C*-disaccharide (Scheme 2-1) is that of Rouzaud and Sinaÿ [14] reported in 1983 which condenses the lithium alkynyl derivative **8** with the protected glucopyranolactone **9**. The mixture of α and β-1-hydroxy-*C*-glycosides **10** so-obtained was reduced first with Et$_3$SiH and BF$_3$·Et$_2$O, followed by double hydrogenation of the acetylene link. A protected form of *C*-gentobiose (β-Glc*p*-CH$_2$(1→6)-Glc*p*) was obtained. Schmidt and Preuss [35] applied a similar concept to generate *C*-(1→4) disaccharides. An alkyllithium derived from 1,6-anhydro-2,3-di-*O*-benzyl-4-deoxy-4-*C*-(iodomethyl)-β-D-galactopyranose by metal-halogen exchange was added to glucopyranolactone **9** and to the corresponding galactopyranolactone.

Table $-1. Summary of the syntheses of C-disaccharides. The C-link is CH_2 or a substituted methylene group.

Type of tredecose, dodecose		Method of chain formation		References
L11	β-D-Glcp-CH$_2$(1→1)-β-D-Glcp (β,β-C-trehalose)	Henry's reaction (7C + 6C)		Martin[22]
	α,α-, α,β-, β,β-C-trehalose	R-C≡CI(Cr^{II}) + aldehyde (9C + 4 C)		Kishi[23]
	β-D-Glcp-CH$_2$(1→1)-β-D-Manp	idem		"
	β-D-Manp-CH$_2$(1→1)-β-D-Glcp	idem		"
	1-Me-β-D-Arap-CH$_2$(1→1)-D-Ara-OH	Cationic dimerisation of 1-deoxy-1-		Nicotra[24]
	1-Me-α-D-Glcp-CH$_2$(1→1)-β-D-Glc-OH	methylidenfuranoses and pyranoses		
B12	α-D-Glcp-CH$_2$(1→2)-β-D-Ribp-OAc	Intermolecular radical glycosidation	(6C+6C)	Giese[25]
	α-D-Glcp-CH$_2$(1→2)-D-Glcp-OAc	idem	(6C+7C)	Giese[27]
	α-D-Manp-CH$_2$(1→2)-D-Glcp-OAc	idem		"
	α-L-Fucp-CH$_2$(1→2)-D-Galp-OAc	idem		"
	α-D-Glcp-CH$_2$(1→2)-L-Altp	idem		Vogel[27]
	β-D-Xylp-CO(1→2)-α-D-Glcp-OEt	Dipolar cycloaddition	(6C + 6C)	Paton[28]
	α-D-Glcp-CH$_2$(1→2)-β-D-Frup (C-sucrose)	2-alkenyl iodide + aldehyde (Cr^{II})	(9C+4C)	Kishi[29]
	α-D-Manp-CH(OH)(1→2)-α-D-Glcp-OBn	Wittig's olefination		Nicotra[30]
		Sm-glycosyl + aldehyde	(6C+7C)	Beau[20]
B13	α-D-Galp-CH$_2$(1→3)-α-D-Galp-OMe	alkenyl iodide + aldehyde	(9C+4C)	Kishi[30]
	α-D-Galp-CH$_2$(1→3)- α-D-Glcp-OMe	idem		Kishi[31]
	β-D-Galp-CH(OH)(1→3)-α-D-Galp-OMe	1-lithio-2-phenylsulfinyl glycal + aldehyde	(6C+7C)	Schmidt[31]
	β-D-Xylp-CO(1→3)- α-D-Manp-OEt	Dipolar cycloaddition	(6C+6C)	Paton[28]
	α-D-Glcp-CH$_2$(1→3)-α-L-Man-OMe	Intermolecular radical glycosidation	(6C+7C)	Vogel[27]
	α-D-Galp-CH$_2$(1→3)-α-D-Man-OMe	idem		Vogel[3h]

B14	α- & β-D-Glcp-CH$_2$(1→4)-α-D-Glcp-OMe	Wittig's or Julia's olefination	(8C+5C)	Kishi[33,34]
	α- & β-D-Manp-CH$_2$(1→4)-α-D-Glcp-OMe	idem		"
	α-D-Glcp-CH$_2$(1→4)-α-D-3-deoxy-Glcp-OMe	R-C≡Cl(Cr^{II}) + aldehyde	(9C+4C)	Kishi[34]
	β-D-Glcp-CH$_2$(1→4)-α-D-Xylp-OMe	Wittig's olefination	(7C+4C)	"
	β-D-Glcp-CH$_2$(1→4)-D-Glcp	alkyllithium + aldonolactone	(7C+6C)	Schmidt[35]
	β-D-Galp-CH$_2$(1→4)-D-Galp	idem		"
	β-D-Galp-CH$_2$(1→4)-D-Glcp (C-lactose)	1-lithio-2-phenylsulfinylglycal+aldehyde	(6C+7C)	Schmidt[36]
	α-D-Glcp-CH$_2$(1→4)-L-Man	Intermolecular radical glycosidation	(6C+7C)	Vogel[27]
	α-D-Galp-CH$_2$(1→4)-D-Man	idem		Vogel[3h]
	α-D-Manp-CH$_2$(1→4)-α-D-Glcp-OMe	Intramolecular radical glycosidation		Sinaÿ[37]
	α-D-Glcp-CH$_2$(1→4)-α-D-Glcp-OMe	idem		Sinaÿ[38,39]
	α- & β-D-Galp-CH$_2$(1→4)-α-D-Glcp-OMe	idem		Sinaÿ[40]
B15	α-D-Glcp-CH$_2$(1→5)-L-TalUf	Intermolecular radical glycosidation	(6C+7C)	Vogel[27]
L16	β-D-Glcp-CH$_2$(1→6)-D-Glcp	R-C≡CLi + aldonolactone	(7C+6C)	Sinaÿ[14]
	α-D-Glcp-CH$_2$(1→6)-D-Glcp (C-isomaltose)	CH$_2$=C(R)I(Cr^{II}) + aldehyde	(9C+4C)	Kishi[41]
	β-D-Glcp-CH$_2$(1→6)-D-Glcp (C-gentiobiose)	idem		"
	β-D-Glcp-CH$_2$(1→6)-α-D-Galp	Henry's condensation	(7C+6C)	Martin[22]
	β-D-Glcp-CH$_2$(1→6)-α-D-Galp	S$_{RN}$1 by nitronate anion +		Martin[42]
	β-D-Galp-CH$_2$(1→6)-α-D-Galp	6-glycosylcobaloxime	(7C+6C)	"
	β-D-Xylp-CH(OH)(1→6)-D-Glc	Dipolar cycloaddition	(6C+6C)	Paton[43]
	α-D-Glcp-CH$_2$(1→6)-D-Glcp-OMe	Intramolecular radical addition	(6C+7C)	Beau[44]
	α-D-GulfCF$_2$(1→6)-D-Glcp	Radical addition	(7C+6C)	Motherwell[45]
	α-D-Glcp-CH$_2$(1→6)-D-Gal, L-Gal	Pyranosyl chain elongation, Wittig olefina-		Armstrong[46]
	α-D-Glcp-CH$_2$(1→6)-D-Ido, L-Ido	tion, quadruple hydroxylation of diene		

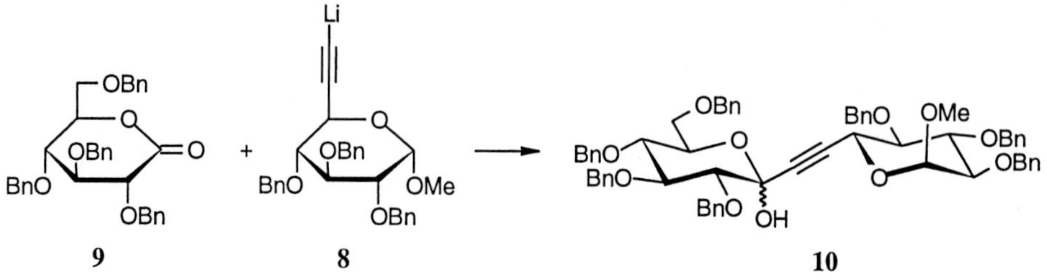

Scheme 2.1

Kishi and his group have prepared a large variety of C-disaccharides including branched-chain systems through the Wittig olefination [33] as exemplified with the reaction **11** + **12** (Scheme 2-2). The (Z)-alkene so-obtained (**13**) can be doubly hydroxylated and converted, after several steps, into the C(1→4) disaccharides **14-17**.

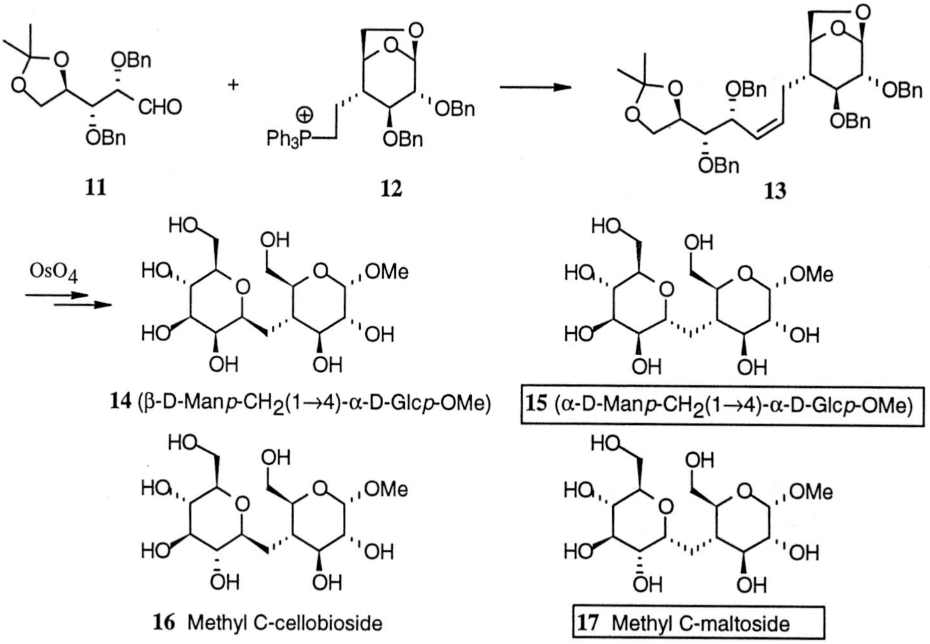

14 (β-D-Man$p$-CH$_2$(1→4)-α-D-Glc$p$-OMe)
15 (α-D-Man$p$-CH$_2$(1→4)-α-D-Glc$p$-OMe)
16 Methyl C-cellobioside
17 Methyl C-maltoside

Scheme 2-2

The same group investigated an alternative route using the nickel(II)/chromium(II) mediated coupling of iodoalkyne [23] or 2-iodoalkene [29, 31] derivatives of pyranosides with protected tetroses. Their synthesis of C-sucrose is out-lined in Scheme 2-3 as an illustration of the method.

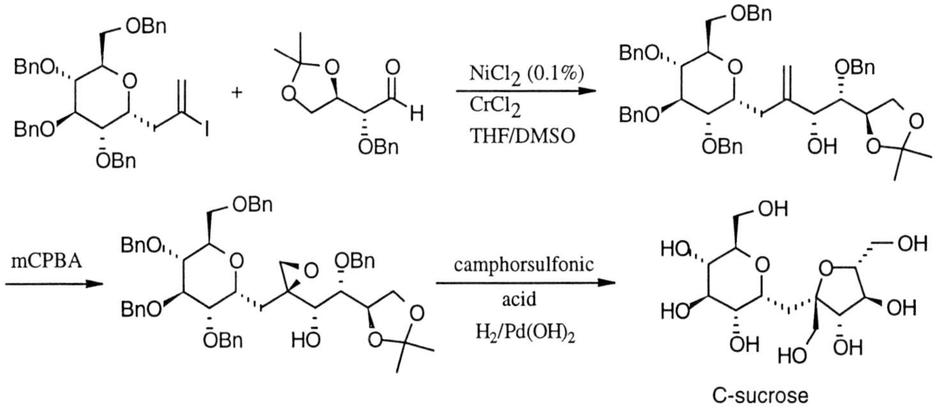

Scheme 2-3

Schmidt and co-workers [32, 36] have used heteroatom-stabilized 1-C-lithiated glycals as nucleophiles and C-formyl derivatives of sugars to construct a variety of C(1→3) and C(1→4) disaccharides. Their synthesis of C-lactose [36] is summarized in Scheme 2-4.

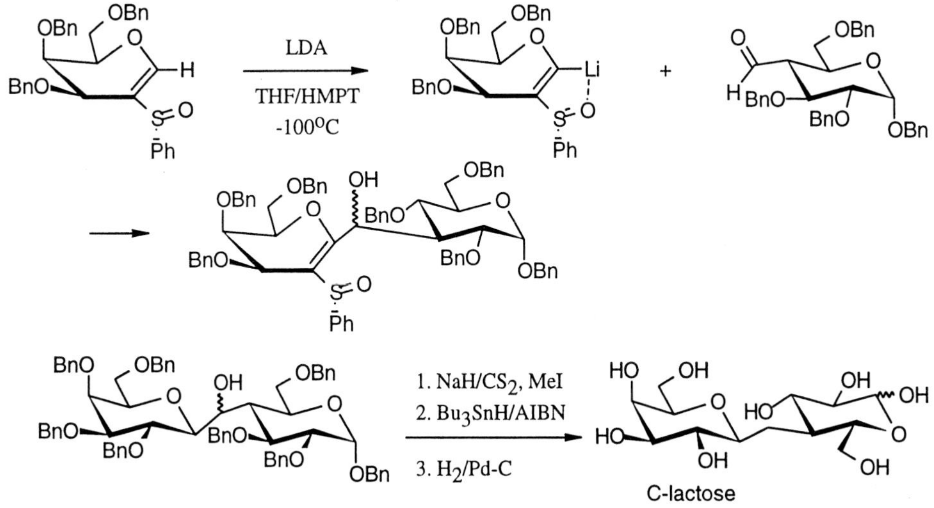

Scheme 2-4

Martin and co-workers [22] have applied the Henry's condensation of 1-deoxy-1-nitromethylpyranoses to sugar derived aldehydes to prepare a variety of C-disaccharides in a short and efficient way. For the moment the method is limited to the synthesis of linear β-C(1→6) and β-C(1→1) disaccharides. The same group [42] has also shown that β-C(1→6) disaccharides can be obtained through $S_{RN}1$ reactions of nitronate anions with sugar derived alkylcobaloxime as illustrated below by the sequence **18 + 19 → 20**.

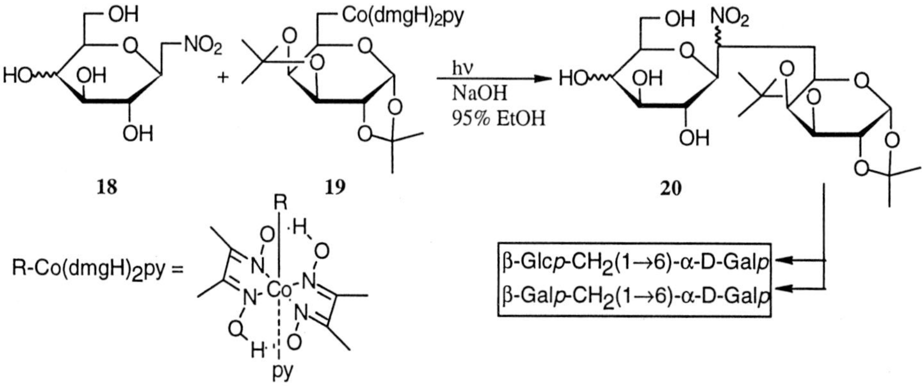

Scheme 2-5

Most of the C-C bond forming processes using ionic intermediate require adequate protection of the sugar derived sub-units that are combined during the synthesis of a *C*-disaccharide. In principle, pericyclic reactions have less severe requirements since they can be carried out under neutral, thermal conditions. Paton and co-workers have used the dipolar-1,3 cycloadditions of pyranose derived nitrile-oxides to unsaturated sugars in which the hydroxy groups are protected as acetates to generate β-CO(1→2), β-CO(1→3) (Scheme 2-6) [28] and β-CH(OH)(1→6) disaccharides [43]. Reductive hydrolysis of the adducts (isooxazolines) generates the corresponding *C*-disaccharides. Whereas the face selectivity can be controlled by the substituents of the eno-pyranoside, the regioselectivity of the cycloaddition is generally low. However, in the case of vinyl substituted sugars, the regioselectivity of their cycloadditions to carbonitrile oxides is good whereas the face selectivity might be low [43].

Provided that stereocontrol (double stereodifferentiation) can be achieved, radical addition to alkenes is a very smooth approach to the generation of C-C bonds which, usually, tolerates a great variety of functionalities.

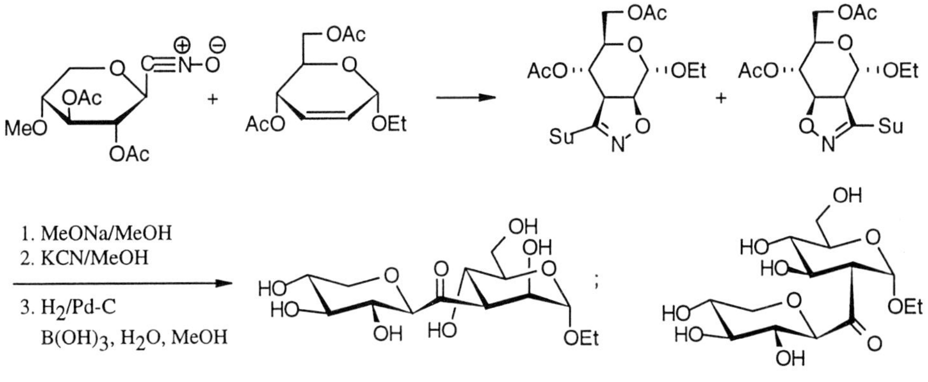

Scheme 2-6

In 1986, Giese and Witzel [25] showed that the D-glucopyranosyl radical generated by bromide abstraction from the commercially available 2,3,4,6-tetra-*O*-acetyl-α-D-glucopyranosyl bromide **21** adds to the unprotected 2-deoxy-2-methylidene-D-*erythro*pyranolactone (**22**) giving a mixture (81%) of the four possible C(1→2) disaccharides **23**. The α:β ratio for the *C*-glucopyranosyl moiety was 10:1 and the 2-deoxy-2-alkylaldonolactone isomeric ratio was 3:2. Somewhat better diastereoselectivities in radical glycosidations of 2-methylidene-aldonolactones bearing benzyl protective groups, obtained from the corresponding C-2-lithiated glycals, were observed [26].

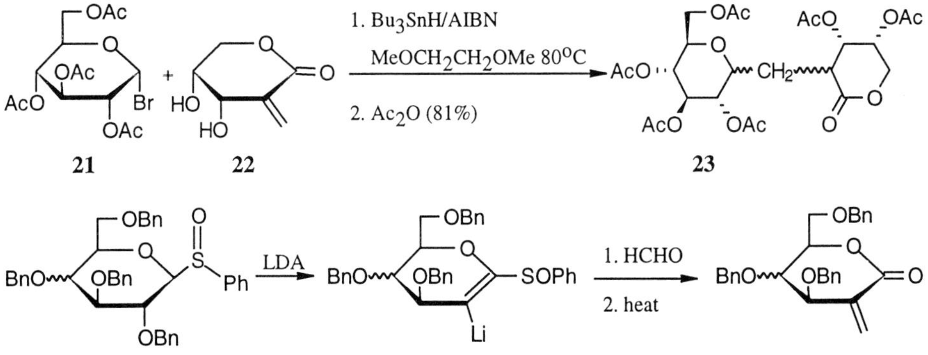

Scheme 2-7

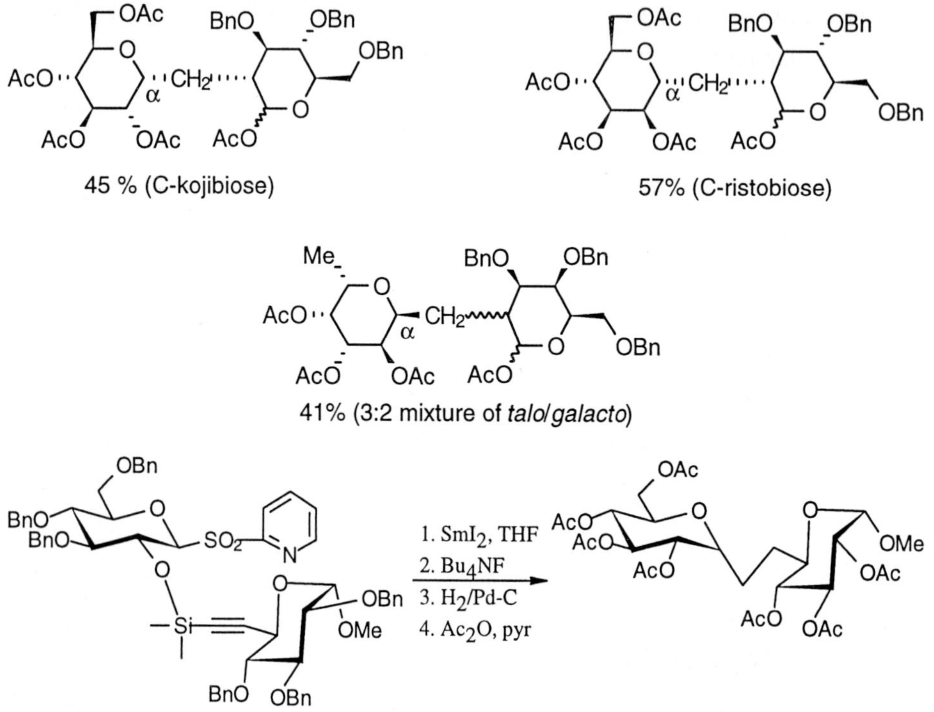
45 % (C-kojibiose)
57% (C-ristobiose)
41% (3:2 mixture of talo/galacto)
1. SmI2, THF
2. Bu4NF
3. H2/Pd-C
4. Ac2O, pyr

Scheme 2-8

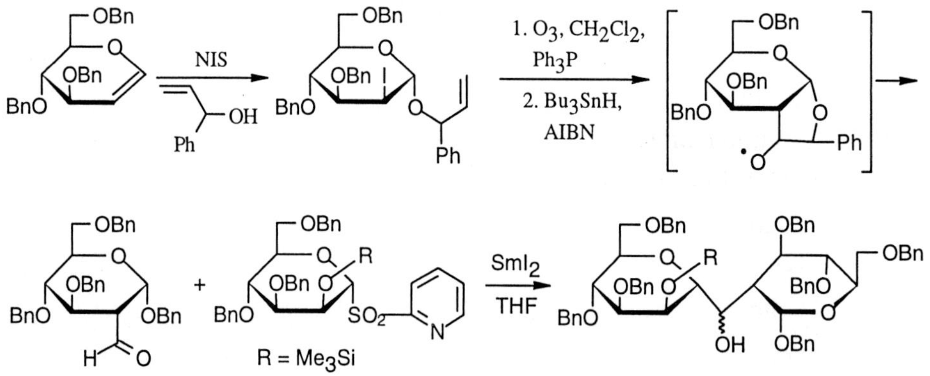
NIS
1. O3, CH2Cl2, Ph3P
2. Bu3SnH, AIBN
R = Me3Si
SmI2
THF

Scheme 2-9

Recently, Sinaÿ and co-workers [47] showed that non-electrophilic methylidenepyranosides can add pyranosyl radicals provided they are temporarily connected through a ketal [37] or a silaketal tether [38-40]. Contrary to the intermolecular radical glycosidation of Giese which often requires the use of a large excess of either the glycosyl precursor or of the electrophilic alkene, this method condenses two carbohydrate sub-units in a 1:1 ratio. The method is expeditious provided that the methylidenepyranosyl derivative and the glycosyl radical precursor are readily available and can be tethered in a few steps. The diastereoselectivity and the yields of the C-C bond forming process can be good; they depend on the nature of the sugars involved and of the tether. Beau and co-workers have applied a similar intramolecular technique to generate a α-C(1→6) [44] disaccharide (Scheme 2-8). The method implies the generation of a (glycosyl)SmI$_2$ intermediate which adds intramolecularly onto an acetylenic moiety. An intermolecular version (Scheme 2-9) in which the (glycosyl)SmI$_2$ intermediate adds to a 2-*C*-formyl-2-deoxy-α-D-glucopyranosyl derivative giving a 2:1 mixture of (*R*) and (*S*) CH(OH)-bridged disaccharides has been described recently by the same group [21].

2.3 The "Naked Sugar" Approach

Optically pure 7-oxabicyclo[2.2.1]hept-5-en-2-yl derivatives (+)-**24**, (+)-**25**, (+)-**26** and (-)-**26** ("naked sugars of the first generation") are obtained readily through the Diels-Alder additions of furan to optically pure 1-cyanovinyl esters [48]. Because of their rigid, bicyclic skeleton, their olefinic moieties, as well as other π systems derived from them (e.g.: ketones, enolates, carbon-centered radicals) are highly *exo* face selective. Furthermore, the regioselectivity of the double substitution of C(5) and C(6) can be controlled by the nature of the functions at C(2) [49]. Polysubstituted 7-oxabicyclo[2.2.1]heptan-2-ones are thus readily obtained with high and predictable stereoselectivity.

These systems can be converted in one step through Baeyer-Villiger oxidations into the corresponding 1,4-anhydrouronolactones (therefore the coinage "naked sugars" for (+)-**24**, (+)-**25**, (+)-**26** and (-)-**26**) [50]), or cleaved into 2,5-anhydrohexonic acids which are direct precursors of pentoses and *C*-nucleosides [51].

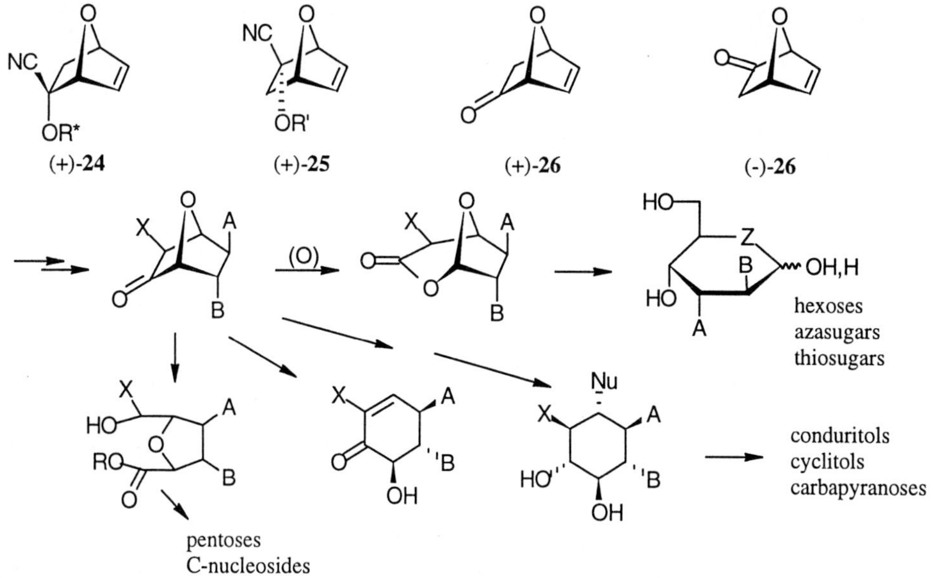

Scheme 2-10

Base-induced [52] or nucleophilic ring opening [53] of the ethereal bridge generates polysubstituted cyclohexene or cyclohexane derivatives, direct precursors of conduritols, cyclitols or carbapyranoses (Scheme 2-10). In all these total, asymmetric syntheses the chiral auxiliaries ((1S)- or (1R)-camphanic acid [48a], SADO(Et)-OH or RADO(Et)-OH [48b]) are recovered at a very early stage of the processes. Furthermore, the two enantiomers of a given target compound can both be obtained enantiomerically pure with the same ease.

The same methods applied to products of Giese's glycosylation of 3-methylidene-7-oxabicyclo[2.2.1]heptan-2-ones or to products of cross-aldolizations of 7-oxabicyclo[2.2.1]heptan-2-ones with sugar-derived aldehydes have allowed us to prepare a variety of α-C(1→2), α-C(1→3), α-C(1→4), α-C(1→5) disaccharides [27, 34], the first example of an aza-C(1→3) disaccharide [54] and the first examples of C-pyranosides of carbapentopyranoses [55] and of aminoconduritols [56].

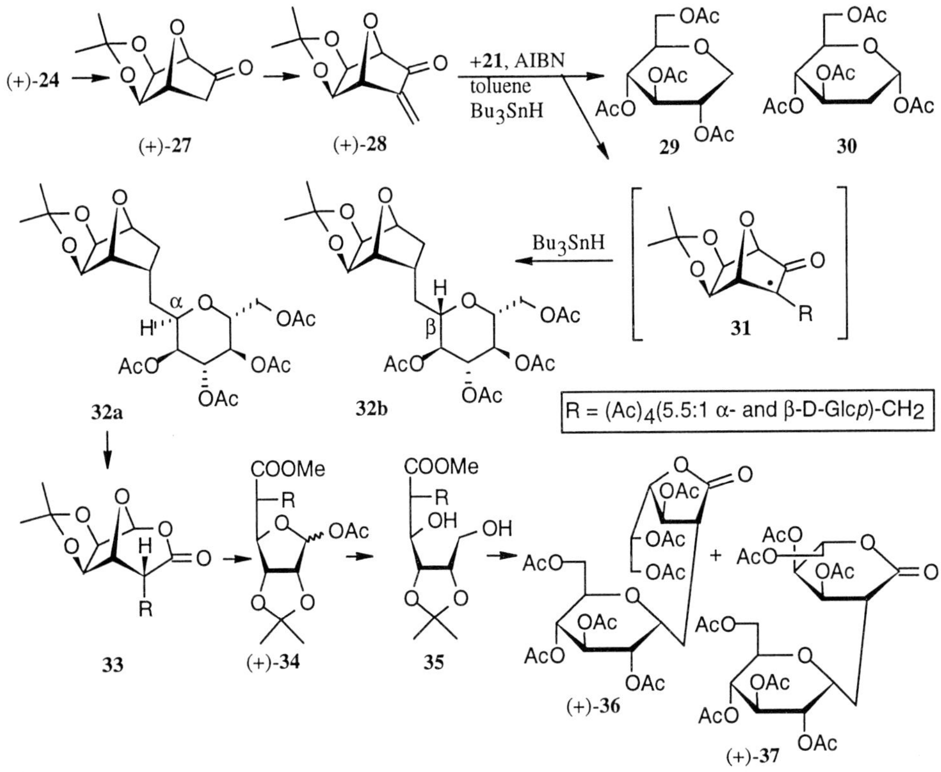

Scheme 2-11

Ketone (+)-**27** derived from (+)-**24** reacted with monomeric formaldehyde and N-methylanilinium trifluoroacetate to give the α-methyleneketone (+)-**28** which, under the conditions recommended by Giese and co-workers [25, 26, 57], was glucosylated with α-acetobromoglucose (**21**, 1.3 equivalent), yielding a mixture of the deoxy-D-glucose derivatives **29** and **30** and a 5.5:1 mixture of the α and β-C-glucopyranoside **32α** and **32β** (68%). No trace of the isomeric *exo*-C-glucosides could be seen, thus demonstrating the high *exo*-face selectivity for the reduction of the intermediate radical **31** by Bu$_3$SnH. When (Me$_3$Si)$_3$SiH was used instead of Bu$_3$SnH as hydrogen donor, the yield in **32α** + **32β** was only 28%. Baeyer-Villiger oxidation (mCPBA/NaHCO$_3$) of **32α** + **32β** was highly regioselective [58] giving the uronolactones **33** (96.5%). Alcaline methanolysis of **33**, followed by acetylation provided (+)-**34** (90%). When the methanolysis was followed by reduction with NaBH$_4$, the diol **35** was formed, which was converted (HCl, then acetylation) into the C(1→2) disaccharide (+)-**36** and (+)-

37, separated by flash chromatography and isolated in 36 and 31% yield respectively. These experiments showed that the Giese's *C*-glycosidation of (+)-**28** can be used to generate in a few synthetic steps a 5.5:1 mixture of α- and β-D-Glc*p*-CH$_2$(1→5)-α,β-L-(*talo*-hexofuranosyl)uronate derivatives, rare examples of C(1→5) disaccharides, and α-D-Glc*p*-CH$_2$(1→2)-L-*altro*-hexono-1,4- and -1,5-lactones derivatives. The same approach has been applied to the synthesis of α-D-(1→3) and α-D-(1→4)-*C*-linked glucopyranosides of L-mannose (Scheme 2-12) [27] and of α-D-(1→3) and α-D-(1→4)-*C*-linked galactopyranosides of D-mannose derivatives (Scheme 2-13) [34].

Under kinetic control, benzeneselenyl chloride adds to enone (+)-**26** to give a single adduct (+)-**38**. The high regioselectivity of this electrophilic addition was attributed to the electron-releasing ability of the homoconjugated carbonyl moiety (n(CO) ↔ σC(2),C(1) ↔ pC(6) interactions [59]). Treatment of the lithium enolate of (+)-**38** with the Eschenmoser's salt (H$_2$C=NMe$_2$I) afforded the methyleneketone (-)-**39** (84%). Its radical glycosidation with **21** led to a mixture of **29**, **30** and *C*-glucosides from which (+)-**40** (β-*C*-glucopyranoside) and (+)-**41** (α-*C*-glucopyranoside) were isolated in 6 and 48.5% yield, respectively. In this case the α/β anomer selectivity was 8:1. Products resulting from the reduction of the benzeneselenyl or/and chloride function were not detected when the excess of Bu$_3$SnH was carefully controlled. In this case also, only 3-*endo*-alkyl-7-oxabicyclo[2.2.1]heptan-2-ones were formed pointing out again the high *exo* face selectivity of the hydrogen atom transfer from Bu$_3$SnH to the 2-oxoalkyl radical intermediate. Reduction of ketone (+)-**41** with NaBH$_4$ furnished the corresponding *endo* alcohol (96%). Oxidative elimination of the PhSe group with mCPBA followed by acetylation gave (+)-**43** (92%). Double hydroxylation of (+)-**43** gave α-hydroxy ketone which was acetylated into (+)-**44** (94%). Baeyer-Villiger oxidation (mCPBA/NaHCO$_3$) provided the uronolactone (+)-**45** (94%). Alkaline methanolysis gave the α-D-Glc*p*-CH$_2$(1→3)-*manno*-furanuronate **46** the reduction of which with NaBH$_4$ (MeOH) yielded a polyol. Treatment of this polyol with HCl followed by acetylation provided (+)-**47** (62%). Under acidic conditions, the methanolysis of (+)-**45** (MeOH/SOCl$_2$) followed by reacetylation afforded (+)-**48** (28%) and (+)-**49** (27%). LiAlH$_4$ reductions of (+)-**48** and (+)-**49**, followed by reacetylation provided the C(1→3) disaccharides (+)-**50** and (+)-**51**, respectively. Thus the same method of *C*-glycosidation has allowed the preparation of α-D-Glcp-CH$_2$(1→4) and α-D-Glcp-CH$_2$(1→3)-L-mannose derivatives.

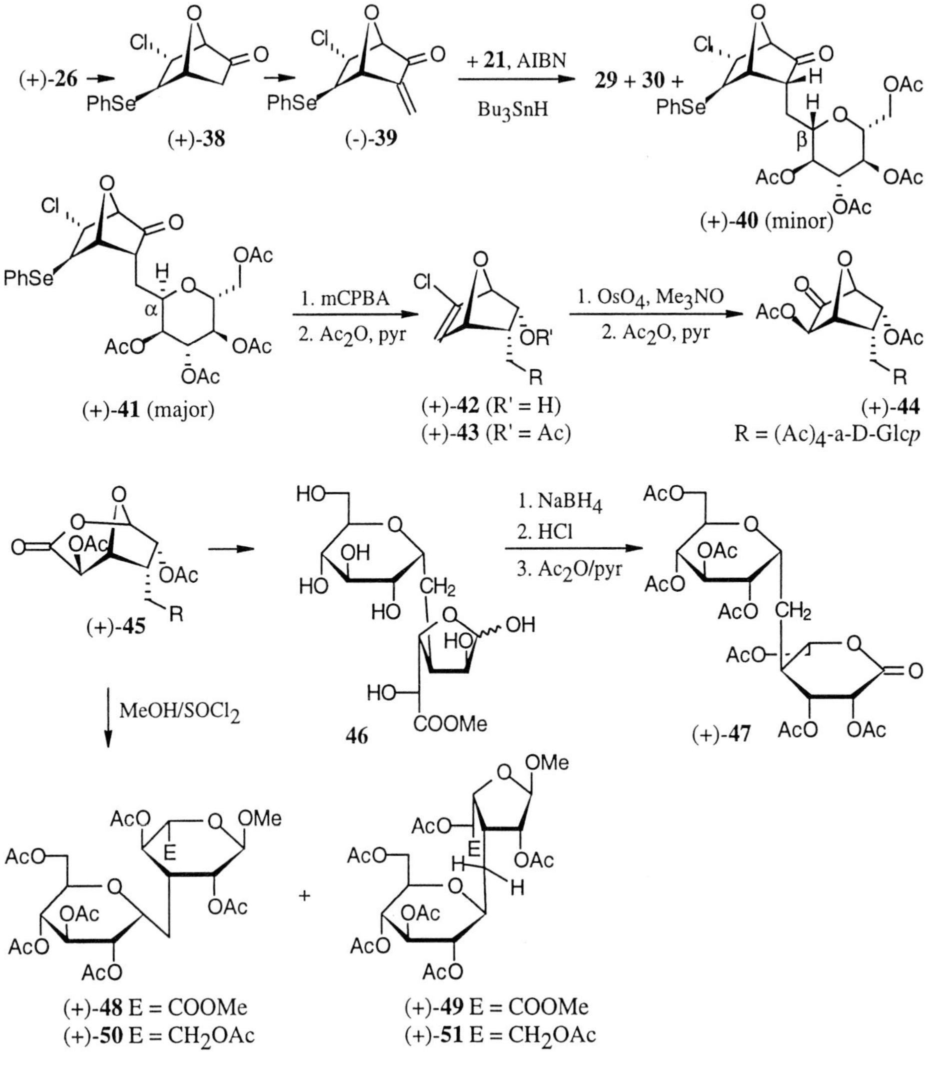

Scheme 2-12

The same approach has been used to prepare a variety of α-C(1→2), α-C(1→3), α-C(1→4) and α-C(1→5) galactosides [3h] and mannosides [60] using the commercially available α-acetobromo-D-galactose and α-acetobromo-D-mannose, respectively, as *C*-pyranosyl precursors. In the case of the Giese's radical *C*-galactosidation, the reaction was carried out with the racemic mixture (±)-**39** giving a 65% yield of a 1:1 mixture of the separable α-galactosides (+)-**52** and (+)-**53** (Scheme 2-13). In this case the α:β ratio was better than 50:1 under optimal conditions (benzene solution 0.5 M in Bu$_3$SnH, 0.38 M in (±)-**39** and 0.006 M in AIBN was added in 90 min to a boiling 0.5 M solution of α-acetobromogalactose in benzene). On lowering the concentration of the Bu$_3$SnH solution, the yield of *C*-galactosidation decreased (58%, 0.2 M in PhH) and the β-*C*-galactosides were formed concurrently (2-4%) with the α-*C*-galactosides! When (Me$_3$Si)$_3$SiH was used instead of Bu$_3$SnH as hydrogen atom donor, the yield in (+)-**52** and (+)-**53** did not exceed 33% (75% conversion of enone (±)-**39**). For unclear reasons, *C*-galactosidation of dienone (±)-**54** led to a 1:1 mixture of the α-galactosides **55** and **56** in a mediocre yield (16%). According to the sequence described in Scheme 2-12, (+)-**53** was transformed into the new α-D-(1→3)-*C*-linked galactosides of D-mannose **57** and **58**, on one hand, and the α-D-(1→4)-*C*-linked galactoside of D-mannono-1,5-lactone **59**, on the other hand.

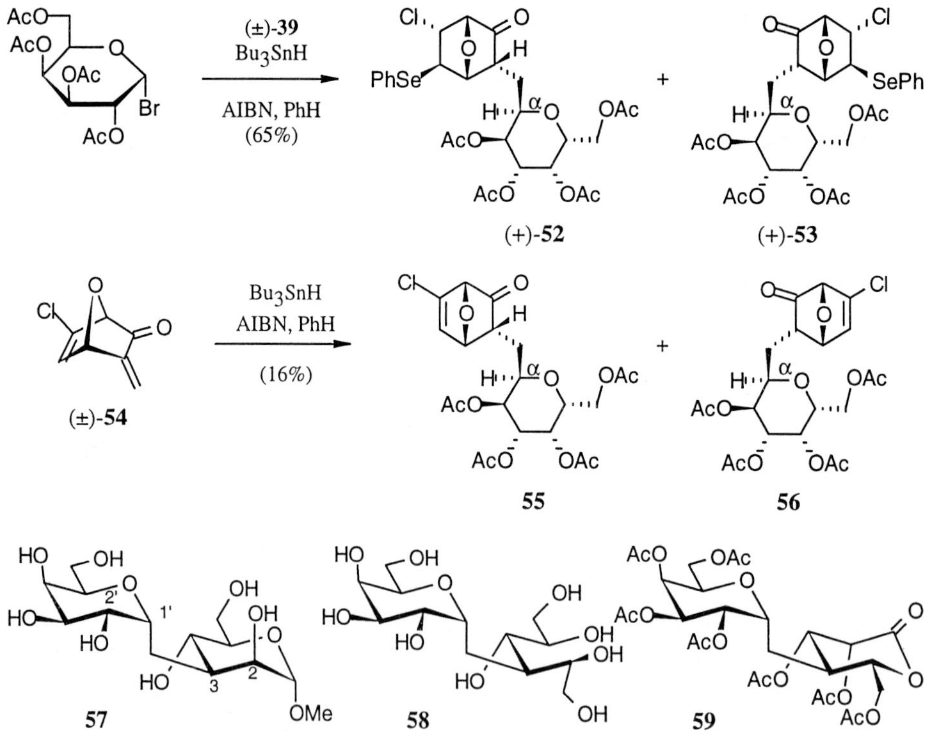

Scheme 2-13

The ^{1}H-NMR spectra of the α-C-galactosides confirmed that their preferred conformations involve antiperiplanar arrangements for the C-linked substrates and bond σC(1'),C(2') of the galactoside unit. (+)-Methyl 3-deoxy-3-[(α-D-galactopyranosyl)methyl]-α-D-mannopyranoside (**57**) adopts a conformation similar to that proposed for methyl 3-O-(α-D-galactopyranosyl)-α-D-mannopyranoside [61].

2.3 C-Glycosides of Carbapyranoses and Analogues

Several antibiotics and compounds of biological interest incorporate glycosides of cyclohexane polyols [62]. Some cyclohexanepolyols have been called sugars [63] or carba-sugars [64]. The replacement of the interglycosidic oxygen atom in a glycoside by a methylene moiety generates the corresponding deoxy-(glycosylmethyl) analogue which may mimic the physical and biological properties of the corresponding O-glycoside but should be inert towards acidic and enzymatic hydrolysis. Our "naked sugar" approach has allowed one to generate the first representatives **60, 61, 62** of C-glycosides of carbasugars, a new class of disaccharide mimics [55, 65]. Meanwhile, Bols and co-workers [66] have reported the synthesis of analogue **63**.

The "naked sugar" (+)-**26** was treated with PhSeCl in MeOH to give an addition product. This ketone was treated with (Me$_3$Si)$_2$NLi in THF (-60°) to generate the lithium enolate which reacted with the Eschenmoser's salt yielding enone (-)-**64** (75%) (Scheme 2-14). Radical glycosidation of (-)-**64** with α-acetobromogalactose gave (+)-**65** in 74% yield. Oxidation of the selenide (+)-**65** with one equivalent of mCPBA at low temperature (-78°C, THF) followed by treatment with Ac$_2$O/AcONa led to (+)-**66** (82%) via a seleno-Pummerer rearrangement [67]. Deselenation of (+)-**66** with Bu$_3$SnH (AIBN, PhH, 80°C) gave (+)-**67** (97%) stereoselectively. As for the reaction (-)-**64** → (+)-**65**, quenching of the intermediate bicyclic radical was highly *exo* face selective. This sequence of reactions gave 7-oxabicyclo[2.2.1]heptan-2-one derivatives with two *endo* hydroxy groups at C(5) and C(6) bearing orthogonal protective groups [68].

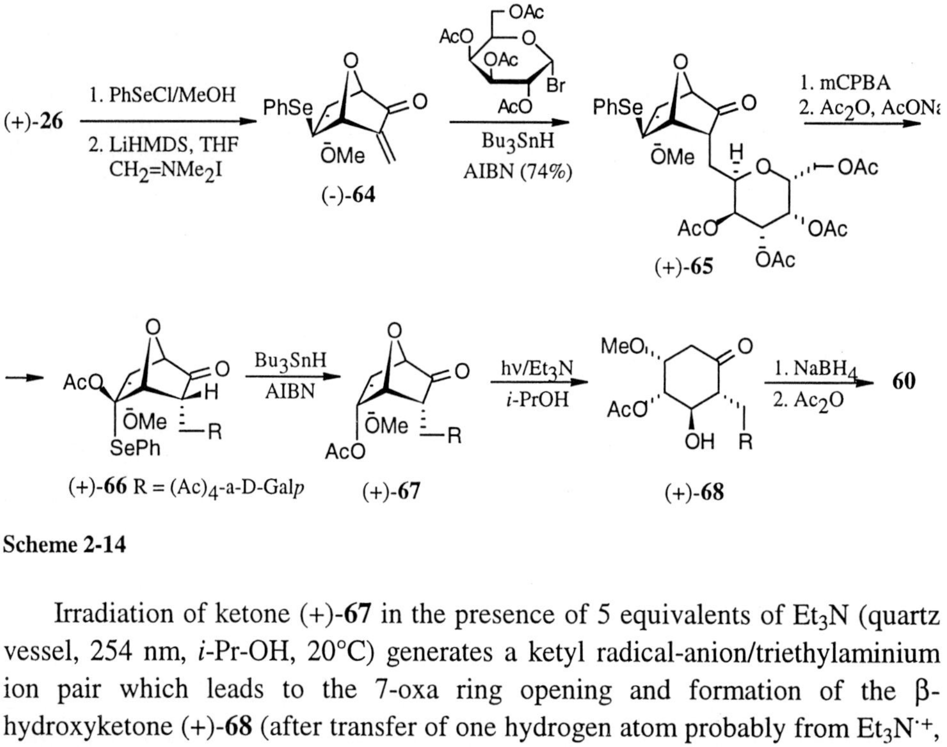

Scheme 2-14

Irradiation of ketone (+)-**67** in the presence of 5 equivalents of Et$_3$N (quartz vessel, 254 nm, *i*-Pr-OH, 20°C) generates a ketyl radical-anion/triethylaminium ion pair which leads to the 7-oxa ring opening and formation of the β-hydroxyketone (+)-**68** (after transfer of one hydrogen atom probably from Et$_3$N$^{\cdot+}$, which is converted into Et$_2$N$^+$=CHMe, and protonation) [69]. Reduction of ketone (+)-**68** with NaBH$_4$/MeOH, followed by acetylation provided **60** (46%) which can be regarded as the dicarba analogue of *O*-α-D-galactopyranosyl-(1→2)-α-D-(xylo-pentodialdo-1,5-pyranose-6-α-methyl acetal) or of *O*-α-D-galactopyranosyl-(1→2)-α-L-(methyl xylopentodialdo-1,5-pyranoside). The same sequence of reactions was performed on the racemic enone (±)-**64** and gave a separable mixture of *C*-galactosidation products. Ketone (+)-**69** so obtained was reduced with NaBH$_4$ (MeOH, 0°C) into a 2.5:1 mixture of alcohols that were acetylated and separated by flash chromatography on silica gel in 40% and 16%

yield, respectively. ¹H-NMR spectra (400 MHz, NOESY and COSY-DQF) confirmed the average conformations shown for **60**, **61** and **62**.

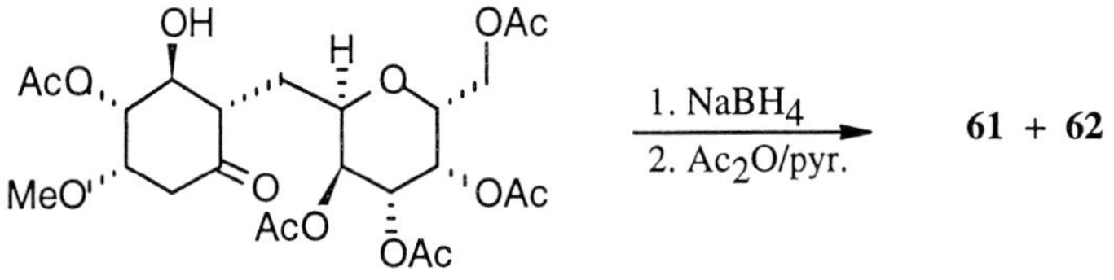

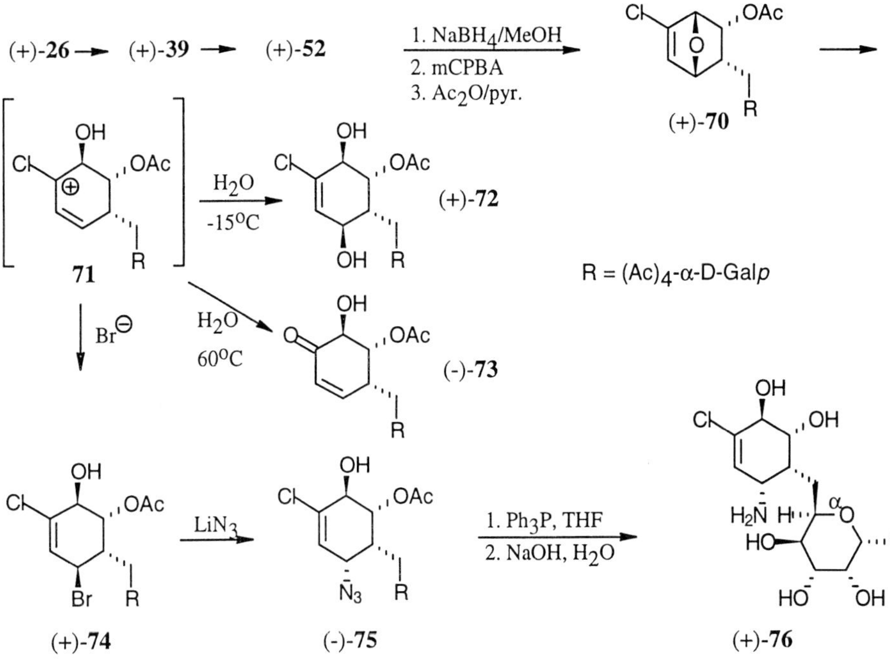

Scheme 2-15

Reduction of the α-*C*-galactoside (+)-**52** with NaBH₄/MeOH gave the corresponding *endo* alcohol which underwent oxidative elimination of selenium on treatment with mCPBA. After acetylation the chloroalkene (+)-**70** was isolated (Scheme 2-15). Under acidic conditions, the 7-oxanorbornene (+)-**70** was ring regioselectively opened into the allylic cationic intermediate **71** which could be reacted with various nucleophiles. For instance, when (+)-**70** was treated with aqueous 70% HClO₄ in CF₃CH(OH)CF₃ (HFIP) at -15°C (18 h), the allyl diol (+)-**72** was formed. The same reaction performed in boiling HFIP/isopropanol 10:1 (60°C, 5 min) led to enone (-)-**73** (65%). Treatment of (+)-**70** with 30% HBr in anhydrous AcOH (20°C, 6h) furnished the bromide (+)-**74** (90%). The high stereoselectivity of the nucleophilic quenchings **71** + H_2O and **71** + Br^- can be attributed to a steric factor, the face *anti* with respect to the bulky (α-D-galactopyranosyl)methyl substituent being preferred by the nucleophiles. In the presence of an excess of LiN_3 in DMF, the bromide (+)-**74** underwent smooth S_N2 displacement (20°C, 15 h) and provided azide (-)-**75** (96%). Reduction of the azide (-)-**75** with Ph_3P in THF (20°C, 8 h), followed by saponification with 5% aqueous NaOH (20°C) and acidification with aqueous 0.7 M HCl gave the hydrobromide of amine (+)-**76**. Filtration on a DOWEX 50W 8X, H^+ column and elution with aqueous NH_3 provided the aminoconduritol α-*C*-galactopyranoside (+)-**76** in 66% yield.

2.4 Branched-Chain Carbohydrates through Cross-Aldolisations

The lithium enolate of 7-oxabicyclo[2.2.1]heptan-2-one (+)-**38** reacted with furfural at -78°C to yield a single aldol **77** isolated in 95% yield (Scheme 2-15). As expected for steric reasons (Zimmerman-Traxler model [70], *like* mode of addition, *exo* face selectivity of the enolate) the *anti* aldol was obtained, the relative configuration of which was proven by the ^{1}H-NMR spectrum of derivative **78** [71] Aldol **77** was protected as the silyl ether and the ketone was reduced with $NaBH_4$ (1:1 THF/MeOH) to give **79** (88%). Oxidation with mCPBA led to the corresponding chloroalkenol, the alcoholic moiety of which was protected as a MOM ether **80** (78%). Double hydroxylation of the chloroalkene **80** with $Me_3N \rightarrow O/NaHCO_3/OsO_4$ gave an α-hydroxyketone which was tosylated into **81** (72%). Baeyer-Villiger oxidation of ketone **81** with $mCPBA/NaHCO_3$ furnished uronolactone **82** (77%). Treatment of tosylate **82** with $NaHCO_3$ in anhydrous MeOH led to a 2:1 mixture of the α- and β-furanose **83** and **84**. With MeONa in THF (-78°C) the **83/84** ratio was 9:1. When treated with anhydrous $K_2CO_3/MeOH/DMF$ (20°C) **82** was converted into a 85:15 mixture of anhydrogalacto and anhydroaltrouronate **85** and **86**, respectively, (69%) from which pure **85** was isolated in 55% yield. Ester **86** could not be equilibrated with **85** in the presence of $K_2CO_3/MeOH/DMF$, thus confirming that epimerization had occurred at C(5) of tosylate **82**. Reduction of methyl uronate **85** with $LiAlH_4$ gave the corresponding alcohol which was protected as di-MOM ether **87** (71%).

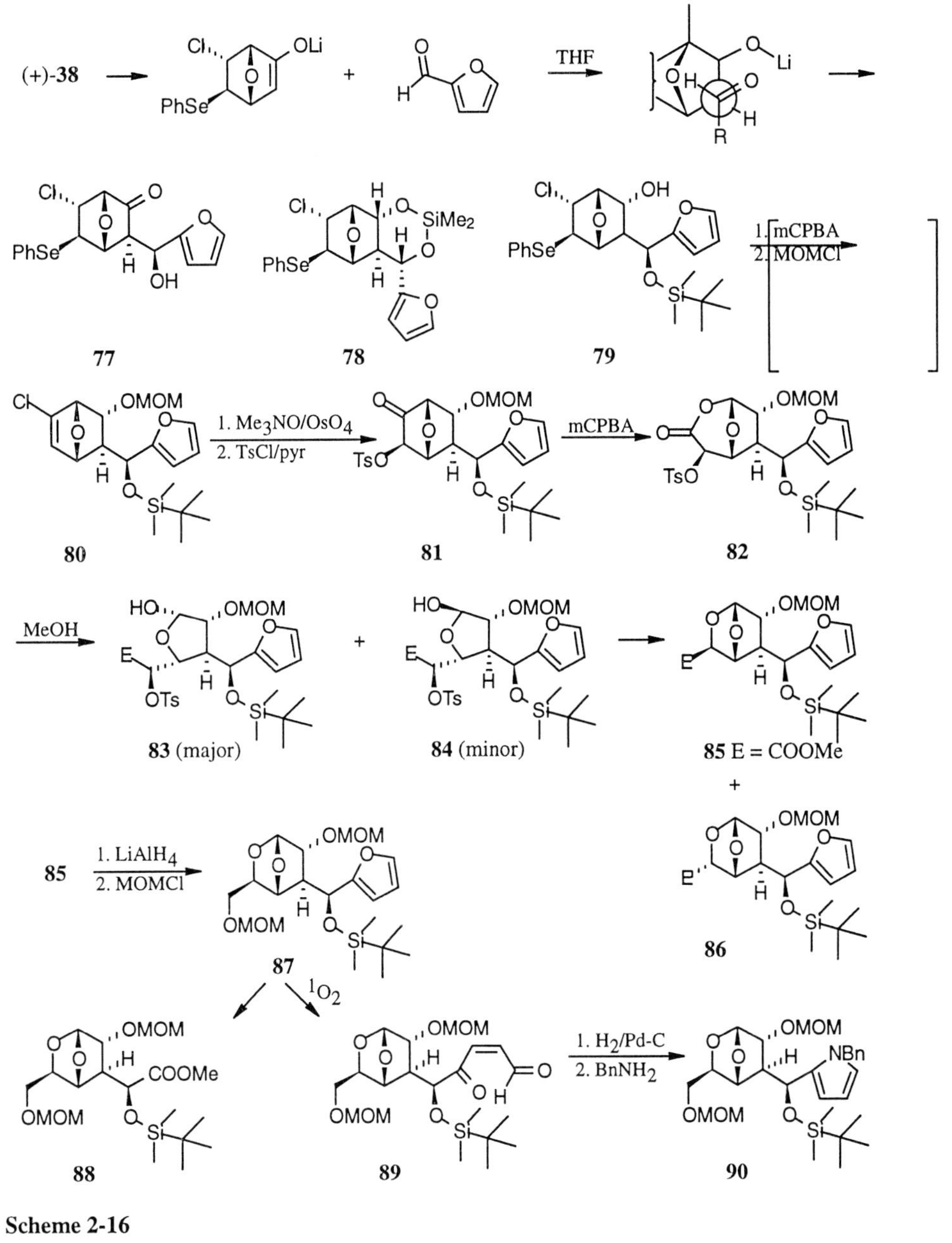

Scheme 2-16

Oxidative cleavage of the furan ring in **87** (RuO₄/NaIO₄), followed by esterification gave the branched-chain carbohydrate **88** (80%). Photooxidation of **87** led to **89** which was hydrogenated and treated with BnNH₂ to give the pyrrole **90**. This work illustrates the possibility to convert our "naked sugars" into 1,4-

anhydro-galactopyranose derivatives. Compounds **88, 89** and **90** are potential precursors for the synthesis of all kinds of disaccharide mimics.

2.5 Synthesis of an Aza-C(1→3)-Disaccharide

The cross-aldolisation of a 2,6,7-trideoxy-2,6-imino-D-*glycero*-L-mannose and a 7-oxabicyclo[2.2.1]heptan-2-one derivative has been used to generate the first example of an aza-C(1→3)-disaccharide. Such disaccharide mimics are azasugars linked to other sugars through non-hydrolysable links; they are potential glycohydrolase inhibitors that are thought to be more selective than simple azasugars (polyhydroxypiperidines and polyhydroxypyrrolidines [72]) because they contain the charge and steric information of the glycosyl moieties, on one hand, and the steric information of the aglycones which they are attached to, on the other hand. A first example of aza-C(1→6)-disaccharide (1,5-dideoxy-1,5-imino-D-mannitol linked at C(6) of D-galactose through a CH_2 unit) has been prepared by Johnson and co-workers [73].

The readily available 2,3:6,7-di-*O*-isopropylidene-D-*glycero*-D-*gulo*-heptono-1,4-lactone ((-)-**91**) [74] was converted into the azidooctulose derivative (+)-**92** (63%) through esterification with $(CF_3SO_2)_2O$/pyridine, displacement of the corresponding triflate with LiN_3 and addition of MeLi (Scheme 2-18). Hydrogenation of (+)-**92** gave the corresponding amine which equilibrated with the corresponding imine resulting from the intramolecular addition onto the ulose moiety. The latter was hydrogenated selectively to give the semi-protected imino-octitol (+)-**93** (95%). For steric reasons, the face *anti* with respect to the neighbouring acetonide was preferred for the hydrogenation. Acidic hydrolysis of (+)-**93** provided the unprotected azasugar (+)-**94**.

Although the piperidine unit of (+)-**94** has the same absolute configuration as β-D-*galacto*-hexopyranosides, neither the *Aspergillus niger* and *Escherichia coli* α-galactosidases, nor the coffee beans, *Aspergillus niger*, *Escherichia coli*, bovine liver and *Asperigillus orizae* β-galactosidases were inhibited by this azasugar. A weak inhibiting activity was detected, however, for the α-galactosidase from coffee bean (34% inhibition at 1 mM concentration of (+)-**94**) and for β-glucosidase from almond (IC_{50} = 98 μM, K_i = 15 μM, pH 4.5, 37°C) and from *Caldocellum saccharolyticum* (IC_{50} = 107 μM, K_i = 41 μM, pH 4.5, 37°C) [75]. Benzylation of (+)-**93** gave (+)-**95**. Protection of the amino group of (+)-**95** with benzyl chloroformate (97%) followed by treatment with 8:1 $AcOH/H_2O$ and $NaIO_4$ (20°C, 15 h) led to selective hydrolysis of the 7,8-*O*-isopropylidine group and oxidative cleavage into aldehyde (-)-**97**. Condensation of (−)-**97** with the lithium enolate of (+)-**38** (−90°C, THF) gave a major aldol (−)-**98** (61%, 27% of unreacted (+)-**38**).

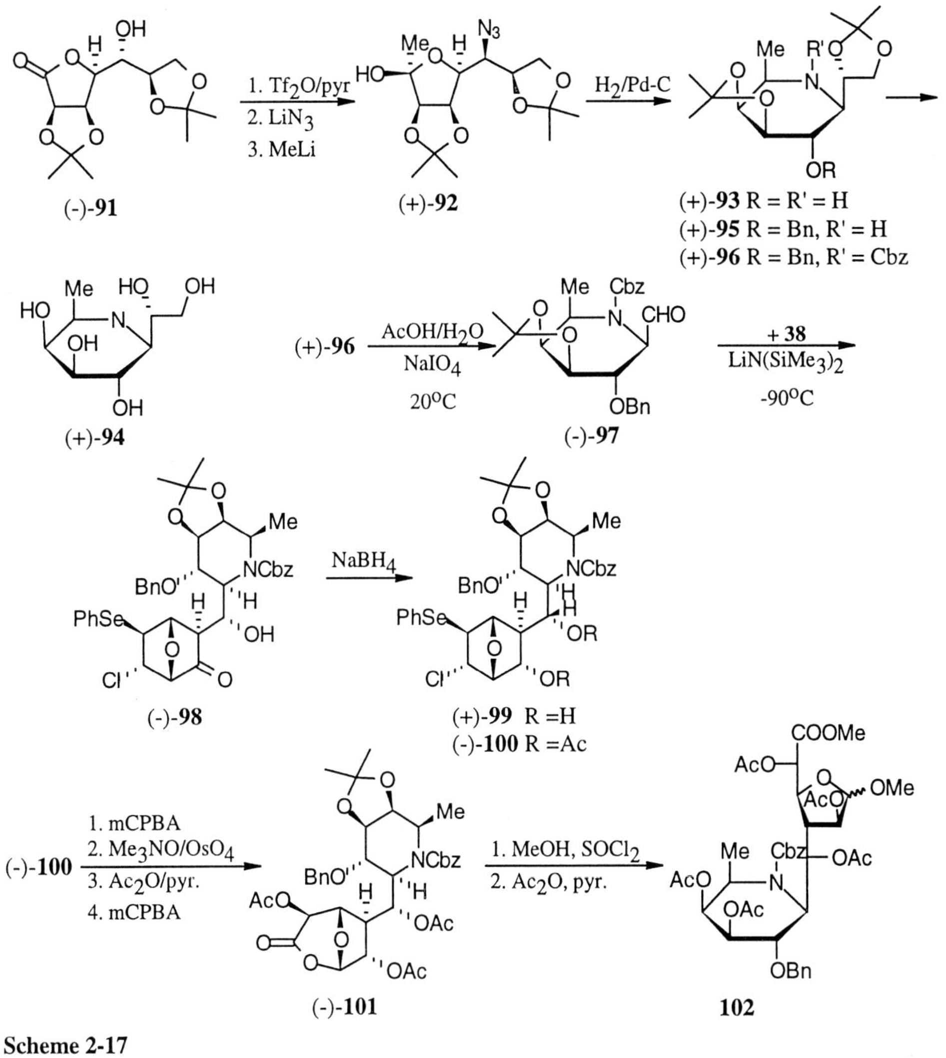

Scheme 2-17

Reduction of (−)-**98** with NaBH$_4$ gave diol (+)-**99** (88%), the acetylation of which afforded (−)-**100** (88%). Oxidation (mCPBA), followed by double hydroxylation, acetylation and Baeyer-Villiger oxidation provided uronolactone (−)-**101** (67%) which was converted into a mixture of the methyl furanosides **102** (73%), that are β-D-(1→3)-C-linked 1,5,6-trideoxy-1,5-iminogalactosides of D-altrose derivatives [54].

2.6 Conclusion

The "naked sugars of the first generation" are powerful tools to construct all kind of disaccharide mimics. At the moment the Giese's radical glycosidation and the cross-aldolisation of 7-oxabicyclo[2.2.1]heptan-2-one derivatives have been used to construct branched-chain tredecose systems with high and predictable diastereoselectivity. These C-C bond forming processes can be run with equimolar amounts of the two sugar-like sub-units. Because of the high radicophilicity of 3-methylidene-7-oxabicyclo[2.2.1]heptan-2-ones, their intermolecular radical glycosidation did not require large excesses of either the glycosyl radical precursor or of the electrophilic alkene. The diversity of the chemistry known for the simple "naked sugars" can be applied to their C-glycosidated derivatives. This allows the construction of different kinds of unusual C-disaccharides including C-glycosides of L-hexoses, C-glycosides of carbasugars and aza-C-disaccharide derivatives.

Acknowledgments

This work would not have been possible without all the efforts put into the development of the "naked sugar" chemistry by co-workers, the names of whom are given in the references. We thank all of these dedicated chemists and also the Swiss National Science Foundation, the "Found Herbette" (Lausanne), Hoffmann-La Roche and Co., AG (Basel), Geigy Jubilaeums Stiftung (Basel), and the European COST chemistry D2 programme (OFES, Bern) for generous financial support.

2.7 References

1. R. Schauer, *Adv. Carbohydr. Chem. Biochem.* **1982**, *40*, 131-234; O. Hindsgaul, D. P. Khare, M. Bach, R. U. Lemieux, *Can. J. Chem.* **1985**, *63*, 2653-2658; R. Kornfeld, S. Kornfeld, *Ann. Rev. Biochem.* **1985**, *54*, 631-664; H. C. Krivan, D. D. Roberts, V. Ginsburg, *Proc. Natl. Acad. Sci. U.S.A.* **1988**, *85*, 6157-6161; T. W. Rademacher, R. B. Parekh, R. A. Dwek, *Annu. Rev. Biochem.* **1988**, *57*, 785-838; G. E. Rice, M. P. Bevilacqua, *Science* **1989**, *246*, 1303-1306; S. Yagel, R. Feinmesser, C. Waghorne, P. K. Lala, M. L. Breitman, J. W. Dennis, *Int. J. Cancer* **1989**, *44*, 685-690; J. Montreuil, In *Compr. Biochem.* A. Neuberger, L. L. M. van Deenen, Eds. Elsevier Scientific: Amsterdam, **1982**, *19B II*, Chapter 1; K. A. Karlsson, *Annu.Rev. Biochem.* **1989**, *58*, 309-350; N. Sharon, H. Lis, *Science* **1989**, *246*, 227-234; T. Feizi, In *Carbohydrate Recognition in Cellular Function*; Ciba Geigy Symposium 145; Wiley: Chichester, 1989; p 62; C. A . Smith, T. Davis, D. Anderson, L. Solam, M. P. Beckmann, R. Jerzy, R. S. K. Dower, D. Cosman, R. G. Goodwin, *Science* **1990**, *248*, 1019-1023; L. A. Lasky, *Science* **1992**, *258*, 964-966; K. Furukawa, A. Kobata, In *Carbohydrates: Synthetic Methods and Applications to Medicinal Chemistry*, H. Ogura, A. Hasegawa, T. Suami, Eds.

VCH Publishers: New York, **1992**; pp 369-384; N. Sharon, H. Lis, *Sci. Am.* **1993**, *268*, 82; T. Feizi, *Curr, Opin. Struct. Biol.* **1993**, *3*, 701-710; A. Varki, A. *Proc. Natl. Acad. Sci. U.S.A.* **1994**, *91*, 7390-7397; K. W. Morenem, R. B. Trimble, A. Herscovics, *Glycobiology*, **1994**, *4*, 113-125.

2. C. P. J. Glaudemans, *Chem. Rev.* **1991**, *91*, 25-33; R. U. Lemieux, *Chem. Soc. Rev.* **1989**, *18*, 347-374; A. Rivera-Sagredo, D. Solis, T. Diaz-Mauriño, J. Jimenez-Barbero, M. Martin-Lomas, *Eur. J. Biochem.* **1991**, *197*, 217-228; F. A. Quiocho, D. K. Wilson, N. K. Vyas, *Nature* **1989**, *340*, 404-407; Y. Bourne, P. Rougé, C. Cambillau, *J. Biol. Chem.* **1990**, *265*, 18161-18165; F. A. Quiocho, *Biochem. Soc. Trans.* **1993**, *21*, 442-448; R. U. Lemieux, In Carbohydrate Antigens; P. J. Garegg, A. A. Lindberg, Eds. ACS Symposium Series 519; American Chemical Society: Washington, DC, **1993**; pp 5-18; R. U. Lemieux, T. C. Wong, H. Thøgersen, *Can. J. Chem.* **1982**, *60*, 81-86; O. P. Srivastava, R. U. Lemieux, M. Shoreibah, M. Pierce, *Carbohydr. Res.* **1988**, *179*, 137-161; S. Sabesan, S. Neira, F. Davidson, J. Ø. Duus, K. Bock, *J. Am. Chem. Soc.* **1994**, *116*, 1616-1634; S. W. Homans, R. A. Dwek, T. W. Rademacher, *Biochemistry* **1987**, *26*, 6571-6578; Z. Y. Yan, C. A. Bush, *Biopolymers* **1990**, *29*, 799-811; J. P. Carver, *Pure & Appl. Chem.* **1993**, *65*, 763-770; i) P. Berthault, N. Birkirakis, G. Rubinstenn, P. Sinaÿ, H. J. Desvaux, *J. Biomolecular NMR*, **1996**, *8*, 23-35; j) J. -F. Espinosa, F. J. Cañada, J. L. Asencio, M. Martin-PAstor, H. Dietrich, M. Martin-Lomas, R. R. Schmidt, J. Jiménez-Barbero, *J. Am. Chem. Soc.*, **1996**, *118*, 10862-10871.

3. a) T.-C. Wu, P. G. Goekjian, Y. Kishi, *J. Org. Chem.* **1987**, *52*, 4819-4823; b) P. G. Goekjian, T.-C. Wu, H.-Y Kang, Y. Kishi, *Ibid.* **1987**, *52*, 4823-4825; c) S. A. Babirad, Y. Wang, P. G. Goekjian, Y. Kishi, *Ibid.* **1987**, *52*, 4825-4827; d) Y. Kishi, *Pure & Appl. Chem.* **1993**, *65*, 771-778; C. A. Duda, E. S. Steven, *J. Am. Chem. Soc.* **1993**, *115*, 8487-8488; A. Wie, Y. Kishi, *J. Org. Chem.* **1994**, *59*, 88-96; e) O. R. Martin, W. Lai, *Ibid.* **1990**, *55*, 5188-5190; **1993**, *58*, 176-185; f) D. J. O'Leary, Y. Kishi, *Tetrahedron Lett.* **1994**, *35*, 5591-5594; g) F. J. López-Herrera, M. S. Pino-González, F. Planas-Ruiz, *Tetrahedron: Asymmetry* **1990**, *1*, 465-475; h) R. Ferritto, P. Vogel, *Ibid.* **1994**, *5*, 2077-2092.

4. A. Wei, K. M. Boy, Y. Kishi, *J. Am. Chem. Soc.* **1995**, *117*, 9432-9436; A. Wei, A. Haudrechy, C. Audin, H.-S. Jun, N. Haudrechybretel, Y. Kishi, *J. Org. Chem.* **1995**, *60*, 2160-2169 and ref. cited therein.

5. see e.g. : A.D. Elbein, R. J. Molyneux, In *Alkaloids: Chemical and Biological Perspectives*; S. W ., Pelletier, Ed. Wiley-Interscience: New York, **1987**; Vol. 5, Ch. 1; A. S . Howard, J. P . Michael, In *The Alkaloids*; Brossi, A., Ed. Academic Press: New York, **1986**; Vol. 28, Ch. 3; A.D. Elbein, *Annu. Rev. Biochem.* **1987**, *56*, 497-534; P. Vogel, *Chimica Oggi* **1992**, 9-15; B. Winchester, G. W. J. Fleet, *Glycobiology* **1992**, *2*, 199-210; G. Legler, *Adv. Carbohydr. Chem. Biochem.* **1990**, *48*, 319-384; L. E . Fellows, *Chem. Br.* **1987**, *23*, 842-844; L. E . Fellows, *New Sci.* **1989**, *123*, 45; M. L Sinnot, *Chem. Rev.* **1990**, *90*, 1171; R. W . Frank, *Bioorganic Chem.* **1992**, *20*, 77-88; A. Tan, L. van den Broek, J. Bolscher, D. J . Vermass, L. Pastoors, C. van Boeckel, H. Ploegh, *Glycobiology* **1994**, *7*, 141-149.

6. N. Ishida, K. Kumagai, T. Niida, T. Tsuruoka, H. Yumoto, *J. Antibiot.*, **1967**, *20*, 66-71.
7. R. Datema, S. Olofsson, P. A . Romero, *Pharmacol. Ther.* **1987**, *33*, 221-286; H.-D. Klenk, R. T Schwarz, *Antiviral Res.* **1982**, *2*, 177-190; B.D. Walker, M. Kowalski, W. C . Goh, K. Kozarsky, M. Krieger, C. Rosen, L. Rohrschneider, W. A . Haseltine, J. Sodroski, *Proc. Natl. Acad. Sci. U.S.A.* **1987**, *84*, 8120-8124.
8. HIV: R. A . Gruters, J. J . Neefjes, M. Tersmette, R. E . de Göde, A. Tulp, H. G . Huisman, F. Miedema, H. L . Plögh, *Nature* **1987**, *330*, 74-77; A. S . Tyms, E. M Berrie, T. A . Ryder,; R. J . Nash,; M. P . Hegarty, D. L . Taylor, M. A . Mobberley, J. M . Davis, E. A . Bell, D. J . Jeffries, D. Taylor-Robinson, L. E . Fellows, *Lancet* **1987**, *2*, 1025-1026; G. W . J . Fleet, A. Karpas, R. A . Dwek, L. E . Fellows, A. S . Tyms, S. Petursson, S. K . Namgoong, N. G . Ramsden, P. W . Smith, J. C . Son, F. Wilson, D. R . Witty, G. S . Jacob, T. W Rademacher, *FEBS Lett.* **1988**, *237*, 128-132; P. S . Sunkara, D. L . Taylor, M. S . Kang, T. L . Bowlin, P. S . Liu, A. S . Tyms, A. Sjoerdsma, *Lancet* **1989**, *1*, 1206; R. M . Ruprecht, S. Mullaney, J. Andersen, R. Bronson, *J. Acquired Immune Defic. Synd.* **1989**, *2*, 149-157.
9. Cytomegalovirus: D. L . Taylor, L. E . Fellows, G. H . Farrar, R. J . Nash, D. Taylor-Robinson, M. A . Mobberley, T. A . Ryder, D. J . Jeffries, A. S . Tyms, *Antiviral Res.* **1988**, *10*, 11-26.
10. M. J . Humphries, K. Matsumoto, S. L . White, K. Olden, *Cancer Res.* **1986**, *46*, 5215-5222; Dennis, J. W . *Cancer Res*, **1986**, *46*, 5131-5136; E. J . Nichols, R. Manger, S. I . Hakomori, L. R . Rohrschneider, *Exp. Cell. Res.* **1987**, *173*, 486-495; J. W . Dennis, S. Laferté, C. Waghorne, M. L . Breitman, R. S Kerbel, *Science* **1987**, *236*, 582-586; G. K . Ostrander, N. K . Scribner, L. R . Rohrschneider, *Cancer Res.* **1988**, *48*, 1091-1094; M. J . Humphries, K. Matsumoto, S. L . White, R. J . Molyneux, K. Olden, *Cancer Res*, **1988**, *48*, 1410-1415; A. Hadwiger-Fangmeier, H. Niemann, T. Tamura, *Arch. Virol.* **1989**, *104*, 339-345Y. ; Nishimura, T. Satoh, H. Adachi, S. Kondo, T. Takeuchi, M. Azetaka, H. Fukuyasu, Y. Iizuka, *J. Am. Chem. Soc.* **1996**, *118*, 3051-3052.
11. See, e.g. : Wassarman, P. M . *Science*, **1987**, *235*, 553-560. *Annu. Rev. Cell. Biol.* **1987**, *3*, 109-142. *Sci. Ann.* **1988**, *259*, 78-84; R. Yanagimachi, In *The Physiology of Reproduction*; Kobil, E. Neill, J.D. Eds. Raven Press: New York, **1988**, Vol. 1, p 35.
12. P. H . Bennet, N. B . Rushforth, M. Miller, P. M . Le Compte, *Rec. Prog. Horm. Res.* **1976**, *32*, 333-376; E. Truscheit, W. Frommer, B. Junge, L. Müller, D.D. Schmidt, W. Wingender, *Angew. Chem. Int. Ed. Engl.* **1981**, *20*, 744-761; Sinnott, M. L . In *Enzyme Mechanisms*; M. I ., Page, A., Williams, Eds. The Royal Society of Chemistry; London, **1987**; p 259-297; P. R . Holt, D. Thea, M. Y . Yang, D. P . Kotler, *Metab. Clin. Exp.* **1988**, *37*, 1163-1170; B. L . Rhinehart, K. M . Robinson, A. J . Payne, M. E . Wheatley, J. L . Fischer, P. S . Liu, W. Cheng, *Life Sci.* **1987**, *41*, 2325-2331; A. J . Reuser, H. A . Wisselaar, *Eur. J. Clin. Invest.* **1994**, *24*, Suppl. 3, 19-24; W. Leonhardt, M. Hanefield, S. Fischer, J. Schulze, *Eur. J. Clin.*

13. *Invest.* **1994**, *24*, Suppl. 3, 45-49; Bischoff, H. *Eur. J. Clin. Invest.* **1994**, *24*, Suppl. 3, 3-10.
14. See, e.g. : W. Puls, H. P . Krause, L. Müller, H. Schutt, R. Sitt, G. Thomas, *Int. J. Obesity* **1984**, *8*, 181-190; P. Layer, G. L . Carlson, E. P . Di-Magno, *Gastroenterology* **1985**, *88*, 1895-1902.
15. D. Rouzaud, P. Sinaÿ, *J. Chem. Soc., Chem. Commun.* **1983**, 1353-1354.
16. T. Haneda, P. G . Goekjian, S. H . Kim, Y. Kishi, *J. Org. Chem.* **1992**, *57*, 490-498.
17. See, e.g. : a) S. J . Danishefsky, M. P . DeNinno, *Angew. Chem. Int. Ed. Engl.* **1987**, *26*, 15-23; J. A . Secrist, K.D. Barnes, S.-R. Wu, in « *Trends in Synthetic Carbohydrate Chemistry* », ACS Symposium Series 386, American Chemical Society, Washington, DC. 1989, Chap. 5, p. 93-106; C.-H. Wong, G. M . Whitesides, *Enzymes in Synthetic Organic Chemistry*, Pergamon, Oxford, **1994**; C.-H. Wong, R. L . Halcomb, Y. Ichikawa, T. Kajimoto, *Angew. Chem. Int. Ed. Engl.* **1995**, *34*, 412-432; **1995**, *34*, 521-546; b) B. Fraser-Reid, B. F . Molino, L. Magdzinski, D. R . Mootoo, *J. Org. Chem.* **1987**, *52*, 4505-4511; Y. Kishi, *Pure Appl. Chem.* **1989**, *61*, 313-324; K. C . Nicolaou, C. K . Hwang, M. E . Duggan, *J. Am. Chem. Soc.* **1989**, *111*, 6682-6690; S. Jarosz, B. Fraser-Reid, *J. Org. Chem.* **1989**, *54*, 4011-4013; S. Jarosz, *Carbohydr. Res.* **1988**, *183*, 209-215; O. Sakanaka, T. Ohmori, S. Kozaki, T. Suami, T. Ishii, S. Ohba, Y. Saito, *Bull. Chem. Soc. Jpn.* **1986**, *59*, 1753-1759; N. Ikemoto, S. L . Schreiber, *J. Am. Chem. Soc.* **1992**, *114*, 2524-2535; J. Ramza, A. Zamojski, *Tetrahedron* **1992**, *48*, 6123-6134; W. Schmid, G. M . Whitesides, *J. Am. Chem. Soc.* **1991**, *113*, 6674-6675; A. G . Myers, D. Y . Gin, K. L . Widdowson, *J. Am. Chem. Soc.* **1991**, *113*, 9661-9663; A. G . Myers, D. Y . Gin, D. H . Rogers, *J. Am. Chem. Soc.* **1994**, *116*, 4697-4718; G. Casiraghi, L. Colombo, G. Rassu, P. Spanu, *J. Org. Chem.* **1991**, *56*, 2135-2139; K. E . McGhie, R. M . Paton, *Tetrahedron Lett.* **1993**, *34*, 2831-2834; J. A . Marshall, S. Beaudoin, K. Lewinski, *J. Org. Chem.* **1993**, *58*, 5876-5879; D. Enders, U. Jegelka, *Tetrahedron Lett.* **1993**, *34*, 2453-2456; J. A . Marshall, S. Beaudoin, *J. Org. Chem.* **1994**, *59*, 6614-6619; W. Karpiesiuk, A. Banaszek, *Tetrahedron* **1994**, *50*, 2965-2974; A. Dondoni, L. Kniezo, M. Martinkova, *J. Chem. Soc., Chem. Commun.* **1994**, 1963-1964; S. Jarosz, *Tetrahedron Lett.* **1994**, *35*, 7655-7658; *Pol. J. Chem.* **1994**, *68*, 1333-1342; R. M . Paton, A. A . Young, *J. Chem. Soc., Chem. Commun.* **1994**, 993-994; A. Fürstner, H. Weidmann, *J. Org. Chem.* **1989**, *54*, 2307-2311; D. P . Sutherlin, R. W . Armstrong, *Tetrahedron Lett.* **1993**, *34*, 4897-4900; A. J . Sinaÿ, P. Fairbanks, *Tetrahedron Lett.* **1995**, *36*, 893-896; c) O. Eyrisch, W.-D. Fessner, *Angew. Chem. Int. Ed. Engl.* **1995**, *34*, 1639-1641; d) F. Emery, P. Vogel, *J. Org. Chem.* **1995**, *60*, 5843-5854 and ref. cited therein; e) G., Casiraghi, F., Zanardi, G., Rassu, G. Spanu, *Chem. Rev.* **1995**, *95*, 1677-1716; S. Knapp, *Chem. Rev.* **1995**, *95*, 1859-1876; J. K ., Cha, N.-S. Kim, *Chem. Rev.* **1995**, *95*, 1761-1795.
18. H. Paulsen, K. Roden, V. Sinnwell, P. Luger, *Liebigs Ann. Chem.* **1981**, 2009-2027.

18. B. Aebischer, J. H . Bieri, R. Prewo, A. Vasella, *Helv. Chim. Acta*, **1982**, *65*, 2251-2272.
19. O. Frey, M. Hoffmann, H. Kessler, *Angew. Chem. Int. Ed. Engl.* **1995**, *34*, 2026-2028 and previous reports.
20. P. De Pouilly, A. Chénedé, J.-M. Mallet, P. Sinaÿ, *Bull. Soc. Chim. Fr.* **1993**, *130*, 256-265.
21. D. Mazéas, T. Skrydstrup, J.-M. Beau, *Angew. Chem. Int. Ed. Engl.* **1995**, *34*, 909-912; see also: O. Jarreton, T. Skrydstrup, J.-M. Beau, *J. Chem. Soc., Chem. Commun.* **1996**, 1661-1662.
22. O. R . Martin, W. Lai, *J. Org. Chem.* **1990**, *55*, 5188-5190; O. R . Martin, W. Lai, *J. Org. Chem.* **1993**, *58*, 176-185; see also: W. R. Koberts, C. R . Bertozzi, M. D Bednarski, *Ibid.* **1996**, *61*, 1894-1897.
23. A. Wei, Y. Kishi, *J. Org. Chem.* **1994**, *59*, 88-96.
24. A. Boschetti, F. Nicotra, L. Panza, G. Russo, L. Zucchelli, *J. Chem. Soc., Chem. Commun.* **1989**, 1085-1086; L. Lay, F. Nicotra, L. Panza, G. Russo, E. Caneva, *J. Org. Chem.* **1992**, *57*, 1304-1306.
25. B. Giese,Witzel, T. *Angew. Chem. Int. Ed. Engl.* **1986**, *25*, 450-451.
26. B. Giese, M. Hoch, C. Lamberth, R. R . Schmidt, *Tetrahedron Lett.* **1988**, *29*, 1375-1388.
27. R. M . Bimwala, P. Vogel, *Tetrahedron Lett.* **1991**, *32*, 1429-1432; R. M . Bimwala, P. Vogel, *J. Org. Chem.* **1992**, *57*, 2076-2083.
28. I. M . Dawson, T. Johnson, Paton, R. M . Rennie, A. C . *J. Chem. Soc., Chem. Commun.* **1988**, 1339-1340.
29. U. C . Dyer, Y. Kishi, *J. Org. Chem*, **1988**, *53*, 3383-3384.
30. M. Carcano, F. Nicotra, L. Panza, G. Russo, *J. Chem. Soc, Chem. Commun.* **1989**, 642-643; L. Lay, F. Nicotra, C. Pangrazio, L. Panza, G. Russo, *J. Chem. Soc., Perkin Trans I* **1994**, 333-338.
31. Y. Wang, P. G . Goekjian, D. M . Ryckman, Y. Kishi, *J. Org. Chem.* **1988**, *53*, 4151-4153.
32. R. R . Schmidt, ; A. Beyerbach, *Liebigs Ann. Chem.* **1992**, 983-986; see also: A. T . Khan, P. Sharma, R. R . Schmidt,*J. Carbohyder. Chem.* **1995**, *14*, 1353-1367.
33. S. A . Babirad, Y. Wang, Y. Kishi, *J. Org. Chem.* **1987**, *52*, 1370-1372.
34. Y. Wang, S. A . Babirad, Y. Kishi, *J. Org. Chem.* **1992**, *57*, 468-481.
35. R. Preuss, R. R . Schmidt, *J. Carbohydr. Chem.* **1991**, *10*, 887-900; see also : R. Preuss, K.-H. Jung, R. R . Schmidt, *Liebigs Ann. Chem.* **1992**, 377-382.
36. H. Dietrich, R. R . Schmidt, *Liebigs Ann. Chem.* **1994**, 975-981.
37. B. Vauzelles, D. Cravo, J. -M. Mallet, P. Sinaÿ, *Synlett* **1993**, 522-524.
38. Y.-C. Xin, J. -M. Mallet, P. Sinaÿ, *J. Chem. Soc., Chem. Commun.* **1993**, 864-864.
39. A. Chénedé, E. Perrin, E.D. Rekaï, Sinaÿ, P. *Synlett* **1994**, 420-422.
40. A. Mallet, J. -M. Mallet, P. Sinaÿ, *Tetrahedron Asym.* **1994**, *5*, 2593-2608.

41. P. G. Goekjian, T.-C. Wu, H.-Y. Kang, Kishi, Y. *J. Org. Chem.* **1991**, *56*, 6422-6434.
42. O. R. Martin, F. Xie, R. Kakarla, R. Benhamza, *Synlett* **1993**, 165-167.
43. R. M. Paton, K. J. Penman, *Tetrahedron Lett.* **1994**, *35*, 3163-3166.
44. D. Mazéas, J. Skrydstrup, O. Doumeix, J.-M. Beau, *Angew. Chem., Int. Ed. Engl.* **1994**, *33*, 1383-1386.
45. W. B. Motherwell, B. C. Ross, M. J. Tozer, *Synlett* **1989**, 68-70; Herpin, T. F. W. B. Motherwell, M. J. Tozer, *Tetrahedron Asymmetry* **1994**, *5*, 2269-2282.
46. R. W. Armstrong, D. P. Sutherlin, *Tetrahedron Lett.* **1994**, *35*, 7743-7746.
47. a) For recent applications, see e.g.: A. J. Fairbanks, E. Perrin, P. Sinaÿ, *Synlett* **1996**, 679-681; b) For an example of carbenoid insertion into a carbaldehyde C-H bond, see: F. Sarabia-Garcia, F. J. Lopez-Herrera, M. S. Pino-Gonzalez, *Tetrahedon* **1995**, *51*, 5491-5500.
48. a) E. Vieira, P. Vogel, *Helv. Chim. Acta* **1983**, *66*, 1865-1871; b) J.-L. Reymond, P. Vogel, *Tetrahedron: Asymmetry* **1990**, *1*, 729-736; c) R. Saf, K. Faber, G. Penn, H. Griengl, *Tetrahedron* **1988**, *44*, 389-392; B. Ronan, H. B. Kagan, *Tetrahedron: Asymmetry* **1991**, *2*, 75-83; E. J. Corey, T.-P. Loh, *Tetrahedron Lett.* **1993**, *34*, 3979-3982.
49. P. Vogel, Y. Auberson, M. Bimwala, E. de Guchteneere, E. Vieira, J. Wagner, In «*Trends in Synthetic Carbohydrate Chemistry*», Eds. Horton, D. Hawkins, L.D. McGarvey, G. J. ACS Symposium Series 386; American Chemical Society, Washington, D. C. **1989**, p. 197-241; P. Vogel, D. Fattori, F. Gasparini, C., Le Drian, *Synlett* **1990**, 173-185; P. Vogel, *Bull. Soc. Chim. Belg.* **1990**, *99*, 295.
50. For the «naked sugars of the second generation», see : P. Kernen P. Vogel, *Tetrahedron Lett.* **1993**, *34*, 2473-2476; A.-F. Sevin, P. Vogel, *J. Org. Chem.* **1994**, *59*, 5920-5926; P. Kernen, P. Vogel, *Helv. Chim. Acta* **1995**, *78*, 301-320; M. Bialecki, P. Vogel, *Ibid.* **1995**, *78*, 325-343.
51. F. Gasparini, P. Vogel, *J. Org. Chem.* **1990**, *55*, 2451-3457; P. Péchy, F. Gasparini, P. Vogel, *Synlett* **1992**, 676-678; V. Jeanneret, F. Gasparini, P. Péchy, P. Vogel, *Tetrahedron* **1992**, *48*, 10637-10644.
52. C. Le Drian, E. Vieira, P. Vogel, *Helv. Chim. Acta*, **1989**, *72*, 338-347; C. Le Drian, J.-P. Vionnet, P. Vogel, *Ibid.* **1990**, *73*, 161-168.
53. See e.g. the first total synthesis of cyclophellitol: V. Moritz, P. Vogel, *Tetrahedron Lett.* **1992**, *33*, 5243-5244.
54. A. Baudat, P. Vogel, *Tetrahedron Lett.* **1996**, *37*, 483-484; see also: E. Frérot, C. Marquis, P. Vogel, *Ibid.* **1996**, *37*, 2023-2026.
55. R. Ferritto,; P. Vogel, *Tetrahedron Lett.* **1995**, *36*, 3517-3518.
56. R. Ferritto,; P. Vogel, *Synlett* **1996**, 281-282
57. See also B. Giese,*Pure Appl. Chem.* **1988**, *60*, 1655-1658; B. Giese, J. Dupuis, M. Nix, *Org. Synth.* **1987**, *65*, 236-242.
58. G. Arvai, D. Fattori, P. Vogel, *Tetrahedron* **1992**, *48*, 10621-10636.

59. P.-A. Carrupt, P. Vogel, *Tetrahedron Lett.* **1982**, *23*, 2563-2566; *Helv. Chim. Acta* **1989**, *72*, 1008-1028; *J. Org. Chem.* **1990**, *55*, 5696-5700.
60. C. Pasquarello, P. Vogel, unpublished results.
61. V. E. Verovskii, G. M . Lipkind, N. K . Kochetkov, *Bioorg. Khim.* **1984**, *10*, 1680-1687.
62. See e.g. : A. A . Higton, A.D. Roberts,« Dictionary of Antibiotics and Related Substances », B. W . Bycroft, Ed. Chapman & Hall, London, **1988**; T. Suami, in « Carbohydrates-Synthetic Methods and Applications in Medicinal Chemistry », H. Ogura, A. Hasegawa, T. Suami, Eds., Kodansha Ltd. & VCH, Tokyo, **1992**; G. Bach, S. Breiding-Mack, S. Grabley, P. Hammann, K. Hütter, R. Thiericke, H. Uhr, J. Wink, A. Zeeck, *Liebigs Ann. Chem.* **1993**, 241-250.
63. G. E . McCasland, S. Furuta, L. J . Durham, *J. Org. Chem.* **1966**, *31*, 1516-1521.
64. T. Suami, Ogawa, S. *Adv. Carbohydr. Chem. Biochem.* **1990**, *48*, 21-90.
65. J. Cossy, J.-L. Ranaivosata, V. Bellosta, J. Ancerewicz, R. Ferritto,; P. Vogel, *J. Org. Chem.* **1995**, *60*, 8351-8359.
66. C. Barbaud, M. Bols, I. Lundt, M. R . Sierks, *Tetrahedron* **1995**, *51*, 9063-9078.
67. See e.g. : N. Ikota, B. Ganem, *J. Org. Chem.* **1978**, *43*, 1607-1608; J. A. Marshall, R. D., Royce, Jr. *Ibid.* **1982**, *47*, 693-698; S. L. Schreiber, C. Santini, *J. Am. Chem. Soc.* **1984**, 106, 4038-4039.
68. F. Emery, P. Vogel, *Synlett*, **1995**, 420-422.
69. J. Cossy, P. Aclinou, V. Bellosta, N. Furet, J. Baranne-Lafont, D. Sparfel, C. Souchaud, *Tetrahedron Lett.* **1991**, *32*, 1315-1316.
70. H. E. Zimmerman, M. D . Traxler, *J. Am. Chem. Soc.* **1957**, *79*, 1920-1923.
71. K. Kraehenbuehl, P. Vogel, *Tetrahedron Lett.* **1995**, *36*, 8595-8598.
72. E. W. Baxter, A. B. Reitz, *J. Org. Chem.* **1994**, *59*, 3175-3185, T. M. Jespersen, W. Dong, M. R. Sierks, T. Skrydstrup, I. Lundt, M. Bols, *Angew. Chem., Int. Ed. Engl.* **1994**, *33*, 1778-1781; C. -H. Wong, L. Provencher, J. A. Porco, S. -H. Jung, Y. -F. Wang, L. Chen, R. Wang, D. H. Steensma, *J. Org. Chem.* **1995**, *60*, 1492-1501 and ref. cited therein.
73. C. R. Johnson, M. W. Miller, A. Golebiowski, H. Sundram, M.B. Ksebati, *Tetrahedron Lett.* **1994**, *35*, 8991-8994; see also: O. R. Martin, L. Liu, F. Yang, *Tetrahdron Lett.* **1996**, *37*, 1991-1994.
74. J. S. Brimacombe, L. C. N. Tucker, *Carbohydr. Res.* **1966**, *2*, 341-348; T. K. M. Shing, H. C. Tsui, Z. -H. Zhou, T. C. W. Mak, *J. Chem. Soc., Perkin Trans. I*, **1992**, 887-893.
75. A. Baudat, S. Picasso, P. Vogel, *Carbohydr. Res.* **1996**, *281*, 277-284.

3 The Nitrile Oxide/Isoxazoline Route to C-Disaccharides

R. M. Paton

3.1 Introduction

Disaccharide analogues in which the glycosidic oxygen is replaced by a carbon atom are the subject of widespread current interest both in their own right as carbohydrate mimics and as potential enzyme inhibitors [1-27]. Typical examples of such carbon-linked disaccharides (C-disaccharides) include C-sucrose **1** - the analogue of natural sucrose with a methylene in place of the glycosidic oxygen [1, 2] - and compound **2** in which two glucopyranose units are β-(1→6)-linked [3]. Although most examples reported to date are methylene-linked some C-disaccharides with functionalized bridges have also been described; thus in compound **3** the two pyranose units are (1→4)-linked by hydroxymethylene [4].

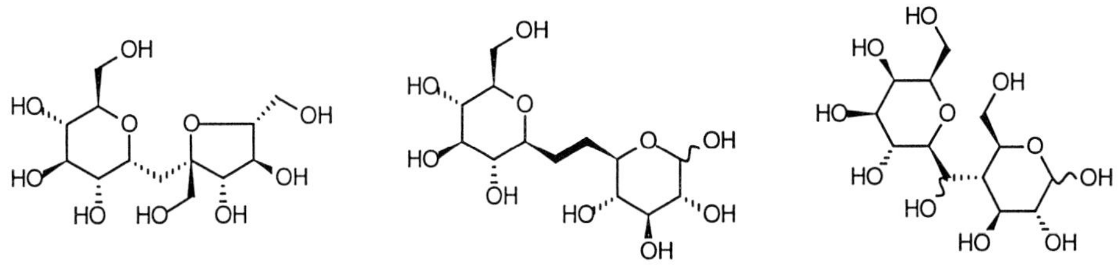

Figure 3-1. Examples of C-disaccharides.

Much effort has been devoted in recent years to developing effective methods for the synthesis of C-disaccharides. The most attractive routes involve the coupling of two monosaccharide units thereby incorporating most of the desired stereochemical features at the outset. Various techniques have been employed for this purpose including radical addition reactions [eg 5-9], aldol condensations [eg 10-12], Wittig methodology [eg 2, 13, 14], nitroaldol-couplings [eg 15-17], and various other carbonyl addition reactions [eg 3, 18-24]. An alternative approach to C-disaccharides is provided by the hetero-Diels-

Alder cycloaddition reactions of suitably substituted dienes with carbohydrate aldehydes [eg 25-27]. We now describe a cycloaddition route, based on nitrile oxide/isoxazoline chemistry [28-32], which is suitable for the preparation from readily accessible precursors of disaccharides possessing a variety of functionalised carbon bridges.

For many years the 1,3-dipolar cycloaddition reactions of nitrile oxides have been used for the construction of five-membered heterocycles incorporating the C=N-O unit [33] but it is only in the last decade or so that the their potential for the synthesis of natural products and analogues has been realised. The utility of nitrile oxides stems from their ease of generation under mild conditions, and the regio- and stereo-chemical control of their cycloaddition to alkenes. The well-established chemistry of the resulting 2-isoxazoline (4,5-dihydroisoxazole) cycloadducts permits rapid accumulation of polyfunctionality in a small molecular framework, and this methodology has been applied successfully for the preparation of a wide variety of natural products including carbocyclics, alkaloids and carbohydrates. The method involves three basic steps (Figure 3-2): firstly, cycloaddition of the nitrile oxide to the alkene; then modification of the resulting isoxazoline which is usually sufficiently robust as to allow the introduction and/or manipulation of substituents; and finally reductive ring cleavage at the N-O bond of the isoxazoline to afford, for example, β-hydroxyketones or γ-aminoalcohols. This route to β-hydroxyketones represents an alternative to the traditional aldol condensation in that the new carbon-carbon bond is adjacent to the carbonyl rather than one bond removed [29].

Figure 3-2. Nitrile Oxide-Isoxazoline approach to β-hydroxyketones and γ-aminoalcohols.

Our aim has been to develop a general method capable of joining sugar units with a variety of functionalized bridges with control of both stereochemistry and position of linkage. We considered that nitrile oxide/isoxazoline chemistry could be well suited for this purpose. The approach, which is illustrated in Figure 3-3, involves as the key steps cycloaddition of a pyranosyl nitrile oxide to a carbohydrate alkene, followed by hydrogenolysis of the resulting isoxazoline.

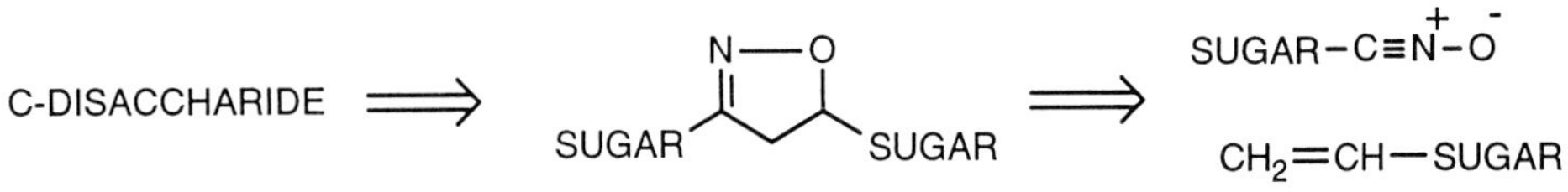

C-DISACCHARIDE $\Longrightarrow$ SUGAR–[isoxazoline]–SUGAR $\Longrightarrow$ SUGAR–C≡N$^+$–O$^-$ + CH$_2$=CH–SUGAR

Figure 3-3. The nitrile oxide-isoxazoline approach to C-disaccharides

3.2 Generation of Nitrile Oxides

Although nitrile oxides have been identified as short-lived intermediates in a variety of reactions [33], for synthetic purposes two methods of generation are in common usage. These involve the dehydration of a nitromethyl compound using, for example, phenyl isocyanate as originally reported by Mukaiyama et al [34], and thermal or base-mediated dehydrochlorination of a hydroximoyl chloride (Figure 3-4). In view of their tendency to dimerize to 1,2,5-oxadiazole-2-oxides (furoxans) they are usually prepared *in situ* in the presence of an excess of the dipolarophile.

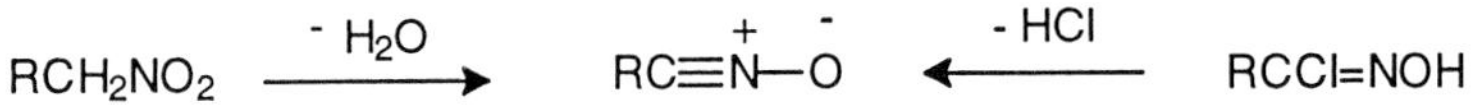

RCH$_2$NO$_2$ $\xrightarrow{- H_2O}$ RC≡N$^+$–O$^-$ $\xleftarrow{- HCl}$ RCCl=NOH

Figure 3-4. Generation of nitrile oxides.

For the present work an efficient source of pyranose 1-carbonitrile oxides was required and, in view of the ready availability by established literature methods of pyranosylnitromethanes, we selected the Mukaiyama dehydration approach. D-Xylose-derived nitrile oxide **4** was chosen for the initial investigations and its precursor, β-xylopyranosylnitromethane derivative **5**, was therefore prepared in three steps (Scheme 3-1) from D-xylose using a modified version of the procedure described by Köll et al [35]. Base-catalyzed condensation of nitromethane with D-xylose afforded a mixture of 1-deoxy-1-nitrohexitols **6** which on heating in boiling water underwent dehydration/cyclization to yield pyranosylnitromethane **7** as the major product; the minor α-isomer was not isolated.

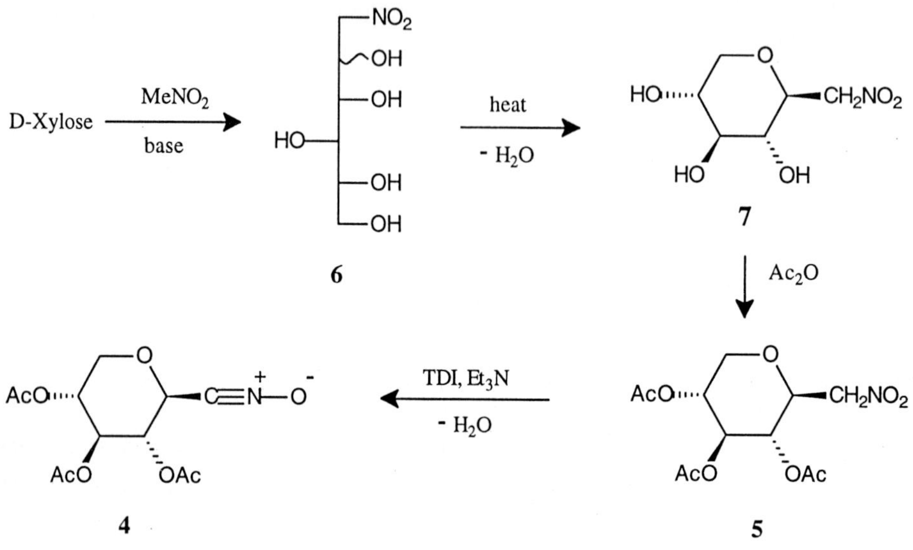

Scheme 3-1

Subsequent acid-catalyzed acetylation afforded triacetate derivative **5** in 49% overall yield from D-xylose as a convenient, crystalline and shelf-stable precursor of the target nitrile oxide.

For the generation of the nitrile oxide the phenyl isocyanate traditionally used as the dehydrating agent was replaced by commercially available tolylene diisocyanate (TDI). In a typical experiment a solution of the nitromethyl compound in dichloromethane was added slowly over 36 hours by means of a motorized syringe pump to a solution of the sugar alkene, TDI and triethylamine, and the mixture heated for a further 24 hours. Work-up of the products was facilitated by quenching the unreacted isocyanate with 1,2-diaminoethane thus forming an insoluble polymeric urea which could be removed by filtration. The same approach is also utilized for the generation of the corresponding pyranosylnitrile oxides derived from D-arabinose, D-galactose and D-mannose [36].

3.3 (1→2)- and (1→3)-Carbonyl-linked Disaccharides

The first targets were disaccharides in which two pyranose units were linked (1→2) and (1→3) by a carbonyl group. It was considered that these should be accessible *via* the regioisomeric isoxazolines formed by cycloaddition of a pyranose 1-carbonitrile oxide to a 2,3-unsaturated pyranoside as illustrated in Figure 3-5.

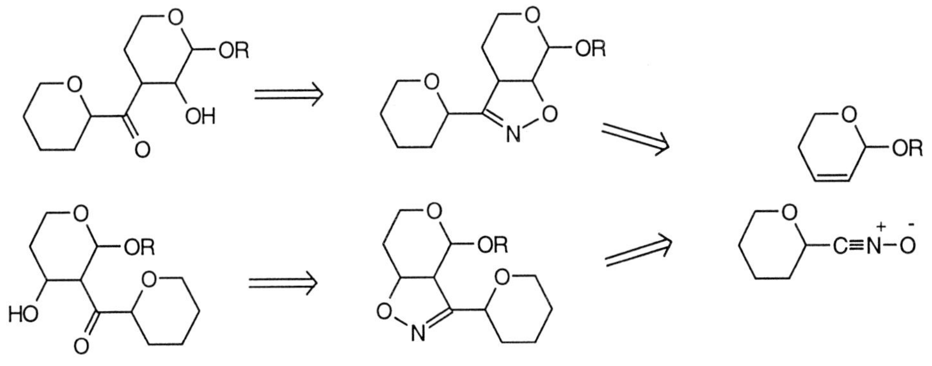

Figure 3-5. The nitrile oxide-isoxazoline approach to carbonyl-linked disaccharides.

2,3-Dideoxy hex-2-enopyranoside derivative **8** was selected as the dipolarophile component. It is readily prepared in multigram quantities from D-glucose by the Ferrier method [37] and it was anticipated that the *cis*-arrangement of the two allylic substituents flanking the alkene moiety might induce a degree of π-facial selectivity to the cycloaddition process [38]. Reaction of xylopyranosyl nitrile oxide **4** with alkene **8** was accomplished using the procedure described above and from the reaction mixture was isolated by chromatography and crystallization a pair of diastereoisomeric isoxazolines **9** and **10**, each in 33% yield (Scheme 3-2).

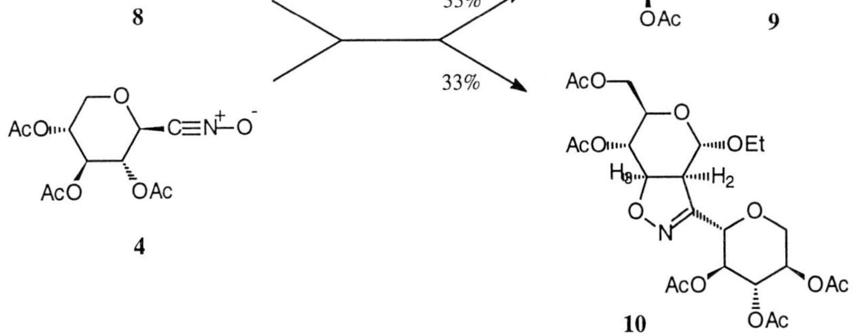

Scheme 3-2

The two adducts can readily be distinguished by the characteristic NMR signals for the isoxazoline ring protons; the CH adjacent to the ring oxygen invariably has the greater chemical shift. Thus in the dipolarophile-derived portion of adduct **9** H(2) resonates at 4.24 ppm in CDCl$_3$ and H(3) at 3.48 ppm, whereas for regioisomer **10** the order is reversed with H(2) at 3.67 and H(3) at 4.78 ppm. The absence in each case of a discernible coupling between the anomeric proton H(1) and the adjacent proton H(2) is indicative of the proposed *trans* stereochemistry [39]. Careful examination of the reaction mixture by TLC and ^{13}C NMR spectroscopy showed no trace of the alternative *cis* products. The cycloaddition is therefore face-specific with the nitrile oxide approaching the alkene, as expected, from the less hindered side *anti* to the allylic substituents. The reaction is, however, non-selective in a regiochemical sense, perhaps not surprisingly in view of the similarity of the groups at C(1) and C(4).

The next stage in the sequence required reductive hydrolytic cleavage of the isoxazoline ring. The technique adopted was that developed by Curran [29] involving hydrogenation with a palladium/charcoal or Raney nickel catalyst in boric acid/methanol/water. Although preliminary experiments performed on peracetylated isoxazoline **9** yielded inseparable mixtures of products, it was found that a clean reaction (97%) could be achieved by initial deacetylation to **11** followed by hydrogenolysis under the standard conditions (Scheme 3-3). That the isolated product was the target carbonyl-bridged analogue **12** of ethyl 3-*O*-β-D-xylopyranosyl-α-D-mannopyranoside was evident from its spectroscopic properties, the presence of the carbonyl group being confirmed by an IR absorption at 1720 cm^{-1} and a characteristic ^{13}C NMR peak at 207 ppm. The ^{1}H NMR spectrum of the hexaacetate derivative **13** showed couplings for the dipolarophile-derived pyranose ring of 1.6, 2.7, 11.0 and 10.6 Hz for H(1)-H(2), H(2)-H(3), H(3)-H(4) and H(4)-H(5) respectively, consistent with the proposed structure in which H(1) and H(2) are equatorial and H(3), H(4) and H(5) all axial. The carbonyl group attached to C(3) thus occupies the sterically less demanding equatorial position. It is noteworthy that, although the original alkene was prepared from D-glucose, this portion of the final product has *manno* configuration; this is a consequence of the concerted nitrile oxide cycloaddition delivering both the incipient carbonyl and the oxygen atom to the same face of the dipolarophile. The structure of compound **13** has also been confirmed by X-ray crystallography [36].

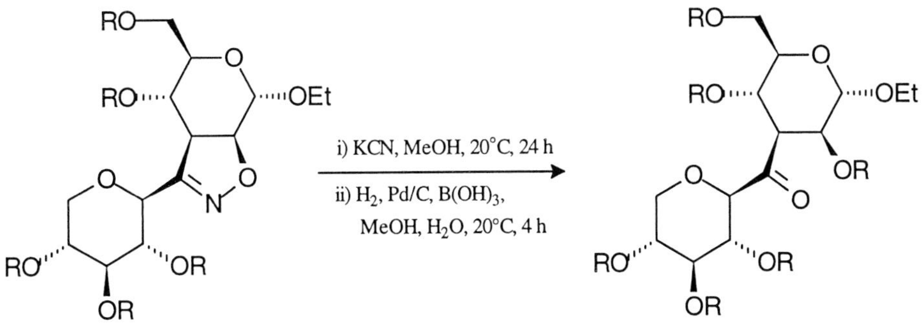

Scheme 3-3

Having successfully prepared a β-1,3-carbonyl-bridged disaccharide from isoxazoline **9** attention was then focussed on the ring-opening of its regioisomer **10**. Similar behaviour would be expected to afford the corresponding β-1,2-linked system **14** (Scheme 3-4). The reaction sequence of deacetylation followed by hydrogenolysis apparently proceded in a similar manner giving in 90% yield a product with ^{13}C NMR and IR peaks at 206 ppm and 1720 cm^{-1} respectively, as expected for the β-hydroxyketone. The ^{1}H NMR data for the hexaacetate derivative were, however, not consistent with this assignment. The diagnostic couplings for the dipolarophile-derived pyranose ring were 3.8, 11.1, 9.3 and 10.3 Hz for H(1)-H(2), H(2)-H(3), H(3)-H(4) and H(4)-H(5) respectively, indicating that it has structure **15** with H(2) axial rather than equatorial. It follows that the isolated product was carbonyl-bridged analogue **16** of ethyl 2-*O*-β-D-xylopyranosyl-α-D-glucopyranoside and not the expected *manno* compound **14**.

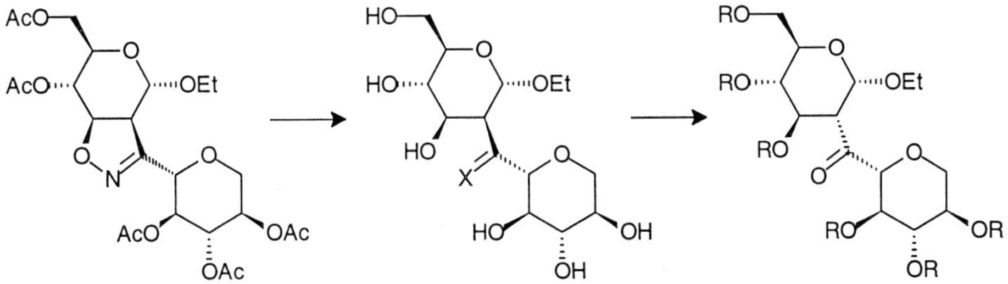

Scheme 3-4

Evidently epimerisation has occured at C(2), either in the first-formed *manno*-ketone or its imine precursor **17**, thus allowing the bulky acyl group to occupy the energetically favoured equatorial position. Epimerization at the position adjacent to an imine function under isoxazoline ring-opening conditions has been observed previously [29].

Reductive ring cleavage of isoxazolines by hydrogenation under non-hydrolytic conditions has become a well-established method for the preparation of γ-aminoalcohols [28]. Such a reaction for the carbohydrate isoxazolines currently under investigation would provide access to a novel class of aminomethylene-linked disaccharides not readily available by other means. To test the viability of this approach isoxazoline **10** was therefore subjected to hydrogenation in acetic acid over a platinum catalyst. From the reaction (Scheme 3-5) was isolated in 90% yield a diastereomeric mixture (5:1) of aminomethylene compounds **18**, which could not be separated nor purified, but were identified spectroscopically and by the preparation of their heptaacetate derivatives. The same mixture of amines (75%, 3:1) was also obtained by reductive amination of carbonyl-linked disaccharide **12**.

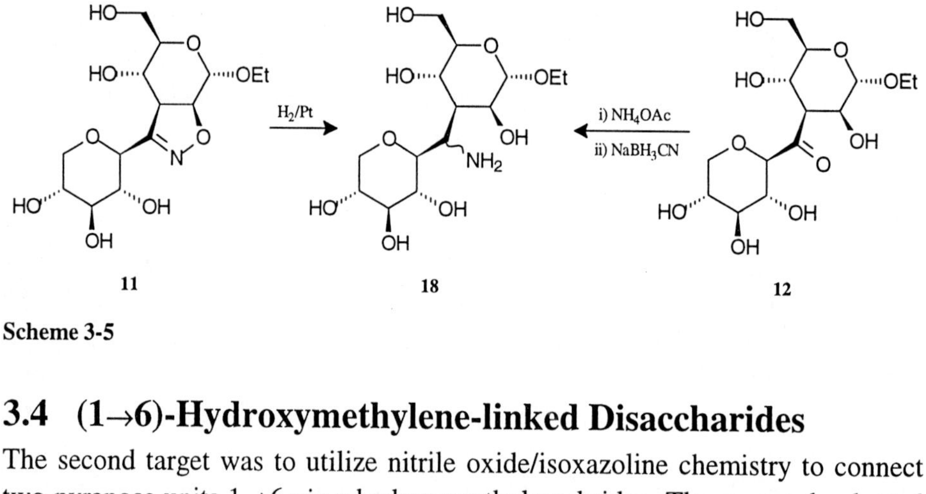

Scheme 3-5

3.4 (1→6)-Hydroxymethylene-linked Disaccharides

The second target was to utilize nitrile oxide/isoxazoline chemistry to connect two pyranose units 1→6 *via* a hydroxymethylene bridge. The approach adopted (Figure 3-6) is based on cycloaddition of a pyranosyl 1-nitrile oxide to a 5,6-dideoxyhex-5-enofuranoside and manipulation of the resulting isoxazoline. To test the strategy D-xylose-derived nitrile oxide **4** was chosen as the dipole component and 5,6-dideoxy-D-*xylo*-hex-5-enofuranoside derivative **19** as the dipolarophile. The latter alkene was prepared from diacetone-D-glucose using established literature methods (Scheme 3-6).

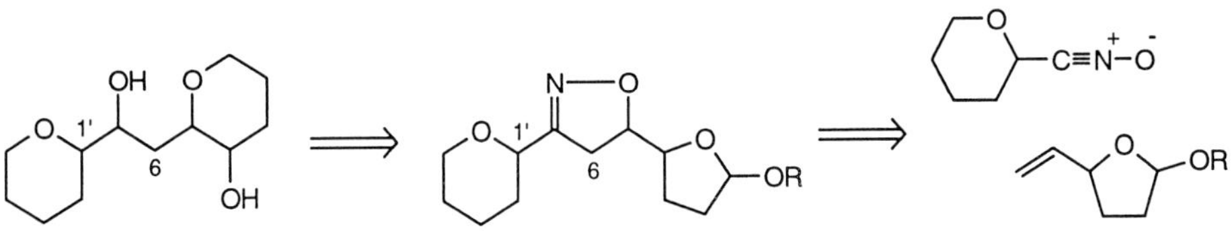

Figure 3-6. Nitrile oxide-isoxazoline route to (1→6)-hydroxymethylene-linked disaccharides.

Benzylation of the hydroxyl at C(3) followed by selective hydrolysis of the 5,6-isopropylidene group afforded 3-*O*-benzyl-1,2-*O*-isopropylidene-D-*gluco*-furanose from which the dimesylate derivative **20** was prepared [40]. The dimesylate, which is a crystalline solid with a good shelf life, was obtained in *ca* 50% overall yield from diacetoneglucose without isolation of the intermediates, and represents a useful source of the required dipolarophile which proved to have limited stability. Alkene **19** was then prepared from **20** as required using a modified Tipsen-Cohen procedure [41] involving treatment with sodium iodide and a zinc/copper couple.

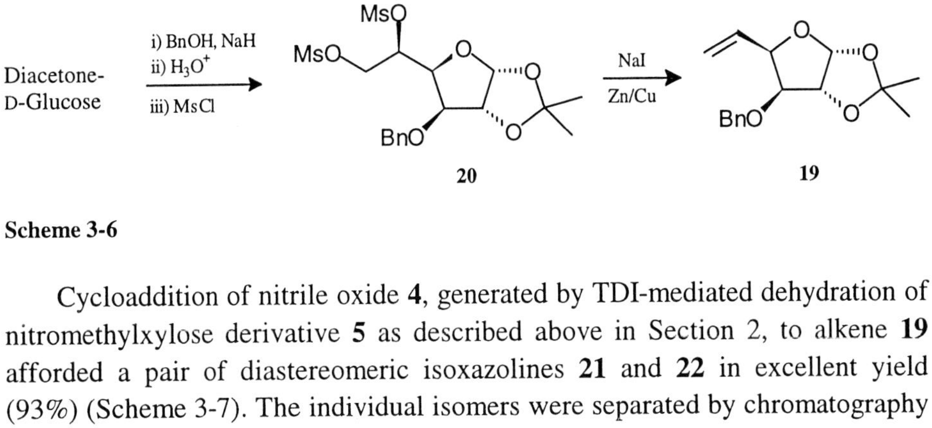

Scheme 3-6

Cycloaddition of nitrile oxide **4**, generated by TDI-mediated dehydration of nitromethylxylose derivative **5** as described above in Section 2, to alkene **19** afforded a pair of diastereomeric isoxazolines **21** and **22** in excellent yield (93%) (Scheme 3-7). The individual isomers were separated by chromatography and their structures provisionally assigned by comparison of their NMR parameters and optical rotations with those of previously reported isoxazolines prepared from the same alkene [42-44]. The structure of the major adduct **21** was confirmed by X-ray crystallography [36], which showed that it has *R*-configuration at the new stereogenic centre C(5); the minor isomer therefore has 5*S* structure **22**. The product ratio **21**:**22** was determined as 78:22 by ^{1}H NMR spectroscopy, which also showed that neither of the other two possible regioisomeric cycloadducts in which the oxygen of the nitrile oxide becomes attached to the terminal methylene of the dipolarophile were formed. The reaction is therefore regiospecific and diastereoselective in favour of the adduct for which there is an *erythro* relationship between C(5) and the adjacent chiral centre C(4). Similar π-facial discrimination has been reported for cycloaddition of various nitrile oxides to chiral allyl ethers [42-47] and has been rationalized in terms of the "inside alkoxy effect" proposed by Houk *et al* [47].

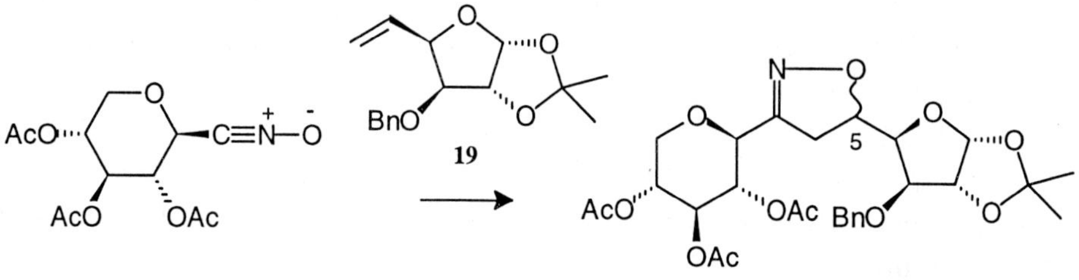

Scheme 3-7

The preferred transition state is considered to have the largest substituent *anti*, the smallest (H) 'outside', and the alkoxy group in the 'inside' position; for alkene **19** the *anti* substituent is linked *via* the five-membered sugar ring to the inside alkoxy group as illustrated in Figure 3-7. There may also be a contribution to the observed *erythro*-selectivity attributable to the homoallylic alkoxy group (R = OBn) at C(3), which in such cyclic systems is known to reinforce the effect of the allylic group when there is a *threo* relationship between these two substituents [42, 43, 46].

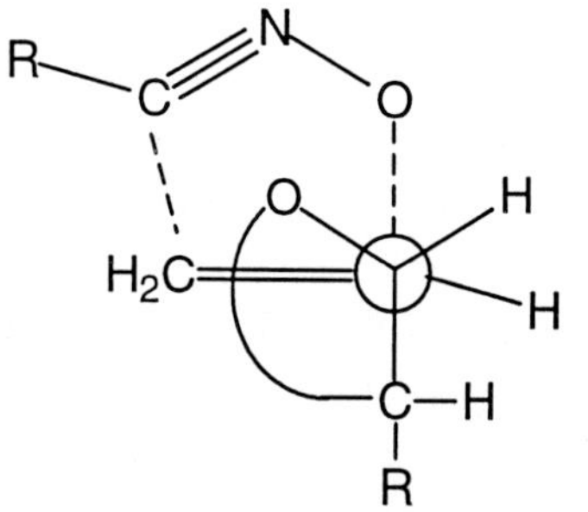

Figure 3-7. Preferred transition state for cycloaddition of nitrile oxides to alkene **19**.

The major adduct **21** was deacetylated and the resulting isoxazoline **23** subjected to reductive hydrolytic ring cleavage in the usual manner using a Raney-nickel, hydrogen and boric acid in methanol-THF-water. From the reaction mixture were isolated the expected β-hydroxyketone/7-osulose **24** in

65% yield, together with lesser amounts (18%) of 7-amino compound **25**. The formation of amine **25** is presumably the result of hydrogenation of the putative intermediate imine **26** competing with its hydrolysis. The presence of the carbonyl group in compound **24** is confirmed by characteristic peaks in its IR and ^{13}C NMR spectra at 1718 cm^{-1} and 207 ppm respectively. The carbonyl group at C(7) was reduced using sodium borohydride in ethanol-water to give in 76% yield a 17:83 mixture of 1,3-diols **27** and **28** which were separated by chromatography. Reversed selectivity was observed using L-Selectride, a 65:35 mixture of **27** and **28** being formed in 79% combined yield. For each isomer the configuration of the new stereogenic centre at C(7) was deduced from its ^{1}H NMR spectrum.

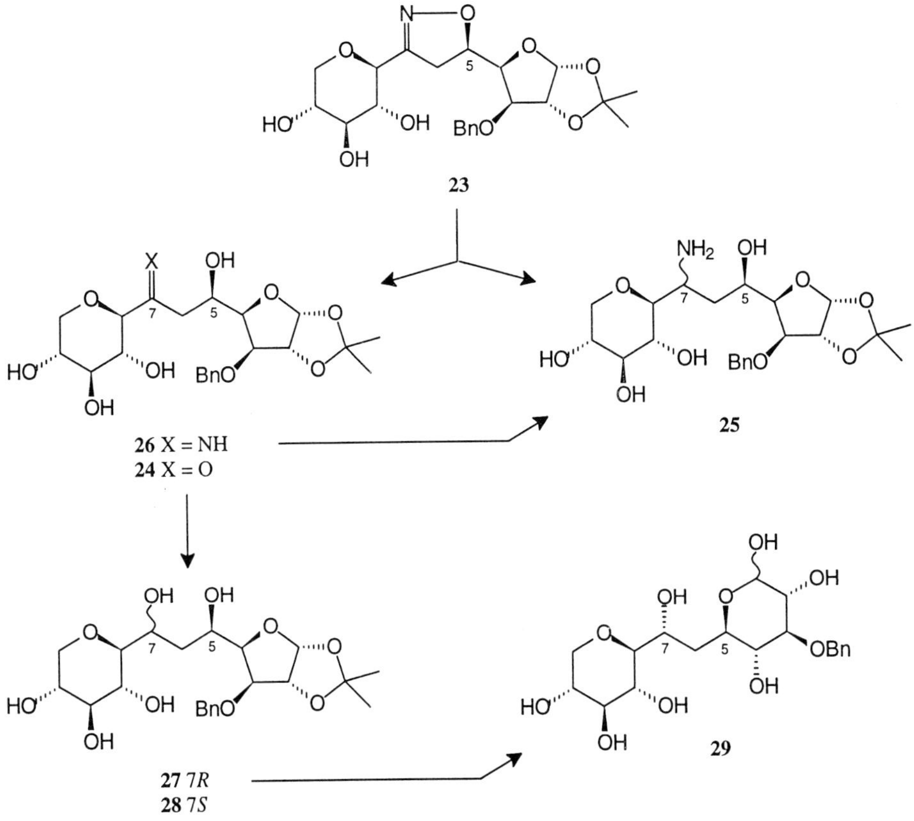

Scheme 3-8

The 5*R*, 7*S* compound **28** adopts a hydrogen-bonded chair-like conformation (Figure 3-8) with the bulky furanosyl (R) and pyranosyl (R') substituents both equatorial. Protons H(5) and H(7) both show typical axial-axial couplings of 10.0 Hz to H(6a) and axial-equatorial couplings (2.4 Hz) to H(6e). In contrast, for diastereoisomer **27** a chair arrangement would require one of the sugar substituents (R or R') to be in a less favoured axial position and there is significant distortion of the chair. A similar rationale has previously been used to assign the configuration of C(7) in 6-deoxydodecadialdose and 6-deoxy-nonose, -decose, -undecose derivatives prepared from the same sugar alkene [44, 48, 49].

In the final stage of the sequence 7R compound **27** was treated with aqueous trifluoroacetic acid which resulted in removal of the 1,2-isopropylidene group followed by furanose to pyranose ring conversion affording the hydroxymethylene-bridged analogue **29** of 6-O-β-D-xylopyranosyl-3-O-benzyl-D-glucopyranose as a 38:62 mixture of α- and β-anomers in 92% combined yield. That the dipolarophile-derived sugar ring adopts the glucopyranose form was evident from the ^{1}H NMR spectrum; the diagnostic data are given in Table 3-1.

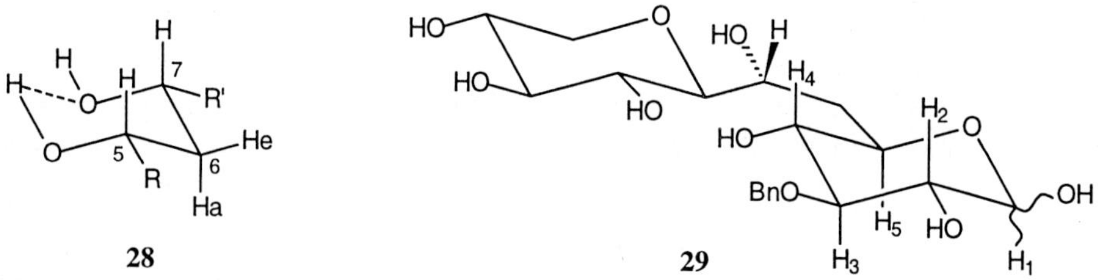

Figure 3-8. Conformation of 1,3-diol **28** and structure of C-disaccharide **29**.

Table 1. Selected ^{1}H NMR data (600 MHz, D$_2$O) for C-disaccharide **29**.

	δ/ppm				J/Hz			
	H(1)	H(2)	H(3)	H(4)	1,2	2,3	3,4	4,5
α	5.11	3.55	3.63	3.22	3.9	9.8	9.2	(a)
β	4.56	3.26	3.42	3.44	8.7	9.3	9.3	9.8

(a) not determined

3.5 Higher-Carbon Dialdoses

Thusfar the disaccharide analogues described herein have all been derived from the cycloaddition reactions of pyranose 1-nitrile oxides, resulting in products with the glycosidic oxygen replaced by carbon. There is, however, also interest in carbohydrates in which sugar units are connected at other positions. In particular, there are higher-carbon sugars in which two sugar rings are joined at their non-reducing termini, *eg* in the eleven-carbon tunicamine unit of the antibiotic tunicamycin. We considered that nitrile oxide/isoxazoline chemistry might also be well suited for the synthesis of such higher-carbon dialdoses as an alternative to traditional methods based on *eg* nitro-aldol reactions [15]. The approach, which is illustrated in Figure 3-9 for a tridecadialdose, involves combination of a hexurononitrile oxide and an ω-unsaturated heptopyranose.

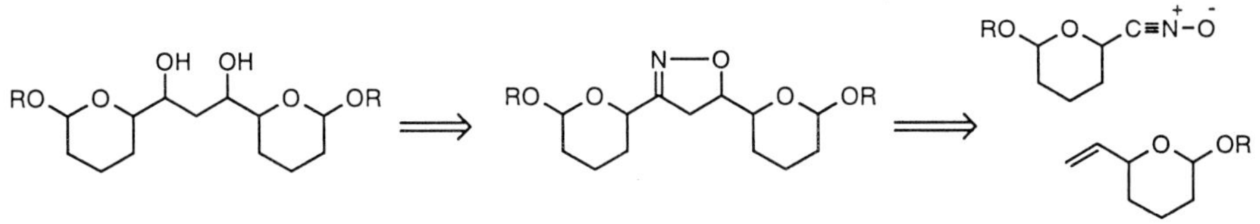

Figure 3-9. Nitrile-oxide-isoxazoline approach to tridecadialdoses.

7-Deoxytridecadialdose derivative **30** was prepared (Scheme 3-9) from hexurononitrile oxide **31** and hept-6-enopyranoside **32**, both of which are accessible from D-galactose. Oxidation with PCC of the 6-hydroxymethyl group of 1,2:3,4-di-*O*-isopropylidene-D-galactose afforded aldehyde **33** from which both the dipolarophile **32** and the oxime precursor **34** of the nitrile oxide were readily prepared by standard procedures. Nitrile oxide **31** was generated *in situ* from oxime **34** by chlorination with NCS followed by addition of base to the resulting hydroximoyl chloride **35**. Reaction in the presence of alkene **32** afforded a 78:22 diastereomeric mixture of isoxazolines **36** and **37** in 91% overall yield. The major adduct was separated by crystallization and its structure determined by comparison of its NMR data with those of other isoxazolines prepared from the same alkene [48-51]; the *R*-configuration at the new stereogenic centre C(6) was also confirmed by X-ray crystallography [42]. The preference for cycloadducts with an *erythro* relationship between C(6) and the adjacent asymmetric centre C(5) can again be explained in terms of the aforementioned inside-alkoxy effect [47].

Reductive hydrolytic cleavage of the isoxazoline ring using hydrogen, palladium/charcoal, boric acid in methanol-water afforded β-hydroxyketone **38** in 72% yield, and in the final stage of the sequence compound **38** was reduced with sodium borohydride to give as the principal product (62%) 7-deoxy-L-*erythro*-D-*gluco*-D-*galacto*-dialdose derivative **30**. The overall yield from oxime **34** was 41%. In order to determine the configuration at the newly-created chiral centre C(8) in the 1,3-diol **30** it was converted into the 6,8-isopropylidene derivative **39** and the NMR spectra of the product compared with those of similar 7-deoxy-nonose and decose derivatives [51].

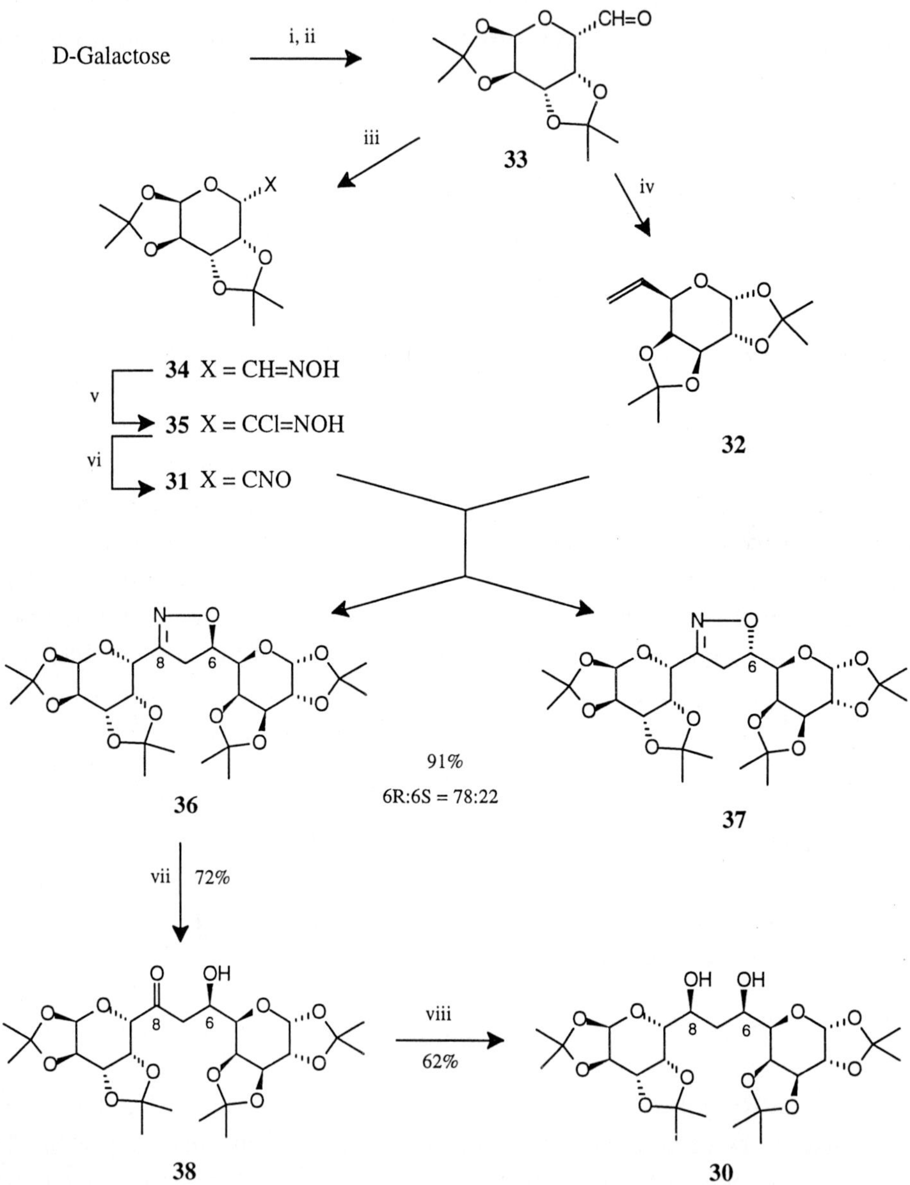

34 X = CH=NOH
35 X = CCl=NOH
31 X = CNO

36 91% 6R:6S = 78:22 **37**

38 62% **30**

Scheme 3-9. Reagents: i) Me$_2$CO, H$^+$; ii) PCC; iii) NH$_2$OH; iv) Ph$_3$P=CH$_2$; v) NCS, pyridine; vi) Et$_3$N; vii) H$_2$, Pd/C; viii) NaBH$_4$.

$3J$-values of 2.5 Hz are observed for both H(6)–H(7e) and H(8)–H(7e), and of 11.6 Hz for H(6)–H(7a) and H(8)–H(7a), indicating that the 1,3-dioxane ring adopts a regular chair conformation as indicated in Figure 3-10. It is concluded that both the galactopyranosyl substituents at the 6- and 8-positions are equatorial and that C(8) therefore has the *S*-configuration.

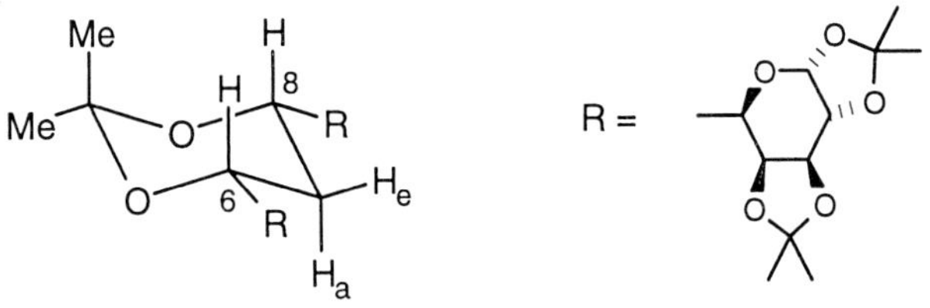

Figure 3-10. 6,8-*O*-Isopropylidene derivative **39** of 1,3-diol **30**.

Using the same general approach the corresponding 6-deoxydodecadialdoses can be prepared by combination of a six-carbon nitrile oxide and a hex-5-enofuranoside. Thus 6-deoxy-L-*erythro*-D-*gluco*-D-*gluco*- and 6-deoxy-L-*erythro*-D-*manno*-D-*gluco*-dialdose derivatives **40** and **41** were synthesized via isoxazoline **42** in 15% overall yield from the hexurononitrile oxide **31** and the D-glucose-derived alkene **19**.

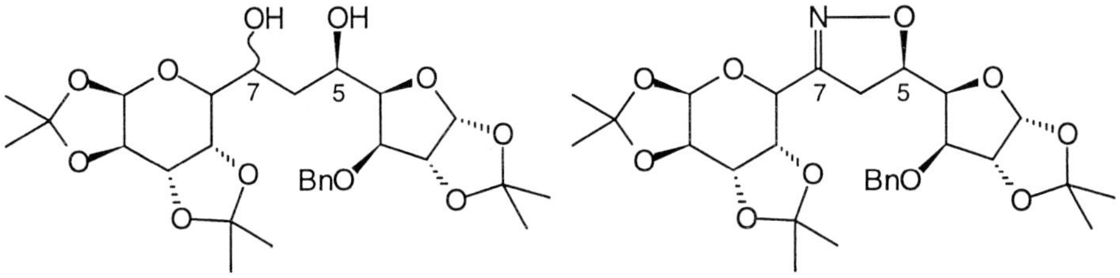

40 7R
41 7S

42

Figure 3-11. Dodecadialdoses and their isoxazoline precursor.

3.6 Conclusion

The results described herein demonstrate that nitrile oxide/isoxazoline chemistry is well suited for coupling two sugar rings and thus creating *C*-disaccharides with functionalized bridges. In the main examples chosen for discussion 1→2, 1→3, 1→6, and 6→6-carbonyl/hydroxymethylene/aminomethylene links were formed between xylo-, gluco-, manno- and galactopyranosyl units. Although most attention has thus been focused on alkenes derived from D-glucose and dipole components for the cycloaddition step in which the nitrile oxide is attached to the 1-position of pyranoses, the method is limited only by the availability of nitromethyl- or hydroxyimino-sugars as precursors of the nitrile oxides and of suitable sugar alkenes for use as the dipolarophile. For example, we have used α-methyl 5,6-dideoxy-2,3-O-isopropylidene-D-*lyxo*-hex-5-enofuranoside, prepared from D-mannose, as the dipolarophile in combination with pyranosyl 1-nitrile oxides derived from D-galactose, D-mannose and D-arabinose [36, 46]; a D-ribofuranosyl 1-nitrile oxide has also been reported [52, 53]. And there is further scope for increasing the range of functionalized bridges by chemical modification, *eg* of the connecting carbonyl group.

Acknowledgements

I am grateful for the contributions to the work described in this article made by the following co-workers: A.J. Blake, I.M. Dawson, W.J. Ferguson, R.O. Gould, T. Johnson, K.J. Penman, D. Reed, R.A.C. Rennie, I.H. Sadler and A.A. Young; and to the SERC/EPSRC, DENI and ICI plc for financial support.

3.7 References

1. D. J. O'Leary, Y. Kishi, *J. Org. Chem.*, **1993**, 58, 304-306.
2. L. Lay, F. Nicotra, C. Pangrazio, L. Panza, G. Russo, *J. Chem. Soc., Perkin Trans. 1*, **1994**, 333-338.
3. D. Rouzaud, P. Sinaÿ, *J. Chem. Soc., Chem. Commun.*, **1983**, 1353-1354.
4. R. R. Schmidt, R. Preuss, *Tetrahedron Lett.*, **1989**, *30*, 3409-3412.
5. B. Giese, T. Witzel, *Angew. Chem., Int. Ed. Engl.*, **1986**, *25*, 450-451.
6. B. Giese, M. Hoch, C. Lamberth, R. R. Schmidt, *Tetrahedron Lett.*, **1988**, *29*, 1375-1378.
7. R. Ferritto, P. Vogel, *Tetrahedron Asymmetry*, **1994**, *5*, 2077-2092.
8. T.F. Herpin, W.B. Motherwell, M.J. Tozer, *Tetrahedron Asymmetry*, **1994**, *5*, 2269-2282.
9. A. Mallet, J. -M. Mallet, P. Sinaÿ, *Tetrahedron Asymmetry*, **1994**, *5*, 2593-2608.
10. T. Haneda, P. G. Goekjian, S. H. Kim, Y. Kishi, *J. Org. Chem.*, **1992**, *57*, 490-498.
11. O. Eyrisch, W. -D. Fessner, *Angew. Chem., Int. Ed. Engl.*, **1995**, *34*, 1639-1641.
12. S. Jarosz, B. Fraser-Reid, *Tetrahedron Lett.*, **1989**, *30*, 2359-2362.
13. R. W. Armstrong, B. R. Teegarden, *J. Org. Chem.*, **1992**, *57*, 915-922.

14. Y. Wang, S.A. Babirad, Y. Kishi *J. Org. Chem.*, **1992**, *57*, 468-481.

15. T. Suami, H. Sasai, K. Matsuno, *Chem. Lett.*, **1983**, 819-822.

16. O. R. Martin, W. Lai, *J. Org. Chem.*, **1990**, *55*, 5188-5190.

17. Z. J. Witczak in *Levoglucosenone and Levoglucosans* (Ed. Z.J. Witczak), ATL Press, Mount Prospect, Il. USA, **1994**, Chapter 1.

18. S. Jarosz, B. Fraser-Reid, *J. Org. Chem.*, **1989**, *54*, 4011-4013.

19. D. P. Sutherlin, R. W. Armstrong, *Tetrahedron Lett.*, **1993**, *34*, 4897-4900.

20. F. Sarabia-Garcia, F. J. Lopez-Herrera, M. S. P. Gonzalez, *Tetrahedron*, **1995**, *51*, 5491-5500.

21. R. R. Schmidt, A. Beyerbach, *Liebigs. Ann. Chem.*, **1992**, 983-986.

22. H. M. Binch, A. M. Griffin, S. Schwidetzky, M. V. J. Ramsay, T. Gallagher, F. W. Lichtenthaler, *J. Chem. Soc., Chem. Commun.*, **1995**, 967-968.

23. B. Aebischer, J. H. Bieri, R. Prewo, A. Vasella, *Helv. Chim. Acta*, **1982**, *65*, 2251-2272.

24. S. Hanessian, M. Martin, R. C. Desai, *J. Chem. Soc., Chem. Commun.*, **1986**, 926-927.

25. A Dondoni, L. Kniezo, M. Martinkova, *J. Chem. Soc., Chem. Commun.*, **1994**, 1963-1964.

26. S. J. Danishefsky, W. H. Pearson, D. F. Harvey, C. J. Maring, J. P. Springer, *J. Am. Chem. Soc.*, **1985**, *107*, 1256-1268.

27. J. Jurczak, T. Bauer, S. Jarosz, *Tetrahedron*, **1986**, *42*, 6477-6486.

28. V. Jäger, I. Müller, R. Schohe, M. Frey, R. Ehrler, B. Häfele, D. Schröter, *Lectures Heterocycl. Chem.*, **1985**, *8*, 79-98.

29. D. P. Curran, *Adv. Cycloaddition*, **1988**, *1*, 129-189.

30. K. B. G. Torssell, *Nitrile Oxides, Nitrones and Nitronates in Organic Synthesis*, VCH Publishers, Weinheim, **1988**.

31. S. Kanemasa, O. Tsuge, *Heterocycles*, **1990**, *30*, 719-736.

32. P. Grünanger, P. Vita-Finzi, *Isoxazoles: The Chemistry of Heterocyclic Compounds*, **1991**, Wiley, New York, Vol. 49, Part 1, p. 572-602.

33. P. Caramella, P. Grünanger in *1,3-Dipolar Cycloaddition Chemistry*, (Ed. A. Padwa), Wiley, New York, **1984**, Chapter 3.

34. T. Mukaiyama, T. Hoshino, *J. Am. Chem. Soc.*, **1960**, *82*, 5339-5342.

35. A. Förtsch, H. Kogelberg, P. Köll, *Carbohydr. Res.*, **1987**, *164*, 391-402.

36. A. J. Blake, R. O. Gould, W. J. Ferguson, T. Johnson, R. M. Paton, K. J. Penman, R. A. C. Rennie unpublished observations.

37. R. J. Ferrier, *Methods Carbohydr. Chem.*, **1972**, *6*, 307-331.

38. A. J. Blake, I. M. Dawson, A. C. Forsyth, T. Johnson, R. M. Paton, R. A. C. Rennie, P. Taylor, *J. Chem. Res. (S)*, **1988**, 328-329.

39. H. Gnichtel, L. Autenreith, P. Luger, K. Vangehr, *Liebigs Ann. Chem.*, **1982**, 1091-1095.

40. H. Paulsen, D. Stoye, *Chem. Ber.*, **1969**, *102*, 820-833.

41. R. S. Tipsen, A. Cohen, *Carbohydr. Res.*, **1965**, *1*, 338-340.

42. A. J. Blake, R. O. Gould, K. E. McGhie, R. M. Paton, D. Reed, I. H. Sadler, A. A. Young, *Carbohydr. Res.*, **1991**, *216*, 461-473 and references therein.

43. F. De Amici, C. De Micheli, A. Ortisi, G. Gatti, R. Gandolfi, L Toma, *J. Org. Chem.*, **1989**, *54*, 793-798.

44. K. E. McGhie, R. M. Paton, *Tetrahedron Lett.*, **1993**, *34*, 2831-2834.

45. U. A. R. Al-Timari, L. Fisera, *Carbohydr. Res.*, **1991**, *218*, 121-127.

46. A. J. Blake, G. Kirkpatrick, K. E. McGhie, R. M. Paton, K. J. Penman, *J. Carbohydr. Chem.*, **1994**, *13*, 409-419.

47. K. N. Houk, S. R. Moses, Y.-D. Wu, N. G. Rondan, V. Jäger, R. Schohe, F. R. Franczek, *J. Am. Chem. Soc.*, **1984**, *106*, 3880-3882 and references therein.

48. R. M. Paton, A. A. Young, *J. Chem. Soc., Chem. Commun.*, **1991**, 132-134.

49. R. M. Paton, A. A. Young, *J. Chem. Soc., Chem. Commun.*, **1994**, 993-994.

50. R. M. Paton, A. A. Young, *J. Chem. Soc., Perkin Trans. 1*, **1997**, in the press.

51. A. J. Blake, R. O. Gould, R. M. Paton, A. A. Young, *J. Chem. Res. (S)*, **1993**, 482-483.

52. A. P. Kozikowski, S. Goldstein, *J. Org. Chem.*, **1983**, *48*, 1139-1141.

53. T. V. RajanBabu, G. S. Reddy, *J. Org. Chem.*, **1986**, *51*, 5458-5461.

4 Synthesis of Glycosyl Phosphate Mimics

F. Nicotra

4.1 Introduction

The interest in "glycomimetics" is growing with the knowledge of the multifarious roles that carbohydrates play in the biological events. It is now well proven that glycoconjugates are the main structures responsible for cell-cell and cell-molecule recognition events and host-pathogen interactions.

Glycomimetics are molecules that mimic a natural sugar or the transition state of an enzymatic transformation. They can, in principle, replace the normal substrate in interactions with receptors and active sites of enzymes.

The modifications introduced in glycomimetics, with respect to the natural sugar, are often more "functional" than structural, in the sense that the functional group mainly involved in the metabolism of the carbohydrates, the acetal function, is modified. This give rise to compounds unable to undergo the normal metabolic transformation of the sugar, interfering in this way in its metabolism.

The metabolism of carbohydrates involves mostly glycosyl phosphates, which are the main natural glycosylating agents. We will describe here our studies devoted to the synthesis of glycosyl phosphates mimics.

4.2 Roles of Glycosyl Phosphates

Glycosyl phosphates play a central role in the metabolism of carbohydrates. They are involved in primary metabolism, for example in glycolysis and gluconeogenesis; they act as glycosyl donors in the biosynthesis of oligosaccharides, polysaccharides and glycoconjugates, and in some cases they are regulators of metabolic processes. β-D-Fructose-2,6-bisphosphate is an interesting example of metabolic regulators. It is a potent inhibitor of fructose 1,6-bisphosphatase (K_i 2.5 µM) and an allosteric activator of phosphofructokinases; and it acts as an activator of the glycolysis and inhibitor of gluconeogenesis [1].

In the glycosylation processes, glycosyl phosphates can be involved in different ways: directly in non Leloir pathways, more commonly as nucleotide

In the glycosylation processes, glycosyl phosphates can be involved in different ways: directly in non Leloir pathways, more commonly as nucleotide diphosphate (NDP) derivatives in the Leloir pathway or finally as dolichyl diphosphate derivatives (e.g. in the biosynthesis of N-linked glycoproteins). Dolichyl diphosphate derivatives are formed by reaction of dolichyl phosphate with the NDP-sugar, which in turn is obtained from NTP and the glycosyl phosphate. In conclusion, the glycosyl phosphate is the general key molecule in the glycosylation processes.

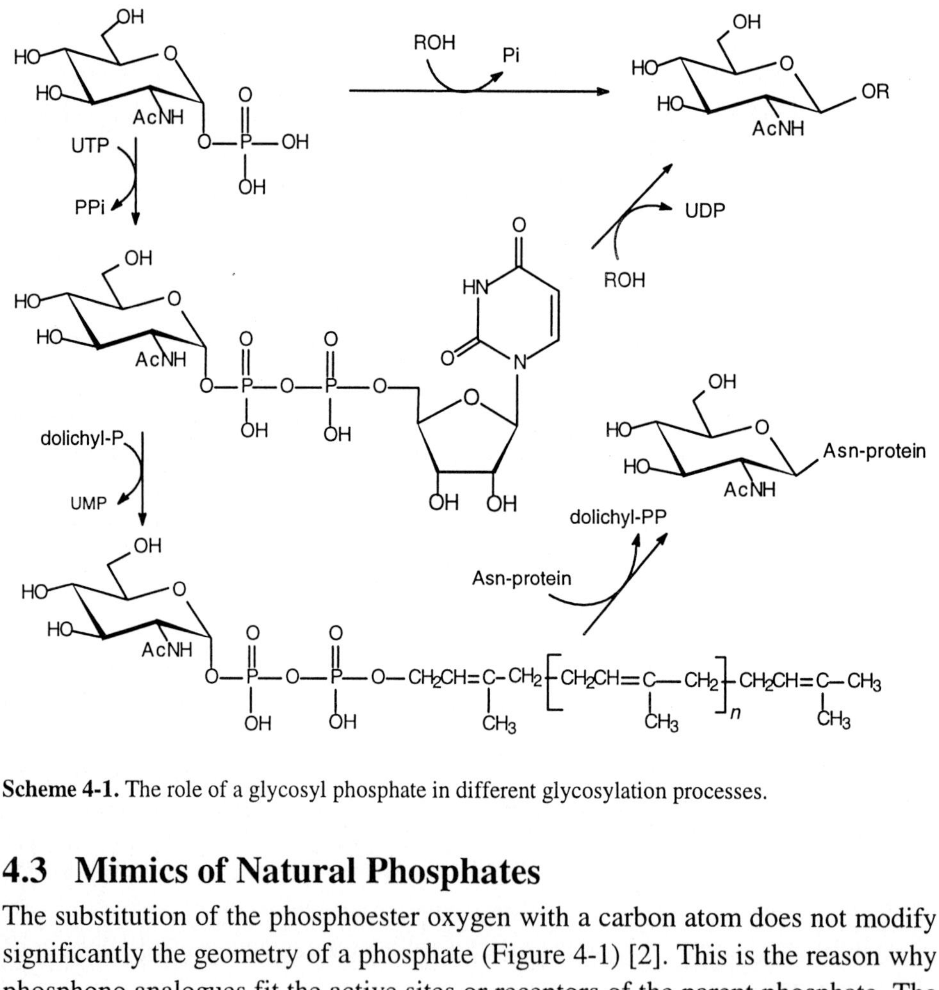

Scheme 4-1. The role of a glycosyl phosphate in different glycosylation processes.

4.3 Mimics of Natural Phosphates

The substitution of the phosphoester oxygen with a carbon atom does not modify significantly the geometry of a phosphate (Figure 4-1) [2]. This is the reason why phosphono analogues fit the active sites or receptors of the parent phosphate. The phosphonates however cannot undergo the cleavage of the phosphoester bond, which is the main metabolic transformation of the natural phosphate, and this often results in an inhibition of the metabolic process.

Figure 4-1. Bond length (in Å) and angles of a phosphate and a phosphonate.

In the light of these observations, different phosphono analogues of natural phosphates have been synthesized, and the biological activity of some of them has been investigated. In general, two different classes can be defined: a) isosteric analogues, in which the phosphoesteric oxygen is substituted by one carbon atom, without any appreciable change of the geometry of the molecule; and b) non-isosteric analogues, in which the phosphoesteric oxygen is simply removed or replaced by two or more carbon atoms. Examples of the two classes are the isosteric analogue of ribose 5-phosphate [3] and the non-isosteric phosphono analogue of glucose 6-phosphate [4] shown in Figure 4-2.

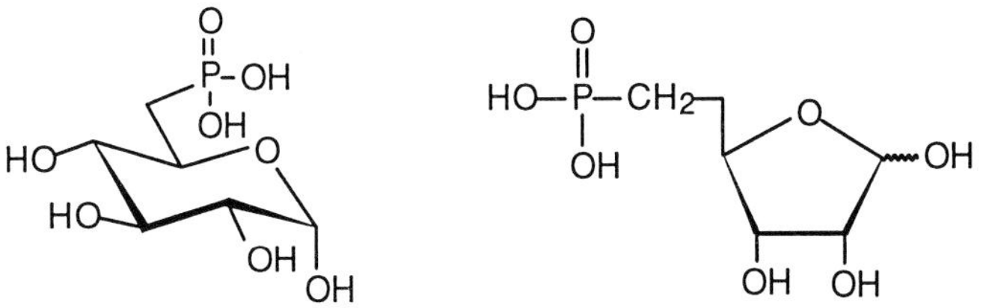

non-isosteric phosphono analogue of glucose 6-phosphate

isosteric phosphono analogue of ribose 5-phosphate

Figure 4-2. Isosteric and non isosteric analogues of natural phosphates.

4.4 Mimics of Glycosyl Phosphates

Despite the biological importance of the glycosyl phosphates, until 1981 no examples of synthesis of phosphono analogues of these molecules were described [5]. Probably one of the reason lies in the chemical and stereochemical difficulties faced in the substitution of the anomeric oxygen of the sugar with a carbon atom, a process requiring C-glycosylation.

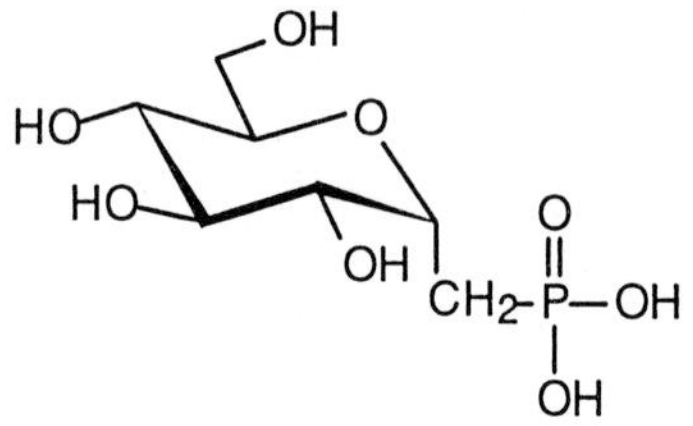

Figure 4-3. Isosteric phosphono analogue of α-D-glucose 1-phosphate.

4.4.1 Direct Methods of Glycosyl Methylphosphonate Synthesis

Following the main C-glycosylation methods, we will first discuss the possibility of direct introduction of a methylenephosphonic group at the anomeric center. C-glycosides can be obtained by nucleophilic substitution on a glycosyl halide using a Grignard reagent [6] or a malonate [7]. The process occurs mainly with inversion of the anomeric configuration, but the initial α or β configuration of the reacting halide depends on different factors including ring size, its preferred conformation and the presence of halide ions in the reaction mixtures. Moreover, an acyl protecting group at C-2 exerts a participating effect directing the attack of the nucleophile from the opposite side. Attempts to use this procedure for the direct introduction of the methylenephosphonic group failed.

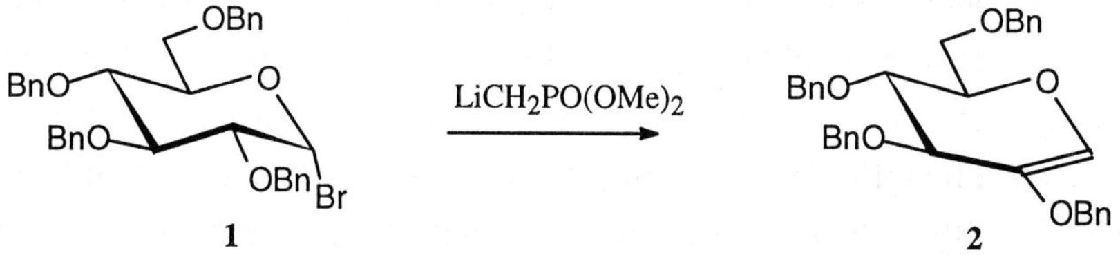

Scheme 4-2

LiCH$_2$PO(OMe)$_2$ [8], a hard nucleophile, reacts with 2,3,4,6-tetra-O-benzyl-α-D-glucopyranosyl bromide **1** affording the elimination product **2** (Scheme 4-2) [9]. Unsatisfactory results have also been obtained with the corresponding cuprate [9].

A different C-glycosylation approach, allowing the use of hard carbon-nucleophiles, involves the addition of the nucleophile to a glyconolactone, followed by the reduction with triethylsilane of the obtained lactol (Scheme 4-3)

[10]. In the case of a glucopyranosyl derivative this process affords products with the β anomeric configuration, the anomeric effect directing the incoming hydride ion from the α face [10].

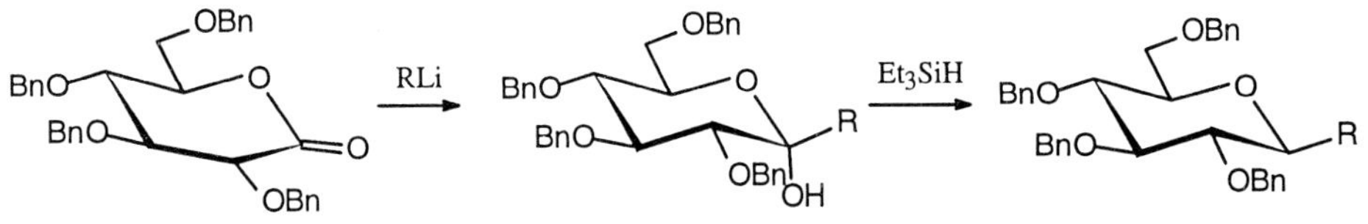

Scheme 4-3

LiCH$_2$PO(OMe)$_2$ reacts with 2,3,4,6-tetra-*O*-benzyl-α-D-gluconolactone **3** to afford the lactol-phosphonate **4**, but the subsequent reduction of **4** with triethylsilane does not occur (Scheme 4-4) [9].

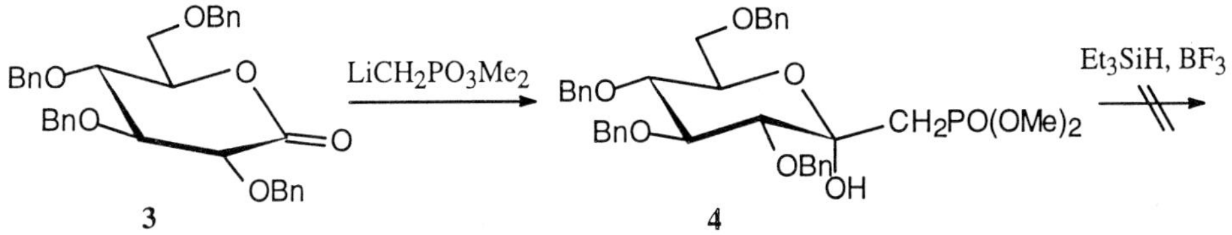

Scheme 4-4

Another interesting C-glycosylation procedure involves the reaction of a stabilized ylide with an aldose, followed by Michael cyclization of the product. This procedure has been employed by McClard who obtained the C-ribosyl phosphonate **6** by refluxing 2,3-*O*-isopropylidene-5-*O*-triphenylmethyl-D-ribose **5** in methanol with diphenyl triphenylphosphoranylidenemethylphosphonate for 40 hours (Scheme 4-5) [11]. The yield however was not reported, and attempts to reproduce the reaction (with dimethyl triphenylphosphoranylidene methylphosphonate) [12] and to extend the reaction to other substrates [1], failed. The use of Horner-Emmons reagents, tetraethyl or tetramethyl methylenediphosphonate in the presence of a base [12, 13], seems to be more general; however, the stereoselectivity of the reaction was very poor (Scheme 4-5).

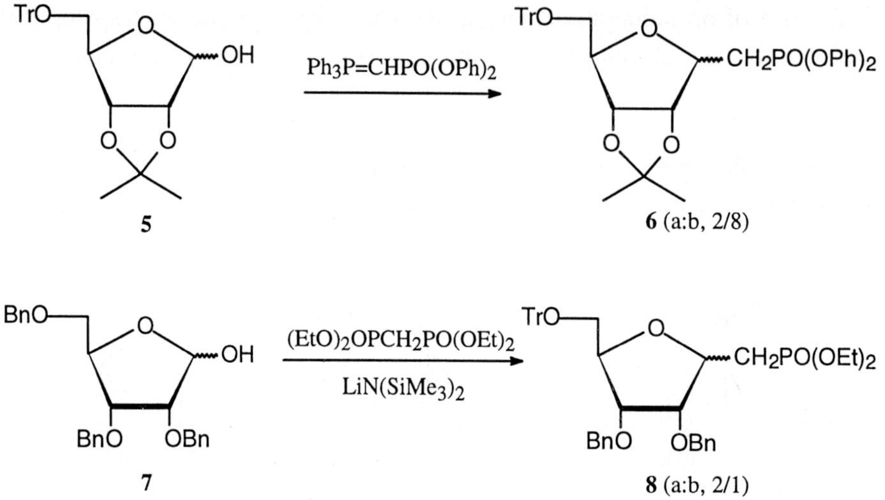

Scheme 4-5. Direct formation of ribofuranosyl phosphonates.

4.4.2 Indirect Methods of Glycosyl Methylphosphonate Synthesis

In view of the difficulties found in the direct introduction of the methylenephosphonic group at the anomeric center of the sugar, indirect methods have been developed. These methods rely upon the formation of a C-glycoside in which the carbon linked to the anomeric center of the sugar, is properly functionalysed for the subsequent introduction of a phosphonic group (Scheme 4-6). Aldehydes, alkyl halides and methanesulfonates have been used.

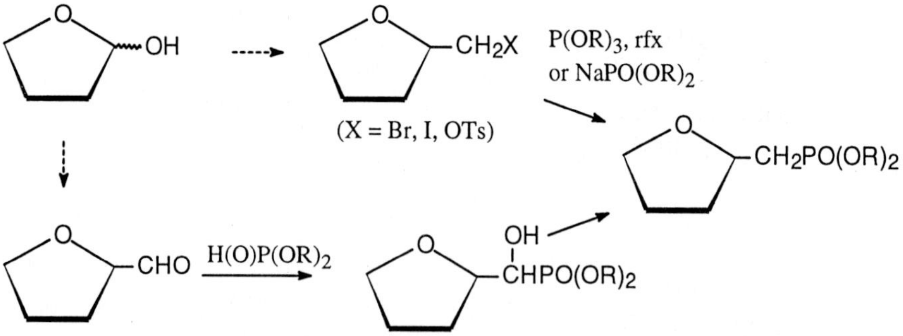

Scheme 4-6. Possible strategies for the synthesis of phosphono analogues of glycosyl phosphates.

The use of an aldehyde has been described for example by Vasella [14] in the synthesis of the phosphono analogue of β-D-fructose 2,6-bisphosphate (Scheme 4-7). The reaction of the aldehyde **9** with diphenyl phosphite in the presence of triethylamine, followed by treatment of the resulting α-hydroxyphosphonate with N,N'-thiocarbonylimidazole, affords **10**. Deoxygenation with Bu₃SnH in refluxing toluene, converts **10** into the phosphonate **11**.

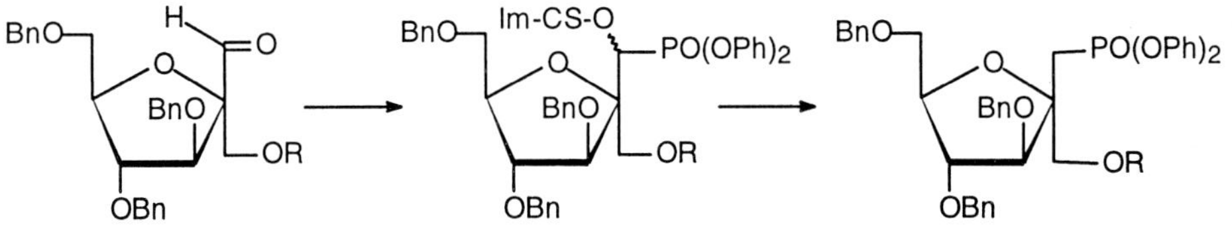

Scheme 4-7. Formation of a C-glycosyl phosphonate by reaction of a C-glycosyl aldehyde with a phosphite.

Most of the reported syntheses involve the use of C-glycosyl halides (or mesylates), which are then treated with a trialkyl phosphite (Arbuzov reaction) to afford the corresponding phosphonate.

4.4.2.1 Synthesis of C-Glycosyl Halides

The stereoselective synthesis of C-glycosyl halides has been achieved by different C-glycosylation methods. Among them, one of the more convenient is the reaction of an aldose with methylenetriphenylphosphorane, followed by the mercurio-cyclization of the obtained glycoenitol [15].

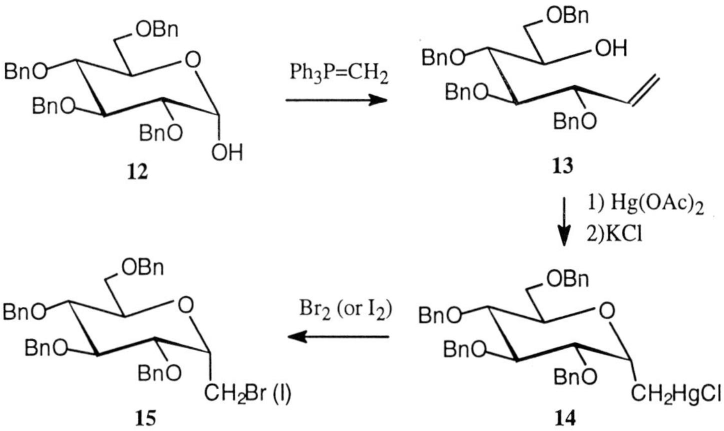

Scheme 4-8. Wittig reaction-mercuriocyclization conversion of an aldose into a C-glycoside.

This process stereoselectively affords C-glycosyl-mercurioderivatives with a 1,2-*cis* relationship [16], which can be easily converted into the corresponding halide by treatment with I_2, Br_2, NBS or NIS.

The direct halocyclization of the intermediate glycoenitol is also possible, but it is less general. 3,4,6-Tri-*O*-benzyl-D-arabinohexenitol **16** reacts with NBS affording the β-*C*-arabinoside **17** in 78% d.e. the stereoselection obtained by mercuriocyclization [13].

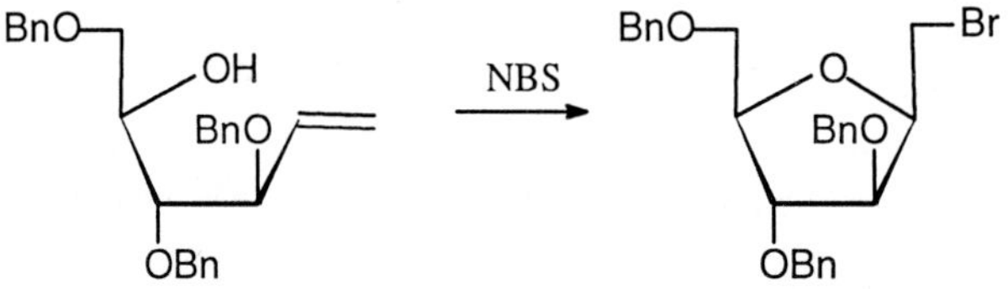

Scheme 4-9. Bromocyclization of 3,4,6-tri-*O*-benzyl-D-arabinohexenitol of **16** to *C*-arabinosyl bromide **17**.

In the case of 3,4,5,7-tetra-*O*-benzyl-D-glucoheptenitol **13**, no reaction occurs by treatment with NIS, or with I_2-$NaHCO_3$-MeCN (Bartlett conditions). On the contrary, treatment with I_2 under slightly acidic conditions (pH 4, oxolane water) results in a debenzylation, affording the more favoured 5-member cyclization product [17].

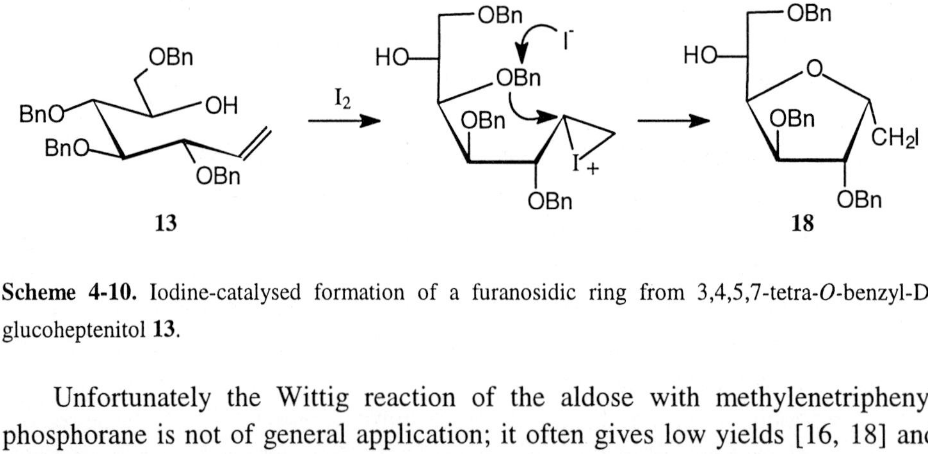

Scheme 4-10. Iodine-catalysed formation of a furanosidic ring from 3,4,5,7-tetra-*O*-benzyl-D-glucoheptenitol **13**.

Unfortunately the Wittig reaction of the aldose with methylenetriphenyl phosphorane is not of general application; it often gives low yields [16, 18] and epimerization at C-2 of the starting sugar [18]. Furthermore, it has never been described on aminosugars. In our hands, any attempt to effect this reaction on differently protected D-glucosamine failed [9].

An interesting alternative to prepare a glycoenitol such as **13** is the vinylation of a "one carbon shorter" aldose, provided that the stereochemistry of the reaction is controlled. This is the case when divinylzinc is used as vinylating agent. It reacts with different tri-*O*-benzyl-D-pentoses to afford stereoselectively the *threo* product (Scheme 4-11) [19].

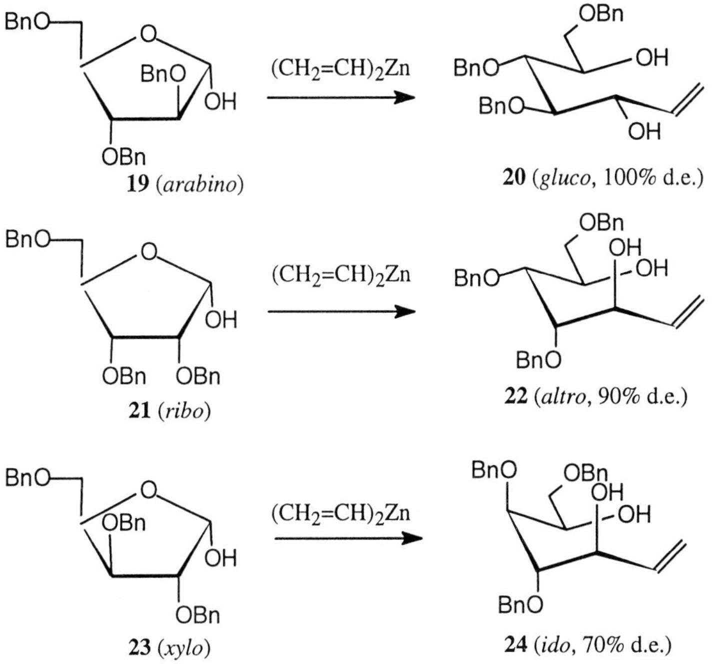

Scheme 4-11. Examples of formation of enitols by stereoselective vinylation of pentoses.

4.4.2.2 Conversion of C-Glycosyl Halides into Phosphonates

The conversion of a C-glycosyl halides into the corresponding phosphonate is usually performed by treatment with a phosphite (Arbuzov reaction). The reaction requires high temperatures, and is usually performed without solvents. The yields are governed by the steric hindrance both of the halide and the phosphite. Phosphites such as tris(trimethylsilyl)phosphite or ethyl diphenyl phosphite allow an easier subsequent conversion of the phosphonic ester into the free phosphonic acid, but their hindrance lowers the yields of the reaction. Hindered halides also result in poor yields. A significant example of the influence of the halide structure in the Arbuzov reaction comes from the synthesis of the phosphono analogue **33** of β-D-fructose 2,6-bisphosphate [20].

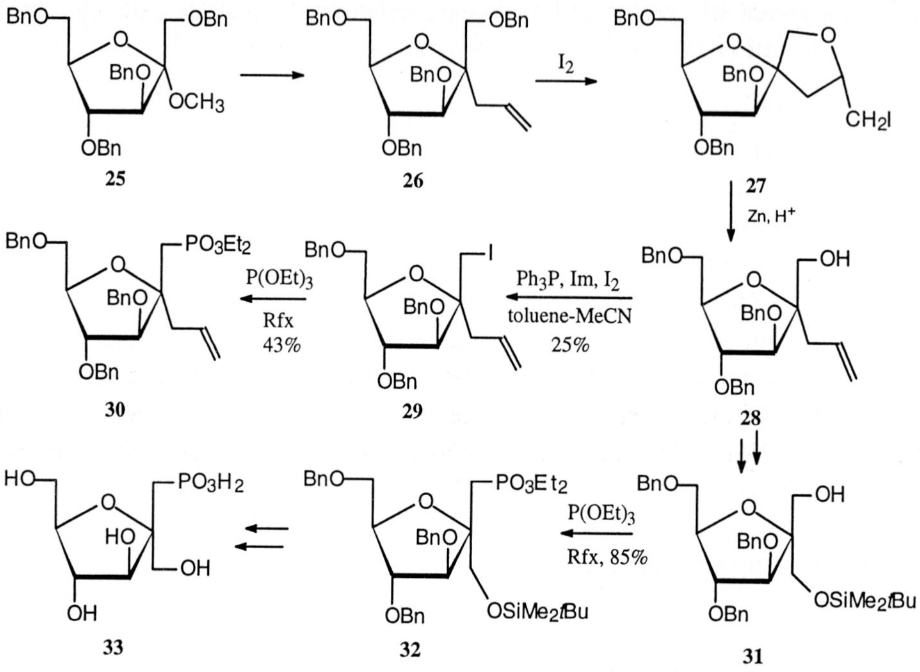

Scheme 4-12. Synthesis of the phosphono analogue of β-D-fructose 2,6-bisphosphate.

In this case both the conversion of the primary hydroxyl-group of **28** into the iodide **29** and the subsequent Arbuzov reaction occur in very poor yields. On the contrary, after the conversion of the allylic substituent of **29** into the protected hydroxymethylene group of **31**, the Arbuzov reaction proceeds in good yields to give **32**.

Following this procedure, involving the formation of a C-glysosyl halide and its reaction with a trialkylphosphite, the phosphono analogues of α-D-glucose 1-phosphate [21], α- and β-D-mannose 1-phosphate [22], α-D-ribose 1-phosphate [23], and β-D-arabinose 1-phosphate [13] have been synthesized. A glycal-1-ylmethyl phosphonate, analogue of the transition state in the glycosylation process catalyzed by a galactosyltransferase, has been obtained by Arbuzov reaction on the corresponding mesylate [24].

4.5 Mimics of *N*-Acetyl-α-D-Glucosamine 1-Phosphate and *N*-Acetyl-α-D-Mannosamine 1-Phosphate

4.5.1 The Biological Roles

Two glycosyl phosphates of particular interest from a biological and pharmacological point of view are *N*-acetyl-α-D-glucosamine 1-phosphate (GlcNAc 1P) and *N*-acetyl-α-D-mannosamine 1-phosphate (ManAc 1P).

N-acetyl-α-D-glucosamine 1-phosphate is the key intermediate in the biosynthesis of the *N*-linked glycoproteins, a class of glycoconjugates involved in many important cell-cell and cell-pathogen recognition phenomena, such as HIV-lymphocyte T adhesion or tumor cells-selectin adhesion. The first step in the biosynthesis of the *N*-linked glycoproteins is the conversion of *N*-acetyl-α-D-glucosamine 1-phosphate into UDP-GlcNAc which in turn is converted into the dolichyl-*N*-acetyl-α-D-glucosamine 1-diphosphate. Further glycosylations afford a dolichyl-diphosphate-oligosaccharide which is then transferred to an asparagine residue of the protein (Scheme 4-13).

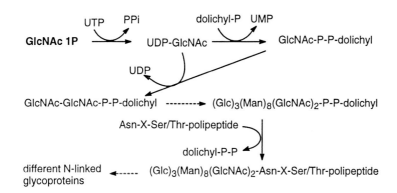

Scheme 4-13. The role *N*-acetyl-α-D-glucosamine 1-phosphate in the biosynthesis of *N*-linked glycoproteins.

Other important processes involving *N*-acetyl-α-D-glucosamine 1-phosphate are the biosynthesis of mureine and theicoic acid, the main components of the bacterial cell walls. Recently it has been shown that *N*-acetyl-α-D-glucosamine 1-phosphate is involved in a glycosylation-deglycosylation of some proteins. This is an abundant and dynamic process the role of which is not clear yet [25], and still under investigation.

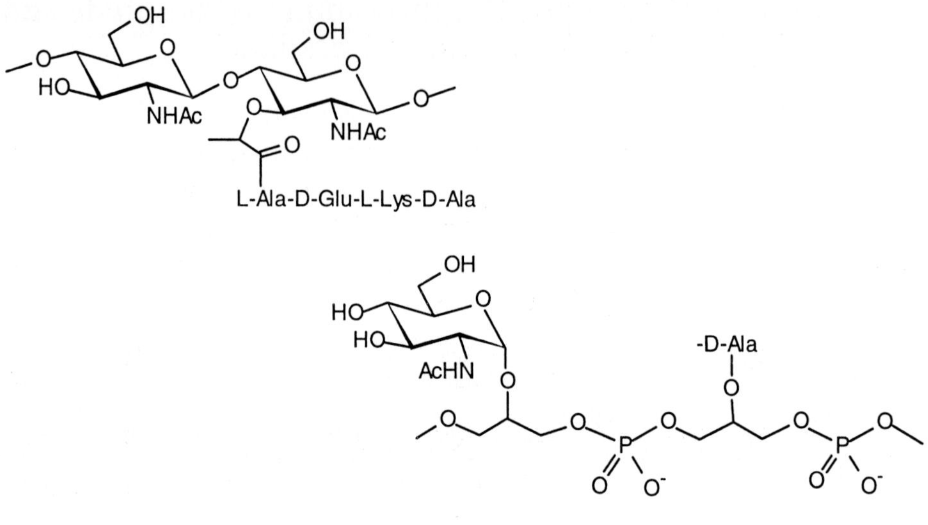

Figure 4-4. Repeating unit of mureine (above) and theicoic acid (below).

N-acetyl-α-D-mannosamine 1-phosphate is involved in the biosynthesis of N-acetylneuraminic acid (Scheme 4-14), a sialic acid present in complex oligosaccharides and, inter alia, in tumor-associated oligosaccharides.

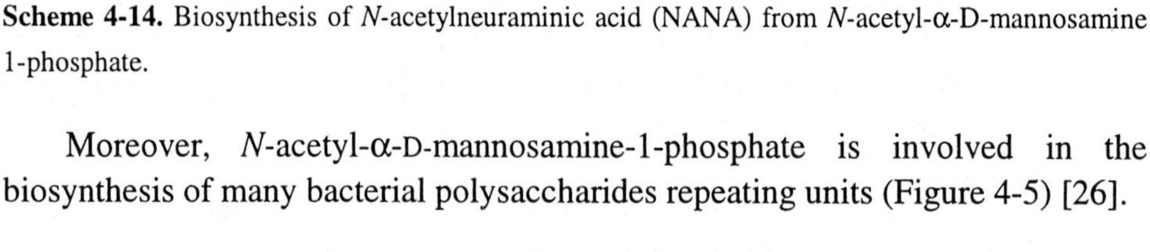

Scheme 4-14. Biosynthesis of N-acetylneuraminic acid (NANA) from N-acetyl-α-D-mannosamine 1-phosphate.

Moreover, N-acetyl-α-D-mannosamine-1-phosphate is involved in the biosynthesis of many bacterial polysaccharides repeating units (Figure 4-5) [26].

2)-β-D-Man$p$-(1,3)-α-**D-Man$p$NAc**-(1,3)-β-L-Rha$p$-(1,4)-α-D-Glc$p$NAc-(1-
S. dysenteriae type 10

-3)-β-**D-Man$p$NAc**-(1,4)β-D-Glc$p$-(1,3)-α-D-Rha$p$-(1,2)-α-D-Rha$p$-(1,5)-β-KDOp-(2-
E. Coli K7 and K56

Figure 4-5. Repeating units of ManNAc containing bacterial polysaccharides

-4)-β-D-Glc*p*NAc-(1,3)-β-**D-Man*p*NAc**-(1-
Haemophilus influenzae type d

-3)-β-D-Glc*p*NAc-(1,4)-β-**D-Man*p*NAc**-(1- -
3
2
β-D-Fuc*f*
Haemophilus influenzae type e

6)-α-**D-Man*p*NAc**-(1-OPO$_3^-$-
Neisseria meningitidis type A

-4)-α-Glp*A*-(1,3)-α-D-Gal*p*-(1,3)-β-**D-Man*p*NAc**-(1,4)-α-D-Glc*p*-(1,4)-β-D-Glc*p*-(1-
S. pneumoniae 9A

-4)-α-Glp*A*-(1,3)-α-D-Gal*p*-(1,3)-β-**D-Man*p*NAc**-(1,4)-α-D-Glc*p*-(1,4)-β-D-Glc*p*-(1-
OAc   OAc
S. pneumoniae 9V

Figure 4-5. (continued) Repeating units of ManNAc containing bacterial polysaccharides

4.5.2 Synthesis

In the light of the biological importance of *N*-acetyl-α-D-glucosamine 1-phosphate and *N*-acetyl-α-D-mannosamine 1-phosphate, there is a great interest in the synthesis of mimics of these 2-acetamido-2-deoxy-glycosyl phosphates. These mimics should interfere in the cell-pathogen adhesion phenomena and in the formation of the bacterial cell walls; and they may shed light on the recently discovered dynamic glycosylation process.

Unfortunately, examples of formation of C-glycosides of D-glucosamine and D-mannosamine are very limited [27, 28], and C-glycosyl halides of these sugars have never been reported. The reaction of methylenetriphenylphosphorane with properly protected D-glucosamino derivatives does not afford the desired product: 2-acetamido-2-deoxy-3,4,6-tri-*O*-benzyl-D-glucopyranose **34** affords the oxazolidine **35**, whereas 2-amino-2-deoxy-3,4,6-tri-*O*-benzyl-D-glucopyranose did not react.

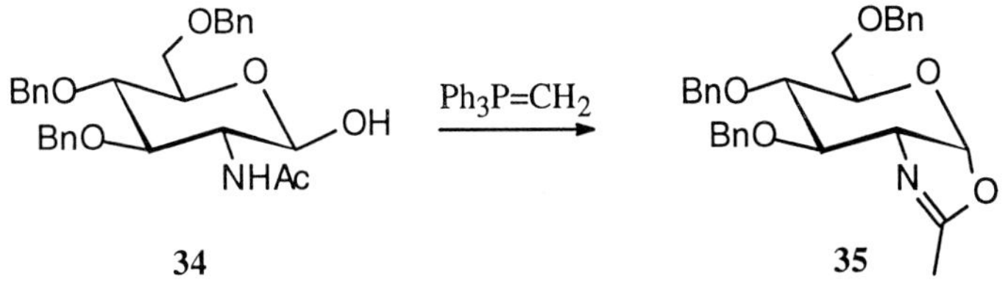

Scheme 4-15

An interesting alternative route to obtain a 2-amino-2-deoxy-glucoenitol is the reaction of N-benzyl(2,3,5-tri-O-benzyl-D-arabinofuranosyl)amine **36** with vinylmagnesium bromide (Scheme 4-16) [29].

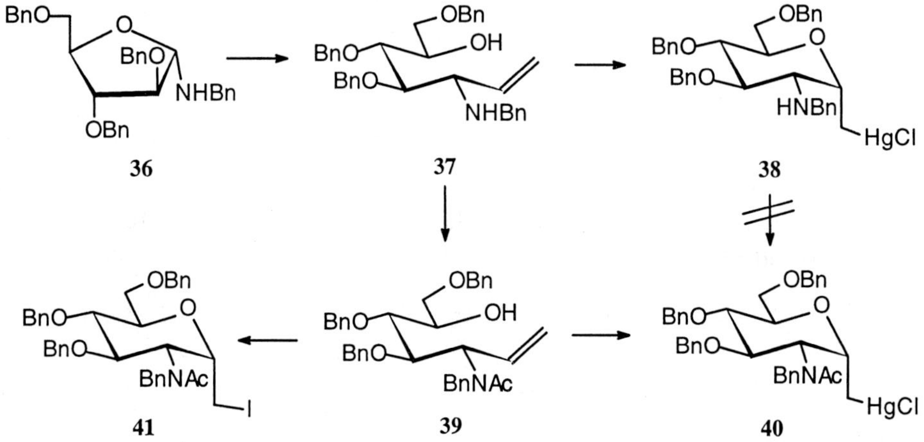

Scheme 4-16

The reaction affords stereoselectively the aminoglucoenitol **37**, which can be cyclized with mercuric trifluoroacetate to the α-C-glucopyranoside **38**. Any attempt to convert the mercurioderivative **38** into the corresponding halide (Br$_2$ or I$_2$ in CH$_2$Cl$_2$) or to halo-cyclize directly **37** (I$_2$, THF-H$_2$O, pH 4) was however unfruitful. In the hypothesis that the nucleophilic character of the amino function of **38**, adjacent to the electrophilic carbon of the desired halide, could interfere in the reaction, its nucleophilicity was lowered by acetylation. This required the selective acetylation of the open chain precursor **37** and the subsequent mercurio-cyclization of the product **39**, as **38** proved inert to acetylation. The N-acetylated mercurio derivative **40** also gave unsatisfactory results in the halodemercuriation; whereas treatment of **39** with iodide in THF at pH 4 afforded the very labile iododerivative **41**, which decomposes under reflux with triethylphosphite.

To overcome all these difficulties, a different synthetic strategy was followed, in which the amino function is introduced in the molecule after the phosphono-function (Scheme 4-17). The strategy requires an α-C-glucopyranoside with a deprotected hydroxyl group at C-2, to be converted into an amino function at the end of the synthesis.

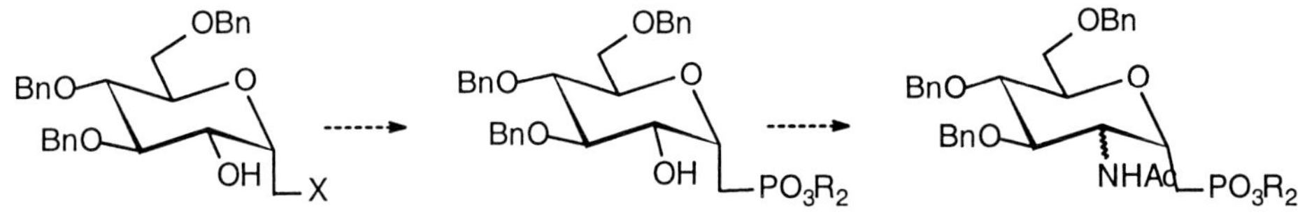

Scheme 4-17

This can be achieved by stereoselective vinylation of 2,3,5-tri-*O*-benzyl-D-arabinose **19** with divinylzinc, as reported in Scheme 4-11, and subsequent stereoselective mercurio-cyclization of the obtained enitol **20**, to afford the α-*C*-glucopyranoside **41** (Scheme 4-18).

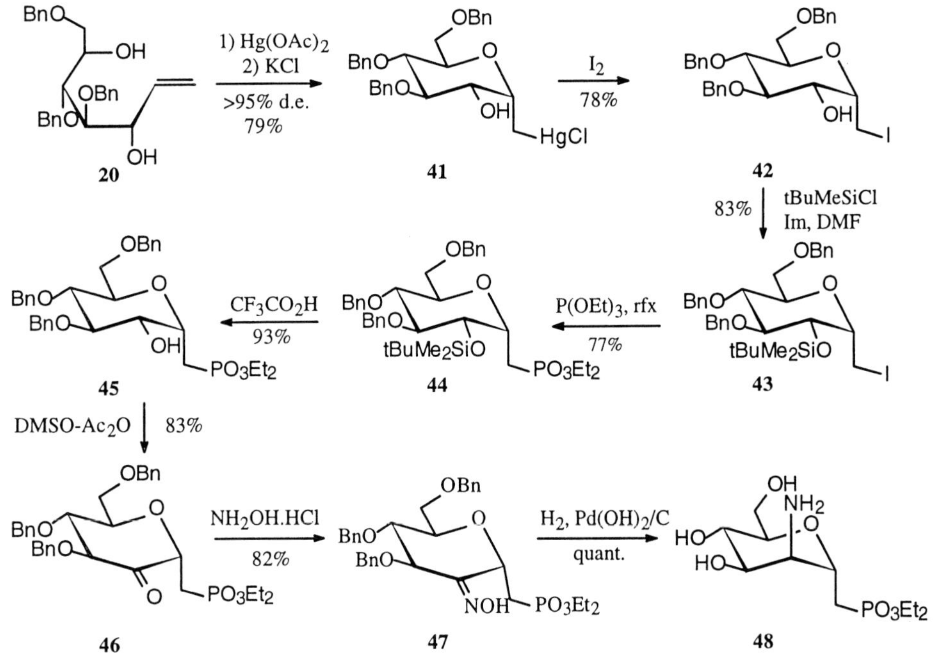

Scheme 4-18. Synthesis of the phosphono analogue of α-D-mannosamine 1-phosphate.

In the synthesis outlined before, the conversion of the mercurioderivative into the corresponding iodide **42** is easy, and the iodide can be converted into the corresponding phosphonate **45** in good yields, provided that the free hydroxyl group of **42** is temporary protected. Arbuzov reaction of iodide **42**, gave the cyclic phosphonate **49** as the only product (Figure 4-6).

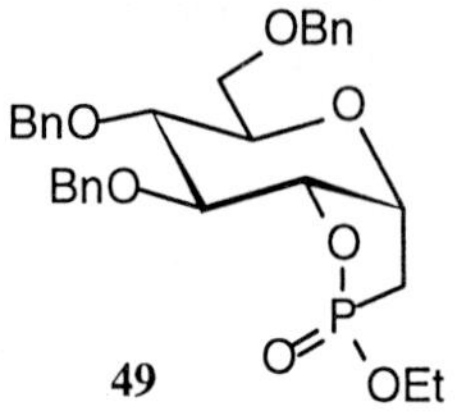

Figure 4-6

The conversion of the free hydroxyl group into an amino group at C-2 of a glucopyranoside is a well established process. Among other procedures, it can be effected by oxidation, oximation, and reduction of the oxime to the corresponding amine by catalytic hydrogenation or treatment with diborane. The process affords stereoselectively a 2-aminosugar with the *manno* configuration if the anomeric configuration of the starting ketone is β; whereas, starting from the α-anomer, the *gluco* isomer is preferentially formed [30].

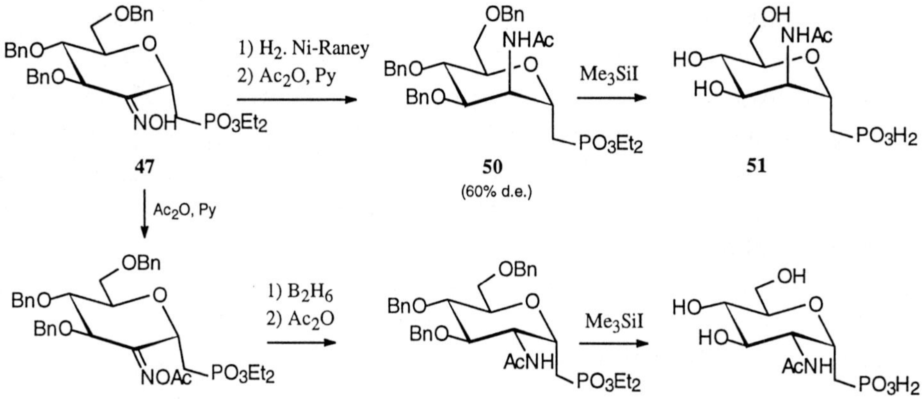

Scheme 4-19. Diastereoselective formation of the phosphono analogues of N-acetyl-α-D-mannosamine 1-phosphate and N-acetyl-α-D-glucosamine 1-phosphate.

In our case, the catalytic reduction of the α-oxime **47** with palladium hydroxide on carbon, unexpectedly affords the *manno*saminoderivative **48** in quantitative yields. Catalytic hydrogenation with Raney nickel also gives the *manno*-isomer (**50**, after acetylation), but with a lower stereoselection (60% d.e.). The use of a methyloxime instead of the oxime **47** does not change the yield and stereoselection of these reductions. These results suggest a coordination of the α-oriented phosphonate group with the metal catalyst, which favors the attack of the hydrogen from the α-face. In this hypothesis, to obtain the *gluco*-isomer, the non-coordinating reducing agent diborane should be used. In fact, the reduction of the acetyloxime **52** with diborane in THF affords, after acetylation, the 2-acetamido-2-deoxy-α-C-*gluco*pyranoside **53** in 64% d.e. (Scheme 4-19).

The synthesis of the phosphono analogues of *N*-acetyl-α-D-mannosamine 1-phosphate **51** and *N*-acetyl-α-D-glucosamine 1-phosphate **54** has been completed by treatment with iodotrimethylsilane which effects at the same time the debenzylation and conversion of the phosphonate ester into the corresponding acid.

4.6 Conclusion

The synthesis of glycosyl phosphate mimics is an area of growing interest as these molecules can widely interfere in the metabolism of carbohydrates. Different synthetic efforts have been reported, some of which directly introduce a methylenephosphonate group at the anomeric center of the sugar, but are not of general application. A more general procedure involves the formation of a C-glycosyl halide (or a mesylate) and the subsequent conversion into a phosphonate. This procedure however does not allow the synthesis of the phosphono analogues of 2-amino-2-deoxy-glycosyl phosphates such as *N*-acetyl-α-D-mannosamine 1-phosphate and *N*-acetyl-α-D-glucosamine 1-phosphate, two molecules of great biological and pharmacological interest. These molecules can be synthesised following a strategy in which the amino function is introduced in the final step of the synthesis.

The availability of the glycosyl phosphate mimics here described opens the way to the preparation of their NDP-derivatives and analogues, and to the evaluation of their biological activity.

4.7 References

1. H. G. Hers, *Biochem. Soc. Trans.* **1984**, *12*, 729-735.
2. R. Hengel, *Chem. Rev.* **1977**, *77*, 349-367.
3. G. H. Jones, J. G. Moffatt, U.S. Patent 2 583 974 (8 June 1971); *Chem. Abs.* **1971**, *75*, 130091q.
4. B. S. Griffin, A. Burgher, *J. Am. Chem. Soc.* **1956**, *78*, 2336-2338.
5. M. Chmielewski, J.N. BeMiller, D. Pat Cerrett, *Carbohydr. Res.* **1981**, *97*, C1-C-4.
6. See for example C.D. Hurd, R. P. Holysz, *J. Am. Chem. Soc.* **1950**, *72*, 1732-1735.
7. See for example S. Hanessian, A. Pernet, *Can. J. Chem.* **1974**, *52*, 1280-1293.
8. E. J. Corey, G. T. Kwiatkowski, *J. Am. Chem. Soc.* **1966**, *88*, 5654-5656.

9. F. Nicotra, unpublished results.
10. M. D. Lewis, J. Kun Cha, Y. Kishi, , *J. Am. Chem. Soc.* **1982**, *104*, 4976-4978.
11. R. McClard, *Tetrahedron Lett.* **1983**, *24*, 2631-2634.
12. B. F. Maryanoff, S. O. Nortey, I. A. Campbell, A. B. Reitz, D. Liotta, *Carbohydr. Res.* **1987**, *171*, 259-278.
13. R. B. Meyer, Jr., T. E. Store, P. K. Jesthi, *J. Med. Chem.* **1984**, *27*, 1095-1097.
14. R. Meuwly, A. Vasella, *Helv. Chim. Acta* **1986**, *69*, 751-760.
15. J.-R. Pougny, M. A. M. Nassr, P. Sinaÿ, *J. Chem. Soc., Chem. Commun.* **1981**, 375-376.
16. F. Nicotra, R. Perego, F. Ronchetti, G. Russo, L Toma, *Gazz. Chim. Ital.* **1984**, *114*, 193-195.
17. F. Nicotra, L. Panza, F. Ronchetti, G. Russo, L. Toma, *Carbohydr. Res.* **1987**, *171*, 49-57.
18. F. Freeman, K. D. Robarge, *Carbohydr. Res.* **1986**, *154*, 270-274.
19. A. Boschetti, F. Nicotra, L. Panza, G. Russo, *J. Org. Chem.* **1988**, *53*, 4181-4185.
20. F. Nicotra, L. Panza, G. Russo, A. Senaldi, N. Burlini, P. Tortora, *J. Chem. Soc., Chem. Commun.* **1990**, 1396-1397.
21. F. Nicotra, F. Ronchetti, G. Russo, *J. Org. Chem.* **1982**,*47*, 4459-4462.
22. F. Nicotra, R. Perego, F. Ronchetti, G. Russo, L. Toma, *Carbohydr. Res.* **1984**, *131* 180-184.
23. F. Nicotra, L. Panza, F. Ronchetti, L. Toma, *Tetrahedron Lett.* **1984**, *25*, 5937-5940.
24. K. Frische, R. R. Schmidt, *Liebigs Ann. Chem.* **1994**, 297-303.
25. G. W. Hart, *Current Op.Cell Biol.* **1992**, *4*, 1017-1023.
26. L. Kenne, B. Lindberg, in *The Polysaccharides*, Ed. by G. O. Aspinall, Academic Press, **1983**, vol 2, 287.
27. a) F. Nicotra, G. Russo, F. Ronchetti, L. Toma, *Carbohydr. Res.* **1983**, *124*, C5-C7; b) G. Grynkiewicz, J. N. BeMiller, *Carbohydr. Res.* **1983**, *112*, 324-327; c) A. Giannis, P. Munster, K. Sandhoff, W. Steglich, *Tetrahedron* **1988**, *44*, 7177-7180; d) R. Grondin, Y. Leblanc, K. Hoogsteen, *Tetrahedron Lett.* **1991**, *32*, 5024-5027; e) H.Vyplel, D. Scholz, I. Macher, K. Schindlmaier, E. Schutze, *J. Med. Chem.* **1991**, *34*, 2759-2767; f) C. Leteux, A. Veyrieres, *J. Chem. Soc., Perkin 1* **1994**, 2647-2655; g) M. Petrušová, F. Fedoronko, L. Petruš. *Chem Papers* **1990**, *44*, 267-271: h) M. Hoffmann, H. Kessler, *Tetrahedron Lett.* **1994**, *35*, 6067-6070.

28. For the synthesis of 2-azido-C-glycosides see: a) M. G. Hoffmann, R. R. Schmidt, *Liebig Ann. Chem.* **1985**, 2403-2419; b) C. R. Bertozzi, M. D. Bednarski, *Tetrahedron Lett.* **1992**, *33*, 3109-3112.

29. (a) M. Carcano, F. Nicotra, L. Panza, G. Russo, *J. Chem. Soc., Chem. Commun.* **1989,** 297-298; (b) L. Lay, F. Nicotra, L. Panza, A. Verani, *Gazz. Chim. Ital.* **1992**, *122***,** 345-348.

30. F. W. Lichtentaler, E, Kaji, *Liebigs Ann. Chem.* **1985**, 1659-1668 and ref. cited therein

31. L. Cipolla, L. Lay, F. Nicotra, L. Panza, G. Russo, *J. Chem. Soc., Chem. Commun.* **1995**, 1993-1994.

5 Synthetic Studies on Glycosidase Inhibitors Composed of 5a-Carba-Sugars

S. Ogawa

5.1 Introduction

Carba-sugars [1], previously known as pseudo-sugars, are a family of sugar mimics currently attracting great interest among chemists and biochemists in various fields. Almost thirty years ago, McCasland et al.[2] synthesised the first example 5a-carba-α-DL-talopyranose **1**, followed by two new isomeric compounds, and called them "pseudo-sugars". Carba-sugars are (hydroxymethyl)-branched-chain cyclitols. They are similar topologically to normal sugars particularly in the arrangement of the hydroxyl and hydroxymethyl groups, but have the oxygen atom of the pyranose ring replaced by a methylene. Their biological features are well exemplified by the fact that humans cannot differentiate carba- from true-glucoses by their taste. Furthermore free carba-sugars exist in structurally stable six-membered ring forms, and chemical modification at anomeric positions may therefore be possible, thus providing biologically interesting derivatives analogous to *C*-glycosides.

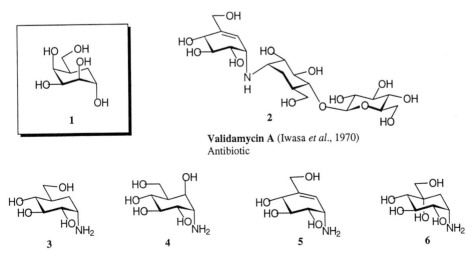

Figure 5-1. A 5a-carba-sugar and related natural and synthetic compounds of biological interest.

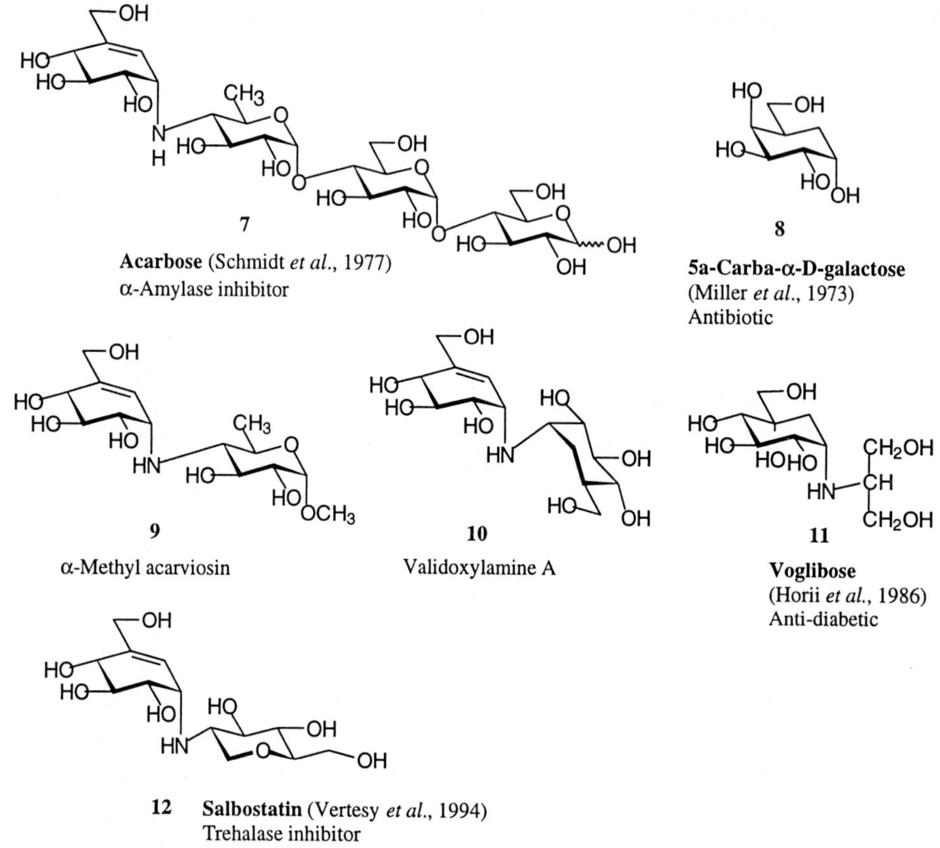

7 Acarbose (Schmidt et al., 1977)
α-Amylase inhibitor

8 5a-Carba-α-D-galactose
(Miller et al., 1973)
Antibiotic

9 α-Methyl acarviosin

10 Validoxylamine A

11 Voglibose
(Horii et al., 1986)
Anti-diabetic

12 Salbostatin (Vertesy et al., 1994)
Trehalase inhibitor

Figure 5-1. continued.

The agricultural antibiotic validamycin A **2**, a naturally occurring biologically active compound composed of carba-sugars, was discovered [3] in 1970 and was shown to possess a unique pseudo-trisaccharide structure. Extensive studies on isolation of minor components of the fermentation broth have led to the discovery [4] of seven homologous antibiotics (validamycins B-H) and four examples of new carba-amino sugar classes of compounds: validamine (**3**), hydroxyvalidamine (**4**), valienamine (**5**), and valiolamine (**6**).

Soon afterwards, the α-amylase inhibitor acarbose (**7**) [5] and homologues, and related pseudo-oligosaccharidic inhibitors [6] were isolated by several research groups between 1976 and 1982. Previously the interesting 5a-carba-α-D-galactopyranose **8** had been isolated as a weak antibiotic from the fermentation broth of *Streptomyces* species in 1973 [7]. Since the enzyme inhibitory potency of **7** has been attributed to the core structure, pseudo-disaccharide α-methyl acarviosin **9** that is itself very strong α-glucosidase inhibitor, a similar survey has been extensively carried out to make clear the mechanism and biological action of validamycins. Finally, the strong potency of validamycin A has been attributed to high trehalase-inhibitory activity [8] (*in vitro*) of the carba-disaccharide validoxylamine A (**10**). This knowledge then opened up the possibility for development of therapeutically useful compounds, resulting in the finding of strong synthetic glycohydrolase inhibitor voglibose **11** [9]. Both acarbose and voglibose are now clinically useful therapeutic agents to control diabetes. Very recently, the new trehalase inhibitor salbostatin (**12**) [10] has been isolated.

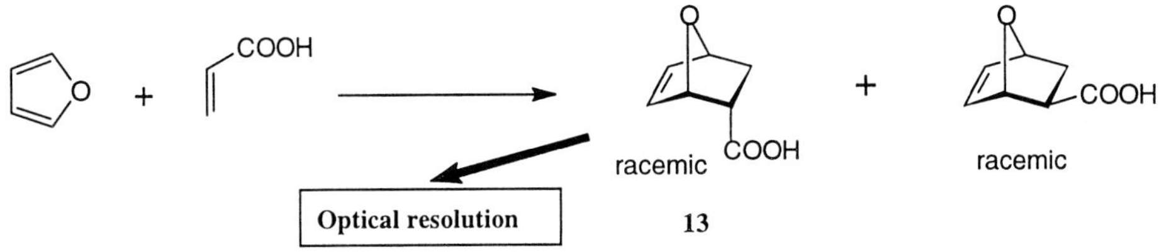

Figure 5-2. *endo*-Adduct of furan and acrylic acid, a versatile synthetic intermediate.

We have so far been interested in synthetic studies on biologically active carba-sugar derivatives, especially 5a-carba series, since our successful preparation of racemic validamine [11] (in 1976), achieved starting from the *endo*-adduct **13** of furan and acrylic acid. The racemic acid was later easily optically resolved to give both enantiomers [12], which were then thought to be versatile intermediates useful for preparation of various kind of carba-sugar derivatives, including validamycins [13] and acarbose [14]. In this chapter are described synthetic efforts to prepare biologically interesting oligosaccharides composed of carba-sugars, and the development of new types of sugar hydrolase inhibitors in the hope of obtaining bio-drugs on the basis of structure-activity relationship of the glycosidase inhibitors.

5.2 Synthesis of Biologically Active Natural Compounds Containing 5a-Carba-Sugars

5.2.1 Synthesis of Validamycins

Construction of imino bridges between two carba-sugar residues may be carried out, for instance, by four synthetic procedures: i) coupling of amines and allyl halides, ii) coupling of amines and epoxides, iii) coupling of amines and α,β-unsaturated epoxides, and iv) reductive amination of ketones with amines. Of these we choose to employ the most simple procedure, i.e. coupling of amines and epoxides. The synthons designed are readily accessible and the reaction products are usually predictable by stereochemical considerations, *e.g.* on the basis of the Fürst-Plattner rule.

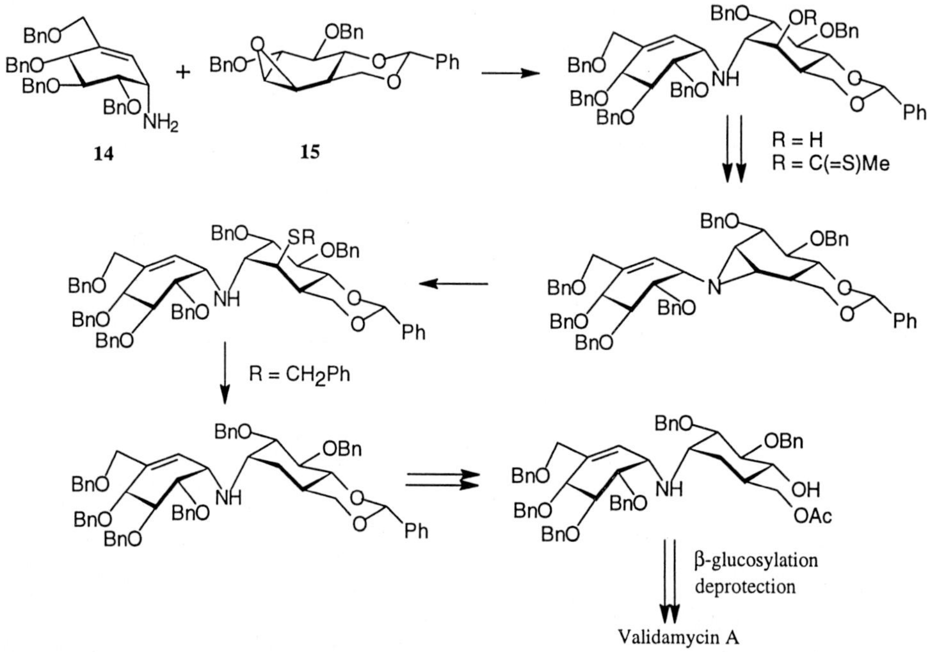

Figure 5-3. Total synthesis of validamycin A.

Complete total synthesis of validamycin A is shown in Figure 5-3 as a typical example [13]. Coupling between the protected valienamine derivative **14** and epoxide **15** was carried out in 2-propanol in a sealed tube at 120°C for 3 to 6 days. The reaction proceeded very slowly, but the coupling products once formed remained unchanged. When the reaction was quenched, the products were isolated and unreacted starting materials easily recovered. The product was transformed by established procedures into the protected validamycin, which gave the natural product was after deprotection.

5.2.2 Synthesis of Acarbose

Synthesis of the pseudo-tetrasaccharide acarbose was also achieved by coupling of the amine **16** and the anhydro derivative **17** of trisaccharide [14]. Protection of the hydroxyl groups of the synthons appreciably hampered and retarded the coupling reaction with the amine. Although a migration of the epoxide ring seemed to be possible under these conditions, the nucleophilic substitution reaction occurred with comparative ease to afford only the two products expected from opening of the original 3",4"-epoxide ring. Deprotection of the diequatorially substituted product **18** was readily carried out conventionally to afford acarbose.

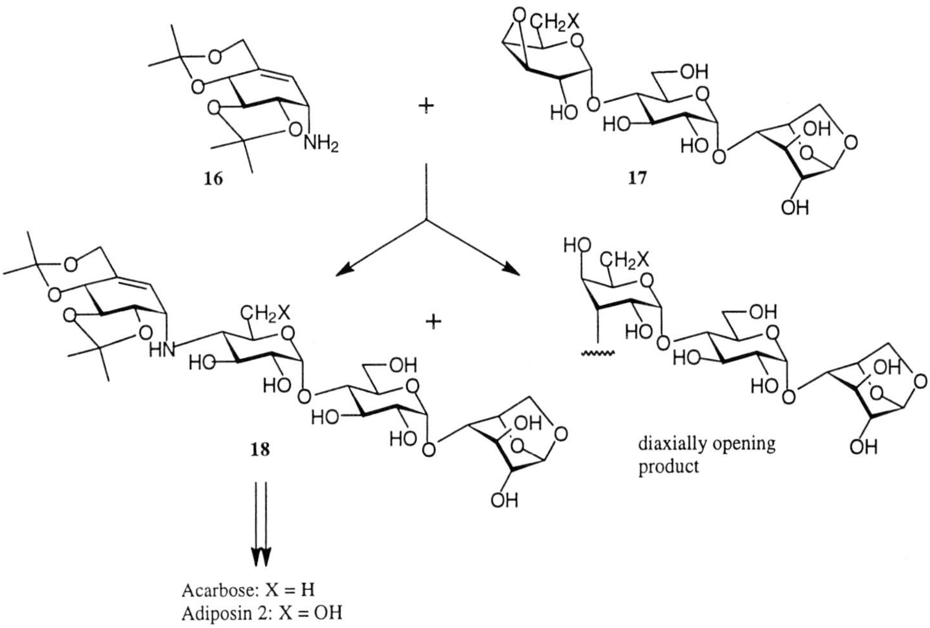

Acarbose: X = H
Adiposin 2: X = OH

Figure 5-4. Total synthesis of acarbose and adiposin 2.

5.2.3 Synthesis of Salbostatin

The recently discovered trehalase inhibitor salbostatin [10] was similarly synthesised [15] via coupling the amine **16** to the sugar epoxide **19**. Interestingly, selective cleavage of the pyranose 2,3-epoxide was shown to give the diaxially opened product **20** in the case when the epoxide adopted the rigid half-chair conformation, due to the presence of a 4,6-O-benzylidene ring. But in the case of the conformationally flexible free epoxide **21** the diequatorially opened product **22** was formed preferentially.

Three naturally occurring carba-sugar-containing products were successfully synthesised by a moderate yielding coupling reaction using the amine and epoxide starting materials, with minimum protections of the hydroxyl groups.

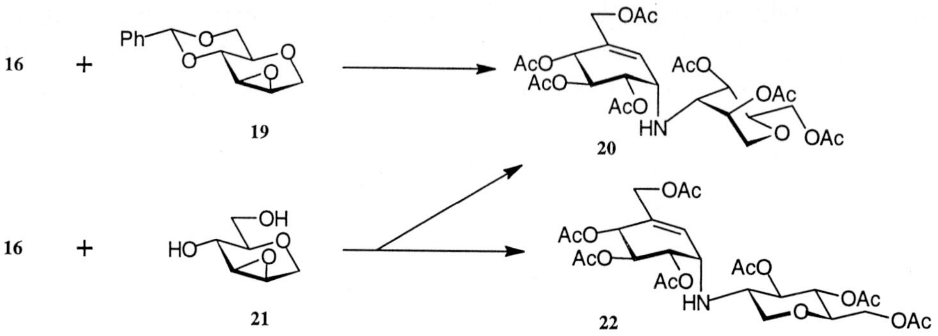

Figure 5-5. Total synthesis of salbostatin.

5.2.4 Structure-Activity Relationship of Naturally Occurring Glycosidase Inhibitors Composed of 5a-Carba-Sugars

Considering the structural similarity of validoxamine and α,α-trehalose, its strong trehalase inhibitory activity seemed to be attributable to the symmetric feature of carba-disaccharide linked by way of an imino bridge. Therefore, three mimics **23**, **24**, and **25**, composed of validamine, valienamine, and validoxylamine residues, respectively, were synthesised [16] and demonstrated to be potent trehalase inhibitors with activity almost comparable with that of validoxylamine A.

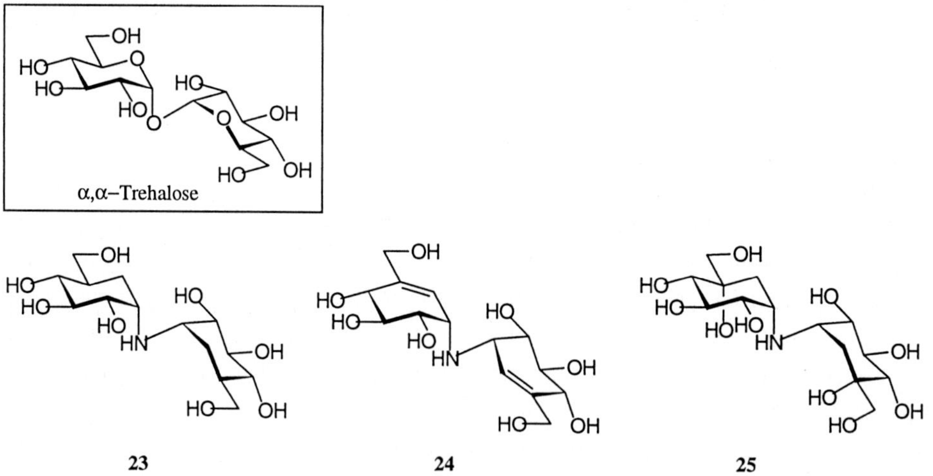

Figure 5-6. Synthetic trehalase inhibitors related to validoxylamine A.

On the other hand, glycosidase inhibitory potentials of acarbose were attributable to the core structure, a carba-disaccharide composed of valienamine and 4-amino-4,6-dideoxy-D-glucose. The valienamine residue plays a role as a mimic of a postulated flattened glucosyl cation when binding to the active site of the enzymes. Methanolysis of acarbose afforded the active core unsaturated carba-disaccharide, α-methyl acarviosin **9** [17]. Therefore, several chemically modified acarviosin analogues were prepared [18]. Concerning the inhibitory activity of the 1,6-anhydro derivatives against Baker's yeast, the more hydrophobic the aglycone parts become, the more the inhibitory activities increase.

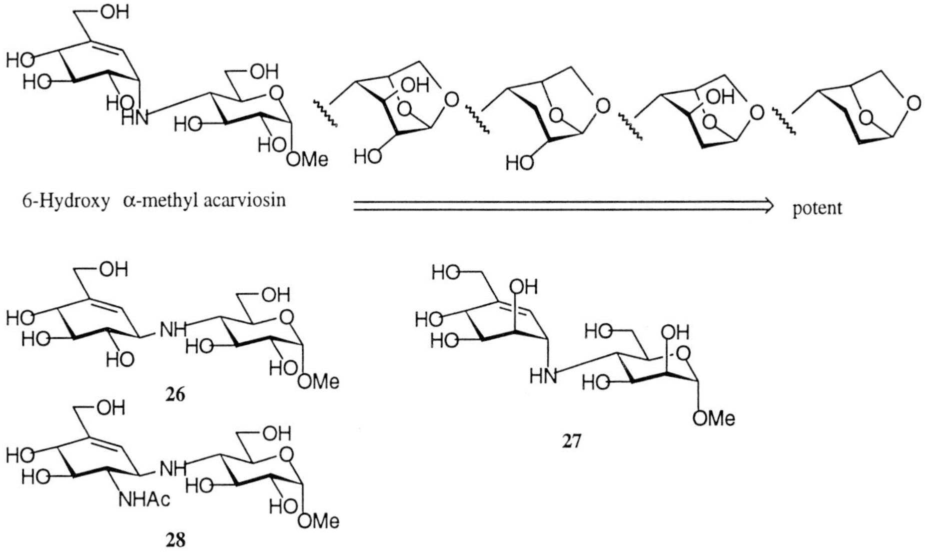

Figure 5-7. Chemical modification of methyl acarviosin.

Furthermore, three glycosidase inhibitors were designed on the basis of branched-chain conduritol-structure. Thus, the valienamine part of **9** was chemically modified to mimic the glycosyl cation intermediates: D-glucosyl, D-mannosyl, and 2-acetamido-2-deoxy-D-glucosyl cations, in the hope that the respective pseudo-disaccharides **26** [19], **27** [20], and **28** [21] would act as inhibitors of β-glucosidase, α-mannosidase, and chitinase. However, our efforts have not been well rewarded. These results suggested that the characteristic structural features of the potent inhibitors in particular case can not always be applied as a lead structure in other related cases.

5.3 Synthesis of Carba-Oligosaccharides of Biological Interests

5.3.1 5a'-Carba-maltoses Linked by Imino, Ether, and Sulfide Linkages

As seen in acarviosin and validoxylamine A the unsaturated carba-sugar residues play important roles in exhibiting inhibitory potency against certain enzymes. We then started to elucidate the biological roles of carba-sugar residues in oligosaccharides. Thus, three 5a'-carbamaltoses **29**, **30**, and **31**, containing an imino, ether, and sulfide bridges, and their corresponding 5'(5a')-unsaturated derivatives were synthesised [22].

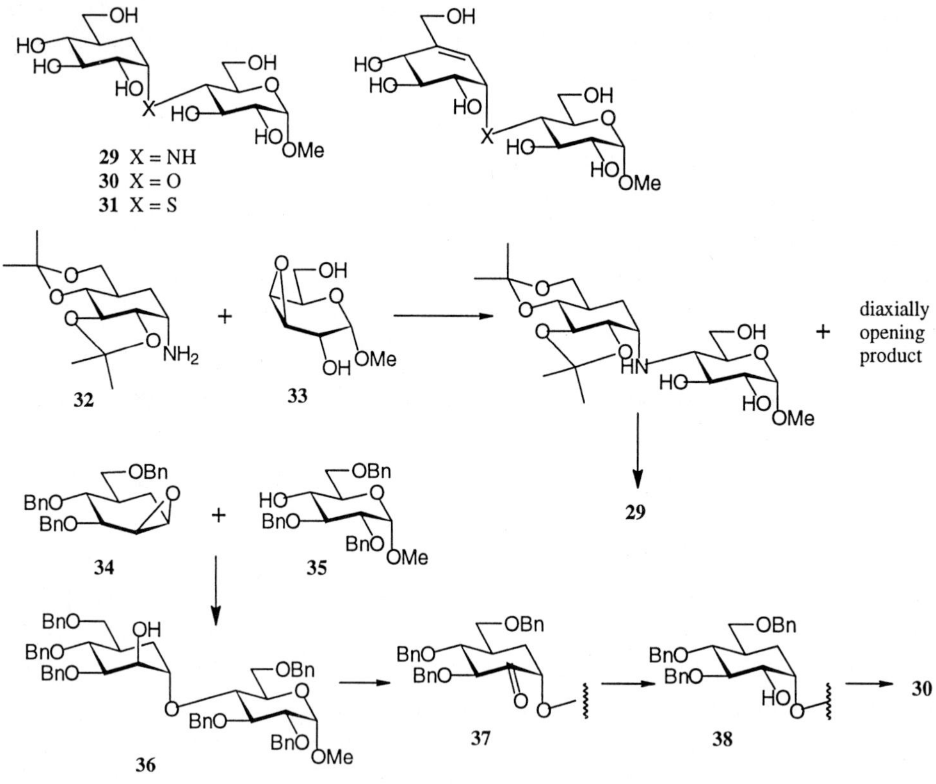

Figure 5-8. Synthesis of 5a'-carbamaltoses.

The imino-linked carba-disaccharide **29** was readily prepared by coupling of the amine **32** and the sugar epoxide **33**, followed by conventional deprotection. Establishing a general procedure to incorporate the carba-sugar residue into the oligosaccharide chain through an ether bridge was considered to be very important for further development of carba-sugar chemistry. However, construction of the ether-linkage proved to be rather difficult, and only one 1,2-anhydro-carba-hexopyranose has so far been found to act as a donor substrate for this purpose. Thus, coupling of the perbenzyl derivative of 1,2-anhydro-5a-carba-β-D-mannopyranose **34** with an oxide anion generated in situ from the 4-OH unprotected derivative of methyl-α-D-glucopyranoside **35** by treatment with sodium hydride in DMF was carried out in the presence of 15-crown-5 ether at 70°C. The coupling product **36** with α-*manno* configuration was obtained (56%). The 2-OH function was then oxidized and the resulting ketone **37** was reduced with L-selectride to give the carbamaltose derivative **38** together with **36**. The 5a'-carbamaltose with a sulfide linkage was synthesised by coupling of the benzoyl derivative of **34** with the 4-deoxy-4-thioacetoxy derivative of methyl α-D-glucopyranoside, followed by deprotection.

The three carbamaltoses were subjected to bioassay on α-glucosidase (Baker's yeast). Only the imino-linked compound **29** showed a weak inhibition against the enzyme. Both the ether or sulfide-linked carbamaltoses have no inhibitory activity, suggesting that they are probably acting as substrate analogues. Although the incorporation of an unsaturation into **29** increased the inhibitory activity greatly, the unsaturated analogues of **30** and **31** remained completely inactive.

5.3.2 Several Carba-Oligosaccharides: Core-Structures of Cell Surface Glycans

In order to elucidate biological influence of ether-linked carba-sugar residues in oligosaccharides, some core structures of cell-surface glycans were synthesised. In the course of this study, attempts were made to subject some anhydro derivatives of carba-hexoses (1,2-anhydro-5a-carba-α-D-gluco and -galacto pyranoses) to nucleophilic substitution by an oxide anion generated from protected sugar derivatives. However, the epoxides were found to undergo a facile elimination reaction, and no coupling products were formed. Therefore, two derivatives **34** and **39** derived from 1,2-anhydro-5a-carba-β-D-mannopyranose were used to prepare effectively a single coupling product with α-*manno* configuration, which were then transformed into a variety of carba-hexopyranose residues by chemical modification. Figure 5-9 shows the preparation of three isomeric 5a'-carba-disaccharides, using 1,6:2,3-dianhydro-β-D-mannopyranose **40** as the glycosyl acceptor [23]. The 2-OH group of the coupling product **41** of **39** and **40** was oxidized and the resulting ketone **42** was epimerized by treatment with a base such as potassium *tert*-butoxide or DBU to afford the β-anomer **43**. Subsequent selective reduction gave the carba-disaccharides **45** and **46** with β-*gluco* and β-*manno* configurations. On the other hand, reduction of **42** gave the α-*gluco* isomer **44** along with **41**. The ratio of the alcohols formed was found to depend on the reducing agent used.

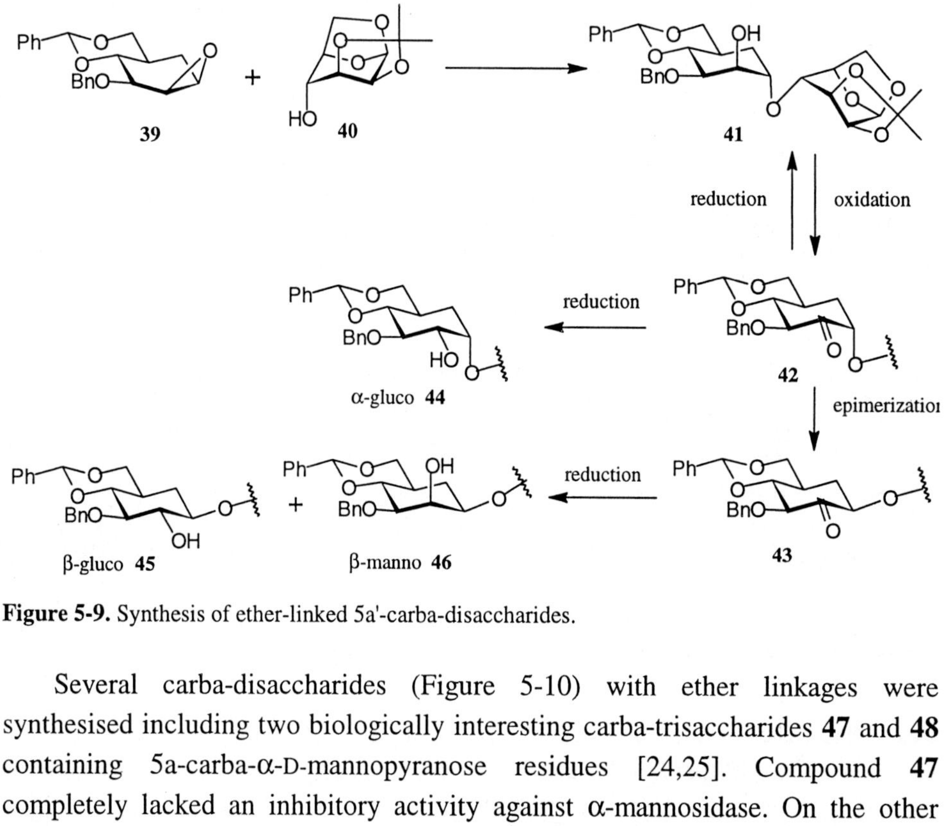

Figure 5-9. Synthesis of ether-linked 5a'-carba-disaccharides.

Several carba-disaccharides (Figure 5-10) with ether linkages were synthesised including two biologically interesting carba-trisaccharides **47** and **48** containing 5a-carba-α-D-mannopyranose residues [24,25]. Compound **47** completely lacked an inhibitory activity against α-mannosidase. On the other hand, compound **48** has been demonstrated [25] to be an effective substrate analogue for GlcNAcT-V. The activity of this enzyme has been shown to correlate with the mutastatics potential of tumor cells. Therefore these results may open up new opportunities for the design of inhibitors of this important enzyme.

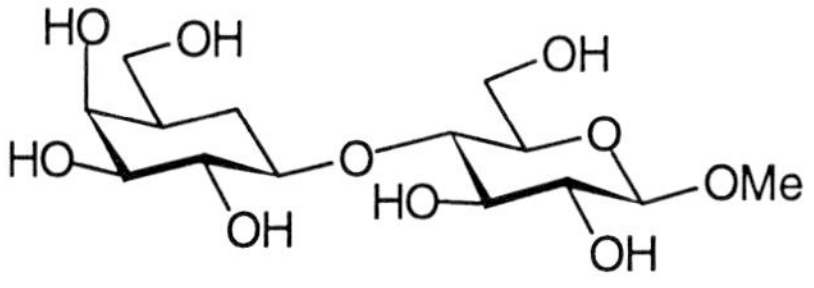

Methyl 5a'-carba-β-lactoside

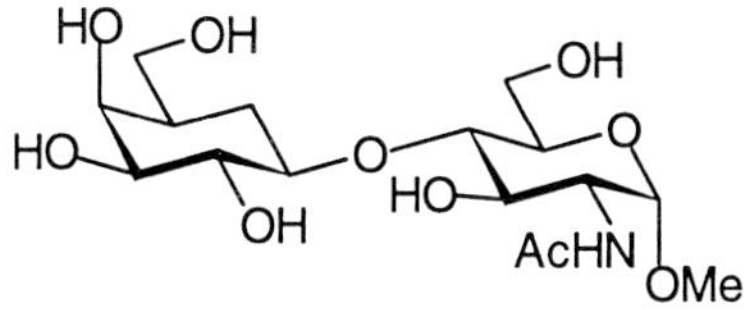

Methyl 5a'-carba-α-lactosaminide

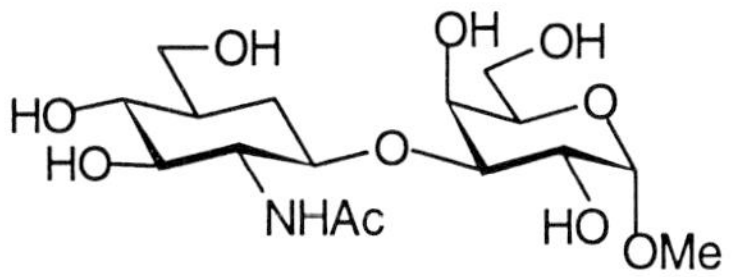

5a-carba-β-D-GlcpNAc-(1 → 3)-α-D-Galp-OMe

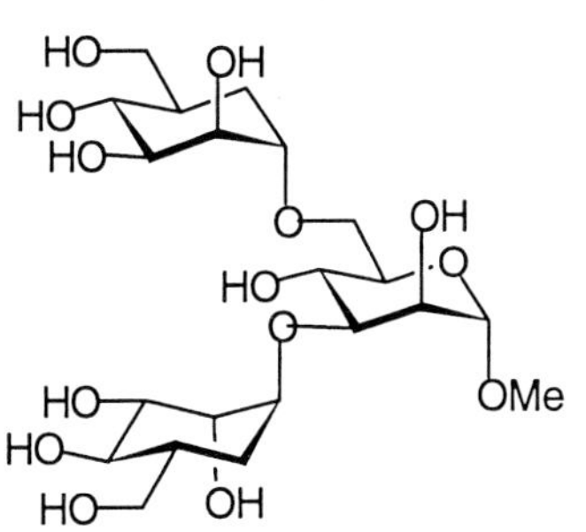

47

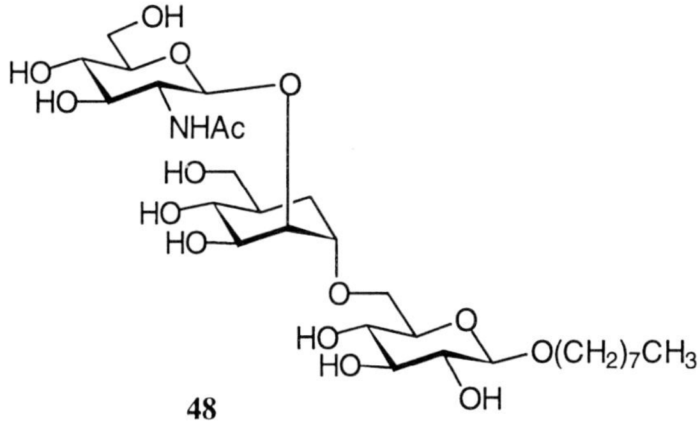

48

Figure 5-10. Ether-linked 5a-carba-di- and trisaccharides of biological interest.

5.4 Synthesis of Glycolipid Analogues Containing 5a-Carba-sugars

5.4.1 Glycosylamide Analogues

Recently, many glycosylamides structurally related to glycosphingolipids and glycoglycerolipids have been synthesised [26] and subjected to several kinds of biological tests. Some of them have been found to be immunomodulators. These studies have stimulated us to synthesize [27] the corresponding carba-sugar analogues, which contain the 5a-carba-hexopyranose residues of α- and β-*gluco*, β-*galacto*, and α- and β-*manno* configurations, respectively. The synthesis was carried out by coupling the carba-sugar epoxides and aliphatic amines and subsequent *N*-acylation. The five carba-sugar analogues thus prepared were subjected to *in vivo* tests. Interestingly, they have been shown to possess potencies as immunomodulators comparable to those of the corresponding true sugar analogues. This suggests that the glycolipid analogues composed of 5a-carba-sugars would possibly be useful model compounds for the elucidation of biological roles and functions of glycolipids.

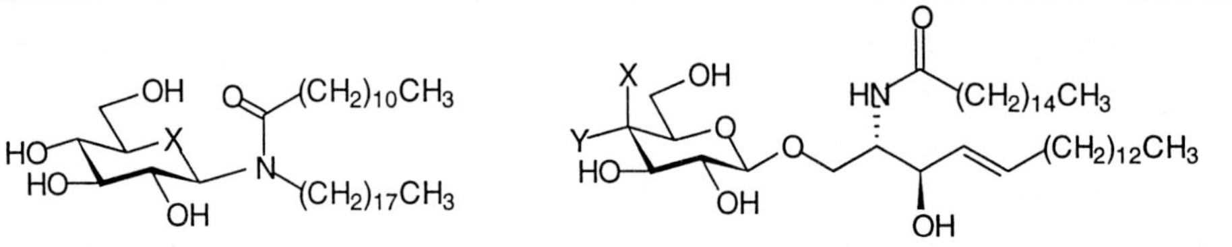

Glucosylamide: X = O
5a-Carba-glucosylamide: X = CH$_2$

Glucosylceramide: X = H, Y = OH
Galactosylceramide: X = OH, Y = H

Figure 5-11. Glycosylamides and 5a-carba sugar analogues.

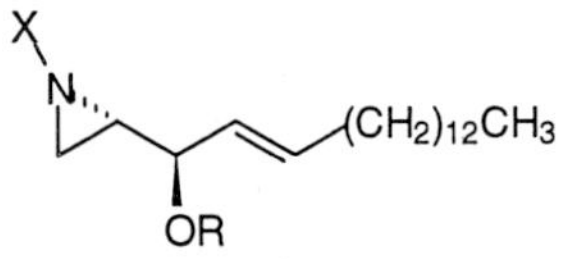

49 R = TBDMS, X = 2,4-dinitrophenyl
50 R = Bn, X = tosyl

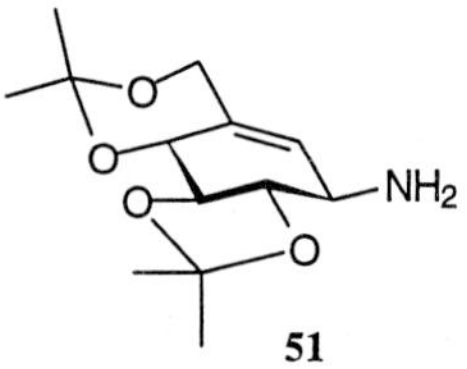

51

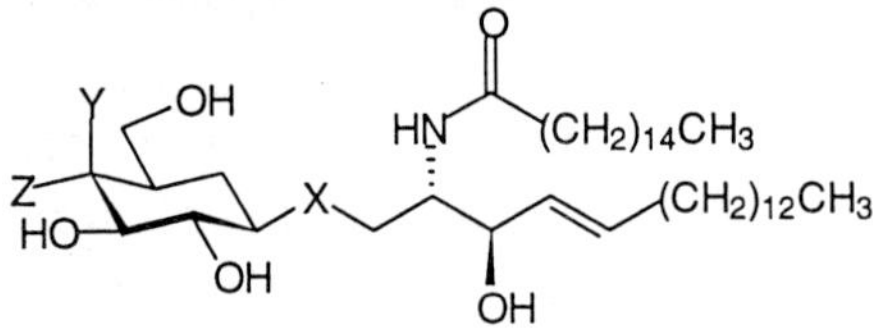

5a-Carba-glycosylceramides

52 X = NH, Y = H, Z = OH
53 X = NH, Y = OH, Z = H
54 X = O, Y = H, Z = OH
55 X = S, Y = H, Z = OH

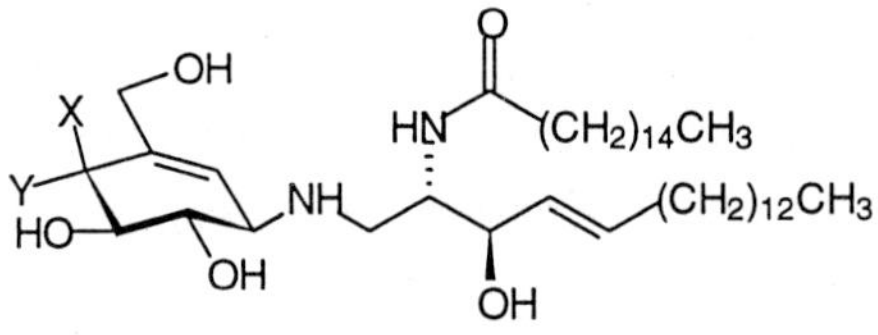

Unsaturated 5a-carba-glycosylceramides

56 X = H, Y = OH
57 X = OH, Y = H

Figure 5-12. Glycosylceramides and 5a-carba-sugar analogues.

5.4.2 Glycosylceramide Analogues

Glycosphingolipids play important roles in biological systems as essential constituents of membranes and cell walls. We have therefore been interested in the preparation of carbocyclic analogues of glycosylceramides in order to get glycohydrolase inhibitors of glycolipid biosynthesis and to shed light on some functions of glycolipids.

The synthetic methods [28] involve coupling of 1,2-aziridines, as sphingosine precursors, with appropriately protected 5a-carba-sugars. The 5a-carba-glycosyl acceptors **49** and **50** were prepared from the known azidosphingosine. Coupling of 4,6-*O*-benzylidene-5a-carba-β-D-glucopyranosyl-amine and **50** afforded a product which after removal of the *N*-protection followed by *N*-acylation, gave the imino-linked 5a-carbaglucosylceramide **52**. The ether **54** and sulfide-linked 5a-carbaglucosyl-ceramides **55** were prepared similarly. The inhibitory activity of the glycosylceramide analogues against β-gluco and β-galactocerebrosidases was assayed. The imino-linked analogues were shown to inhibit moderately both β-gluco and β-galactocerebrosidases, but the *galacto* isomer **53** had very weak activity. In contrast, the other analogues show no marked inhibitory activity against both enzymes. None of these analogues were inhibitors either of glucocerebroside synthase or of galactocerebroside synthase.

5.4.3 Glycocerebrosidase Inhibitors

The above results have prompted us to synthesize the 5a-carba-unsaturated sugar analogues [29] **56** and **57**, as possible more potent and specific glycocerebrosidase inhibitors. The coupling product obtained from the aziridine **49** and the protected β-valienamine **51** was conventionally transformed into the unsaturated analogue **56**. The β-*galacto* type carba-sugar donor was also introduced into the ceramide chain, giving the β-galactosylceramide analogue **57**. The corresponding Z-isomers were also prepared. The inhibitory activities against glycocerebrosidases (rat liver) of the four glycosylceramide analogues were assayed. The glucocerebroside analogues almost completely inhibited β-glucocerebrosidase activity (IC_{50} 3 x 10^{-7} M) selectively. The galactocerebrosidase analogues specifically inhibited β-galactocerebrosidase (IC_{50} 2.7 x 10^{-6} M) but they did not inhibit glucocerebrosidase at all.

We have thus demonstrated that the unsaturated analogues of 5a-carba-sugar inhibit β-glucocerebrosidase or β-galactocerebrosidase more strongly and selectively than the saturated compounds. These results indicate that the unsaturated analogues should more efficiently mimic the transition state of the natural substrate during the enzyme reaction. This has so far been the first demonstration that the structures of carba-sugar moieties designed to mimic the substrate molecules that are likely to be correlated to the inhibitory action. The analogues have not shown any inhibitory potential against glycocerebroside synthase.

Attempts were made to develop a similar type of inhibitors by transforming β-valienamine into the derivatives with more simple structures. The specific glucosylceramide synthase inhibitor PDMP [30] (D-*threo*-1-phenyl-2-decanoyl-amino-3-morpholino-1-propanol) and a homologous series of compounds has been studied extensively in order to extend these potential from biological tools to therapeutic agents.

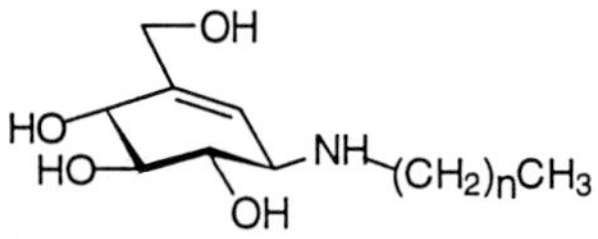

Figure 5-13. *N*-Alkyl-β-valienamines and its PDMP analogues.

58: *N*-Alkyl β-valienamine (n = 7) (n = 3, 5, 9, 13, 17)

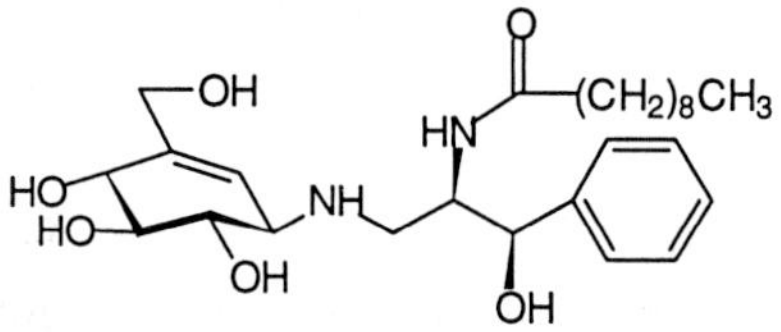

59: PDMP analog

It has become of interest to prepare PDMP analogues in which the morpholine units are replaced by carba-sugar residues and to test their inhibitory activity. Several *N*-acyl derivatives of **51**, prepared in the usual way, were reduced with LAH and deprotected to afford a homologous series [31] of *N*-alkyl-β-valienamines. Among the six homologs tested, *N*-octyl derivative **58** possessed the highest inhibitory potency (IC_{50} 3 x 10^{-8} M) against rat liver β-glucocerebrosidase; this is almost 10-fold potent than the related 5a-carbaglucosylceramide analogue. The PDMP analogues [32] containing β-valienamine residues have also been shown to inhibit not β-glucocerebroside synthase but β-glucocerebrosidase (D-*threo* isomer **59**, IC_{50} 7 x 10^{-7} M). Therefore, it still remains unresolved what kind of structural functions can modulate the inhibitory action against each two types of enzymes.

5.5 Development of New-Types of Glycosidase Inhibitors

5.5.1 Chemical Modification of the Trehalase Inhibitor Trehazolin

In 1990 trehazolin (**60**), a potent trehalase inhibitor was isolated from the culture broth of *Amycolatopsis trehalostatica* [33] and *Micromonospora* strain [34]. The proposed structure was finally established by total synthesis [35] to be a unique pseudo-disaccharide composed of α-D-glucopyranosylamine and 5-amino-1-*C*-hydroxymethyl-1,2,3,4-cyclopentanetetraol bonded by way of a cyclic isourea function. In connection with our studies on glycosidase inhibitors, especially trehalase inhibitor, we became interested in the structure-activity relationship of the trehazolin structure.

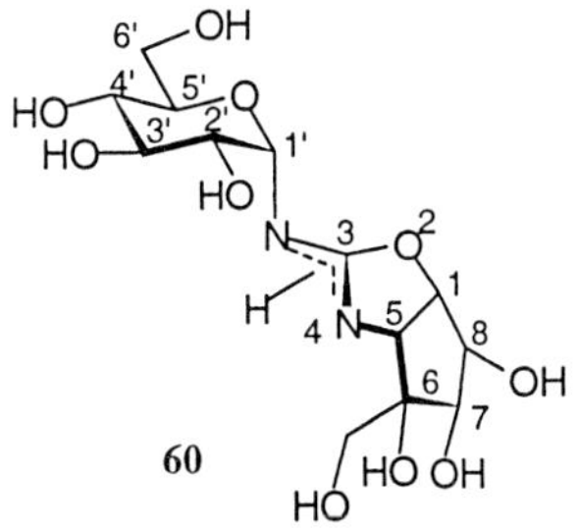

60 Trehazolin (4.9×10^{-8})

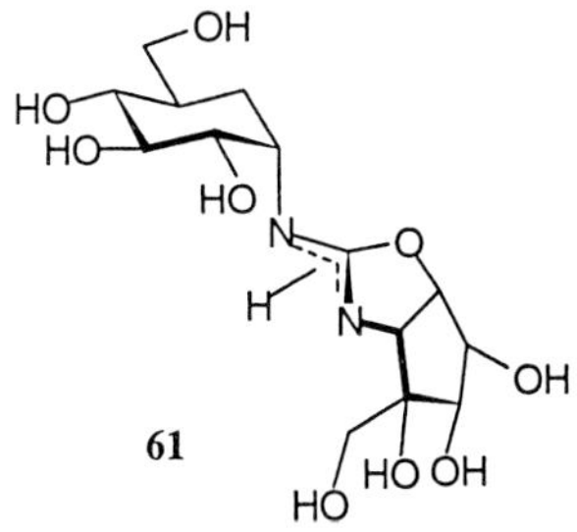

61 5a'-Carbatrehazolin (4.9×10^{-8})

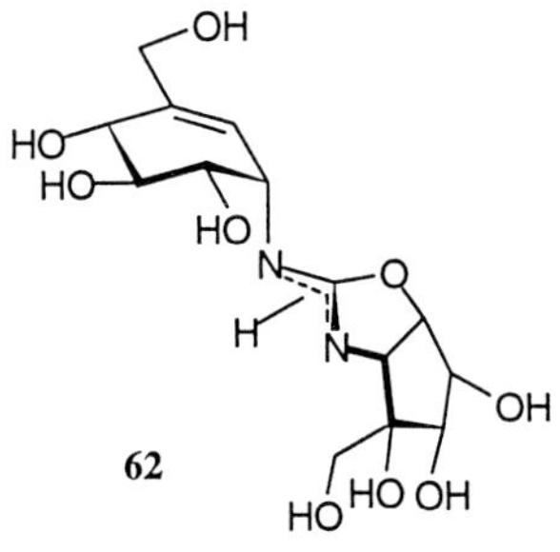

62 5a'-Carba-5'(5a')-eno-trehazolin (3.1×10^{-7})

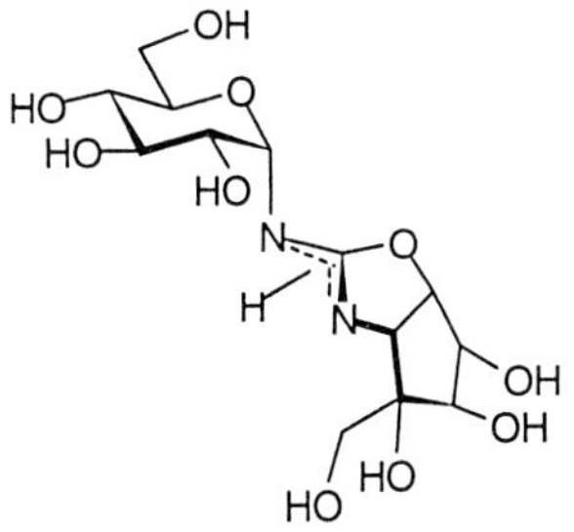

7-Epitrehazolin No inhibition

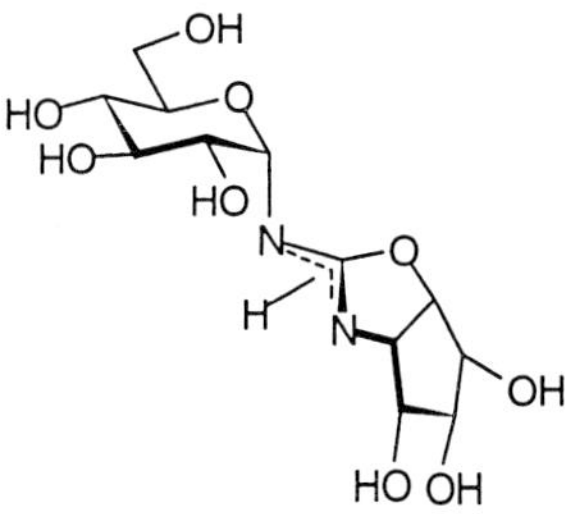

6-De(hydroxymethyl)-trehazolin (7.8×10^{-6})

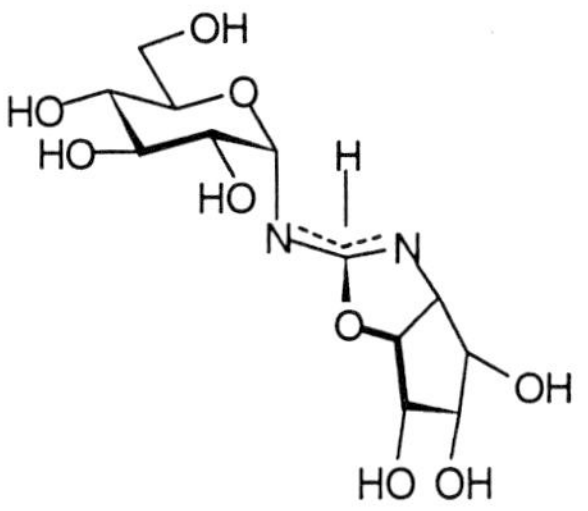

6-Dehydroxymethyl trehazolin (isomer) (1.6×10^{-7})

Figure 5-14. Trehalase inhibitor trehazolin and its chemical modification [Inhibitory activity (IC_{50}/M) against silkworm trehalase].

Systematic chemical modification of each sugar [36] and aminocyclopentanepolyol moieties [37, 38] was carried out. The inhibitory activities against silkworm trehalase of 17 stereoisomers were assayed. Except for 5a'-carbatrehazolin **61**, all showed decreased potency compared with **60**. Interestingly, the unsaturated 5a'-carbatrehazolin **62** shows rather low activity probably due to a conformational deformation of the 5a-carbaglucopyranose ring, thus indicating the importance of the D-glucopyranose residue. The pseudo-disaccharide structure of **60** may be a close mimic of the substrate α,α-trehalose or its intermediate which is involved in the substrate recognition and hydrolytic steps. The aminocyclopentanepolyol moiety with a cyclic isourea group would constitute charge-distribution part (glycone part) for binding to the active site of the enzymes. All hydroxyl groups on the cyclopentane ring are considered to be

5.5.2 Development of 5-Amino-1,2,3,4-cyclopentanetetraol Derivatives as New Potent Glycosidase Inhibitors

According to the results obtained from modifications of trehazolin, some stereoisomers of the cyclic isourea derivatives of this type were expected to be lead compounds for elaboration to new glycosidase inhibitors. Trehazolamine **D-63**, the aminocyclitol part of **60**, has already been shown [39] to exhibit moderate inhibitory activity against β-glucosidase.

When the structures of **D-63**, **D-64**, and **L-64** were compared, **L-64** seemed to have the closest similarity to the flattened half-chair conformation of the glucosyl cation probably formed during glucoside hydrolysis. As had been expected, it was demonstrated that **L-64** is a strong inhibitor (IC_{50} 4 x 10^{-7} M) of α-glucosidase (Baker's yeast) but had a mild activity against almond β-glucosidase [40]. On the other hand, **63** was shown to be a weak α-glucosidase inhibitor (IC_{50} 2.6 x 10^{-4} M). Next, the readily available aminocyclopentanepolyols **D** and **L-65**, and **66** which mimic the half-chair conformation of the galactosyl cation were chosen as potential galactosidase inhibitors. As had been expected, only **D-65** was shown to be a moderate inhibitor of β-galactosidase (*E. coli*). Thus, 5-aminocyclopentanepolyols such as **L-64** and **D-65** may belong to a new type of 4a-carba-amino sugars of biological interest.

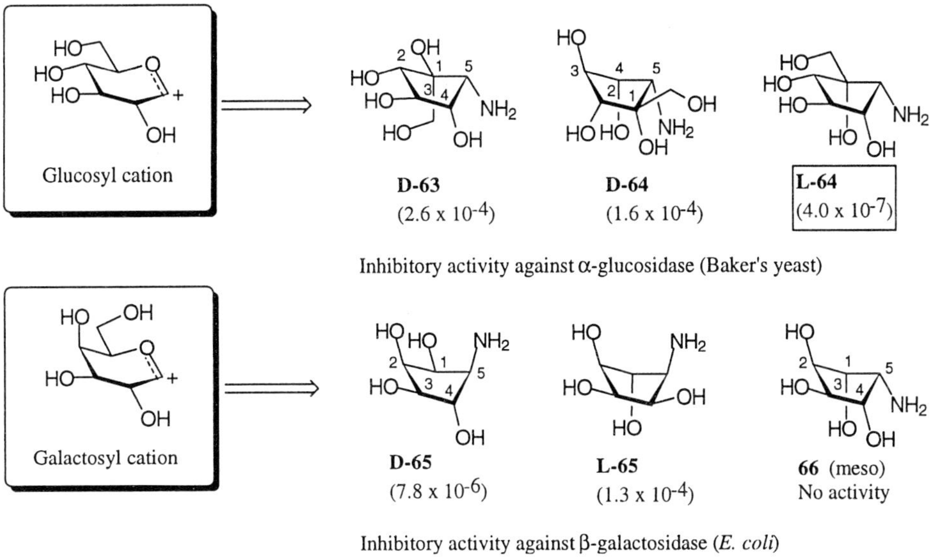

Figure 5-15. Glycosidase inhibitory activity of 5-amino-1,2,3,4-cyclopentanetetraols [Inhibitory activity (IC$_{50}$/M)].

5.5.3 N-Alkyl and N-Phenyl Cyclic Isourea Derivatives of 5-Aminocyclopentanepolyol Derivatives: New Types of Potent α-Glucosidase Inhibitors

The above observations have prompted us to improve the inhibitory activity of the lead compounds by modification of charge-distributions and conformations. We chose first to introduce N-phenyl cyclic isourea groups in order to add hydrophobic functions and/or stacking effects to the inhibitors. Thus, nine derivatives were prepared [41] by coupling of aminocyclitols with phenyl isothiocyanate, followed by treatment with yellow mercury(II) oxide.

Significant improvements in specific inhibitory potential against α-glucosidase were observed for derivatives **L-67** and **L-68,**. The configuration of the hydroxyl groups, along with the oxygen atoms of the isoureas, proved important for obtaining specific activity in this series of inhibitors. In order to further increase potency, N-butyl and N-butyl cyclic isourea functions were introduced into **L-64**. The latter **L-69** had also a strong potency against Baker's yeast glucosidase I (yeast) at a level almost comparable with that of N-butyl deoxynojirimycin [42].

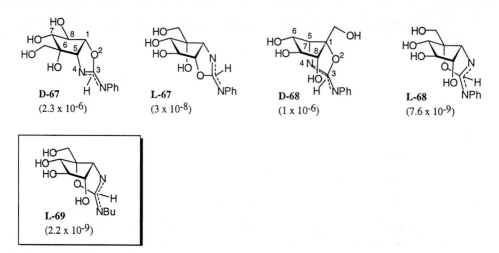

Figure 5-16. Development of potent α-glucosidase inhibitors [Inhibitory activity (IC$_{50}$/M) against Baker's yeast].

In summary, the present results have demonstrated that the *N*-substituted 3-amino-6-hydroxymethyl-2-oxa-4-azabicyclo[3.3.0]oct-3-ene-6,7,8-triols would be model compounds for designing new potential sugar-hydrolase inhibitors.

5.6 References

1. T. Suami, S. Ogawa, *Advan. Carbohydr. Chem.* **1990**, Vol. 48, p. 21.

2. G. E. McCasland, S. Furuta, L. J. Durham, *J. Org. Chem.* **1966**, *31*, 1516-1521; ibid. **1968**, *33*, 2835-2841 and 2841-2844.

3. T. Iwasa, H. Yamamoto, M. Shibata, *J. Antibiot.* **1970**, *32*, 595-602.

4. S. Horii, T. Iwasa, E. Mizuta, Y. Kameda, *J. Antibiot.* **1971**, *24*, 59-63; Y. Kameda, N. Asano, M. Yoshikawa, M. Takeuchi, T. Yamaguchi, K. Matsui, S. Horii, H. Fukase, *J. Antibiot.* **1984**, *37*, 1301-1307.

5. D. D. Schmidt, W. Frommer, B. Junge, L. Müller, W. Wingender, E. Truscheit, D. Schafer, *Naturwissenschafter* **1977**, 64, 535-536.

6. D. D. Schmidt, W. Frommer, B. Junge, L. Müller, W. Wingender, E. Truscheit, *Proceedings First International Symposium on Acarbose* (Ed. : W. Creutzfeldt), Excerpta Medica, Amsterdam-Oxford-Princeton, **1982**, P. 5.

7. T. W. Miller, B. H. Arison, G. Albers-Schonberg, *Biotech. Bioeng.* **1973**, *15*, 1075-1080.

8. N. Asano, M. Takeuchi, Y. Kameda, K. Matsui, Y. Kono, *J. Antibiot.* **1990**, *43*, 722-726.

9. S. Horii, H. Fukase, T. Matsuo, Y. Kameda, N. Asano, K. Matsui, *J. Med. Chem.* **1986**, *29*, 1038-1046.

10. L. Vertésy, H-W. Fehlhaber, A. Schulz, *Angew. Chem. Int. Ed. Engl.* **1994**, *33*, 1844-1845.

11. T. Suami, S. Ogawa, K. Nakamoto, I. Kasahara, *Carbohydr. Res.* **1977**, *58*, 240-244.

12. S. Ogawa, Y. Iwasawa, T. Nose, T. Suami, S. Ohba, M. Ito, Y. Saito, *J. Chem. Soc., Perkin Trans. 1* **1985**, 903-906.

13. Y. Miyamoto, S. Ogawa, *J. Chem. Soc., Perkin Trans. 1* **1989**, 1013-1018.

14. Y. Shibata, S. Ogawa, *Carbohydr. Res.* **1989**, *189*, 309-322.

15. T. Yamagishi, C. Uchida, S. Ogawa, *Bioorg. Med. Chem. Lett.* **1995**, *5*, 487-490; *Chem. Eur. J.* **1995**, *1*, 634-636.

16. S. Ogawa, K. Sato, Y. Miyamoto, *J. Chem. Soc., Perkin Trans. 1* **1993**, 691-696.

17. B. Junge, F. R. Heiker, J. Kunz, J. Müller, D. D. Schmidt, C. Wunsche, *Carbohydr. Res.* **1984**, *128*, 235-268.

18. Y. Shibata, Y. Kosuge, T. Mizukoshi, S. Ogawa, *Carbohydr. Res.* **1992**, *228*, 377-398; S. Ogawa, D. Aso, *ibid.* **1993**, *250*, 177-184.

19. S. Ogawa, C. Uchida, Y. Shibata, *Carbohydr. Res.* **1992**, *223*, 279-286.

20. S. Ogawa, Y. Nakamura, *Carbohydr. Res.* **1992**, *226*, 79-89.

21. S. Ogawa, H. Tsunoda, *Liebigs Ann. Chem.* **1993**, 755-769.

22. H. Tsunoda, S.-i. Sasaki, T. Furuya, S. Ogawa, *Liebigs Ann. Chem.* **1996**, 159-165.

23. S. Ogawa, K. Hirai, M. Ohno, T. Furuya, S-i Sasaki, H. Tsunoda, *Liebigs Ann. Chem.* **1996**, 673-677.

24. S. Ogawa, S.-i. Sasaki, H. Tsunoda, *Carbohydr. Res.* **1995**, *274*, 183-196.

25. S. Ogawa, T. Furuya, H. Tsunoda, O. Hindsgaul, K. Stangier, M. M. Palcic, *Carbohydr. Res.* **1995**, *271*, 197-205.

26. O. Lockhoff, *Angew. Chem. Int. Ed. Engl.* **1991**, *30*, 1611-1620.

27. H. Tsunoda, S. Ogawa, *Liebigs Ann. Chem.* **1994**, 103-107.

28. H. Tsunoda, S. Ogawa, *Liebigs Ann. Chem.* **1995**, 267-277.

29. H. Tsunoda, J.-i. Inokuchi, K. Yamagishi, S. Ogawa, *Liebigs Ann. Chem.* **1995**, 279-284

30. J.-i. Inokuchi, N. S. Radin, *Lipid Res.* **1987**, *28*, 565-571; N. S. Radin, J.-i. Inokuchi *Trends Glycosci. Glycotechnol.* **1991**, *3*, 200-213.

31. S. Ogawa, M. Ashiura, C. Uchida, S. Watanabe, C. yamazaki, K. Yamagishi, J.-i. Inokuchi, *Bioorg. Med. Chem. Lett.* 1996, 6, 929-932.

32. S. Ogawa, T. Mito, E. Taiji, K. Yamagishi, J.-i. Inokuchi, unpublished results.

33. S. Murao, T. Sakai, H. Gibo, T. Nakayama, T. Shin, *Agric. Biol. Chem.* **1991**, *55*, 895-897.

34. O. Ando, H. Satake, K. Itoi, A. Sato, M. Nakajima, S. Takahashi, H. Haruyama, Y. Okuma, T. Kinoshita, R. Enokita, *J. Antibiot.* **1991**, *44*, 1165-1168.

35. S. Ogawa, C. Uchida, *J. Chem. Soc., Perkin Trans. 1* **1992**, 1939-1942; C. Uchida, T. Yamagishi, S. Ogawa, *ibid.* **1994**, 589-602.

36. C. Uchida, H. Kitahashi, T. Yamagishi, Y. Iwaisaki, S. Ogawa, *J. Chem. Soc., Perkin Trans. 1* **1994**, 2775-2785.

37. C. Uchida, H. Kitahashi, S. Watanabe, S. Ogawa, *J. Chem. Soc., Perkin Trans. 1* **1995**, 1707-1717.

38. C. Uchida, T. Yamagishi, H. Kitahashi, Y. Iwaisaki, S. Ogawa, *Bioorg. Med. Chem.* **1995**, *3*, 1605-1624.

39. O. Ando, M. Nakajima, K. Hamano, K. Itoi, S. Takahashi, Y. Takamatsu, A. Sato, R. Enokita, T. Okazaki, H. Haruyama, T. Kinoshita, *J. Antibiot.* **1993**, *46*, 1116-1125.

40. C. Uchida, H. Kimura, S. Ogawa, *Bioorg. Med. Chem. Lett.* **1994**, *4*, 2643-2648.

41. C. Uchida, J.-i. Inokuchi, S. Ogawa, unpublished results.

42. R. A. Gruters, J. J. Neefjes, M. Tersmette, R. E. de Goede, A. Tulp, H. G. Huisman, A. F. Miedem, H. L. Ploegh, *Nature* **1987**, *330*, 74-77.

6 The Dithiane Route from Carbohydrates to Carbocycles

K. Krohn, G. Börner and S. Gringard

6.1 Introduction

Carbohydrates are the most abundant source of chiral natural products [1] and Nature uses this wealth for the conversion not only of sugar derivatives but also for other carbocyclic and heterocyclic products (e.g. shikimic acid, cyclitols etc [2]). The last two decades have seen an increasing interest of the synthetic community to use sugars as chiral templates for the chemical conversion to other natural products [3] (for further related references see ref. [4, 5]).

The use of carbohydrates in the construction of functionalized carbocyclic ring systems is an emerging new field in carbohydrate chemistry. Ferrier and Middleton [6], in a recent comprehensive review on the conversion of sugars to cyclopentanes and cyclohexanes, state that more than 80% of the papers relevant to the subject were published in the last decade. The use of sugars for incorporation in condensed ring systems [7] or in radical cyclizations [8] has also recently been reviewed.

The products prepared have more or less resemblance to the starting materials. Very often the overfunctionalization with hydroxy groups presents a problem. On the other hand, new functional groups have to be introduced such as double bonds to merge the sugar part into more complex ring systems for instance by Diels-Alder reaction [9]. If the oxygen in the furanose or pyranose ring systems is merely replaced by a methylene group the term "pseudo sugar" has been coined.

In many cases the carbocyclic framework is constructed by cycloaddition reactions and the carbohydrate serves merely as a chiral template. Methods for the conversion of a carbohydrate to a carbocycle involving predominantly the carbon atoms of the starting sugar are rare [10] and only a few examples are known where the nucleophile centre was generated from nitro sugars [11], dicarbonyl derivatives [12], or vinyl ethers, as in the Ferrier reaction [13].

In our investigations we have explored two alternative approaches outlined in Scheme 1. In the first case the aldehyde function of 2-deoxysugars such as 2-deoxyribose present in the predominant hemiacetal form is converted into the 1,3-dithiane. Alternatively, protected sugars (mostly acetonides, see below) are reacted with lithiated 1,3-dithiane. Deoxygenation of the newly created hydroxyl group is required to prevent β-elimination. In both cases the common next steps are appropriate protection and activation of the hydroxyl groups. This step determines the ring size if activation is performed by tosylation. After conversion into epoxides, two cyclization modes of the lithiated dithiane are possible in principle.

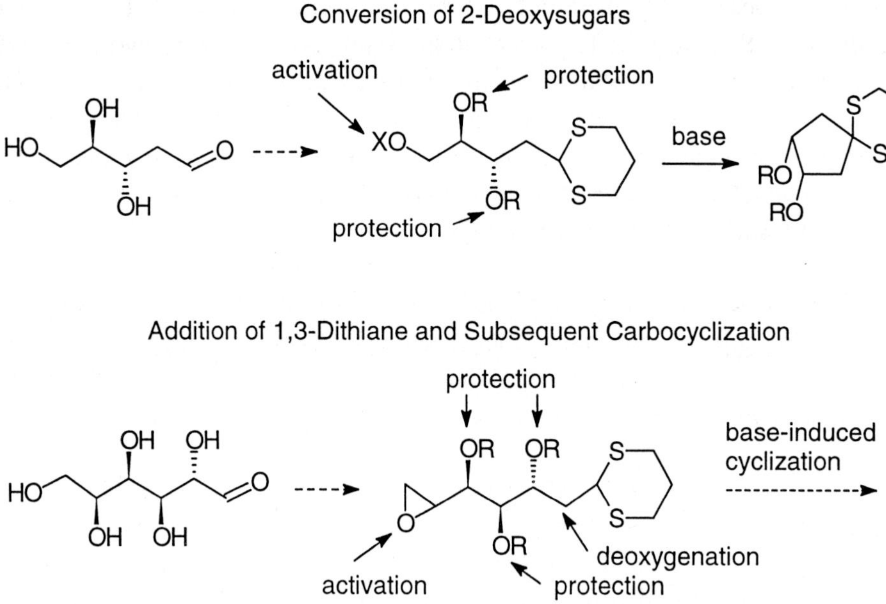

Scheme 6-1

It must be stressed at this point that in spite of the principal feasibility of the cyclization with 1,3-dithianes explored by Seebach et al [14, 15]. with simple hydrocarbons it was by no ways certain the cyclizations would work with sugars. The increased difficulty of nucleophilic substitution with sugars is a well known phenomenon [16] and the reasons are partly steric and partly electronic (nucleophiles have to pass close to electronegative oxygens).

We shall first discuss the conversion of 2-deoxy-sugars followed by the addition of 1,3-dithiane to sugars. In each case the material is presented according to ring size. Finally, examples for application in natural product syntheses are given.

6.2 Three- to Five-membered Carbocyclic Rings from 2-Deoxy ribose

A prerequisite for the realization of this synthetic scheme is the absence of leaving groups at the neighboring position at C-2 to avoid 1,2-eliminations during anion formation. For that reason we selected the commercially available 2-deoxy-D-ribose **1** as the starting material.

6.2.1 Five-membered Ring

2-Deoxy-D-ribose **1** was converted in 90% yield into the known [17] dithioacetal **2** as shown in Scheme 6-2. The selective tosylation at the primary hydroxy group was achieved by the addition of dibutyltin oxide to yield 59% of **3**. The monotosylate **3** was converted into the acetonide **4** to avoid epoxide formation and Payne rearrangements in the subsequent base treatment. A clean cyclization to give the achiral cyclopentane derivative **5** in 71% yield occurred upon treatment of **4** with *n*-butyl lithium at −30°C in THF.

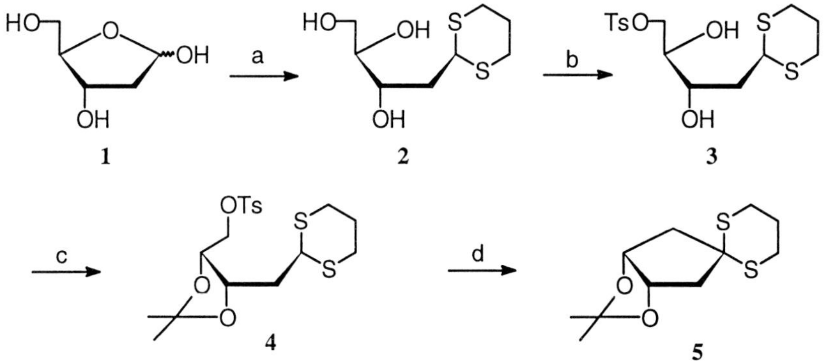

Scheme 6-2. Reagents: a) CHCl$_3$, 6 N HCl, 1,3-Propanedithiol; b) TsCl, *n*-Bu$_2$SnO, Py; c) DMP, TsOH; d) THF, 1 eq *n*-BuLi, −30 °C

6.2.2 Three-membered rings

Three-membered carboyclic rings have mainly been constructed from sugars by external addition of diazomethane, sulfur ylids or carbenes to unsaturated sugars (for a review see [18]). In our plans we wanted to create an epoxide next to the nucleophilic dithiane center. Stork and Cohen [19] have shown that cyclobutanes were formed in an epoxynitrile cyclization in preference to cyclopentanes and it was interesting to see whether three- or four-membered rings would result from the intramolecular dithiane-cyclization.

Dithiane **2** was tritylated at the primary position to afford **6** in 81% yield as shown in Scheme 6-3. The monotosylation of the two secondary hydroxy groups in **6** gave a 3 : 1 mixture of **7** and **8** (69%) that could be separated by layer chromatography on silica gel. Treatment of the α-hydroxytosylates **7** and **8** gave the enantiomeric epoxides **9** and **11** in essentially quantitative yield. Over 70% of the cyclopropane-dithioacetals **10** and **12** were isolated upon cyclization of the epoxides. The products **10** and **12** are enantiomers as are the parent epoxides **9** and **11**.

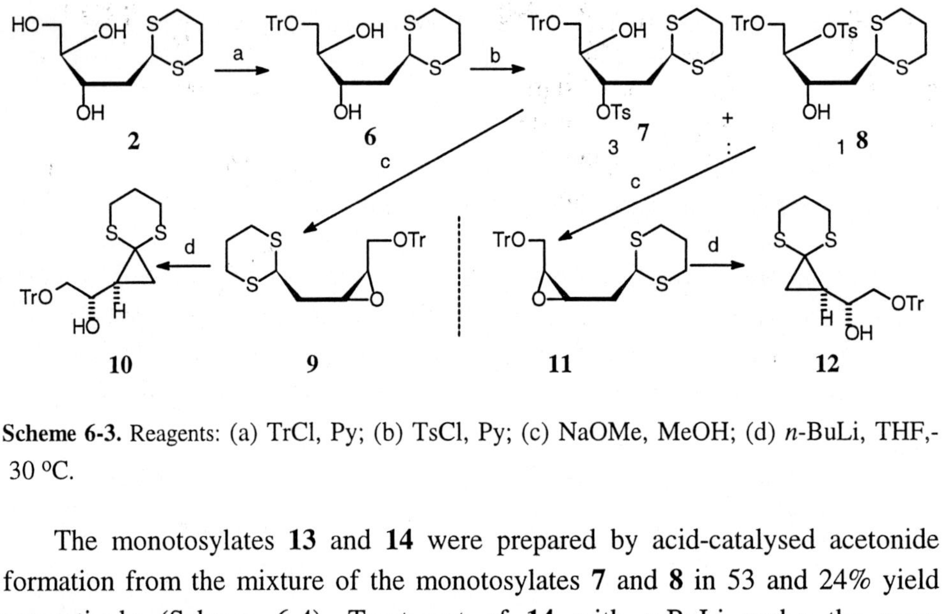

Scheme 6-3. Reagents: (a) TrCl, Py; (b) TsCl, Py; (c) NaOMe, MeOH; (d) *n*-BuLi, THF,-30 °C.

The monotosylates **13** and **14** were prepared by acid-catalysed acetonide formation from the mixture of the monotosylates **7** and **8** in 53 and 24% yield respectively (Scheme 6-4). Treatment of **14** with *n*-BuLi under the same conditions as described before gave a clean cyclization to afford the cyclopropane acetonide **15**. However, no cyclobutane formation could be induced by *n*-BuLi treatment of **13** in agreement with model studies confirming that, for steric reasons, no S_N2 displacement reaction was possible with **13**.

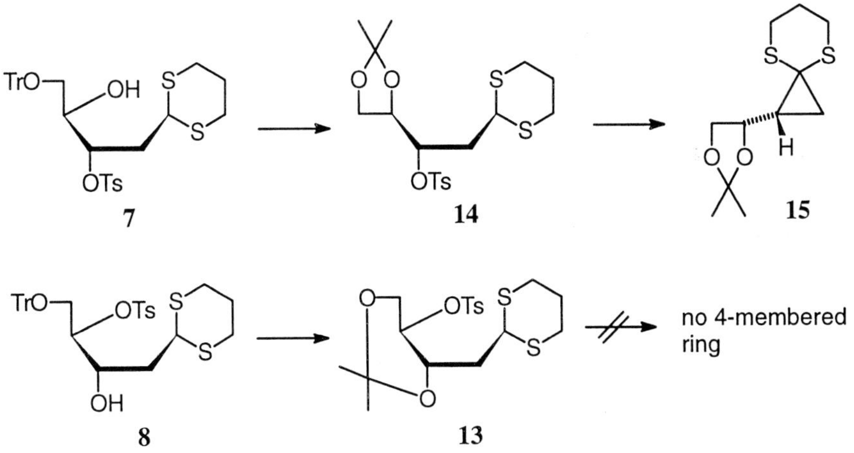

Scheme 6-4

6.2.3 Four-membered Ring

A more flexible system had to be chosen to investigate the formation of four-membered rings. For that purpose the anomeric mixture **16a/b** of the methyl glycosides derived from 2-deoxy-D-ribose **1** was converted into the methyl ethers **17a/b** (72% overall yield from **1**). Thioacetalization of **17a/b** gave the 1,3-dithiane **18** in 81% yield and tosylation of the free hydroxy group at C-4 afforded the desired monotosylate **19** that can be cyclized to the cyclobutane **20** (18%).

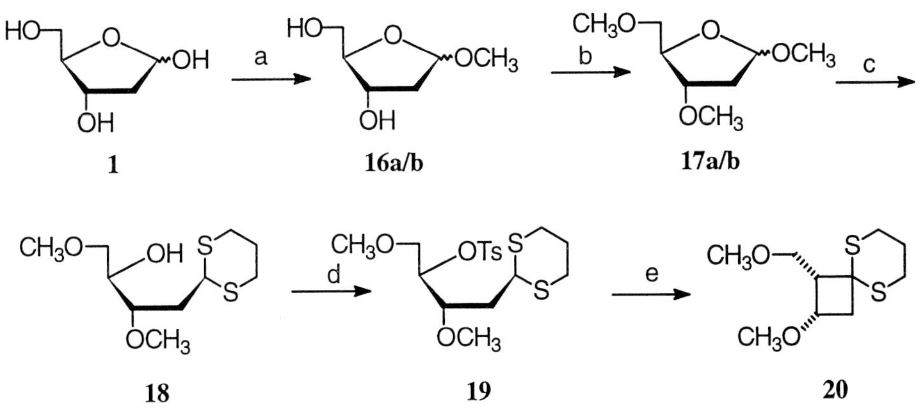

Scheme 6-5. Reagents: a) MeOH, H_2SO_4; b) MeI, NaH, THF; c) $HS(CH_2)_3SH$, SiO_2/HCl, $ZnCl_2$, CH_2Cl_2; d) TsCl, Py; e) n-BuLi, –50 °C.

The examples presented so far show that three-, five- and to a lesser extent four-membered carbocyclic rings can be prepared from 2-deoxy sugars by intramolecular nucleophilic displacement reactions.

6.3 Three- to Seven-membered Carbocyclic Rings from Mannose

The addition of 1,3-dithiane to protected sugars is an extension of the methodology shown in the first section that expands the scope and flexibility of the reaction. Simple starting materials such as 2,3:5,6-di-*O*-isopropylidene-D-mannofuranose (**21**) can be used for addition with 1,3-dithiane to afford **22**. Transacetalization of the known alcohol **22** [20] to the triacetonide **23** was achieved by treatment with acetone and acid (84%). The next two steps can be performed in one reaction vessel: elimination to the transient thioketene acetal **24** is initiated by treatment with 2 equiv of *n*-BuLi and subsequent reduction to the saturated alcohol **25** with 1.2 equiv of LAH[21] (Scheme 6-6, 90% from **23**).

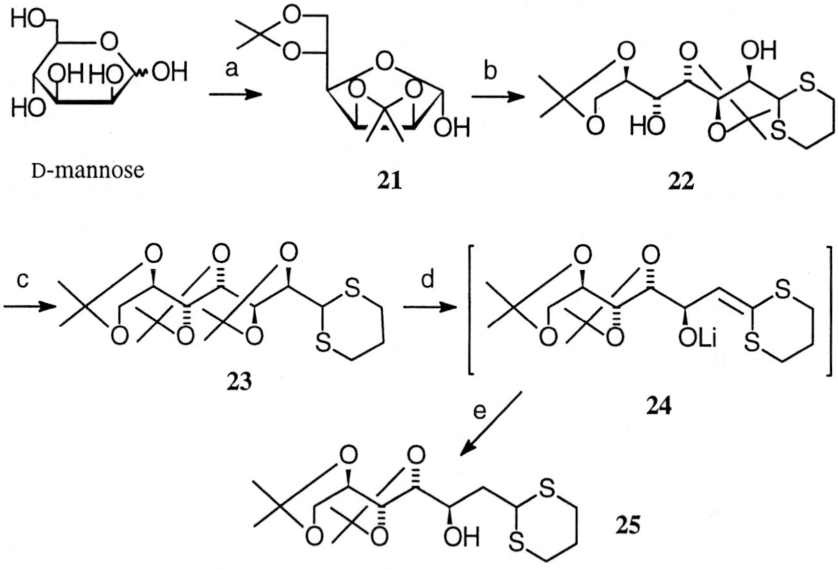

Scheme 6-6. Reagents: a) ref. [22]; b) ref. [23]; c) CH$_3$COCH$_3$, H$_2$SO$_4$ (84%); d) *n*-BuLi, −40 to −20°C; e) LAH, 0°C.

6.3.1 Cyclopropane Derivatives

Cyclopropane derivatives are accessible by our strategy in a very short reaction sequence (6 steps from mannose, Scheme 6-7). Thus, the LAH reduction product

25 is tosylated to the monotosylate **26** (96%) and then cyclized by treatment with one equiv of *n*-BuLi to the crystalline cyclopropane derivative **27** in 77% yield (Scheme 6-7). Desulfurization of **27** with Raney nickel was successful and two products could be isolated. The major compound (51%) was identified as the chiral cyclopropane derivative **28**. The minor product (44%) could be assigned the open chain structure **29** (Scheme 6-7).

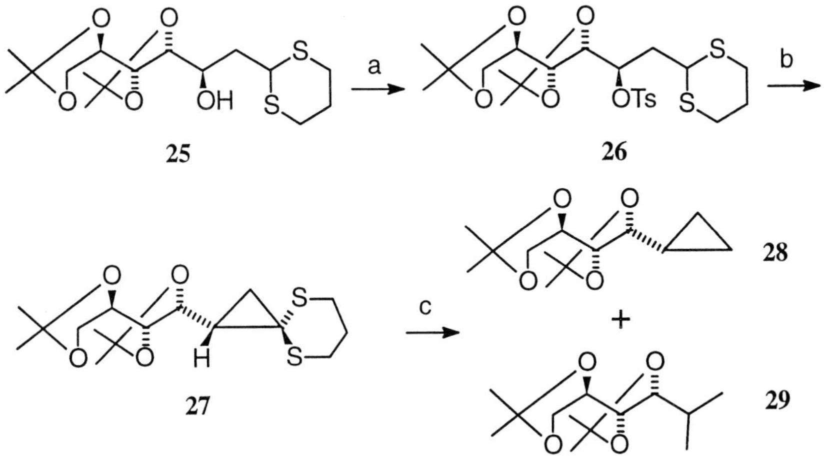

Scheme 6-7. Reagents: a) Py/TsCl/DMAP, 24 h, 20°C (96%); b) THF, *n*-BuLi, −40°C to 20°C, 2 h, (77%); c) Ra-Ni, EtOH, 76°C, 30 min (51% **28**, 44% **29**).

6.3.2 Cyclobutane Derivatives

In our initial studies (see above) the yield of the produced cyclobutane was low (18%). The required epoxide **36** for an alternative approach to this ring system was prepared starting from the common intermediate **25** (Scheme 6-8). The hydroxy group at C-3 was protected as the methyl ether **30** (96%) and the primary acetonide of **30** cleaved selectively to afford the diol **31** (68%). Protection as the methyl ether **32** (96%), cleavage of the acetonide to the diol **33** (90%), and tosylation afforded a mixture of the 5-*O*-monotosylate **34** (25%) and the 4-*O*-monotosylate **35** (56%). The major tosylate **35** was converted to the epoxide **36** by treatment with NaOMe in MeOH (95%). Employing the usual conditions for deprotonation of the 1,3-dithiane (1 equiv of *n*-BuLi in THF, −40°C) a cyclobutane derivative **37** was isolated in 34% yield. No cyclopentane derivates could be detected in the non-polar fraction, in agreement with Stork's observation

[24]. Thus, cyclobutanes are still formed in relatively low yield using the dithiane methodology.

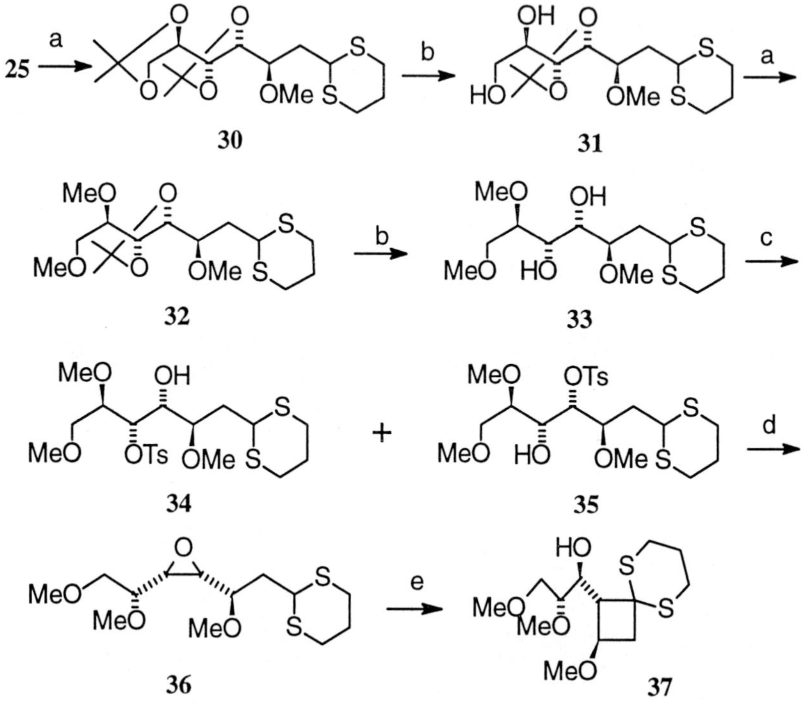

Scheme 6-8. Reagents: a) THF, NaH, MeI, 18 h, 20°C; b) MeOH, HCl; c) Py/TsCl/DMAP, 2 h, 20°C; d) NaOMe/MeOH, 18 h, 20°C (95%); e) THF, n-BuLi, −40 to 20°C, 18 h (34%).

6.3.3 Cyclohexanes and Cycloheptanes

The diol **31** can also be used to selectively activate the positions at C-6 and C-7 for the anticipated nucleophilic ring closure. A selective activation of 6-OH can be achieved by protection of the primary hydroxy group by tritylation to afford the trityl ether **38** in 82% yield. The remaining secondary hydroxy group was then tosylated to provide **39** (86%). However, even under prolonged reaction times no cyclization products could be observed. Inspection of models shows that a 1,3-diaxial interaction in the transition state may prevent the cyclization reaction (see below).

Consequently, we investigated the behaviour of the primary epoxide **41**. The synthesis of **41** started from the diol **31**, which was selectively tosylated at the primary position (73%) and subsequently treated with NaOMe in MeOH to afford the epoxide **41** quantitatively. The reaction of this epoxide with *n*-BuLi gave two crystalline products (55 and 12%), that could be identified as the six- and seven-membered ring systems **42** and **43** (Scheme 6-9). Again, the result is in agreement with the Baldwin rules since 6-*exo-tet* and *-trig* as well as 7-*endo trig* processes are possible [25] (Scheme 6-9).

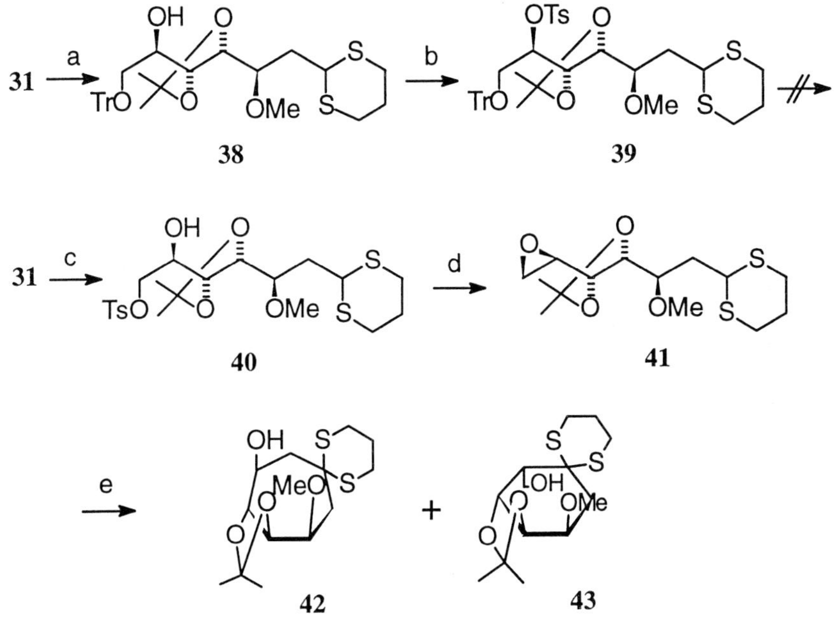

Scheme 6-9. Reagents: a) TrCl/Py/DMAP, 2 d, 20°C, (82%); b) TsCl/Py/DMAP, 2 d, 20°C (86%); c) Py/TsCl, 18 h, 3°C (73%); d) NaOMe/MeOH, 2 h, 20°C, (97%); e) THF, *n*-BuLi, 4 h, –40°C to 20°C.

6.4 Application to Natural Product Synthesis

6.4.1 Validatol

A few examples in natural product synthesis demonstrates the scope of the dithiane route from sugars to carbocycles. Analysis of the stereochemistry of compound **43** revealed that it is epimeric at C-3 to validatol (**44**), the product of hydrogenolysis[26] of the aminoglycoside antibiotic validamycin A (**45**)[27] (Scheme 6-10).

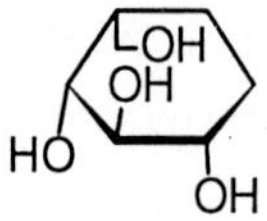

validatol (**44**)

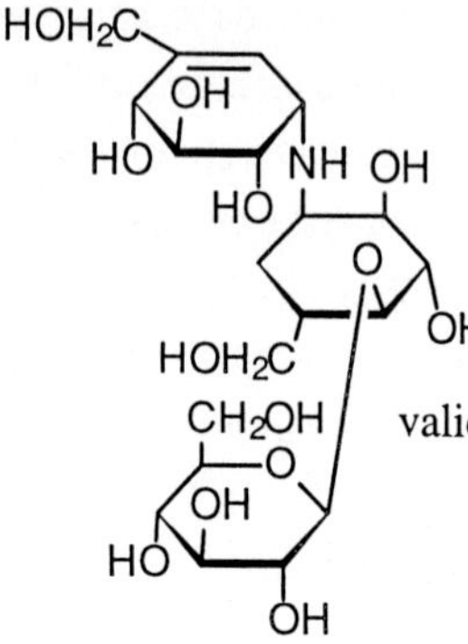

validamycin A (**45**)

Scheme 6-10

In our synthesis the center at C-3 in **43** had to be inverted at an early stage of the synthesis. The desulfurization of **43** with Raney nickel (80%) afforded 4-*epi*-validatol after deprotection. The correct configuration was obtained by inversion of the intermediate **46** to the alcohol **47** and the series of transformations described above led to validatol **44**. Alternatively, one could start from glucose. In this case a mixture of epimeric alcohols resulted from the 1,3-dithiane addition to the pyranoside bisacetonide of glucose that was converted to the triacetonides **48** and **49**. However, this was of no disadvantage since the newly generated hydroxy group had to be removed in the course of the synthetic sequence (Scheme 6-11).

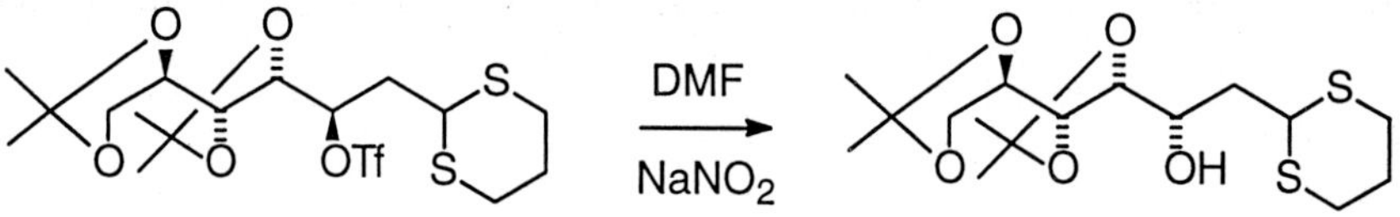

46 **47**

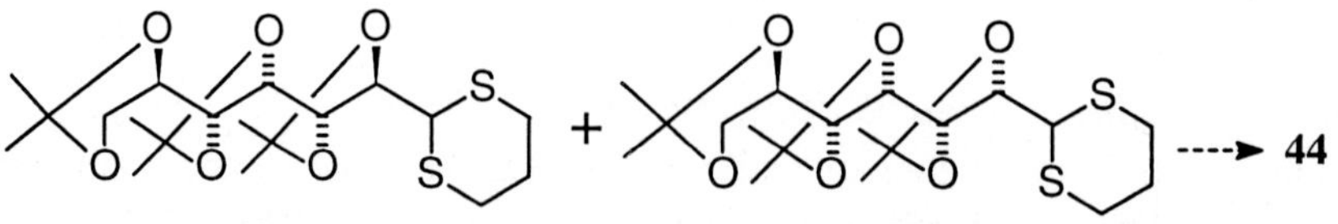

48 **49**

Scheme 6-11

Thus, the synthesis of validatol (**44**) carried out as described above proceeded in 12 steps and 6% overall yield starting from glucose.

6.4.2 Calcitriol

Calcitriol (1α,25-dihydroxy vitamin D_3, **I**) is the physiologically active form of vitamin D_3 (calciol) and is formed by hydroxylation in the liver and kidney. The interesting biological properties have stimulated a great number of syntheses of calcitriol (**I**) (review 28]). The first total synthesis based on the disconnection method of Lythgoe [28] has been achieved by a Hoffman LaRoche group [29] (retrosynthetic analysis see Scheme 6-12).

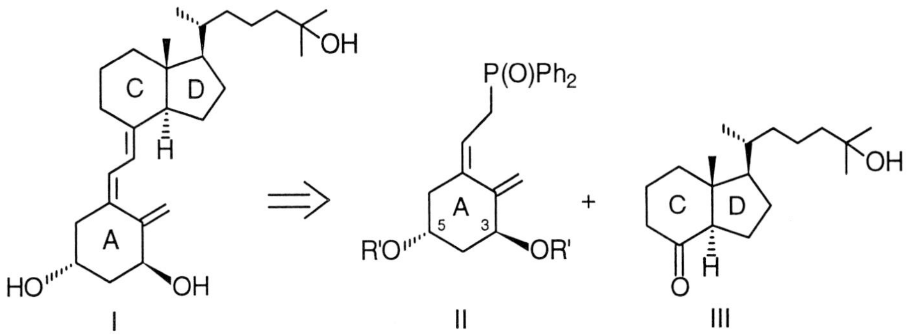

Scheme 6-12

Building block **II** can be derived from D-glucose (Scheme 6-13). The chain-elongated 1,3-dithiane derived from D-glucose has the correct configuration at carbon atoms C-3 and C-5. The synthetic scheme involves intermediate epoxides and deoxygenations at C-2 and C-6 along with the dithiane addition and the cyclization step. However, it had to be shown whether six- or seven-membered ring systems would result from the cyclization of the epoxy-1,3-dithiane intermediate.

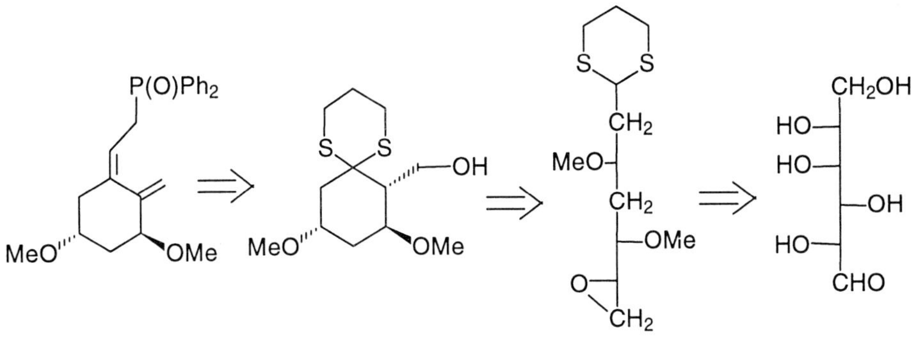

Scheme 6-13

In contrast to the previous methods, the lithiated 1,3-dithiane was added at C-6 of the sugar derivative **50**. The dithiane addition to the epoxide **50** [30] showed that only half the amount (1.1 equiv.) of lithiated 1,3-dithiane was required compared to the original procedure to afford the adduct **51** (89%). The diol **51** was then methylated to produce the dimethyl ether **52** (95%). The relatively stable acetonide **52** was cleaved by treatment with THF/2 N HCl (1:1.2) at 50°C (95%) and the oily furanose **53** (anomeric mixture) was reduced with sodium borohydride in EtOH to afford the triol **54** (80%) and the terminal hydroxy groups were protected as the acetonide **55** (79%) (Scheme 6-14).

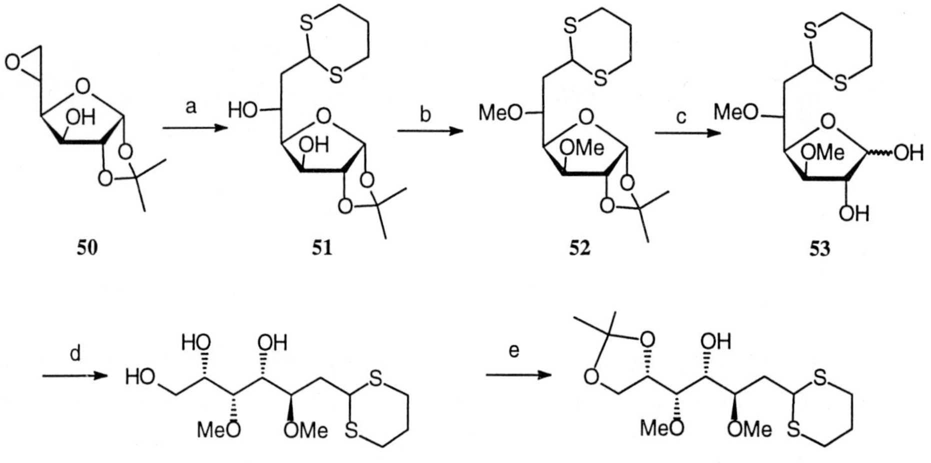

Scheme 6-14. Reagents: a) 1,3-dithiane, 2 eq. *n*-BuLi, THF/HMPT, –60°C (89%); b) NaH, MeI, THF, rt (95%); c) THF/ 2 N HCl, 50°C, (95%); d) NaBH$_4$, EtOH, rt (80%); e) DMP, H$_2$SO$_4$, 0°C, (79%)

To get the ring-A building block of calcitriol the hydroxy group at C-4 must be removed. The monoalcohol **55** was deoxygenated using the Barton-McCombie procedure [31] to afford the 2,4-dideoxy heptose derivative **56** (81%). The acetonide of compound **56** was easily cleaved using silica gel/6 N HCl (Huet procedure [32]) (Scheme 6-15).

Next, the terminal hydroxy groups had to be converted into potent electrophiles. A differentiation of the secondary and primary hydroxy groups by selective tosylation was not possible yielding only the ditosylate or product mixtures. A very convenient and generally applicable stereospecific conversion into the terminal epoxide was finally developed. In this procedure the diol **57** was converted with thionyl chloride into the mixture of the diastereomeric cyclic sulfites **58a/b** (ratio 1:2) in nearly quantitative yield [33]. The sulfites **58a/b** were then treated with sodium iodide in DMF to selectively afford the C-7 iodide **59** by attack on the less hindered primary carbon atom. The epoxide **60** was then generated under mild conditions by NaOMe treatment of the iodide **59** in 85% overall yield (Scheme 6-15). We believe that this procedure presents a general solution for the conversion of vicinal diols into epoxides if chemical differention of the hydroxy groups by selective activation (e.g. tosylation) is difficult to achieve.

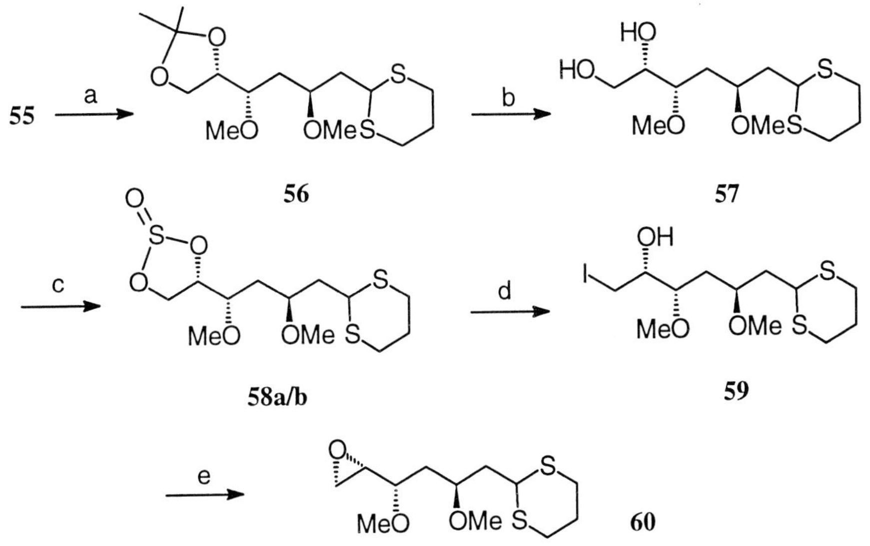

Scheme 6-15. Reagents: a) NaH, CS_2, MeI, THF, rt, then TBTH, toluene, reflux (81%); b) HCl, SiO_2, diethyl ether (80%); c) $SOCl_2$, CH_2Cl_2, 0°C (98%); d) NaI, DMF, 100°C; e) NaOMe, MeOH, 0°C (85% from **58**)

With the epoxide in hand the cyclization experiment was performed as described previously using *n*-BuLi as the base for the dithiane deprotonation. Surprisingly, the experiment did not lead to the expected six-membered ring **61**, but in a slow but clean reaction to the cycloheptane derivative **62** (64%). The structure of **62** was unambiguously established by the observation of a signal at 73 ppm in the ^{13}C NMR-spectrum for a tertiary carbon atom (CHOH) and the absence of a signal for a secondary carbon (CH_2OH), expected for six-membered rings at ca. 60 ppm [5, 34] (Scheme 6-16).

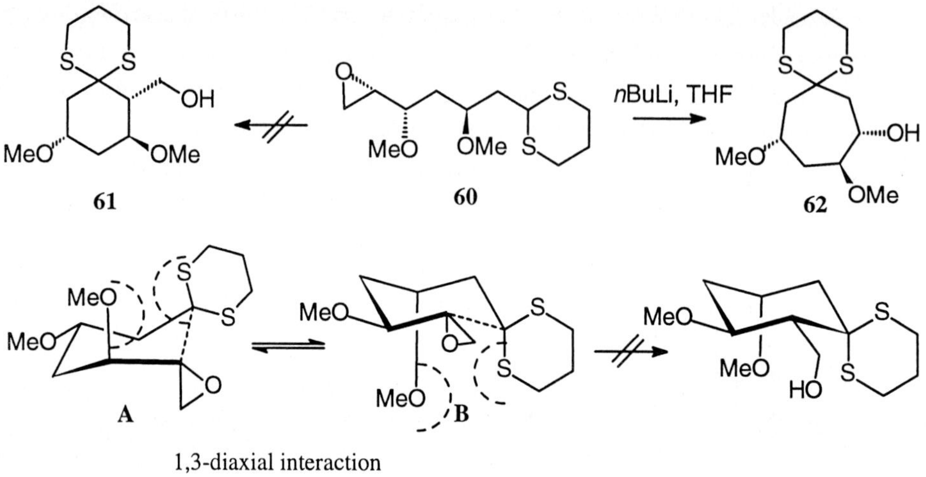

1,3-diaxial interaction

Scheme 6-16

Evidently, formation of the six-membered ring **61** is prevented by severe sterical hindrance of the 1,3-diaxial substituents in both transitions states A and B (see Scheme 6-16). Inspection of models of the more flexible transition states leading to the seven-membered ring revealed the absence of such sterical hindrance. Enantiomerically pure highly-substituted seven-membered rings are much more difficult to synthesize than their six-membered counterparts and, with this respect, the initially disappointing failure in the vitamin D_3 building block synthesis turned out to be a fortunate event. It is reasonable to assume that deliberate use of 1,3-diaxial strain in six-membered transition states (e.g. **A** or **B**) will generally allow the construction of seven-membered carbocycles using the dithiane methodology.

6.5 References

1. P. Collins, R. Ferrier, *Monosaccharides, Their Chemistry and Their Roles in Natural products*, John Wiley & Sons, New York, **1995**.

2. J. Thiem, *Hoechst High Chem.* **1989**, 8, 27–32.

3. S. Hanessian, *Total Synthesis of Natural Products: The 'Chiron' approach*, Pergamon, Oxford, **1983**.

4. K. Krohn, G. Börner, *J. Org. Chem.* **1991**, 56, 6038–6043.

5. K. Krohn, G. Börner, S. Gringard, *J. Org. Chem.* **1994**, 59, 6069–6074.

6. R. J. Ferrier, S. Middleton, *Chem. Rev.* **1993**, 93, 2779–2831.

7. K. J. Hale, *Rodd's Chemistry of Carbon Compounds*, 2nd Supplement, vol. 1 Parts E, F, G, Elsevier, Amsterdam, **1993**, p. 315.

8. B. Fraser-Reid, R. Tsang, *Strategies and Tactics in Organic Synthesis* (Ed.: T. Lindberg), vol. 2, Academic Press, New York, **1989**, pp. 123–162.

9. S. Chew, R. J. Ferrier, *J. Chem. Soc., Chem. Commun.* **1984**, 911–912.

10. G. D. Vite, R. Alfonso, B. Fraser-Reid, *J. Org. Chem.* **1989**, 54, 2268–2271.

11. H. H. Baer, *New Approaches to the Synthesis of Nitrogenous and Deoxy sugars and Cyclitols* in *Trends in Synthetic Carbohydrate Chemistry* (Eds.: D. Horton, L. D. Hawkins, G. J. McGarvey), chap. 2, American Chemical Society, Washington, **1989**, pp. 22–44.

12. S. Achab, J.-P. Cosson, B. C. Das, *J. Chem. Soc., Chem. Commun.* **1984**, 1040–1041.

13. D. H. R. Barton, J. Camara, P. Dalko, S. T. Géro, B. Quiclet-Sire, P. Stütz, *J. Org. Chem.* **1989**, 54, 3764–3766.

14. D. Seebach, N. R. Jones, E. J. Corey, *J. Org. Chem.* **1968**, 33, 300–305.

15. D. Seebach, E.-M. Wilka, *Synthesis* **1976**, 476–477.

16. R. W. Binkley, *Modern Carbohydrate Chemistry*, Marcel Dekker, New York, **1988**.

17. A. M. Sepulchre, G. Vass, S. D. Gero, *Tetrahedron Lett.* **1973**, 3619–3620.

18. J. Salaün, *Chem. Rev.* **1989**, 89, 1247–1270.

19. A. A. Kandil, K. N. Slessor, *J. Org. Chem.* **1985**, 50, 5649–5655.

20. H. Paulsen, M. Schüller, M. A. Nashed, A. Heitmann, H. Redlich, *Tetrahedron Lett.* **1985**, 26, 3689–3692.

21. M. Y. Wong, G. R. Gray, *J. Am. Chem. Soc.* **1978**, 100, 3548–3553.

22. O. Th. Schmidt, *Isopropylidene Derivatives* in *Methods in Carbohydrate Chemistry* (Ed.: R. L. Whistler), vol. II, Academic Press, New York, **1963**, pp. 319–325.

23. H. Paulsen, M. Schüller, M. A. Nashed, A. Heitmann, H. Redlich, *Liebigs Ann. Chem.* **1986**, 675–686.

24. G. Stork, L. D. Cama, D. R. Coulson, *J. Am. Chem. Soc.* **1974**, 96, 5268–5270.

25. J. E. Baldwin, *J. Chem. Soc., Chem. Commun.* **1976**, 734–736.

26. S. Horii, T. Iwasa, E. Muizuta, Y. Kameda, *J. Antibiot.* **1971**, 24, 59–63.

27. S. Horii, T. Iwasa, Y. Kameda, *J. Antibiot.* **1971**, 24, 57–58.

28. O. Isler, *Die D-Vitamine* in *Fettlösliche Vitamine* (Ed.: G. Brubacher), Georg Thieme Verlag, Stuttgart, **1982**, pp. 90–125.

29. E. G. Baggiolini, J. A. Iacobelli, B. M. Hennessy, M. R. Uskokovic, *J. Am. Chem. Soc.* **1982**, 104, 2945–2948.

30. A.-M. Sepulchre, G. Lukacs, G. Vass, S. D. Gero, *Bull. Soc. Chim. Fr.* **1972**, 10, 4000–4007.

31. D. H. R. Barton, S. W. McCombie, *J. Chem. Soc., Perkin Trans.1* **1975**, 1574–1585.

32. F. Huet, A. Lechevallier, M. Pellet, J. M. Conia, *Synthesis* **1978**, 63, 63–65.

33. K. Nymann, J. S. Svendsen, *Acta Chem. Scand.* **1994**, 48, 183–186.

34. K. Krohn, G. Börner, *J. Org. Chem.* **1994**, 58, 6063–6068.

7 New Reductive Carbocyclisations of Carbohydrate Derivatives Promoted by Samarium Diiodide

J. L. Chiara

7.1 Introduction

The range of methods available for the conversion of carbohydrate derivatives into highly functionalized, chiral carbocyclic derivatives has expanded dramatically in recent years [1]. A significant contribution to this development has come from the field of radical cyclisation reactions [2]. When the attacking radical in the cyclisation is α-heterosubstituted, the net loss of functionality which often limits the usefulness of classical tin-mediated radical reactions is avoided. Ketyl radical reactions belong to this type. This chapter presents some new advances in the application of ketyl radical cyclisations for the transformation of carbohydrates into highly functionalized carbocycles using the single electron transfer reducing agent samarium diiodide [3].

The molecules that inspired this work are shown in Figure 7-1. Inositol phosphates, such as 1,4,5-IP_3, and their derivatives are crucial for the transduction of information in living organisms [4]. Other inositol derivatives, the glycosylphosphatidylinositol (GPI) membrane anchors, are found on many eukariote plasma membrane proteins [5] and have been reported to be implicated in the signal transduction mechanism of growth factors and hormones such as insulin [6]. For the synthesis of these molecules it is important to have a ready access to selectively protected and enantiomerically pure inositol derivatives [7]. The aminocyclopentitol derivatives trehazolin [8], mannostatin A [9], and allosamidin [10] are potent glycosidase inhibitors of natural origin. While inhibitors of glycosidases have traditionally been molecules with structural homology to the natural substrates, usually polyhydroxylated 6-membered rings, these cyclopentane derivatives are the strongest competitive inhibitors known for their respective enzymes. The superiority of 5-membered rings over 6-membered analogues is intriguing. Besides their remarkable biological activity, all these natural products are challenging synthetic targets that offer an opportunity to develop new methodologies for the construction of polyoxygenated and enantiomerically pure cyclopentanes and cyclohexanes.

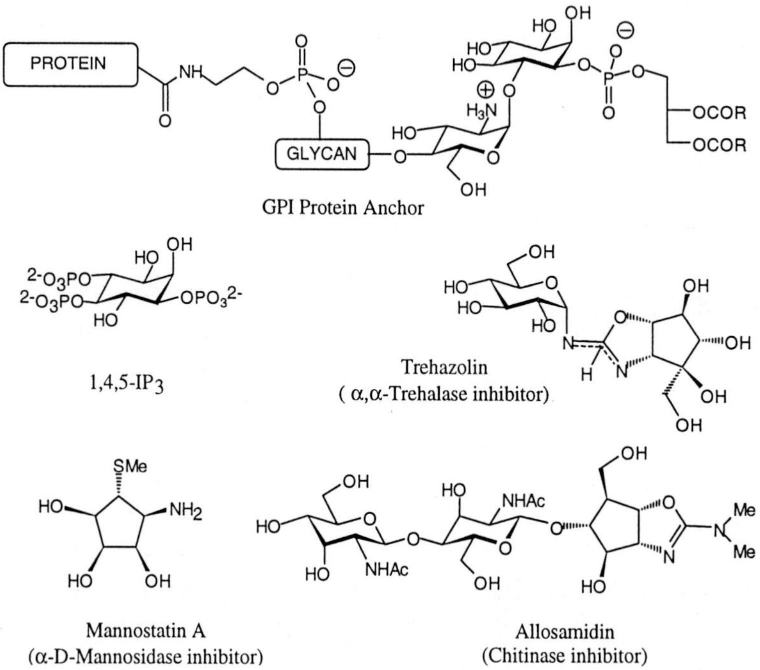

Figure 7-1. Biologically active natural products containing polyhydroxylated cyclohexanes and cyclopentanes.

With these targets in mind, three general routes were devised starting from carbohydrate derivatives and using a reductive carbocyclisation reaction promoted by SmI$_2$ (Figure 7-2) [11]. In the first route (a), the carbocyclisation step is an intramolecular pinacol coupling reaction promoted by SmI$_2$ that gives a carbocyclic vicinal diol stereoselectively. The dicarbonyl compound can be easily prepared by oxidation of a carbohydrate-derived 1,5- or 1,6-diol. In the second route (b), the carbocyclisation step consists of an intramolecular reductive cross-coupling of an oxime ether with a carbonyl group. This substrate can be readily prepared from a conveniently protected sugar lactol in two high-yielding steps. The resultant carbocyclic hydroxylamine ether can be further reduced in situ with excess samarium reagent to the corresponding aminocyclitol. Finally, in the last sequence (c), treatment of a 6-deoxy-6-iodopyranoside with 4 equivalents of SmI$_2$ starts up a cascade reaction that produces eventually a contraction of the pyranose ring via reductive elimination followed by 5-*exo-trig* ketyl-olefin radical cyclisation and trapping with an electrophile. A highly functionalized cyclopentitol is thus obtained in a single synthetic operation.

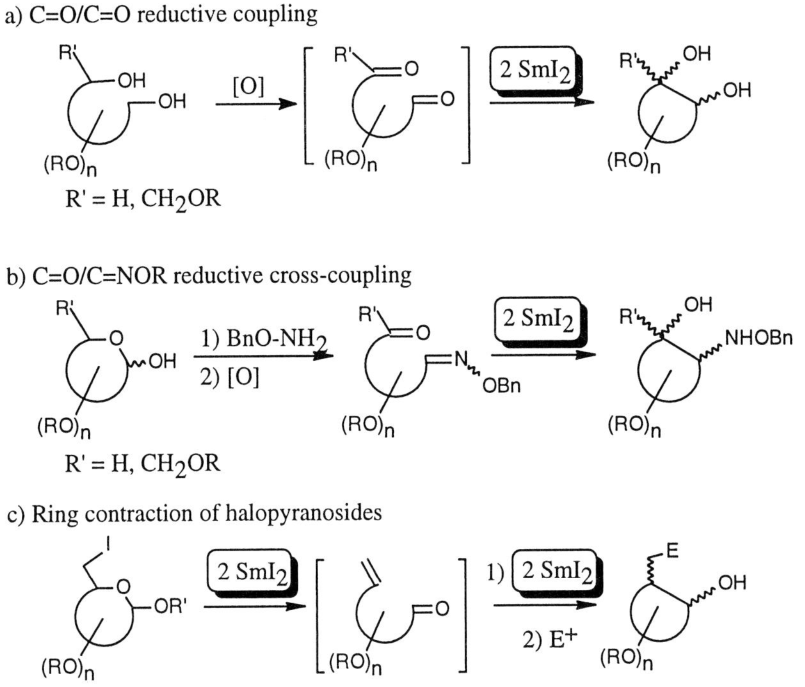

a) C=O/C=O reductive coupling

b) C=O/C=NOR reductive cross-coupling

c) Ring contraction of halopyranosides

Figure 7-2. Reductive carbocyclisations of carbohydrate derivatives using samarium diiodide.

7.2 Results and Discussion

7.2.1 Intramolecular C=O/C=O Reductive Coupling

As the first objective, a new synthesis of selectively protected and enantiomerically pure *myo*-inositol derivatives was developed. When designing a carbocyclisation approach to inositols, the most direct disconnection is based on an intramolecular pinacol coupling reaction, that gives the carbocycle and a vicinal diol in a single step [12]. Since pinacol coupling reactions produce mainly *cis*-diols for ring size 8 or smaller [13], there are two possible disconnections for the *myo*-inositol ring, as shown in Figure 7-3. Due to the symmetry of *myo*-inositol, each of these disconnections will lead to enantiomeric linear dialdehydes that can be prepared eventually from enantiomeric C_2-symmetric linear polyols of *ido* configuration. Our route [14] followed disconnection A and the required L-iditol derivative was easily prepared from D-mannitol, also C_2-symmetric, by inversion of the configuration of carbons 2 and 5. This task was performed following a literature procedure that has been improved recently [15] (Scheme 7-1).

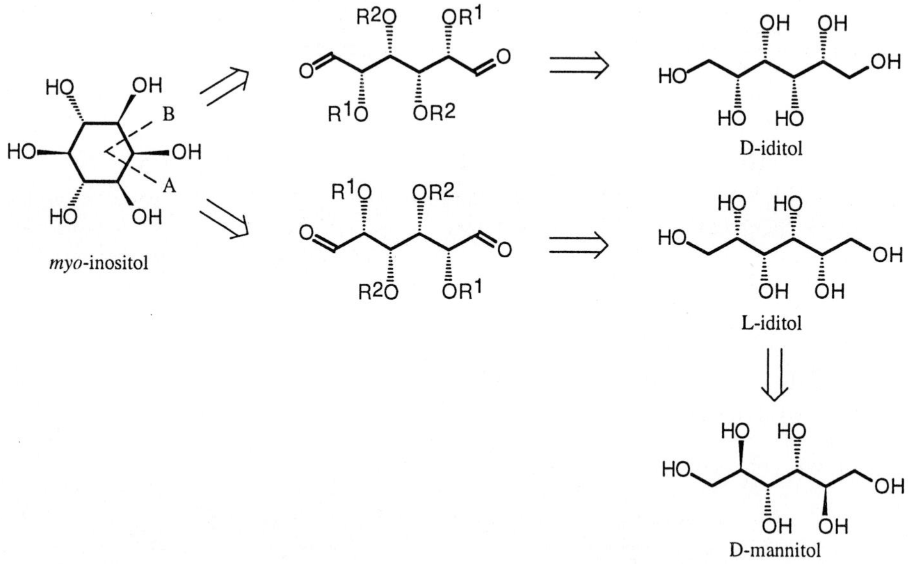

Figure 7-3. Retrosynthesis of *myo*-inositol using a pinacol coupling cyclisation strategy.

The transformation of D-mannitol into the ido-derivative **5** could be easily performed on a multigram scale, and required only two chromatographic purifications. Simple protecting group manipulations generated the required 1,6-diol **7** in high yield. Swern oxidation of **7** in THF produced the C_2-symmetric dialdehyde **8**, which was not isolated. The crude reaction mixture was directly added to samarium diodide in THF containing *t*-BuOH, at low temperature. The reductive cyclisation took place in excellent yield and with very good diastereoselectivity giving the expected *cis*-diol, the *myo*-inositol derivative **9**, as the major product and a small amount of the corresponding C_2-symmetric *trans*-diol, that was shown to be the *scyllo*-isomer **10** by ^{1}H NMR. Both diastereoisomers could be readily separated by column chromatography. The other possible diastereoisomer, the D-*chiro*-inositol **11** could not be detected in the ^{1}H NMR of the crude. This synthesis provides an enantiomerically pure and selectively protected *myo*-inositol derivative from D-mannitol in 10 steps and 22% overall yield. The survival of the *trans*-isopropylidene group shows the mildness of the reductive coupling conditions.

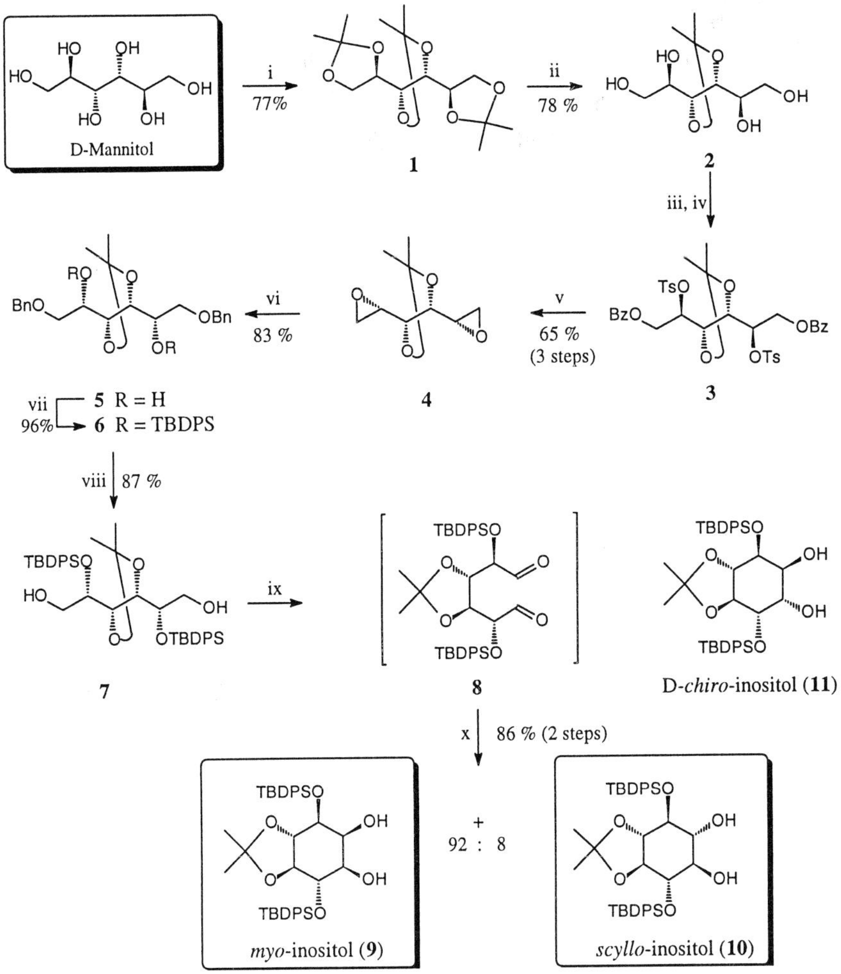

Scheme 7-1. Reagents: i) Me$_2$CO, H$_2$SO$_4$; ii) AcOH, H$_2$O, 40°C, 2h: iii) PhCOCl, pyr. CH$_2$Cl$_2$, -80°C to 0°C; iv) TsCl, Et$_3$N, DMAP, CH$_2$Cl$_2$, 0°C to 23°C; v) K$_2$CO$_3$, CH$_2$Cl$_2$/MeOH; vi) PhCH$_2$OH, NaH, DMF; vii) *t*Bu(Ph)$_2$SiCl, imid. DMAP, DMF; viii) Pd/C, H$_2$, *i*PrOH,/AcOEt; ix) (COCl)$_2$, DMSO, Et$_3$N, THF, -78°C to 23°C; x) 2 SmI$_2$, THF, *t*BuOH, -50°C to 23°C.

The structure of the major product was further confirmed by its transformation into the known 1,2;4,5-di-O-isopropylidene derivative **15** (Scheme 7-2) by standard protecting group manipulations. Its physical data agreed with those expected on the basis of the values reported for its enantiomer (**16**) [16]. Simple transformations provided other useful synthons (**13, 14**) for the preparation of chiral inositol polyphosphates. The strategy described takes advantage of the symmetry in *myo*-inositol for its preparation from a readily available C_2-symmetric precursor. The C_2 symmetry has been conserved throughout the route up to the cyclisation step. This simplifies the number of synthetic operations allowing reactions to be performed simultaneously at two homotopic centers, and also reduces the number of possible diastereoisomers resulting from the carbocyclisation. After cyclisation, the previously homotopic groups become diastereotopic and could, in principle, be differentiated.

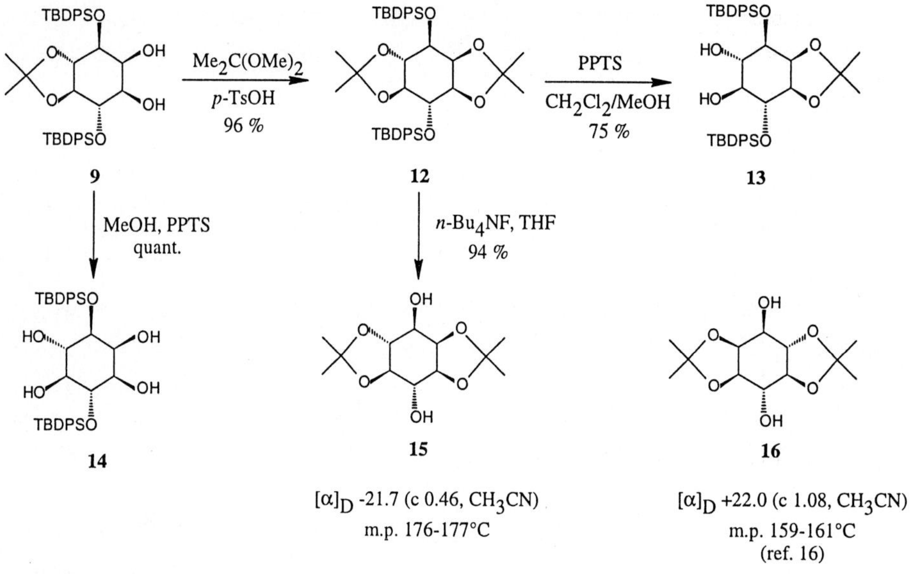

[α]$_D$ -21.7 (c 0.46, CH$_3$CN)
m.p. 176-177°C

[α]$_D$ +22.0 (c 1.08, CH$_3$CN)
m.p. 159-161°C
(ref. 16)

Scheme 7-2. Selective transformations of compound **9**.

The same strategy can also be applied to other alditols. For instance, using a parallel route, D-sorbitol can be readily transformed into an L-*chiro*-inositol derivative (Scheme 7-3) [17]. Selective benzoylation of tetrol **18**, followed by silylation of the remaining secondary hydroxyls gave the fully protected sorbitol derivative **20**. The benzoate groups were selectively removed by reduction with DIBALH. These conditions prevented migration of the silyl groups that took place under the more common basic solvolysis. With the selectively protected 1,6-diol **21** in hand, the one-pot sequence of Swern oxidation and reductive cyclisation was studied. A readily separable mixture of *cis* and *trans* diols was obtained in good overall yield and with good diastereoselectivity. The major diasteroisomer was again a *cis*-diol. Of the two possible *cis*-diols, only the one that is *trans* oriented with respect to the two α-alkoxy groups, i.e. L-*chiro*-diastereoisomer **24**, was obtained. The *trans*-diol was shown to be the *myo*-inositol derivative **23** by ^{1}H NMR.

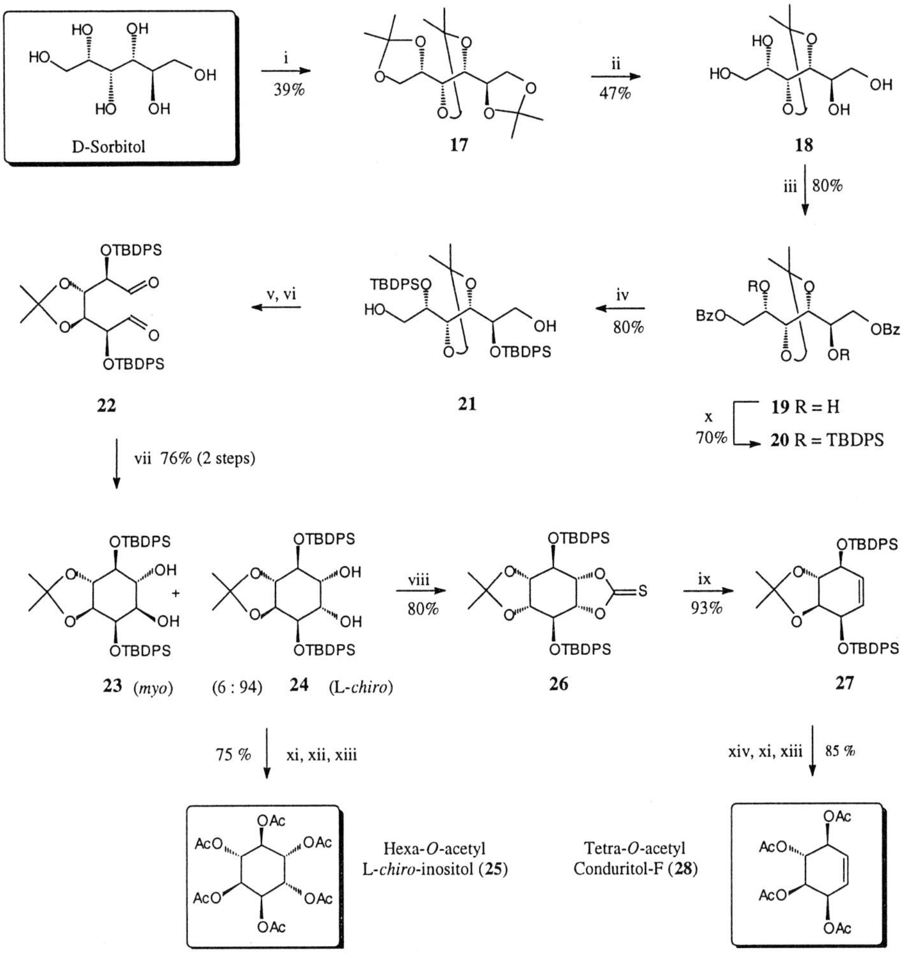

Scheme 7-3. Reagents: i) Acetone, H_2SO_4; ii) AcOH, H_2O, 50°C; iii) PhCOCl, pyr, CH_2Cl_2, -78°C to 0°C; iv) DIBALH, tol., -78°C; v) $(COCl)_2$, DMSO, THF, -78°C; vi) Et_3N, -78°C to 23°C; vii) $2SmI_2$, *t*BuOH, THF, -50°C to 23°C; viii) Im_2CS, tol, 100°C; ix) $(MeO)_3P$, 110°C; x) *t*-$BuPh_2SiCl$, imidazole, DMAP, DMF; xi) TBAF, THF; xii) MeOH, *p*TsOH; xiii) Ac_2O, pyr; xiv) PPTS, CH_2Cl_2-MeOH.

The major diastereoisomer was fully deprotected to give L-*chiro*-inositol that was characterized as its hexa-acetate **25**. Its C_2-symmetry was readily apparent in the ^{1}H and ^{13}C NMR spectra. Dehydroxylation of the *cis*-diol via the cyclic thiocarbonate **26** provided a (-)-conduritol F derivative (**27**) which was again fully deprotected and characterized as its tetraacetate **28**.

After showing that 1,6-diols derived from sugar alditols can be cyclized efficiently and stereoselectively to give enantiomerically pure inositols, the cyclisation protocol was extended to 1,5-diols that would afford polyhydroxylated cyclopentanes. Scheme 7-4 shows a short, simple and highly efficient route to branched cyclopentitols that could be transformed into a number of derivatives by further manipulations [18]. The synthesis started with the protected lactol **29** [19]. Reduction with sodium borohydride afforded the 1,5-diol **30**. Again, the one-pot sequence of Swern oxidation followed by treatment with two equivalents of SmI$_2$ at low temperature provided an equimolar mixture of two inseparable cyclopentane diols (**32**) in an excellent overall yield. Both diols were shown to have the *cis* stereochemistry by transformation into the corresponding isopropylidene derivatives (**33, 34**) and the cyclic thiocarbonates (**35, 36**), that were now readily separable by flash chromatography.

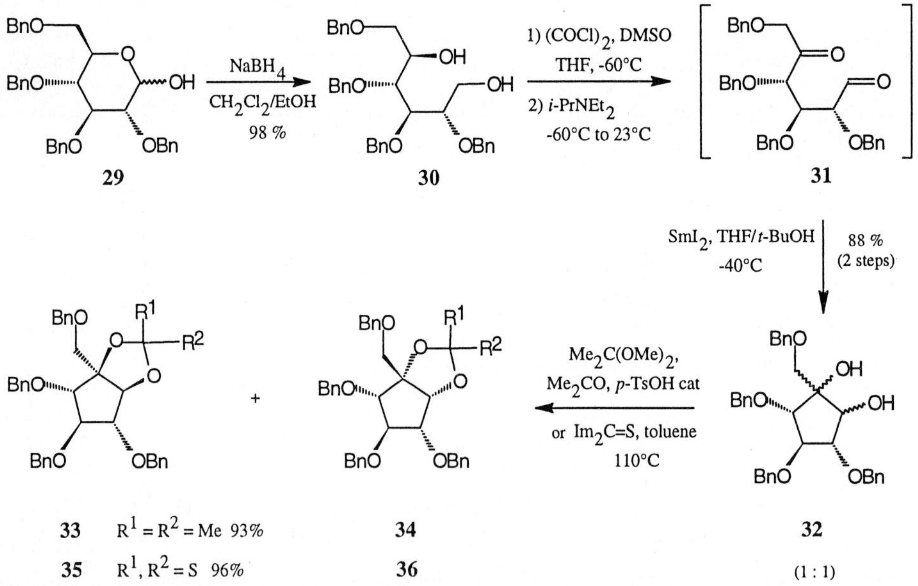

Scheme 7-4. Synthesis of cyclopentitols by pinacol coupling cyclisation of 1,5-diols.

Other groups have reported on the use of samarium diiodide for the stereoselective synthesis of cyclitols from sugar-derived diols via an intramolecular pinacol coupling reaction [11b, c]. Additionally, the same type

Reductive Carbocyclisations of Carbohydrate Derivatives.. 131

of reaction proved to be very useful for the construction of other complex carbocyclic structures [20]. From this experience, samarium diiodide can be considered today as the reagent of choice to implement an intramolecular pinacol coupling reaction in a synthetic scheme due to its ready availability, mildness, high efficiency, usually high stereoselectivity, and simple work-up procedures. Moreover, the one-flask sequence of Swern oxidation followed by reductive carbocyclisation avoids the isolation of the intermediate dicarbonyl compound thus providing greatly improved overall yields.

7.2.2 Intramolecular C=O/C=N-OR Reductive Cross-Coupling

The first reductive cross-coupling between an oxime ether and a carbonyl compound was reported by Corey in 1983 using zinc metal [21]. More recently, it has been shown that this coupling could be carried out electrochemically [22] or using tributyltin hydride [23]. Samarium diiodide was also shown to promote the intermolecular cross-coupling of ketones with the O-benzyloxime of formaldehyde [24]. In this case, the reaction requires the presence of HMPA as a cosolvent [25]. The related reductive cyclisation of diphenylhydrazones using samarium diiodide has been reported recently and needs also the presence of HMPA [26]. Although the reported intermolecular cross-coupling of ketones and oxime ethers using samarium diiodide only worked for oxime ethers of formaldehyde, an intramolecular version could probably be feasible.

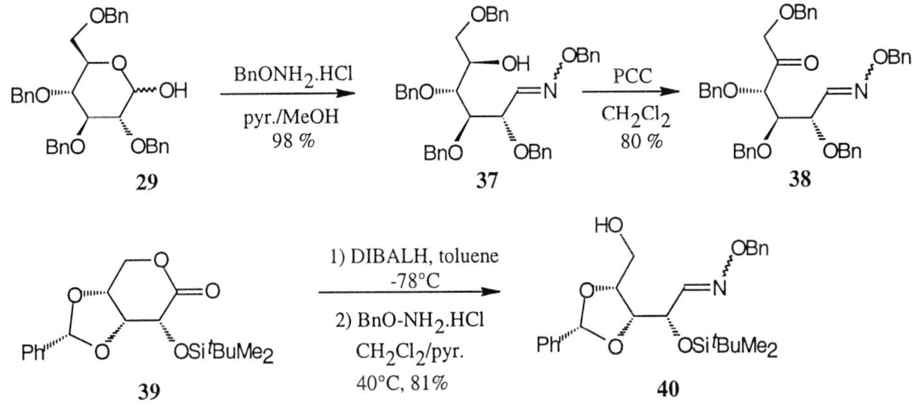

Scheme 7-5. Synthesis of precursors for the preparation of aminocyclopentitols via intramolecular C=O/ C=N-OR reductive coupling.

The precursors needed for this study were readily obtained from appropriate sugar derivatives as shown in Scheme 7-5. Condensation of a protected sugar lactol **29** with *O*-benzylhydroxylamine produced a mixture of syn and anti oximes **37** liberating the hydroxyl group at C-5. In the case of hexoses, the hydroxyl was then oxidized to a ketone with PCC. In the case of pentoses, the lactol was prepared by DIBALH reduction of the corresponding protected lactone **39**, but the oxidation step, that will generate an aldehyde, was preferably performed within a one-pot oxidation-cyclisation sequence, as in the intramolecular pinacol coupling case.

The ketone/oxime ether reductive coupling was studied first [27]. Scheme 7-6 shows the reactions of keto-oxime ethers derived from D-glucose **38** and D-mannose **42** that differ only in the configuration at C-2. Confirming our expectations, the reductive coupling was in fact very efficient and did not require the presence of HMPA, in contrast to the intermolecular case [24] and to the corresponding coupling with diphenylhydrazones [26]. The stereochemistry of the products was independent of the sterochemistry of the oxime ether, syn and anti isomers giving the same mixtures of cyclopentane diastereosiomers [28].

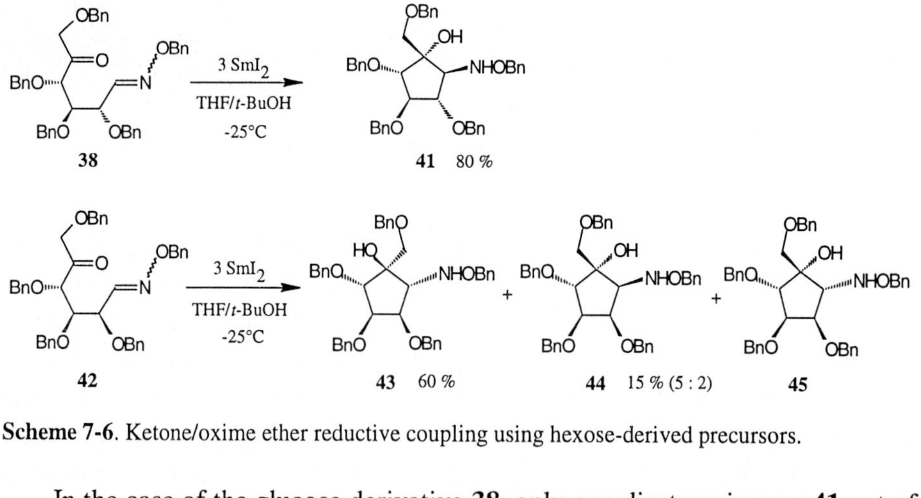

Scheme 7-6. Ketone/oxime ether reductive coupling using hexose-derived precursors.

In the case of the glucose derivative **38**, only one diastereoisomer **41** out of the four possible was formed. In the case of mannose, however, three diastereoisomers **43-45** were formed. The major one could be readily separated from the others by column chromatography. In both cases, the major diasteroisomer had a *trans* disposition between the hydroxyl and the hydroxylamine functions, as observed by Fallis in the similar coupling with diphenylhydrazones [26]. This is contrary, however, to the general tendency found for the intramolecular pinacol coupling reactions. The hydroxylamine group is also *trans* to the vicinal alkoxy group in the major diastereoisomer.

It should be mentioned that cyclopentitols **41**, **43-45** can be easily transformed by simple reduction into epimers of the aminocyclopentitol moiety of the trehalase inhibitor trehazolin. This route constitutes the shortest synthetic approach described so far to these complex carbocycles [29], involving only five steps from the starting unprotected hexose to the fully functionalized cyclopentane derivative. A range of analogues could thus be readily prepared from different hexoses following this strategy.

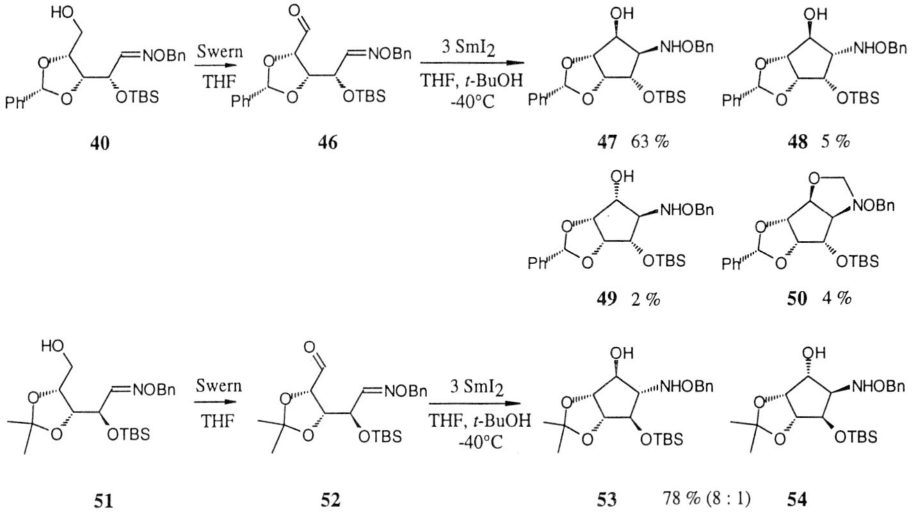

Scheme 7-7. Aldehyde/oxime ether reductive coupling using pentose-derived precursors.

Aldehydes can also be used in this reductive coupling [27], providing access to unbranched aminocyclopentitols. In this case, as in the pinacol coupling reactions, the yields were greatly improved by performing the oxidation and the reductive coupling steps in a one-pot sequence thus avoiding the isolation of the intermediate aldehyde. Scheme 7-7 shows two examples derived from D-ribose (**40**) and from D-arabinose (**51**). As in the previous case, both substrates differ only in the configuration at C-2, the center vicinal to the oxime moiety. In both cases, the two-step one-pot sequence took place in good overall yield giving mixtures of cyclopentanes with good diasteroselectivity. The major isomer **47** in the cyclisation of the precursor prepared from D-ribose has the hydroxyl and hydroxylamine groups in *cis* disposition. This is the only example where a *cis*-selective coupling has been observed. In this case, apart from the two minor *trans*-diastereoisomers **48, 49**, a bicyclic oxazolidine (**50**) was also isolated and was shown to have the same stereochemistry as the major product. Thus, compound **50** was independently prepared from **47** by treatment with aqueous formaldehyde in the presence of citric acid, confirming its structure. We found that this minor product only appeared when the crude mixture of the Swern oxidation was not allowed to warm above -10°C. Although its origin is not clear at the moment, we speculate that it could be formed by reaction of a radical intermediate with carbon monoxide that is released in the preparation of the Swern reagent and subsequent reduction. Alternatively, this oxazolidine could also result from an intermediate methylthiomethyl ether produced by reaction of the major cyclitol product with excess Swern reagent. In the case of substrate **51**, derived from D-arabinose, two inseparable *trans*-diastereoisomers **53, 54** were formed.

Scheme 7-8. One-pot C=O/C=N-OR reductive carbocyclisation and N-O reductive cleavage for the synthesis of aminocyclopentitols.

The resultant carbocyclic hydroxylamines could be reduced to the free amines by a number of reducing agents, including samarium diiodide itself [30]. Most interestingly, the reductive cleavage reaction can be carried out in a one-pot sequence following cyclisation if an excess of samarium diiodide is used (Scheme 7-8) [27]. This reduction is faster in the presence of water [31]. The overall yields for the one-pot cyclisation-reductive cleavage sequence are excellent, and benzyl groups are unaffected. The resultant aminoalcohols were conveniently isolated after acetylation. In the case of compound **40**, the reductive cleavage reaction promoted a partial scrambling of the silyl group between the hydroxyl and the amino groups (**57**). Desilylation and peracetylation gave finally the protected aminocyclopentitol **58** in excellent yield.

7.2.3   Ring Contraction of 6-Deoxy-6-Iodo-Hexopyranosides

The treatment of 6-deoxy-6-iodohexopyranosides with excess samarium diiodide in the presence of HMPA promotes a cascade of reactions that leads to cyclopentitol derivatives; this process constitutes a novel mode of contraction of the pyranose ring [32,33]. The sequence (Figure 7-2, route c) requires four single electron transfer steps (i.e. 4 mol equiv of SmI_2) and consists of (1) a reductive dealkoxyhalogenation [34] that opens the pyranose ring and liberates the carbonyl function at the anomeric center; (2) a ketyl-olefin reductive cyclisation [11a, 35] affording a ring-contracted organosamarium intermediate, and (3) the intermolecular trapping of this organosamarium with an appropriate electrophile [35] to produce finally a branched cyclopentitol derivative.

A series of 6-deoxy-6-iodohexopyranosides of different configuration and substitution pattern were studied (Table 7-1). Treatment of these compounds with 6 equivalents of samarium diiodide in THF containing HMPA (5 equivalents with respect to samarium reagent) at room temperature gave a mixture of methyl cyclopentitols, that are the expected ring-contracted products, and the corresponding dehalogenated pyranosides. In the case of the benzyl derivative **62**, the latter was however the only product that could be isolated. Probably in this case, the initial primary radical that is formed at C-6 in the first step of the reaction is quickly quenched by a 1,5-hydrogen atom transfer from the benzyl group at O-4. There are precedents for this hydrogen transfer in other pyranose systems [36]. The cyclisation reaction was more selective for the silyl-protected pyranosides (**63-65**), and only two of the four possible diastereoisomers were formed. However, the amount of dehalogenated pyranoside was also higher in these cases. The most interesting case is shown in the first entry of Table 7.1. Reaction of compound **59** afforded a separable mixture of three cyclopentitols (**66-68**), isolated after acetylation, and a small amount of the dehalogenated pyranoside (**69**). The total yield of ring-contracted products was remarkably good (89%). It should be emphasized that these cyclitols are obtained in just three steps from D-galactose.

Table 7-1. Reaction of 6-deoxy-6-iodo-hexopyranosides with SmI$_2$ in THF-HMPA.

Entry	Substrate	Products (isolated yield)			
1	**59**	**66** (65%)[a]	**67** (20%)[a]	**68** (4%)[a]	**69** (9%)
2	**60**	**70** (47%)	**71** (17%)[b]		**72** (11%)
3	**61**	**73a-c** (76%) (1.8 : 1.7 : 1)[c]			**74** (11%)
4	**62**	—			**75** (50%)
5	**63**	**76** (42%)	**77** (11%)		**78** (29%)
6	**64**	**79** (51%)	**80** (8%)		**81** (31%)
7	**65**	**82** (43%)	**83** (27%)		**84** (21%)

[a] Products obtained after acetylation of the crude reaction mixture; [b] Stereochemistry of **71** was further confirmed by its transformation into **66** by desilylation (TBAF in THF followed by in situ acetylation (Ac$_2$O, pyr.); [c] Inseparable mixture. Ratio determined by ^{1}Hnmr. Reprinted with

permission from *The Journal of Organic Chemistry*, vol 61, p 6488. Copyright 1996 American Chemical Society.

The diastereoselectivity of the cyclisation reaction was also high. Cyclopentane **67** is probably formed by nucleophilic attack of the final organosamarium intermediate on the acetone molecule that is liberated after ring-opening of the pyranose. In most cases, the major diastereomer from the cyclisation has the methyl and the hydroxyl groups in *trans* relationship, except in the reaction of the gluco derivative **63**. There is also a *trans* relation between the hydroxyl and the vicinal silyloxy group, with the exception of the galacto derivatives **60** and **65**. Changes in protecting groups can produce important changes in the diastereoselectivity of the cyclisation (cf. Table 7.1, entries 1 and 2; 3 and 5).

Trapping experiments with D_2O (Scheme 7-9) confirmed the intermediacy of unstable alkyl anions both in the reductive elimination step and in the carbocyclisation reaction [32]. The formation of a different mixture of cyclopentanes under these conditions is probably a consequence of protonation of the O-2 samarium(III) alkoxide in the ketyl radical anion intermediate. Chelation of samarium(III) in this radical intermediate with the hydroxyl group at C-2 or intramolecular hydrogen bonding could explain the predominant formation of cyclopentanes with a *cis* relative orientation of the hydroxyls at C-1 and C-2 (**85, 86**).

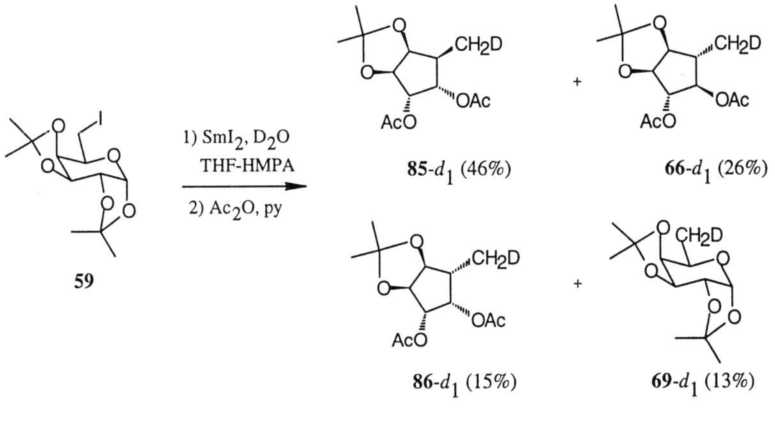

Scheme 7-9. Reaction of compound **59** with SmI_2 in THF-HMPA-D_2O.

The effect of temperature was also studied using compound **59**. Very low temperatures favor the competing dehalogenation reaction giving the reduced pyranose **69** as the major product (63% at -78 °C) [32, 37].

7.3 Conclusions

Samarium diiodide is a very versatile reagent for the stereoselective transformation of readily available carbohydrate derivatives into highly functionalized 5- and 6-membered cyclitols under very mild conditions (Figure 7-4). Starting from a selectively protected hexose **A**, a 1,6- (**B**) or 1,5-dicarbonyl derivative (**D**) can be easily prepared and transformed in situ into an inositol (**C**) or a cyclopentitol (**E**) via a stereoselective intramolecular pinacol coupling reaction in excellent yield. Alternatively, a keto-oxime ether (**F**) can be prepared and converted into a carbocyclic hydroxylamine that can be chemoselectively reduced *in situ* with excess reagent into an aminocyclitol (**G**), by reductive cleavage of the N-O bond. Finally, a new ring contraction reaction of halopyranosides (**H**) has been uncovered. A methyl cyclopentitol (**I**) is formed in a cascade process under mild conditions.

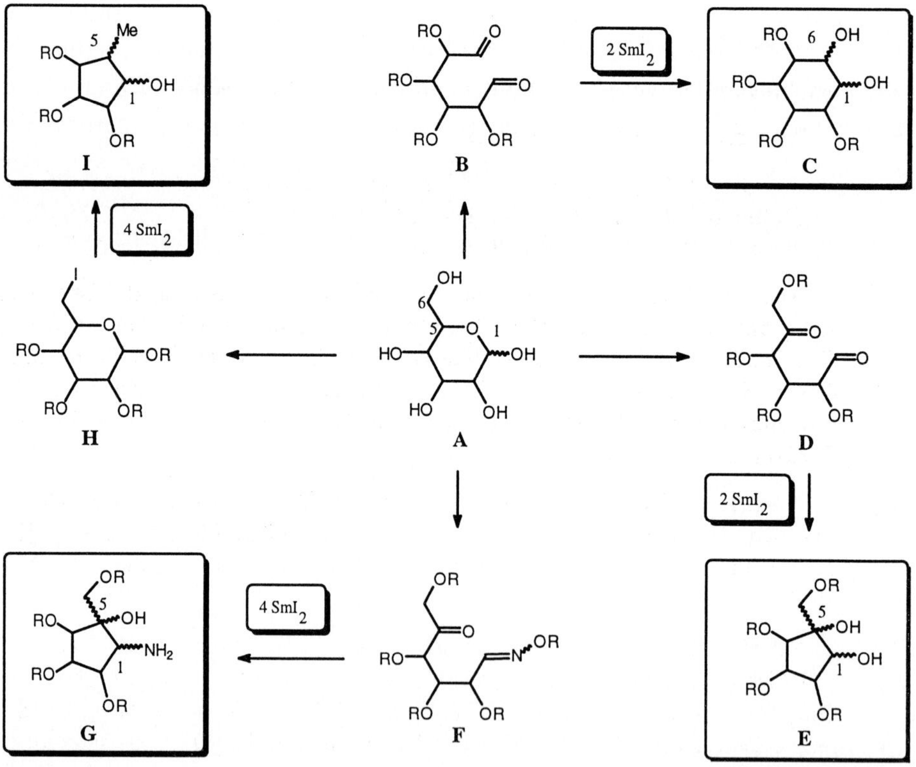

Figure 7-4. Transformation of hexoses into polyhydroxylated cyclopentanes and cyclohexanes using SmI$_2$.

Acknowledgements

This work was possible only through the skillful participation of the group of coworkers whose names are cited in the references from our laboratory. Financial support by DGICYT (grant PB93-0127-C02-01) is also gratefully acknowledged.

7.4 References

1. R. J. Ferrier, S. Middleton, *Chem. Rev.* **1993**, *93*, 2779-2831.

2. B. Giese, B. Kopping, T. Göbel, G. Thoma, K. J. Kulicke, F. Trach, *Organic Reactions* **1996**, *48*, 301-856.

3. P. Girard, J. L. Namy, H. B. Kagan, *J. Am. Chem. Soc.* **1980**, *102*, 2693-2698. Molander, G. A. *Chem. Rev.* **1992**, *92*, 29-68. Imamoto, T. *Lanthanides in Organic Synthesis*; Academic Press: London, 1994. Molander, G.A. *Org. React.* **1994**, *46*, 211-367. Molander, G. A. *Chem. Rev.* **1996**, *96*, 307-338.

4. M. J. Berridge. *Nature* **1993**, *361*, 315-325.

5. M. J. McConville, M. A. J. Ferguson, *Biochem. J.* **1993**, *294* 305-324.

6. G. Romero, J. Larner, *Adv. Pharmacol.* **1993**, *24*, 21-50.

7. B. V. L. Potter, D. Lampe, *Angew. Chem. Int. Ed. Engl.* **1995**, *34*, 1933-1972. D. C. Billington, *The Inositol Phosphates: Chemical Synthesis and Biological Significance, VCH, Weinheim*, **1993**.

8. O. Ando, H. Satake, K. Itoi, A. Sato, M. Nakajima, S. Takahashi, H. Haruyama, *J. Antibiot.* **1991**, *44*, 1165-1168. T. Nakayama, T. Amachi, S. Murano, T. Sakai, T. Shin, R. T. M. Kenny, T. Iwashita, M. Zagorski, H. Komura, K. Nomoto, *J. Chem. Soc., Chem. Commun.* **1991**, 919-921. O. Ando, M. Nakajima, K. Hamano, K. Itoi, S. Takahashi, H. Haruyama, T. Konishita, *J. Antibiot.* **1993**, *46*, 1116-1125.

9. T. Aoyagi, T. Yamamoto, K. Kojiri, H. Morishima, M. Nagai, M. Hamada, T. Takeuchi, H. Umezawa, *J. Antibiot.* **1989**, *42*, 883-889. H. Morishima, K. Kojiri, T. Yamamoto, T. Aoyagi, H. Nakamura, Y. Iitaka, *J. Antibiot.* **1989**, *42*, 1008-1011.

10. S. Sakuda, A. Isogai, S. Matsumoto, A. Suzuki, *Tetrahedron Lett.* **1986**, *27*, 2475-2478. S. Sakuda, A. Isogai, S. Matsumoto, A. Suzuki, *J. Antibiot.* **1987**, *40*, 296-300. S. Sakuda, A. Isogai, S. Matsumoto, K. Koseki, H. Komada, A. Suzuki, *Agric. Biol. Chem.* **1987**, *51*, 3251-3259. S. Sakuda, A. Isogai, S. Matsumoto, A. Suzuki, K. Koseki, H. Komada, Y. Yamada, *Agric. Biol. Chem.* **1988**, *56*, 1615-1617. Y. Nishimoto, S. Sakuda, S. Takayama, Y. Yamada, *J. Antibiot.* **1991**, *44*, 716-722.

11. For other transformations of carbohydrate derivatives into carbocycles using samarium diiodide, see: (a) J. J. C. Grové, C. W. Holzapfel, D. B. G. Williams, *Tetrahedron Lett.* **1996**, *37*, 1305-1308; (b) E. Perrin, J.-M. Mallet, P. Sinaÿ, *Carbohydr. Lett.* **1995**, *1*, 215-216; (c) J. P. Guidot, T. Le Gall, C. Mioskowski, *Tetrahedron Lett.* **1994**, *36*, 6671-6672; (d) K.-i. Tadano, Y. Isshiki, M. Minami, S. Ogawa, *J. Org. Chem.* **1993**, *58*, 6266-6279, and references cited therein; (e) E. J. Enholm, A. Trivellas, *J. Am. Chem.Soc.* **1989**, *111*, 6463-6465; (f) E. J. Enholm, H. Satici, A. Trivellas, *J. Org. Chem.***1989**, *54*, 5841-5843; see also ref. 33.

12. Y. Watanabe, M. Mitani, S. Ozaki, *Chem. Lett.* **1987**, 123-126.

13. J. E. McMurry, J. G. Rico, *Tetrahedron Lett.* **1989**, *30*, 1169-1172. J. E. McMurry, N. O. Siemers, *Tetrahedron Lett.* **1993**, *34*, 7891-7894.

14. J. L. Chiara, M. Martín-Lomas. *Tetrahedron Lett.* **1994**, *35*, 2969-2972.

15. Y. Le Merrer, A. Duréault, C. Greck, D. Micas-Languin, C. Gravier, J.-C. Depezay, *Heterocycles* **1987**, *25*, 541-548. A. Duréault, M. Portal, J. C. Depezay, *Synlett* **1991**, 225-226.

16. M. Jones, K. K. Rana, J. G. Ward, R. C. Young, *Tetrahedron Lett.* **1989**, *30*, 5353. For (±)-**15**, m.p. 171-173°C (J. Gigg, R. Gigg, S. Payne, R. Conant, *Carbohydr. Res.* **1985**, *142*, 132-134).

17. J. L. Chiara, N. Valle, *Tetrahedron: Asymmetry* **1995**, *6*, 1895-1898.

18. H. Dietrich, J. L. Chiara, unpublished results.

19. E. Decoster, J.-M. Lacombe, J.-L. Strebler, B. Ferrari, A. A. Pavia, *J. Carbohydr. Chem.* **1983**, *2*, 329-341.

20. H. M. R. Hoffmann, I. Münnich, O. Nowitzki, H. Stucke, D. J. Williams, *Tetrahedron* **1996**, *52*, 11783-11798. C. S. Swindle, W. Fan, *Tetrahedron Lett.* **1996**, *37*, 2321-2324. C. Anies, A. Pancrazi, J.-Y. Lallemand, T. Prangé, *Tetrahedron Lett.* **1994**, *35*, 7771-7774. T. Kan, S. Hosokawa, S. Nara, M. Oikawa, S. Ito, F. Matsuda, H. Shirahama, *J. Org. Chem.* **1994**, *59*, 5532-5534. S. Arseniyadis, D. V. Yashunsky, R. Pereira de Freitas, M. Muñoz-Dorado, E. Toromanoff, P. Potier, *Tetrahedron Lett.* **1993**, *34*, 1137-1140.

21. E. J. Corey, S. G. Pyne, *Tetrahedron Lett.* **1983**, *24*, 2821-2824.

22. T. Shono, N. Kise, T. Fujimoto, A. Yamanami, R. Nomura, *J. Org. Chem.* **1994**, *59*, 1730-1740.

23. T. Kiguchi, K. Tajiri, I. Ninomiya, T. Naito, H. Hiramatsu, *Tetrahedron Lett.* **1995**, *36*, 253-256. T. Naito, K. Tajiri, T. Harimoto, I. Ninomiya, T. Kigushi, *Tetrahedron Lett.* **1994**, *35*, 2205-2206.

24. T. Hanamoto, J. Inanaga, *Tetrahedron Lett.* **1991**, *32*, 3555-3556.

25. For the effect of added HMPA on SmI$_2$ reactions, see: Z. Hou, Y. Wakatsuki, *J. Chem. Soc., Chem. Commun.* **1994**, 1205-1206, and references cited therein.

26. C. F. Sturino, A. G. Fallis, *J. Org. Chem.* **1994**, *59*, 6514-6516. C. F. Sturino, A. G. Fallis, *J. Am. Chem. Soc.* **1994**, *116*, 7447-7448.

27. J. L. Chiara, J. Marco-Contelles, N. Khiar, P. Gallego, C. Destabel, M. Bernabé, *J. Org. Chem.* **1995**, *60*, 6010-6011.

28. P. A. Bartlett, K. L. McLaren, P. C. Ting, *J. Am. Chem. Soc.* **1988**, *110*, 1633-1634.

29. For other synthetic approaches to trehazolin five-membered ring, see: B. E. Ledford, E. M. Carreira, *J. Am. Chem. Soc.* **1995**, *117*, 11811-11812, and references cited therein. See also Chapter 7 by B. Ganem.

30. G. E. Keck, S. F. McHardy, J. A. Murry, *J. Am. Chem. Soc.* **1995**, *117*, 7289-7290. G. E. Keck, S. F. McHardy, T. T. Wager, *Tetrahedron Lett.* **1995**, *36*, 7419-7422. J. L. Chiara, C. Destabel, P. Gallego, J. Marco-Contelles, *J. Org. Chem.* **1996**, *61*, 359-360.

31. For the effect of added H$_2$O on SmI$_2$ reactions, see: E. Hasegawa, D. P. Curran, *J. Org. Chem.* **1993**, *58*, 5008-5010.

32. J. L. Chiara, S. Martínez, M. Bernabé, *J. Org. Chem.* **1996**, *61*, 6488-6489.

33. For a recent highlight on ring contractions in carbohydrates, see: H. Redlich, *Angew. Chem., Int. Ed. Engl.* **1994**, *33*, 1345-1347. For a related ring contraction of carbohydrate derivatives using samarium diiodide, see: A. Chénedé, P. Pothier, M. Sollogoub, A. J. Fairbanks, P. Sinaÿ, *J. Chem. Soc., Chem. Commun.* **1995**, 1373-1374.

34. Originally described with Zn or butyllithium by: B. Bernet, A. Vasella, *Helv. Chim. Acta* **1979**, *62*, 1990-2016. For previous ring scissions of cyclic β-halo ethers promoted by samarium diiodide, see: L. Crombie, L. J. Rainbow, *J. Chem. Soc., Perkin Trans. 1* **1994**, 673-687; *Tetrahedron Lett.* **1988**, *29*, 6517-6520.

35. G. A. Molander, C. R. Harris, *J. Am. Chem. Soc.* **1996**, *118*, 4059-4071, and references cited therein.

36. L. J. Liotta, R. C. Bernotas, D. B. Wilson, B. Ganem, *J. Am. Chem. Soc.* **1989**, *111*, 783-785. O. R. Martin, F. Xie, R. Kakarla, R. Benhamza, *Synlett* **1993**, 165-167. C. Barbaud, M. Bols, I. Lundt, M. R. Sierks, *Tetrahedron* **1995**, *51*, 9063-9078.

37. A similar effect of temperature on the extend of competing dehalogenation vs dealkoxyhalogenation has been previously observed in the reaction of halosugars with metal-graphite reagents: A. Fürstner, U. Koglbauer, H. Weidmann, *J. Carbohydr. Chem.* **1990**, *9*, 561-570. A. Fürstner, D. Jumbam, J. Teslic, H. Weidmann, *J. Org. Chem.* **1991**, *56*, 2213-2217.

8 A Concise Route to Carba-Hexopyranoses and Carba-Pentofuranoses from Sugar Lactones

F. Chrétien, M. Khaldi and Y. Chapleur

8.1 Introduction

The existence of some natural structures in which Nature itself has replaced the sugar moiety by a close analogue, a so-called pseudosugar [1] is intriguing (Figure 8-1). For example, valienamine [2], which is a component of a natural aminoglycoside antibiotic, may be regarded as an unsaturated form of a carba-glucosamine [3]. Gabosines, another interesting class of natural compounds, also involve a carbasugar-like key structure [4]. Carbocyclic pentoses analogues are also known, the most representatives being the nucleoside analogues aristeromycin [5] and neplanocin-A [6]. Another intriguing structure furnished by Nature is mannostatin-A [7], an inhibitor of glycoprotein processing enzymes, which is of interest as a biological tool. Other oxygenated five-membered ring are also included in biologicaly active non-carbohydrate compounds such as the potent antitumor neocarzinostatin [8].

The existence of these highly oxygenated carbocyclic structures as well as the recognition and the use of sugar analogues as potential drugs and pharmacological tools has prompted the search for new routes to these compounds [9,10]. The biosynthesis of naturally occurring "carbocyclic" sugars is not fully elucidated but in most cases proceeds via the transformation of a sugar into a carbocycle using an intramolecular aldol-like reaction triggered by the presence of a phosphate group [11]. For example Nature uses an hexose, D-glucose, to construct the carbocyclic part of neplanocin-A. The biomimetic chemical approach to carba-analogues of glucose or ribose using an aldol reaction appears to be attractive but needs to add an extra carbon to the corresponding hexose or pentose and to form the carbocycle.

We have investigated such a route to carba-hexopyranoses on this basis and we found that, en route to six membered rings, it was also possible to prepare five membered rings along the same lines by a slight modification of the reaction conditions.

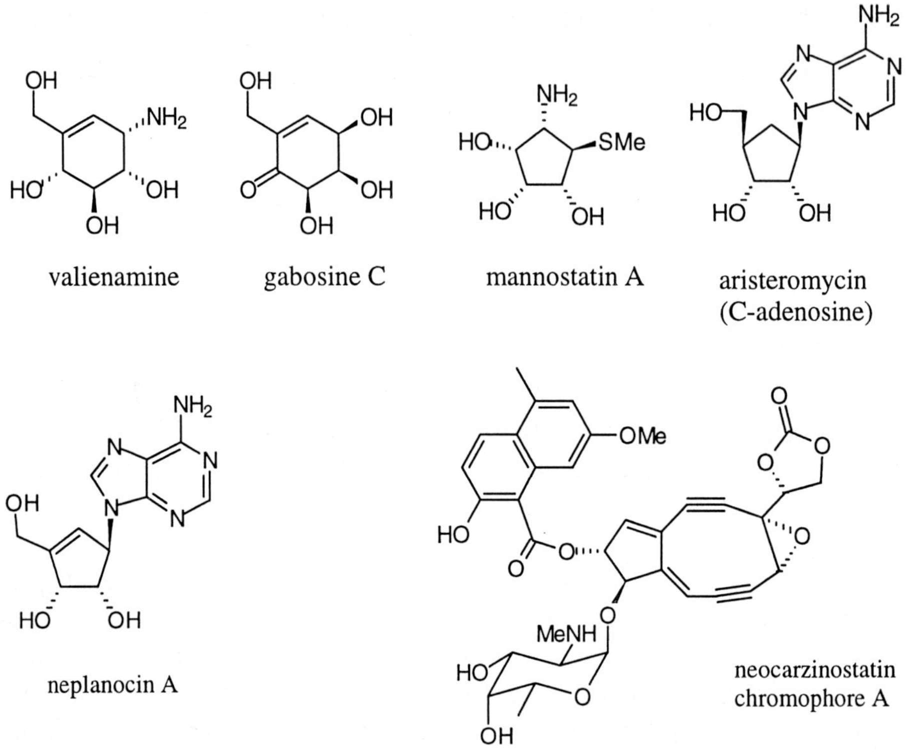

Figure 8-1: Structure of some natural oxygenated carbocycles.

8.2 Results

8.2.1 Carbahexopyranoses

A general strategy for the synthesis of carbasugars relies on the transformation of a chiral branched-chain cyclohexenone (see **2**, scheme 8-4) which, by reduction, should give the expected polyhydroxylated cyclohexane **1**. A number of solutions for the synthesis of chiral cyclohexanes from sugars have been proposed during the last twelve years using, for example, the Ferrier carbocyclisation [9]. However the synthesis of chiral branched-cyclohexanes from sugar precursors is less well documented. Some examples of branched-cyclohexene ring synthesis based on carbon-carbon double bond formation have been reported. Intramolecular Wittig reactions have been exploited by Fleet *et al.* [12], by Vasella [13] and by Mirza [14] in the synthesis of shikimate derivatives (Scheme 8-1).

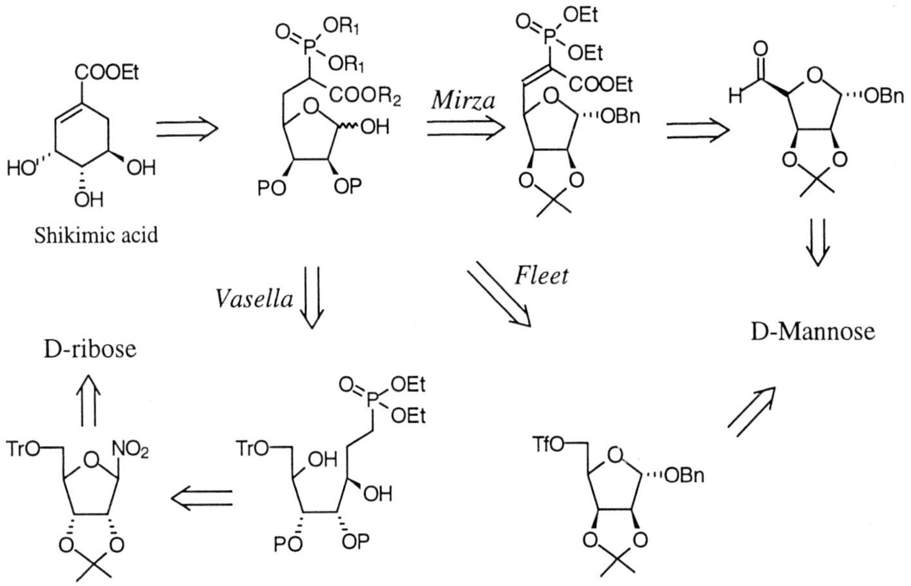

Scheme 8-1

Altenbach *et al.* proposed an alternative route to cyclohexenone-based structures from sugar lactones using an intramolecular Horner olefination (Scheme 8-2) [15].

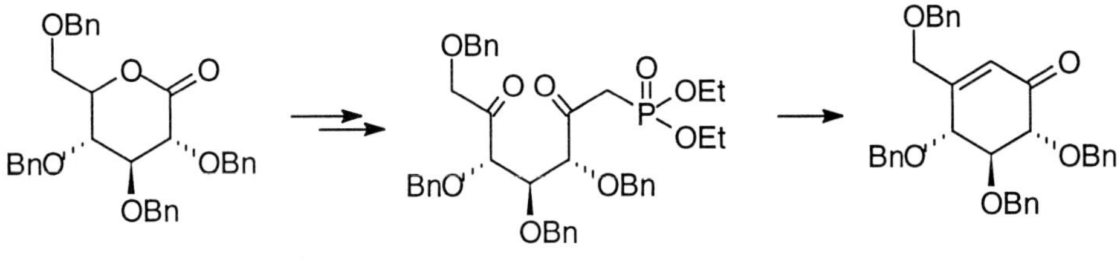

Scheme 8-2

Aldol type cyclisations have been proposed also by Vasella *et al.* via condensation of an enolate onto a pseudolactone (Scheme 8-3) [16].

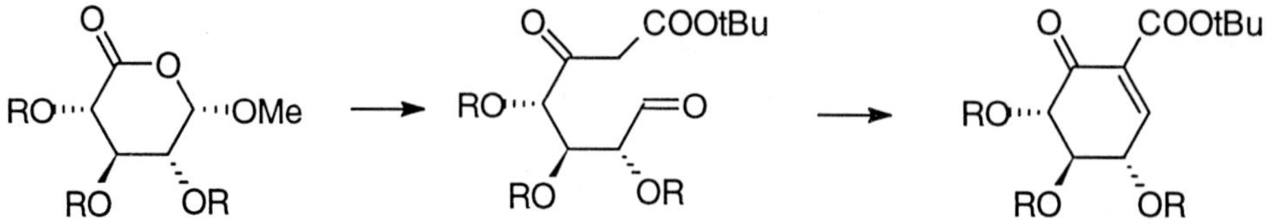

Scheme 8-3

Our strategy is summarized on Scheme 8-4. We have investigated the use of a sugar lactone as starting compound. Accordingly, compound **2**, a suitable precursor of carbasugars, might be prepared by an intramolecular aldol reaction followed by dehydration of aldol **3**. Thus synthon **4**, derived from disconnection of the double bond, should incorporate a precursor of an aldehyde. We chose to use a protected diol which could be deprotected and cleaved to yield the required aldehyde group. The second key feature of our retrosynthetic analysis was the formation of the active methylene group needed in synthon **4**. For that purpose we needed to create a carbon-carbon bond between an anion and a suitably activated carbonyl group. The use of a five membered ring lactone was by far the most suitable unit for use in this process. Indeed the lactonic carbonyl of compound **6** should be reactive enough to react with an enolate anion (synthon **7**). Furthermore, the well known tendancy of these lactones to undergo a single nucleophilic addition would led to a mono condensation of anion **7**. Concerning this synthon, the electron-withdrawing group required for the formation of the lithium anion must be carefully chosen in order to be in turn transformed into an hydroxymethyl, aldehyde or carboxylic acid group, (Z group in Scheme 8-4). Activating groups such as nitrile or N,N-diethylamido, seemed to fulfil the above requirements. Oxygenated ortho-toluamides were also investigated, given the possibility of subsequent Birch reduction followed by oxidative cleavage of the aromatic ring.

The first experiments were conducted on the readily available D-gulonolactone bis-acetonide **8**. Several lithiated anions were prepared by treatment of acetonitrile or benzophenone with lithium diisopropylamide or by treatment of N,N-diethylacetamide or an aromatic amide with s-butyllithium at low temperature. These anions were allowed to react with lactone **8**. In all cases, reasonable to good yields of the expected adducts **9**, **10**, **11**, **12** and **13** were obtained (see Table 8-1).

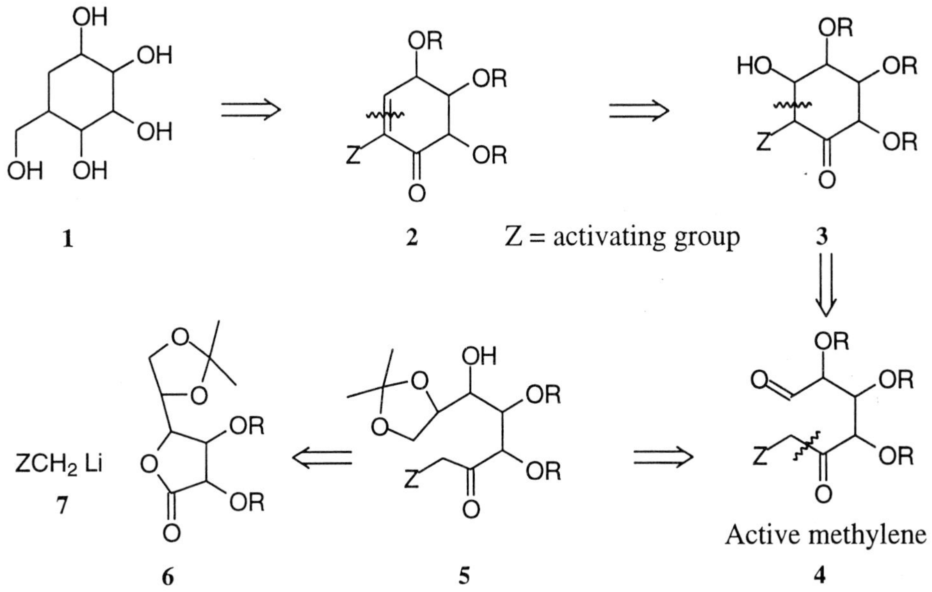

Scheme 8-4

Although nitrile **9** seemed to be the more suitable group for the required subsequent transformation into an hydroxymethyl group, we have been unable to use this compound for carbocycle formation (see below). The tertiary amide group seemed to be the most appropriate for this purpose, and we undertook a series of reactions of the anion derived from N,N-diethyl acetamide and several lactones such as 2,3:5,6-di-*O*-isopropylidene-L-gulono-1,4-lactone (**14**), 2,3:5,6-di-*O*-isopropylidene-D-mannono-1,4-lactone (**15**), 2,3-*O*-isopropylidene-D-ribono-1,4-lactone (**16**) and 5,6-O-isopropylidene-2,3-O-trimethylsilyl-L-galactono-1,4-lactone (**17**). The results are summarized on Table 8-2. The relatively poor results obtained with the *ribo* (entry 3) and *galacto* derivatives (entry 4) could be explained in terms of steric hindrance on both faces of the carbonyl group. This was not the case with other substrates in which only one face of the lactone ring is hindered by the substituents.

Table 8-1. Condensation of selected carbanions on lactone **8**.

Entry	Base	Anion equivalent	T °C	Z	Products	Yield %
1	LDA	1.2	0	–CN	9	72
2	LDA	1.2	0	–C(=O)–Ph	10	51
3	LDA	1.2	0	–C(=O)–NEt₂	11	60
4	sBuLi	2	-78	2-OMe-6-Me-C₆H₃–C(=O)NEt₂	12	80
5	sBuLi	2	-78	3,5-(MeO)₂-2-OMe-6-Me-C₆H–C(=O)NEt₂	13	65

Table 8-2. Condensation of N,N-diethyl acetamide.

Entry	Starting lactone	Product	Yield %
1	14	18	90
2	15	19	50
3	16	20	55
4	17	21	20

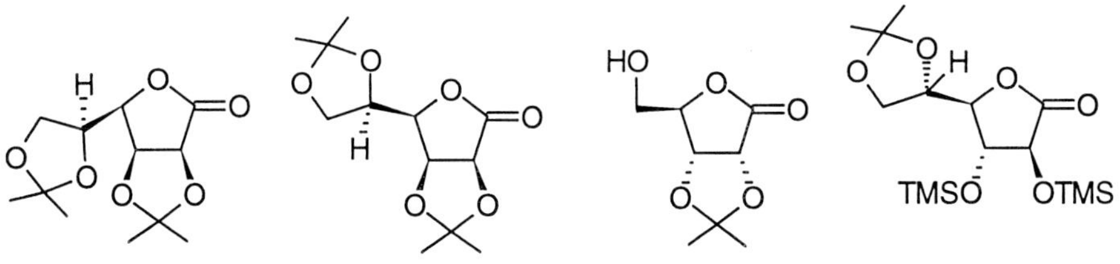

A Concise Route to Carba-hexopyranoses... 149

14, **15**, **16**, **17**, 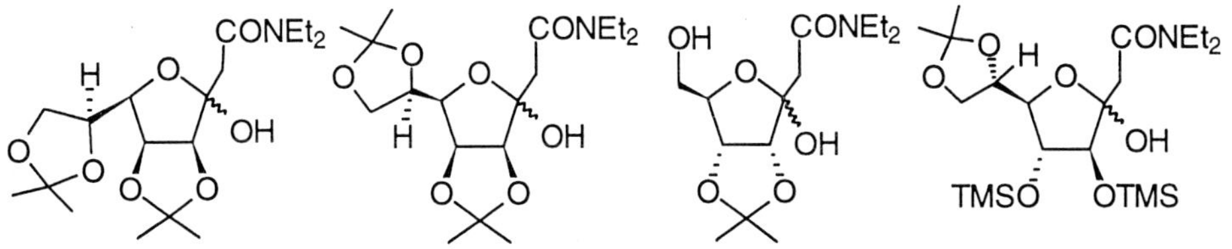 **18**, **19**, **20**, **21**

The next key step was the formation of the carbocycle, i.e. removal of the 5,6-isopropylidene unit, glycol cleavage and aldol reaction as summarized in Scheme 8-5. The acetonide of the adduct **22** was removed upon treatment with aqueous acetic acid (9/1:v/v) at 60°C. The resulting diol **23**, after purification, was treated with one molar equivalent of sodium periodate in a methanol/water mixture to give the expected aldehyde **24** which was used, without further purification, after filtration of the residual salts and evaporation of the solvent.

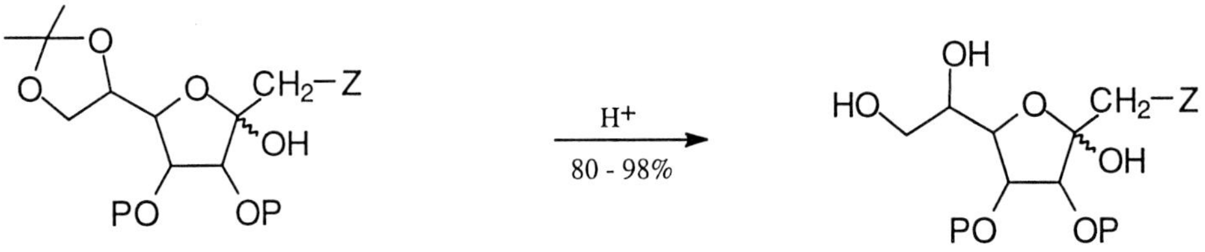

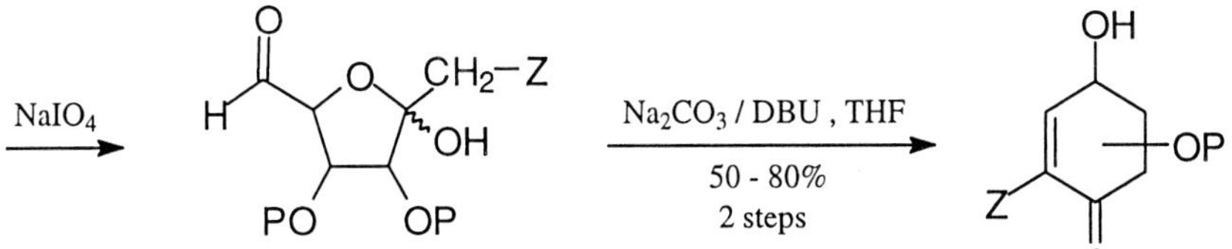

Scheme 8-5

26   **27**   **28**

Some preliminary experiments were needed to find the appropriate aldolisation conditions. We observed that a basic medium such as ethanolic sodium hydroxide, or LDA invariably led to decomposition. Fortunately, sodium carbonate suspended in dry tetrahydrofuran cleanly transformed the aldehyde **24** into the cyclohexenone **25** *provided that a catalytic amount of DBU* was added to the reaction mixture. Use of DBU alone caused extensive decomposition of the starting material. Under these conditions no intermediate aldol product was formed. This sequence was applied to several substrates of general formula **22** including **11**, **12**, **18** and **19** which gave **26**, **27**, **28** respectively (Table 8-3). We should mention that in some experiments cyclohexenone was accompanied by a minor side product which explained the lower yield observed. No cyclisation was observed when starting from compounds **9** and **10**.

Table 8-3. Carbocyclisation of compounds **9, 10, 11, 12, 18** and **19**.

Entry	Starting compound	Product	Yield %
1	11	26	58
2	12	27	80
3	18	28	60
4	19	28	52
5	9	no	-
6	10	no	-

The second crucial step was the chemical manipulation of the functionalities of the branched-chain cyclohexenones. We first attempted to reduce the keto group, (with and without reduction of the double bond) and subsequently the amide group. Transformation of the oxygenated aromatic ring of **21** was also explored.

Before reduction of the keto group of cyclohexenones **25** it was of interest to protect the free OH group. Treatment of **28** with dihydropyran under acidic conditions yielded the corresponding tetrahydropyranyl ether (THP) **29** in 70% yield. Upon reduction of this compound under Luche conditions [17], the corresponding allylic alcohol **30** was obtained as a mixture of diastereoisomers (due to the THP group). However standard acetylation followed by removal of the THP group through careful acid hydrolysis gave the acetate **31** as a single isomer. The stereochemistry of the newly created chiral centre was found to be (*R*) by ^{1}H nmr, in agreement with an hydride attack occurring *trans* to the vicinal dioxolane ring.

Complete reduction of the α,β unsaturated ketone was cleanly achieved on treatment of **28** with sodium borohydride. A single isomer **32** was isolated. The stereochemistry at the newly created centres was again found to be *cis* as in the case of 1,2-reduction. The absolute configuration at C-2 was tentatively assigned as *R* on the basis of the ^{1}H nmr spectrum although the complex pattern of H-2 did not allow complete analysis in terms of coupling constants.

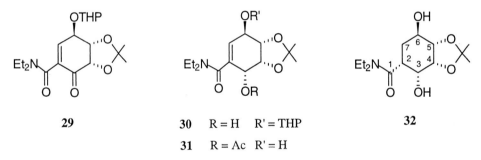

Figure 8-2: Reductions products of α,β unsaturated ketones **28** and **29**.

Finally the reduction of the amide group was attempted. Although a number of methods are, in principle, available for the transformation of tertiary amides into aldehydes or alcohols, [18-21] it proved impossible to reduce, in a reproducible and efficient way, the amide group of **28**, **29** or **32**. Reducing agents such as DIBAH, LiAlH$_4$, with or without additives, failed to give any detectable amount of aldehyde or alcohol. The inertness toward reduction of tertiary amides, which are often used in metalation processes, is a real problem which we have encountered in a related work in the alkaloid field [22] together with other authors [23].

8.2.2 Carbapentofuranoses

As mentioned above the oxidative cleavage of the diol **23** (Scheme 8-5) followed by cyclisation to **25** was, in several experiments, poor yielding. In these cases a by-product was isolated, the structure of which was very similar to that of **25**. Careful examination of the ^{1}H and ^{13}C nmr spectra showed that a CH group was absent and led us to the conclusion that a cyclopentenone could be formed in this process. We suspected that the formation of this by-product might be related to the oxidative cleavage conditions. The use of an excess of sodium periodate or extended reaction times resulted in the extensive formation of cyclopentenone **36**. This reaction was easily rationalized. Given its hemiketalic structure, the intermediate aldehyde **33** is in equilibrium with the open hydroxy-aldehyde form **34** which can be further oxidatively cleaved to **35**. Subsequent treatment of this compound under our cyclisations conditions gave **36** in 52% yield for the two steps.

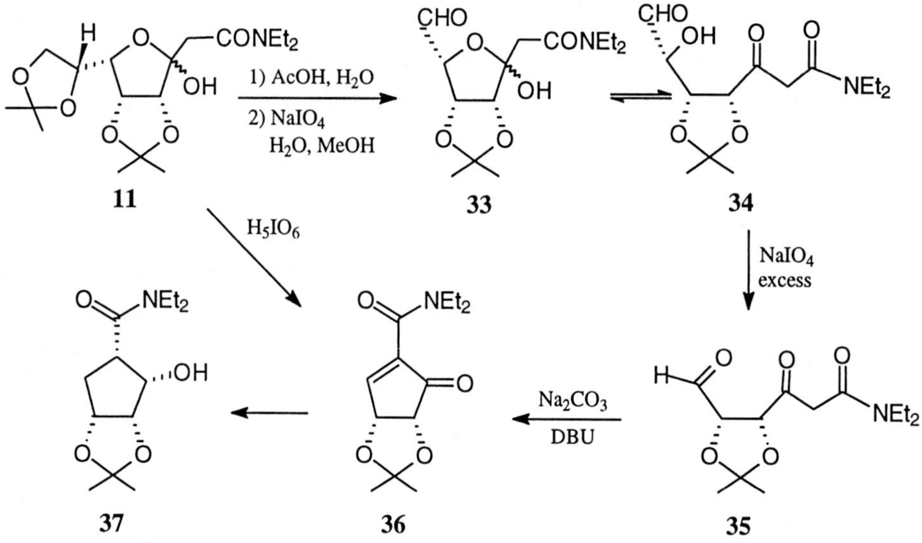

Scheme 8-6.

In an effort to shorten our synthesis we became interested by new methodology proposed by Wu and Wu [24]. By using periodic acid in ether, these authors were able to transform, in one pot, the 5,6-O-isopropylidene acetal of some sugars, including those bearing acid-sensitive protecting groups to an aldehyde with one carbon less.

Applied to amide **11**, this method gave the expected aldehyde which upon treatment under our cyclisation conditions gave cyclopentenone **36** in 52% overall yield (Scheme 8-6). It should be noted that Wu's procedure is slow probably because of the heterogeneous nature of the medium (H_5IO_6 is only sparingly soluble in ether). In order to accelerate the process, the reaction vessel was placed, from time to time, in an ultrasonic bath for 20 minutes. This operation removed solids from the reactor walls. To our surprise, extended reaction times according to the above procedure led to the formation of **36** in signifiant amount. Thus, it became clear that transformation of an adduct such as **11** into cyclopentenone **36** can be performed in a single one-pot operation in the acceptable yield of 59%. This procedure worked equally well with L-gulonolactone derivative **18** which gave *ent*-**36** in 50% yield. Using D-ribonolactone derivative **21**, compound **36** can be also prepared, albeit in a lower yield (20%).

The reduction of **36** was also investigated and, not unexpectedly, gave, after reduction with $NaBH_4$, alcohol **37** with an all *cis* arrangement of the substituents. In this case attack of the hydride on the double bond cleanly proceeds *anti* to the dioxolane ring, further protonation of the intermediate enolate also proceeds from this face, explaining the relative stereochemistry in **37**.

8.2.3 Epilogue

With the basic principles of our approach being established, the last (but not least) problem to solve was the chemical reduction of the tertiary amide group. More recently we have tested Weinreb acetamide (N-methyl,N-methoxy acetamide) as a surrogate of the N,N-diethylamide group. The condensation of the lithium anion of N-methyl, N-methoxy-acetamide and lactone **8** proceeded well and gave the expected adduct **38** in 40% yield. The relatively low yield could be explained by a less efficient enolisation procedure.

The cleavage of the acetonide of **38** and the cleavage of the resulting diol with $NaIO_4$ proceeded uneventfully. The cyclisation of the resulting aldehyde **39** gave the expected cyclohexenone **40** in 25% yield. Obviously, this sequence needs to be improved but will probably open the way to highly functionnalised intermediates en route to carbahexopyranoses by controlled reduction of the keto, amide and olefin groups.

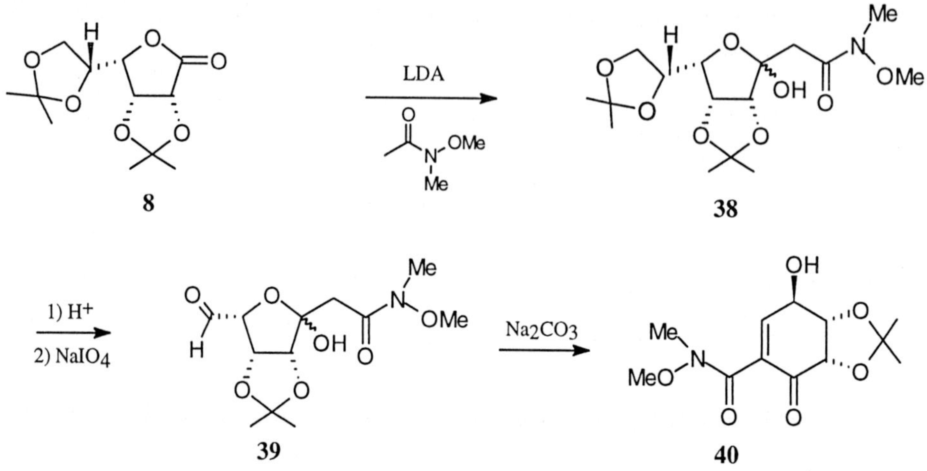

Scheme 8-7

8.3 Conclusion

In conclusion we have demonstrated that tertiary amides are suitable for condensation onto sugar lactones. This addition results in the formation of an active methylene group. These structural features are suitable for an intramolecular Knoevenagel-type reaction, (the aldehyde needed in this reaction coming from cleavage of a diol) to form cyclohexane-based structures. Over-oxidation of the intermediate aldehyde can led to the loss of two carbons allowing the formation of cyclopentane-based structures. Thus, by careful choice of the experimental conditions, it is possible to obtain either six or five membered oxygenated rings from a single chiral progenitor. Adjustment of the amide structure, using for example Weinreb amides, should complete this approach, allowing regioselective reduction of the amide group into the corresponding alcohol or aldehyde as needed for the synthesis of sugar mimics.

8.4 References

1. This term has been proposed first by McCasland: see G. E. McCasland, S. Furuta, L. J. Durham, *J. Org. Chem.* **1966**, *31*, 1516-1521.

2. T. Iwasa, H. Yamamoto, M. Shibata, *J. Antibiot.* **1970**, *23*, 595-602.

3. Sugar having a methylene instead of the ring oxygen have been named more precisely carba-sugars by Ogawa and Suami: see T. Suami, S. Ogawa, *Adv. Carbohydr. Chem. Biochem.* **1990**, *48*, 22-90. See also Chapter 5 of this book.

4. G. Bach, S. Breidingmack, S. Grabley, P. Hammann, K. Hutter, R. Thiericke, H. Uhr, J. Wink, A. Zeeck, *Liebigs Ann. Chem.* **1993**, 241-250.

5. T. Kusaka, H. Yamamoto, M. Shibata, M. Muroi, T. Kishi, K. Mizuno, *J. Antibiot.* **1968**, *21*, 255-263.

6. Y. F. Shealy, J. D. Clayton, *J. Am. Chem. Soc.* **1966**, *88*, 3885-3887.

7. T. Aoyagi, T. Yamamoto, K. Kojiri, H. Morishima, M. Nagai, M. Hamada, T. Takeuchi, H. Umezawa, *J. Antibiot.* **1989**, *42*, 883-889 ; H. Morishima, K. Kojiri, T. Yamamoto, T. Aoyagi, H. Nakamura, Y. Iitaka, *J. Antibiot.* **1989**, *42*, 1008-1011.

8. N. Ishida, K. Miyazaki, K. Kumagai, M. Rikimaru, *J. Antibiot.* **1965**, *18*, 68-76.

9. R. J. Ferrier, S. Middleton, *Chem. Rev.* **1993**, *93*, 2779-2831.

10. L. Agrofoglio, E. Suhas, A. Farese, R. Condom, S. R. Challand, R. A. Earl, R. Guedj, *Tetrahedron* **1994**, *50*, 10611-10670.

11. G. N. Jenkins, N. J. Turner, *Chem. Soc. Rev.* **1995**, *24*, 169-176.

12. G. W. J. Fleet, T. K. M. Shing, S. M. Warr, *J. Chem. Soc., Perkin Trans. I* **1984**, 905-908.

13. S. Mirza, A. Vasella, *Helv. Chim. Acta* **1984**, *67*, 1562-1569.

14. S. Mirza, J. Harvey, *Tetrahedron Lett.* **1991**, *32*, 4111-4114.

15. H.-J. Altenbach, W. Holzapfel, G. Smerat, S. Finkler, *Tetrahedron Lett.* **1985**, 6329-6332.

16. S. Mirza, L.P. Molleyres, A. Vasella, *Helv. Chim. Acta* **1985**, *68*, 988-996.

17. J. L. Luche, A. L. Gemal, *J. Am. Chem. Soc.* **1979**, *101*, 5848-5849.

18. V. M. Micovic, M. L. Mihailovic, *J. Org. Chem.* **1953**, *18*, 1190-1200 ; F. Weygand, G. Eberhardt, H. Linden, F. Schäfer, I. Eigen, *Angew. Chem.* **1953**, *65*, 525-531 ; H. C. Brown, A. Tsukamoto, *J. Am. Chem. Soc.* **1961**, *83*, 2016-2017 ; ibid **1961**, *83*, 4549-4552.

19. H. C. Brown, A. Tsukamoto, *J. Am. Chem. Soc.* **1964**, *86*, 1089-1095 ; J. Malek, M. Cerny, *Synthesis* **1972**, 217-234.

20. N. S. Ramegowa, M. N. Modi, A. K. Koul, J. M. Bora, C. K. Narang, N. K. Mathur, *Tetrahedron* **1973**, *29*, 3985-3986.

21. S. Tsay, J. A. Robl, J. R. Hwu, *J. Chem. Soc., Perkin Trans. I* **1990**, 757-759.

22. M. Khaldi, F. Chrétien, Y. Chapleur, *Tetrahedron Lett.* **1995**, *36*, 3003-3006.

23. T. K. Park, S. J. Danishefsky, *Tetrahedron Lett.* **1995**, *36*, 195-196 ; X. R. Tian, T. Hudlicky, K. Konigsberger, *J. Am. Chem. Soc.* **1995**, *117*, 3643-3644.

24. W. L. Wu, Y. L. Wu, *J. Org. Chem.* **1993**, *58*, 3586-3588.

9 Asymmetric Synthesis of Aminocyclopentitols *via* Free Radical Cycloisomerization of Enantiomerically Pure Alkyne-Tethered Oxime Ethers Derived from Carbohydrates

J. Marco-Contelles

9.1 Introduction

Free radical inter- and intramolecular carbon-carbon bond forming reactions are of paramount importance in organic synthesis.[1] In recent years, complex and densely functionalized carbocycles have been efficiently prepared from chiral precursors, using free radical based methodologies [1,2].

In this chapter we describe the first examples of the free radical cycloisomerization of enantiomerically pure, polyoxygenated alkyne-tethered oxime ethers [3]. This strategy has resulted in a new and stereospecific method [4] for the asymmetric synthesis of aminocyclopentitols [5]. These compounds are key intermediates for the preparation of carbonucleosides [6] and cyclopentane type glycosidase inhibitors [7], such as trehazolin (**1a**) and trehalostatin (**1b**) (Figure 9-1) [8]. These glycoside processing enzymes play critical roles in intra- and intercellular processes including cell adhesion and recognition, membrane transport and signal transduction. Consequently, the use of such compounds and their analogs as glycosidase inhibitors has important implications in virology and oncology [7a].

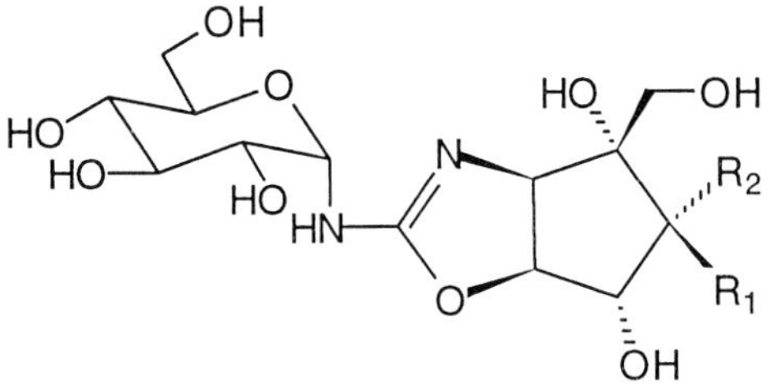

1a R_1 = OH, R_2 = H Trehazolin
1b R_1 = H, R_2 = OH, Trehalostatin

Figure 9-1. Structures of Trehazolin and Trehalostatin.

In addition to their remarkable biological properties, these natural products are challenging synthetic targets and offer an opportunity to develop reaction methodology for the construction of substituted cyclopentanes.

Three general strategies have been employed for the synthesis of such five-membered rings:

(1) Heterocyclization reactions of substituted cyclopentadienes,[7b] (Figure 9-2).

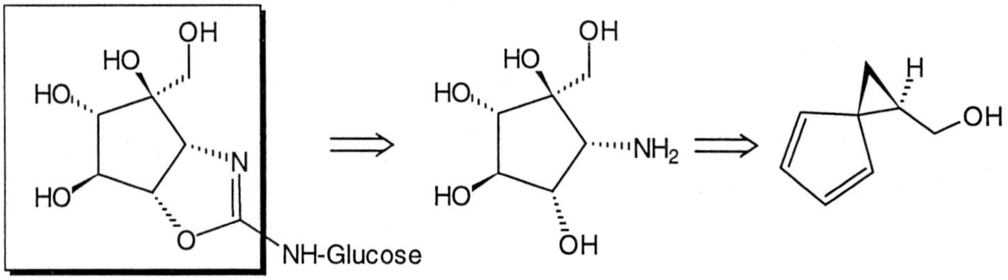

Trehazolin

E. M. Carreira et al, J. Am. Chem. Soc. **1995**, *117*, 11811-11812
B. Ganem et al, J. Am. Chem. Soc. **1994**, *116*, 11811-11812

Figure 9-2. Retrosynthetic analysis of Trehazolin.

(2) Desymmetrization reactions of substituted *meso*-cyclopent-2-en-1,4-diols[7c] (Figure 9-3) and

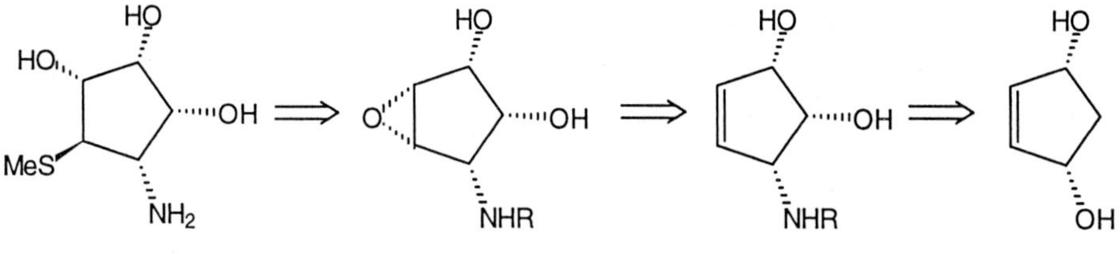

Mannostatin A

B. M. Trost et al. J. Am. Chem. Soc. **1991**, *113*, 6317-6318

Figure 9-3. Retrosynthetic analysis of Mannostatin A.

(3) Fragmentation and refunctionalization reactions of carbohydrates [7d,e, l,m,n].

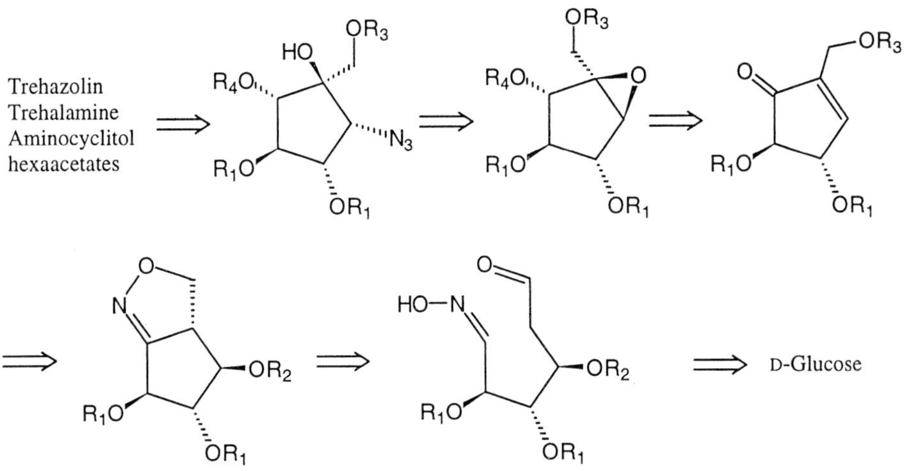
Trehazolin
Trehalamine
Aminocyclitol
hexaacetates
D-Glucose

M. Shiozaki *et al. J. Org. Chem.* **1994**, *59*, 813-822

Figure 9-4. Retrosynthetic analysis of aminocyclopentitols.

9.2 Results and Discussion

For years we have been interested in the development of new strategies for the synthesis of substituted cyclopentanes [2]. Our approach to the aminocyclopentitol derived natural products is described in Figure 9-5.

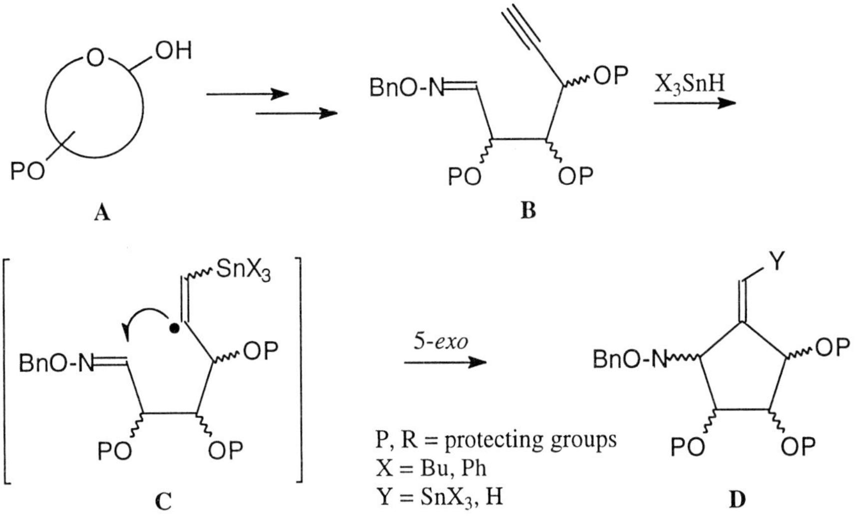
X3SnH
5-exo
A
B
C
D
P, R = protecting groups
X = Bu, Ph
Y = SnX3, H

Figure 9-5. Synthesis of aminocyclopentitols **D** via free radical cyclization.

The essential aspects of this scheme include nucleophilic attack of an ethynyl anion to an aldose **A** [9] followed by selective protection and activation to afford the conveniently functionalized precursor **B**. This compound, upon attack by the appropriate tin hydride reagent, provides vinyl radical [10] species **C** which was expected to lead to aminocyclopentitol **D**. Compounds of type **D** are designed for further synthetic manipulation.

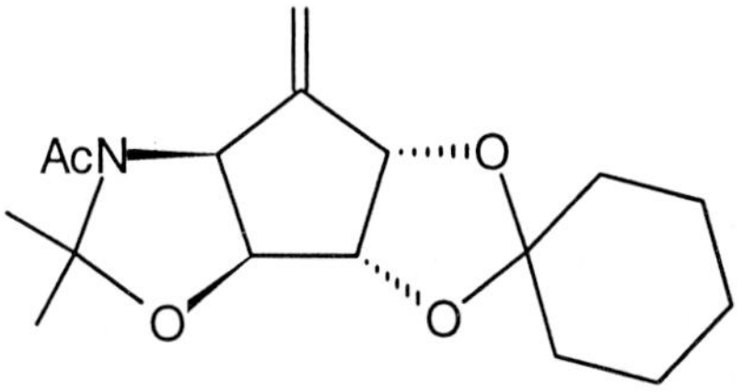

Figure 9-6

In fact, an example of a compound of this type, 1,2-*O*-cyclohexylidene-3,4-*N,O*-isopropylidene derivative of D and L-(1,2:3,4)-4-acetamido-5-methylenecyclopentane-1,2,3-triol (**2**) (Figure 9-6), has been prepared from 1,2-*O*-cyclohexylidene-*myo*-inositol in poor yield, in a time consuming process (an additional step was required for resolution of the racemate), and it has been finally transformed into trehazolin (**1a**) and trehalostatin (**1b**) (Figure 9-1) [11]. In view of these precedents, it is clear that compounds of type **D** are of high synthetic interest and that new, more efficient and asymmetric synthetic routes are needed. Our present approach largely fulfills these requirements and can be considered as a new and significant solution to this problem. Coming back to our scheme, we have to highlight two points. First, although the free radical cyclization of terminal alkynes with oxime ethers has been described by Enholm and coworkers some years ago, its high synthetic potential has been neglected [12]. Second, as the starting aldoses are readily available materials, it is possible to synthesize a large spectrum of compounds of type **D**, natural or analogues, having different configurations at the carbons bearing the protected hydroxyl groups; in addition, the particular stereodirecting properties and conformational bias of the acyclic sugar derivatives offer potentially high degrees of stereocontrol in the formation of the new stereocenters.

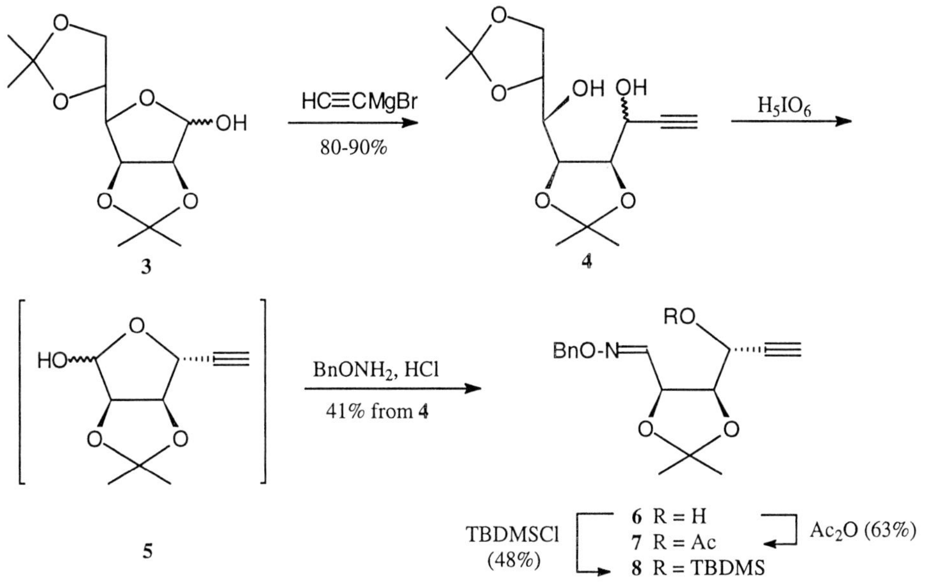

Scheme 9-1. Synthesis of the radical precursors **6-8**.

9.2.1 Synthesis and Radical Cyclization of D-*manno* Derivatives 6-8

The synthesis was carried out starting from 2,3:5,6-di-*O*-isopropylidene-D-mannofuranose (**3**) [13] (Scheme 9-1). Treatment of **3** with ethynylmagnesium bromide [14] gave compound **4** (C6 *R*)) as the only isolated isomer [15]. Sequential "one-pot" acid hydrolysis and triol cleavage [16] gave lactol **5**, which after oxime ether formation, afforded the radical precursor **6**, in 41% overall yield from **4**. Using standard manipulations the acetyl (**7**) and the *t*-butyldimethylsilyl (**8**) derivatives were also prepared. These compounds were isolated as inseparable mixtures of *E* and *Z* isomers in a 2:1 ratio, respectively, as determined by integration of the well resolved H-1 signal of each isomer in the ^{1}H NMR spectrum of the mixture [(*E*) δ H-1:≈ 7.60 ppm; *J* 7.3 Hz; (*Z*) δ H-1: ≈ 7.00 ppm; *J* 4.2 Hz)]. Since no cyclization was observed under the conditions described by Enholm [12], we turned our attention to the triethylborane-triphenyltin hydride-mediated carbocyclization of enynes, as described by Oshima [17]. Under these conditions, precursors **6-8** gave the vinyltin derivatives **9-11** in good yield (Scheme 9-2). These products were obtained as mixtures of *Z* and *E* isomers that could be separated and isolated by flash chromatography (for compound **10** only the Z isomer was detected and isolated). Each geometrical isomer was stereochemically homogeneous at the new stereocenter formed during the cyclization (C-4) and, in all cases, only the isomer having the *R*-configuration was detected. This was determined by analysis of the corresponding ^{1}H NMR spectra.

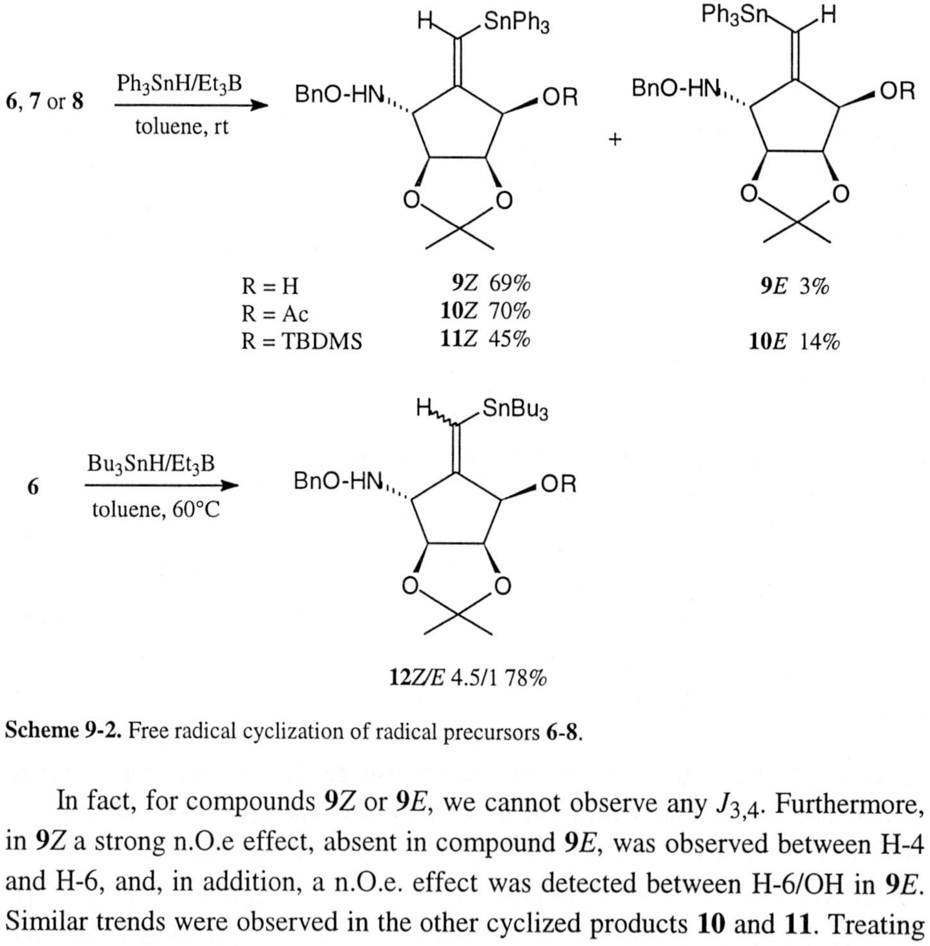

Scheme 9-2. Free radical cyclization of radical precursors **6-8**.

In fact, for compounds **9Z** or **9E**, we cannot observe any $J_{3,4}$. Furthermore, in **9Z** a strong n.O.e effect, absent in compound **9E**, was observed between H-4 and H-6, and, in addition, a n.O.e. effect was detected between H-6/OH in **9E**. Similar trends were observed in the other cyclized products **10** and **11**. Treating **6** with triethylborane and tributyltin hydride gave carbocycle **12** in good yield and high stereoselectivity, as an inseparable mixture of *Z:E* isomers (Scheme 9-2).

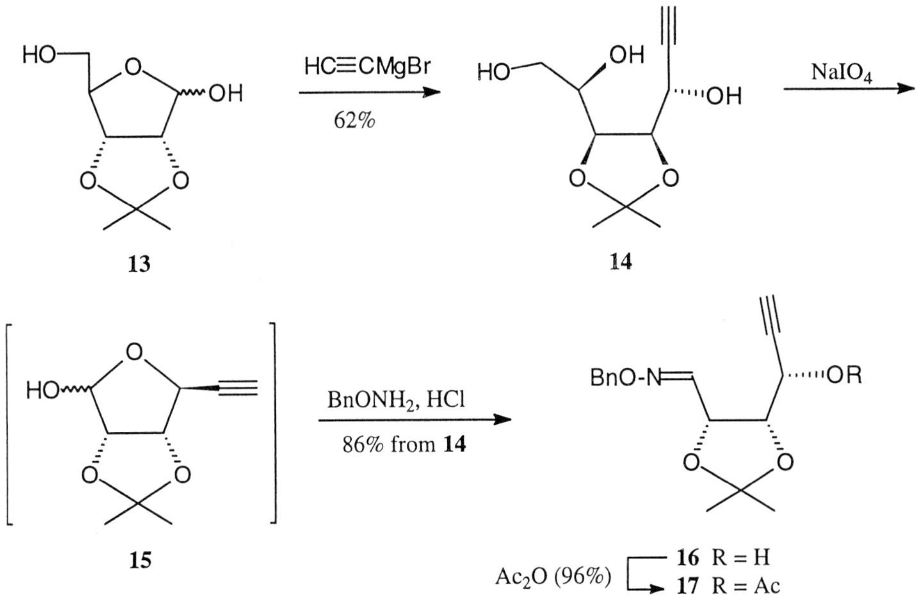

Scheme 9-3. Synthesis of the radical precursors **16-17**.

9.2.2 Synthesis and Radical Cyclization of D-*ribo* Derivatives 16 and 17.

In order to broaden the scope of the methodology, the synthesis of the D-*ribo* radical precursors **16** and **17** was examined. These products were prepared from 2,3-*O*-isopropylidene-D-ribose (**13**) [18] following the same synthetic sequence as for the synthesis of **6** and **8** (see above) and were isolated as mixtures of *E* and *Z* isomers, in a 2:1 ratio, respectively (Scheme 9-3). The radical cyclization of these precursors, under the same experimental conditions, gave in high yield the corresponding aminocyclopentitol derivatives **18**, **19** (as a separable mixture of *E/Z* isomers) and **20** (as the exclusive *Z* isomer) (Scheme 9-4). The detailed analysis of the ^{1}H NMR spectrum of these compounds showed that all of them were also stereochemically homogeneous and had the *S*-configuration at the new stereocenter. In fact, the ^{1}H and ^{13}C NMR spectra of compounds **18Z** {$[\alpha]_D^{25}$ -76.3 (*c* 1.19, CHCl$_3$)} (Scheme 9-4) and **9Z** {$[\alpha]_D^{25}$ +80.1 (*c* 1.08, CHCl$_3$)}(Scheme 9-2) were superimposable. These two compounds differ only in their chiroptical properties. Thus, starting from two readily available and cheap D-sugars, we have prepared an enantiomeric pair of aminocyclopentitols in a short synthetic sequence, with complete diastereoselection.

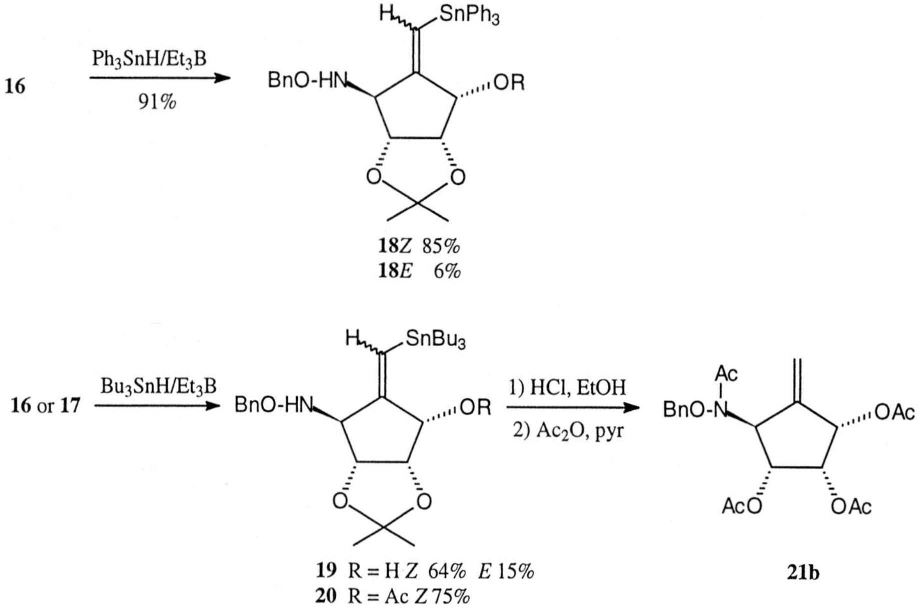

Scheme 9-4. Free radical cyclization of precursors **16** and **17**.

The high degree of stereochemical control observed in the cyclization of precursors **16** and **17** (the same argument is also valid for compounds **6-8**) can be explained according to Beckwith's analysis, [1c] assuming that, in the early transition state, the favored vinyl radical species is in a chair-like conformation with most of the substituents in the preferred pseudoequatorial orientation (see radical species **E** in Figure 9-7).

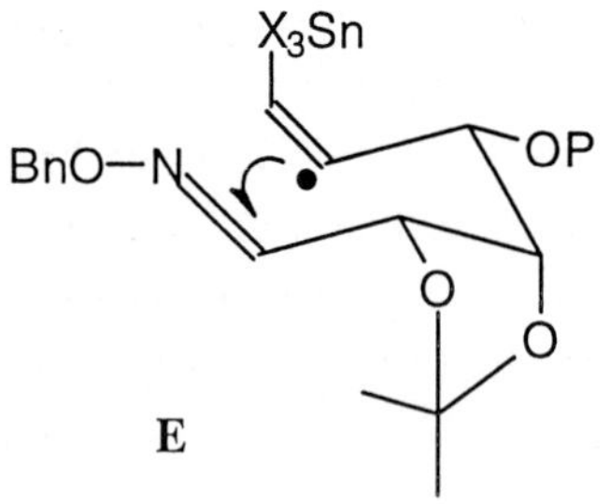

Figure 9-7. Transition state model for the cyclization of *ribo* radical precursors.

The formation of mixtures of geometrical isomers in these vinyltin intermediates is in agreement with the results of Oshima [17], but has no detrimental influence for synthetic purposes (see below).

9.2.3 Synthetic Elaboration of the Carbocyclic Vinyltin Derivatives. Synthesis of Aminocyclopentitols.

Different synthetic protocols for transforming the vinyltin intermediates [19] into useful aminocyclopentitol derivatives were investigated. First, we tested different conditions for the protodestannylation of the vinyl intermediates. After some experimentation, we found that acid hydrolysis promoted by ethanol saturated with hydrogen chloride was the method of choice for this transformation [20]. For instance, when **20Z** was submitted to these conditions, followed by acetylation, the corresponding peracetylated aminocyclitol **21b** (Scheme 9-4) was obtained in 73% overall yield.

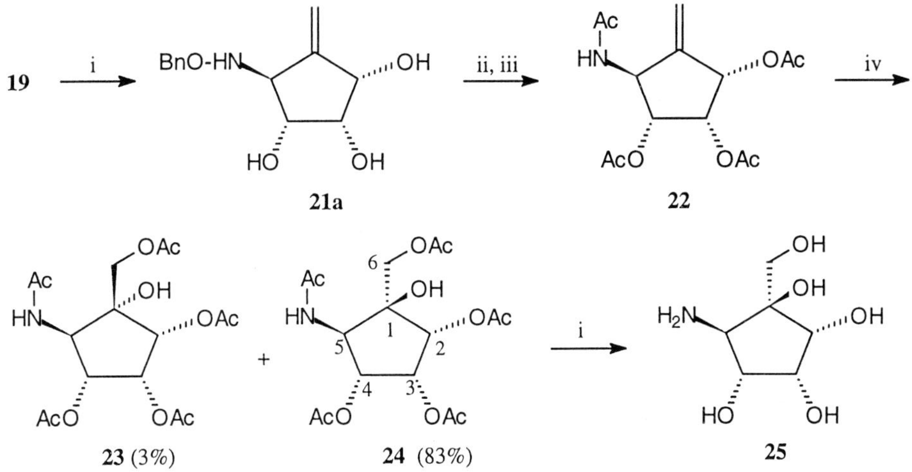

Scheme 9-5. Reagents: i) HCl, EtOH, 92%; ii SmI$_2$; iii) Ac$_2$O, 79% 2 steps; iv OsO$_4$, Ac$_2$O.

The synthetic usefulness of this methodology was demonstrated by further transforming compound **19** (or **20**) into **24** (Scheme 9-5). This compound is the C-1,2,4-tri-epimer of the aminocyclitol moitey of trehazolin (**1a**) and the C-1,4-di-epimer of the aminocyclitol moitey of trehalostatin (**1b**); the C-4 epimer of aminocyclitol **24** has been prepared previously and converted into an analogue of these glycosidase inhibitors [11a]. The transformation (**19→24**) was accomplished as follows. Acid hydrolysis of compound **19** (**21a**) was followed by samarium diiodide-mediated [21a] cleavage of the nitrogen-oxygen bond in the *O*-benzyl hydroxylamine **21a**, then by acetylation in a "one-flask" operation, thus affording peracetate **22** in good overal yield (79%) from **21a**. Treatment of this allylic acetamide with osmium tetroxide, NMO in 80% aqueous acetone [12a] gave almost exclusively, after partial acetylation, the aminocyclopentitol **24** in excellent yield; only traces of the minor isomer (**23**), obtained by reaction from the α face, were isolated. A similar high *syn*-stereoselectivity has also been observed in the osmylation of some related substrates [22]. The configuration at the new stereocenter (C-1) was established by ^{1}H NMR analysis. Thus, the 2D-NOESY spectrum of **24** shows crosspeaks for H-5/H-6a, and also OH/H-2, indicating that the methylene C-H6, is *cis* to H-5, and that the OH group is *cis* to H-2. On the other hand, the 2D-NOESY spectrum of aminocyclopentitol **23** exhibits crosspeaks for H-2/H-6a and H-2/H-6b, thus indicating that C-H6 is now *cis* to H-2. Finally, acid hydrolysis of compound **24** gave in good yield the fully deprotected aminocyclopentitol **25**.

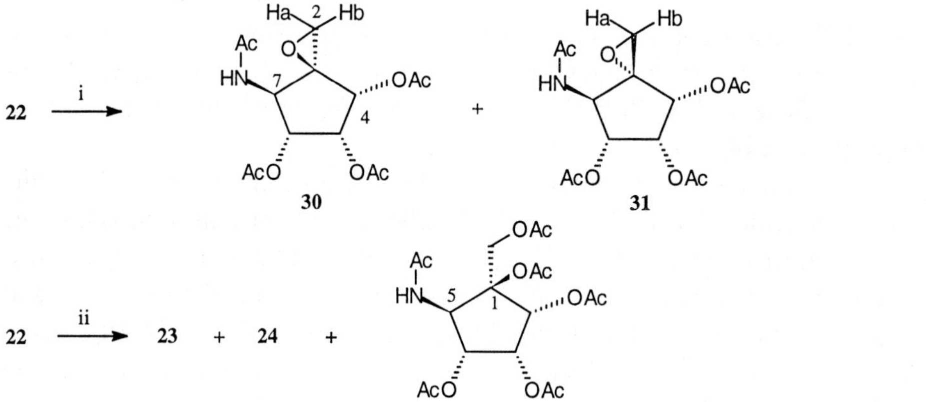

Scheme 9-6. Reagents: i) mCPBA (**30**, 58%, **30+31**, 20% 2.8:1); ii) mCPBA then NaOAc, (isolated **23+24**, 16%, **24**, 37% **32**, 14%); iii) O$_3$, MeOH

The elaboration of intermediate **22** with other oxidizing agents was also considered. For instance, epoxidation (*m*-CPBA) of **22** gave the expected epoxides **30** and **31** (Scheme 9-6), in a 10:1 ratio, as determined by ^{1}H NMR. After careful chromatography, the major compound **30** and a mixture of **30+31** (2.8:1) were isolated in 58% yield and in 20% yield, respectively (combined overall yield: 78% yield). The configuration at C-3 (the new stereocenter) in the major isomer (**30**) was assigned by n.O.e. experiments; a strong n.O.e. effect was detected between H-2a and H-7 and a very weak one between H-2b and H-4, and between H-2a and NH. This assignment was also definitively established by chemical correlation. When compound **22** was epoxidized and the resulting mixture treated with sodium acetate in DMF [11a], followed by acetylation, compounds **23**, **24** and a new compound **32** (Scheme 9-6) were detected. After purification by column chromatography, product **24** (37% yield), a mixture of **23** and **24** (in 1.7:1 ratio; 16% yield) and peracetate **32** (14% yield) were isolated. The configuration at C-1 in aminocyclitol **32** was determined by detailed ^{1}H NMR analysis. In its NOESY spectrum, a strong crosspeak H-5/H-6a was observed, which allowed us to assign the *S*-configuration at C-1, as shown in Scheme 9-6. Thus, this product comes from the peracetylation of the major product **24**.

The ozonolysis of products **21a** and **22** was then investigated. Under the usual conditions (O$_3$, MeOH at -78_C, followed by reaction with DMS) we obtained ketones **33** and **34** (Scheme 9-6) in 42% and 23% yield, respectively. Straightforward assignment of the structures for the major and minor isomers to **33** and **34**, respectively, was deduced from the corresponding ^{1}H NMR spectra. A similar substrate **33a**; (Scheme 9-6) was recently obtained and characterized by Ogawa [11a]. The spectroscopic data of **33a** are in full agreement with those observed for **33**. These compounds are extremely interesting substrates for further development of a synthesis of mannostatin A analogues [11].

9.3 Conclusion

From the results reported here, we can conclude that (a) the tributyl or triphenyltin hydride mediated free radical cycloisomerization of polyoxygenated, enantiomerically pure alkyne tethered oxime ethers is possible; (b) the presence of triethylborane is absolutely necessary for the success of the process; (c) the process gives highly functionalized vinyltin [20] derivatives with large synthetic potential and (d) by selecting the appropiate radical precursor a very high degree of diastereoselection was achieved, independent of the type of substituents at the propargylic position.

Acknowledgements.

The author thanks DGICYT (SAF94-0818-C02-02), CICYT (CE93-0023), Comunidad Autónoma de Madrid-Consejería de Educación y Cultura-(AE-0094/94) and EU (Human Capital and Mobility; Contract n_ ERBCHRXCT 92-0027) for generous financial support. The author thanks also to Dr José Luis Chiara and Dr Manuel Bernabé for the fruitful collaboration during this project and Dr Christine Destabel and Dr Pilar Gallego for the experimental work described here.

9.4 References

1. (a) B. Giese, *Radicals in Organic Synthesis: Formation of Carbon-Carbon Bonds*, Pergamon Press, New York, **1986**. (b) D. P. Curran, *Synthesis* **1988**, 417-439, 489-513. (c) A. L. J. Beckwith, C. H. Schiesser, *Tetrahedron* **1985**, *41*, 3925-3941. (d) W. B. Motherwell, D. Crich, *Free Radical Chain Reactions in Organic Synthesis*; Academic Press, London, **1992**.

2. (a) J. Marco-Contelles, L. Martínez, A. Martínez-Grau, C. Pozuelo, M. L. Jimeno, *Tetrahedron Lett.* **1991**, *32*, 6437-6440. (b) J. Marco-Contelles, C. Pozuelo, M. L. Jimeno, L. Martínez, A. Martínez-Grau, *J. Org. Chem.* **1992**, *57*, 2625-2631. (c) J. Marco-Contelles, B. Sánchez, C. Pozuelo, *Nat. Pro. Lett.* **1992**, *1*, 167-170. (d) J. Marco-Contelles, B. Sánchez, *J. Org. Chem.* **1993**, *58*, 4293-4297. (e) J. Marco-Contelles, M. Bernabé, D. Ayala, B. Sánchez, *J. Org. Chem.* **1994**, *59*, 1234-1235. (f) J. Marco-Contelles, B. Sánchez, C. Pozuelo, *Tetrahedron: Asymmetry* **1992**, *3*, 689-692. (g) J. Marco-Contelles, L. Martínez, A. Martínez-Grau, *Tetrahedron: Asymmetry* **1991**, *2*, 961-964.

3. S. E. Booth, P. R. Jenkins, C. J. Swain, J. B. Sweeney, *J. Chem. Soc., Perkin Trans 1* **1994**, 3499-3508, and references cited therein.

4. (a) J. Marco-Contelles, C. Destabel, J. L. Chiara, M. Bernabé, *Tetrahedron: Asymmetry* **1995**, *6*, 1547-1550. (b) J. Marco-Contelles, C. Destabel, P. Gallego, J. L. Chiara, M. Bernabé, *J. Org. Chem.* **1996**, *61*, 1354-1362.

5. For some recent synthetic approaches, see reference 7 and the following: (a) T. Kiguchi, K. Tajiri, I. Ninomiya, T. Naito, H. Hiramatsu, *Tetrahedron Lett.* **1995**, *36*, 253-256. (b) J. K. Gallos, E. G. Goga, A. E. Koumbis, *J. Chem. Soc., Perkin Trans. 1* **1994**, 613-614. (c) J. Marco-Contelles, M. Bernabé, *Tetrahedron Lett.* **1994**, *35*, 6361-6364. (d) J. Marco-Contelles, P. Ruiz, L. Martínez, A. Martínez-Grau, *Tetrahedron* **1993**, *49*, 6669-6694. (e) R. P. Elliot, A. Hui, A. J. Fairbanks, R. J. Nash, B. G. Winchester, G. Way, C. Smith, R. B. Lamont, R. Storer, G. W. J. Fleet, *Tetrahedron Lett.* **1993**, *34*, 7949-7952. (f) J. L. Chiara, J. Marco-Contelles, N. Khiar, C. Destabel, P. Gallego, M. Bernabé, *J. Org. Chem.* **1995**, *60*, 6010-6011.

6. A. D. Borthwick, K. Biggadike, *Tetrahedron* **1992**, *48*, 571-623.

7. **Mannostatins:** (a) S. Knapp, T. G. M. Dhar, *J. Org. Chem.* **1991**, *56*, 4096-4097. (b) B. Ganem, S. B. King, *J. Am. Chem. Soc.* **1991**, *113*, 5089-5090; B. E. Ledford, E. M. Carreira, *J. Am. Chem. Soc.* **1995**, *117*, 11811-11812 (c) B. M. Trost, D. L. Van Vranken, *J. Am. Chem. Soc.* **1991**, *113*, 6317-6318. (d) S. Ogawa, Y. Yuming, *J. Chem. Soc., Chem. Commun.* **1991**, 890-891. (e) C. Li, P. L. Fuchs, *Tetrahedron Lett.* **1994**, *35*, 5121-5124. (**Allosamizoline**-the cyclopentane component of **Allosamidine**): (f) B. K. Goering, B. Ganem, *Tetrahedron Lett.* **1994**, *35*, 6997-7000. (g) N. S. Simpkins, S. Stokes, A.Whittle, *Tetrahedron Lett.* **1992**, *33*, 793-796. (h) M. Nakata, S. Akazawa, S. Kitamura, K. Tatsuta, *Tetrahedron Lett.* **1991**, *32*, 5363-5366. (i) S. Takahashi, H. Terayama, H. Kuzuhara, *Tetrahedron Lett.* **1991**, *32*, 5123-5126. (j) B. M. Trost, D. L. Van Kranken, *J. Am. Chem. Soc.* **1993**, *115*, 444-445. (k) D. A. Griffith, S. Danishefsky, *J. Am. Chem. Soc.* **1991**, *113*, 5863-5864. (**Trehalostatin**): (l) C. Uchida, T. Yamagashi, S. Ogawa, *Chem. Lett.* **1993**, 971-974. (**Trehazolin**): (m) Y. Kobayashi, H. Miyazaki, M. Shiozaki, *J. Am. Chem. Soc.* **1992**, *114*, 10065-10066. (n) S. Knapp, A. Purandare, K. Rupitz, S. G. Withers, *J. Am. Chem. Soc.* **1994**, *116*, 7461-7462. (**Merrell-Dow cyclopentylamine**): (o) R. A. Farr, N. P. Peet, M. S. Kang, *Tetrahedron Lett.* **1990**, *31*, 7109-7112.

8. O. Ando, H. Satake, K. Itoi, M. Nakajima, S. Takahashi, H. Haruyama, *J. Antibiot.* **1991**, *44*, 1165-1168.

9. The incorporation of a triple bond into a carbohydrate precursor can be achieved by Wittig-type reagents and dehydrobromination: (a) M. Mella, L. Panza, F. Ronchetti, L. Toma, *Tetrahedron* **1988**, *44*, 1673-1678. (b) M. McIntosh, S. Weinreb, *J. Org. Chem.* **1993**, *58*, 4823-4832, or by simple Grignard addition to the corresponding hemiacetals: (c) J. J. Gaudino, C. S. Wilcox, *J. Am. Chem. Soc.* **1990**, *112*, 4374-4380.

10. For some free radical cyclization of vinyl radicals, see: (a) D. L. J. Clive, H. W. Manning, *J. Chem. Soc., Chem. Commun.* **1993**, 666-667. (b) O. Moriya, M. Okawara, Y. Ueno, *Chem. Lett.* **1984**, 1437-1440. (c) N. Moufid, Y. Chapleur, *Tetrahedron Lett.* **1991**, *32*, 1799-1802.

11. (a) C. Uchida, T. Yamagashi, S. Ogawa, *J. Chem. Soc., Perkin Trans. 1* **1994**, 589-602; (b) C. Uchida, H. Kitahashi, T. Yamagashi, Y. Iwaisaki, S. Ogawa, *J. Chem. Soc., Perkin Trans. 1* **1994**, 2775-2785. (c) S. Ogawa, C. Uchida, Y. Yuming, *J. Chem. Soc., Chem. Commun.* **1992**, 886-888. (d) S. Ogawa, C. Uchida, *Chem. Lett.* **1993**, 173-176. (e) Y. Kobayashi, H. Miyazaki, M. Shiozaki, *Tetrahedron Lett.* **1993**, *34*, 1505-1506. (f) Y. Kobayashi, H. Miyazaki, M. Shiozaki, *J. Org. Chem.* **1994**, *59*, 813-822.

12. (a) E. J. Enholm, J. A. Burroff, L. M. Jaramillo, *Tetrahedron Lett.* **1990**, *31*, 3727-3230. For the analogous cyclization of β-allenic oxime ethers, see: (b) J. Hatem, C. Henriet-Bernard, J. Grimaldi, R. Maurin, *Tetrahedron Lett.* **1992**, *33*, 1057-1058.

13.   B. Lee, T. J. Nolan, *Tetrahedron* **1967**, 2789-2794.

14.   J. G. Buchanan, A. D. Dunn, A. R. Edgar, *J. Chem. Soc., Perkin Trans. 1* **1975**, 1191-1200.

15.   T. K. H. Shing, D. A. Elsley, J. G. Gillhouley, *J. Chem. Soc., Chem. Commun.* **1989**, 1280-1282.

16.   W.-L. Wu, Y.-L. Wu, *J. Org. Chem.* **1993**, *58*, 3586-3588.

17.   K. Nozaki, K. Oshima, K. Utimoto, *J. Am. Chem. Soc.* **1987**, *109*, 2547-2549.

18.   (a) C. J. F. Bichard, A. J. Fairbanks, G. W. J. Fleet, N. G. Ramsden, K. Vogt, O. Doherty, L. Pearce, D. J. Watkin, *Tetrahedron : Asymmetry* **1991**, *2*, 901-912. (b) S. B. Mandal, B. Achari, *Synth. Commun.* **1993**, *23*, 1239-1244.

19.   M. Pereyre, J. P. Quintard, A. Rahm, *Tin in Organic Synthesis*, Butterworths, London, **1986**.

20.   J. Ardisson, J. P. Férézou, M. Julia, A. Pancrazi, *Tetrahedron Lett.* **1987**, *28*, 2001-2004.

21.   Although the reduction of oximes to amines *via* SmI_2 is known (T. Mukaiyama, K. Yorozu, K. Kato, T. Yamada, *Chem. Lett.* **1992**, 181-183), the specific reduction of *O*-alkyl hydroxylamines to amines *via* SmI_2 seems to be unprecedented. For the first very recent reports on the samarium diiodide-mediated cleavage of the nitrogen-oxygen bond, see ref. 5, 6g, and the following: (a) J. L. Chiara, C. Destabel, P. Gallego, J. Marco-Contelles, *J. Org. Chem.* **1996**, *61*, 359-360. (b) G. E. Keck, S. F. McHardy, J. A. Murry, *J. Am. Chem. Soc.* **1995**, *117*, 7289-7290. (c) G. E. Keck, S. F. McHardy, T. T. Wager, *Tetrahedron Lett.* **1995**, *36*, 7419-7422 (personal communication; we thank Professor Keck for sending us a copy of this manuscript prior to publication).

22.   (a) B. Ganem, S. B. King, *J. Am. Chem. Soc.* **1994**, *116*, 562-570. (b) B. M. Trost, G.-H. Kuo, T. J. Benneche, *J. Am. Chem. Soc.* **1988**, *110*, 621-622. (c) G. Poli, *Tetrahedron Lett.* **1989**, *30*, 7385-7388. However, it has been shown that the complexation of amides with OsO_4 does not take place in the hydroxylation of certain acyclic sulfoxide substituted allylic amides: F. M. Hauser, S. R. Ellenberger, J. C. Clardy, L. S. Bass, *J. Am. Chem. Soc.* **1984**, *106*, 2458-2459.

10 Carbohydrates as Sources of Chiral Inositol Polyphosphates and their Mimics

D. J. Jenkins, A. M. Riley and B. V. L. Potter*

10.1 Introduction

Intracellular Ca^{2+} mobilization mediated by the second messenger D-*myo*-inositol 1,4,5-trisphosphate [Ins(1,4,5)P_3 **1**] is the prime response to phosphoinositidase C activation *via* stimulation of an extracellular G-protein coupled receptor in a vast array of cell types [1]. Understanding the subtleties of the polyphosphoinositide signaling pathway has been a fundamental biological aim since the discovery of the Ca^{2+} releasing activity of Ins(1,4,5)P_3 in 1983 [2]. Since 1986 there has been an intensive chemical focus upon the synthesis of inositol polyphosphates and on understanding the structure-recognition parameters at the Ins(1,4,5)P_3 receptor and other binding proteins [3]. The synthesis of structurally-modified Ins(1,4,5)P_3 analogs offers the prospect of pharmacological intervention in such signalling pathways. We have been active in this area for several years [4–7], primarily with the synthesis of analogues and mimics of Ins(1,4,5)P_3. In this chapter we review several areas of current interest where the use of carbohydrates as starting materials can lead to the synthesis of chiral cyclitol and even carbohydrate based polyphosphates for potential pharmacological intervention in the polyphosphoinositide pathway of cellular signalling.

1 $R_1=PO_3^{2-}$; $R_2=OH$
2 $R_1=R_2=H$
3 $R_1=R_2=PO_3^{2-}$
4 $R_1=H$; $R_2=PO_3^{2-}$

5

Ins(1,4,5)P_3 is metabolised by two pathways: hydrolysis of the phosphate at position 5 by a low affinity, high capacity Ins(1,4,5)P_3 5-phosphatase giving Ins(1,4)P_2 **2**; or phosphorylation at position 3 by a high affinity, low capacity Ins(1,4,5)P_3 3-kinase giving Ins(1,3,4,5)P_4 **3**. Ins(1,3,4,5)P_4 was discovered in

1985 [8], and the kinase was subsequently characterised [9,10]. Its metabolism has been studied and the main route of production is from $Ins(1,4,5)P_3$ by 3-kinase; it has also been isolated *in vitro* as a product of $Ins(1,3,4)P_3$ 5/6-kinase action on D-*myo*-inositol 1,3,4-trisphosphate, $Ins(1,3,4)P_3$ **4** [11], but this route is not believed to be physiologically significant [11]. The main route of $Ins(1,3,4,5)P_4$ catabolism is dephosphorylation by 5-phosphatase giving $Ins(1,3,4)P_3$ [12]; it is also dephosphorylated by a 3-phosphatase, regenerating $Ins(1,4,5)P_3$ [13–15]. It has been noted [12] that $Ins(1,3,4,5)P_4$ is probably not the physiological substrate of 3-phosphatase as the enzyme is inhibited by physiological concentrations of $Ins(1,3,4,5,6)P_5$ and $InsP_6$. However, the activity of the phosphatase may be significant in experimental situations which involve broken and permeabilised cells. It has been suggested that $Ins(1,3,4,5)P_4$ is involved in Ca^{2+} homeostasis at the plasma membrane, helping to control entry of extracellular Ca^{2+} into the cell [16]. In support of this hypothesis, an $Ins(1,3,4,5)P_4$-sensitive Ca^{2+}-permeable channel has been characterised from endothelial cells [17]; $Ins(1,4,5)P_3$ failed to induce an increase in this channel's activity.

Intracellular sites that specifically bind $Ins(1,3,4,5)P_4$ with high affinity [and with high selectivity over $Ins(1,4,5)P_3$] have been noted in HL60 cells [18], bovine adrenal cortex [19], bovine parathyroid [20], rat [21] and pig [22] cerebellum and human platelets [23]. $Ins(1,3,4,5)P_4$-binding proteins have been purified from pig [24] and rat [25,26] cerebellum and porcine platelets [27]. The last example has received particular recent interest: this protein has been tentatively proposed as an $Ins(1,3,4,5)P_4$ receptor [27–29], has demonstrated a high specificity for $Ins(1,3,4,5)P_4$ over all other inositol tetrakis- [28] and various polyphosphates [27], and *in vitro* $Ins(1,3,4,5)P_4$-stimulated GAP activity against the oncogene *ras* has been demonstrated [29]. The $Ins(1,3,4,5)P_4$ binding region of this protein has been identified [29]. Evidence for the physiological significance of this protein and involvement of $Ins(1,3,4,5)P_4$ is awaited with interest. Consistent with their proposed Ca^{2+}-entry function, $Ins(1,3,4,5)P_4$-binding proteins are located on the inner leaflet of the plasma membrane. However, it has yet to be unambiguously demonstrated that these proteins are not in fact receptors for phosphatidylinositol-3,4,5-trisphosphate another putative second messenger located in the plasma membrane [30], of which $Ins(1,3,4,5)P_4$ essentially represents the polar headgroup.

Early studies investigating the ability of $Ins(1,3,4,5)P_4$ to mobilise intracellular Ca^{2+} directly *via* the $Ins(1,4,5)P_3$ receptor produced conflicting results. Thus, it was reported that $Ins(1,3,4,5)P_4$ did not mobilise intracellular

Ca^{2+} in Swiss 3T3 cells [31], permeabilised Jurkat T-lymphocytes [32] or from aortic sarcoplasmic reticulum vesicles [33]. However, other studies reported direct Ca^{2+} release in cerebellar [34] and adrenal [35] microsomes, *Xenopus* oocytes [36] and permeabilised SH-SY5Y neuroblastoma cells [37].

To help investigate potential biological roles for Ins(1,3,4,5)P_4 we decided to synthesise the novel analogue D-2-deoxy-Ins(1,3,4,5)P_4 **5**. This compound is potentially useful to determine the importance of the hydroxyl at position 2 of Ins(1,3,4,5)P_4 with respect to binding to its putative receptor, similarly to 2-deoxy-Ins(1,4,5)P_3 at the Ins(1,4,5)P_3 receptor [38,39]. Deletion of hydroxyl groups adjacent to phosphomonoesters has also produced phosphatase inhibitors [*e.g.* 2-deoxy-Ins(1)P and 6-deoxy-Ins(1,4,5)P_3 inhibited inositol monophosphatase [40] and Ins(1,4,5)P_3 5-phosphatase [41] respectively], so 2-deoxy-Ins(1,3,4,5)P_4 **5** is also of interest as a potential inhibitor of Ins(1,3,4,5)P_4-3-phosphatase.

10.2 Synthesis of D-2-deoxy-*myo*-inositol 1,3,4,5-tetrakis-phosphate

A practical route to chiral deoxyinositols is *via* Ferrier carbocyclisation [42] of hex-5-enopyranoside precursors. The Ferrier carbocyclisation is included in an extensive recent review of the conversion of carbohydrates to cyclopentane and cyclohexane derivatives [43]. It has found many imaginative applications, notably allowing efficient access to modified aminocyclitol components of aminoglycoside antibiotics [44–49], and to pseudo-sugars [50–55]. Additionally, it has been used to prepare various deoxy and non-deoxy inositol phosphates and precursors. Sato *et al.* [56] unambiguously prepared L-1,2,4-tri-*O*-benzyl-*myo*-inositol, a phosphorylation precursor for Ins(1,4,5)P_3, from D-glucose. Several 6-deoxy-*myo*-inositol polyphosphates have been prepared from methyl β-D-galactopyranoside, by way of olefin (**6**) [57]. A 2:1 mixture of the axial and equatorial products **7** and **8** respectively were obtained on Ferrier carbocyclisation. The equatorial product **8** was reduced stereoselectively with lithium borohydride to L-6-*O*-benzyl-1,2-*O*-cyclohexylidene-4-deoxy-*myo*-inositol, a versatile intermediate which was elaborated to 6-deoxy-Ins(1)P **9**, 6-deoxy-Ins(1,5)P_2 (**10**), 6-deoxy-Ins(1,4,5)P_3 (**11**), 6-deoxy-Ins(1,3,4,5)P_4 (**12**) and 6-deoxy-Ins(1,2-cyclic-4,5)P_3 (**13**). D-Glucose was used as starting material in a preparation of 3-deoxy-Ins(1,4,5)P_3 (**14**) and D-3-deoxy-*muco*-Ins(1,4,5)P_3 (**15**) [58], the latter synthesis being noteworthy for incorporating allyl ethers in this mercury(II)-catalysed carbocyclisation.

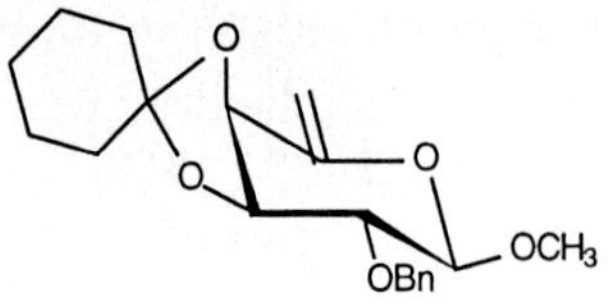

6

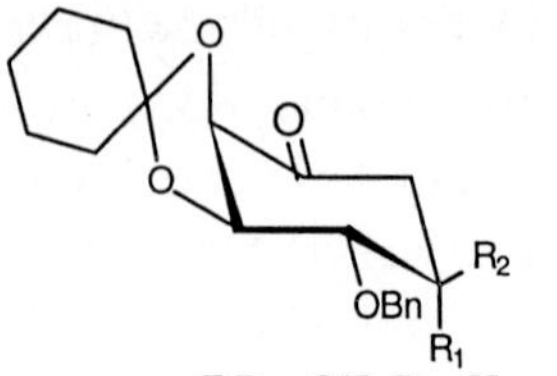

7 $R_1=OH; R_2=H$
8 $R_1=H; R_2=OH$

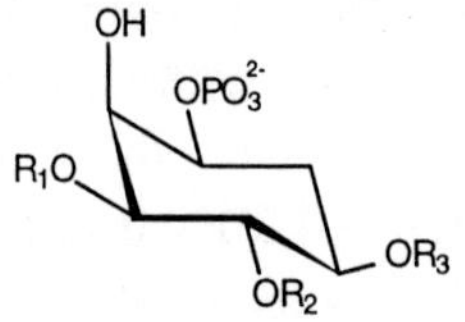

9 $R_1=R_2=R_3=H$
10 $R_1=R_2=H; R_3=PO_3^{2-}$
11 $R_1=H; R_2=R_3=PO_3^{2-}$
12 $R_1=R_2=R_3=PO_3^{2-}$

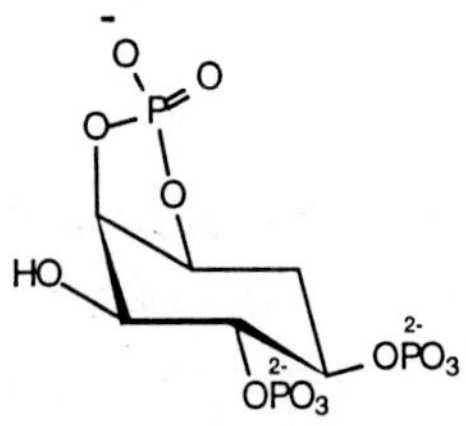

13

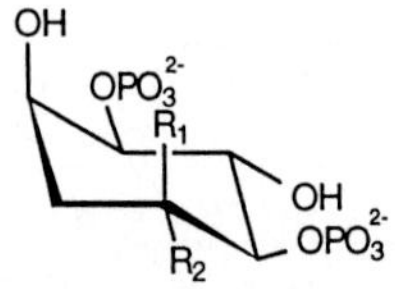

14 $R_1=H; R_2=OPO_3^{2-}$
15 $R_1=OPO_3^{2-}; R_2=H$

Bender and Budhu [59] devised an elegant biomimetic route to optically active *myo*-inositol derivatives. They oxidised partially protected glucose intermediates **16** to aldehydes **17**, which were trapped as enolates by acetic anhydride to give enol acetates **18**. These underwent Ferrier carbocyclisation with high stereoselectivity for the D-*myo*-inos-6-oses **19**, which could be reduced stereospecifically with sodium triacetoxyborohydride to give *myo*-inositol derivatives **20**.

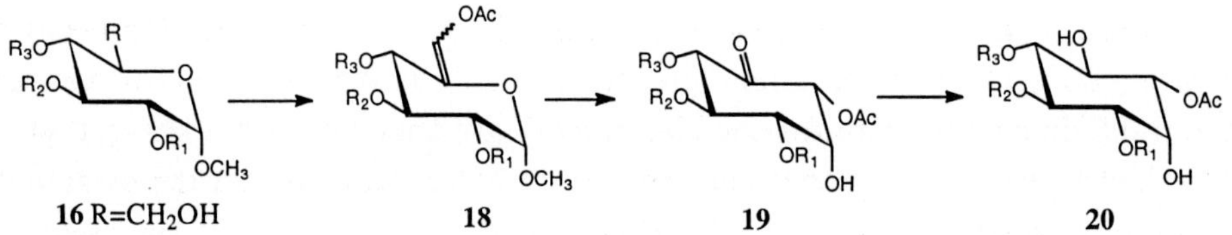

16 $R=CH_2OH$
17 $R=CHO$

18

19

20

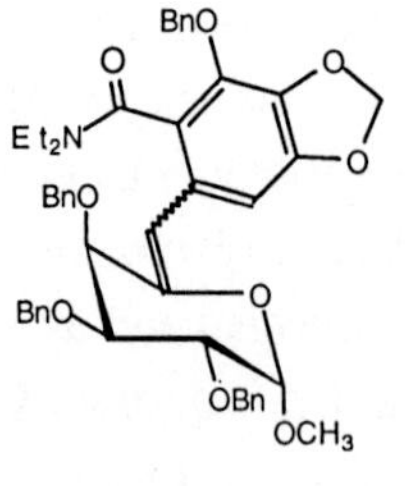

21

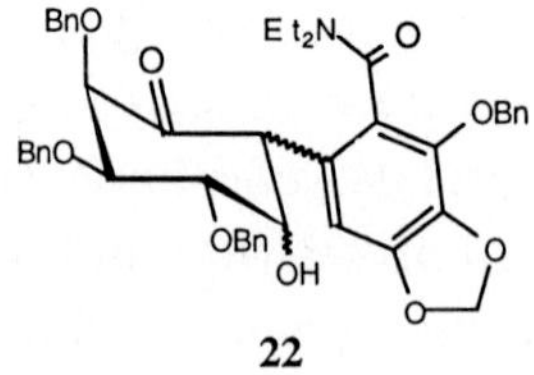

22

Of related interest is the recent report by Park and Danishefsky [60] in which the substituted enol ether **21** was subjected to classical Ferrier conditions to give **22**. The axial or equatorial orientation of the aromatic group was not described. Bender and Budhu's modification of the Ferrier carbocyclisation has also been applied by Estevez and Prestwich [61] to prepare a chiral, P-1-tethered Ins(1,3,4,5)P_4 affinity label, which has been used to purify Ins(1,3,4,5)P_4 binding sites [25,26]. Similar methodology has been used to prepare 1-*O*-alkyl and 1-*O*-acyl analogues of PIP$_3$ [62] and a P-5-tethered photoaffinity label of Ins(1,2,5,6)P_4 [63].

The Ferrier carbocyclisation was therefore potentially useful to prepare optically pure 2-deoxy-Ins(1,3,4,5)P_4 from an appropriate carbohydrate. To synthesise D-2-deoxy-Ins(1,3,4,5)P_4 the phosphorylation precursor L-1-*O*-benzyl-3-deoxy-*scyllo*-inositol **23** is required. In **23** all substituents are equatorial. This arrangement may be obtained from reduction of the ketone of a Ferrier product in which all substituents are equatorial (*i.e.* **24**) to the corresponding equatorial alcohol **25**, followed by removal of protecting groups R, which must be orthogonal with respect to benzyl ethers.Cyclohexanones of type **24** are obtained as the minor products of the mercury(II) reaction on D-*xylo*-hex-5-enopyranosides, ultimately derived from D-glucose. To prepare **5**, a methyl α-D-*xylo*-hex-5-enopyranoside derivative was required with a benzyl ether at position 2 and an alternative protecting group at positions 3 and 4. Benzoate esters are easily introduced in high yield, orthogonal with respect to benzyl ethers, compatible with the conditions of the Ferrier carbocyclisation and often facilitate interpretation of ^{1}H NMR spectra. They were therefore chosen for this route.

The initial route envisaged to prepare **5** stereospecifically from methyl α-D-glucopyranoside **26** is outlined in Scheme 10-1. This route involved benzylation of the 4,6-*O*-benzylidene derivative of **26**, selection of the 2-*O*-benzyl product **27** and conversion *via* **28** to methyl 3,4-di-*O*-benzoyl-2-*O*-benzyl-6-bromo-6-deoxy-α-D-glucopyranoside **29** using standard carbohydrate chemistry. Elimination of HBr from **29** would give the olefin **30**, which on Ferrier carbocyclisation should give a mixture of C-5 epimeric cyclohexanones, **31** and **32**. Inversion of stereochemistry at position 5 of the expected major product **32** and acylation would provide an intermediate with the stereochemistry and selective protection of positions 2, 3, 4 and 5 consistent with positions 4, 5, 6 and 1 respectively of Ins(1,4,5)P_3. This intermediate could potentially furnish several useful phosphorylated compounds. In particular, stereoselective equatorial reduction of the ketone followed by ester hydrolysis, phosphorylation and deprotection should give **5**.

10.2.1 Improved Preparation of Methyl 4,6-*O*-Benzylidene-α-D-Glucopyranoside and Methyl 4,6-*O*-Benzylidene-α-D-Manno-pyranoside

The benzylidene acetal is a commonly used protecting group for the 4,6-diol of hexopyranosides. Many methods have been described for the preparation of methyl 4,6-*O*-benzylidene-α-D-glucopyranoside **33**, including the reaction of methyl α-D-glucopyranoside **26** with zinc chloride-benzaldehyde [64,65] (and note an important modification [66]); with benzaldehyde dimethyl acetal in DMF in the presence of *p*-toluenesulphonic acid [67–69] or pyridinium *p*-toluenesulphonate [69] or tetrafluoroboric acid [70]; with benzaldehyde dimethyl acetal in 10% methanolic H_2SO_4 [71]; and with benzaldehyde diethyl acetal in DMF in the presence of HCl [72], or in dioxane in the presence of a strong cation exchange resin [73], or in chloroform in the presence of camphorsulphonic acid [74]. Although excellent yields are reported for most of these techniques, suitability to large-scale preparation (*i.e.* 0.5 mole or more) is rarely discussed. The zinc chloride-benzaldehyde method, which has been used on a 0.4 mole scale [66], is messy.

We applied the method of Evans [67,68] modified by Horton and Weckerle[75] to methyl α-D-glucopyranoside, using 1.05 molar equivalents of benzaldehyde dimethyl acetal on a 0.5 mole scale. The product was obtained in 90% yield as fine white needles by crystallisation from 2%$^{w/v}$ aqueous sodium hydrogen carbonate solution. This improved preparation worked well on a 0.5 mole scale and required neither dried solvent, nor purified reagents.

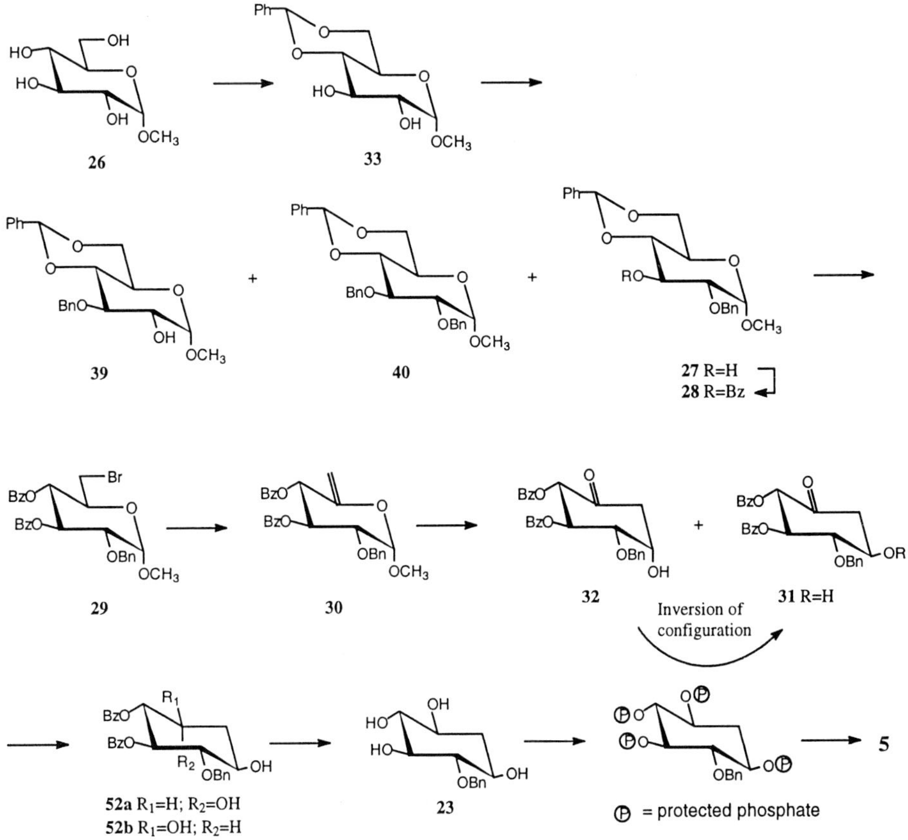

Scheme 10-1

This method was applied to methyl α-D-galactopyranoside **34** and methyl α-D-mannopyranoside **35**. The 4,6-*O*-benzylidene galactoside **36** could not be conveniently freed from the contaminating 3,4-*O*-benzylidene derivative. However, the mannoside isomer **37** was easily separated from the dibenzylidene derivative **38** by crystallisation and was obtained in 48% yield (100g scale) [76].

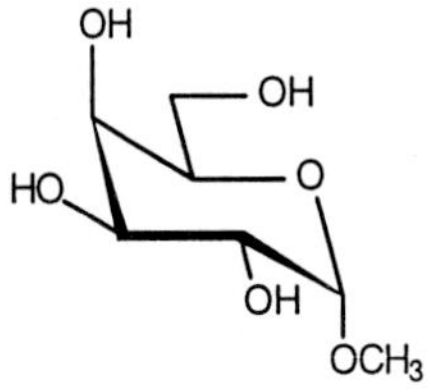

34 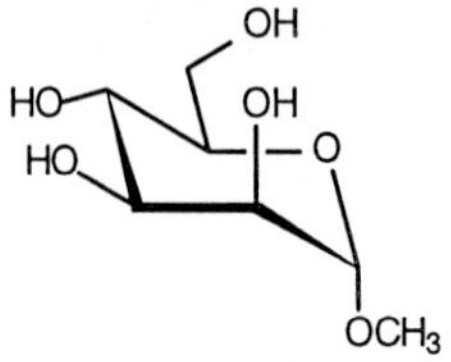 **35** 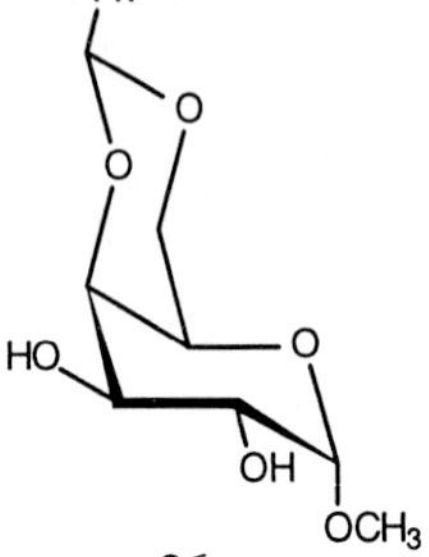 **36**

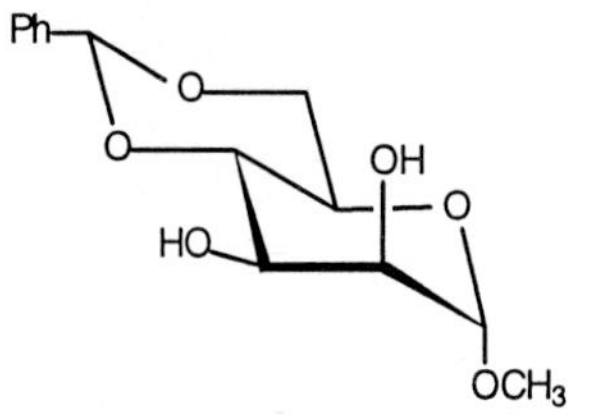

37 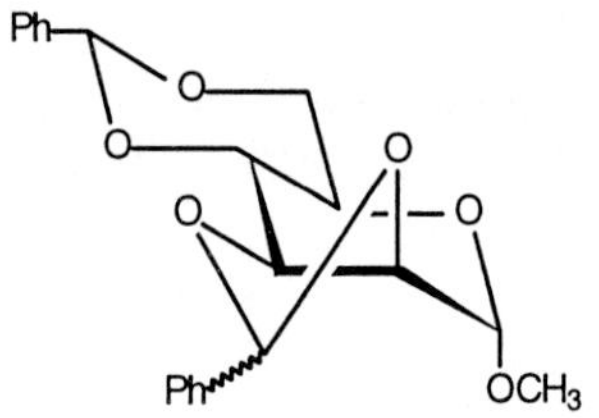 **38**

10.2.2 Benzylation of Methyl 4,6-*O*-Benzylidene-α-D-Glucopyranoside and Related Reactions

With **33** in hand, attention turned to preparing the 2-*O*-benzyl derivative **27**. Treatment of a solution of **33** in DMF with 1 equiv. of sodium hydride and 1.05 equiv. of benzyl bromide at room temperature gave 39% of **27**, together with its 3-substituted isomer **39** (8%) and a relatively large proportion of the disubstituted derivative **40** (22%). As the yield of **27** was rather disappointing, an alternative benzylation method was sought. The monobenzyl ethers **27** and **39** have been prepared with little or no disubstitution using 1.1 equivalents of sodium hydride in neat benzyl chloride [77], *via* phase transfer catalysis [78], copper and mercuric chelates [79], stannyl ethers [80] and stannylene acetals [80]. Of these, the last gave the highest reported yield of **27** and was selected for further investigation. After completion of this work, Dasgupta and Garegg [81] reported a refinement of the stannyl ether method providing the best selectivity for, and yield of, **27** yet described. The selective manipulation of hydroxyl groups using organotin derivatives has been reviewed [82-84]. The dibutylstannylene intermediate **41** was formed by boiling a mixture of **33** and dibutyltin oxide in toluene for 2–3 h with azeotropic removal of water. The residue from this process was then heated with benzyl bromide in dry DMF to give **27** (46%) and **39** (19%). No disubstitution occurred. Slightly improved selectivity (and yield of **27**) was observed when the reaction was carried out in refluxing acetonitrile in the presence of quaternary ammonium salts, a procedure employed by Corrie [85] to *o*-nitrobenzylate **33**.

41 R_1,R_2=Bu_2Sn
42 R_1=Bz; R_2=H
43 R_1=Ts; R_2=H
45 R_1=H; R_2=Bz

44

46

A claim by Boons *et al.* [86] that benzylation of **41** (obtained from **33** and dibutyltin dimethoxide) in toluene at 50°C gave exclusively the 3-substituted product **39** could not be reproduced. Instead, an approximate ratio of 2.7:1, with **39** as the minor product, was obtained [87]. In addition, treatment of the cooled solution of **41** with benzoyl chloride or tosyl chloride was found to give exclusively the 2-substituted product (**42** and **43** respectively) in either case, contrary to the above report [86]. Our results are consistent with those of a previous report [88], which employed dioxane as solvent. However, conditions that did reverse benzylation selectivity involved treating the stannylene **41** with benzyl bromide in dry DMF at room temperature in the presence of caesium fluoride [89], a mild method frequently employed in cyclitol chemistry. This procedure gave **27** and **39** in an approximate ratio of 1:2, a surprising result as it is well established that, in the absence of steric effects, the 2-hydroxyl group is the more reactive in methyl α-D-glucopyranoside derivatives [82, 90, 91].

Some generalisations have been observed in the properties of 2- and 3-*O*-substituted derivatives of **33**: thus, 2-substituted derivatives tend to be less polar by TLC, possess lower melting points, exhibit sharper OH stretching bands in their IR spectra and have smaller $^3J_{H,OH}$ values than their 3-substituted isomers. These could be useful qualitative pointers for preliminary identification of similar compounds.

Attempts to isolate **27** from the purified reaction mixture by *crystallisation* met with limited success, there being little consistency between successive experiments. However, after benzoylation of the purified mixture, the corresponding 3-benzoate ester **28** could be selectively obtained by fractional crystallisation from ethanol in 30–35% yield, a result reproducible on a 0.5 mole scale. Compound **28** was identified by its ^{1}H NMR spectrum, which displayed a deshielded one-proton triplet at 5.85 ppm corresponding to the methine at position 3.

10.2.3 Preparation of Methyl 3,4-Di-*O*-benzoyl-2-*O*-benzyl-6-deoxy-α-D-*xylo*-hex-5-enopyranoside 30

The conversion of 4,6-*O*-benzylidene hexopyranosides to 4-*O*-benzoyl-6-bromo-6-deoxyhexopyranosides by *N*-bromosuccinimide (NBS) in carbon tetrachloride containing excess barium carbonate was first described by Hanessian [92] and a modified procedure subsequently by Hullar and co-workers [93]. This transformation has been widely used in carbohydrate chemistry and works well on large scales [94]. Thus, a mixture of **28**, NBS (1.2 equiv.) and barium carbonate (1.5 equiv.) was heated under reflux in dry carbon tetrachloride. On purification, only 28% of the required bromodibenzoate **29** was obtained; the known methyl 3,4-di-*O*-benzoyl-6-bromo-6-deoxy-α-D-glucopyranoside [95] (**44**; 16%) and methyl 3-*O*-benzoyl-4,6-*O*-benzylidene-α-D-glucopyranoside [96] (**45**; 10%), representing *O*-debenzylated product and starting material respectively, were also obtained, together with unreacted starting material (16%).

A more efficient route from **28** to **29** required acidic hydrolysis of the benzylidene acetal to give methyl 3-*O*-benzoyl-2-*O*-benzyl-α-D-glucopyranoside **46**, followed by bromination with triphenylphosphine-carbon tetrabromide in THF and conventional benzoylation. Elimination of HBr from **29** to provide the Ferrier precursor **30** was first attempted using silver (I) fluoride in dry pyridine [97]. However, for complete reaction at least six equivalents of the expensive silver salt were required, and a poor yield was obtained (31% after crystallisation). Iodide exchange in acetone followed by treatment with DBU in refluxing toluene failed to effect complete conversion, but iodide exchange at 100°C in DMSO containing molecular sieves, followed by addition of DBU [98] worked efficiently, giving **30** in 63% yield after crystallisation. Olefin **30** crystallised readily from ethanol and was perfectly stable at room temperature in its crystalline form.

10.2.4 Ferrier Carbocyclisation of 30

When a stoichiometric amount of mercury (II) trifluoroacetate was added to a solution of **30** in aqueous acetone containing 1% acetic acid, the starting material was rapidly consumed. However, no clearly defined product was apparent until the addition of sodium chloride. On purification, the expected C-5 epimeric cyclohexanones **31** and **32** were isolated, in yields of 5% and 50% respectively. The proportion of minor product **31** in this case is much lower than in those reported for the corresponding 2,3,4-tri-*O*-benzyl (3:1) [44] and 2,4-di-*O*-allyl-3-*O*-benzyl (6:1) [58] D-*xylo*- precursors and the 2-*O*-benzyl-3,4-*O*-cyclohexylidene-L-*arabino*- precursor (2:1) [57]. It is noteworthy that lower proportions of minor product (if any) have been isolated in all previous examples in which esters, rather than ethers, have been employed at positions 3 and 4. This may indicate the subtle influence of protecting groups on the stereochemical outcome of the mercury (II)-catalysed carbocyclisation.

A catalytic Ferrier carbocyclisation of **30** using mercury (II) trifluoroacetate increased the overall yield obtained, but did not alter the product ratio. Neither did carbocyclisation using palladium (II) chloride [99,100], which had significantly increased the proportion of the minor product in a previous example [100].

10.2.5 Attempted Inversion of Stereochemistry at Position 5 of 32

Attention now turned to inversion of stereochemistry at position 5 of **32**; a literature search found no previous attempt at inverting this position of Ferrier products. Ferrier products readily undergo β-elimination to give β-enones [42,45,46,101]. To minimise this side reaction it was decided to protect **32** as ethylene acetal **47**. Ferrier and Haines [101] examined several methods for introducing an ethylene acetal to a related cyclohexanone and optimised their yield by reaction with ethylene glycol in dioxane-benzene with catalytic H_2SO_4. In the case of **32** this method (substituting toluene for benzene) provided 54% of **47**, but also 33% of the eliminated product **48**.

Mitsunobu conditions [102] [DEAD-Ph$_3$P or 1,1'-(azodicarbonyl)-dipiperidine–Bu$_3$P] failed to give any reaction, in keeping with previous inositol examples [103,104]. Attempted S_N2 inversions of the 5-*O*-triflate with caesium acetate in DMF [105], or of the 5-*O*-mesylate with caesium acetate in refluxing toluene in the presence of crown ethers [106] provided only 12–18% of the required acetate **49**, together with 50–55% of **48**. The tendency of **47** to eliminate despite protection of the ketone of **32** may be explained by the antiperiplanar arrangement of the axial H-6 proton and the leaving group at C-5.

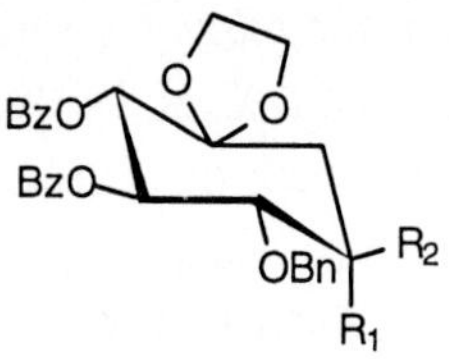

47 R$_1$=OH; R$_2$=H
49 R$_1$=H; R$_2$=OAc

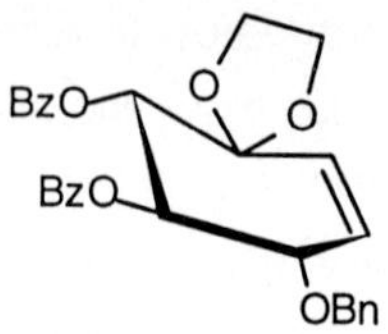

48

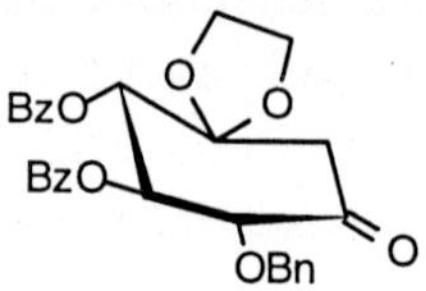

50

As the yield of inverted products using the above methodology was unsatisfactory, attention turned to an oxidation-reduction procedure which should avoid the elimination problem. Oxidation of **47** using oxalyl chloride and DMSO in dichloromethane [107] gave a highly crystalline, strongly dextrorotatory product in high yield which was not, however, the required product **50**, but enone **51**, i.e. elimination of the position 3 benzoate had occurred.

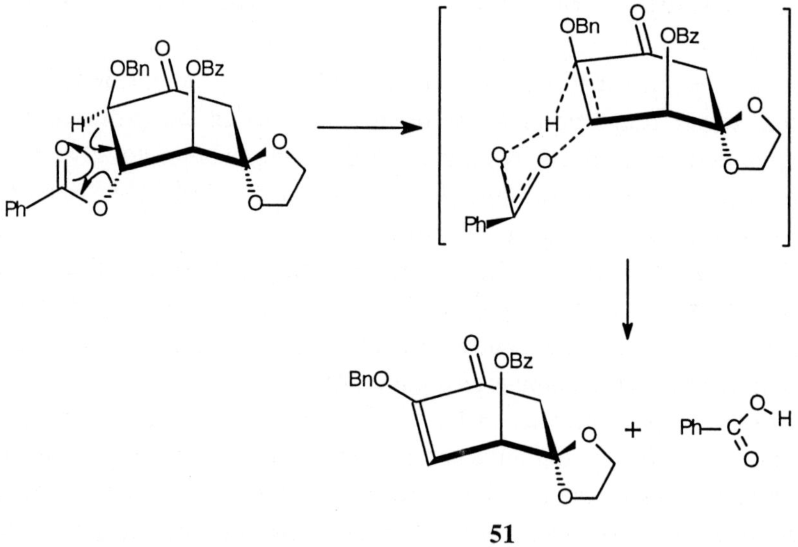

51

The driving force for this elimination is presumably the resultant conjugation, but the mechanism is less obvious. An antiperiplanar arrangement of the axial H-4 proton and the *cis*-C-3-benzoate cannot occur and triethylamine is therefore unlikely to abstract H-4. An alternative possibility is an entirely intramolecular mechanism involving anchimeric assistance from the C-3-benzoate to give a six-membered ring transition state. The required ketone **50** was eventually obtained by oxidation of **47** with the acidic pyridinium chlorochromate [108] Compound **50** was obtained in 53% yield by crystallisation from ethanol and further quantities (total 62%) were available by flash chromatography of the mother liquors. Reduction of **50** with sodium borohydride in dioxane or with sodium borohydride-cerium trichloride in methanol gave exclusively the axial alcohol **47**. Jaramillo and Martín-Lomas [109] experienced similar difficulty in attempting to reduce a cyclohexanone derived from D-*chiro*-inositol to a *myo*-inositol derivative. The only reagents which gave their required product in appreciable yield were (*R*)- and (*S*)-Alpine hydride. Reduction of **50** with (*R*)-Alpine hydride at −78°C in THF also gave only **47**. In a Dreiding model of **50** it is clear that equatorial approach of the reducing agent (leading to **47**) is much less hindered than axial approach. At this stage, inversion attempts were not pursued further and the minor Ferrier carbocyclisation product **31** was used to prepare the target tetrakisphosphate **5**.

10.2.6 Synthesis of D-2-Deoxy-*myo*-inositol 1,3,4,5 Tetrakisphos-phate 5

Conversion of **32** to the target tetrakisphosphate required reduction of the ketone to the equatorial alcohol, benzoate hydrolysis to provide tetrol **23**, phosphorylation and complete deblocking. Treatment of **32** with sodium borohydride in dioxane gave L-1,6-di-*O*-benzoyl-5-*O*-benzyl-3-deoxy-*myo*-inositol **52a** and L-1,2-di-*O*-benzoyl-3-*O*-benzyl-5-deoxy-*scyllo*-inositol **52b**, in a ratio of *ca.* 1:3.3 respectively (Scheme 10-1).

The benzoate esters of **52b** were smoothly removed with methanolic sodium hydroxide to give L-1-*O*-benzyl-3-deoxy-*scyllo*-inositol **23** in 86% yield. Phosphitylation of **23** was carried out at room temperature using the P(III) reagent bis(*p*-chlorobenzyloxy)(diisopropylamino)phosphine with 2 equivalents of phosphitylating reagent per hydroxyl group and 3 equivalents of tetrazole, in a small volume of dry dichloromethane. After the mixture of phosphitylating reagent and tetrazole had been stirred for 15min, the ^{31}P NMR spectrum indicated a peak at 126.2 ppm, corresponding to phosphitylating agent-tetrazolide. Thirty minutes after addition of the tetrol, the ^{31}P NMR spectrum indicated a complex overlapping pattern of phosphite signals at 139.7–140.0 and 141.5–142.1 ppm. The complexity of the spectrum arises from multiple $^5J_{PP}$ coupling [110]. After oxidation with MCPBA and work-up, the ^{31}P NMR spectrum revealed four phosphate signals at −1.59, −1.63, −1.68 and −2.15ppm.

Deblocking of the nine benzyl/*p*-chlorobenzyl protecting groups was achieved using sodium in liquid ammonia. The product was purified using ion-exchange chromatography and phosphate-containing fractions were detected and then quantified after combining using a modification of the Briggs phosphate assay [5]. The structure of the product was identified as the required tetrakisphosphate **5** on the basis of its ^{1}H-coupled ^{31}P NMR spectrum which showed four doublets, and its ^{1}H NMR spectrum in D$_2$O which showed the distinct methylene protons and a triplet corresponding to the ring methine geminal to the sole unphosphorylated alcohol at position 6 (*myo*-inositol numbering).

With the current interest in recently purified Ins(1,3,4,5)P$_4$-binding proteins, it is hoped that **5** will prove a useful ligand in determining the relative importance of the hydroxyl at position 2 of Ins(1,3,4,5)P$_4$ in ligand-protein binding. The full synthesis of **5** has recently been published [111] and its biological properties being evaluated.

10.3 Synthesis of Ring-contracted Analogues of Ins(1,4,5)P$_3$

All previously reported approaches to the structural modification of Ins(1,4,5)P$_3$ producing agonists have focused upon modifications at phosphorus or hydroxyl group deletion, reorientation, alkylation or replacement with isosteres and other groups in the six-membered ring. Despite numerous single and multiple modifications, the fundamental requirement of a six-membered ring had not been addressed. Even in the adenophostins (*vide infra*), which differ in many respects from Ins(1,4,5)P$_3$, the important 3,4-bisphosphate / 2-hydroxyl triad, analogous to the 4,5-bisphosphate / 6-hydroxyl arrangement of Ins(1,4,5)P$_3$ is contained within the six-membered pyranoside ring.

Since many studies demonstrated that positions 2 and 3 of Ins(1,4,5)P$_3$ are tolerant to extensive modification, it was reasoned that a contracted structure such as **53a**, obtained essentially by deletion of the 2-position carbon of Ins(1,4,5)P$_3$ with its associated hydroxyl group, should also fulfil the recognition requirements of the Ins(1,4,5)P$_3$ receptor. A carbohydrate based method to such a "pentagon IP$_3$"was therefore devised [112]. Many methods for converting carbohydrates into cyclopentane derivatives have been described [43], but the disclosure by Ito *et al.* [113] that treatment of methyl 2,3,4-tri-*O*-benzyl-6,7-dideoxy-α-D-*gluco*-hex-6-enopyranoside(1,5) **54** with zirconocene ("Cp$_2$Zr") followed by boron trifluoride etherate produced vinylcyclopentane **55** (see ref. 114 for a review) was of particular interest, as positions 1,2,3 and 4 of **55** possess the same relative stereochemistry as the equivalent positions in **53a**, and therefore as positions 4,5,6 and 1 respectively in Ins(1,4,5)P$_3$. It was decided to apply this methodology in an attempt to prepare (1*R*, 2*R*, 3*S*, 4*R*, 5*S*)-

3-hydroxy-1,2,4-trisphospho-5-vinylcyclopentane **53b**, an analogue of **53a** which would allow the viability of cyclopentane-based Ins(1,4,5)P$_3$ mimics to be assessed.

Trisphosphate **53b** would require a ring-contracted intermediate in which the protecting groups at positions 1 and 2 are different from that at position 3, to allow selective removal to provide the appropriate triol for phosphorylation. It was decided to employ *p*-methoxybenzyl ethers at positions 1 and 2, and a benzyl ether at position 3 (*i.e.* giving intermediate **56**), as this combination was as close as possible to the tribenzyl arrangement used by Ito *et al.* [113]. The carbohydrate precursor to **56** is heptoside **57**. The 5-vinyl moiety of **57** would be introduced by successive Swern oxidation and Wittig methylenation [115] of methyl 2-*O*-benzyl-3,4-di-*O*-(*p*-methoxybenzyl)-α-D-glucopyranoside **58**. We first examined methods to prepare primary alcohol **58**.

10.3.1 Preparation of Methyl 2-*O*-Benzyl-3,4-di-*O*-(*p*-methoxybenzyl)-α-D-Glucopyranoside 58

p-Methoxybenzylidene acetals have the property that either of the acetal C—O bonds may be selectively cleaved to furnish a *p*-methoxybenzyl ether. The direction of cleavage depends upon steric and electronic factors as well as upon the choice of cleavage reagent. Various methods [116,117] have been used to cleave the acetal of 2,3-disubstituted derivatives of methyl 4,6-*O*-(*p*-methoxybenzylidene)-α-D-glucopyranoside **59** resulting in selective formation of 4-*O*-(*p*-methoxybenzyl) ethers.

59 $R_1=R_2=H$
60 $R_1=Bn; R_2=H$
61 $R_1=H; R_2=Bn$
62 $R_1=Bn; R_2=PMB$
63 $R_1=R_2=Bn$
69 $R_1=R_2=CH_3$

64 $R_1=Bn; R_2=PMB; R_3=H$
65 $R_1=Bn; R_2=H; R_3=PMB$
66 $R_1=R_3=PMB; R_2=H$
67 $R_1=R_3=PMB; R_2=Bz$
68 $R_1=R_2=PMB; R_3=Bz$

70

The known [116] **59** was prepared in an analogous procedure to **33**, *i.e.* by reaction of methyl α-D-glucopyranoside with *p*-methoxybenzaldehyde dimethyl acetal at 70°C in DMF containing a catalytic quantity of *p*-toluenesulfonic acid, with continuous removal of methanol. Benzylation of **59** with dibutyltin oxide and benzyl bromide in acetonitrile in the presence of tetrabutylammonium iodide and 4Å molecular sieves gave a major and a minor product by TLC. These were easily separated by column chromatography and were identified as the 2-benzyl ether **60** (48%) and the 3-benzyl isomer **61** (11%). The major product **60** was smoothly *p*-methoxybenzylated using sodium hydride and *p*-methoxybenzyl chloride in DMF at room temperature, to furnish **62**. With **62** in hand, attention turned to cleavage of the acetal. When Johansson and Samuelsson [116] treated methyl 2,3-di-*O*-benzyl-4,6-*O*-(*p*-methoxybenzylidene)-α-D-glucopyranoside **63** with sodium cyanoborohydride and trimethylsilyl chloride in acetonitrile, they obtained the 4-*O*-substituted *p*-methoxybenzyl ether **64** in 76% yield, together with 13% of the 6-substituted isomer **65**.

When **62** was treated with these reagents, the chromatographically separable methyl 2-*O*-benzyl-3,6-di-*O*-(*p*-methoxybenzyl)-α-D-glucopyranoside **66** and the required 4-substituted isomer **58** were obtained in a ratio of *ca.* 1:1.7 . The structures of **66** and **58** were established by comparison of chemical shifts of their position 6 carbon atoms (identifiable by 135DEPT experiments), as observed in their ^{13}C NMR spectra in CDCl$_3$. The C-6 of **66** resonates at lower field (69.18ppm) than that of **58** (61.84ppm) due to the α-effect of alkylation [118]. These assigned structures were confirmed by preparation of benzoate esters **67** and **68**. The ^{1}H NMR spectrum of **67** displayed a deshielded triplet at

5.28ppm, corresponding to the position 4 methine; the position 6 methylene protons of **68** were deshielded to 4.46–4.97ppm.

As the selectivity of cleavage using the above method was disappointing, alternative conditions were explored. Joniak *et al.* [117] used LiAlH$_4$-AlCl$_3$ to convert methyl 4,6-*O*-(*p*-methoxybenzylidene)-2,3-di-*O*-methyl-α-D-glucopyranoside **69** exclusively to the 4-*O*-(*p*-methoxybenzyl) ether **70**. Reaction of **62** with LiAlH$_4$-AlCl$_3$ in refluxing THF gave exclusively the required product **58** in 73% yield. This result supports the suggestion [117] that a substituent at position 3 of the glucopyranoside ring hinders access of the reagent to the oxygen lone pair at position 4, thereby directing exclusive coordination to the oxygen at position 6. Cleavage of the acetal C—O-6 bond then results in exclusive formation of the 4-substituted ether. Similar effects have been reported for the cleavage of related 4,6-*O*-benzylidene acetals in 3-substituted glucose derivatives [119]. Why the electrophile in the Me$_3$SiCl-NaCNBH$_3$ experiment should coordinate to the position 4 acetal oxygen more easily than does the LiAlH$_4$-AlCl$_3$ reagent is not clear.

10.3.2 Preparation of (1*R*, 2*R*, 3*S*, 4*R*, 5*S*)-3-Hydroxy-1,2,4-trisphospho-5-vinyl Cyclopentane 53b

Swern oxidation of **58** with DMSO/oxalyl chloride gave aldehyde **71**. This compound was anticipated to be somewhat unstable, and so was quickly converted by Wittig methylenation to the required vinyl carbohydrate **57**. The ring contraction was carried out as described [113] where a vinyl carbohydrate is treated with Cp$_2$Zr(*n*-Bu)$_2$ (prepared *in situ* from zirconocene dichloride and 2 equivalents of *n*-butyllithium), followed by boron difluoride etherate in THF.

The authors propose a reaction mechanism based on an NMR study of the nine-membered zirconocycle intermediate, suggesting that the boron trifluoride etherate functions to accelerate the decomposition of this hemiacetal-zirconate by coordinating to the methoxy oxygen [113]. Elimination of the methoxyl group can give one of two alternative oxonium ion transition states, which can react intramolecularly to give the ring-contracted carbocycle. One of these transition states is of higher energy due an unfavourable steric interaction involving the protecting group at position 4 of the starting vinylpyranoside and a cyclopentadiene ring. The diastereoselectivity of the ring contraction is thought to originate in this energy difference. The stereochemistry of the new chiral centres in the product is thus strongly influenced by the stereochemistry at position 4 and the nature of the substituent. In this respect it was expected that a *p*-methoxybenzyl protecting group at this position would exert a similar influence to a benzyl group. One complication we encountered, however, was that gradual loss of *p*-methoxybenzyl protecting groups after the addition of boron trifluoride etherate became a competing reaction. Nevertheless, the desired vinylcyclopentane **56** was obtained in fair (46%) yield as a waxy solid, together with a small amount of the kinetically disfavoured product **72**, which was crystalline. In the event, the isolation of this unwanted diastereoisomer **72** was fortuitous because, with both compounds available, it was possible to compare the two. As might be expected, the two diastereoisomers had nearly identical IR spectra, but very different NMR properties (and optical rotations). The relative stereochemistries of **72** and of **56** were confirmed by phase-sensitive 2D-NOESY and NOE difference NMR spectroscopy.

Removal of the *p*-methoxybenzyl protecting groups from **56** gave the triol **73**. Molecular modelling of this rather simple structure yielded predictions of the vicinal coupling constants between protons in the cyclopentane ring. These agreed well with those measured from the 400 MHz ^{1}H NMR spectrum of **73**, further confirming the stereochemistry deduced at the previous stage. Phosphitylation using bis(benzyloxy)diisopropylaminophosphine, followed by oxidation of phosphites with *m*-chloroperoxybenzoic acid (MCPBA) gave the fully protected trisphosphate **74**. ^{31}P NMR spectroscopy of the intermediate trisphosphite triester showed an unusually high $^5J_{PP}$ coupling of 6.7 Hz [*cf.* 2.9 Hz and 3.4 Hz for precursors of Ins(4,5)P$_2$ [110] and Ins(1,4,5)P$_3$ [120] respectively], presumably reflecting the altered geometry of the P(III)-P(III) interaction in a five-membered ring. Deprotection using sodium in liquid ammonia removed the seven benzyl protecting groups, leaving the vinyl group intact. The logic of retaining this unusual feature in the target molecule was to produce a five-membered ring Ins(1,4,5)P$_3$ analogue containing the essential recognition elements in the minimum number of synthetic steps. Should this prototype prove to be active, the vinyl group of **56** would provide a convenient starting-point for a range of modifications at this position, while maintaining the desired stereochemistry. Purification by ion-exchange chromatography of the crude product on Sepharose Q Fast Flow resin gave the target trisphosphate **53b**, which was isolated as the triethylammonium salt and quantified by phosphate assay.

Trisphosphate **53b** was examined for Ca^{2+} mobilising activity at the platelet Ins(1,4,5)P$_3$ receptor using fluorescence techniques, and also using saponin-permeabilised platelets loaded with ^{45}Ca^{2+}. It was found to be a full agonist, although with an EC$_{50}$ some 65-fold higher than Ins(1,4,5)P$_3$. The effect was inhibited by addition of heparin, and **53b** was also active in Jurkat T-lymphocytes. These results demonstrated for the first time that Ins(1,4,5)P$_3$ receptor mediated Ca^{2+} mobilisation does not necessarily require a cyclohexyl (or equivalent) structural motif. A smaller ring phosphate which retains crucial recognition elements of Ins(1,4,5)P$_3$, *i.e.* three appropriately orientated phosphates and a pseudo-6-hydroxyl group, can still exhibit agonistic activity.

10.3.3 Preparation of (1*R*, 2*R*, 3*S*, 4*R*, 5*S*)-3-Hydroxy-5-hydroxymethyl-1,2,4 trisphospho Cyclopentane 75

Noting that DL-3-*O*-ethyl-Ins(1,4,5)P$_3$ and DL-3-*O*-propyl-Ins(1,4,5)P$_3$ both have EC$_{50}$ values of greater than 100μM in SH-SY5Y neuroblastoma cells [*cf.* Ins(1,4,5)P$_3$ 0.18μM] [121], it seemed reasonable that replacement of the vinyl group of **53b** with a hydroxyl-containing side chain should markedly increase its potency. An obvious target was **75**, in which the vinyl substituent is replaced by hydroxymethyl. Initially it was intended to prepare **75** by *p*-methoxybenzylation of **56** followed by treatment with OsO$_4$–NaIO$_4$ and reduction of the intermediate aldehyde to give **76**. Benzylation of the primary alcohol followed by removal of *p*-methoxybenzyl ethers would furnish triol **77**, which on phosphorylation and deprotection would provide **75**. However, a recent report by Chénedé *et al.* [122] offered the possibility of a simpler route. These workers treated aldehyde **78** with samarium (II) iodide in the presence of *t*-butanol and HMPA, and obtained cyclopentane **79**. The structure of **79** was established by conversion to known compounds, and a mechanism for the ring contraction has been proposed [122]. Applying this carbocyclisation to aldehyde **71** should give **80**. Benzylation of the primary hydroxyl group and acidic hydrolysis should furnish triol **77**, which could be elaborated to **75**.

73 R_1=H; R_2= CH=CH$_2$
74 R_1=P(O)(OBn)$_2$; R_2= CH=CH$_2$
76 R_1=PMB; R_2=CH$_2$OH
77 R_1=H; R_2=CH$_2$OBn
83 R_1=P(O)(OBn)$_2$; R_2=CH$_2$OBn

75

78 R=Bn
71 R=PMB

79 R=Bn
80 R=PMB

81 R_1=Bn; R_2=H
82 R_1=H; R_2=Bn

Aldehyde **71** was prepared by Swern oxidation of **58** as described above. Although sufficiently stable to allow full characterisation, **71** gradually rehydrated after standing in air for several days, and was therefore best prepared freshly. Treatment of **71** with samarium (II) iodide in THF in the presence of *t*-butanol and HMPA with rigorous exclusion of air and moisture gave two products after 1 h. These were identified as alcohol (**58**; 19%), arising from reduction of the aldehyde, and (**80**; 37%).

Benzylation of **80** at 0°C with 1 equiv. sodium hydride and 1.1 equiv. benzyl bromide in dry DMF gave a mixture of **81** and **82** in a ratio of 3:2 respectively. The structures of **81** and **82** were distinguished between by comparison of chemical shifts of the methylene carbon atoms of the C-5 side chain in the ^{13}C NMR spectra: the α-effect of alkylation causes this signal in the spectrum of **82** (67.9 ppm) to be deshielded relative to that of **81** (60.9 ppm). Acidic hydrolysis of **82** with 2:1 ethanol–1M aqueous HCl provided triol **77** in 72% yield. Phosphitylation of **77** with tetrazole-activated bis(benzyloxy) (diisopropylamino)phosphine in dichloromethane provided a trisphosphite intermediate, the ^{31}P NMR spectrum of which exhibited a singlet at 140.3 ppm and an AB system at 140.2 and 139.9 ppm with the characteristically large $^5J_{PP}$ coupling constant of 7.0 Hz. Oxidation of this intermediate with MCPBA provided the fully protected intermediate **83**. Although compound **83** contained some non-phosphorylated impurities, the crude material was deprotected using sodium in liquid ammonia. The target trisphosphate **75** was obtained pure as its triethylammonium salt after ion-exchange chromatography.

The ^{1}H-coupled ^{31}P NMR spectrum of **75** comprised three doublets, confirming the presence of three ring phosphates. The ^{1}H NMR spectrum of **75** contained a high-field quintet at 2.29ppm corresponding to the methine at position 5; an ABX system at 3.65 and 3.71ppm corresponding to the methylene protons; a triplet at 3.94ppm corresponding to the methine at position 3; and a deshielded three-proton multiplet corresponding to the ring methines geminal to phosphate esters.

Preliminary evaluation of the Ca^{2+} mobilising activity of **75** in Jurkat T-lymphocytes has demonstrated an EC_{50} value only –4-fold higher than that of Ins(1,4,5)P$_3$ itself (D. J. Jenkins, A. M. Riley, B. V. L. Potter, *J. Org. Chem.*, **1996**, in press).

10.4 Synthesis of a Carbohydrate-based Inositol Polyphosphate Mimic Based on Adenophostin A

Although most of the active inositol polyphosphates and related compounds tested for Ca^{2+} release at the Ins(1,4,5)P$_3$ receptor during 1987–93 had been full agonists, few exhibited a potency comparable to the natural ligand. However, in late 1993 a Japanese group reported the isolation of two potent trisphosphates from a culture broth of *Penicillium brevicompactum* [123]. These were named adenophostins A and B, and identified as 3'-*O*-(α-D-glucopyranosyl)-adenosine-2',3",4"-trisphosphate **84** and its 6"-*O*-acetyl derivative **85** respectively [124]. The structure of **84** has recently been confirmed by total synthesis [125].

The adenophostins have been demonstrated to be heparin-sensitive full agonists in rat cerebellar microsomes, with potencies 100-fold higher than Ins(1,4,5)P$_3$ [EC_{50} values 1.4nM **84** and 1.5nM **85**; *cf.* 170nM Ins(1,4,5)P$_3$] [126], these relative potencies are consistent with binding data [126]. In another study involving the purified Ins(1,4,5)P$_3$ type 1 receptor, **85** was found to be 10-fold more potent than Ins(1,4,5)P$_3$ [EC$_{50}$ values 11nM **85**; *cf.* 100nM Ins(1,4,5)P$_3$] [127]. In this study, **85** demonstrated a positive cooperativity in binding to the Ins(1,4,5)P$_3$ receptor not exhibited by Ins(1,4,5)P$_3$. Both **84** and **85** are resistant to the metabolic enzymes 5-phosphatase and 3-kinase and, as expected, produce a sustained Ca^{2+} release [126].

The high potency of the adenophostins is intriguing, as they bear little apparent resemblance to Ins(1,4,5)P$_3$. However, the structures of **84** and **85** show consistencies with many of the features known to be important for agonism. The glucose 3,4-bisphosphate moiety possesses D-*threo* stereochemistry and the position 2 hydroxyl group may be regarded as analogous to position 6. Indeed, molecular modelling studies [126,128] demonstrate similarity of positions 4, 3 and 2 of **84** with positions 4, 5 and 6 respectively of Ins(1,4,5)P$_3$. In addition, the adenophostins possess a third phosphate, which, similarly to Ins(1,4,5)P$_3$, is essential for high potency: 3'-*O*-(α-D-glucopyranosyl)-adenosine-3",4"-bisphosphate **86**, in which this third phosphate is removed, possessed a 1000-fold lower binding affinity than **84** [126]. However, although the broad basis for the activity of **84** and **85** is clear, a full structural rationalisation for their exceptional potency is lacking. An interesting combination of compounds to compare, for example, would be **84**, methyl 3-*O*-(α-D-glucopyranosyl)-β-D-ribofuranoside-2,3',4'-trisphosphate **87** and (2-hydroxyethyl) α-D-glucopyranoside 2',3,4-trisphosphate **88**. Such a study would establish the relative importance of the adenine and adenosine components of **84**. The work described here represents initial attempts at preparing selectively protected glucose-based intermediates for synthesis of adenophostin A and related targets, and at preparing glucose-based polyphosphates. Adenophostin A synthesis requires deprotection of a fully protected, phosphorylated intermediate such as **89**, derived from triol **90**. Triol **90** could be obtained from a derivative in which positions 2', 3" and 4" are protected with a group removable in the presence of benzyl ethers, benzamides and glycosidic linkages. The *p*-methoxybenzyl ether fulfils this requirement. 6-*N*-Benzoyl-2'-*O*-(*p*-methoxybenzyl)-adenosine **91** is known [129] and benzylation of the primary 5'-hydroxyl in favour of the secondary 3'-hydroxyl to give **92** ought to be straightforward. Selective coupling of **92** to position 1 of 2,6-di-*O*-benzyl-3,4-di-*O*-(*p*-methoxybenzyl)-D-glucopyranose **93** to provide the α-anomeric derivative **94** should be possible using trichloroacetimidate [130] methodology commonly employed in oligosaccharide synthesis.

89 R=P(O)(OBn)$_2$
90 R=H
94 R=PMB

93 R$_1$=PMB; R$_2$=H
95 R$_1$=Bn; R$_2$=H
96 R$_1$=Bn; R$_2$=All
100 R$_1$=H; R$_2$= α-Me
105 R$_1$=H; R$_2$= α-All
106 R$_1$=Bz; R$_2$= α-All
108 R$_1$=PMB; R$_2$= α-All
109ab R$_1$=PMB; R$_2$= α-prop-1-enyl
110 R$_1$=H; R$_2$= α-CH$_2$CH$_2$OH
111 R$_1$=P(O)(OBn)$_2$; R$_2$= α-CH$_2$CH$_2$OP(O)(OBn)$_2$

91 R=H
92 R=Bn

The β-trichloroacetimidate derivative of **93** could also act as a glycosyl donor to other alcohols, such as 2-(*p*-methoxybenzyloxy)-ethanol, giving a precursor to **88**, or methyl 5-*O*-benzyl-2-*O*-(*p*-methoxybenzyl)-β-D-ribofuranoside, giving a precursor to **87**. Clearly, compound **93** is an important intermediate and, as part of an effort to synthesise adenophostin A analogues, we first sought a method to prepare **93**

10.4.1 Synthesis of 2,6-Di-*O*-benzyl-3,4-di-*O*-(p-methoxybenzyl) -D-Glucopyranose 93

Preparation of 2,3,4,6-tetra-*O*-benzyl-D-glucopyranose **95** from the corresponding allyl glycosides has been described [131]. Allyl ethers have been widely used as protecting groups in carbohydrate and inositol chemistry because, while relatively stable themselves, they can be induced to isomerise to prop-1-enyl ethers by a variety of reagents [132, 133]. These enol ethers are hydrolysed by mildly acidic (and other) conditions. Thus, Gigg and Gigg prepared an anomeric mixture of allyl 2,3,4,6-tetra-*O*-benzyl-D-glucopyranosides **96**, the allyl groups were isomerised using potassium *t*-butoxide in dry DMSO at 100°C for 15 min and **95** was obtained by refluxing a solution of the products in 9:1 acetone-0.1M aqueous HCl for 30 min [131]. Using essentially this methodology, allyl ethers have been removed in the presence of *p*-methoxybenzyl ethers [134], making this an ideal route for the preparation of **93**.

97ab 7:3 α, β anomeric mixture
97a α-anomer

101

102 R=Bz
103 R=H
104 R=Bn

98 R=CH$_2$OH
99 R=CH=CH$_2$

107

Selective protection of positions 2 and 6 of the allyl glycoside(s) is required in this route. As position 2 bears the most reactive secondary hydroxyl group in α-alkyl glycosides, whereas position 3 bears the most reactive in the β-isomers [82,90,91] the preparation of allyl α-D-glucopyranoside **97a** was required. In addition, use of the α-allyl glycoside potentially allowed preparation of (2-hydroxyethyl) α-D-glucopyranoside 2',3,4-trisphosphate **88** directly: Ferrier and Stütz [52] had prepared **98** from **99** by reaction of the latter with OsO$_4$–NaIO$_4$ followed by reduction of the intermediate aldehyde with sodium borohydride. It followed that an allyl glycoside ought to be similarly convertible to a (2-hydroxyethyl) glycoside.

Glycoside **97a** was first obtained pure by reaction of D-glucose with dry allyl alcohol containing dissolved HCl gas (*i.e.* a classical Fischer glycosidation) followed by dry acetone extraction and fractional crystallisation [135] although a recent report [136] found this general method unsatisfactory. A more commonly used method is a modified Fischer glycosidation in which a mixture of D-glucose, allyl alcohol and a strong ion-exchange resin is heated under reflux [136–138]. When the latter procedure was used, an orange syrup was obtained after cooling, filtration and evaporation. In agreement with Lee and Lee [137] no crystallisation could be induced at this stage. Column chromatography of this syrup gave a white solid **97ab**. ^{1}H NMR spectroscopy of **97ab** in D$_2$O revealed a *ca.* 7:3 α:β anomeric mixture of allyl glycosides, estimated from the integral ratio of the anomeric protons, resonating at 4.92ppm

(*J* 3.7Hz) and 4.46ppm (*J* 7.9Hz) respectively. A single crystallisation from ethyl acetate-ethanol typically gave a 9:1 α:β mixture. Fractional crystallisation gave the pure α-anomer **97a**, but only in poor yield (14%).

Attention now turned to selective protection of positions 2 and 6. Methyl 2,6-di-*O*-benzyl-α-D-glucopyranoside **100** was produced in 30% yield when methyl α-D-glucopyranoside was treated with 1.5 equiv. of bis(tributyltin) oxide, followed by neat benzyl bromide [118], presumably by formation of the 2,6-bis(tributylstannyl) diether. A possible method of disubstitution of **97a** would therefore be to react it with 2 equiv. of bis(tributyltin) oxide and to benzylate it as above. An alternative possibility was to use stannylene acetals.

Reaction of methyl α-D-glucopyranoside with 1 equiv. of dibutyltin oxide results in formation of the 2,3-*O*-dibutylstannylene derivative, with selective esterification [88] or alkylation [139] occurring at position 2. In addition, benzyl 2,3-di-*O*-benzyl-α-D-glucopyranoside has been regioselectively methoxymethylated at position 6 *via* its 4,6-*O*-dibutylstannylene derivative [140]. Therefore, it seemed reasonable that reacting an alkyl glucopyranoside with more than 2 equiv. of dibutyltin oxide ought to result first in the formation of the 2,3-*O*-dibutylstannylene, followed by the 2,3:4,6-di-*O*-dibutylstannylene. The latter derivative of an α-anomer would be expected to direct substitution at positions 2 and 6. As the α-anomer **97a** had been more difficult than expected to isolate from the glycosidation reaction, an ideal situation would be the isolation of a 2,6-disubstituted α-anomeric derivative directly from **97ab**. Such methodology would be potentially generally useful to carbohydrate chemists and consequently, both esterification and alkylation were attempted using a bis-stannylene approach. When **97ab** was reacted with 2.5 equiv. of dibutyltin oxide in toluene and the reaction mixture cooled, a precipitate formed which could not be redissolved in toluene or dioxane. Hoping that gradual dissolution and reaction would occur, this suspension (in toluene) was stirred for several days with 2.1 equiv. of benzoyl chloride, but no reaction occurred. After completion of these studies, Qin and Grindley [141] noted the poor solubility of methyl 2,3:4,6-di-*O*-dibutylstannylene-α-D-glucopyranoside in chloroform and suggested oligomerisation.

The stannylation product of **97ab** with 1.05–1.2 equiv. of dibutyltin oxide did not precipitate on cooling. Treatment of the cooled solution with 2.1 equiv. of benzoyl chloride gave a mixture of products by TLC, from which the known [138] allyl 2,6-di-*O*-benzoyl-α-D-glucopyranoside **101** was isolated in 34% yield by a combination of column chromatography and crystallisation. Presumably the 2,3-*O*-dibutylstannylene derivative of the α-anomer formed, was soluble in cold toluene and directed substitution at position 2, while the primary hydroxyl was selectively benzoylated over the remaining free secondary hydroxyl at position 4 as would be expected. This two-step preparation of **101** from D-glucose represents an improvement on the five steps of a previous report [138]. Reaction of **101** with 2-methoxypropene gave fully protected **102**. It is noteworthy that best yields of 3,4-O-isopropylidene derivatives were obtained with a short reaction time. Subsequently, a variation of this method was described on the enantiomer of **101** using THF as solvent over a longer period [142]. The benzoate esters of **101** were easily replaced with benzyl ethers in two-steps: basic methanolysis provided diol **103** which was benzylated with sodium hydride and benzyl bromide in DMF to provide allyl 2,6-di-O-benzyl-3,4-O-isopropylidene-α-D-glucopyranoside **104**.

Attention now turned to potential direct benzylation. Alkylation is carried out at higher temperatures than esterification, at which the bis stannylene intermediate might be soluble. As benzylation tends to be less selective than benzoylation, the pure α-anomer **97a** was used initially to test the viability of this method. Treatment of the stannylated product with benzyl bromide in refluxing toluene or acetonitrile (both containing quaternary ammonium salts) resulted in sluggish reactions providing many products by TLC, none of which dominated and which were not completely separable by column chromatography. However, when stirred with neat benzyl bromide at 100–110°C, the stannylated product dissolved within one hour and after two days a major product and several minor products (of similar R_f value) were obtained. After work up and column chromatography to remove excess benzyl bromide, attempts were made to crystallise the major product. Crystallisation in diisopropyl ether yielded a single dibenzyl derivative in 44% yield (two crops). This was identified as the required allyl 2,6-di-O-benzyl-α-D-glucopyranoside **105** by preparation of its 3,4-dibenzoate **106**, the ^{1}H NMR spectrum of which displayed deshielded triplets at 5.52 and 5.98ppm corresponding to H-4 and H-3 respectively, and by reaction with 2-methoxypropene to provide **104**.

In the hope that further quantities of **105** could be obtained as the less polar **104**, the mother liquor of the benzylation mixture was reacted with 2-methoxypropene. A single new product was formed, which was easily isolated by flash chromatography. This compound was not **104**, however, but a dibenzyl derivative possessing an isopropylidene acetal in a 1,3-dioxane chair conformation, as apparent from the characteristic [143] chemical shifts of the isopropylidene axial methyl (19.18ppm), equatorial methyl (29.22ppm) and quaternary (99.34ppm) carbons in the ^{13}C NMR spectrum in CDCl$_3$. Its structure was therefore assigned as allyl 2,3-di-*O*-benzyl-4,6-*O*-isopropylidene-α-D-glucopyranoside **107**, thereby demonstrating that all of **105** had been isolated by crystallisation. With appropriate benzylation methodology and a convenient isolation procedure established, the technique was attempted on **97ab** on a 35g scale. In this case **105** could still be crystallised from the (more complicated) product mixture, but only in poor yield (*ca.* 15%). Reaction of the mother liquor with 2-methoxypropene in this case gave a mixture of several acetal-containing species which could not be separated further. Although the yield of crystalline **105** from the anomeric mixture was rather disappointing, this method has been found to be the most convenient overall to produce it in multigram quantities.

While this work was in progress, Qin and Grindley [141] reported similar benzylation selectivity on methyl α-D-glucopyranoside, confirming the general applicability of this method. In their studies, conducted on a 1mmol scale, only the 2,6-dibenzyl derivative was obtained, and an improved yield was reported when using only 1.5 equiv. of dibutyltin oxide. This has not been attempted on **97a**. Another interesting result was the product mixture obtained from methyl β-D-glucopyranoside: the 2,6-di-*O*-benzyl derivative was the major product (56%) and not the 3,6-di-*O*-benzyl derivative (16%). This selectivity is the reverse of that observed with methyl 4,6-*O*-benzylidene β-D-glucopyranoside which gave a *ca.* 1:2 ratio of the 2- and 3-*O*-benzyl ethers [144], although it is consistent with the ratio of monoalkylated products from use of 1 equiv. of dibutyltin oxide with methyl β-D-glucopyranoside [139]. The conversion of **105** to target **93** was achieved in three steps. *p*-Methoxybenzylation of **105** with sodium hydride and *p*-methoxybenzyl chloride gave fully protected **108**, which was isomerised to the corresponding prop-1-enyl glycoside **109ab** using potassium *t*-butoxide in dry DMSO. Finally, hydrolysis with refluxing 10:1 acetone-1M aqueous HCl for 20 min gave **93** in 77% yield after crystallisation.

10.4.2 Synthesis of (2-Hydroxyethyl)-2',3,4-Trisphosphate-α-D-Glucopyranoside 88

Attention now turned to direct preparation of **88**, which represents the derivative of adenophostin A in which most of the adenosine moiety has been deleted. As both benzyl ethers and isopropylidene acetals should be stable to osmium tetraoxide, **104** was chosen as an appropriate selectively protected intermediate for the synthesis of **88**. Reaction of **104** with 0.3 equiv. of osmium tetraoxide and excess sodium metaperiodate for 5 h produced a polar product which was not isolated but reduced with sodium borohydride to furnish (2-hydroxyethyl)

2,6-di-*O*-benzyl-α-D-glucopyranoside **110** in 56% overall yield from **104**. The structure of **110** was assigned chiefly on the basis of its ^{13}C NMR spectrum, which showed loss of isopropylidene and allyl signals and the presence of methylene carbons at 61.47 and 70.28 ppm, typical [145] of the 2' and 1' positions respectively of an α-(2-hydroxyethyl) glucoside.

Loss of the isopropylidene acetal in this reaction was unexpected. It could not have occurred in the presence of sodium metaperiodate, as the resulting diol would have been oxidised to a dialdehyde. Noting the higher solubility of osmium tetraoxide in organic solvents compared to water [146] it presumably partitioned into the organic layer during purification of the intermediate aldehyde and was subsequently reduced to osmic acid (H_2OsO_4), an acid sufficiently strong to remove the labile *trans* ketal. Nevertheless, the loss of the acetal was advantageous, as it directly provided the triol required for phosphorylation. Phosphorylation of **110** with tetrazole-activated bis(benzyloxy) (diisopropylamino)phosphine gave a trisphosphite triester, in which the vicinal bisphosphite presented as an AB quartet and the primary phosphite as a singlet in the ^{31}P NMR spectrum. After oxidation with MCPBA and purification, the ^{1}H-coupled ^{31}P NMR spectrum of the product **111** exhibited a septet corresponding to the phosphorylated primary alcohol and two sextets corresponding to the protected ring phosphates. Deprotection of **111** with sodium in liquid ammonia, followed by purification by ion-exchange chromatography, provided the required trisphosphate **88**, isolated as the triethylammonium salt and quantified by Briggs phosphate assay. The full synthesis of **88** has recently been published [147].

Trisphosphate **88** has also been prepared by Gigg and co-workers from D-galactose [148,149]. We examined **88** for Ca^{2+}-mobilising activity at the platelet Ins(1,4,5)P$_3$ receptor. It was found to be a full agonist with a potency *ca.* 10-fold lower than Ins(1,4,5)P$_3$ [EC$_{50}$ 0.6μM; *cf.* Ins(1,4,5)P$_3$ 0.05μM], a result consistent with binding data and with another study in SH-SY5Y neuroblastoma cells [128]. The sample of **88** prepared by Gigg *et al.* was found to be resistant to hydrolysis by 5-phosphatase; it was not turned over by 3-kinase, being neither a substrate nor an inhibitor of this enzyme.

Two related studies are of interest to this discussion. First, DL-6-deoxy-6-hydroxymethyl-*scyllo*-inositol-1,2,4-trisphosphate **112** demonstrated a Ca^{2+}-mobilising potency approximately equal to that of Ins(1,4,5)P$_3$ at the platelet Ins(1,4,5)P$_3$ receptor [EC$_{50}$ 0.1μM; *cf.* Ins(1,4,5)P$_3$ 0.11μM] [150]. Assuming that one enantiomer is inactive, and noting that L-*scyllo*-Ins(1,2,4)P$_3$ has approximately equal Ca^{2+}-mobilising potency to Ins(1,4,5)P$_3$, implies that the

CH$_2$OH component is at least tolerated by the Ins(1,4,5)P$_3$ receptor, and may even give rise to a modest increase in potency. The observation that **88** is less active than either **112** or Ins(1,4,5)P$_3$ suggests that the conformationally rather mobile 2'-phosphate group of **88** is not a good mimic of the 2'-phosphate in adenophostin A, nor of the 1-phosphate of Ins(1,4,5)P$_3$. Thus, all or part of the adenosine moiety in adenophostin A may be necessary to orientate the 2'-phosphate group in a particularly favourable way at the receptor binding site.

The second finding of interest is the relative Ca^{2+}-mobilising potencies of a series of both α and β-anomeric (2-hydroxyethyl) and (3-hydroxypropyl) xylopyranoside trisphosphates [152]. It is noteworthy that **113**, which represents the derivative of **88** in which the hydroxymethyl group has been deleted, and which of the four xylopyranosides most closely resembles the adenophostins, is the most potent. Increasing the aglycon chain length approximately halved potency, while inversion of configuration at the anomeric centre reduced it approximately four-fold. Given that the hydroxymethyl group of **88** may engender a favourable ligand-receptor interaction (*vide ultra*), it would be interesting to compare **88** and **113** in the same system.

10.4.3 Preparation of a Fluorescent Label Based upon (2-Hydroxy-ethyl)-2',3,4-trisphosphate-α-D-Glucopyrano-side

The metabolic resistance of **88** made it an ideal compound onto which to attach a fluorescent label. Of particular interest was the possibility of preparing a probe to carry out fluorescence resonance energy transfer (FRET) experiments to study the purified Ins(1,4,5)P$_3$ receptor. It was reasoned that, as the 3,4-bisphosphate/2-hydroxy triad of **88** is essential for binding, the primary phosphate would be the ideal place onto which to append the label. Collaborators had previously used the BODIPY FL label, and consequently it was decided to react iodoacetamide **114** (Scheme 10-2) with (2-hydroxyethyl) α-D-glucopyranoside 3,4-bisphosphate-2'-phosphorothioate **115** in order to prepare the required **116**. Allyl 2,6-di-*O*-benzyl-α-D-glucopyranoside **105** was phosphitylated with tetrazole-activated bis(*p*-chlorobenzyloxy)(diisopropyl-amino)-phosphine, then oxidised with MCPBA to give fully protected **117**. The *p*-chlorobenzyl phosphate diester was chosen because of its increased acid stabilty compared to the benzyl diester [153]. This was considered important as the following step had produced acid strong enough to remove an isopropylidene acetal (*vide ultra*). Intermediate **117** was converted to the corresponding (2-hydroxyethyl) glycoside **118** using OsO$_4$–NaIO$_4$ followed by sodium borohydride as described above. In this case a much longer reaction time was required and the intermediate aldehyde was purified by flash chromatography in an attempt to remove osmium tetraoxide. Compound **118** was phosphitylated with tetrazole-activated bis(p-chlorobenzyloxy) (diisopropylamino)phosphine in the usual way. The solvents were then evaporated and the phosphite was treated with elemental sulphur in DMF-pyridine (2:1) [154] to give **119**, the sulphoxidation being complete within 30 min. The ^{1}H-coupled ^{31}P NMR spectrum of **119** in CDCl$_3$ exhibited the two protected ring phosphates as sextets at −2.33 and −1.89 ppm, and the protected phosphorothioate as a septet at 68.7 ppm. Deprotection of **119** with sodium in liquid ammonia provided **115**, which was purified by ion-exchange chromatography and isolated as its triethylammonium salt. The ^{1}H-coupled ^{31}P NMR spectrum of **115** showed a triplet at 45.5 ppm, corresponding to the primary phosphorothioate and two doublets at 1.43 and 0.94 ppm, corresponding to the ring phosphates [76].

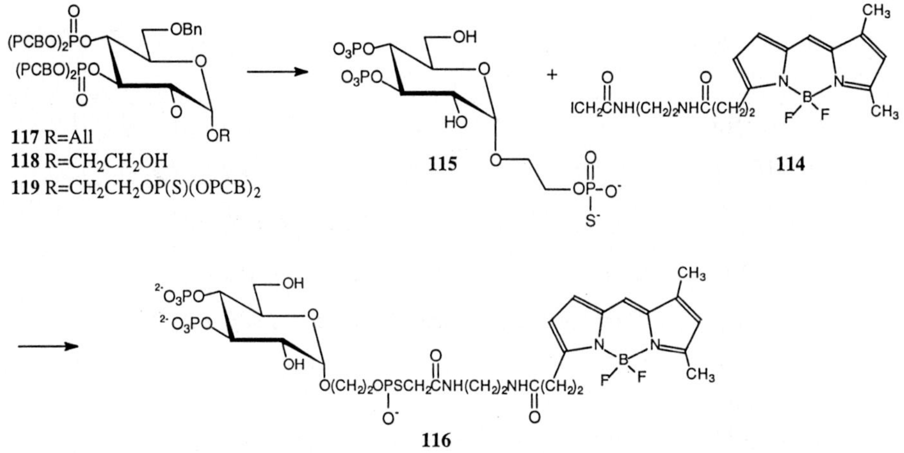

Scheme 10-2

Phosphorothioate **115** was stirred with 1.2 equiv. of **114** in ethanol at room temperature. After two hours ^{31}P NMR spectroscopy indicated that the orange suspension contained a coupled product, judging by signals at δ$_P$ 16.5, 1.3 and 0.9 ppm. After ion-exchange chromatography, the ^{1}H-coupled ^{31}P NMR spectrum of **116** demonstrated the presence of the two ring phosphates, presenting as doublets, and the coupled phosphorothioate, presenting as a quintet. In addition, the ^{19}F NMR spectrum showed a peak at δ$_F$ −12.6 ppm and fluorescence spectra were consistent with presence of the BODIPY label. Biological exploitation of this probe is in progress.

10.5 Conclusion

Carbohydrates have served as useful starting materials in the synthesis of structurally diverse carbocycles using different strategies. The enantiomerically pure carbocyles thus obtained are suitable precursors in the preparation of new chiral phosphoinositide ligands which are mimics of *myo*-inositol 1,4,5-trisphosphate.

Acknowledgements

We thank the BBSRC Intracellular Signalling Programme and the Wellcome Trust for support.

10.6 References

1. M. J. Berridge, *Nature (London)* **1993**, *361*, 315–325.
2. H. Streb, R. F. Irvine, M. J. Berridge, I. Schulz, *Nature (London)* **1983**, *206*, 67–69.
3. B. V. L. Potter, D. Lampe, *Angew. Chem. Int. Ed. Engl.* **1995**, *34*, 1933–1972.
4. D. Lampe, S. J. Mills, B. V. L. Potter, *J. Chem. Soc., Perkin Trans. 1* **1992**, 2899-2906.
5. D. A. Sawyer, B. V. L. Potter, *J. Chem. Soc., Perkin Trans. 1* **1992**, 923–932.
6. D. Lampe, C. Liu, B. V. L. Potter, *J. Med. Chem.* **1994**, *37*, 907–912.
7. A. M. Riley, R. Payne, B. V. L. Potter, *J. Med. Chem.* **1994**, *37*, 3918–3927.
8. I. R. Batty, S. R. Nahorski, R. F. Irvine, *Biochem. J.* **1985**, *232*, 211–215.
9. R. F. Irvine, A. J. Letcher, J. P. Heslop, M. J. Berridge, *Nature (London)* **1986**, *320*, 631–634.
10. K. Y. Choi, H. K. Kim, S. Y. Lee, K. H. Moon, S. S. Kim, J. W. Kim, H. K. Chung, S. K. Rhee, *Science* **1990**, *248*, 64–66.
11. M. Abdullah, P. J. Hughes, A. Craxton, R. Gigg, T. Desai, J. F. Marecek, G. D. Prestwich, S. B. Shears, *J. Biol. Chem.* **1992**, *267*, 22340–22345.
12. S. B. Shears, *Advances in Second Messenger and Phosphoprotein Research Vol. 26*, Ed. J. W. Putney Jr., Raven Press Ltd., New York **1992**, 63–92.
13. P. J. Hughes, S. B. Shears, *J. Biol. Chem.* **1990**, *265*, 9869–9875.
14. A. Höer, E. Oberdisse, *Biochem. J.* **1991**, *278*, 219–224.
15. K. Nogimori, P. J. Hughes, M. C. Glennon, M. E. Hodgson, J. W. Putney Jr., S. B. Shears, *J. Biol. Chem.* **1991**, *266*, 16499–16506.
16. R. F. Irvine, *FEBS Lett.* **1990**, *263*, 5–9.

17. A. Lückhoff, D. E. Clapham, *Nature (London)* **1992**, *355*, 356–358.
18. P. G. Bradford, R. F. Irvine, *Biochem. Biophys. Res. Commun.* **1987**, *149*, 680–685.
19. P. Enyedi, G. H. Williams, *J. Biol. Chem.* **1988**, *263*, 7940–7942.
20. P. Enyedi, E. Brown, G. Williams, *Biochem. Biophys. Res. Commun.* **1989**, *159*, 200–208.
21. A. B. Theibert, S. Supattapone, P. F. Worley, J. M. Baraban, J. L. Meek, S. H. Snyder, *Biochem. Biophys. Res. Commun.* **1987**, *148*, 1283–1289.
22. F. Donié, G. Reiser, *FEBS Lett.*, **1989**, *254*, 155–158.
23. P. J. Cullen, Y. Patel, V. V. Kakkar, R. F. Irvine, K. S. Authi, *Biochem. J.* **1994**, *298*, 739–742.
24. F. Donié, G. Reiser, *Biochem. J.* **1991**, *275*, 453–457.
25. A. B. Theibert, V. A. Estevez, C. D. Ferris, S. K. Danoff, R. K. Barrow, G. D. Prestwich, S. H. Snyder, *Proc. Natl. Acad. Sci. USA* **1991**, *88*, 3165–3169.
26. A. B. Theibert, V. A. Estevez, R. J. Mourey, J. F. Marecek, R. K. Barrow, G. D. Prestwich, S. H. Snyder, *J. Biol. Chem.* **1992**, *267*, 9071–9079.
27. P. J. Cullen, A. P. Dawson, R. F. Irvine, *Biochem. J.* **1995**, *305*, 139–143.
28. P. J. Cullen, S.-K. Chung, Y.-T. Chang, A. P. Dawson, R. F. Irvine, *FEBS Lett.* **1995**, *358*, 240–242.
29. P. J. Cullen, J. J. Hsuan, O. Truong, A. J. Letcher, T. R. Jackson, A. P. Dawson, R. F. Irvine, *Nature (London)* **1995**, *376*, 527–530.
30. L. Stephens, *Biochem. Soc. Trans.* **1995**, *23*, 207–221.
31. R. F. Irvine, A. J. Letcher, D. J. Lander, M. J. Berridge, *Biochem. J.* **1986**, *240*, 301–304.
32. A. H. Guse, E. Roth, F. Emmrich, *Biochem. J.* **1992**, *288*, 489–495.
33. B. E. Ehrlich, J. Watras, *Nature (London)* **1988**, *336*, 583–586.
34. S. K. Joseph, C. A. Hansen, J. R. Williamson, *Mol. Pharmacol.* **1989**, *36*, 391–397.
35. J. A. Ely, L. Hunyady, A. J. Baukal, K. J. Catt, *Biochem. J.* **1990**, *268*, 333–338.
36. I. Parker, I. Ivorra, *J. Physiol.* **1991**, *433*, 207–227.
37. D. J. Gawler, B. V. L. Potter, R. Gigg, S. R. Nahorski, *Biochem. J.* **1991**, *276*, 163–167.
38. M. Hirata, Y. Watanabe, T. Ishimatsu, T. Ikebe, Y. Kimura, K. Yamaguchi, S. Ozaki, T. Koga, *J. Biol. Chem.* **1989**, *264*, 20303–20308.

39. M. Hirata, F. Kanaga, T. Koga, T. Ogasawara, Y. Watanabe, S. Ozaki, *J. Biol. Chem.* **1990**, *265*, 8404–8407.

40. R. Baker, J. J. Kulagowski, D. C. Billington, P. D. Leeson, I. C. Lennon, N. Liverton, *J. Chem. Soc., Chem. Commun.* **1989**, 1383–1385.

41. S. T. Safrany, R. J. H. Wojcikiewicz, J. Strupish, J. McBain, A. M. Cooke, B. V. L. Potter, S. R. Nahorski, *Mol. Pharmacol.* **1991**, *39*, 754–761.

42. R. J. Ferrier, *J. Chem. Soc., Perkin Trans 1* **1979**, 1455–1458.

43. R. J. Ferrier, S. Middleton, *Chem. Rev.* **1993**, *93*, 2779–2831.

44. D. Semeria, M. Philippe, J.-M. Delaumeny, A.-M. Sepulchre, S. D. Gero, *Synthesis* **1983**, 710–713.

45. R. Blattner, R. J. Ferrier, P. Prasit, *J. Chem. Soc., Chem. Commun.* **1980**, 944–945.

46. I. Pintér, J. Kovács, A. Messmer, G. Tóth, S. D. Géro, *Carbohydr. Res.* **1983**, *116*, 156–161.

47. I. Pelyvás, F. Sztaricskai, R. Bognár, *J. Chem. Soc., Chem. Commun.* **1984**, 104–105.

48. M. Mádi-Puskás, I. Pelyvás, R. Bognár, *J. Carbohydr. Chem.* **1985**, *4*, 323–331.

49. R. J. Ferrier, A. E. Stütz, *Carbohydr. Res.* **1990**, *200*, 237–245.

50. N. Sakairi, H. Kuzuhara, *Tetrahedron Lett.* **1982**, *23*, 5327–5330.

51. R. Blattner, R. J. Ferrier, *J. Chem. Soc., Chem. Commun.* **1987**, 1008–1009.

52. R. J. Ferrier, A. E. Stütz, *Carbohydr. Res.* **1990**, *205*, 283–291.

53. D. H. R. Barton, S. Augy-Dorey, J. Camara, P. Dalko, J. M. Delauměny, S. D. Géro, B. Quiclet-Sire, P. Stütz, *Tetrahedron* **1990**, *46*, 215–230.

54. D. F. Corbett, D. K. Dean, S. R. Robinson, *Tetrahedron Lett.* **1993**, *34*, 1525–1528.

55. D. F. Corbett, D. K. Dean, S. R. Robinson, *Tetrahedron Lett.* **1994**, *35*, 459.

56. K.-i. Sato, M. Bokura, M. Taniguchi, *Bull. Chem. Soc. Jpn.* **1994**, *67*, 1633–1640.

57. J. Cleophax, D. Dubreuil, S. D. Géro, A. Loupy, M. V. de Almeida, A. D. Da Silva, G. Vass, E. Bischoff, E. Perzborn, G. Hecker, O. Lockhoff, *Bioorg. Med. Chem. Lett.* **1995**, *5*, 831–834.

58. E. Poirot, H. Bourdon, F. Chrétien, Y. Chapleur, B. Berthon, M. Hilly, J.-P. Mauger, G. Guillon, *Bioorg. Med. Chem. Lett.* **1995**, *5*, 569–572.

59. S. L. Bender, R. J. Budhu, *J. Am. Chem. Soc.* **1991**, *113*, 9883–9885.

60. T. K. Park, S. J. Danishefsky, *Tetrahedron Lett.* **1995**, *36*, 195–196.

61. V. A. Estevez, G. D. Prestwich, *J. Am. Chem. Soc.* **1991**, *113*, 9885–9887.

62. T. Sawada, R. Shirai, Y. Matsuo, Y. Kabuyama, K. Kimura, Y. Fukui, Y. Hashimoto, S. Iwasaki, *Bioorg. Med. Chem. Lett.* **1995**, *5*, 2263–2266.

63. A. Chaudhary, G. Dormán, G. D. Prestwich, *Tetrahedron Lett.* **1994**, *35*, 7521–7524.

64. K. Freudenberg, H. Toepffer, C. C. Andersen, *Chem. Ber.* **1928**, *61*, 1750–1760.

65. N. K. Richtmyer, *Methods Carbohydr. Chem.* **1962**, *1*, 107–113.

66. D. M. Hall, *Carbohydr. Res.* **1980**, *86*, 158–160.

67. M. E. Evans, *Carbohydr. Res.* **1972**, *21*, 473–475.

68. M. E. Evans, *Methods Carbohydr. Chem.* **1980**, *8*, 313–315.

69. J. J. Patroni, R. V. Stick, B. W. Skelton, A. H. White, *Aust. J. Chem.* **1988**, *41*, 91–102.

70. R. Albert, K. Dax, R. Pleschko, A. E. Stütz, *Carbohydr. Res.* **1985**, *137*, 282–290.

71. V. I. Grishkovets, A. E. Zemyakov, V. Ya. Chirva, *Khim. Prir. Soedin.* **1982**, 119. *Chem. Abstr.* **1982**, *96*: 200035j.

72. K. D. Philips, J. Zemlicka, J. P. Horowitz, *Carbohydr. Res.* **1973**, *30*, 281–286.

73. A. Yu. Romanovich, A. F. Sviridov, S. V. Yarotskii, *Bull. Acad. UDSSR Div. Chem. Sci.* **1977**, 2002–2003.

74. V. Ferro, M. Mocerino, R. V. Stick, D. M. G. Tilbrook, *Aust. J. Chem.* **1988**, *41*, 813–815.

75. D. Horton, W. Weckerle, *Carbohydr. Res.* **1975**, *44*, 227–240.

76. D. J. Jenkins, Ph.D. Thesis, University of Bath, **1995**.

77. J. M. Küster, I. Dyong, *Liebigs Ann. Chem.* **1975**, 2179–2189.

78. P. J. Garegg, T. Iversen, S. Oscarson, *Carbohydr. Res.* **1976**, *50*, C12–C14.

79. R. Eby, K. T. Webster, C. Schuerch, *Carbohydr. Res.* **1984**, *129*, 111–120.

80. T. Ogawa, T. Kaburagi, *Carbohydr. Res.* **1982**, *103*, 53–64.

81. F. Dasgupta, P. J. Garegg, *Synthesis* **1994**, 1121–1123.

82. S. David, S. Hanessian, *Tetrahedron* **1985**, *41*, 643–663.

83. M. Pereyre, J. P. Quintard, A. Rahm. In: *Tin in Organic Synthesis*, Butterworths, London **1987**, pp.261–323.

84. T. B. Grindley. In: *Synthetic Oligosaccharides: Indispensible Probes for the Life Sciences* (P. Kovác, Ed.) ACS, Washington **1994**, pp.51–76.

85. J. E. T. Corrie, *J. Chem. Soc., Perkin Trans. 1* **1993**, 2161–2166.

86. G.-J. Boons, G. H. Castle, J. A. Clase, P. Grice, S. V. Ley, C. Pinel, *Synlett* **1993**, 913–914. *Corrigendum*: *Synlett* **1994**, 764.

87. D. J. Jenkins, B. V. L. Potter, *Carbohydr. Res.* **1994**, *265*, 145–149.

88. R. M. Munavu, H. H. Szmant, *J. Org. Chem.* **1976**, *41*, 1832–1936.

89. N. Nagashima, M. Ohno, *Chem. Pharm. Bull.* **1991**, *39* 1972–1982.

90. J. M. Sugihara, *Adv. Carbohydr. Chem.* **1953**, *8*, 1.

91. R. C. Chalk, D. H. Ball, L. Long Jr., *J. Org. Chem.* **1966**, *31*, 1509.

92. S. Hanessian, *Carbohydr. Res.* **1966**, *2*, 86–88.

93. D. L. Failla, T. L. Hullar, S. B. Siskin, *J. Chem. Soc., Chem. Commun.* **1966**, 716–717.

94. P. J. Kocienski, *Protecting Groups*, Georg Thieme Verlag, Stuttgart **1994**, p.100.

95. J. Thiem, J. Elvers, *Carbohydr. Res.* **1978**, *60*, 63–73.

96. E. J. Bourne, A. J. Huggard, J. C. Tatlow, *J. Chem. Soc.* **1953**, 735–741.

97. T. M. Cheung, D. Horton, W. Weckerle, *Carbohydr. Res.* **1977**, *58*, 139–151.

98. K.-i. Sato, N. Kubo, R. Takada, S. Sakuma, *Bull. Chem. Soc. Jpn.* **1993**, *66*, 1156–1165.

99. S. Adam, *Tetrahedron Lett.* **1988**, *29*, 6589–6592.

100. D. H. R. Barton, J. Camara, P. Dalko, S. D. Géro, B. Quiclet-Sire, P. Stütz, *J. Org. Chem.* **1989**, *54*, 3764–3766.

101. R. J. Ferrier, S. R. Haines, *Carbohydr.Res.* **1984**, *130*, 135–146.

102. O. Mitsunobu, *Synthesis* **1981**, 1–28.

103. M. F. Boehm, G. D. Prestwich, *Tetrahedron Lett.* **1988**, *29*, 5217–5220.

104. C. Jaramillo, J.-L. Chiara, M. Martín-Lomas, *J. Org. Chem.* **1994**, *59*, 3135–3141.

105. D. Lampe, B. V. L. Potter, *Tetrahedron Lett.* **1993**, *34*, 2365–2368.

106. Y. Torisawa, H. Okabe, S. Ikegami, *Chem. Lett.* **1984**, 1555–1556.

107. K. Omura, D. Swern, *Tetrahedron* **1978**, *34*, 1651–1660.

108. G. Piancatelli, A. Scettri, M. D'Auria, *Synthesis* **1982**, 245–258.

109. C. Jaramillo, M. Martín-Lomas, *Tetrahedron Lett.* **1991**, *32*, 2501–2504.

110. M. R. Hamblin, B. V. L. Potter, R. Gigg, *J. Chem. Soc., Chem. Commun.* **1987**, 626–627.

111. D. J. Jenkins, D. Dubreuil, B. V. L. Potter, *J. Chem. Soc., Perkin Trans. 1* **1996**, 1365–1372.

112. A. M. Riley, D. J. Jenkins, B. V. L. Potter, *J. Am. Chem. Soc.* **1995**, *117*, 3300–3301.

113. H. Ito, Y. Motoki, T. Taguchi, Y. Hanzawa, *J. Am. Chem. Soc.* **1993**, *115*, 8835–8836.

114. Y. Hanzawa, H. Ito, T. Taguchi, *Synlett* **1995**, 299–305.

115. K. Tatsuta, Y. Niwata, K. Umezawa, K. Toshima, M. Nakata, *J. Antibiot.* **1991**, *44*, 456–458.

116. R. Johansson, B. Samuelsson, *J. Chem. Soc., Perkin Trans. 1* **1984**, 2371–2374.

117. D. Joniak, B. Košíková, L. Kosáková, *Collect. Czech. Chem. Commun.* **1978**, *43*, 769–773.

118. T. Ogawa, Y. Takahashi, M. Matsui, *Carbohydr. Res.* **1982**, *102*, 207–215.

119. A. Lipták, I. Jodál, P. Nánási, *Carbohydr. Res.* **1975**, *44*, 1–11.

120. A. M. Cooke, B. V. L. Potter, R. Gigg, *Tetrahedron Lett.* **1987**, *28*, 2305–2308.

121. R. A. Wilcox, S. T. Safrany, C. Liu, S. R. Nahorski, B. V. L. Potter, unpublished results.

122. A. Chénedé, P. Pothier, M. Sollogoub, A. J. Fairbanks, P. Sinaÿ, *J. Chem. Soc., Chem. Commun.* **1995**, 1373–1374.

123. M. Takahashi, T. Kagasaki, T. Hosoya, S. Takahashi, *J. Antibiot.* **1993**, *46*, 1643–1647.

124. S. Takahashi, T. Kinoshita, M. Takahashi, *J. Antibiot.* **1994**, *47*, 95–100.

125. H. Hotoda, M. Takahashi, K. Tanzawa, S. Takahashi, M. Kaneko, *Tetrahedron Lett.* **1995**, *36*, 5037–5040.

126. M. Takahashi, K. Tanzawa, S. Takahashi, *J. Biol. Chem.* **1994**, *269*, 369–372.

127. J. Hirota, T. Michikawa, A. Miyawaki, M. Takahashi, K. Tanzawa, I. Okura, T. Furuichi, K. Mikoshiba, *FEBS Lett.* **1995**, *368*, 248–252.

128. R. A. Wilcox, C. Erneux, W. U. Primrose, R. Gigg, S. R. Nahorski, *Mol. Pharmacol.* **1995**, *47*, 1204–1211.

129. H. Takaku, K. Kamaike, *Chem. Lett.* **1982**, 189–192.

130. R. R. Schmidt, *Angew. Chem. Int. Ed. Eng.* **1986**, *25*, 212–235.

131. J. Gigg, R. Gigg, *J. Chem. Soc. (C)* **1966**, 82–86.

132. P. J. Kocienski, *Protecting Groups*, Georg Thieme Verlag, Stuttgart **1994**, pp.61–68.

133. T. W. Greene, P. G. M. Wuts, *Protecting Groups in Organic Synthesis (Second Edition)*, John Wiley, Sons, Inc. **1991**, pp.42–46.

134. T. Desai, J. Gigg, R. Gigg, E. Martín-Zamora, *Carbohydr. Res.* **1994**, *262*, 59–77.

135. E. A. Talley, M. D. Vale, E. Yanovsky, *J. Am. Chem. Soc.* **1945**, *67*, 2037–2039.

136. S. A. Nepogod'ev, L. V. Backinowsky, B. Grzeszczyk, A. Zamojski, *Carbohydr. Res.* **1994**, *254*, 43–60.

137. R. T. Lee, Y. C. Lee, *Carbohydr. Res.* **1974**, *37*, 193–201.

138. I. Pelyvás, T. Lindhorst, J. Thiem, *Liebigs Ann. Chem.* **1990**, 761–769.

139. M. E. Haque, T. Kikuchi, K. Yoshimoto, Y. Tsuda, *Chem. Pharm. Bull.* **1985**, *33*, 2243–2255.

140. P. J. Kocienski, *Protecting Groups*, Georg Thieme Verlag, Stuttgart **1994**, p.71.

141. H. Qin, T. B. Grindley, *J. Carbohydr. Chem.* **1994**, *13*, 475–490.

142. J. Cai, B. E. Davison, C. R. Ganellin, S. Thaisrivongs, *Tetrahedron Lett.* **1995**, *36*, 6535–6536.

143. J. G. Buchanan, A. R. Edgar, D. I. Rawson, P. Shahidi, R. H. Wightman, *Carbohydr. Res.* **1982**, *100*, 75–86.

144. K. Takeo, K. Shibata, *Carbohydr. Res.* **1984**, *133*, 147–151.

145. J.-P. Praly, G. Descotes, M.-F. Grenier-Loustalot, F. Metras, *Carbohydr. Res.* **1984**, *128*, 21–35.

146. *The Merck Index (Eleventh Edition)*, Merck, Co., Inc., Rahway, New Jersey, USA **1989**, p.1090.

147. D. J. Jenkins, B. V. L. Potter, *Carbohydr. Res.* **1996**, *287*, 169–182.

148. T. Desai, J. Gigg, R. Gigg, *Aust. J. Chem.* **1996**, *49*, 305–309.

149. T. Desai, J. Gigg, R. Gigg, *Abstr.*, 8th Europ. Carbohydr. Symp., Seville, Spain, 2–7 July **1995**, A-14.

150. A. M. Riley, C. T. Murphy, C. J. Lindley, J. Westwick, B. V. L. Potter, *Bioorg. Med. Chem. Lett.* **1996**, *6*, 2197–2200.

151. D. Lampe, C. Liu, M. F. Mahon, B. V. L. Potter, *J. Chem. Soc., Perkin Trans. 1* **1996**, 1717–1727.

152. N. Moitessier, F. Chrétien, Y. Chapleur, C. Humeau, *Tetrahedron Lett.* **1995**, *36*, 8023–8026.

153. H. B. A. de Bont, J. H. van Boom, R. M. J. Liskamp, *Recl. Trav. Chim. Pays-Bas* **1990**, *109*, 27–28.

154. S. J. Mills, A. M. Riley, C. T. Murphy, A. J. Bullock, J. Westwick, B. V. L. Potter, *Bioorg. Med. Chem. Lett.* **1995**, *5*, 203–208.

11 Chemo-enzymatic Total Synthesis of Some Conduritols, Carbasugars and (+)-Fortamine

L. Dumortier, P. Liu, J. Van der Eycken and M. Vandewalle*

11.1 Introduction

In the context of our ongoing programme directed towards enantioselective synthesis of natural products and analogues containing a pivotal cyclohexane ring we have studied the formation of enantiopure building blocks *via* the action of hydrolytic enzymes [1]. Three types of chiral building blocks suitable for the synthesis of the title compounds will be described.

11.2 Results

11.2.1 Substrates and Enantioselective Enzymatic Hydrolysis

The synthesis of chiral templates derived from the 1,2,3-cyclohexane triols **2** and **5** is described in Schemes 11-1 and 11-2. The substrate (±)-**3** was prepared *via* initial epoxidation of **1**.

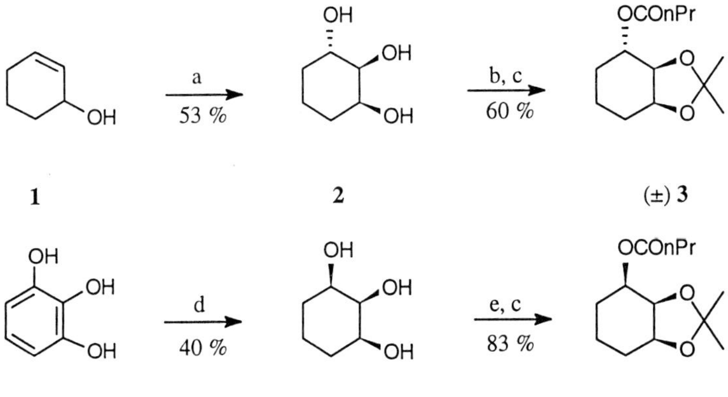

Scheme 11-1. Reagents: a) $HO_2CCH_2CH_2CO_3H$, H_2O, r.t., 2 h; b) Me_2CO, PTSA, r.t., 12 h; c) n-PrCOCl, Et_3N, DMAP, CH_2Cl_2, 0°C, 2 h; d) Raney-Ni, H_2, EtOH, 100 atm, 100°C, 3 h; e) $Me_2C(OMe)_2$, PTSA, r.t., 12 h.

Hydrolytic epoxide opening under neutral conditions, acetonide formation and esterification gave butyrate (±)-**3**. The all-*cis* isopropylidene derivative (±)-**6** with an *endo* oriented acyloxy group was obtained in a facile multi-gram scale preparation consisting of catalytic hydrogenation of unexpensive pyrogallol (**4**) and subsequent ketalisation and esterification. To the best of our knowledge hydrogenation of **4** has not yet been described in the literature. It is evident that the all-*cis* isomer (±)-**6** is the most easily available one and therefore by far the preferred substrate.

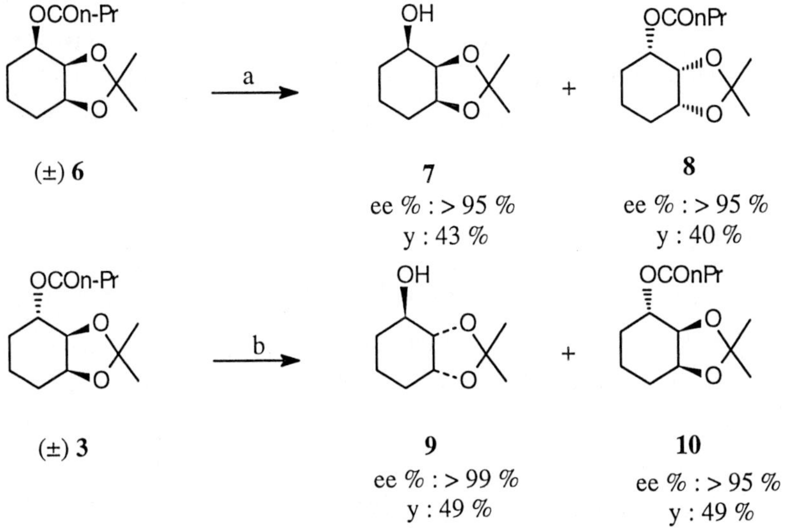

Scheme 11-2. Reagents: a) SAM II lipase, pH 7, r.t., 30 min, NaOH; b) PGL, pH 7, r.t., 45 min, NaOH.

The enzymatic hydrolysis of (±)-**3** and (±)-**6** was studied in the presence of several hydrolases [1a]. The best results are shown in Scheme 11-2 and lead to 1(*R*)-alcohols **7** and **9** [2]. The optical purity of **7, 8** and **10** was determined by ^{1}H NMR at 500 MHz in the presence of the chiral shift reagent Eu(hfc)$_3$, only one enantiomer could be detected. In the case of the *exo*-alcohol **9** determination was possible with GC on a Chirasil-Val column.

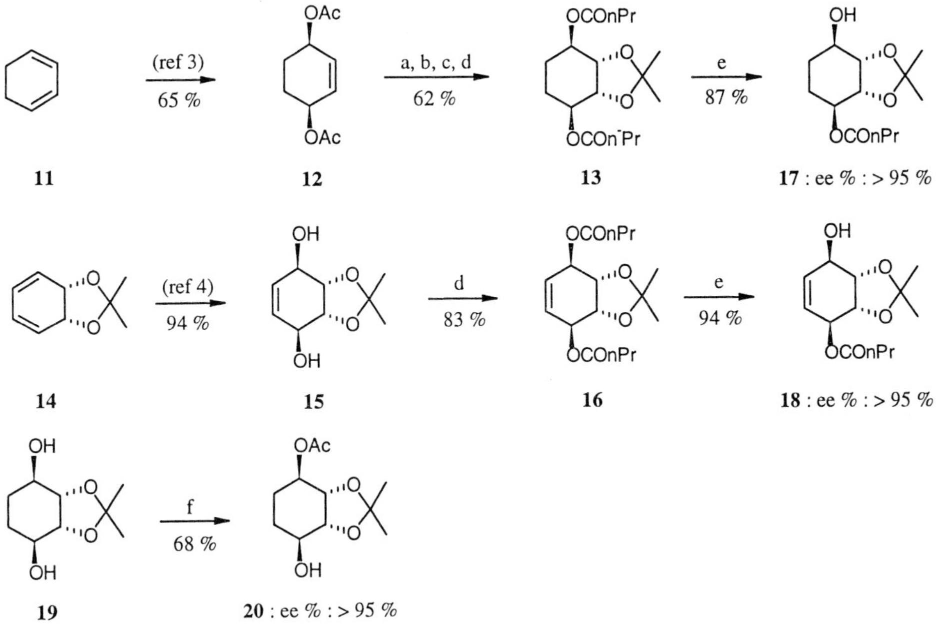

Scheme 11-3. Reagents: a) OsO$_4$, KClO$_3$, THF, H$_2$O, r.t., 48 h; b) acetone, PTSA, r.t., 5 h; c) K$_2$CO$_3$, MeOH, r.t., 1 h; d) n-PrCOCl, NEt$_3$, DMAP, CH$_2$Cl$_2$, r.t., 4 h; e) PGL, pH 7, 35°C, NaOH; f) vinyl acetate, lipase PS, 25°C, 4 d.

The synthesis of the tetrol derivatives **13** and **16** and their enantiotoposelective hydrolysis is described in Scheme 11-3 [1d]. Diacetate **12** can be easily obtained from **11** by the method described by Bäckvall et al. [3]. Syn-dihydroxylation of **12** yielded exclusively a single diol, which was transformed to the acetonide. Because of its high melting point, the resulting diacetate was not a suitable substrate and was therefore transformed in the dibutyrate **13** (steps c and d).

Substrate **16** is a protected form of conduritol A, and was prepared from **14** according to the procedure described by Balci et al. [4]. Enzyme-catalyzed hydrolysis of **13** and **16** led respectively to the (R)-alcohols **17** and **18**. The optical purities of **17** and **18** were determined by ^{1}H NMR at 500 MHz in the presence of Eu(hfc)$_3$. Johnson et al. [5] have described the enzymatic acetylation of **15** in an organic solvent. By analogy we performed the transformation of **19** into **20**, leading to the enantiomeric series of **17**.

11.2.2 The Synthesis of Some Advanced Intermediates

With the highly optically pure alcohols in hand we turned our attention towards the construction of some advanced intermediates. In order to introduce a higher functionalisation we studied regioselective dehydration of **7** and **9** (Scheme 11-4). Surprisingly *endo*-alcohol **7** resisted all classical *syn*-eliminations [1f]. A solution to this problem was found when we observed that flash-vacuum pyrolysis [6] of the corresponding xanthate **22** led in high yield to **21**. In contrast *ent*-**9** could be transformed into **21** upon Mitsunobu type *anti*-elimination [7]. Epoxidation of **21** gave exclusively the *exo*-epoxide **23**. Base mediated regioselective opening of the epoxide and deprotection led to pseudo α–D-xylopyranose or (1S)-dihydroconduritol F **24**.

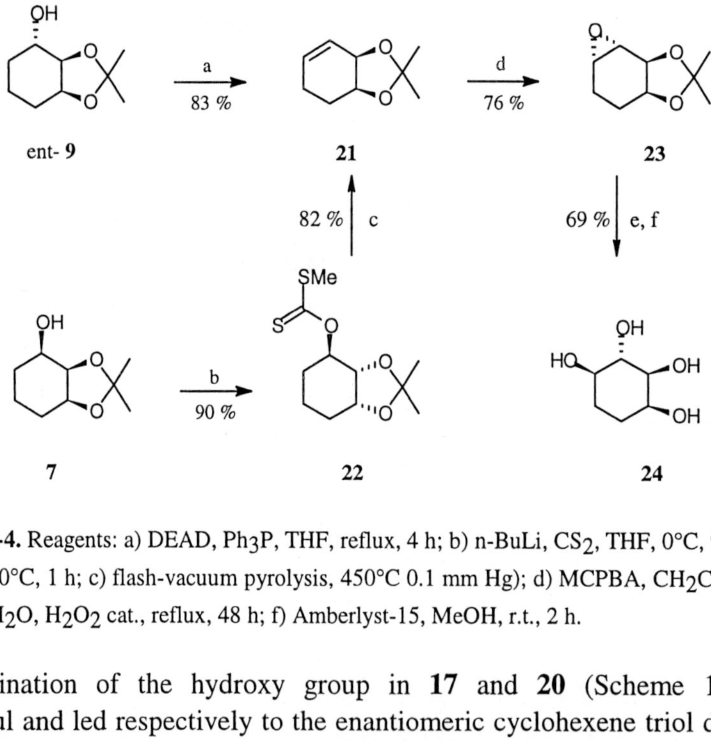

Scheme 11-4. Reagents: a) DEAD, Ph$_3$P, THF, reflux, 4 h; b) n-BuLi, CS$_2$, THF, 0°C, 90 min, then MeI, THF, 0°C, 1 h; c) flash-vacuum pyrolysis, 450°C 0.1 mm Hg); d) MCPBA, CH$_2$Cl$_2$, 0°C, 2 h; e) NaOH, H$_2$O, H$_2$O$_2$ cat., reflux, 48 h; f) Amberlyst-15, MeOH, r.t., 2 h.

Elimination of the hydroxy group in **17** and **20** (Scheme 11-5) was uneventful and led respectively to the enantiomeric cyclohexene triol derivatives 1(S)-**26** and 1(R)-**26**. It is noteworthy that Mitsunobu inversion of **17** and **20** failed.

20 1-(S)- **26** b 85 % **25** ; R = n-PrCO → 1-(*R*)-**26** ; R = H **17**

Scheme 11-5. Reagents: a) DEAD, Ph_3P, THF, reflux, 16 h; b) K_2CO_3, MeOH, r.t., 3 h.

In contrast to this, inversion carried out on cyclohexenol **18** and subsequent selective hydrolysis of the resulting p-nitrobenzoate gave **27** in high yield (Scheme 11-6). An inherent feature of optically active derivatives of *meso*-compounds is the fact that also the enantiomeric series is easily available upon protective group manipulation as demonstrated by the formation of **28a,b** and **29a,b**.

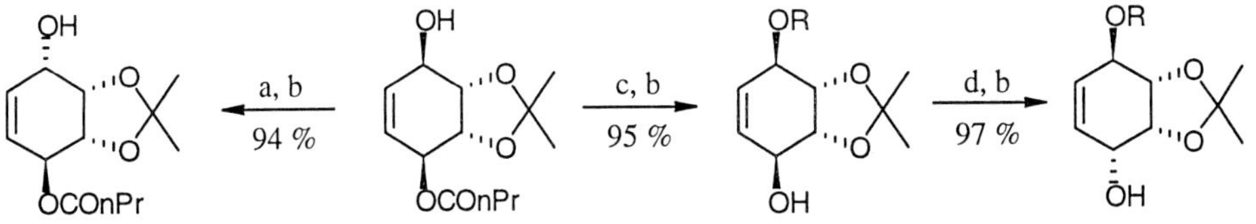

27 **18** **28a** ; R = TBDPS **28b** ; R = MPM **29a** ; R = TBDPS **29b** ; R = MPM

Scheme 11-6. Reagents: a) p-$NO_2C_6H_4COOH$, DEAD, Ph_3P, THF, r.t., 3 h; b) $KHCO_3$, MeOH, r.t., 4 h; c) **28a** : TBDPSCl, imidazole, DMF, r.t., 5 h; **28b** : MPMOC(=NH)CCl_3, CSA, CH_2Cl_2, r.t., 3 h; d) $PhCO_2H$, DEAD, Ph_3P, THF, r.t., 3 h.

11.2.3 Synthesis of Conduritols

Cyclohexenetetrols are known as the conduritols. The 6 diastereoisomers are designated from A to F; they are potential glycosidase inhibitors. Only conduritols A and (+)-F occur in nature. There has been considerable interest in the synthesis of enantiopure conduritols as they are useful precursors for cyclitols and carbasugars [8]. The synthesis of conduritols (-)-C, (-)-E, (-)-F and A is described in Scheme 11-7 [1d].

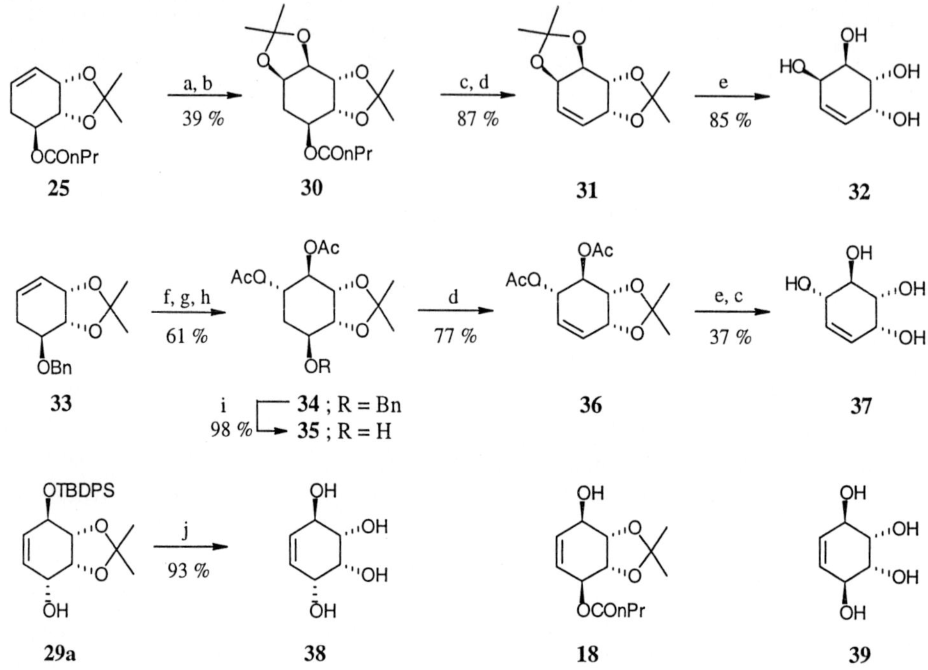

Scheme 11-7. Reagents: a) OsO$_4$, KClO$_3$, THF, H$_2$O, r.t., 5 days; b) DMP, PTSA, r.t., 5 h; c) K$_2$CO$_3$, MeOH, r.t.; d) Ph$_3$P, DEAD, THF, reflux, 12 h; e) MeOH, PTSA, 40°C, 16 h; f) MCPBA, CH$_2$Cl$_2$, r.t., 48 h; g) NaOH, H$_2$O, H$_2$O$_2$ (cat.), 80°C, 16 h; h) Ac$_2$O, Py, DMAP (cat.), r.t., 3 h; i) Pd/C, H$_2$ (1 atm), EtOH, 4 h; j) Amberlyst-15, MeOH, r.t., 12 h.

Syn-dihydroxylation of **25** followed by protection of the resulting diol as an acetonide yielded **30** as a single isomer. Ester solvolysis and anti-dehydration resulted in **31** which after deprotection led to (-)-conduritol E **32**.

The synthesis of (-)-conduritol F (**37**) starts with the benzyl ether **33** easily obtained from 1-(R)-**26**. Epoxidation afforded exclusively the exo-epoxide which upon base-catalyzed hydrolysis and subsequent acetylation resulted in diacetate **34**. Hydrogenolytic cleavage of the benzyl ether and selective anti-elimination of the hydroxy group in **35** led to **36**, which after deprotection afforded (-)-conduritol F **37**. (-)-Conduritol C (**38**) is directly available upon deprotection of **29a**. Compound **18** is an optically active derivative of conduritol A (**39**) and is interesting as a synthetic intermediate.

11.2.4 Synthesis of Carba-sugars

Carba- or pseudo-sugars [9] are carbocyclic analogues of carbohydrates. 2,3,4,5-Tetrahydroxy-1-(hydroxymethyl)-cyclohexanes or 5a-carba-hexopyranoses, are thus related to hexopyranoses in which the ring oxygen has been replaced by a methylene group. A number of them are found as components of important antibiotics [10]. The structural resemblance to true-sugars endows them with interesting biological activities in the area of enzyme inhibitors, sweeteners and antibiotic, antiviral and anticancer therapy.

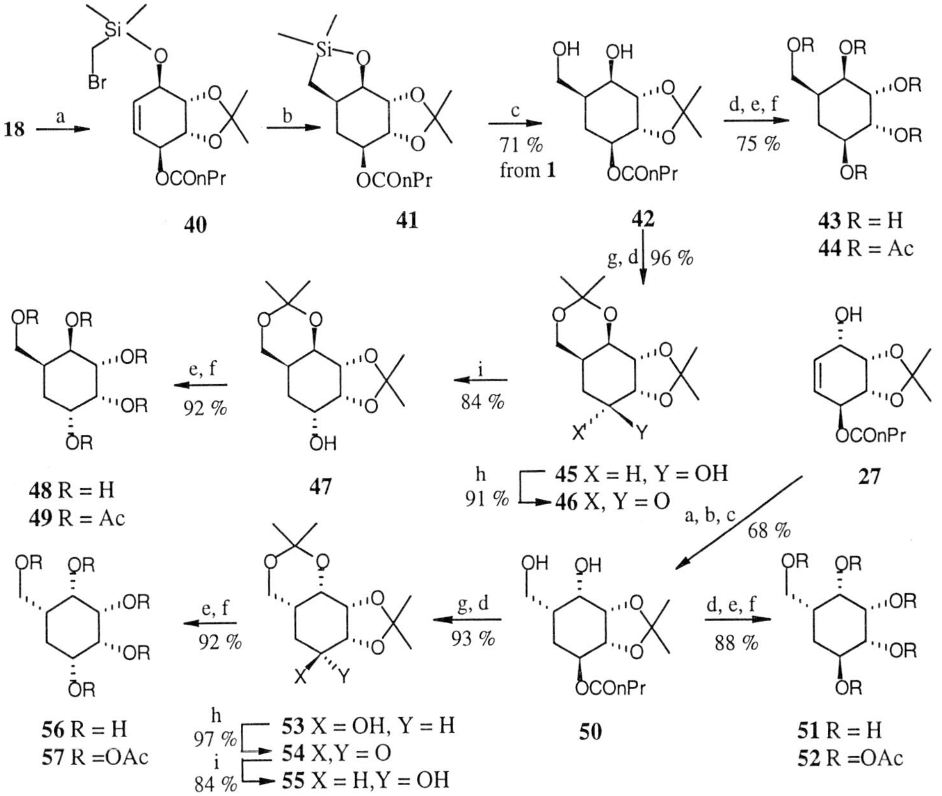

Scheme 11-8. Reagents: a) (bromomethyl)chlorodimethylsilane (1.1 eq), Et$_3$N (1.1 eq), DMAP (0.1 eq), CH$_2$Cl$_2$, 0°C, 2 h; b) n-Bu$_3$SnH (1.5 eq), AIBN (0.1 eq), C$_6$H$_6$, reflux 5 h, then r.t., 5 h; c) KF (2 eq), KHCO$_3$ (1 eq), H$_2$O$_2$ (35 %, 12 eq), THF/MeOH (1/1), 2.5 h, r.t., then Na$_2$SO$_3$ (12 eq), 0°C; d) KHCO$_3$ (1 eq), MeOH, r.t.; e) PTSA, MeOH, r.t.; f) Ac$_2$O, r.t., 24 h; g) 2,2-dimethoxypropane, DMF, PPTS, r.t., 24 h; h) oxalylchloride (2 eq), DMSO (4 eq), Et$_3$N (5 eq), CH$_2$Cl$_2$, -78°C, 3 h; i) NaBH$_4$, THF/MeOH (1:1), -78°C.

For the synthesis of these target molecules we need to introduce a functionalized 1-C-substituent on one of the sp^2-carbon atoms of either **18**, **27**, **28** or **29**. As a consequence all D- and L-5a-carba-hexoses, with a 2,3-*cis* substitution pattern (carbohydrate numbering), can be obtained [1g,1h].

For the synthesis (Scheme 11-8) of 5a-carba-gulopyranoses **43** and **48** and 5a-carba-talopyranoses **51** and **56** the Stork radical cyclisation approach is ideally suited [11]. Accordingly, the bromomethyl dimethyl silyl ether **40** was transformed into **41**. Ether **40** is rather unstable and had to be used directly after rapid filtration on celite and solvent evaporation. Oxidative cleavage of the carbon-silicon [12] bond in crude **41** led to **42**, a protected form of 5a-carba-β-L-gulopyranose **43**. Methanolysis of **42** gave **43**, as a hygroscopic syrup, which was characterized as its penta-acetate **44**.

The α-anomer was obtained from **42**, *via* the bis-acetonide **45**. Not surprisingly, Mitsunobu [7] inversion failed on this highly congested alcohol. Alternatively, an oxidation-reduction sequence led to **47**. Swern [13] oxidation gave ketone **46** in high yield, while methods based on PDC, PCC and Collins oxidation failed. The subsequent reduction of **46** afforded an easily separable mixture of **47** and **45** in 9:1 ratio. Solvolysis of **47** led to 5a-carba-α-L-gulopyranose **48**, characterized as **49**.

For the synthesis of the 5a-carba D-talopyranoses **51** and **56**, the same reaction sequence can be used starting from **27**. The Stork procedure gave the key intermediate **50** which was transformed into the penta-acetate **52** of 5a-carba-α-D-talopyranose **51**. The β-anomer **56** and the penta-acetate **57** were obtained *via* ketone **54**; in this case the reduction was completely diastereoselective.

For the synthesis of the 5a-carba-mannopyranoses **62** and **67** and the 5a-carba-allopyranoses **73** and **76** the hydroxymethyl group was introduced *via* 2,3-Wittig rearrangement, subsequent to inversion of one of the allylic oxy-substituents in **18**. Allylic alcohol **29b** was selected as the starting material (Scheme 11-9). Formation of stannane **58** and [2,3]-sigmatropic shift, using Still's conditions [14] cleanly led to the cyclohexene **59**. Hydroboration and oxidative work-up [15] gave, next to 5 % of unidentified isomers, alcohol **61** which was transformed into 5a-carba-α-D-mannopyranose **62** and characterized as its penta-acetate **63**.

On the other hand, hydration of the double bond in **60**, allowed chemo-selective inversion of the pseudo-anomeric hydroxyl group in **64**. This led to 5a-carba-β-D-mannopyranose **67** and its penta-acetate **68**. The 5a-carba-D-allopyranoses **73** and **76** were synthesized in essentially the same way after interchanging the destiny of the allylic oxy-groups in **29b**. The benzyl ether **69** now served as the starting material for Still's procedure. After protection of the hydroxy group in **70**, hydration of the double bond in **71**, led exclusively to **72**; full deprotection gave 5a-carba-β-D-allopyranose **73** and subsequently penta-acetate **74**. The α-anomer **76** and its penta-acetate **77** were obtained *via* **75** as described for **67** from **64**.

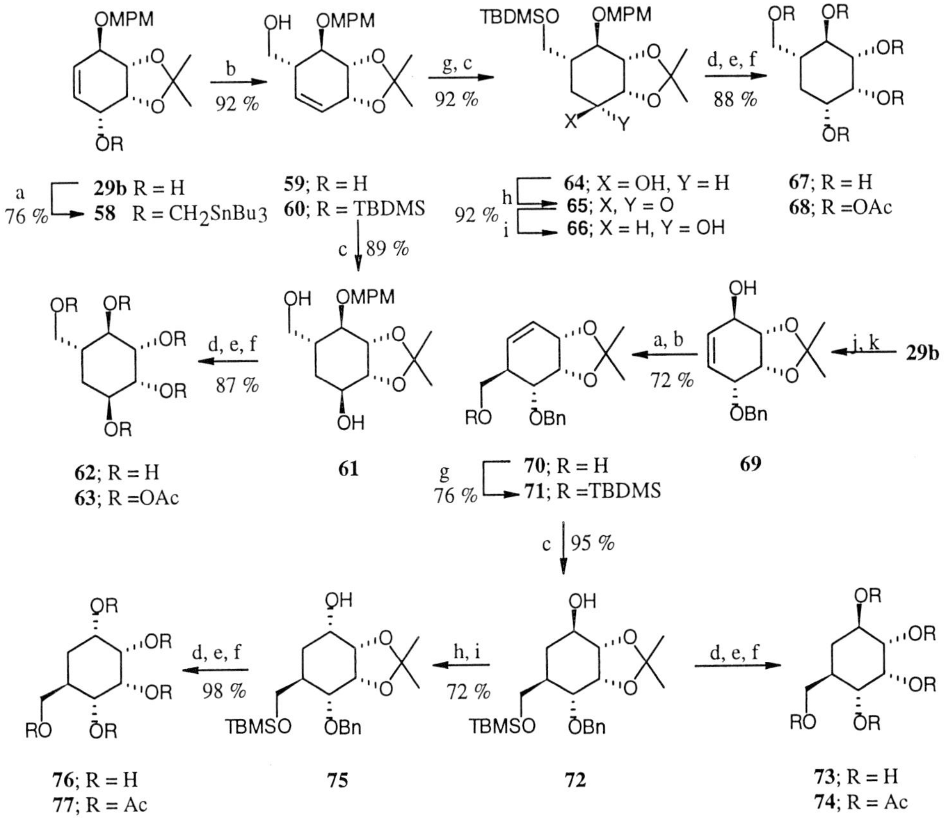

a 76%	**29b** R = H **58** R = CH$_2$SnBu$_3$	**59**; R = H **60**; R = TBDMS	**64**; X = OH, Y = H **65**; X, Y = O **66**; X = H, Y = OH	**67**; R = H **68**; R = OAc

62; R = H
63; R = OAc

61

70; R = H
71; R = TBDMS

69

76; R = H
77; R = Ac

75

72

73; R = H
74; R = Ac

Scheme 11-9. Reagents: a) KH, ICH$_2$SnBu$_3$, THF, 0°C, r.t.; b) n-BuLi, THF, -78°C; c) BH$_3$, THF, -78°C, then H$_2$O$_2$, NaOH, 0°C; d) Pd/C (10 %), MeOH, H$_2$ (1 atm.); e) PTSA, MeOH, r.t.; f) Ac$_2$O, py, r.t., 24 h; g) TBDMS-Cl, imidazole, DMF; h) oxalyl chloride, DMSO, Et$_3$N, -78°C; i) NaBH$_4$, MeOH, 0°C; j) NaH, THF, INBu$_4$, BnBr, r.t.; k) DDQ, CH$_2$Cl$_2$/H$_2$O (5 %), 0°C → r.t.

Except for the 5a-carba-mannopyranoses, these studies led, to the best of our knowledge, to the first reported synthesis of the 5a-carba-L-gulo (**43** and **48**), 5a-carba-D-talo- (**51** and **56**) and of the 5a-carba-D-allopyranoses (**73** and **76**). Moreover, because of the *meso*-nature of the common starting chiral building block **18**, it is obvious that the respective enantiomers are also available.

11.2.5 Synthesis of (+)-Fortamine (78)

Amongst the diaminocyclitols (+)-fortamine (**78**) takes a special position, the 1,4-relative position of the amino functions is a unique feature as normally the 1,3-disposition is observed. It is a component of fortimycin A (**79**) and B (**80**) which are 6-*epi*-purpurosamine glycosides of **78** or of its glycylamide [16]. Its unique structure and intrinsic properties make it an attractive target; next to several total syntheses of racemic **78** [17], one enantioselective route has been described [18].

Our strategy centers around the introduction of the amino functions at C–1 and C–4 in **29b** (fortamine numbering). Upon comparing structures **29b** and **78**, it is obvious that the regioselectivity for substitution at C-1 (*versus* C-6) in **29b** will be critical. However, due to the "*meso*-nature" of **18**, also selective substitution in **27** at C-4 by a methylamine precursor, would open a viable approach. After exploring several alternatives a successful route was found in which the C-atoms of **29b** correspond to those of fortamine as indicated by the numbering of **29b** and **78**.

We envisioned to carry out the crucial nucleophilic substitution at C-1 on cyclic sulfate **82** because this involves less functional group interconversions than the chemoselective formation of a leaving group at C-1 and a protected C-6 oxy-function in diol **81**. The cyclic 1,2-diol was transformed directly into the sulfate upon treatment with sulfuryl chloride [19]. Reaction of cyclic sulfate **82** with lithium azide led in a moderate regioselectivity of 3:1 to the desired **83** as the major product; **83** and **84** are easily separated by column chromatography. This indicates a preference for the axial C-O bond of the cyclic sulfate to act as the leaving group [20].

Stereoselective functionalization of the double bond in intermediate **83**, can be performed using the directing ability of allylic hydroxy functions. Deprotection of the 2-hydroxy group and *syn*-epoxidation gave **85**. Protection of the two hydroxy functions as benzyl ethers and cleavage of the MPM ether afforded the alcohol **86**, essential for the regioselective opening of the epoxide ring. Epoxy-urethane **87**, upon treatment with sodium hydride, underwent intramolecular displacement at C-4 [21]; subsequent *in situ* methylation of the oxy-anion gave **88** in virtually quantitative yield.

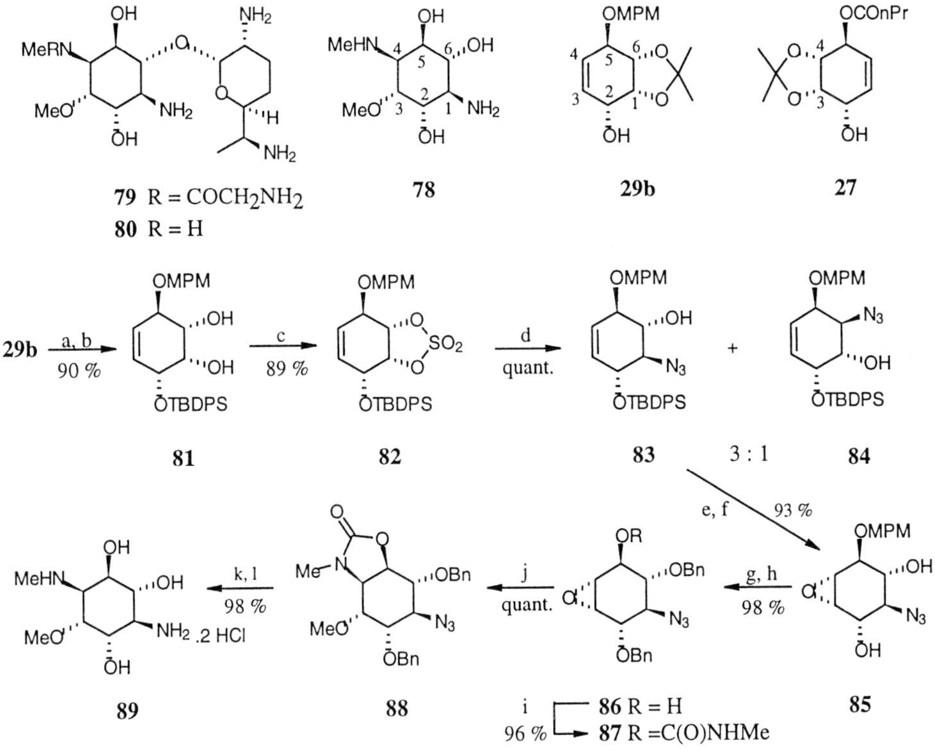

Scheme 11-10. Reagents: a) TBDPSCl, imidazole, DMF, 45°C; b) MeOH, PPTS, 45°C, 12 h; c) SO$_2$Cl$_2$, Et$_3$N, CH$_2$Cl$_2$, r.t., 12 h; d) LiN$_3$, DMF, 10°C, 24 h, then THF, H$_2$SO$_4$ (1 eq), H$_2$O (0.2 eq), r.t. 30 min then NaHCO$_3$ (s), r.t.; e) TBAF, THF, r.t.; f) mCPBA, CH$_2$Cl$_2$, r.t., 72 h; g) NaH, BnBr, THF, r.t.; h) DDQ, CH$_2$Cl$_2$/H$_2$O 20:1, r.t.; i) Me-N=C=O, Et$_3$N, CH$_2$Cl$_2$, r.t., 24 h; j) NaH, THF, r.t., 30 min then MeI, r.t.; k) 5 % HCO$_2$H in EtOH, 10 % Pd/C, H$_2$, r.t.; l) 6 N HCl, reflux.

Concomitant hydrogenolysis of the benzyl ethers during the hydrogenation of the azide in **88** was possible only when an acid solution was used. Finally, acid catalyzed hydrolysis of the oxazolidinone ring afforded (+)-fortamine dihydrochloride **89** which was purified on an Amberlyst IR-120 (H$^+$) column. Salt **89** exhibits $[\alpha]_D^{20}$ = +3.96 (c = 1.0, H$_2$O) and spectral data identical with those of (+)-fortamine dihydrochloride obtained from degradation of natural fortimicin B [22].

Since fortamine dihydrochloride has been converted to the free base form **78** [22] and since natural **78**, obtained from degradation of fortimicin A (**79**), has been converted back to **80** [23], our work also constitutes a formal synthesis of fortimicin B.

11.3 References

1. (a) L. Dumortier, J. Van der Eycken, M. Vandewalle, *Tetrahedron Lett.* **1989**, *30*, 3201-3204; (b) M. Carda, J. Van der Eycken, M. Vandewalle, *Tetrahedron: Asymm.* **1990**, *1*, 17-20; (c) L. Dumortier, M. Carda, J. Van der Eycken, G. Snatzke, M. Vandewalle, *Tetrahedron:Asymm.* **1991**, *2*, 789-792; (d) L. Dumortier, P. Liu, S. Dobbelaere, J. Van der Eycken, M. Vandewalle, *Synlett* **1992**, *3*, 243-245; (e) L. Dumortier, J. Van der Eycken, M. Vandewalle, *Synlett* **1992**, *3*, 245-246; (f) X. Deruyttere, L. Dumortier, J. Van der Eycken, M. Vandewalle, *Synlett* **1992**, *3*, 51-52; (g) P. Liu, M. Vandewalle, *Synlett* **1994**, *4*, 228-230; (h) P. Liu, M. Vandewalle, *Tetrahedron* **1994**, *50*, 7061-7074.

2. PGL (recombinant cutinase from *Fusarium solani pisi*), produced by Plant Genetic Systems, J. Plateaustraat, 22, B-9000 Gent, Belgium.

3. J. E. Bäckvall, E. B. Styrbjörn, R. E. Nordberg, *J. Org. Chem.* **1984**, *49*, 4619-4631.

4. Y. Sütbeyaz, H. Seçen, M. Balci, *J. Chem. Soc., Chem. Comm.* **1988**, 1330-1331.

5. C. R. Johnson, P. A. Plé, J. P. Adams, *J. Chem. Soc., Chem. Comm.* **1991**, 1006-1007.

6. U. E. Wiersum, *Aldrichim Acta* **1984**, *17*, 31-40.

7. O. Mitsunobu, *Synthesis* **1981**, 1-28.

8. For a review see : M. Balci, Y. Sütbeyaz, H. Seçen, *Tetrahedron* **1990**, *46*, 3715-3742.

9. G. E. Mc Casland, S. Furata, L. S. Durham, *J. Org. Chem.* **1966**, *31*, 1516-1521. The term pseudo-sugar, first suggested in this reference, is now better replaced by the term composed of the prefix "carba" and the name of the corresponding carbohydrate; see ref. 10.

10. For a review see : T. Suami, *Top. Curr. Chem.* **1990**, *154*, 257-283.

11. G. Stork, M. Kahn, *J. Am. Chem. Soc.* **1985**, *107*, 500-501; G. Stork, M. J. Sofia, *J. Am. Chem. Soc.* **1986**, *108*, 6826-6828.

12. K. Tamao, K. Maeda, *Tetrahedron Lett.* **1986**, *27*, 65-68.

13. K. Omura, D. Swern, *Tetrahedron* **1978**, *34*, 1651-1660.

14. W. C. Still, *J. Am. Chem. Soc.* **1978**, *100*, 1481-1487; W. C. Still, A. Mitra, *Ibid.* **1978**, *100*, 1927-1928.

15. M. Nakatsuka, J. A. Ragan, T. Sammakia, D. B. Smith, D. E. Uehling, S. L. Schreiber, *J. Am. Chem. Soc.* **1990**, *112*, 5583-5601.

16. (a) T. Nara, M. Yamamoto, I. Kawamoto, K. Takayama, R. Okachi, S. Takasawa, T. Sato, S. Sato, *J. Antibiot.* **1977**, *30*, 522; (b) R. Okachi, S. Takasawa, T. Sato, M. Yamamoto, I. Kawamoto, T. Nara, *ibid.* **1977**, *30*, 541.

17. (a) Y. Honda, T. Suami, *Bull. Chem. Soc. Jpn.* **1982**, *55*, 1156-1162; (b) S. Knapp, M. J. Sebastian, H. Ramanathan, *J. Org. Chem.* **1983**, *48*, 4786-4788; (c) S. Knapp, M. J. Sebastian, H. Ramanathan, P. Bharadwaj, J. A. Potenza, *Tetrahedron* **1986**, *42*, 3405-3410; (d) J. Schubert, R. Schwesinger, H. Prinzbach, *Angew. Chem.* **1984**, *96*, 162-163; (e) J. Schubert, R. Schwesinger, L. Knothe, H. Prinzbach, *Liebigs Ann. Chem.* **1986**, 2009-2052; (f) C. H. Kuo, N. L. Wendler, *Tetrahedron Lett.* **1984**, *25*, 2291-2294.

18. S. Kobayashi, K. Kamiyama, M. Ohno, *J. Org. Chem.* **1990**, *55*, 1169-1169.

19. T. J. Tewson, *J. Org. Chem.* **1983**, *48*, 3507-3510.

20. For a precedent see : M. S. Berridge, M. P. Franceschini, E. Rosenfeld, T. J. Tewson, *J. Org. Chem.* **1990**, *55*, 1211-1217.

21. W. R. Roush, M. A. Adam, *J. Org. Chem.* **1985**, *50*, 3752-3757.

22. H. Sano, T. Sakaguchi, Y. Mori, *Bull. Chem. Soc. Jap.* **1979**, *52*, 2727-2728.

23. W. Rosenbrook, Jr.; J. S. Fairgrieve, *J. Antibiot.* **1981**, *34*, 681.

12 Chemo-enzymatic Approaches to Enantiopure Carbasugars and Carbanucleosides

J. Baumgartner and H. Griengl

12.1 Introduction

Formal replacement of the ring oxygen of the hemiacetal group of carbohydrates by a methylene group results in carba analogs [1]. In the beginning of research in this area these compounds were termed pseudosugars [2]. In later years, however, the use of the prefix carba before the parent name of these compounds became generally accepted practice in accordance with IUPAC recommendations. The same holds for carbocyclic nucleosides. Therefore cyclopentane and cyclohexane rings bearing hydroxy groups can be considered as carbasugars. Of course, some resemblance to a carbohydrate structure must still exist to make these considerations reasonable. However, no strict guidelines can be applied.

Regarding nomenclature, besides other suggestions [2c], earlier practice used the Cahn-Ingold-Prelog *R/S* system. The advantage of being generally applicable is counterbalanced by the lack of intuition to have a quick correlation between structure and name. The α,ß-system of Chemical Abstracts has also been applied for racemates. Several years ago the application of the carba nomenclature, provided by IUPAC rules for those compounds where a heteroatom of a heterocyclic structure, having a trivial or semitrivial name to define structure and stereochemistry, is replaced by a carbon atom, was suggested for carbocyclic analogs of sugars and nucleosides [2a]. This proposal seems to be accepted by most of the research groups being active in this area [2]. As the only difference to IUPAC rule F-4.12 (see Scheme 12-1) the methylene group which replaces the ring oxygen is not numbered 1a but rather 4a or 5a in case of furanoses and pyranoses, resp., since this is in better accordance with chemistry. If this carbon atom is also substituted, thereby creating a new center of chirality, the stereochemistry is given by *R* or *S* as provided by the IUPAC carbohydrate rules. In Scheme 12-1 both examples for synthetic carbasugars and also for naturally occurring compounds having trivial names are included.

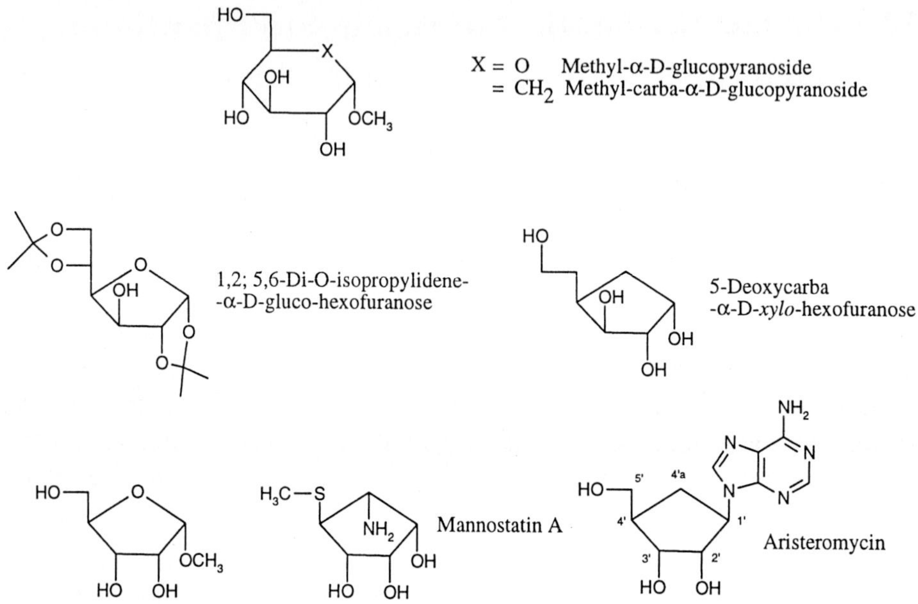

Scheme 12-1 structures

X = O   Methyl-α-D-glucopyranoside
= CH$_2$ Methyl-carba-α-D-glucopyranoside

1,2; 5,6-Di-O-isopropylidene-
-α-D-gluco-hexofuranose

5-Deoxycarba
-α-D-*xylo*-hexofuranose

Mannostatin A

Aristeromycin

IUPAC-Rule

F-4.12: If the parent structure is heterocyclic, its analogs in which one or more of the heteroatomes are replaced by carbon may be named by use of the prefix 'carba'. The original numbering is retained. If in the parent structure the heteroatom is un-numbered the carbon atom replacing it will be numbered by affixing the letter 'a' to the locant of the lower-numbered of the immediately adjacent atoms.

Scheme 12-1. Carbasugars and carbanucleosides and the nomenclature according to IUPAC-rules.

Besides the interest in the synthesis of naturally occurring carbasugars and carbanucleosides these compounds are of pharmacological importance. Since the labile acetal moiety is replaced by a more stable C-C bond a better biostability is to be expected [3]. Nucleoside analogs are of particular importance in the field of antiviral chemotherapy, in particular regarding Herpes viruses and HIV [1a, d]. It was hoped that the high doses to be applied, caused by short half lives of the drugs in the organism, leading to severe side effects, might be reduced by application of the carba analogs. This stimulated intense research efforts in this area. Some carbanucleosides revealed considerable pharmacological activity but soon it was learned that the structure/activity correlations in the carba field seem to be considerably different from those of the normal nucleosides [4].

12.2 Enantiopure Carbahexo- and Pentofuranoses

Carbanucleosides are chiral compounds. Taking into consideration that, as a rule, only one enantiomer has to be regarded as the biologically active one - yet exceptions from this assumption are known - much work was devoted to the development of approaches to obtain these compounds enantiomerically pure [1c,d].

Besides chiral pool methodology, classical resolutions chemical asymmetric syntheses and chemoenzymatic approaches are known. In the latter case enzymatic asymmetrisation of *meso*-compounds [5], enzymatic resolutions [6] and enzymic asymmetric syntheses [7] were applied in this field. Surprisingly, recently it was published [8] that one of these resolutions did not work reproducibly while other authors stated the contrary [9].

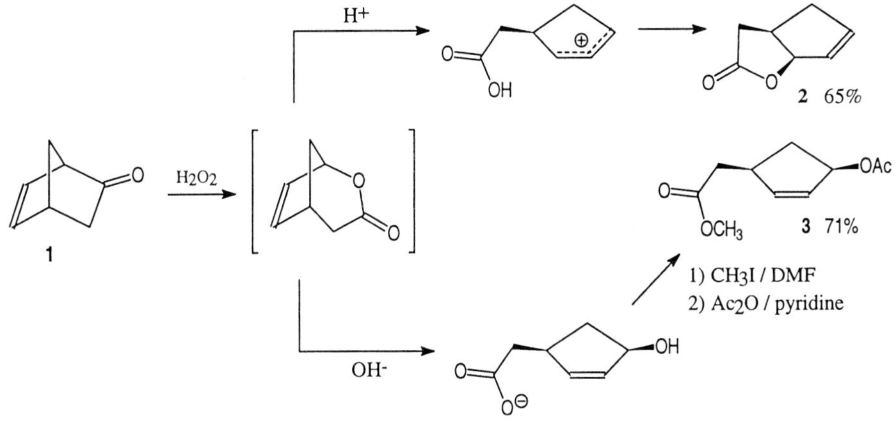

Scheme 12-2. Bayer-Villiger reaction of (+)-5-norbornen-2-one (**1**). In acidic medium 3,6-anhydro-2,5-dideoxy-3a-carba-D-*threo*-hex-5-enono-1,4-lactone (**2**) is formed, in alkaline medium the reaction leads to methyl 1-O-acetyl-2,3,5-trideoxy-4a-carba-β-D-*glycero*-hex-2-enofuranuronoate (**3**) [10].

Some years ago a procedure for the preparation of (+)-5-norbornen-2-one (**1**) with up to 98% enantiomeric excess was developed in our laboratory [2a, 11]. Using this compound as starting material is particularly effective for the synthesis of carbasugars since both the double bond and the oxo group enable the cleavage of the bicyclic system to give differently substituted cyclopentanes. Thus, carbafuranoses are easily accessible. Synthetic use of (+)-*endo*-5-norbornenyl acetate employing ozonolysis of the double bond was reported earlier from our laboratory [2a, 4, 12]. Therefore, in this report attention will be paid to Baeyer-Villiger reaction of the oxo moiety where, depending on the conditions, unsaturated carbahexofuranuronic acid derivatives are obtained (see

Scheme 12-2). The double bond offers the opportunity of stereospecific functionalisation with hydroxyl groups to provide access to all stereoisomers of 5-deoxycarbahexofuranoses and carbapentofuranoses.

The synthetic strategy to be applied was twofold: [2]
1) stereocontrolled introduction of the hydroxyl groups and
2) side chain degradation to obtain the carbapentofuranoses

For the stereocontrol during the formation of the hydroxyl groups three approaches were used.

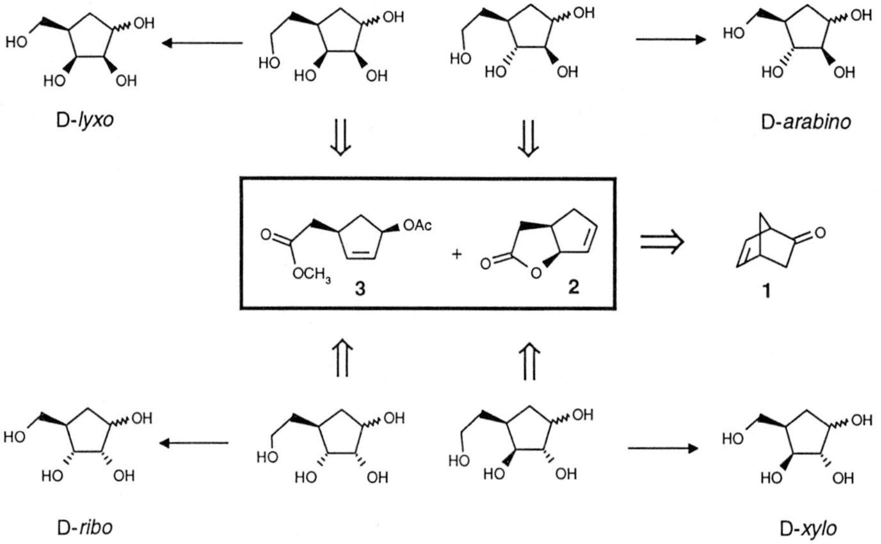

Scheme 12-3. Strategy for carbapentofuranoses.

a) 2,3-cis-diols (Scheme 12-4) were obtained by cis-dihydroxylation of the 2,3-double bond of methyl 1-O-acetyl-2,3,4-trideoxycarba-ß-D-*glycero*-hex-2-enofuranuronate (3) using OsO$_4$/NMNO. The discrimination between both faces of the plane of the double bond was effected by the C-1 substituent. In case of D-*glycero*-stereochemistry (3), the ß-D-*ribo*-product 5 was formed by an approach *anti* to the substituents on C-1 and C-4. If the configuration at C-1 is inverted by a Mitsunobu reaction using diethyl azodicarboxylate, triphenylphosphine and benzoic acid, from the 1-O-benzoate, by attack *anti* to the benzoyloxy group, exclusively the α-D-*lyxo*-compound 8 is obtained. α/ß-Isomerisation was easily achieved by the sequence: Swern oxidation/sodium borohydride reduction, leading in both the *ribo*- and *lyxo*-series to diastereomers 6 and 9 resulting from attack from the sterically more accessible side.

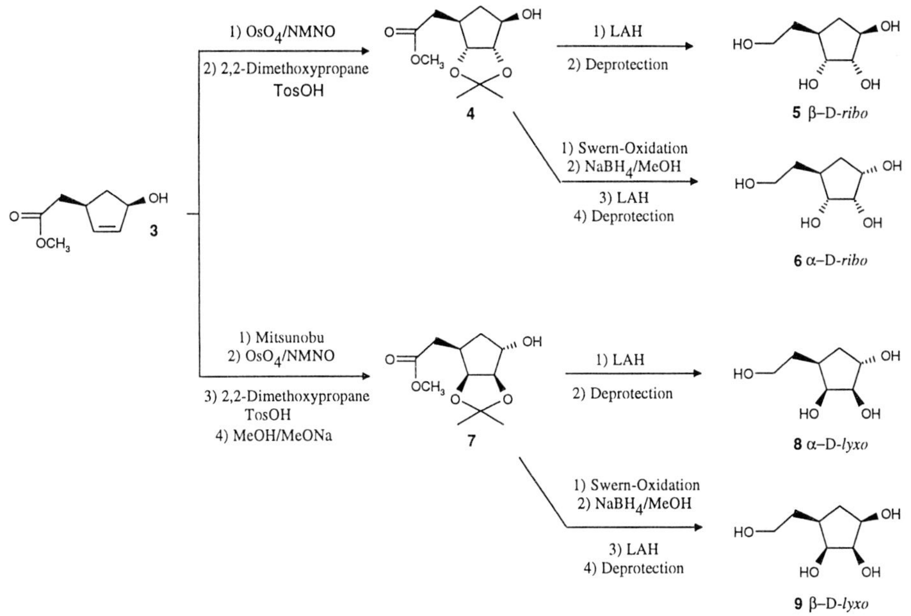

Scheme 12-4. Synthesis of 5-deoxycarba-D-*ribo*- and D-*lyxo*-hexofuranoses.

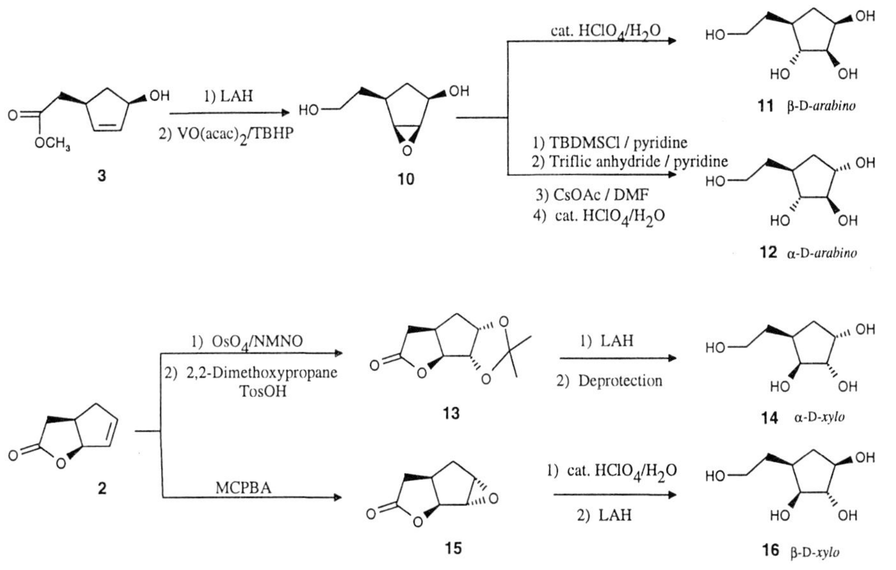

Scheme 12-5. Synthesis of 5-deoxycarba-D-*arabino*- and -D-*xylo*-hexofuranoses.

b) To obtain *2,3-trans*-diols, acid catalysed opening of a 2,3-epoxide appeared to be the method of choice (Scheme 12-5). Therefore, the same starting material as used for the preparation of 2,3-*cis*-diols was reduced to the diol to have the hydroxyl groups available for a vanadium catalysed hydroxyl directed epoxidation which gave the all-*cis*-epoxide **10**. Opening of this epoxide using perchloric acid yielded the ß-D-*arabino* configuration (**11**). For an access to the α-stereoisomer in intermediate **10** the configuration of C-1 was inverted after protection of the primary hydroxyl group as TBDMS ether by reaction of the secondary triflate with cesium acetate.

c) The best approach for steering the direction of the introduction of the hydroxyl groups in the *xylo*-series was to start with the product of the acid catalysed Baeyer-Villiger reaction of norbornenone (**2**). *cis*-Dihydroxylation with OsO_4/NMNO resulted in *anti*-addition with respect to the lactone ring to give after reduction of the lactone 5-deoxycarba-α-D-*xylo*-hexofuranose (**14**), while epoxidation - again *anti* to the lactone ring - followed by oxirane ring opening was the synthetic route to the ß-D-*xylo* stereoisomer **16**.

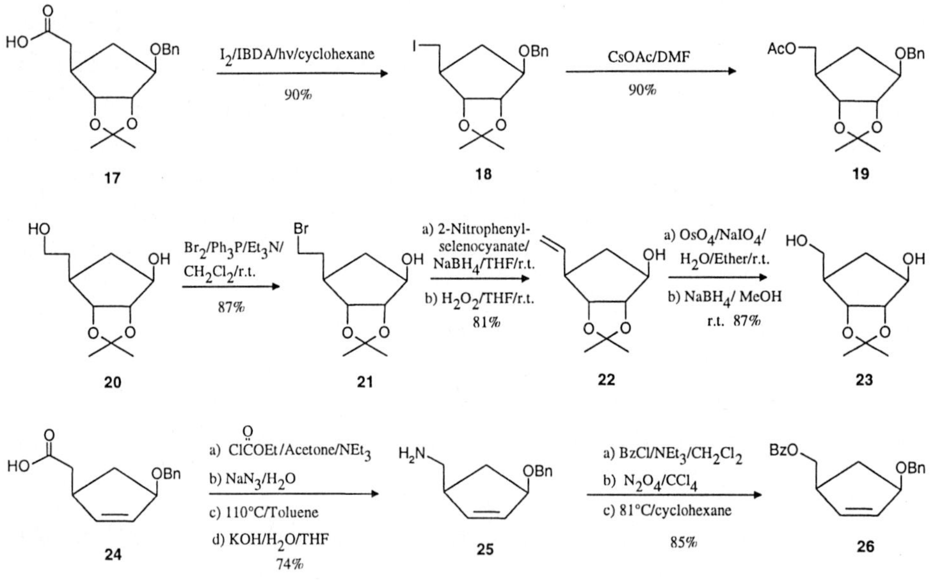

Scheme 12-6. Examples for side chain degradation to convert carbahexofuranoses into carbapentofuranoses.

To switch over into the pentose series, side chain degradation by one carbon atom is necessary. In Scheme 12-6 examples for those methods are given which worked best in our hands. A Hunsdiecker type reaction employing iodine/IBDA [13] proved to be excellent in the carba-ß-D-*ribo* series.

Alternatively, although with additional steps and slightly lower yields, selective mild bromination of the hydroxyl group on C-6 followed by treatment with 2-nitrophenylselenocyanate/NaBH$_4$ and reaction with H$_2$O$_2$ gave the olefin which was cleaved by OsO$_4$/NaIO$_4$ and then reduced with NaBH$_4$ to the protected carba-ß-D-ribofuranose. Still another method, applied in the 2,3-eno-series, was a Curtius degradation to the 5-amino derivative which was converted into the 5-O-benzoyl compound by benzoylation, nitrosation and thermal cleavage [14].

12.3 Introduction of the Base

Methods to link the purine or pyrimidine base to the sugar moiety are well established in nucleoside chemistry [15]. The situation is more complicated in the carba series since the carba analogs of the halogenoses are less reactive. The main approaches used are shown in Scheme 12-7. In principle two strategies are possible. Either the purine or pyrimidine ring is constructed by using an amino function at C-1. As an example from our laboratory the synthesis of protected 2-deoxycarbauridine is shown [16] following the procedure of Shaw and Warrener [17].

Alternatively, displacement reactions using the base as a nucleophile can be applied, as in conventional nucleoside synthesis. Sufficient reactivity must, however, be secured by either having allylic substrates, e.g. as applied by Marquez [18] or by using Mitsunobu techniques, e.g. by Benner [19]. Further possibilities are the opening of an epoxide by the alkali salt of the base, see e.g. Roberts [20] or, in case of a neighbouring double bond, by employing palladium chemistry [21]. In addition, Michael type reactions have been applied [22]. Recently, we used AlEt$_3$ for the formation of the nucleophile from the base in palladium assisted reactions which resulted in good isolated yields (Table 12-1) [23]. An advantage of this method is the good solubility of AlEt$_3$ in organic solvents, which results also in a good solubility of the metallated base. Reaction conditions are comparably mild and allow the use of THF as a solvent for all the common purine bases.

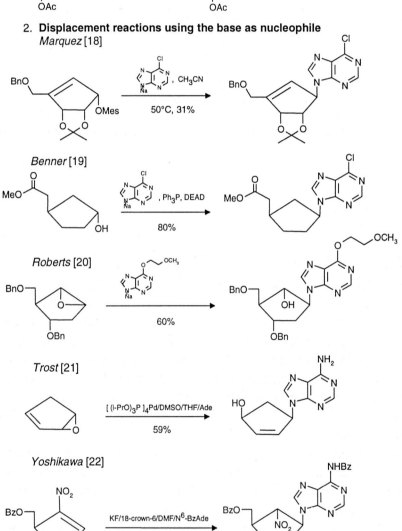

Scheme 12-7. Synthetic approaches to carbocyclic nucleosides.

Table 12-1. Indroduction of the base using AlEt$_3$.

Isolated yields [%]

	Ura	Thy	Cyt	CytN^{4}Bz	Ade	AdeN^{6}Bz	6-Cl-Purine	GuaNAc
28	17	97	69	98	N^7 20	7	92	16
					N^9 44	80		60
29	37	40	-	-	N^7			10
					N^9			42

12.4 Carbauloses

Contrary to the large amount of work devoted to carbaaldoses much less is known about carbauloses [24]. A recent synthesis of α and ß-D-*ribo*-carbahex-2-ulofuranose from this laboratory is shown in Scheme 12-8 [25]. Starting from 2,3-O-isopropylidene-5-O-trityl-DL-carbariburono-1,4-lactone (**30**) 1,2-anhydro-3,4-O-isopropylidene-6-O-trityl-α-DL-*ribo*-carbahex-2-ulofuranose (**31**) was formed in the reaction with CH$_2$Br$_2$/n-BuLi. As it was to be expected the reaction occurred by an attack *anti* to the isopropylidene ring. Nucleophilic opening of the epoxide and deprotection gave ß-DL-*ribo*-carbahex-2-ulofuranose (**32**).

Surprisingly, when using instead the sulfur ylide formed from (CH$_3$)$_3$SOI/NaH *syn* attack with respect to the isopropylidene ring was observed leading to product **33**. Besides the NMR data, evidence for the correct assignment of the stereochemistry came from the formation of the same product from 2,5-anhydro-1-deoxy-3,4-O-isopropylidene-6-O-trityl-DL-*ribo*-carbahex-1-enitol (**34**) by epoxidation with dimethyldioxirane where, as expected, attack from the stereochemically less hindered side occurred.

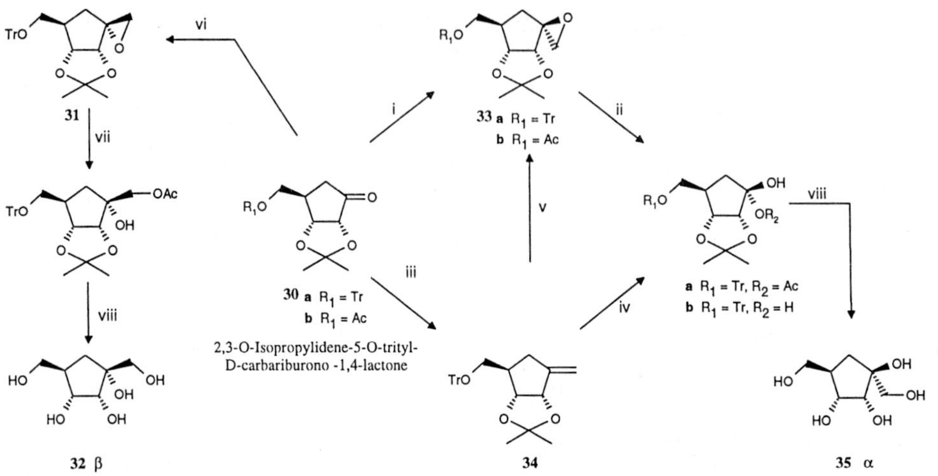

Scheme 12-8. Reagents: i) NaH/(CH$_3$)$_3$SOI/DMSO/THF/r.t.; ii) NaOAc/DMF/140°C or CsOAc/DMF/80°C iii) Cp$_2$Ti(CH$_3$)$_2$/toluene/60-70°C; iv) OsO$_4$/NMNO/acetone/r.t. v) oxone/acetone/18-crown-6/NaHCO$_3$/H$_2$O/CH$_2$Cl$_2$/r.t. or MCPBA/benzene/reflux; vi) CH$_2$Br$_2$/n-BuLi/THF/-80°C-r.t.; vii) CsOAc/DMF/90°Cviii a) Amberlite IR-120/CH$_3$CN/H$_2$O/50°C b) Ac$_2$O/pyridine/DMAP/CH$_2$Cl$_2$ c) CH$_3$OH/CH$_3$ONa/r.t.

Apparently, there must be some special interaction between the sulfur ylide and the neighbouring oxygen of the isopropylidene ring. The stereoselectivities observed are again summarised in Scheme 12-9 which also contains information on related reactions such as with NaBH$_4$, TMSCH$_2$MgCl, (CH$_3$)$_2$SCH$_2$ and CH$_2$N$_2$. To obtain further information on the factors responsible for this unexpected steric course a more detailed investigation was performed [25].

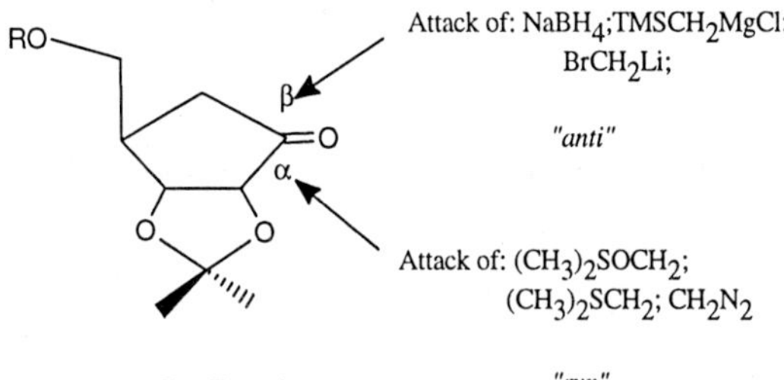

Scheme 12-9. Nucleophilic attack to α-oxygenated cyclic ketones.

Various uloses and carbauloses were reacted with (CH$_3$)$_2$SCH$_2$, (CH$_3$)$_2$SOCH$_2$, BrCH$_2$Li, and the sequence Cp$_2$Ti(CH$_3$)$_2$/MCPBA or dimethyldioxirane. The results are given in Table 12-2. The reason for the predominance of *syn* attack in the case of the sulfur ylides seems to be an electrostatic attraction between the positively charged sulfur atom and the

Chemoenzymatic Approaches to Enantiopure.. 233

neighbouring oxygen atom. A similar explanation was given as rationalisation of a related behaviour of diazomethane [26].

Table 12-2. The reaction of some α-oxygenated cyclic ketones with sulfur ylides. For conditions A, B, C, D see foonote at the bottom of the Table.

Entry	Substrate	Method and side of attack	Product	Yield
1[25b]		R = Tr A syn B syn D anti		A 74 % B 86 % D 73 %
2		R = Ac C anti A syn B syn		C 81 % A 92 % B 88 %
3		A anti C syn		A 70 % C 51 %
		B syn		B 64 %
4		A B		A 68 % B 72 %
5 [24b]		A anti		A 48 %

Table 12-2. The reaction of some α-oxygenated cyclic ketones with sulfur ylides, continued.

Entry	Substrate	Method and side of attack	Product	Yield
6[27]		A syn B syn D anti		A 82 % B 83 % D 73 %
		C anti		C 72 %
7		B anti D anti		B 81 % D 83 %
		C anti A syn : anti 2:1		C 47 % A 35 %
8[28]		B syn : anti 2:1 C anti		B 44 % C 67 %
		D anti A anti		D 68 % A 53 %

Table 12-2. The reaction of some α-oxygenated cyclic ketones with sulfur ylides, continued..

Entry	Substrate	Method and side of attack	Product	Yield
9		B *syn*		B 94 %
		A *syn : anti* 2:1 C *syn : anti* 2:1 D *anti*		A 45 % C 98 % D 47 %

Conversions of ketones with α-oxygen substituents into epoxides using either reactions with dimethylsulfonium methylide (A), dimethyloxosulfonium methylide (B), bromomethyllithium (C), or synthesis of the isosteric olefin and subsequent epoxidation with MCPBA or dimethyldioxirane (D).

Acknowledgements

This work could not have been done without the skill and committment of a number of gratuate students especially C. Marschner, who did most of the work, H. Kapeller and H. Baumgartner. Financial support from the Austrian Science Foundation and the Christian Doppler Society is gratefully acknowledged.

12.5 References

1. Recent reviews: (a) V. E. Marquez, M.-I. Lim, *Med. Res. Rev.* **1986**, *6*, 1-40. (b) T. Suami, *Top. Curr. Chem.* **1990**, *154*, 258-283. (c) T. Suami, S. Ogawa, *Adv. Carbohydr. Biochem.* **1990**, *48*, 21-90. (d) A. D. Borthwick, K. Biggadike, *Tetrahedron* **1992**, *48*, 571-623.(e) L. Agrofoglio, E. Suhas, A. Farese, R. Condom, S. R. Challand, R. A. Earl, R. Guedj, *Tetrahedron* **1994**, *50*, 10611-10670.

2. (a) J. Balzarini, H. Baumgartner, M. Bodenteich, E. De Clercq, H. Griengl, *Nucleosides & Nucleotides* **1989**, *8*, 855-858. (b) C. Marschner, J. Baumgartner, H. Griengl, *J. Org. Chem.* **1995**, *60*, 5224-5235. (c) G. E. McCasland, S. Furuta, L. J. Durham, *J. Org. Chem.* **1966**, *31*, 1516-1521.

3. L. L. J. Bennett, W. M. Shannon, P. W. Allan, G. Arnett, *Ann. N.Y. Acad. Sci.* **1975**, *255*, 342-358.

4. C. Desranges, G. Razaka, M. Rabaud, H. Bricaud, P. Herdewijn, E. De Clerq, *Biochem. Pharmacol.* **1983**, *32*, 3583-3586.

5. (a) M. Arita, K. Adachi, Y. Ito, H. Sawai, M. Ohno, *J. Am. Chem. Soc.* **1983**, *105*, 4049-4055. (b) J. R. Medich, K. B. Kunnen, C.R. Johnson, *Tetrahedron Lett.* **1987**, *28*, 4131-4134. (c) D. R. Deardorff, S. Shambayati, D.C. Myles, D. Heerding, *J. Org. Chem.* **1988**, *53*, 3614-3615. (d) M. Tanaka, M. Yoshioka, K. Sakai, *Tetrahedron: Asymmetry* **1993**, *4*, 981-996.

6. (a) M. Bodenteich, H. Griengl, *Tetrahedron Lett.* **1986**, *27*, 4291-4292. (b) S. Sicsic, M. Ikbal, F. Le Goffic, *Tetrahedron Lett.* **1987**, *28*, 1887-1888. (c) M. S. Levitt, R.F. Newton, S. M. Roberts, A. Willetts, *J. Chem. Soc., Chem. Commun.* **1990**, 619-620. (d) S. J. C. Taylor, A.G. Sutherland, C. Lee, R. Wisdom, S. Thomas, S. M. Roberts, C. Evans, *J. Chem. Soc., Chem. Commun.* **1990**, 1120-1121.

7. (a) T. Hudlicky, H. Luna, G. Barbieri, L. D. Kwart, *J. Am. Chem. Soc.* **1988**, *110*, 4735-4741. (b) H. A. J. Carless, *Tetrahedron: Asymmetry* **1992**, *3*, 795-826.

8. R. Csuk, P. Dürr, *Tetrahedron* **1995**, *20*, 5799-5805.

9. (a) M. Ikbal, C. Cerceau, F. Le Goffic, S. Sicsic, *Eur. J. Med. Chem.* **1989**, *24*, 415-420. (b) M. Mahmoudian, B. S. Baines, M. J. Dawson, G. C. Lawrence, *Enzyme Microbiol. Technol.* **1992**, *14*, 911-916. (c) A. Haupt, private communication.

10. (a) J. Meinwald, M. C. Seidl, B. C. Cadoff, *J. Am. Chem. Soc.* **1958**, *80*, 6303-6309. (b) C. J. Harris, *J. Chem. Soc., Perkin Trans I* **1980**, 2497-2502. (c) P. T. W. Cheng, S. Mc Lean, *Can. J. Chem.* **1989**, *67*, 261-267.

11. (a) G. Eichberger, G. Penn, K. Faber, H. Griengl, *Tetrahedron Lett.* **1986**, *27*, 2843-2844. (b) T. Oberhauser, M. Bodenteich, K. Faber, G. Penn, H. Griengl, *Tetrahedron* **1987**, *43*, 3931-3944. (c) C. Marschner, Ph.D. Thesis, Technical University Graz, **1992**.

12. (a) M. Bodenteich, K. Faber, G. Penn, H. Griengl, *Nucleosides & Nucleotides* **1987**, *6*, 233-237. (b) M. Bodenteich, H. Griengl, *Nucleic Acids Res. Symp. Ser.* **1987**, *18*, 13-16. (c) M. Bodenteich, H. Griengl, *Tetrahedron Letters* **1987**, *28*, 5311-5312. (d) H. Baumgartner, M. Bodenteich, H. Griengl, *Tetrahedron Lett.* **1988**, *29*, 5745-5746. (e) J. Balzarini, H. Baumgartner, M. Bodenteich, E. De Clerq, H. Griengl, *J. Med. Chem.* **1989**, *32*, 1861-1865.

13. (a) J. I. Concepcion, C. G. Francisco, R. Freire, R. Hernandez, J. A. Salazar, E. Suarez, *J. Org. Chem.* **1986**, *51*, 402-404. (b) L. Ötvös, J. Beres, G. Sagi, I. Tömösközi, L. Gruber, *Tetrahedron Lett.* **1987**, *28*, 6381-6384. (c) H. Kapeller, Diploma Thesis, Technical University Graz, **1991**.

14. (a) E. H. White, *J. Med. Chem. Soc.* **1955**, *77*, 6011-6014. (b) E. Neufellner, Diploma Thesis, Technical University Graz, **1994**.

15. H. Vorbrüggen, K. Krolikiewicz, B. Bennua, *Chem. Ber.* **1981**, *114*, 1234-1255.

16. J. Balzarini, H. Baumgartner, M. Bodenteich, E. De Clerq, H. Griengl, *Nucleosides & Nucleotides* **1989**, *8*, 855-858.

17. G. Shaw, R. N. Warrener, *J. Chem. Soc.* **1958**, 153-156 and 157-161.

18. (a) V. E. Marquez, M. -I. Lim, C. K. H. Tseng, A. Markovaz, M. A. Priest, M. S. Khan, B. Kaskar, *J. Org. Chem.* **1988**, *53*, 5709-5714. (b) C. K. H. Tseng, V. E. Marquez, *Tetrahedron Lett.* **1985**, *26*, 3669-3672.

19. (a) T. F. Jenny, N. Previsani, S. A. Brenner, *Tetrahedron Lett.* **1991**, *32*, 7029-7032. (b) T. F. Jenny, J. Horlacher, S. A. Brenner, *Helv. Chim. Acta.* **1992**, *75*, 1944-1954. (c) for a review, see: O. Mitsunobu, *Synthesis*, **1981**, 1. (d) J. Nokami, H. Matsura, H. Takahashi, M. Yamashita, *Synlett* **1994**, 491-493.

20. (a) K. Biggadike, A. D. Borthwick, A. M. Exall, B. E. Kirk, S. M. Roberts, P. Youds, *J. Chem. Soc., Chem. Commun.* **1987**, 1083-1084. (b) G. V. B. Madhavan, D. P. C. McGee, R. M. Rydzewski, R. Boehme, J. C. Martin, E. J. Prisbe, *J. Med. Chem.* **1988**, *31*, 1798-1804.

21. (a) B. M. Trost, G. H. Kuo, T. Benneche, *J. Am. Chem. Soc.* **1988**, *110*, 621-622. (b) B. M. Trost, L. Li, S.D. Guille, *J. Am. Chem. Soc.* **1992**, *114*, 8745-8749.

22. (a) M. Yoshikana, T. Nakae, B. C. Cha, Y. Yokokama, I. Kitagawa, *Chem. Pharm. Bull.* **1989**, *37*, 545-551. (b) M. Yoshikana, Y. Okaichi, B. C. Cha, I. Kitagawa, *Tetrahedron* **1990**, *46*, 7459-7470. (c) I. Kitagawa, B. C. Cha, T. Nakae, Y.Okaichi, Y. Takinami, M. Yoshikana, *Chem. Pharm. Bull.* **1989**, *37*, 542-544.

23. C. Marschner, H. Kapeller, M. Weissenbacher, unpublished results.

24. (a) C. S. Wilcox, J. J. Gaudino, *J. Am. Chem. Soc.* **1986**, *108*, 3102-3104 and *Carbohydr. Res.* **1990**, *206*, 233-250. (b) M. Bodenteich, V. E. Marquez, *Tetrahedron Lett.* **1990**, *31*, 5977-5980.

25. (a) C. Marschner, J. Baumgartner, H. Griengl, *Liebigs Ann. Chem.* **1994**, 999-1004. (b) C. Marschner, G. Penn, H. Griengl, *Tetrahedron* **1993**, *49*, 5067-5078.

26. (a) J. Yoshimura, K. Sato, K. Kobayashi, C. Shin, *Bull. Chem. Soc. Jpn. 1973*, *46*, 1515-1519. (b) B. Flaherty, S. Nahar, W. G. Overend, N. R. Williams, *J. Chem. Soc., Perkin Trans I* **1973**, 632-638. (c) J. Yoshimura, K. Sato, *Bull. Chem. Soc. Jpn.* **1978**, *51*, 2116-2121. (d) J. Yoshimura, K. Sato, *Carbohydr. Res.* **1979**, *73*, 75-84 and **1982**, *103*, 221-238.

27. (a) I. I. Cubero, *Carbohydr. Res.* **1983**, *114*, 311-316. (b) K. Sato, K. Suzuki, M. Ueda, M. Katayama, Y. Kajihara, *Chem. Lett.* **1991**, 1469-1472.

28. M. Funabashi, N. Hong, H. Kodama, J. Yoshimura, *Carbohydr. Res.* **1978**, *67*, 139-145.

13 Glycomimetics that Inhibit Carbohydrate Metabolism

Bruce Ganem

13.1 Introduction

Glycosyl group transfer reactions are among the most important processes in carbohydrate biochemistry, and are catalyzed by two families of enzymes: glycosidases and glycosyl transferases [1]. Overall, the process involves cleavage of the glycoside bond linking the anomeric carbon of a sugar with an oligo or polysaccharide or a nucleoside diphosphate group. The liberated glycosyl group may then be transferred to water (by glycosidases) or to some other nucleophilic acceptor (by transferases).

Glycosidases play many fundamental roles in biochemistry and metabolism [2]. *Exoglycosidases*, which remove sugars one at a time from the non-reducing end of an oligo or polysaccharide, are involved in the breakdown of starch and glycogen, the processing of eucaryotic glycoproteins, the biosynthesis and modification of glycosphingolipids, and the catabolism of peptidoglycans and other glycoconjugates. *Endoglycosidases* are capable of cleaving interior glycosidic bonds within polysaccharides. Besides being involved in the catabolism of aged glycoproteins, they also catalyze the alteration of bacterial and plant cell wall, and the hydrolysis of highly insoluble, structural polysaccharides like chitin and cellulose. Given the fact that cellulose and its derivatives constitute two-thirds of the biosphere's carbonaceous material, and a large-scale source of fuel ethanol [3], exo and endocellulases perform one of the world's most important biochemical reactions.

A major goal of our research has been to design and synthesize new sugar mimics that might function as glycosidase inhibitors [4]. Using such agents, chemists might exert more potent and effective control over glycoside hydrolysis. Aside from their potential value in basic biochemicl research, several synthetic carbohydrate mimics have already demonstrated promising therapeutic applications, both in the area of diabetes management [5] and antiviral chemotherapy [6].

Glycoside hydrolysis may be accomplished either with retention or inversion of configuration (Figure 13-1) [1,2,7]. While the catalytic apparatus of both retaining and inverting enzymes typically features bilateral carboxylic acid groups, the different stereochemical outcomes are achieved by mechanistically distinct pathways. Inverting enzymes use a combination of carboxylic acid and carboxylate groups to achieve acid and base catalysis of direct attack by a water molecule at the anomeric center, whereas in retaining enzymes, the carboxylate functions as a nucleophile in a process involving a glycosyl-enzyme intermediate. X-ray crystallographic studies of several glycosidases further establish that the two carboxyl residues are more widely separated in inverting enzymes than in retaining enzymes, presumably to create enough space for the participating water molecule [8].

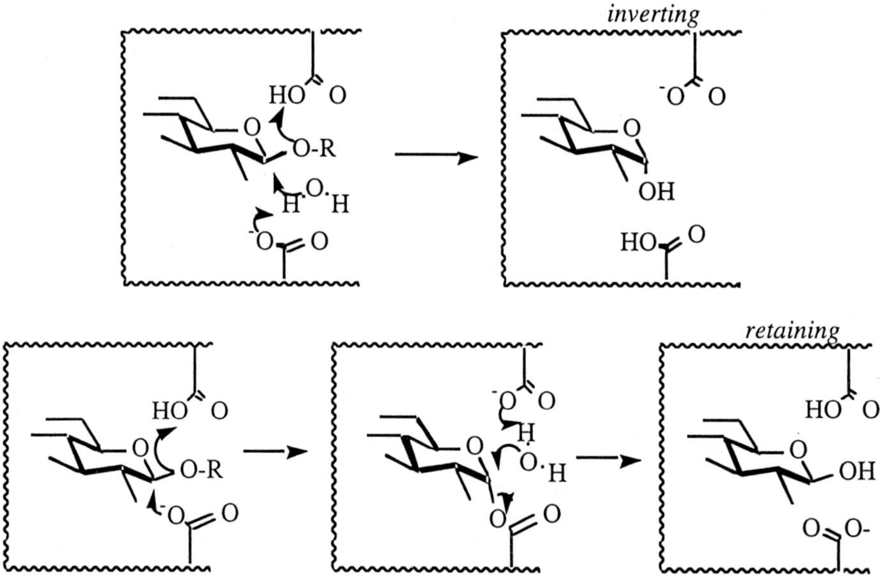

FIGURE $.1

Figure 13-1

Several families of azasugars, in which the pyranose oxygen is replaced by NH, have been widely investigated as glycomimetics. Likewise, the discovery of polyhydroxylated aminocyclohexanes (i.e. "aminocyclohexitols") such as streptomycin, gentamycin, and tobramycin over fifty years ago triggered considerable interest in the assembly of stereochemically well-defined pseudosugar analogs. Quite recently, a new family of aminocyclopentitol-containing natural products has been discovered whose representatives display potent and selective effects on a variety of biologically important glycosidases. Examples (Scheme 13-1) include mannostatins A and B (**1-2**), which are selective mannosidase inhibitors [9], allosamizoline (**3**), derived from the allosamidin family of pseudotrisaccharide chitinase inhibitors [10], and the related trehalase inhibitor trehazolin (**4**) [11]. Inhibitors **1-4** have aroused considerable interest among synthetic chemists [12-22]. Several laboratories have reported total synthesis endeavors, either involving (a) use of an enantiomerically pure starting material (typically carbohydrate-based) drawn from the chiral pool [12-19], or (b) elaboration of di- and trisubstituted cyclopentenes, via osmylation and other vicinal addition reactions, to functionalize the cycloalkene bond stereoselectively [20-25].

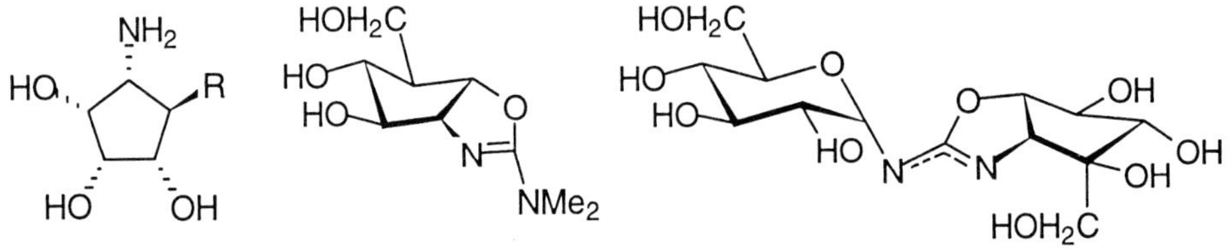

1 R = SCH$_3$
2 R = SOCH$_3$

Scheme 13-1

13.2 Studies on Mannostatin

In 1989, extracts of the soil microorganism *Streptoverticillium verticillus* were found to contain an unusual pentasubstituted cyclopentane named mannostatin A (**1**, Scheme 13-2) [9]. A corresponding sulfoxide of mannostatin A was also isolated and designated mannostatin B (**2**). Both **1** and **2** were potent competitive inhibitors of rat epididymal α-mannosidase, with inhibition constants (K_I) of 48 nM. Mannostatin A also competitively inhibited jackbean, mung bean, and rat liver lysozomal α-mannosidases with IC$_{50}$ values of 70 nM, 450 nM, and 160 nM, respectively [26]. Mannostatin A was also a potent inhibitor of Golgi processing mannosidase II (IC$_{50}$ =10-15 nM), but was inactive against processing mannosidase I [26]. Studies on cell cultures demonstrated that **1** blocked glycoprotein processing with concomitant increases in hybrid oligosaccharides, as would be consistent with mannosidase II inhibition.

The structure of mannostatin A was first established by nuclear magnetic resonance and mass spectrometry [9a]. This assignment was later confirmed by

X-ray diffraction, and the absolute stereochemistry of the inhibitor was determined to be that shown in **1** [9b]. X-ray analysis also established the R-configuration of the additional stereocenter in mannostatin B, as indicated in **2**.

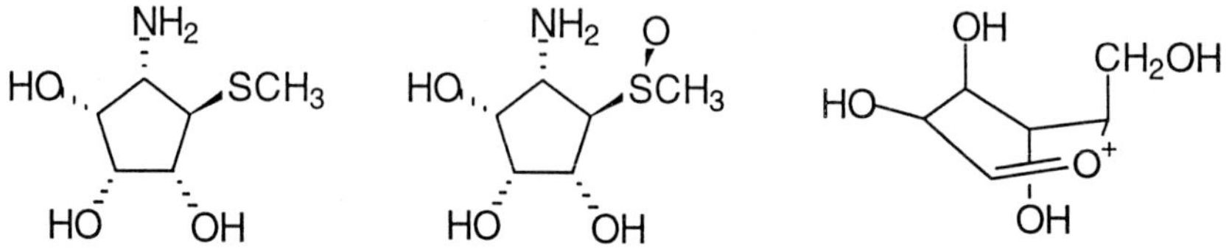

Scheme 13-2

The potent activity of **1** is intriguing since its carbocyclic structure represents a significant departure from known alkaloid-based inhibitors, and bears little resemblance either to D-mannose or to the mannopyranosyl cation **5** the purported intermediate in hydrolysis [27]. It is even more intriguing to note that the mirror image of **1** (not shown) more closely resembles the putative transition structure for mannopyranoside hydrolysis than does enantiomer **1** itself [28].

From a biochemical perspective, interest in mannostatin A was heightened by preliminary reports that glycosidase inhibitors exerted immunoregulatory effects and showed promising activity in a variety of antitumor and antiviral screens [29,30]. Together with its noteworthy biological activity, the unusual and complex stereostructure of **1** made mannostatin A a worthwhile target for synthesis.

The first total syntheses of **1** were reported simultaneously in 1991 by us [23] and by Knapp and Dhar at Rutgers [28]. The Cornell route employed an acylnitroso cycloaddition to prepare both enantiomerically pure and racemic **1**, while the Rutgers team synthesized the individual enantiomers of **1** separately from D- or L-ribonolactone. Both efforts confirmed the assignment of absolute stereochemistry and further demonstrated that the levorotatory enantiomer of **1** had no inhibitory effect on mannosidases [23,28]. Since then, additional syntheses of racemic mannostatin A have been reported [31,32].

Stereocontrolled assembly of the cyclopentane ring in mannostatin A, with its five chiral centers and imposing array of functionality, poses a significant challenge to the synthetic chemist. While a variety of {3+2} cycloaddition processes are known to provide access to functionalized cyclopentanes [33]. We preferred a {4+2} cycloaddition strategy that would generate three of the five chiral centers in **1** while simultaneously installing two of the heteroatoms as part of an appropriate N=O dienophile. The cycloaddition of acylnitroso compounds with dienes was especially attractive [34], not only because the high reactivity of the nitroso dienophile would accommodate densely functionalized or relatively unreactive dienes, but also because a successful asymmetric version of the diene/nitroso cycloaddition reaction has been developed for enantioselective synthesis [35].

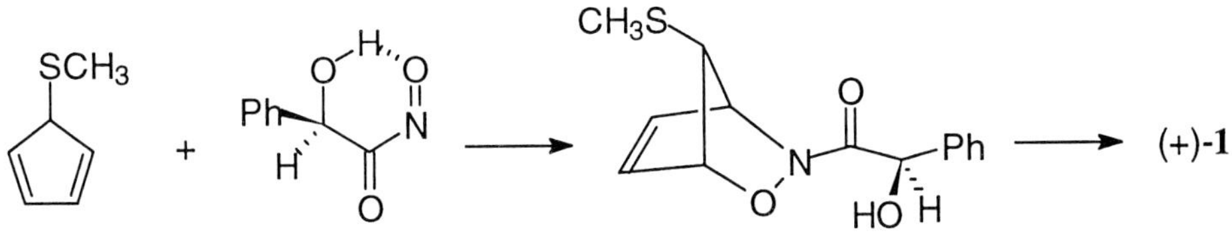

Scheme 13-3

Asymmetric cycloaddition of the acylnitroso compound **7** derived from R-mandelic acid [35a] with the known 1-(methylthio)cyclopenta-2,4-diene (**6**) (Scheme 13-3) [36], should afford predominantly bicyclic adduct **8** with the correct absolute configuration for (+)-mannostatin A. Both relative and absolute stereocontrol in forming the three contiguous O,S and N-containing chiral centers would be assured by (a) *anti* orientation of the methylthio group in **6** as it approaches the heterodienophile, and (b) intramolecular hydrogen bonding in **7** which would direct face-selective *endo* cycloaddition *anti* to the bulky phenyl group in **7**. The former effect is well-precedented in the chemistry of 5-substituted cyclopentadienes, while the latter has been documented in prior cycloadditions of **7** itself [35a]. Subsequent osmylation of the remaining double bond was expected to complete the efficient assembly of substituents in **1**.

Both **6** and **7** are highly reactive species, which complicated the desired cycloaddition. The reported synthesis of **6** involved addition of methylsulfenyl chloride to a solution of cyclopentadienyl-thallium in CCl$_4$ at room temperature, whereupon the precipitated salts were filtered and the product distilled at reduced pressure [36]. Acylnitroso compound **7** was customarily generated by *in situ* oxidation of the corresponding (R)-mandelohydroxamic acid [37] which was then immediately trapped by diene. A 2.6:1 ratio of diastereomers was formed, with **8** as the major component. Several experimental observations have led to a significantly improved procedure.

Preparation of **6** by the published method [36] also led to bis-sulfenylated byproducts; however, by using CH$_3$SCl as the limiting reagent (typically 0.8 equiv), bis-sulfenylation could be minimized. Nevertheless yields of cycloadduct **8** never exceeded 35%. Since commercial samples of cyclopentadienylthallium (the best quality available is listed as 97% pure by Aldrich) were slightly colored, the material was purified by sublimation [38]. Gratifyingly, the yields of **8** rose using pure **6**, and by further limiting the quantity of CH$_3$SCl to 0.6 equiv, multigram scale cycloaddition reactions routinely afforded **8** and its diastereomer (not shown; now as a 3.3:1 ratio) in 45-50% overall yield. Flash chromatography and recrystallization gave pure **8** (mp 89-90°C). With an efficient route to key intermediate **8** at hand, the synthesis of **1** could be developed on a preparative scale.

It was anticipated that mannostatin's two remaining *cis*-hydroxyl groups might be introduced stereoselectively by vicinal hydroxylation of **8** from the less hindered *endo* face of the bicyclic alkene. However, this reaction proved more difficult than expected. For example, attempted catalytic osmylation using the published procedure with N-methylmorpholine-N-oxide [39] formed mixtures of the corresponding sulfoxides and sulfones of **8**. In fact, S-oxidation in the presence of co-oxidant has been observed previously [40]. On the other hand, stoichiometric amounts of OsO$_4$ led to preferential oxidation of the chiral auxiliary group, producing α-ketoamide **9** (Scheme 13-4). Clearly, *endo*-approach to the heterobicyclic {2.2.1} ring system was less favorable than several competing undesirable oxidations.

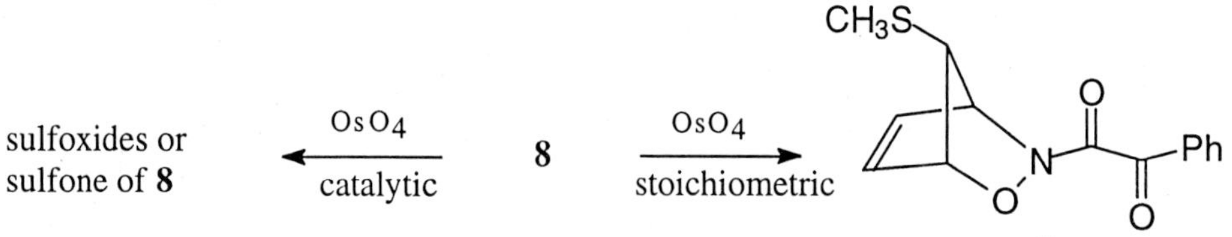

$$\text{sulfoxides or sulfone of 8} \xleftarrow[\text{catalytic}]{\text{OsO}_4} \textbf{8} \xrightarrow[\text{stoichiometric}]{\text{OsO}_4} \textbf{9}$$

Scheme 13-4

In an alternative approach that seemed ideally suited to the system at hand, we decided to exploit the high *syn*-stereoselectivity which has recently been observed in the vicinal hydroxylation of certain bis-allylically substituted cyclopentenes. In 1988, Trost *et al.* reported that osmylation of nitrosulfone **10** (Scheme 13-5) occurred with complete *syn*-stereoselectivity [41]. A subsequent investigation by Poli further noted a marked preference for *syn*-hydroxylations of the monoallylically substituted nitrosulfone **11**, cyanosulfone **12** and malonate **13** (Scheme 13-5) [42]. *syn*-Stereoselectivity leading to **14**, which was highly solvent-dependent (nonpolar solvents favored *syn*-osmylation), was documented in both stoichiometric and catalytic osmylations.

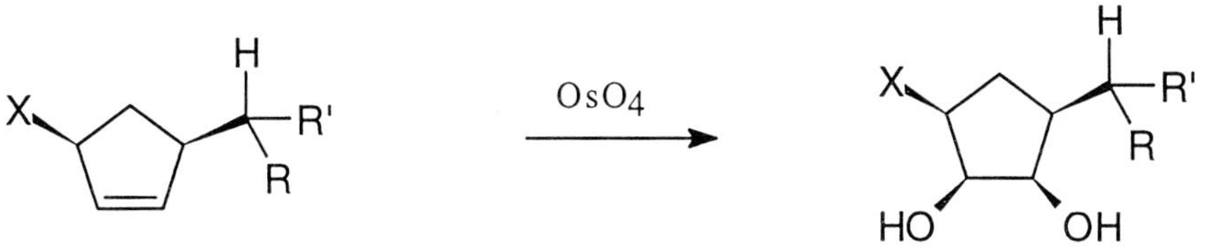

10 X = OR or N-Z-Adenyl
R = NO$_2$, R' = SO$_2$Ph
11 X = H, R = NO$_2$, R' = SO$_2$Ph
12 X = H, R = CN, R' = SO$_2$Ph
13 X = H, R, R' = CO$_2$Et

Scheme 13-5

While directed osmylations have been noted previously [40b,40c], no clearcut mechanistic explanation of this phenomenon has yet emerged. Poli has suggested that conformational effects along the reaction coordinate leading to a late transition state may be responsible for the preferential *syn*-osmylation of **10-13** [42]. Poli noted that neither ^{1}H, ^{13}C and ^{17}O NMR nor IR measurements indicate any complexation between OsO$_4$ and either nitro or sulfonyl groups. Furthermore, when the corresponding methylated analogs of **11-13** (R'= CH$_3$) were osmylated, predominantly *anti*-selectivity was observed [42].

The desired cyclopentene **16** (Scheme 13-6) was prepared by reductive cleavage of the key bicyclic intermediate **8** using aluminum amalgam in THF-H$_2$O [43]. Reduction smoothly formed cyclopentenol **15**, and the yield (originally 41%) [23a] was improved to 78% by using a large excess of amalgam (30-40 equiv). Complete characterization of **15** was facilitated by acetylation to diester **16** (94%). As with Trost's N,O-bis-allylically substituted cyclopentene **10**, stoichiometric osmylation of **16** (1.5 equiv OsO$_4$, pyridine, rt, 24 h) proceeded with high facial selectivity to afford diols **17** and **19** in a 20:1 ratio (60%). The mixture was acetylated and the resulting tetraacetates **18** and **20** were easily separable by silica gel flash column chromatography.

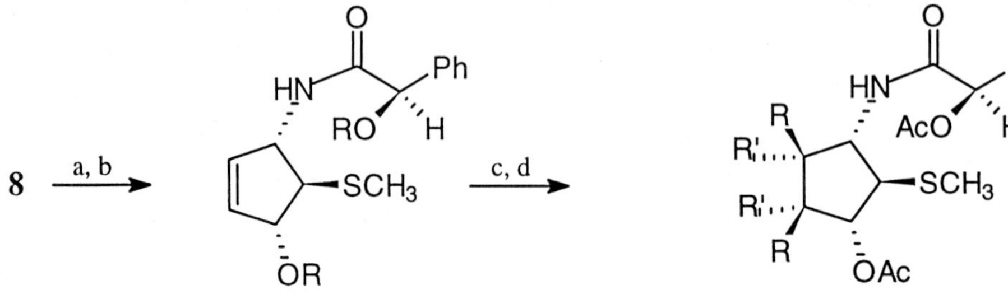

Structures

15 R = H
16 R = Ac

17 R = H, R' = OH
18 R = H, R' = OAc
19 R = OH, R' = H
20 R = OAc, R' = H

21, **22**

Scheme 13-6: Reagents: a) Al(Hg); b) Ac$_2$O, pyr. DMAP; c) OsO$_4$, pyr; d) Ac$_2$O, pyr. DMAP.

Exhaustive deacylation of **18** by exposure to acid (0.4 M HCl-CH$_3$OH, 60°C, 24 h) quantitatively formed optically active mannostatin A hydrochloride (100%), whose physical properties were identical in every respect with those of an authentic sample [44]. With several improvements in the synthesis as noted, enantiomerically pure (+)-**1** could now be obtained in about 10% yield from **4**.

Two new mannostatin congeners, bis-epi analogue **21** and dideoxy alkene **22**, were also synthesized at this stage. Acid methanolysis of **20**, the peracetylated minor product of osmylation (Scheme 13-6), required 5 days to go to completion. Apparently the rate-determining amide methanolysis in **18** was anchimerically assisted by a *cis*-hydroxyl group, an observation that would prove significant in later work. Nevertheless, more vigorous conditions (6N HCl-CH$_3$OH, reflux, 2-3 h) smoothly transformed **20** into 3,4-bis-epi-mannostatin A **21** in 77% yield. The synthesis of **22** by hydrolysis of mandelamide **15** proceeded smoothly in refluxing sodium hydroxide to afford the target aminoalkene in 71% yield.

In tests against various glycosidases, synthetic and naturally-occurring (+)-**1** exhibited the same inhibitory activity against Golgi processing mannosidase II. Moreover a synthetic sample of racemic **1** possessed one-half the potency of (+)-**1**, indicating that the unnatural antipode of mannostatin A had no effect on the activity of processing mannosidase II. In addition, synthetic (+)-mannostatin A was a potent competitive inhibitor of jackbean α-mannosidase, with K_I = 25 ±4 nM. As reported, natural (+)-**1** was a very weak inhibitor of almond β-glucosidase and had no effect on amyloglucosidase.

13.3 Electrophilic Additions to Substituted Cyclopentenes

As has already been noted, the osmylation of di- and trisubstituted cyclopentenes seems to follow no consistent stereochemical trend. Nevertheless it was of interest to ascertain whether other alkene addition reactions would display any stereochemical selectivity in functionalized cyclopentenes. Of particular note was the report that additions of HOBr and HOCl to 3-methoxy and 3-alkylcyclohexenes proceed via *syn*-halonium ions [45-47]. In fact, hypobromous acid additions to 2-cyclopenten-1-ol (**23**) and cyclopentene-*cis*-1,4-diol (**26**) also occurred with good stereoselectivity (Scheme 13-7), presumably by the intermediacy of cyclic *syn*-bromonium ions **24** and **26**, respectively.

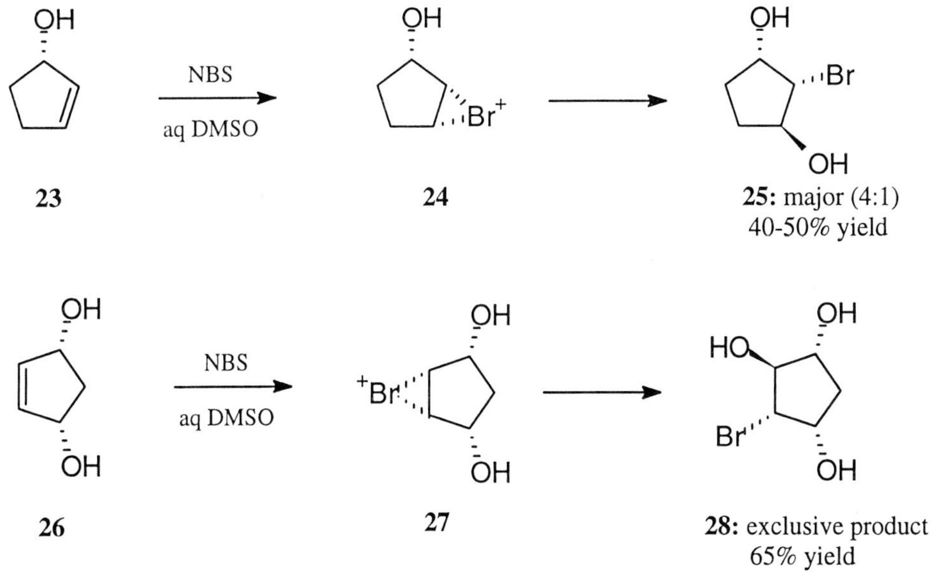

Scheme 13-7

Such three-membered bromonium ions were first proposed in as intermediates in the electrophilic addition of Br_2 to alkenes [48] and have been implicated in electrophilic additions of other Br^+-generating reagents [49]. Some relatively stable mono- and bicyclic bromonium ions have also been observed and characterized by NMR spectroscopy [50] and X-ray crystallography [51,52].

Because of the size of the molecules we wished to consider, high level *ab initio* calculations were not practical. Instead PM3 semiempirical molecular orbital calculations [53] were carried out on bromonium ion intermediates using the code in MOPAC6 [54]. The key words employed were PM3 Charge = 1 Precise GNorm = 0.02. The starting geometry for optimization was obtained by running PCMODEL [55] on the epoxide corresponding to each bromonium ion.

Table 13-1. PM3 Calculations on Cyclopentene Bromonium Ions.

Structure		$E_{anti} - E_{syn}$ (kcal/mol)	
		calculated	observed
24 OH, Br^+ $\mu = 7.0$ D $q_{Br} = 0.22$	**29** OH, Br^+ $\mu = 5.3$ D $q_{Br} = 0.16$	0.27	0.84
27 HO, OH, Br^+ $\mu = 5.9$ D $q_{Br} = 0.27$	**30** HO, OH, Br^+ $\mu = 3.1$ D $q_{Br} = 0.18$	2.30	> 1.8

For comparison, an optimized structure for the ethylene bromonium ion was obtained in which the C-Br bond length was determined to be 2.129 Å. A structure for the 2-bromoethyl cation gave a C-Br bond length of 1.925 Å. Both values were consistent with *ab initio* results. Based on the PM3 models, however, the ethylene cyclic bromonium ion was more stable than the open 2-bromoethyl cation by only 3.7 kcal/mol, far below the *ab initio* results. Because of the poor energy result for this model system, *syn/anti* energy differences for a particular cyclic bromonium ion might well be suspect; nevertheless, it was hoped that the general pattern of energy differences might offer insights into the origins of the experimentally observed stereoselectivities. PM3 calculations were therefore performed on the four cyclic bromonium ions depicted in Table 13-1.

In each cyclopentene, the *syn*-bromonium ion is favored over the *anti*-isomer. Also noteworthy about the calculated structures is the fact that in each *syn/anti* pair, the *syn*-structure displays the higher atomic charge on the bromine as well as the higher dipole moment.

An attractive rationale for the differences in dipole moments, bromine charges, and overall energy differences may be developed by representing the charge distribution as a sum of three components. The first component arises from the dipoles associated with the allylic hydroxyl groups; the second is a unit positive charge centered on the bromine atom; the third is a dipole originating from the double bond and pointing towards the bromine atom. This third component describes the electron donation from the double bond to the bromine cation.

To a first approximation, the dipole moments and bromine charges are the same for both the *syn-* and *anti*-bromonium ions formed from a given hydroxylated cyclopentene. The third component of charge distribution, the net C-Br dipole, varies markedly in the *syn-* and *anti*-bromonium ions. In the *anti*-structures, the C-Br dipole is relatively large, reflecting the greater transfer of positive charge from bromine to the attached carbons. This larger, anti-dipole moment is oriented in opposition to the dipoles of the substituent hydroxyl groups, thus reducing the molecular dipole moment.

By contrast, in both *syn*-bromonium ions, the C-Br dipole component is smaller, with relatively less charge transfer to the ethylene carbons and a greater net positive charge on the bromine atoms. In both *syn*-bromonium ions, the substituent dipoles are oriented with their negative poles towards the bromine atom, which reinforces the C-Br dipole and leads to a larger molecular dipole moment. The reinforcing dipoles of the *syn*-ions, combined with the greater positive charge on the bromine atom, represent a favorable interaction that lowers the molecular energy relative to the *anti*-ions.

13.4 Studies on Allosamizoline: an Aminocyclopentitol from Allosamidin

To test the generality of synthesizing pentasubstituted cyclopentanes by the stereocontrolled addition of charged electrophiles to polysubstituted cyclopentenes, we turned our attention to the assembly of other naturally-occurring aminocyclopentitol structures. One of nature's most abundant polysaccharides is chitin, the β-1,4-linked polymer of N-acetylglucosamine [56]. Chitin is the main component of insect cuticle. It is also a constituent of fungal cell wall, from which approximately 3×10^4 metric tons of the polysaccharide is derived annually [57]. In addition, chitin is the principal structural macromolecule in crustacean shells, with an estimated 1.2×10^5 metric tons being produced worldwide annually as a major waste product of the seafood processing industry. Not surprisingly, interest in research on chitinases, which can degrade chitin to soluble mono- and oligosaccharides, has grown enormously over the past five years [58]. Chitinase inhibitors are also of potential import as insecticides or fungicides [59].

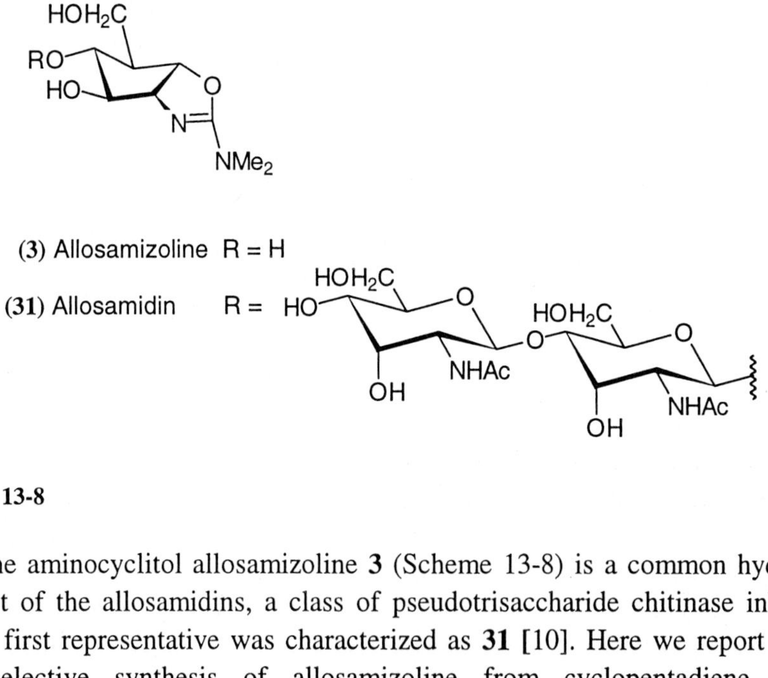

Scheme 13-8

The aminocyclitol allosamizoline **3** (Scheme 13-8) is a common hydrolysis product of the allosamidins, a class of pseudotrisaccharide chitinase inhibitors whose first representative was characterized as **31** [10]. Here we report a short stereoselective synthesis of allosamizoline from cyclopentadiene by an alkylation/cycloaddition route recently developed in our laboratory [23b]. Several groups have achieved total syntheses of allosamidins [20,22,60], and several routes to **31** have been developed from carbohydrate precursors [13,14,19a,61]. Retrosynthetic analysis of allosamizoline focused on construction of the (dimethylamino)oxazoline late in the overall plan, and this tactic revealed an inherent symmetry which we hoped to exploit. Aminotriol **37** (Scheme 13-9), the precursor of thiooxazolidinone **38**, might be prepared by azidolysis of *meso*-epoxydiol **35**, which in turn seemed accessible by an appropriate functionalization of the *meso*-trisubstituted cyclopentene **33**.

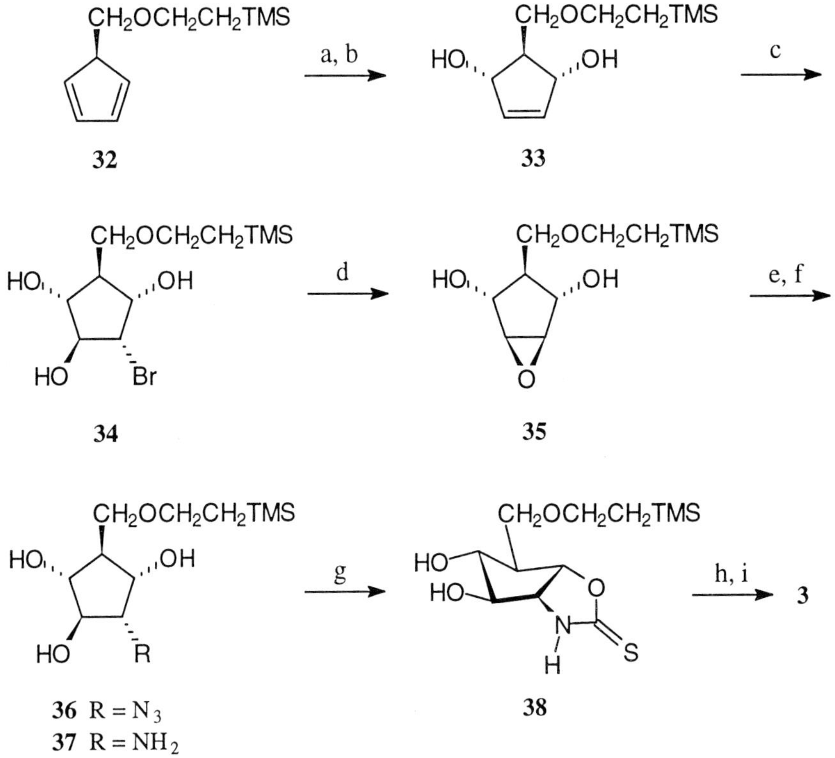

Scheme 13-9: Reagents: a) O_2, hv, sens; b) thiourea; c) NBS, aq. DMSO; d) Na_2CO_3, CH_3OH; e) NaN_3; f) H_2, Pd/C; g) $Im_2C=S$; h) Me_2NH; i) $LiBF_4$-CH_3CN.

A suspension of thallous cyclopentadienide (Aldrich, freshly sublimed) in CH_2Cl_2 was reacted with β-trimethylsilylethoxymethyl chloride (SEM-Cl) to afford monosubstituted cyclopentadiene **32**, which without purification was immediately subjected to singlet oxygen generated using methylene blue as sensitizer. Reduction *in situ* of the transient endoperoxide afforded *meso*-cyclopentenediol **33** in 35% overall yield [62].

Epoxidation of **33** with peracids gave exclusively the undesired *syn*-epoxydiol. We were unable, even using *bis*-trimethylsilyl-**33** [63], to obtain better than a 2:1 facial selectivity for the *anti*-epoxydiol **35**. However reaction of diol **33** at room temperature with NBS in wet DMSO furnished a single bromohydrin

34 (60% yield), in line with earlier predictions from semi-empirical calculations described above. The structure of that bromohydrin was confirmed by cyclization (Na_2CO_3-CH_3OH) exclusively to **35**. Addition of HOCl to **33** likewise gave **35** via the corresponding chlorohydrin.

Epoxydiol **35** underwent smooth ring opening with NaN_3 (5 equiv, 8:1 $CH_3OCH_2CH_2OH$:H_2O) to afford racemic azidotriol **36** (66%) which was then quantitatively reduced to the corresponding aminotriol **37**. Cyclization with thiocarbonyldiimidazole produced thiooxazolidinone **38** in 81% yield. From **38**, a one-step construction of the (dimethylamino)oxazoline ring was achieved by heating with $(CH_3)_2NH$-CH_3OH in a sealed tube. The SEM group was deprotected using $LiBF_4$ to afford racemic allosamizoline **3** in 85% yield for 2 steps.

13.5 Studies on Trehazolin and Kerrufaride

A separate cycloaddition strategy has been developed to construct the trehalose inhibitor trehazolin (**4**) [8], as well as the keruffarides [64] and crasserides (Scheme 13-10) [65], two families of marine-derived glycolipids having general structure **39** embodying a core cyclopentanepentaol.

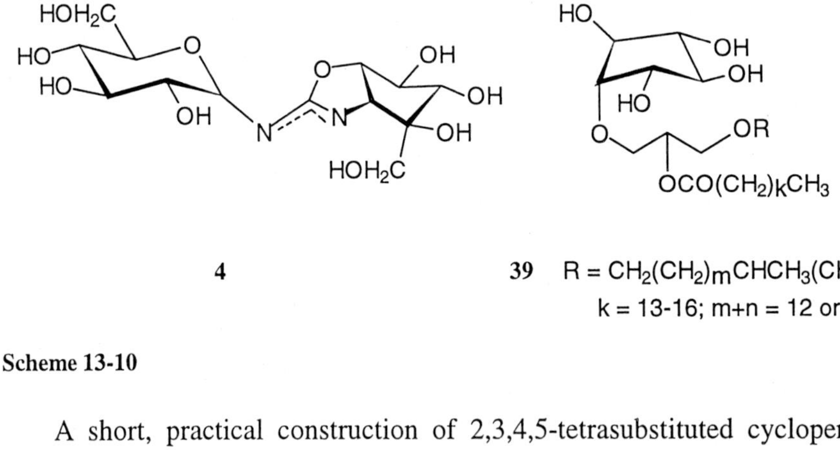

4

39 R = $CH_2(CH_2)_mCHCH_3(CH_2)_nCH_3$
k = 13-16; m+n = 12 or 13

Scheme 13-10

A short, practical construction of 2,3,4,5-tetrasubstituted cyclopentanones has been achieved from 6,6-dimethylfulvene (**40**). We have also explored highly stereoselective reductions and reductive aminations of such polyoxygenated cyclopentanones. In so doing, we have developed a significantly improved formal total synthesis of trehazolin (**4**) as well as a convergent route to the pentahydroxy-cyclopentane ring of keruffaride, thus serving to illustrate the utility of this approach in constructing biologically relevant pentasubstituted cyclopentanes.

Although Diels-Alder reactions of fulvenes have been extensively investigated, few examples of {4+2}-heterocycloadditions are known. Complex product mixtures are usually produced, affording little or none of the desired adducts. For example, sensitized photooxygenation of dimethylfulvene **40** (Scheme 13-11) gave two 1,4-ketols as well as three other rearrangement products [66]. By using thiourea, to reduce the initially formed 1,4-epidioxide, and sodium acetate, to trap traces of acid, the Rose-Bengal sensitized photooxygenation of commercially available **40** (CH_3OH, -30 °C, 2 h) afforded isopropylidene-cyclopentenediol **41** in 78% yield after chromatography. Presumably diol **41** adopted the pseudo-diaxial conformation, thus relieving $A^{(1,3)}$ strain, so that catalytic osmylation (1.1 equiv N-methylmorpholine-N-oxide, 10:1 acetone:water, rt, 22 h, 50%) gave tetraol **42** which was subsequently protected as the cyclohexylidene acetal **43** (1,1-diethoxycyclohexane, p-TsOH, DMF, 95 °C, 62%).

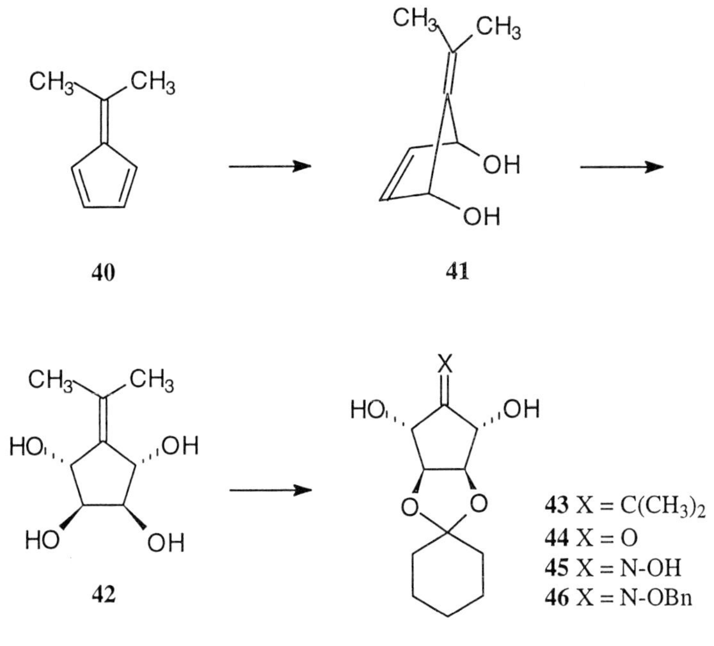

43 X = C(CH$_3$)$_2$
44 X = O
45 X = N-OH
46 X = N-OBn

Scheme 13-11

To prepare the requisite aminocyclopentitol for trehazolin, alkene **43** was first ozonized at -78 °C in a Rubin apparatus [67]. Existing exclusively in its hydrated form, the resulting ketone **44** could not be purified or characterized. Ketone **44** also decomposed during attempts at reductive amination (NH_4OAc,

NaBH3CN, CH3OH). However, the corresponding oxime **45** was obtained in good yield by workup of **44** with NH2OH.

Exhaustive hydrogenolysis of **45** (H2, PtO2, CH3CO2H, rt, 45 psi) produced a 1:1 mixture of aminodiols characterized as triacetyl derivatives **47a** and **48a** (Scheme 13-12, 74% combined yield). Hydroboration of the corresponding O-benzyl oxime **46** using BH3-THF (4 equiv in THF, 0 °C to rt) was stereoselective, giving a 1:12 mixture of **47a** and **48a** after acetylation (64% overall) [68]. Unfortunately, the stereoisomer needed for trehazolin was the minor product, probably resulting from hydroxyl-directed intramolecular reduction. Therefore, oxime ether **46** was silylated (3 equiv TBDMSCl, 5 equiv imidazole, DMF, 2 d, rt, 92%) to furnish triether **49**.

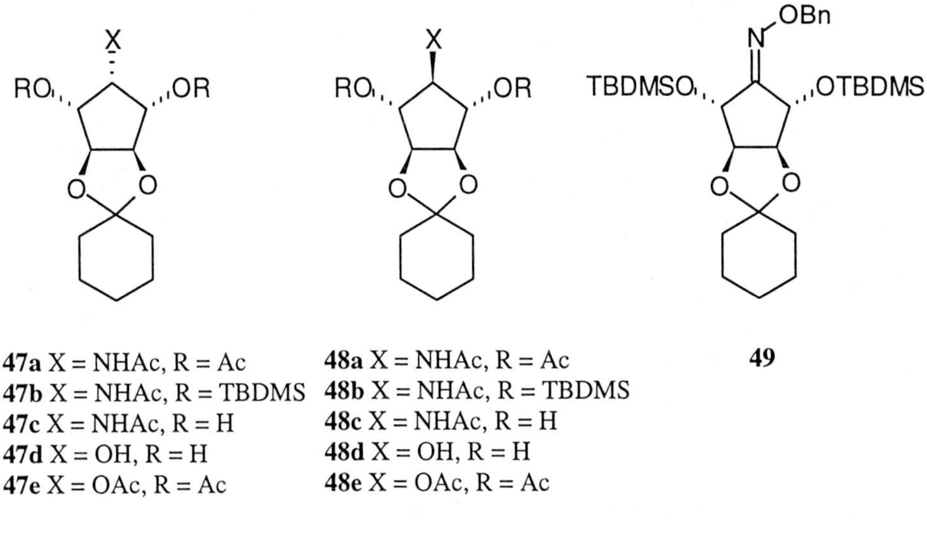

47a X = NHAc, R = Ac
47b X = NHAc, R = TBDMS
47c X = NHAc, R = H
47d X = OH, R = H
47e X = OAc, R = Ac

48a X = NHAc, R = Ac
48b X = NHAc, R = TBDMS
48c X = NHAc, R = H
48d X = OH, R = H
48e X = OAc, R = Ac

49

Scheme 13-12

When **49** was reacted with BH3-THF (15 equiv in THF, 70 °C, 18 h), a much slower reduction occurred with the expected reversal of stereoselectivity leading to a 6:1 mixture of **47b** and **48b** in 89% yield. Desilylation of **47b** (1.8 equiv Bu4NF, THF, rt, 20 min, 97%) afforded meso-diol **47c**, which has previously been carried on, with a subsequent resolution, to (+)-trehazolin (**4**). Compared to the published route to **47c** (6 steps from *myo*-inositol; 0.1% yield), the present 7-step stereoselective route (15.3% overall yield) constitutes a significant improvement in the overall synthesis of trehazolin. Routes to new trehazaloids

[12b] from other aminocyclopentitols available using this methodology are currently being developed.

Although ketone **44** was unstable, ozonolysis of **43** followed by direct reduction using NaBH$_4$ in CH$_3$OH did furnish a single *meso*-triol **48d** (35% yield; no trace of **47d**) which was characterized spectroscopically as **48e**. In further support of this assignment, proton and carbon NMR spectra of the alternative *meso*-reduction product **47e**, an authentic sample of which was prepared by osmylation and ketalization of *cis*-3,4,5-triacetoxycyclopentene [69], were distinctly different from spectra of **48e**. With triol **48d** in hand, convergent approaches to members of the keruffaride and crasseride families may now be pursued.

Acknowledgment

Sincere thanks are owed to the creative and hard-working postdoctoral fellows, graduate and undergraduate students associated with this research program in my laboratory over the past several years; their names appear in the references. Continuous support of this project by the National Institutes of Health (GM 35712) is gratefully acknowledged.

13.6 References

1. *Glycoconjugates: Composition, Structure, and Function,* (Ed.: H. J. Allen, E. C. Kisailus) Dekker, NY, **1992**.

2. (a) M. L. Sinnott, *Enzyme Mechanisms,* (Eds.: M. I. Page, A. Williams), The Royal Society of Chemistry, London, 1987, p. 259; (b) G. Legler, *Adv. Carbohydr. Chem. Biochem.* **1990,** *48,* 319.

3. L. R. Lynd, J. H. Cushman, R. J. Nichols, C. E. Wyman, *Science* **1991,** *251,* 1318.

4. B. Ganem, *Proceedings of the Robert A. Welch Foundation Conference on Chemical Research* **1991,** *25,* 207.

5. (a) P. B. Anzeveno, L. J. Creemer, J. K. Daniel, C.-H. R. King, P. S. Liu, *J. Org. Chem.* **1989,** *54,* 2539; (b) A. D. Elbein, *Annu. Rev. Biochem.* **1987,** *56,* 497.

6. G. W. J. Fleet, A. Karpas, R. A. Dwek, L. E. Fellows, A. S. Tyms, S. Petursson, S. K. Namgoong, N. G. Ramsden, P. W. Smith, J. C. Son, F. Wilson, D. R. Witty, G. S. Jacob, T. W. Rademacher, *FEBS Lett.* **1988,** *237,* 128.

7. M. L. Sinnott, *Chem. Rev.* **1990,** *90,* 1171.

8. Q. Wang, R. W. Graham, D. Trimbur, R. A. J. Warren, S. G. Withers, *J. Am. Chem. Soc.* **1994,** *116,* 11594.

9. (a) T. Aoyagi, T. Yamamoto, K. Kojiri, H. Morishima, M. Nagai, M. Hamada, T. Takeuchi, H. Umezawa, H. *J. Antibiot.* **1989**, *42,* 883; (b) H. Morishima, K. Kojiri, T. Yamamoto, T. Aoyagi, H. Nakamura, Y. Iitaka, *J. Antibiot.* **1989**, *42,* 1008.

10. (a) S. Sakuda, A. Isogai, S. Matsumoto, A. Suzuki, K. Koseki, H. Kodama, A. Suzuki, *Agric. Biol. Chem.* **1987**, *51,* 3251; (b) S. Sakuda, A. Isogai, T. Makita, S. Matsumoto, A. Suzuki, K. Koseki, H. Kodama, Y. Yamada, *Agric. Biol. Chem.* **1988**, *52,* 1615.

11. C. Uchida, T. Yamagishi, S. Ogawa, S. *J. Chem. Soc., Perkin Trans. 1* **1994**, 589.

12. S. Knapp, A. Purandare, K. Rupitz, K.; S. G. Withers, *J. Am. Chem. Soc.* **1994**, *116,* 7461.

13. M. Nakata, S. Akazawa, S. Kitamura, K. Tatsuka, *Tetrahedron Lett.* **1991**, *32,* 5363.

14. (a) N. S. Simpkins, S. Stokes, A. J. Whittle, *Tetrahedron Lett.* **1992**, *33,* 793; (b) idem, *J. Chem. Soc., Perkin Trans. I* **1992**, 2471.

15. Y. Kobayashi, H. Miyazaki, M. Shiozaki, *J. Org. Chem.* **1994**, *59,* 813.

16. C. Uchida, T. Yamagishi, J. Ogawa, *J. Chem. Soc., Perkin Trans. I* **1994**, 589.

17. C. Li, P. L. Fuchs, *Tetrahedron Lett.* **1994**, *35,* 5121.

18. T. Kiguchi, K. Tajiri, L. Minomiya, T. Naito, H. Hiramatsu, H. *Tetrahedron Lett.* **1995**, *36,* 253.

19. (a) S. Takahashi, H. Terayama, H. Kuzuhara, *Tetrahedron Lett.* **1991**, *32,* 5123; (b) S. Takahashi, H. Terayama, H. Kuzuhara, *Tetrahedron Lett.* **1992**, *33,* 7565; (c) S. Takahashi, H. Inoue, H. Kuzuhara, *J. Carbohydr. Chem.* **1995**, *14,* 273.

20. D. A. Griffith, S. J. Danishefsky, *J. Am. Chem. Soc.* **1991**, *113,* 5863.

21. (a) B. M. Trost, D. L. van Vranken, *J. Am. Chem. Soc.* **1990**, *112,* 1261; (b) idem *J. Am. Chem. Soc.* **1993**, *115,* 444.

22. (a) J.-L. Maloisel, A. Vasella, B. M. Trost, D. L. van Vranken, *J. Chem. Soc., Chem. Commun.* **1991**, 1099; (b) idem, *Helv. Chim. Acta* **1992**, *75,* 1515.

23. (a) S. B. King, B. Ganem, *J. Am. Chem. Soc.* **1991**, *113,* 5089; (b) S. B. King, B. Ganem, *J. Am. Chem. Soc.* **1994**, *116,* 562.

24. B. K. Goering, B. Ganem, B. *Tetrahedron Lett.* **1994**, *35,* 6997.

25. B. K. Goering, J. Li, B. Ganem, *Tetrahedron Lett.* **1995**, *36,* 8905.

26. J. E. Tropea, G. P. Kaushal, I. Pastuszak, M. Mitchell, T. Aoyagi, R. J. Molyneux, A. D. Elbein, *Biochemistry* **1990**, *29,* 10062.

27. D. A. Winkler, G. Holan, *J. Med. Chem.* **1989**, *32,* 2084.

28. S. Knapp, T. G. M. Dhar, *J. Org. Chem.* **1991**, *56,* 4096.

29. (a) L. E. Fellows, *New Scientist* **1989**, *123,* 45; (b) L. Fellows, *Chemistry in Britain* **1987**, *23,* 842.

30. (a) G. Trugan, M. Rousset, A. Zweibaum, *FEBS Lett.* **1986**, *195*, 28; (b) V. V. Sazak, J. M. Ordovas, A. D. Elbein, R. W. Berninger, *Biochem. J.* **1985**, *232*, 759; (c) M. J. Humphries, K. Matsumoto, S. L. White, R. J. Molyneux, K. Olden, *Cancer Res.* **1988**, *48*, 1410; (d) J. W. Denis, *Cancer Res.* **1986**, *46*, 5131.

31. B. M. Trost, D. L. Van Vranken, *J. Am. Chem. Soc.* **1991**, *113*, 6317.

32. S. Ogawa, Y. Yuming, *J. Chem. Soc., Chem. Commun.* **1991**, 890.

33. See, inter alia, (a) R. L. Danheiser, D. J. Carini, A. Basak, A. *J. Am. Chem. Soc.* **1981**, *103*, 1604; (b) R. L. Danheiser, C. Martinez-Davila, R. J. Auchus, J. T. Kadonaga, *ibid.* **1981**, *103*, 2443.

34. G. W. Kirby, *Chem. Soc. Rev.* **1977**, *6*, 1.

35. (a) G. W. Kirby, M. Nazeer, M. *Tetrahedron Lett.* **1988**, *29*, 6173; (b) A. Miller, T. M. Paterson, G. Procter, *Synlett* **1989**, *1*, 32; (c) A. Brouillard-Poichet, A. Defoin, J. Streith, *Tetrahedron Lett.* **1989**, *30*, 7061; (d) A. Miller, G. Procter, *Tetrahedron Lett.* **1990**, *31*, 1041, 1043; (e) S. F. Martin, M. Hartmann, J. A. Josey, *Tetrahedron Lett.* **1992**, *33*, 3583; (e) G. W. Kirby, M. Nazeer, *J. Chem. Soc., Perkin Trans. 1* **1993**, 1397.

36. K. Hartke, H.-G. Zerbe, H.-G. *Arch Pharm. (Weinheim)* **1982**, *315*, 406.

37. The specific rotation we measured for (R)-mandelohydroxamic acid (-60°, c=1, H_2O) differed considerably from Kirby's published value of -164° (c=2.5, H_2O; Ref 35a). Professor Kirby has informed us that this value was erroneous, and that upon re-determination, a value of -63° (c=1.6, H_2O) was obtained.

38. F. A. Cotton, L. T. Reynolds, *J. Am. Chem. Soc.* **1958**, *80*, 269.

39. V. VanRheenan, R. C. Kelly, D. Y. Cha, *Tetrahedron Lett.* **1976**, 1973.

40. (a) H. B. Henbest, S. A. Khan, *J. Chem. Soc., Chem. Commun.* **1968**, 1036; (b) F. M. Hauser, S. R. Ellenberger, J. C. Clardy, L. S. Bass, *J. Am. Chem. Soc.* **1984**, *106*, 2458; (c) D. A. Evans, S. W. Kaldor, T. K. Jones, J. Clardy, T. J. Stout, *J. Am. Chem. Soc.* **1990**, *112*, 7001.

41. B. M. Trost, G.-H. Kuo, T. Benneche, *J. Am. Chem. Soc.* **1988**, *110*, 621.

42. G. Poli, *Tetrahedron Lett.* **1989**, *30*, 7385 and references cited therein.

43. G. E. Keck, S. Fleming, S.; D. Nickell, P. Weider, P. *Synth. Commun.* **1979**, *9*, 281.

44. We are grateful to Professor Alan D. Elbein (University of Arkansas School of Medicine) for providing an authentic sample of (+)-**1**.

45. R. A. B. Bannard, A. A. Casselman, L. R. Hawkins, *Can. J. Chem.* **1965**, *43*, 2398.

46. E. J. Langstaff, E. Hamanaka, G. A. Neville, R. Y. Moir, *Can. J. Chem.* **1967**, *45*, 1907.

47. G. Bellucci, G. Berti, G. Ingrosso, E. Mastrorilli, E. *Tetrahedron Lett.* **1973**, 3911.

48. I. Roberts, G. E. Kimball, *J. Am. Chem. Soc.* **1937**, *59*, 947.

49. R. C. Fahey, *Top. Stereochem.* **1968,** *3,* 237.
50. G. Olah, "Halonium Ions," Wiley-Interscience, New York, 1975.
51. (a) H. Slebocka-Tilk, R. G. Ball, R. S. Brown, R. *J. Am. Chem. Soc.* **1985,** *107,* 4504.
52. R. S. Brown, R. W. Nagorski, A. J. Bennet, R. E. D. McClung, G. H. M. Aarts, M. Klobukowski, R. McDonald, B. D. Santarsiero, B. D. *J. Am. Chem. Soc.* **1994,** *116,* 2448.
53. (a) J. J. P. Stewart, *J. Comput. Chem.* **1989,** *10,* 209; (b) idem, *ibid.* 221.
54. J. J. P. Stewart, QCPE #504, *Quantum Chemical Program Exchange*, Indiana University, Bloomington, IN 47405.
55. Serena Software, Box 3076, Bloomington, IN 47402-3076.
56. E. R. Pariser, *Chitin Sourcebook: A Guide to the Research Literature* **1989,** Wiley, New York.
57. D. Knorr, Food Technology **1991,** 45, 114.
58. J. Flach, P.-E. Pilet, P. Jollés, Experientia **1992,** 48, 701.
59. E. Cabib, A. Sburlati, B. Bowers, S. J. Silverman, J. Cell. Biol. **1989,** 108, 1665.
60. S. Takahashi, H. Terayama, H. Kuzuhara, *Tetrahedron Lett.* **1992,** 33, 7565.
61. T. Kitahara, N. Suzuki, K. Koseki, K. Mori, *Biosci. Biotech. Biochem.* **1993,** 57, 1906.
62. Satisfactory ^{1}H, ^{13}C-NMR, IR and mass spectrometric data were obtained for all new compounds.
63. (a) C. G. Chavdarian, C. H. Heathcock, *Synth. Commun.* **1976,** 6, 277; (b) R. Schlessinger, A. Lopes, A. *J. Org. Chem.* **1991,** 46, 5252.
64. M. Ishibashi, C.-M. Zeng, J. Kabayashi, J. *J. Natural Prod.* **1993,** 56, 1856.
65. V. Costantino, E. Fattorusso, A. Mangoni, A. *J. Org. Chem.* **1993,** 58, 186.
66. (a) W. Skorianetz, K. H. Schulte-Elte, G. Ohloff, *Angew Chem. Int. Ed. Engl.* **1972,** 11, 330; (b) N. Harada, H. Uda, H. Ueno, S. Utsumi, *Chem. Lett.* **1973,** 1173.
67. M. B. Rubin, J. Chem. Ed. **1964,** 41, 388.
68. A. K. Ghosh, S. P. Mckee, W. M. Sanders, *Tetrahedron Lett.* **1991,** 32, 711.
69. G. Wolczunowicz, L. Bors, F. Cocu, Th. Posternak, *Helv. Chim. Acta* **1970,** 53, 2288.

14 Toward Azaglycoside Mimics: Aza-*C*-glycosyl Compounds and Homoazaglycosides

O. R. Martin**

14.1 Introduction

Azasugars[‡] [1] constitute undoubtedly one of the most fascinating class of carbohydrate mimics. The replacement of the ring oxygen atom by nitrogen in both pyranoid and furanoid carbohydrates, a structural modification that has been designed by Nature itself, confers remarkable biological activities to the resulting analogs, most prominently as glycosidase inhibitors [2-4].

The high affinity of azasugars for glycosidases results from their ability to become protonated and to form a cation which can interact strongly with an anionic group (carboxylate) at the enzyme active site [5]; in addition, the spatial arrangement of the hydroxyl groups in piperidine azasugars resemble closely that of the natural substrates in their ground state conformation, whereas the flattened structure of pyrrolidine azasugars is believed to mimic the half-chair structure of the glycosyl cation involved as an intermediate (or a transition state) in the mechanism of hydrolysis [3]. Consistent with this interpretation are the observations that the topology of the hydroxyl groups is less important in pyrrolidine than in piperidine azasugars and that pyrrolidine azasugars have a broader spectrum of inhibition [3].[†] For example, 2,5-dideoxy-2,5-imino-D-glucitol **1** [6] and its D-*manno* epimer **2** [7] inhibit not only α- and β-glucosidases, but also an α-galactosidase (**1**), an α-mannosidase (**1**) and a mammalian β-galactosidase (**2**). On the other hand, 1-deoxynojirimycin **3** is primarily an α- and β-glucosidase inhibitor [3,7].

---

[*] Correspondance address: Institut de Chimie Organique et Analytique, Université d'Orléans, B.P. 6759 F-45067 Orléans, France.

[‡] Strictly, azasugars should be named as amino sugars or imino alditols. Since it is commonly used and conveniently characterize amino sugars having nitrogen in the ring, the term azasugar will be used throughout the review.

[†] There are some exceptions, see for example, ref. [8].

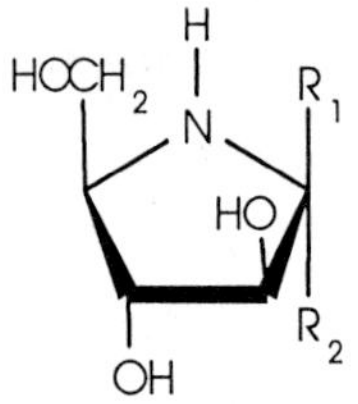

Figure 14-1

1 $R_1 = CH_2OH$, $R_2 = H$
2 $R_1 = H$, $R_2 = CH_2OH$

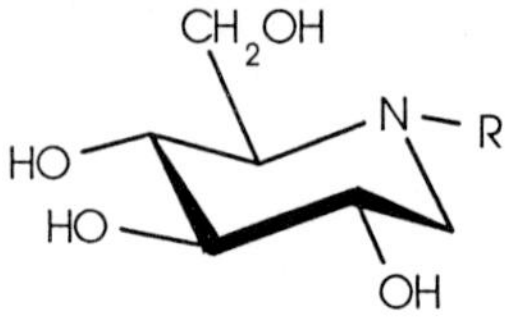

3 $R = H$
4 $R = n\text{-}C_4H_9$

Azasugars have evolved into most useful probes for the study of glycosidases and a wide variety of derivatives have been prepared [4, 9]. Since glycosidases are involved in a number of essential biological processes (absorption of carbohydrates, catabolism of glycoconjugates, post-translational modification of glycoproteins), azasugars have the potential of becoming useful agents for the treatment of metabolic diseases (diabetes), cancer and viral diseases [2]; for example, N-butyl-1-deoxynojirimycin **4** is the azasugar exhibiting the strongest inhibition of HIV-induced cytopathogenicity [4,7].

There is, however, a fundamental limitation in the class of azasugars as carbohydrate mimics: as a result of the lability of the O,N-acetal function [10], free azaglycosides (e.g. compounds of type **I**), which include aza-analogs of disaccharides and most glycoconjugates, would undergo rapid cleavage under hydrolytic conditions [1] and could not be used as biological probes. In fact, most piperidine azasugars reported so far are derivatives of 1-deoxynojirimycin [9].

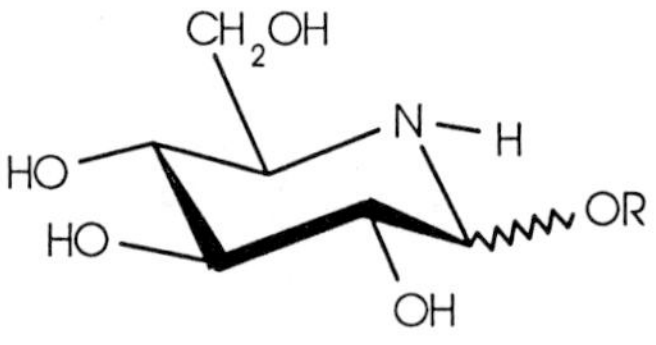

Figure 14-2

In order to generate stable structures simulating aza-analogs of glycoconjugates, chemists have proposed a few ingenious solutions: for example, analogs of elusive aza-disaccharides have been prepared by replacing the interglycosidic oxygen atom by sulfur (e.g. **5**) [11], by linking the aglycone directly to the nitrogen atom (e.g. **6**) [12], or by shifting the position of the nitrogen atom (e.g. **7**) [13].

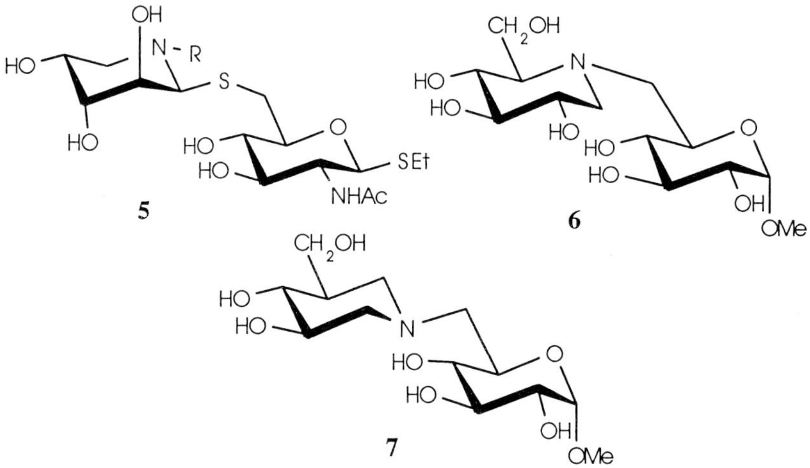

Figure 14-3

Conscious of the enormous potential of aza-analogs of glycoconjugates as selective inhibitors of targeted glycosidase-mediated processes, we have been engaged in the search for effective, second-generation carbohydrate mimics that might replace such elusive compounds. Stable analogs of aza-pyranosides can be generated in one of the following three ways (Scheme 14-1):

(a) by replacing the oxygen atom of the *O,N*-acetal function by a methylene group, to form "aza-*C*-glycosides."

(b) by inserting a methylene group into the C-O bond of the *O,N*-acetal function, to form "homoazaglycosides."

(c) by inserting a methylene group into the C-N bond of the *O,N*-acetal function, to form another type of homoazaglycosides which could be designated as "1-deoxyazaseptanoses."

In this chapter, a brief review of existing synthetic methodologies towards each of these three types of modified azasugars will be provided as well as a description of our own contributions to this field. Since most glycoconjugates contain carbohydrate residues in the pyranoid form, and since our goal is the design of inhibitors exhibiting the greatest possible degree of selectivity, our efforts have concentrated primarily on the preparation of azasugars in the piperidine series.

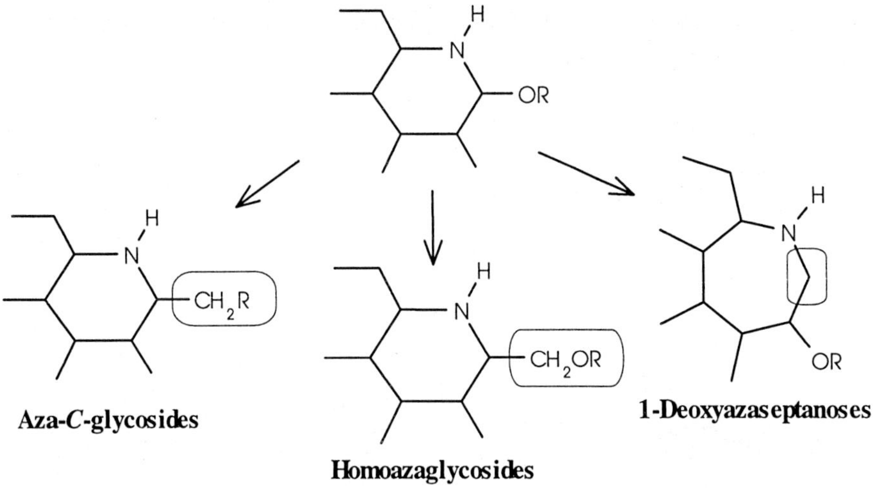

Scheme 14-1

Aza-*C*-glycosides

Homoazaglycosides

1-Deoxyazaseptanoses

14.2 Aza-*C*-glycosyl Compounds

Aza-*C*-glycosyl compounds form an emerging class of carbohydrate mimics. If one excludes "homoazasugars" (discussed in Section 3), very few examples of aza-*C*-glycosyl compounds have been reported so far and only one example of a true aza-*C*-analog of a biologically significant glycoside has been described, namely aza-*C*-disaccharide **8** (Johnson et al. [14]).

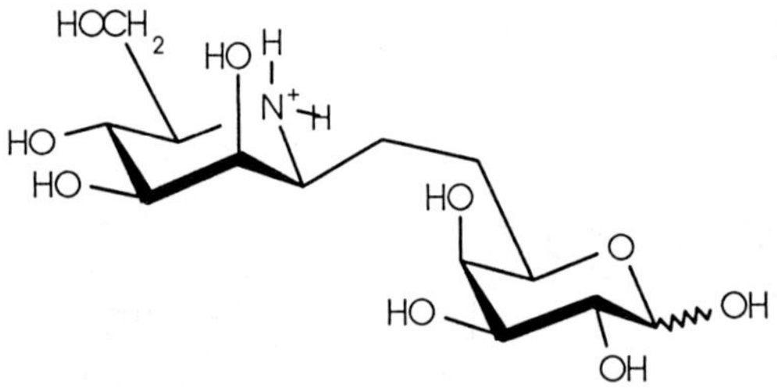

8
(aza-β-D-Man-*C*-(1-6)-D-Gal)

Figure 14-4

Most aza-*C*-glycosyl compounds including homoazasugars have been prepared by way of cyclization procedures in which the chain extension (C–C bond formation, if any) precedes the formation of the C–N bond(s). However, the creation of a C–C bond α to the nitrogen atom in a preformed piperidine derivative is possible if advantage is taken of the intermediacy of an iminium cation [15]: thus, Johnson [16] and, very recently, Schmidt [17] have shown that the azasugar-derived glycoside **9** and glycosyl fluoride **11**, respectively, could be used as precursors of aza-C-glycosyl compounds (**10** and **12** respectively) under Lewis acidic conditions.

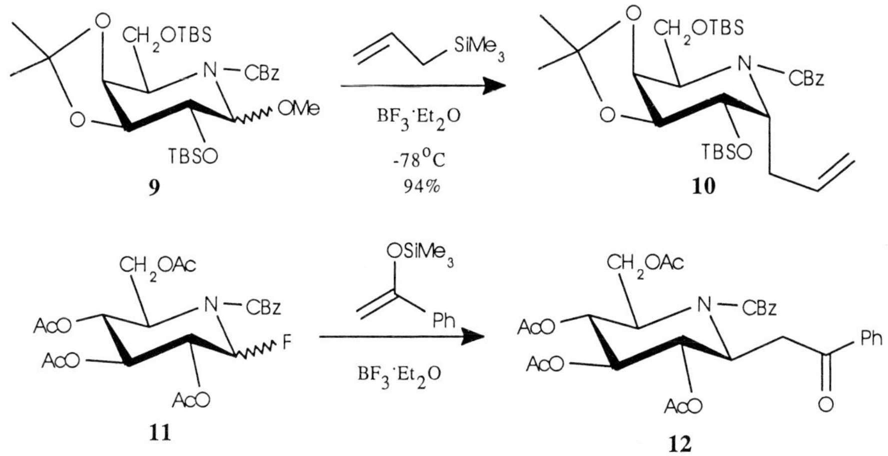

Scheme 14-2

The methods most commonly used for the preparation of C-glycosides [18, 19] appear therefore to be applicable to azaglycoside derivatives provided that the nitrogen atom carries a strongly deactivating substituent. The recently reported α-alkoxy piperidine derivatives **13** [17] and **14** [20], as well as α-acetoxy pyrrolidine **15** [21], constitute three more potential substrates for this C-glycosidation methodology.

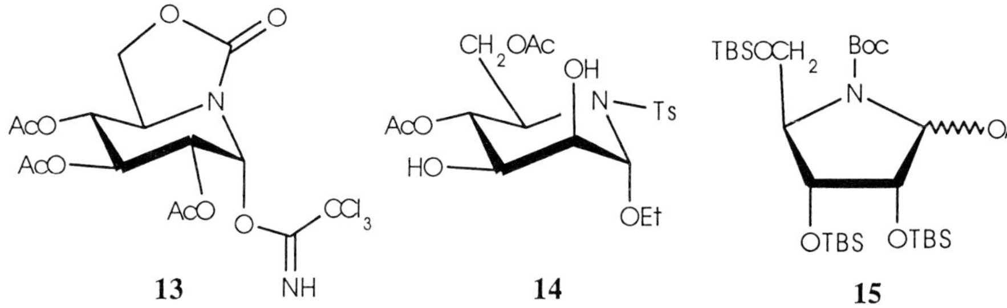

Figure 14-5

Using a well-known method of piperidine chemistry [15], Junge & coworkers [22] have shown that the cyano group could act as a leaving group in 1-cyano-1-deoxynojirimycin derivatives *unsubstituted* at nitrogen, on reaction with a Grignard reagent, e.g.:

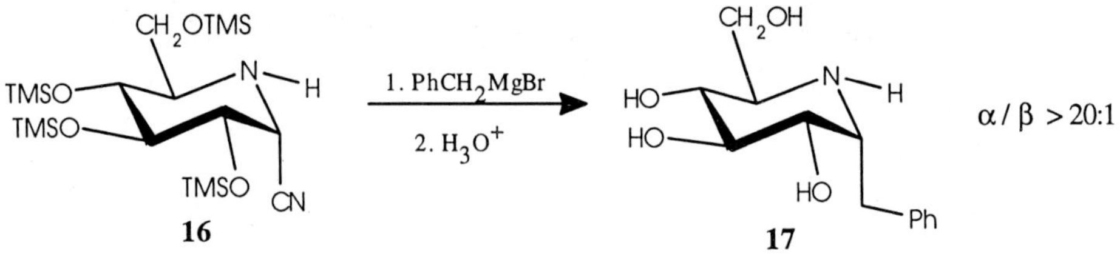

16 **17**

Scheme 14-3

Such azaglycosyl cyanides are readily available from the bisulfite adduct of the corresponding 5-amino-5-deoxy-hexoses and have been used as precursors of a variety of homoazasugars [23,24] and aza-*C*-glycosyl compounds [25]. The benzotriazolyl group was also shown to be an appropriate leaving group for alkylation α to nitrogen in *N*-benzylated polyhydroxypiperidines [26].

In significant examples of *C*-alkylated piperidine derivatives that were obtained by a cyclization procedure, Ganem and coworkers [27] have prepared 1-epoxyalkyl-1-deoxynojirimycin derivatives **18a** and **18b** as potential glycosidase inactivators by way of the following retrosynthetic approach:

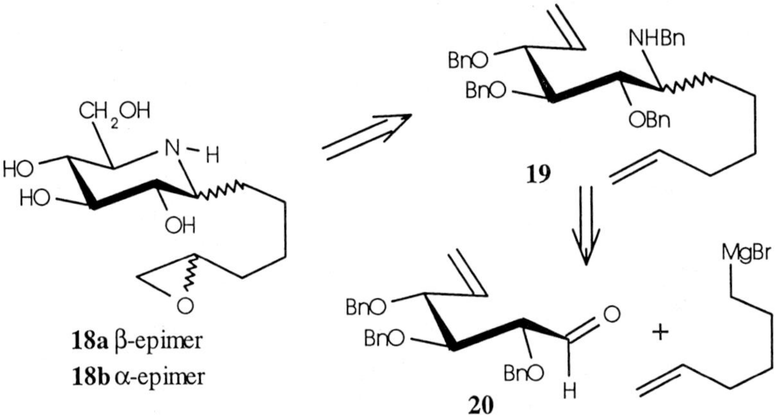

18a β-epimer
18b α-epimer
19
20

Scheme 14-4

Continuing their studies on the highly stereoselective addition of Grignard reagents to the imino form of glycosylamines [28], Nicotra *et al.* have shown that the resulting adduct can be cyclized efficiently to the corresponding 2-*C*-alkylated pyrrolidine or piperidine derivative on reaction with triflic anhydride [29], for example from glucosylamine **21**:

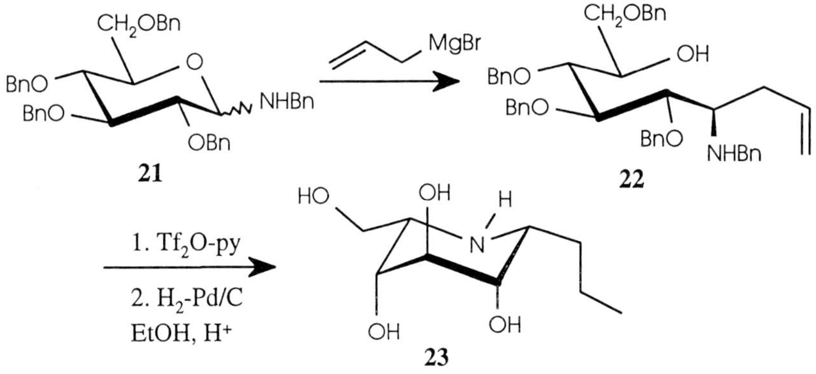

Scheme 14-5

The fact that the cyclization promotes inversion at C-5 of the original substrate and leads from most substrates to a product in the L-series should be noted. Finally, the *C*-formyl piperidine derivative **24** was prepared in seven steps from 2,3:6,7-di-*O*-isopropylidene-D-*glycero*-D-*gulo*-heptono-1,4-lactone and used in the synthesis of the precursor of a β-(1→3)-linked aza-*C*-disaccharide, compound **25** [30]:

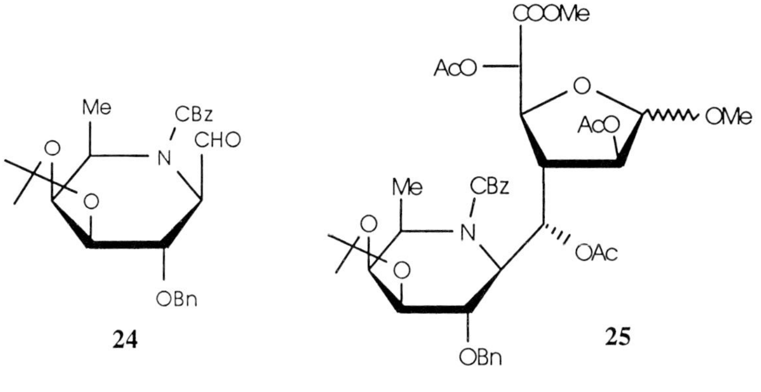

Figure 14-6

In our own investigations on aza-*C*-glycosyl compounds we have developed a strategy designed to generate building blocks (type **27**) for the synthesis of aza-*C*-disaccharides according to the following plan:

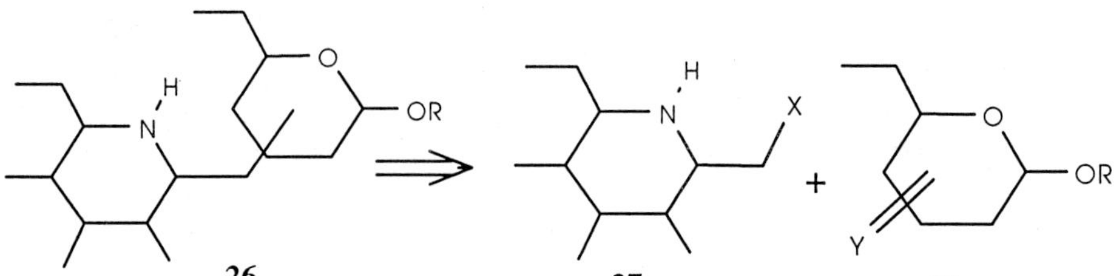

Scheme 14-6

Considering the very high degree of stereoselectivity of the electrophile-mediated cyclization of heptenitols of type **30** [31,32] and on the basis of a favorable precedent reported by Liu [33], we adopted an approach to α-linked aza-*C*-glycosyl compounds based on a chain-extension (I)-amination (II)-cyclization (III) sequence.

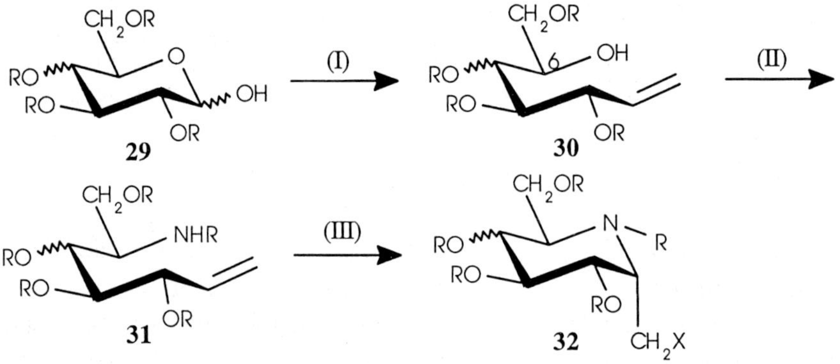

Scheme 14-7

The amination step turned out to be the most challenging step of this sequence! In particular, the conversion of the OH-group at C-6 in heptenitols such as **30** (R=Bn) into a good leaving group (sulfonate) was accompanied by a rapid cyclization involving the benzyloxy group at C-3 and leading to *C*-vinyl furanosides [34]. In Liu's approach to α-homonojirimycin [33], the amino group at C-6 was introduced by way of the reduction of an oxime. However, the separation of the resulting epimers was found to be tedious and the reported ratio of D- vs. L-product difficult to reproduce. Although not highly stereoselective, reductive amination provided a more convenient approach to the desired 6-amino-6-deoxy heptenitols [35]: the D-*gluco* epimer **35** was thus obtained predominantly from the L-*xylo* 2-heptulose **34**.

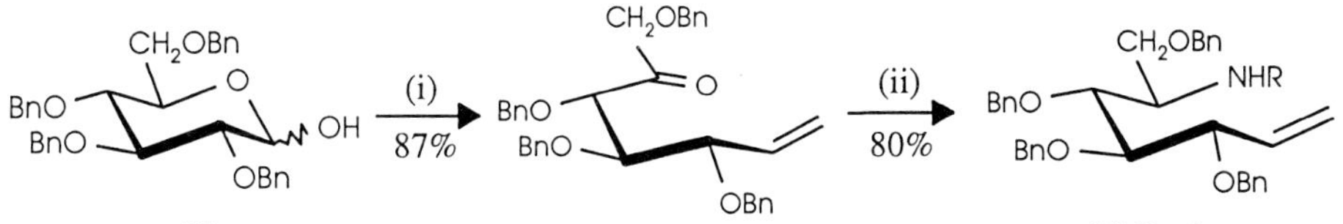

Scheme 14-8. Reagents: (i) a. Ph₃P=CH₂; b. Swern oxid. (ii) BnNH₂-AcOH, NaBH₃CN, MeOH

The NIS-mediated cyclization of **35** gave exclusively the α-linked aza-*C*-glycosyl compound **37**, a key intermediate in our studies on aza-*C*-analogs of glycosides now available in four steps (35-40% yield) from tetra-*O*-benzyl-D-glucose.

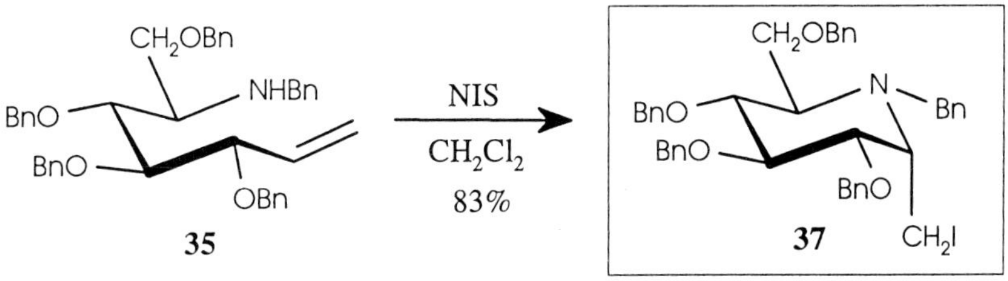

Scheme 14-9

Reductive amination of the L-*lyxo* epimer of **34**, prepared under the same conditions from tetra-*O*-benzyl-D-galactose, gave the L-*altro* and D-*galacto* 6-amino-6-deoxy heptenitols in a 2:1 ratio. On reaction with NIS, both epimers gave the corresponding 2,3-*cis*-disubstituted piperidine derivatives highly stereoselectively (i.e. with the β-D-*altro* and α-D-*galacto* configuration, resp.) [35].

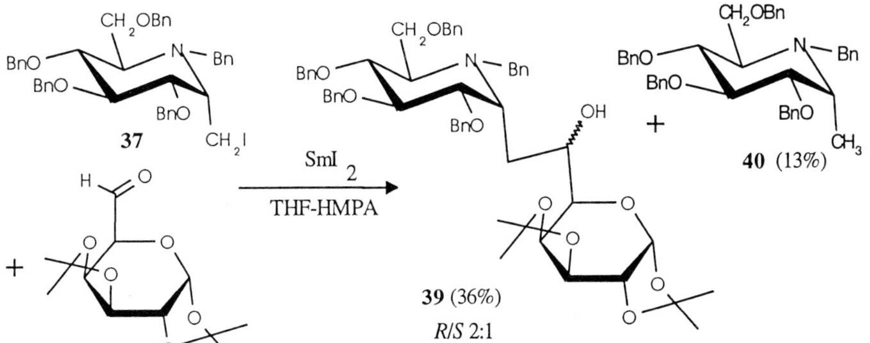

Scheme 14-10

Following the plan set forth for the synthesis of aza-C-disaccharides, the conversion of **37** into an organometallic species and its reaction with carbonyl compounds was investigated. While the organolithium species derived from **37** was not sufficiently stable to be used as a C-nucleophile, the reaction of **37** with SmI$_2$ and the *aldehydo* sugar **38** under samarium-Barbier conditions [36, 37] afforded coupling product **39** in 36% yield (yield based on consumed **37**, ~50%). Compound **39** is the precursor of the aza-C-analog of D-Glc-α-(1→6)-D-Gal.

The reaction of **37** with *keto*-sugars under similar conditions has been, so far, unsuccessful. Compound **37** was also used in the synthesis of aza-C-analogs of glycosyl amino acids. Thus, its reaction with cyanide ion followed by reduction of the resulting nitrile **41** gave the labile aldehyde **42**; compound **42** was converted into the α,β-unsaturated amino acid derivative **43** by a Wittig reaction. Saturation of the double bond of **43** and deprotection should should give **44** ("aza-α-C- glucosylserine").

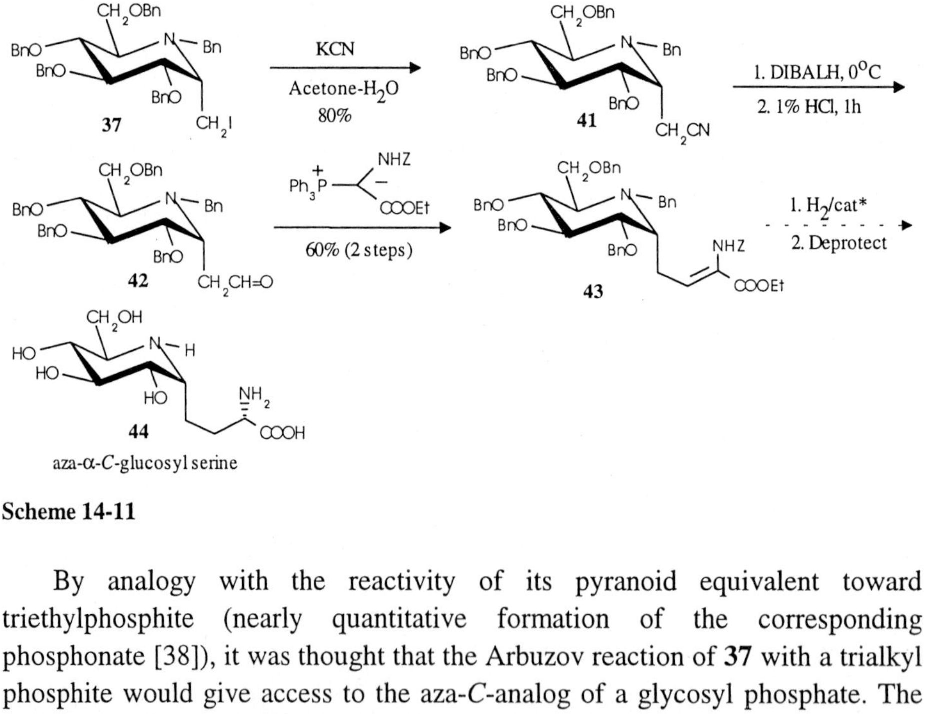

Scheme 14-11

By analogy with the reactivity of its pyranoid equivalent toward triethylphosphite (nearly quantitative formation of the corresponding phosphonate [38]), it was thought that the Arbuzov reaction of **37** with a trialkyl phosphite would give access to the aza-C-analog of a glycosyl phosphate. The reaction took, however, a completely unexpected path leading to the bicyclic anhydro compound **45** (participation of the benzyloxy group at C-5) and none of the desired phosphonate was formed. This outcome may be due to the formation of an intermediate aziridinium cation (**46**) which decreases the reactivity of the

C-1 center toward the soft phosphorus nucleophile and favors the reaction with the internal oxygen nucleophile.

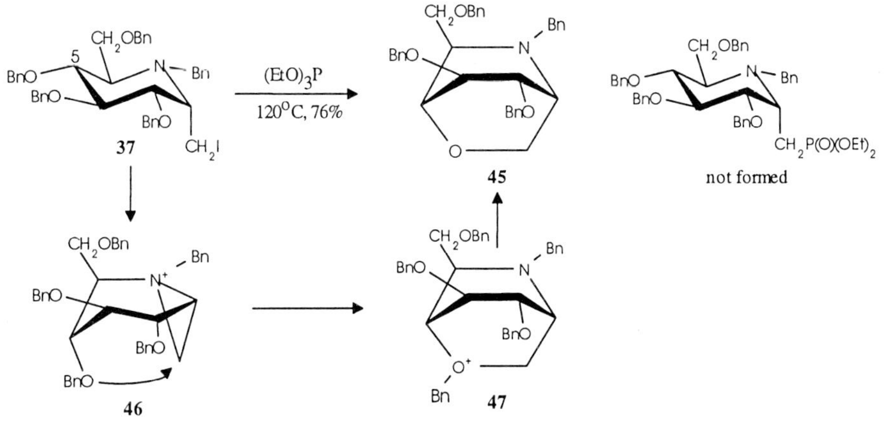

Scheme 14-12

As described in Section 3, compound **37** was also used as a precursor of homoazaglycosides and homoazadisaccharides. In related studies, we have investigated the reductive amination of ketone **51** derived from heptenitol **50** which carries a *benzoyloxy* group instead of a benzyloxy group at the allylic position.

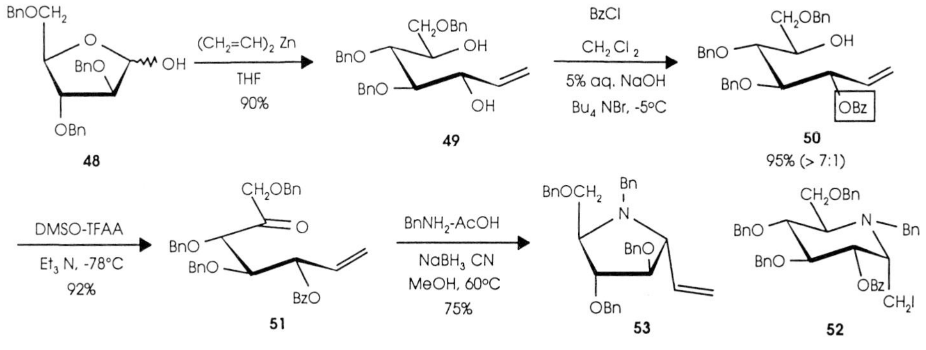

Scheme 14-13

Compound **50** was prepared by way of the highly stereoselective addition of divinylzinc to tri-O-benzyl-D-arabinofuranose **48** [39] followed by the regioselective benzoylation of the resulting diol **49** under phase-transfer catalysis conditions. When conducted at low temperature (10-20°C), the reductive amination of **51** gave the corresponding amino-heptenitol **52** having the D-*gluco* configuration only: the improvement of the stereoselectivity of this process resulting from the change of protecting group at the allylic position is quite remarkable (compare with **34**) and its origin remains to be explained. At higher temperature, the reductive amination of **51** was accompanied by a most interesting ring closure resulting from the internal substitution of the allylic benzoate by the secondary amino group, leading to a *C*-vinyl pyrrolidine derivative (**53**). The inversion/retention ratio at the allylic position was most favorable (~15:1) at a temperature of about 60°C. Under these conditions, compound **51** was converted directly into the D-*manno* pyrrolidine derivative **53** in high yield [40]. This compound is a useful precursor of aza-*C*-glycosyl compounds and homoazasugars in the pyrrolidine series and could also be used in a synthesis of the natural product *australine*.

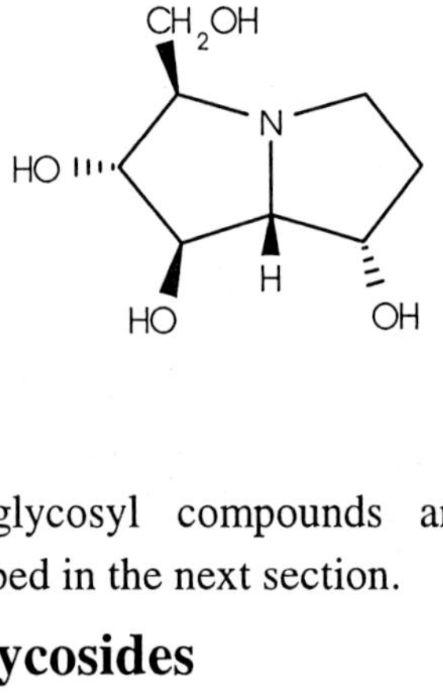

Figure 14-7

β-linked aza-*C*-glycosyl compounds are accessible by way of the methodologies described in the next section.

14.3 Homoazaglycosides

"Homoazaglycosides" are derivatives of so-called "homoazasugars," i.e. azasugars carrying an additional hydroxymethyl substituent α to the nitrogen atom (2,5-dideoxy-2,5-imino-hexitols or 2,6-dideoxy-2,6-imino-heptitols); piperidine homoazasugars are stable homologs of 5-amino-5-deoxy-hexoses in their pyranoid hemiaminal form. They constitute the simplest and also, so far, the most common type of aza-*C*-glycosyl compounds. Interest in homoazasugars was heightened by the discovery of α-homonojirimycin **54** in a neotropical liana (*Omphalea diandra* L.) [41] and in a larva feeding on that plant (*Urania fulgens*) [42], as well as by its potent activity as an α-glucosidase inhibitor [41].

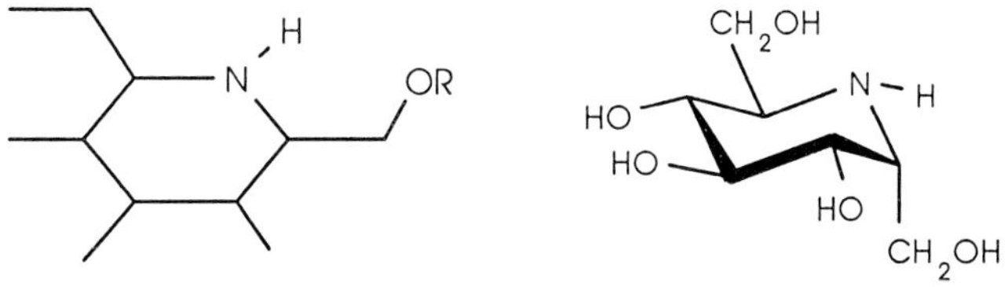

Figure 14-8

homoazasugars: R = H
homoazaglycosides: R = aglycone

54

The first synthesis of **54** (in protected form) had been reported [33] by Liu shortly before it was found to occur as a natural product. The synthesis involved, in the key steps, a Wittig chain extension, amination by way of an oxime, a highly stereoselective mercury(II)-mediated cyclization, and oxidative demercuration:

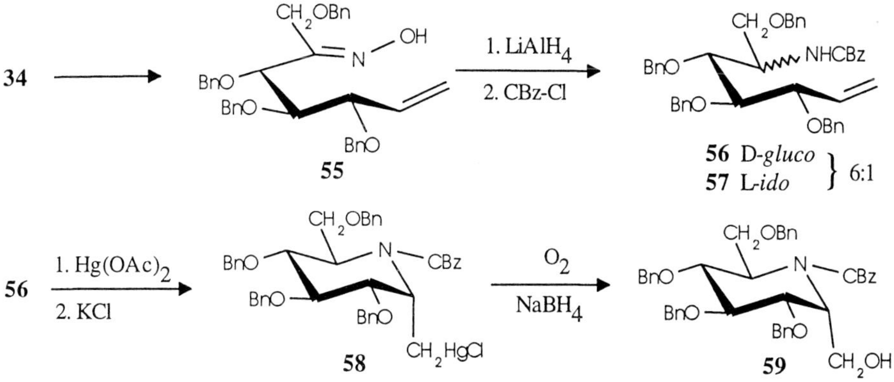

34 → **55** → (1. LiAlH$_4$, 2. CBz-Cl) → **56** D-*gluco* / **57** L-*ido* } 6:1

56 → (1. Hg(OAc)$_2$, 2. KCl) → **58** → (O$_2$, NaBH$_4$) → **59**

Scheme 14-14

The same sequence of reactions was applied recently by the same authors to the preparation of a pyrrolidine homoazasugar, namely 2,5-dideoxy-2,5-imino-D-mannitol [43]. A number of "homo" analogs of piperidine azasugars have been reported since Liu's pioneering studies: α- and β-L-homofuconojirimycin [44,45], and the α-homo analog of mannojirimycin [46] from seven-carbon sugar derivatives by way of internal reductive amination or displacement procedures, β-homomannojirimycin [47,48] and β-homonojirimycin [48] by way of aldolase-mediated coupling processes followed by internal reductive aminations.

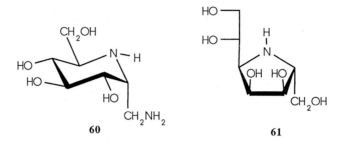

Figure 14-9

A total synthesis of α-homonojirimycin **54** has also been described [49]. More recently, azaglycosyl cyanides have been exploited in a more efficient synthesis of **54** by the Merrell Dow group [23] and for the preparation of amino homoazasugars such as **60** [24]. Finally, Fleet [50] and Wong [51] have recently reported the first examples of homoaza analogs of hexoses in the furanose form (e.g. **61** [50]).

The higher homologs retain most of the activity of the parent azasugars as glycosidase inhibitors and exhibit in some cases greater selectivity [46,50]; for example, α-homomannojirimycin is a much more selective inhibitor of α-mannosidases than the parent compound, 1-deoxymannojirimycin [50]. However, further improvement of the selectivity towards certain family of glycosidases (e.g., sequence-specific inhibition of oligosaccharidases) can be expected if the complete aglycone is incorporated into the inhibitor: indeed, as the first example of a compound of this kind, the homoazadisaccharide **62** ("homoaza-α,β-trehalose") prepared by Liu and coworkers [23,33] was shown to be a very potent competitive inhibitor of intestinal sucrase and α-glucanases.

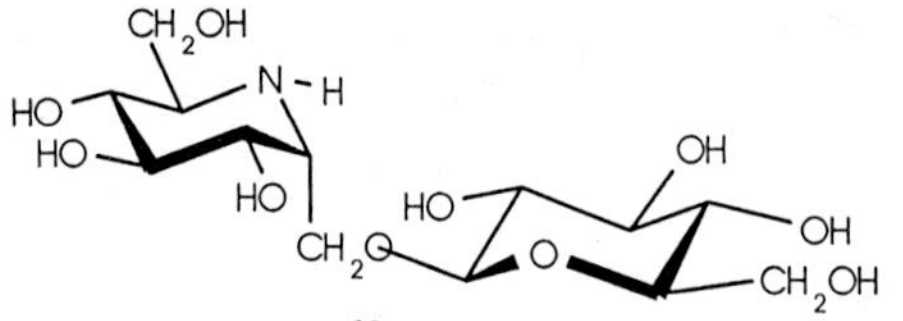

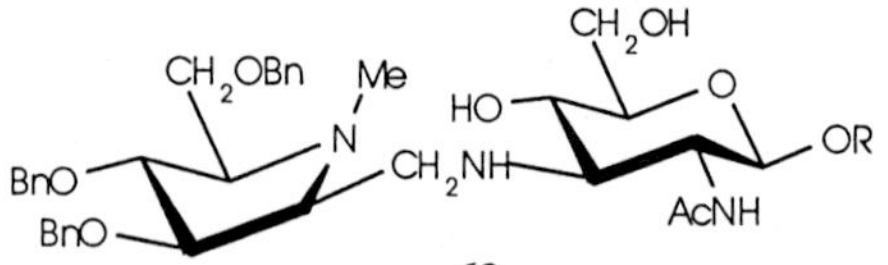

Figure 14-10

As it was able to significantly reduce postprandial hyperglycemia after a sucrose or starch load, this compound was identified as a candidate drug for antidiabetic therapy [52]. In spite of this very promising lead, few other compounds of this kind have been reported: the *N*-linked disaccharide analog **63**

[24] and the related *O*-linked derivative **64** [53]. Wong has also reported the synthesis of a homoaza analog of a nucleoside (**65**) [24].

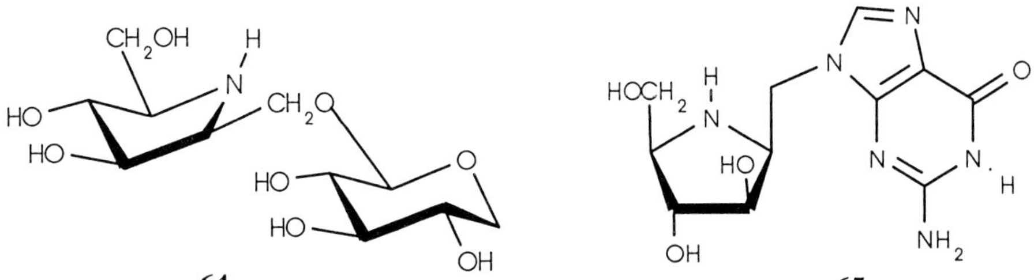

Figure 14-11

As one of our goals was the development of more efficient and selective inhibitors of glucan-processing enzymes, our own investigations on homoazaglycosides have dealt primarily with the synthesis of homoazadisaccharides. The homoaza analogs of both maltose and cellobiose have thus been prepared for the first time.

The maltose analog **69** was obtained in two steps from precursor **37** by way of its reaction with methyl 2,3,6-tri-*O*-benzyl-α-D-glucopyranoside (**66**); this reaction led not only to homoazadisaccharide **67**, but also to the azepane derivative **68** (ratio **67/68** ~ 1:1) which indicated that the apparent displacement process might in fact occur by way of an aziridinium cation intermediate (**70**). Related ring expansion reactions in azasugar chemistry have recently been reported by Depezay [54].

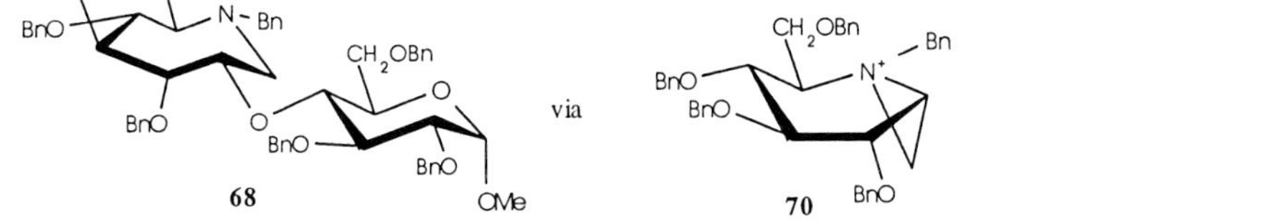

Scheme 14-15

Compound **69** is a most interesting azamaltose mimic and could be a useful tool for the study of α-glucan-processing enzymes. The azepane-containing pseudodisaccharide **68** constitutes the first example of a new and intriguing class of azadisaccharide mimics, with a -CH$_2$- group inserted into the endocyclic C–N bond of the *O,N*-acetal function (see Section 4).

Our synthetic approach to β-**linked** aza-*C*-glycosyl compounds as well as homoazaglycosides, including the analog of cellobiose, is based on the highly stereoselective double reductive amination of 2,6-heptodiuloses.

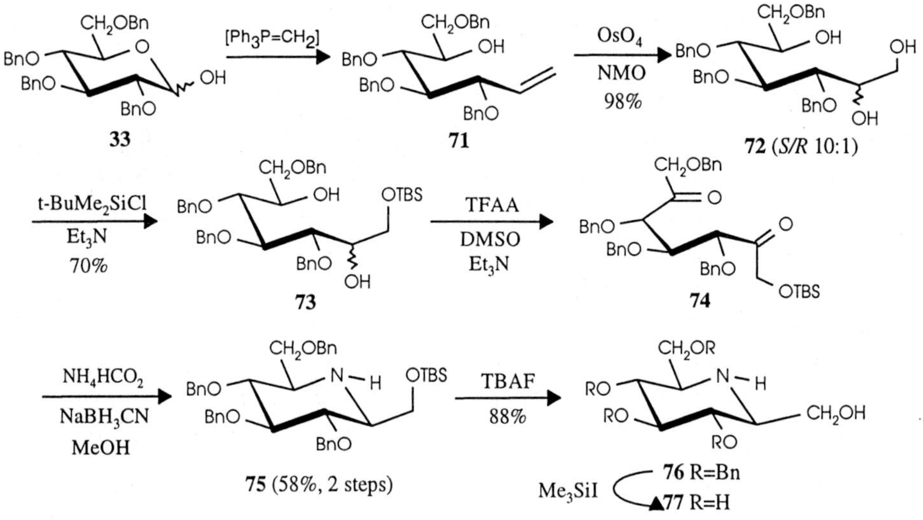

Scheme 14-16

Thus, β-homonojirimycin **77** and derivatives, precursors of azaglycoside mimics in the β-D-*gluco* series, were prepared by two related reaction sequences, as recently published [55]; the most efficient sequence is shown above and involves, in the key step, the reductive amination of D-*xylo* diketosugar **74** to give exclusively the all-equatorial piperidine derivative **75**. Such stereocontrol in this reaction was expected on the basis of the results of extensive studies on the synthesis of 1-deoxynojirimycin by internal reductive amination [3, 56] as well as from dicarbonyl sugars [57]. We have shown very recently that the same degree of stereocontrol was operative in the reductive amination of the L-*arabino* 2,6-heptodiulose **79**, thus giving access to β-homogalactostatin **81** and derivatives [58].

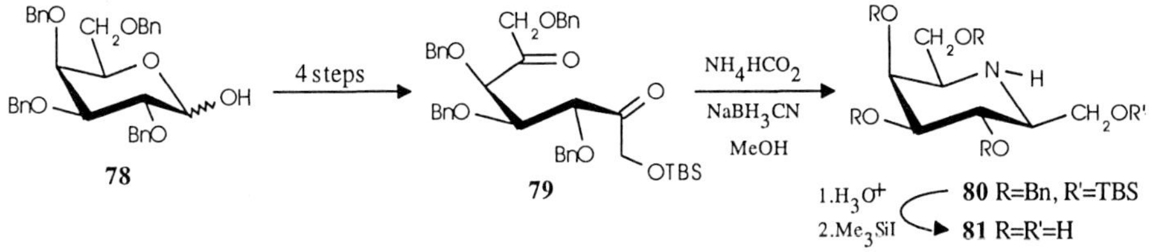

Scheme 14-17

It is worth noting that the nitrogen atom in these 2,6-disubstituted piperidine derivatives is highly hindered and that N-alkylation requires unusual conditions [55]. Remarkably, the reaction of amino alcohol **76** with acylating agents led only to the corresponding O-acylated product: for example, its sulfonylation gave sulfonate **82**, a most useful precursor of β-homoazaglucosides. The reaction of **82** with **66**, followed by deprotection of the resulting coupling product (**83**) gave the homoaza analog of methyl α-cellobioside, compound **84**, a potential inhibitor of β-glucan processing enzymes. No rearranged product was observed which suggested that the reaction occured by direct displacement only in this case. On treatment with DBU, the amino tosylate **82** gave the corresponding aziridine **85**, which could be deprotected under dissolving-metal reduction conditions to 1,N-anhydro homoazasugar **86**; this compound is a potential inactivator of glucosidases [59].

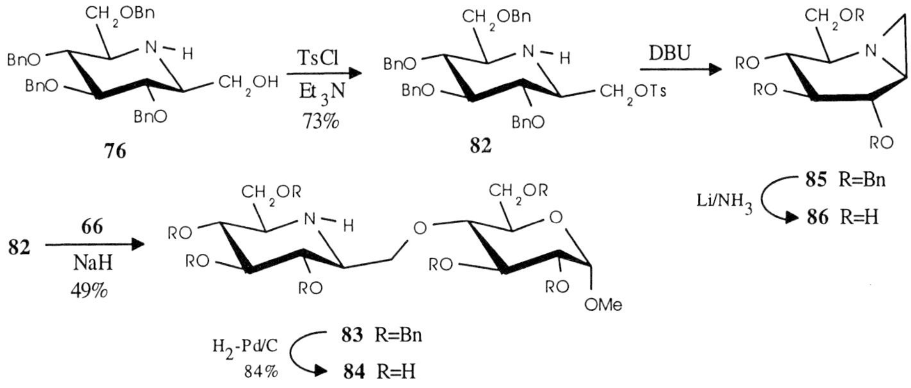

Scheme 14-18

Finally, precursors of α-linked homoazagalactosides as well as α-homogalactostatin were prepared by way of the chain extension-amination cyclization procedure [60] (see Section 2). Double inversion at C–6 of **87** (prepared from **78** by a Wittig reaction) under Mitsunobu conditions* afforded compound **88**. The corresponding benzylcarbamate **89** was cyclized efficiently and highly stereoselectively using mercury (II) trifluoroacetate [61].

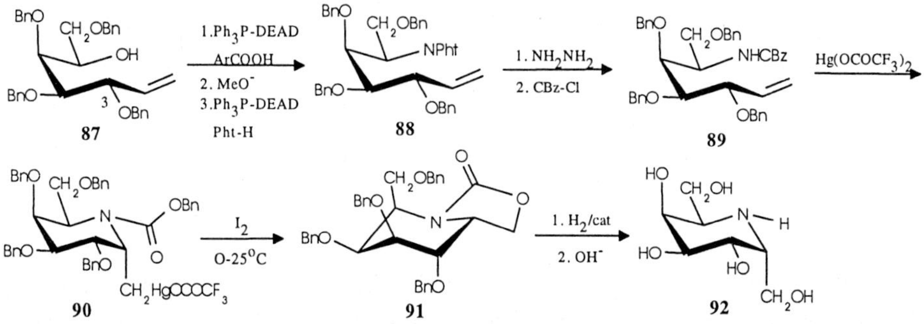

Scheme 14-19

On treatment with iodine, the resulting organomercurial (**90**, not isolated) underwent an unexpected but most useful reaction, namely the participation of the benzyl carbamate function leading to the formation of a cyclic carbamate (**91**) [62]. Deprotection of **91** gave α-homogalactostatin **92**. Removal of the cyclic carbamate function from **91** would lead to the D-*galacto* epimer of **76**, a compound which could be used as the precursor of homoaza analogs of various α-galactosides.

Preliminary studies of the activity of **92** as a glycosidase inhibitor revealed that this compound is a much more selective inhibitor of α-galactosidases than galactostatin (5-amino-5-deoxy-D-galactose) and its 1-deoxy derivatives [63].

The homoazasugar derivatives described in this and the previous section provide access to a wide variety of homoaza analogs of glycosides: on reaction with appropriate nucleophiles, they could thus lead to mimics of glycosyl phosphates, glycosyl amino acids, glycolipids, etc.

14.4 1-Deoxyazaseptanoses

In pioneering studies on sugar analogs with nitrogen in the ring, Paulsen and coworkers [64] have shown that, although they exist almost exclusively in the pyranose form in aqueous solution, 6-amino-6-deoxy-D-glucose and D-galactose

---

* The participation of the benzyloxy group at C–3, which occurs spontaneously upon formation of the triflate of **87** [34], was not observed when displacements were performed at C–6 under Mitsunobu conditions in the D-*galacto*/L-*altro* series. In the D-*gluco*/L-*ido* series, however, mixtures of products were obtained under these conditions.

give the corresponding 1,6-dideoxy-1,6-iminohexitols on catalytic hydrogenation (e.g.: **96**).

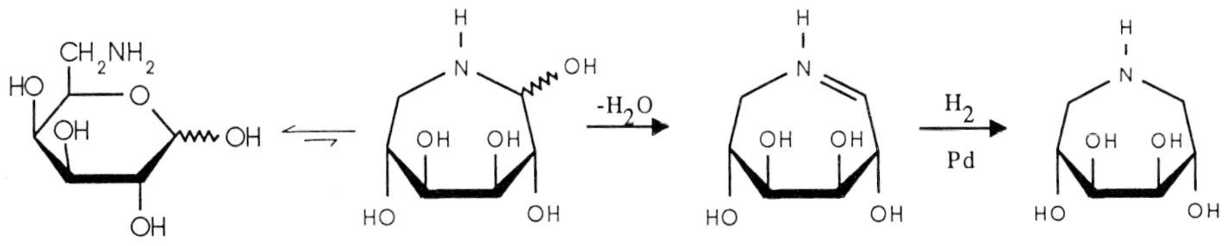

Scheme 14-20

Other examples of polyhydroxylated azepanes ("azaseptanoses" or "1-deoxyazaseptanoses") have appeared only very recently: the 1,6-imino-D-mannitol derivative **97** was prepared as a mimic of the mannopyranosyl cation in the "flap up" half-chair conformation [65] and a series of epimeric 1,6-dideoxy-1,6-imino-hexitols (e.g. **98**) have been obtained by opening 1,2:5,6-dianhydrohexitols with a primary amine [66, 67] or by internal reductive amination of 6-azido-6-deoxy-hexopyranose derivatives [68]. Several members of this family were shown to exhibit significant activities as glycosidase inhibitors [68].

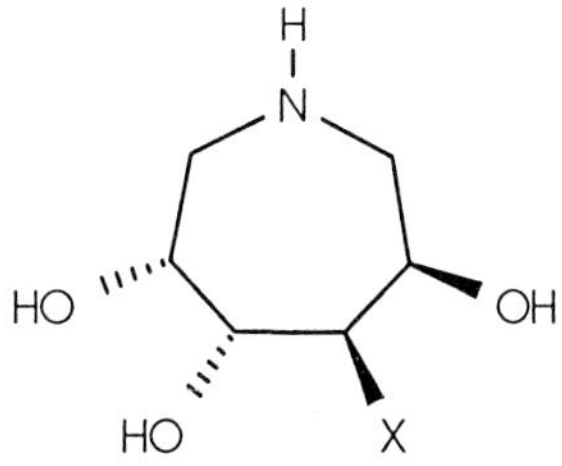

97 X=NH$_2$
98 X=OH

Figure 14-12

It appears that no examples of 1,6-dideoxy-1,6-iminoheptitols have been described. With the goal of developing an approach to this new class of azasugar and azaglycoside mimics, we have investigated the internal reductive amination of a 7-amino-7-deoxy-2-heptulose derivative, compound **101**. Compound **101** was obtained by way of the highly stereoselective hydroxylation of the alkenyl function of the unsaturated ketone **34**, followed by a functional group exchange.

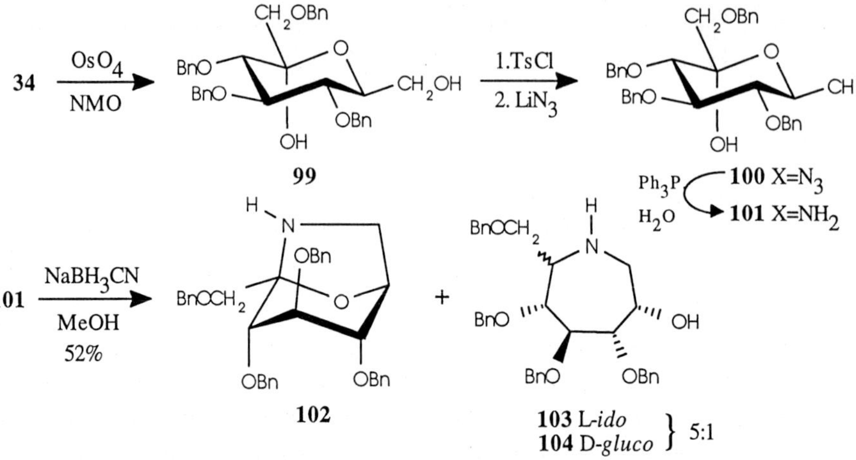

Scheme 14-21

The reaction of **101** with NaBH$_3$CN in methanol gave a mixture of the internal *O,N*-acetal **102** and the desired 1,6-dideoxy-1,6-imino-heptitol derivatives, the L-epimer **103** (L-*glycero*-D-*gulo*, equivalent to "β-L-*ido*") being predominant. Compound **102** was slowly converted into **103/104**, which suggested that **102** is not an intermediate in the reaction of **101** to **103/104** but rather a competing product resulting from an internal reaction of the intermediate cyclic imine (see ref. [64]) (final overall yield: ~52%).

Compounds **103** and **104** carry a free OH group at C-3 and therefore constitute useful precursors for the synthesis of a wide variety of azaglycoside mimics of the 1-deoxyazaseptanose type. As described in the previous section, one such compound (**68**) has already been obtained by way of a ring enlargement process in the course of our synthesis of the homoaza analog of maltose. Interestingly, the NMR parameters of **68** indicate that the seven-membered ring adopts a *TC* conformation close to the $^{4,5}C_1$ form, thus making it resemble the structure of a β-glucoside more than that of an α-glucoside:

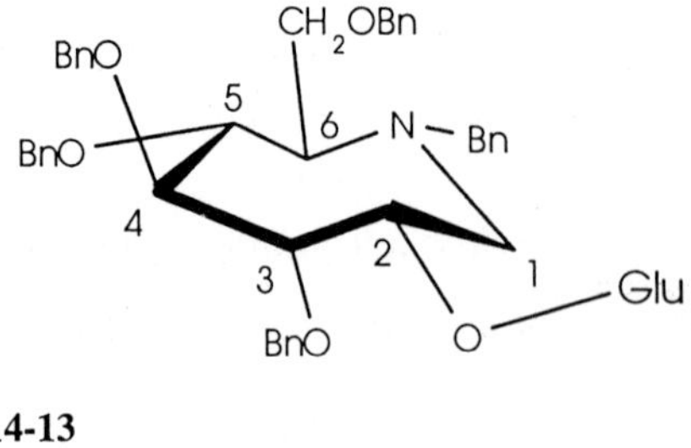

68 $^{1,N}TC_{2,3}$-$^{4,5}C_1$-$^{1,2}TC_{6,N}$

Figure 14-13

Further investigations on the synthesis of this intriguing new type of azaglycoside mimics as well as on their biological activity are in progress in our laboratory.

14.5 Conclusion

Aza-*C*-glycosides and homoazaglycosides form an emerging class of second-generation carbohydrate mimics. By providing stable analogs of potentially most interesting but hydrolytically labile azaglycosides, the preparation and study of these compounds may lead to the finding of novel potent and highly selective glycosidase inhibitors; such compounds would not only constitute most useful biochemical probes but might also be endowed with potential therapeutic value, for example as new antiviral agents. Few members of this new class of carbohydrate mimics have been described so far. We have presented in this review methodologies as well as a number of precursors that could be used for the preparation of a wide range of azaglycoside analogs. It is this author's hope that this review will promote further work in this area.

Acknowledgment

Support of our research activities by a grant from the National Institutes of Health (DK35766) is gratefully acknowledged.

14.6 References

1. H. Paulsen, K. Todt, *Adv. Carbohydr. Chem.* **1968**, *23*, 115-232.
2. B. Winchester, G. W. J. Fleet, *Glycobiology* **1992**, *2*, 199-210, and ref. cited.
3. G. C. Look, C. H. Fotsch, C.-H. Wong, *Acc. Chem. Res.* **1993**, *26*, 182-190.
4. (a) L. A. G. M. Van den Broek, D. J. Vermaas, B. M. Heskamp, C. A. A. van Boeckel, M. C. A. A. Tan, J. G. M. Bolscher, H. L. Ploegh, F. J. van Kemenade, R. E. Y. De Goede, F. Miedema, *Recl. Trav. Chim. Pays-Bas* **1993**, *112*, 82-94. (b) L. A. G. M. van den Broek, in "Carbohydrate in Drugs," Witczak, Z. J. Nierfoth, K. A. Eds., Dekker: New York, 1996, pp.
5. G. Legler, *Adv. Carbohydr. Chem. Biochem.* **1990**, *48*, 319-384.
6. K. K.-C. Liu, T. Kajimoto, L. Chen, Z. Zhong, Y. Ichikawa, C.-H. Wong, *J. Org. Chem.* **1991**, *56*, 6280-6289.
7. N. Asano, K. Oseki, H. Kizu, K. Matsui, *J. Med. Chem.* **1994**, *37*, 3701-3706.
8. Y.-F. Wang, Y. Takaoka, C.-H. Wong, *Angew. Chem. Int. Ed. Engl.* **1994**, *33*, 1242-1244.
9. A. B. Hughes, A. Rudge, *J. Nat. Prod. Rep.* **1994**, 135-162.

10. P. A. S. Smith, *The Chemistry of Open-chain Organic Nitrogen Compounds*, Benjamin: New York, 1965. In most α-alkoxyamino compounds, the nitrogen atom is part of an amide, carbamate, or equivalent function. See: Houben-Weyl, *Methoden der Organischen Chemie*, Bd. E14a/2 (O/N-Acetale), Thieme Verlag: Stuttgart, 1991.

11. K. Suzuki, H. Hashimoto, *Tetrahedron Lett.* **1994**, *34*, 4119-4122.

12. K. M. Robinson, M. E. Begovic, B. L. Rhinehart, E. W. Heineke, J. B. Ducep, P. R. Kastner, F. N. Marshall, C. Danzin, *Diabetes* **1991**, *40*, 825-830.

13. T. M. Jespersen, W. Dong, M. R. Sierks, T. Skrydstrup, I. Lundt, M. Bols, *Angew. Chem. Int. Ed. Engl.* **1994**, *33*, 1778-1779.

14. C. R. Johnson, M. W. Miller, A. Golebiowski, H. Sundram, M. B. Ksebati, *Tetrahedron Lett.* **1994**, *35*, 8991-8994.

15. M. Rubiralta, M. Giralt, A. Diez, "Piperidine," *Studies in Organic Chemistry,* Vol. 43, Elsevier: Amsterdam, 1991.

16. C. R. Johnson, A. Golebiowski, H. Sundram, M. W. Miller, R. L. Dwaihy, *Tetrahedron Lett.* **1995**, *36*, 653-654.

17. T. Fuchss, R. R. Schmidt, XIXth International Carbohydrate Symposium, Milano, Italy, July 1996; Abstr. No. BP040.

18. M. H. D. Postema, *C-Glycoside Synthesis*, CRC Press: Boca Raton, FL, 1995.

19. D. E. Levy, C. Tang, *The Chemistry of C-Glycosides*, Pergamon Press: Oxford, 1995.

20. H.-J. Altenbach, R. Wischnat, *Tetrahedron Lett.* **1995**, *36*, 4983-4984.

21. G. Rassu, L. Pinna, P. Spanu, F. Ulgheri, G. Casiraghi, *Tetrahedron Lett.* **1994**, *35*, 4019-4022.

22. H. Böshagen, W. Geiger, B. Junge, *Angew. Chem. Int. Ed. Engl.* **1981**, *20*, 806-807.

23. P. B. Anzeveno, L. J. Creemer, J. K. Daniel, C.-H. R. King, P. S. Liu, *J. Org. Chem.* **1989**, *54*, 2539-2542 and ref. cited.

24. C.-H. Wong, L. Provencher, J. A., Porco, Jr. S.-H. Jung, Y.-F. Wang, L. Chen, R. Wang, D. H. Steensma, *J. Org. Chem.* **1995**, *60*, 1492-1501.

25. P. S. Liu, R. S. Rogers, M. S. Kang, P. S. Sunkara, *Tetrahedron Lett.* **1991**, *32*, 5853-6.

26. B. B. Shankar, M. P. Kirkup, S. W. McCombie, A. K. Ganguly, *Tetrahedron Lett.* **1993**, *34*, 7171-7174.

27. L. J. Liotta, J. Lee, B. Ganem, *Tetrahedron* **1991**, *47*, 2433-2447.

28. L. Lay, F. Nicotra, A. Paganini, C. Pangrazio, L. Panza, *Tetrahedron Lett.* **1993**, *34*, 4555-4558.

29. L. Cipolla, L. Lay, F. Nicotra, C. Pangrazio, L. Panza, *Tetrahedron* **1995**, *51*, 4679-4690.

30. A. Baudat, P. Vogel, *Tetrahedron Lett.* **1996**, *37*, 483-484.

31. J. -R. Pougny, M. A. M. Nassr, P. Sinaÿ, *J. Chem. Soc. Chem. Commun.* **1981**, 375-376.

32. F. Nicotra, L. Panza, F. Ronchetti, G. Russo, L. Toma, *Carbohydr. Res.* **1987**, *171*, 49-57.

33. P. S. Liu, *J. Org. Chem.* **1987**, *52*, 4717-4721.

34. O. R. Martin, F. Yang, F. Xie, *Tetrahedron Lett.* **1995**, *36*, 47-50.

35. O. R. Martin, L. Liu, F. Yang, *Tetrahedron Lett.* **1996**, *37*, 1991-1994.

36. D. P. Curran, T. L. Fevig, C. P. Jasperse, M. J. Totleben, *Synlett* **1992**, 943-961.

37. P. de Pouilly, A. Chénedé, J. -M. Mallet, P. Sinaÿ, *Bull. Soc. Chim. Fr.* **1993**, *130*, 256-265.

38. F. Nicotra, F. Ronchetti, G. Russo, *J. Org. Chem.* **1982**, *42*, 4459-4462.

39. A. Boschetti, F. Nicotra, L. Panza, G. Russo, *J. Org. Chem.* **1988**, *53*, 4181-4185.

40. O. R. Martin, J. Yuan, unpublished results.

41. G. C. Kite, L. E. Fellows, G. W. J. Fleet, P. S. Liu, A. M. Scofield, N. G. Smith, *Tetrahedron Lett.* **1988**, *29*, 6483-6486.

42. G. C. Kite, J. M. Horn, J. T. Romeo, L. E. Fellows, D. C. Lees, A. M. Scofield, N. G. Smith, *Phytochemistry* **1990**, *29*, 103-105.

43. M. S. Chorghade, C. T. Cseke, P. S. Liu, *Tetrahedron: Asymmetry* **1994**, *5*, 2251-2254.

44. D. M. Andrews, M. I. Bird, M. M. Cunningham, P. Ward, *Bioorg. Med. Chem. Lett.* **1993**, *3*, 2533-2536.

45. G. W. J. Fleet, S. K. Namgoong, C. Barker, S. Baines, G. S. Jacob, B. Winchester, *Tetrahedron Lett.* **1989**, *30*, 4439-4442.

46. I. Bruce, G. W. J. Fleet, I. Cenci di Bello, B. Winchester, *Tetrahedron* **1992**, *48*, 10191-10200.

47. I. Henderson, K. Laslo, C.-H. Wong, *Tetrahedron Lett.* **1994**, *35*, 359-362.

48. K. E. Holt, F. J. Leeper, S. Handa, *J. Chem. Soc. Perkin Trans. I.* **1994**, 231-234.

49. S. Aoyagi, S. Fujimaki, C. Kibayashi, *J. Chem. Soc. Chem. Commun.* **1990**, 1457-1459.

50. P. M. Myerscough, A. J. Fairbanks, A. H. Jones, I. Bruce, S. S. Choi, G. W. J. Fleet, S. S. Al-Daher, I. Cenci di Bello, B. Winchester, *Tetrahedron* **1992**, *48*, 10177-10190.

51. S. Hiranuma, T. Shimizu, T. Nakata, T. Kajimoto, C.-H. Wong, *Tetrahedron Lett.* **1995**, *36*, 8247-8250.

52. B. L. Rhinehart, K. M. Robinson, P. S. Liu, A. J. Payne, M. E. Wheatley, S. R. Wagner, *J. Pharmacol. Exp. Ther.* **1987**, *241*, 915-920.

53. G. Mikkelsen, T. V. Christensen, M. Bols, I. Lundt, *Tetrahedron Lett.* **1995**, *36*, 6541-6544.

54. L. Poitout, Y. Le Merrer, J. -C. Depezay, *Tetrahedron Lett.* **1996**, *37*, 1609-1612.

55. O. M. Saavedra, O. R. Martin, *J. Org. Chem.* **1996**, *61*, 6987-6993.

56. A. Straub, F. Effenberger, P. Fischer, *J. Org. Chem.* **1990**, *55*, 3926-3932.

57. E. W. Baxter, A. B. Reitz, *J. Org. Chem.* **1994**, *59*, 3175-3185.

58. O. M. Saavedra, L. Connell, O. R., Martin, unpublished results.

59. M. K. Tong, B. Ganem, *J. Am. Chem. Soc.* **1988**, *110*, 312-313.

60. O. R. Martin, F. Xie, L. Liu, *Tetrahedron Lett.* **1995**, *36*, 4027-4030.

61. R. C. Bernotas, B. Ganem, *Tetrahedron Lett.* **1985**, *26*, 1123-1126.

62. For a related example, see ref. 43.

63. P. Vogel, S. Picasso-Bourdenet, L. Liu, O. R. Martin, unpublished results.

64. H. Paulsen, K. Todt, *Chem. Ber.* **1967**, *100*, 512-520.

65. R. A. Farr, A. K. Holland, E. W. Huber, N. P. Peet, P. M. Weintraub, *Tetrahedron* **1994**, *50*, 1033-1044.

66. L. Poitout, Y. Le Merrer, J.-C. Depezay, *Tetrahedron Lett.* **1994**, *35*, 3293-3296.

67. B. B. Lohray, Y. Jayamma, M. Chatterjee, *J. Org. Chem.* **1995**, *60*, 5958-5960.

68. F. Moris-Varas, X.-H. Qian, C.-H. Wong, *J. Am. Chem. Soc.* **1996**, *118*, 7647-7652.

15 Total Synthesis and Chemical Design of Useful Glycosidase Inhibitors

K. Tatsuta

15.1 Introduction

In recent years, much attention has focused on the synthesis and development of glycosidase inhibitors because of an increasing awareness of the vital role played by carbohydrates in biological processes. Glycosidase inhibitors have become interesting as antiobesity drugs, antidiabetes, antifungals, insecticides, and antivirals, including substances active against the human immunodeficiency virus (HIV) and metastasis. Therefore, the chemical and biochemical studies on glycosidase inhibitors may lead to understand the molecular basis of intractable diseases such as diabetes mellitus, cancer and AIDS, and may also provide us with therapeutic approaches to these diseases. As part of an ongoing program to clarify the mode of action of glycosidase inhibitors, we have synthesized cyclophellitol, nagstatin and gualamycin, and their analogues having different configurations and functionalities. These syntheses feature general methods of entry into carba-sugars and nitrogen-containing carbohydrates.

1: Cyclophellitol   20: Nagstatin   47: Gualamycin

The synthesis of cyclophellitols including the aziridine and thiirane analogues was mainly based on the stereospecific intramolecular [3+2] cycloaddition of a nitrile oxide to an olefin [1 - 5]. Nagstatins including a variety of hydroxyl and triazole analogues were synthesized by inter- and intramolecular nucleophilic reaction of imidazole and triazole moieties [6 - 9]. Their glycosidase inhibiting activities were quite substrate-specific, indicating that the glycosidases recognize especially each carbons and configurations of the glycosidase inhibitors, and consequently, the inhibitors serve as antagonists of the corresponding glycopyranosides. Total synthesis of gualamycin was accomplished by glycosylation of a thiophenol derivative of the disaccharide portion with a pyrrolidine aglycon. The anti-mite activity of gualamycin was suggested to be due to its maltase inhibiting activity [10 - 11].

15.2 Results

15.2.1 Cyclophellitols

Cyclophellitol (**1**) is a novel β-D-glucosidase inhibitor isolated from culture filtrates of the mushroom *Phellinus* sp.. Structurally, cyclophellitol (**1**) is a fully oxygenated cyclohexane corresponding to a 5a-carba analogue of D-glucopyranose [12]. To clarify their mode of action in glycosidase inhibition, compound **1** and its analogues **2 - 7** have been enantiospecifically synthesized from carbohydrates in our laboratories [1 - 5].

1 X = O (Cyclophellitol)
4 X = NH
6 X = S

2 X = O
5 X = NH
7 X = S

3

Recently, elegant syntheses of **1** have been reported by several groups [13]. Our strategy for construction of these highly oxygenated compounds is based on an intramolecular cycloaddition of a nitrile oxide to an alkene [1 - 3].

Swern oxidation of **8**, which was derived from L-glucose, afforded the unstable aldehyde, which was subjected to Wittig alkenation with salt-free methylidene-triphenylphosphorane to afford the alkene **9** (Scheme 15-1). This compound was hydrolysed with aqueous HCl in dioxane to an idopyranose derivative, which was treated with hydroxylamine hydrochloride in pyridine to give the oxime **10**. Intramolecular cycloaddition of **10** was performed *via* the intermediary nitrile oxide to afford the isoxazoline **11** as a single product in 70%

yield. The stereochemistry was confirmed by ^{1}H nmr studies of compounds **11** - **14** and, finally by completion of the synthesis presented next.

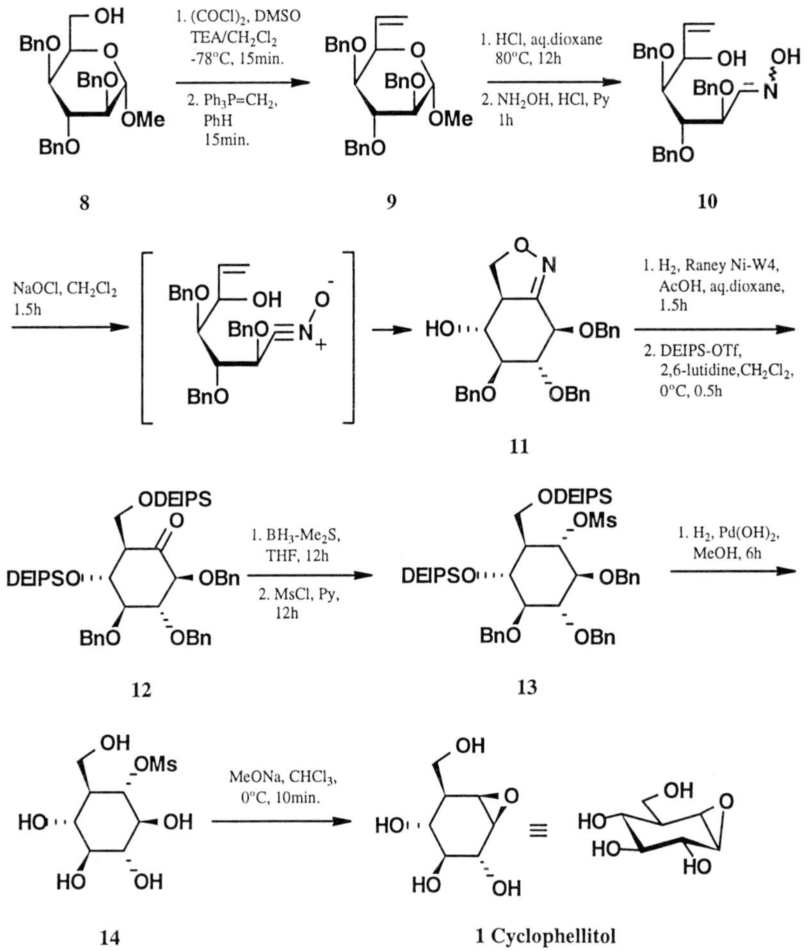

Scheme 15-1

The isoxazoline ring opening was achieved by treatment of **11** with H$_2$ and Raney Ni-W4 in aqueous. dioxane in the presence of AcOH to afford the keto-diol, which was silylated with diethylisopropylsilyl triflate to give the protected ketone **12**. Reduction with BH$_3$-Me$_2$S afforded the desired α-alcohol, which was converted into the labile mesylate **13**. The diethylisopropylsilyl (DEIPS) group was developed in our laboratories and effectively used as an *O*-protecting group, because this silyl is readily removed under hydrogenolytic conditions using

Pd(OH)$_2$ [14]. Accordingly, the mesylate **13** was subjected to hydrogenolysis with Pd(OH)$_2$ in MeOH to give the deprotected **14**, epoxidation with MeONa gave cyclophellitol (**1**).

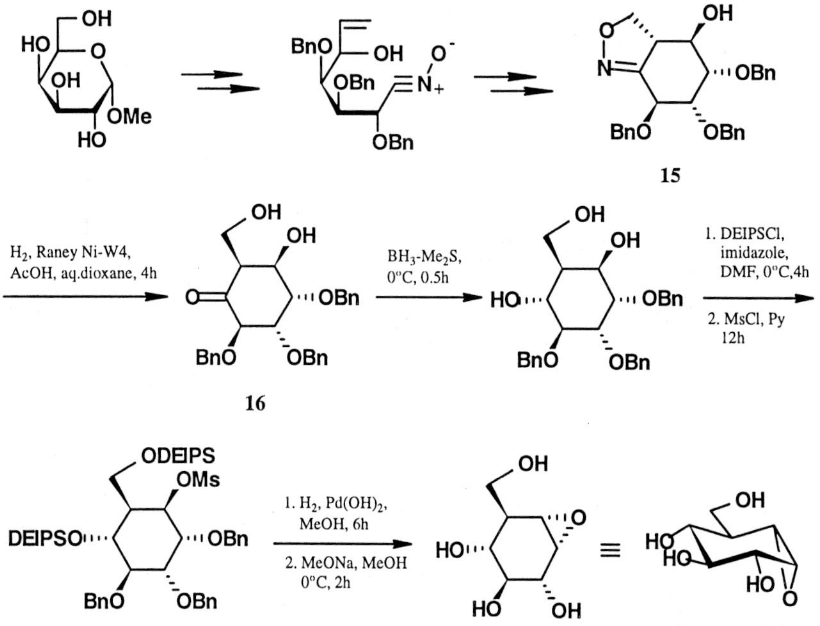

Scheme 15-2

In order to provide additional insight into the mode of action of cyclophellitol (**1**), the unnatural epoxide diastereomers (**2-3**) and heteroatom-containing analogues (**4-7**) were synthesized. From the fact that cyclophellitol (**1**) exhibits a very high β-D-glucosidase inhibiting activity, we expected that 1,6-*epi*-cyclophellitol (**2**) and α-*manno* analogue **3** inhibit α-D-glucosidase and α-D-mannosidase activities, respectively.

1,6-*Epi*-cyclophellitol (**2**) was similarly synthesized from methyl 2,3,4-tri-*O*-benzyl-α-D-galactopyranoside through the isoxazoline **15**, which was subjected to acidic hydrogenolysis with Raney Ni-W4 to afford the desired keto-alcohol **16** with epimerization at the C-1 position (Scheme 15-2).

The α-*manno* analogue **3** was synthesized from **15** without epimerization at C-1 position. Hydrogenolysis of **15** was conducted using Raney Ni and B(OH)$_3$ to afford the keto-alcohol **17** in a quantitative yield, which was converted into **3** by the above route (Scheme 15-3) [4].

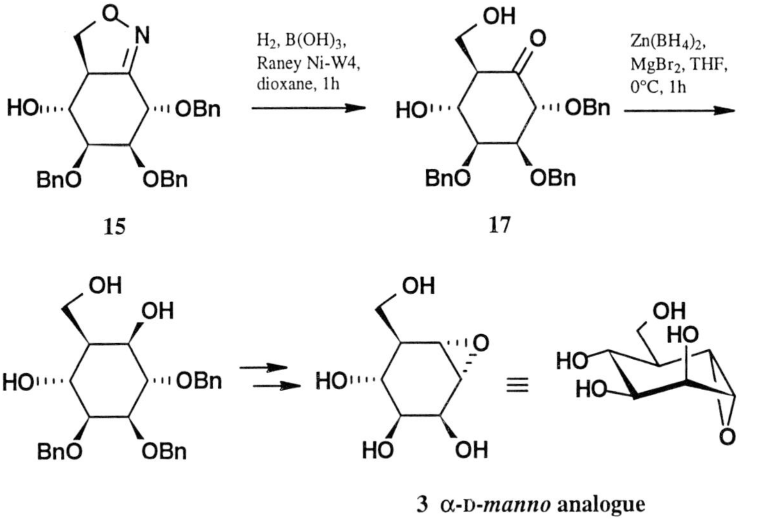

Total Synthesis of Useful Glycosidase Inhibitors... 287

Scheme 15-3

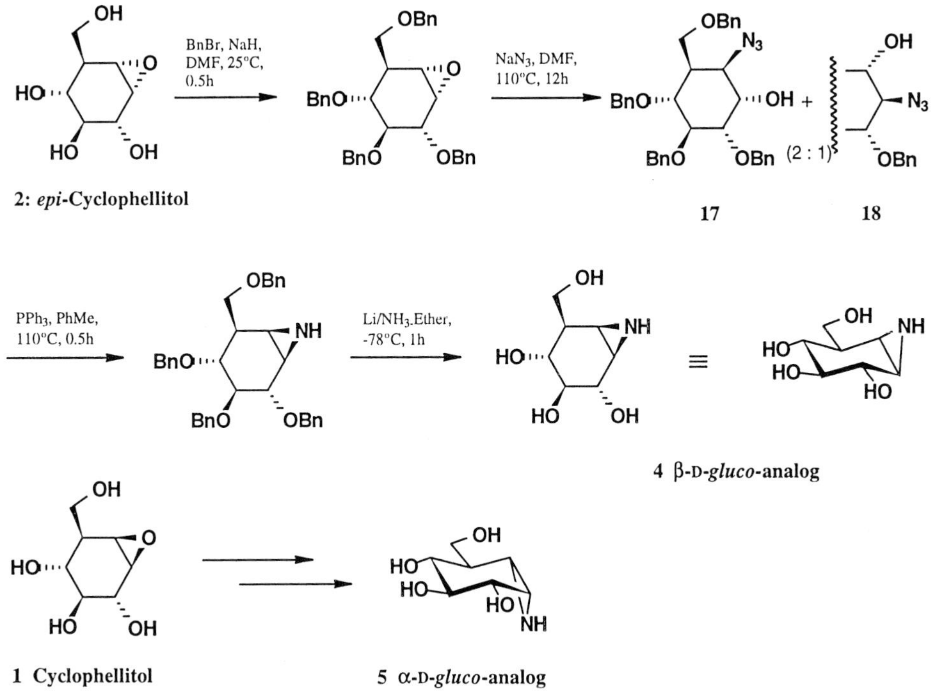

Scheme 15-4

The aziridine analogue **4** was synthesized from 1,6-*epi*-cyclophellitol (**2**). The tetra-*O*-benzyl derivative of **2** was treated with NaN_3 to afford a mixture of **17** and **18**, which was subjected to reduction with PPh_3 to give a single aziridine. De-*O*-benzylation gave the β-aziridine analogue **4** (Scheme 15-4) [4 - 5]. Similarly, the α-aziridine **5** was derived from cyclophellitol (**1**) [4 - 5].

The thiirane analogues **6** and **7** were prepared from **2** and **1**, respectively, on treatment of their *O*-MPM derivatives with $Ph_3P{=}S$ and trifluoroacetic acid, followed by de-*O*-methoxybenzylation with DDQ (Scheme 15-5) [5].

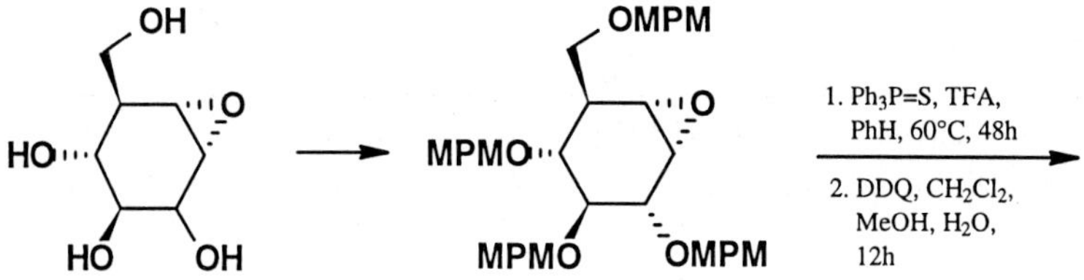

2 ***epi*-Cyclophellitol**

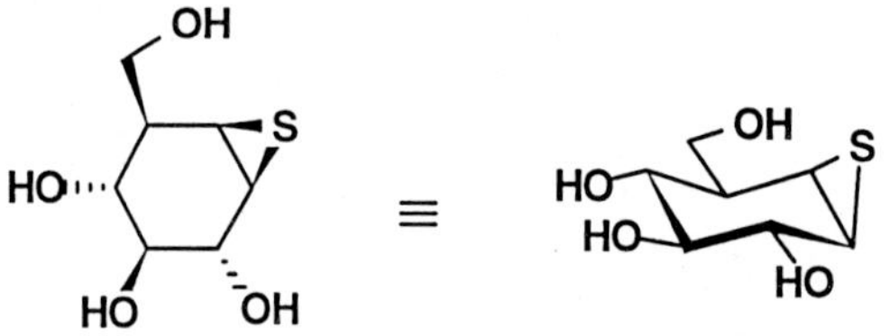

6 β-D-*gluco*-analog

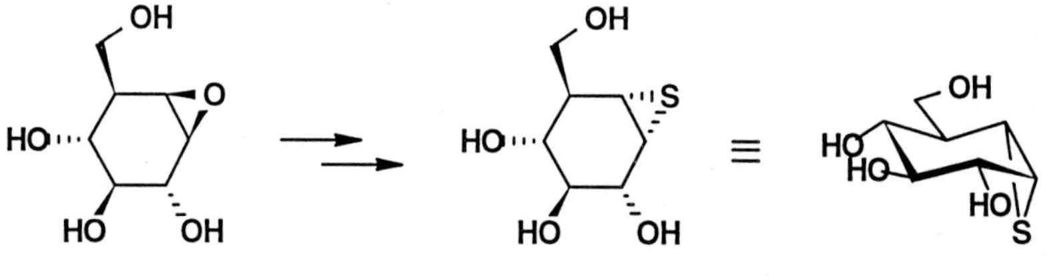

1 Cyclophellitol **7 α-D-*gluco*-analog**

Scheme 15-5

The glycosidase inhibiting activities of cyclophellitol (**1**), 1,6-*epi*-cyclophellitol (**2**), and their analogues **3** - **7** were generally assayed according to the method reported by Saul *et al.* [12] and are shown in Tables 15-1 and 2.

Table 15-1. Inhibitory activities of cyclophellitol, nagstatin (**20**) and their analogues (**2-4** and **21-27**) against glycosidases (IC$_{50}$: μg/ml).

	Inhibitors											
Glycosidases	1	2	3	4	20	21	22	23	24	25	26	27
α-D-Glc[a]		10										
β-D-Glc[b]	0.8			0.22				0.14				
α-D-Man[c]			19									
β-D-Man[d]										0.023		0.078
β-D-Gal[e]							0.0016				0.081	
NAc-β-D-Glc[f]					0.004	0.0015			0.0017			

a: Baker's Yeast α-D-glucosidase; b: Almond β-D-glucosidase; c: Jack beans α-D-mannosidase; d: Snail β-D-mannosidase; e: Escherichia coli β-D-galactosidase; f: Bovine kidney N-acetyl β-D-glucosaminidase

Table 15-2. Glycosidase inhibitory activities of cyclophellitol, nagstatin and their analogues.

Glycosidases	Inhibitors (IC$_{50}$: μg/ml)
α-D-Glucosidase	(Cyclophellitol) (0.8) — (0.22) — (0.14)
β-D-Glucosidase	(10)
β-D-Galactosidase	(10) — (0.0016) — (0.081)
α-D-Mannosidase	(19)
β-D-Mannosidase	(0.0023) — (0.078)
N-acetyl-β-D-Glucosaminidase	(Nagstatin) (0.004) — (0.0015) — (0.0017)

In contrast to natural cyclophellitol (**1**) which inhibited only β-D-glucosidase activity for 50% at 0.8 mg/ml, the *epi*-epoxide **2** exhibited an inhibiting activity only against α-D-glucosidase at IC_{50} 10 mg/ml. The α-*manno* analogue **3** as expected showed inhibitory activity against α-mannosidase of IC_{50} 19 mg/ml. The β-aziridine analogue **4** showed very high inhibitory activity against β-glucosidase at IC_{50} 0.22 mg/ml, while the α-aziridine **5** showed little α-glucosidase inhibiting activities.

Remarkably, both thiirane analogues **6** and **7** showed no significant activities. Structurally, cyclophellitol (**1**) and its aziridine analogue **4** have *quasi*-equatorially oriented C1-O and C1-N bonds, which correspond to the equatorial C1-O bond of β-D-glucopyranosides, whereas *epi*-cyclophellitol (**2**) and α-*manno* analogue **3** have *quasi*-axial C1-O bonds corresponding to the axial C1-O bond of α-D-glycopyranosides. Their glycosidase inhibiting activities emphasized that the α- and β-glycosidase recognized especially the C-1 positions and the residual portions as corresponding to those of α- and β-glycopyranosides. Consequently, these glycosidase inhibitors **1 - 4** serve as antagonists of the corresponding α- and β-D-glycopyranosides.

15.2.2 Nagstatins

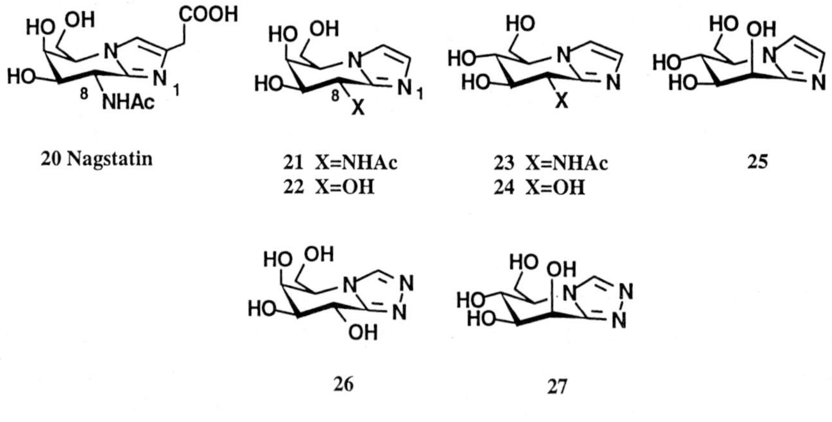

Figure 15-2

Nagstatin (**20**) is an *N*-acetyl-β-D-glucosaminidase inhibitor isolated from fermentation broth of *Streptomyces amakusaensis* (Figure 15-2) [15]. High *N*-acetyl-β-D-glucosaminidase activity in serum has been associated with several diseases such as diabetes mellitus, leukaemia and cancer. Nagstatin (**20**) and a variety of its analogues (**21 - 27**) have been synthesized from carbohydrates through the inter- and intramolecular nucleophilic reactions with the imidazole and triazole moieties to clarify their structure - activity relationships [6 - 9]. These compounds were expected to serve as antagonists of the corresponding β-glycopyranosides from the aforesaid findings.

First of all, de-branched nagstatin (**21**) and its hydroxyl analogue **22** were effectively synthesized from 2,3,5-tri-*O*-benzyl-L-ribofuranose (**28**) (Scheme 15-6). Reaction with lithiated *N*-tritylimidazole which was prepared from *N*-tritylimidazole and *n*-BuLi, gave the L-*allo* **29** and L-*altro* derivatives **30** in a ratio of approximately 1 : 1.

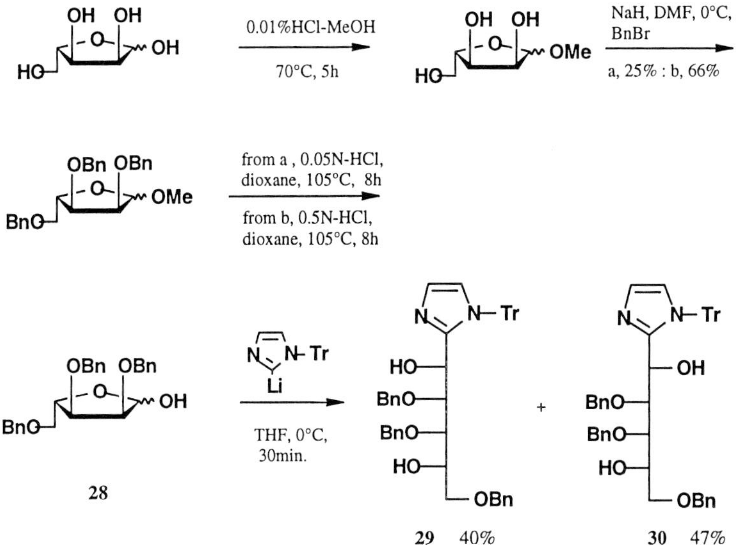

Scheme 15-6

This lack of selectivity was expected from chelation of **28** as shown in Fig 15-3. The chelation occurred preferentially between the cis-oriented oxygen atoms at C-2 and C-3. Therefore, in the nucleophilic reaction, the C-1 hydroxyl group was not configurationally biased to give **29** or **30**. On the other hand, reaction of tri-*O*-benzyl-L-xylofuranose (**35**) with lithiated *N*-tritylimidazole gave the L-gulo analogue **36** as the major product, because the nucleophilic approach was reasonably controlled by the *cis* Li-chelation between the C-1 and C-2 as shown in Fig. 15-3 and Scheme 15-9. However, both imidazoles **29** and **30** were efficiently transformed to useful analogues **21** and **22** as well as nagstatin (**20**) as follows.

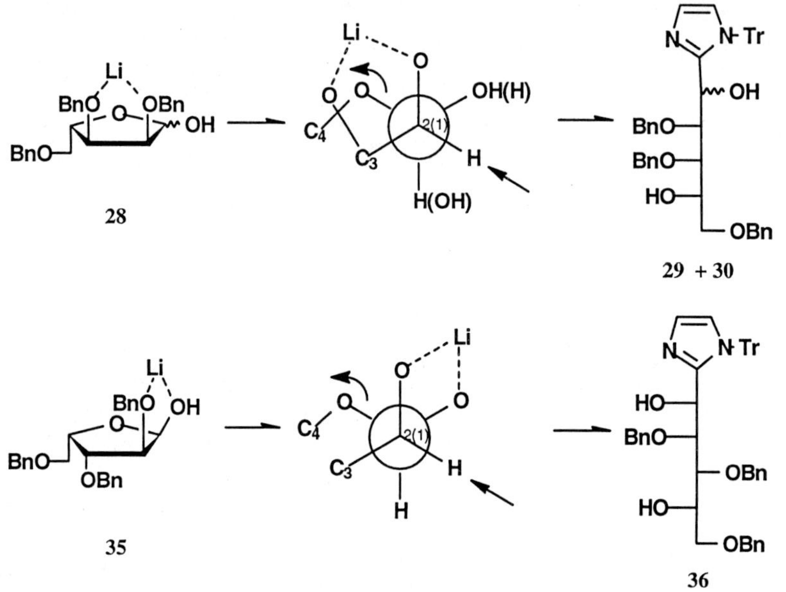

Figure 15-3

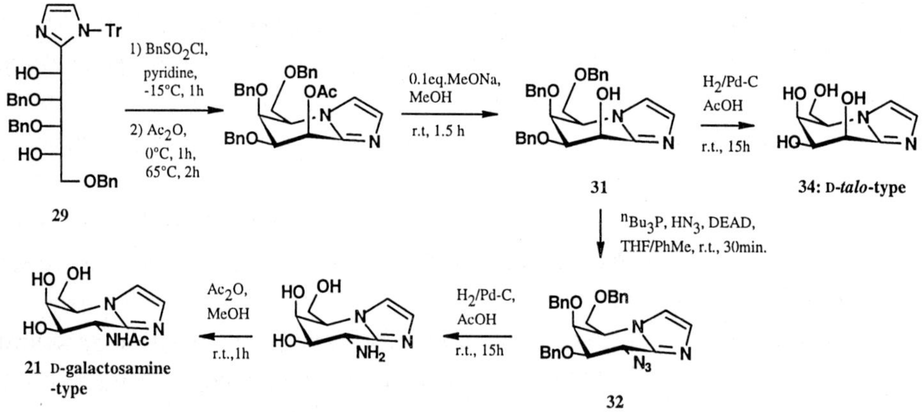

Scheme 15-7

De-*N*-tritylation and the S_N2-type intramolecular cyclisation of **29** were effectively achieved in a one-pot reaction with $BnSO_2Cl$ in pyridine to give preferentially the 5-*O*-sulfonate followed by treatment with Ac_2O to give the desired acetate, which was de-*O*-acetylated to the nitrogenous D-talose analogue **31** (Scheme 15-7). The effective de-*N*-tritylation seemed to be affected by the producing pyridinium acetate. Inversion of the hydroxyl group in **31** using HN$_3$, n-Bu$_3$P and DEAD afforded the azido derivative **32**, which was subjected to hydrogenolysis with Pd-C and N-acetylation with Ac$_2$O in MeOH, leading to the N-acetyl-D-galactosamine analogue **21**, which was related to de-branched nagstatin (Scheme 15-7).

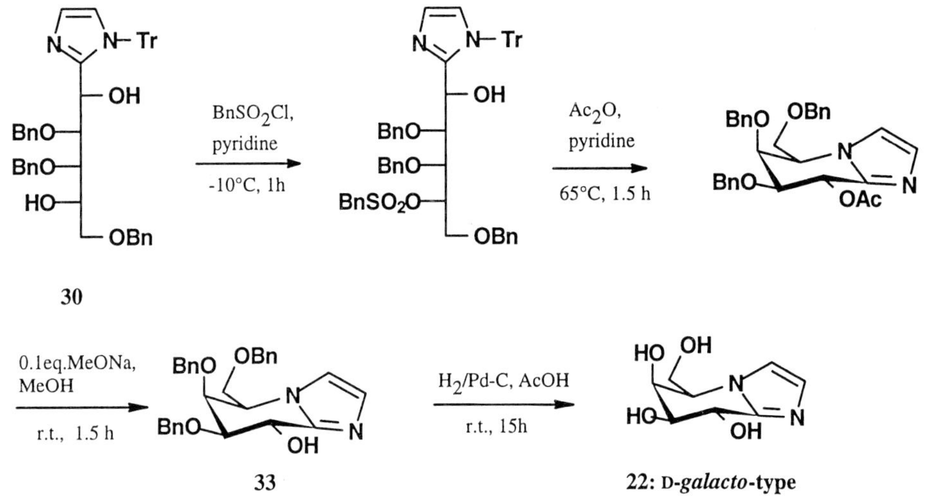

Scheme 15-8

Alternatively, **32** was prepared from the other isomer **30** through **33** (Scheme 15-8). Reaction of **33** with HN$_3$, n-Bu$_3$P and DEAD gave **32** with the expected retention of the C-8 configuration. The formal substitution at C-8 is most probably following an elimination/addition mechanism, and the diastereoselectivity is governed mainly by the axial C6-O substituent and the ring nitrogen. Similar observations have been made in carbohydrate chemistry. The S$_N$2 replacement of equatorial groups at C-2 of glycopyranosides, which is corresponding to C-8 in **33**, is notoriously difficult because of the combined influence of the ring oxygen, the anomeric substituent and dipolar effects [6]. Moreover, in this case, an intermediary carbonium ion would be stabilized by the resonance with the imidazole ring, as shown in Fig. 15-4.

Figure 15-4

This retention was confirmed by the fact that **33** was treated with benzoic acid, *n*-Bu3P and DEAD to give the corresponding benzoate, which was deacylated to the starting **33**.

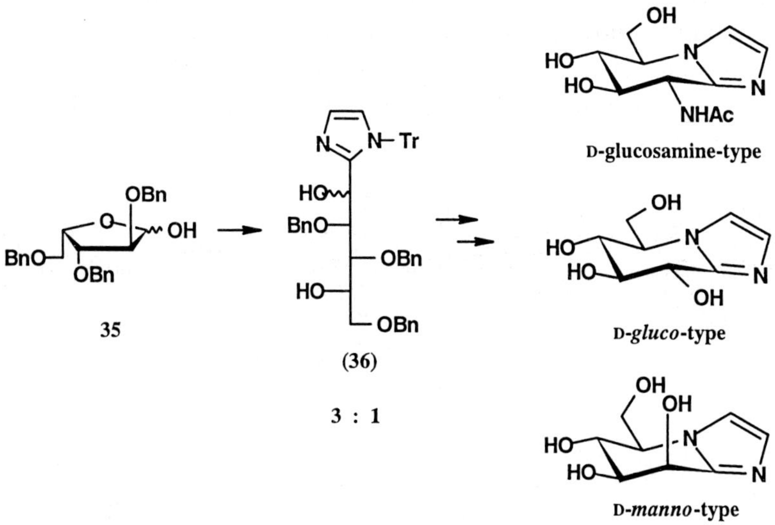

Scheme 15-9

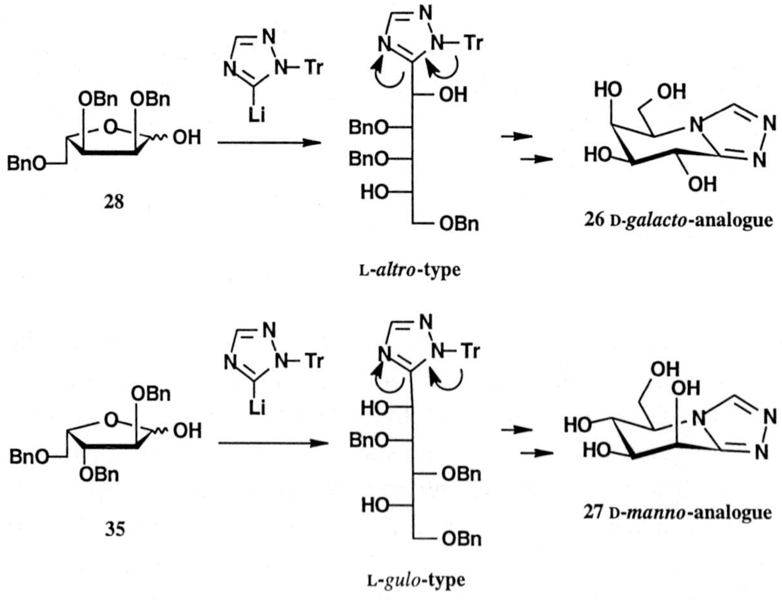

Scheme 15-10

Hydrogenolysis of **31** and **33** afforded the nitrogenous D-talose and D-galactose nitrogenous analogues **34** and **22**, respectively. Similarly, the *N*-acetyl-L-galactosamine analogue **21'** nitrogenous and L-galactose nitrogenous analogue **22'** were prepared from D-ribofuranose derivative by the above procedures.

Furthermore, nitrogenous *N*-acetyl-D-glucosamine, nitrogenous D-glucose nitrogenous and D-mannose nitrogenous analogues (**23**, **24** and **25**) were efficiently prepared from L-xylofuranose derivative **35** by the similar fashion as described above (Scheme 15-9).

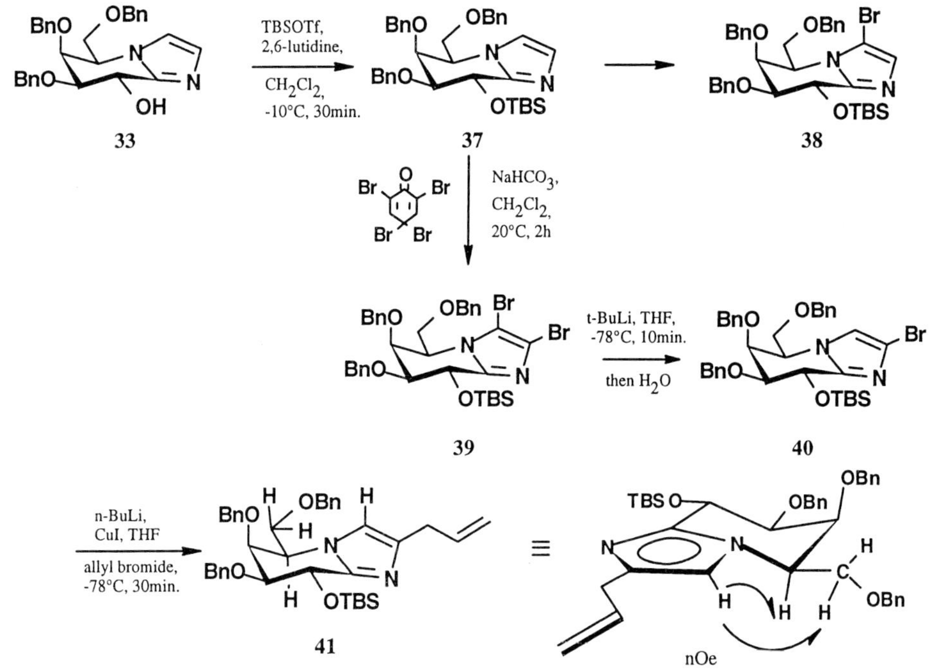

Scheme 15-11

Also, the triazole analogues **26** and **27** were predominantly synthesized from **28** and **35** by reaction with lithiated triazole, while their *N*-regio isomers were little observed (Scheme 15-10). Now, the next step is set for the enantiospecific synthesis of nagstatin (**1**). The rational starting point is the aforesaid isomers **28** and **30** [12]. The regioselective introduction of an allyl group on their C-2 positions was investigated under various conditions. The C-2 position is generally known to be less reactive than the C-3. In fact, selective bromination of **37** gave the undesired C-3 bromo compound **38** (Scheme 15-11).

Accordingly, **37** was fully brominated with 2,4,4,6-tetrabromo-2,5-cyclohexadien-1-one to the dibromo compound **39**. Selective debromination of this compound was then assayed. The best result was obtained by regioselective lithiation with *t*-BuLi in THF followed by quenching with H_2O to give the desired monobromo compound **40**. The structure was confirmed by the ^{1}H-NMR nOe studies of the corresponding allyl derivative **41**.

Dihydroxylation of **41** with OsO_4 and NMO afforded **42**, which was oxidized by using modified Fetizon's conditions using Ag_2CO_3 to give the keto-alcohol **43** (Scheme 15-12). Periodate oxidation followed by esterification with $TMSCHN_2$ provided the methyl ester **44**. Direct ozonolysis of **41**, or periodate oxidation of **42** caused concomitant oxidation at C-9 position.

Conversion of **44** to the azido compound **45** was carried out by de-*O*-silylation followed by treatment with HN_3, *n*-Bu_3P and DEAD. Expected retention of the configuration was observed at the C-8 position as described above.

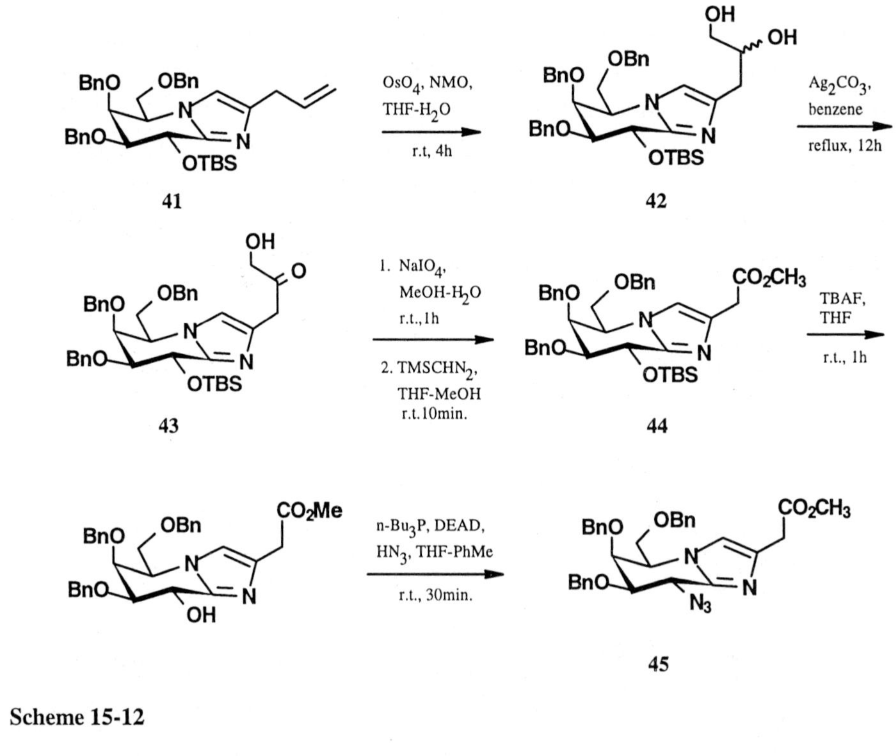

Scheme 15-12

In a similar manner, but with inversion of the configuration, the other C-8 *axial* isomer **31** was converted into the azido compound **45** in a ten-step sequence through the allyl derivative **46** (Scheme 15-13). Hydrogenolysis of **45** followed by successive *N*-acetylation and saponification with aqueous NaOH provided nagstatin (**20**) (Scheme 15-14), which was identical with the natural product in all respects including glycosidase inhibiting activities. The completion of the synthesis confirmed the absolute structure **20**.

The glycosidase inhibiting activities were assayed as described above [7] and summarized in Tables 15-1 and 15-2. *N*-Acetyl-D-galactosamine analogue **21** exhibited the same strong activity even against *N*-acetyl-β-D-glucosaminidase as nagstatin (**20**) and, consequently, was expected to inhibit *N*-acetyl-β-D-galactosaminidase, although this glycosidase was not available now. *N*-Acetyl-D-glucosamine analogue **23** strongly inhibited *N*-acetyl-β-D-glucosaminidase activity and weakly β-D-glucosidase activity. The D-*galacto*, D-*gluco* and D-*manno* analogues (**22**, **24** and **25**) showed very much stronger inhibiting activities against β-D-galactosidase, β-D-glucosidase and β-D-mannosidase, respectively, than against the corresponding α-D-glycosidases.

The triazole analogues **26** and **27** (Figure 15-2) having *galacto* and *manno* configurations also showed strong inhibitory activities against β-D-galactosidase and β-D-mannosidase, respectively [9]. Remarkably, the L-galactose analogues **21'** and **22'** showed no significant glycosidase inhibitory activities. All analogues possess a *quasi*-equatorially oriented C8a-N1 bond, which corresponds to an equatorial C1-O bond of β-glycopyranosides, due to the fused imidazole and triazole rings. The configurations from C8a to C5 of the analogues parallel the alignment from C1 to C5 of the corresponding glycopyranosides.

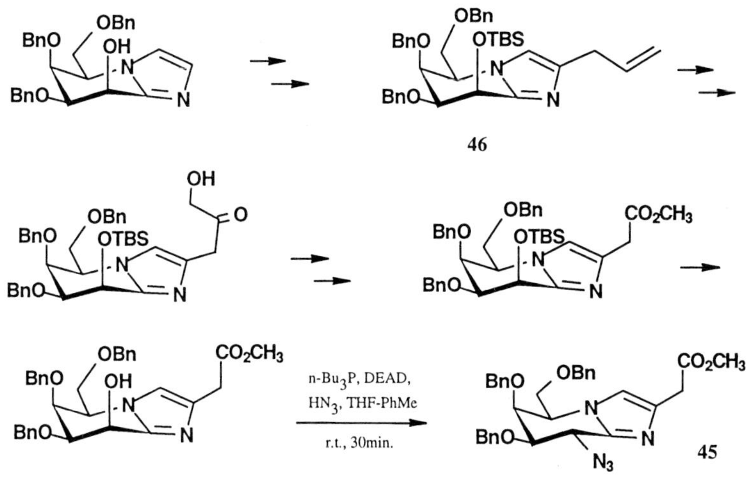

Scheme 15-13

The strong β-D-glycosidase inhibiting activities of the analogues **21 - 27** indicated that the β-D-glycosidases including *N*-acetyl-β-D-glucosaminidase recognized especially their C-8a positions as the C-1 position of β-D-glycopyranosides. Furthermore, their substrate-specific activities emphasized that the analogues serve essentially as the antagonists of the corresponding stereochemically oriented β-D-glycopyranosides. These findings are similar with those of the aforesaid cyclophellitol (**1**) and its analogues.

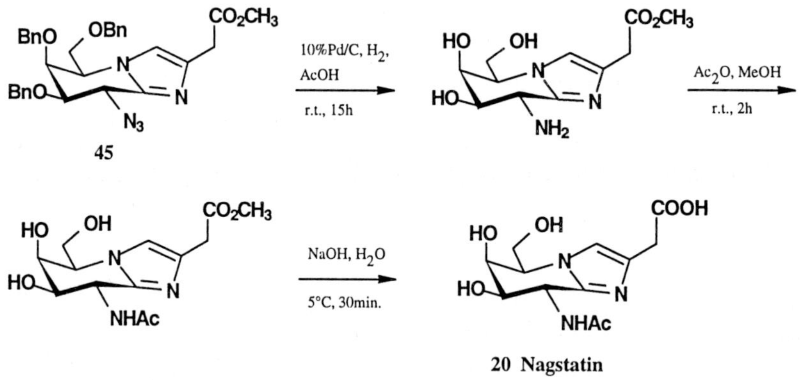

Scheme 15-14

15.2.3 Gualamycin

Gualamycin (**47**) is a novel water-soluble acaricide (anti-mite substance) isolated from the culture broth of *Streptomyces* sp. NK11687 (Fig. 15-5) [16]. The absolute structure was mainly confirmed by enantiospecific syntheses of its amino-disaccharide and pyrrolidine-aglycone portions (**48** and **49**) in our laboratories [10, 17]. The structural complexity, as well as the goal of studying structure-activity relationships, prompted us to explore the total synthesis, which was expected to confirm the absolute structure **47** and elucidate the origin of the acaricidal activity. The synthesis is mainly based on glycosylation of the glycosyl acceptor **63** with the donor **67** [11].

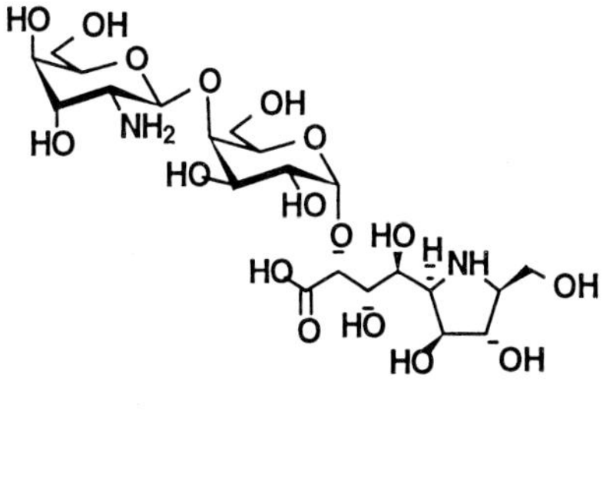

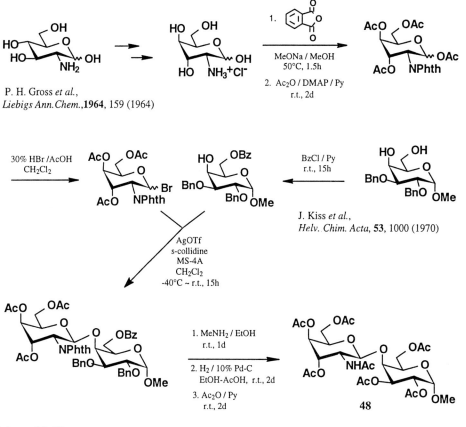

47 Gualamycin

48

49

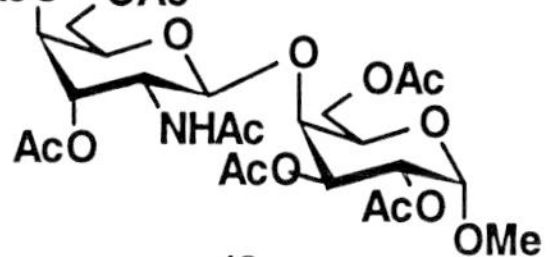

Figure 15-5

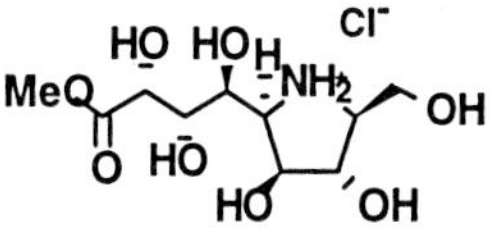

The amino-disaccharide **48** was prepared by the route described in Scheme 15-15.

48

Scheme 15-15

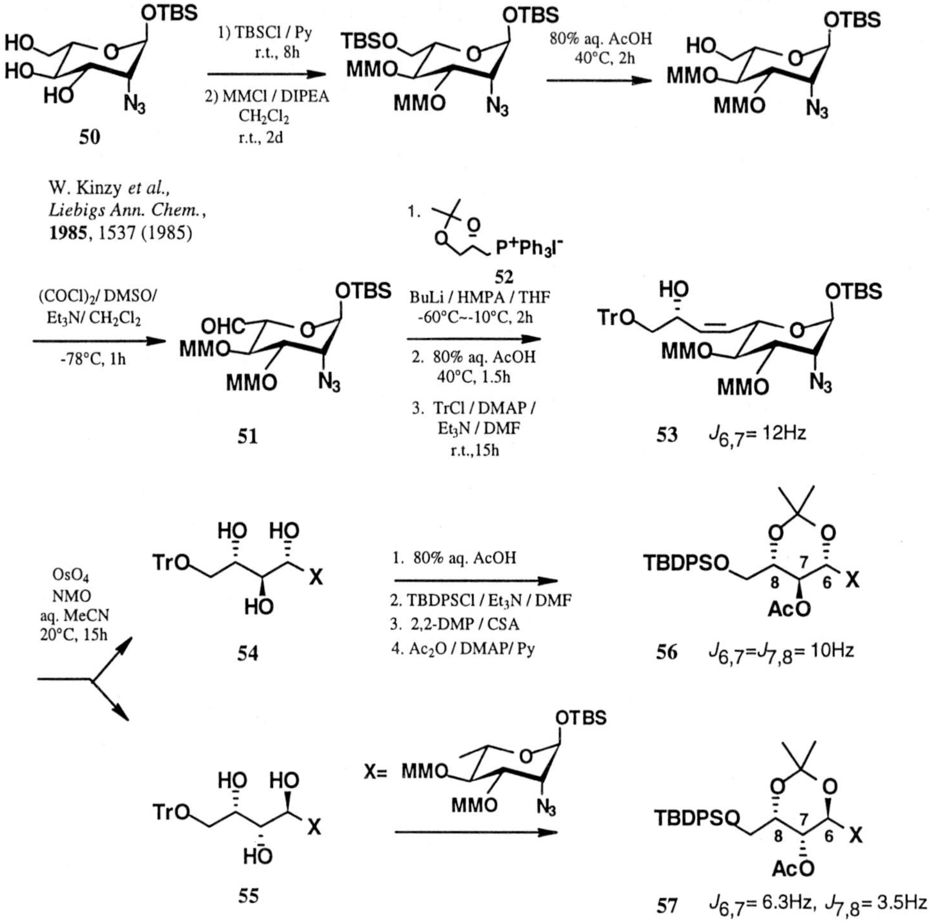

Scheme 15-16

The pyrrolidine-containing aglycon unit **49** was synthesized from the azido sugar **50**, which was prepared by de-*O*-acetylation of *t*-butyldimethylsilyl 3,4,6-tri-*O*-acetyl-2-azido-2-deoxy-α-L-mannopyranoside (Scheme 15-16). The azide **50** was selectively silylated with TBDMSCl and methoxymethylated to the fully protected product. Selective desilylation gave the corresponding alcohol. Swern oxidation gave the labile aldehyde **51**. This compound was treated with the Wittig reagent prepared from (4*S*)-(2,2-dimethyl-1,3-dioxolan-4-yl)methyl-triphenyl-phosphonium iodide (**52**) and *n*-BuLi. Removal of the isopropylidene group and tritylation of the resulting primary alcohol yielded the *cis* olefin **53**. The *cis* dihydroxylation of **53** by OsO$_4$ gave two triols **54** and **55** in 34% and 46% yields, respectively. Their configurations were determined by the ^{1}H-NMR studies of their corresponding isopropylidene derivatives **56** and **57** showing that **55** was the desired triol for the natural product. Compound **55** was methoxymethylated, followed by successive treatment with Ph$_3$P in PhMe and with aqueous THF in refluxing to give the amino compound, which was tosylated to give **58** (Scheme 15-17).

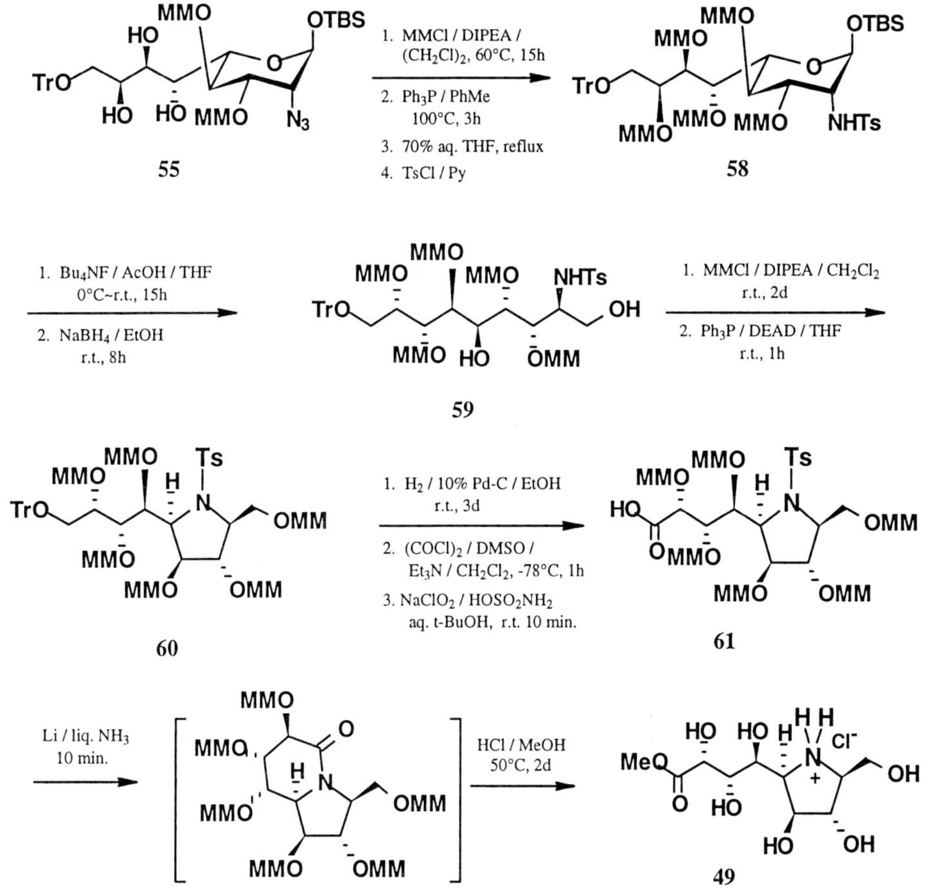

Scheme 15-17

Desilylation with *n*-Bu$_4$NF and hydride reduction with NaBH$_4$ gave the alcohol **59**. This was selectively methoxymethylated and then submitted to the S$_N$2-type cyclisation using Ph$_3$P and DEAD to give the pyrrolidine derivative **60**. After detritylation by hydrogenolysis, the resulting alcohol was oxidized stepwise by Swern's conditions to give the aldehyde and by sodium chlorite/H$_2$NSO$_3$H to give the carboxylic acid **61**. De-*N*-tosylation under Birch's conditions with Li in liq. NH_3 followed by esterification with 5% HCl-MeOH gave the hydrochloride of the pyrrolidine hydrochlorid aglycon **49**, which was identical with the naturally derived sample in every respects.

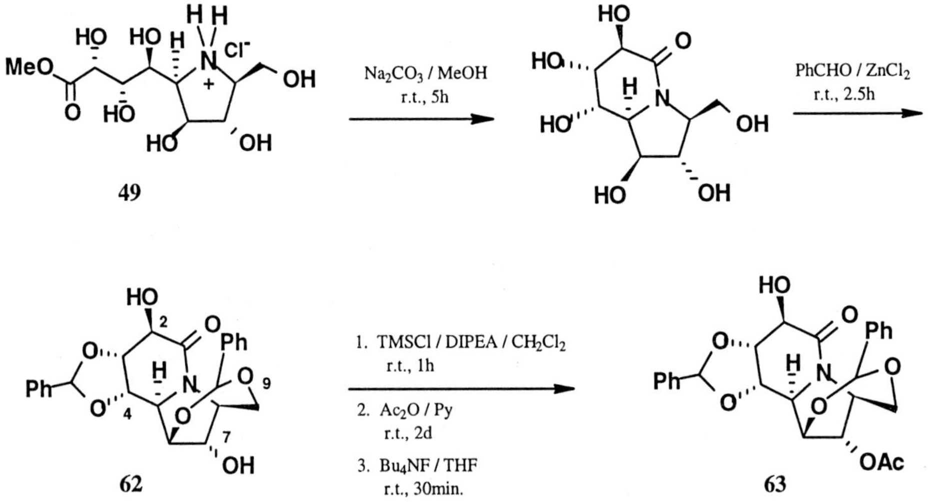

Scheme 15-18

The glycosyl acceptor, di-*O*-benzylidene derivative **63** was prepared in a 5-step sequence from the aglycon **49** (Scheme 15-18). Thus, treatment of **49** with Na_2CO_3 to give the Δ–lactam whose *O*-benzylidenation with PhCHO and $ZnCl_2$ provided **62** in 54% overall yield. The two sets of two hydroxyl groups at the C-3 and 4 positions and at the C-6 and 9 positions were effectively protected by *O*-benzylidene groups as expected from the Dreiding model examination. Protection of the hydroxyl group at C-7 in **62** proceeded non-selectively under a variety of conditions, while the glycosylation of **62** gave unexpectedly the undesired 7-*O*-glycosyl derivative by using the glycosyl donor **67**. Accordingly, **62** was silylated with TMSCl and DIPEA, acetylated and desilylated to give the monoacetate **63** in 70% yield.

The glycosyl donor **67** was prepared from phenyl-1-thio-galactoside **64** in five steps through **65** (Scheme 15-19). The alcohol **65** was subjected to the reaction with the protected gulosaminyl bromide **66** in the presence of AgOTf and *s*-collidine to give the glycosyl donor **67**.

Coupling of **67** with the acceptor **63** was accomplished by using a modified Fraser-Reid's conditions using NIS and TfOH at -40°C for 1 hour to provide exclusively the desired α-glycoside **68** (Scheme 15-20). Hydrogenolysis of **68** followed by treatment with 40% MeNH$_2$ in MeOH furnished the corresponding δ–lactam.

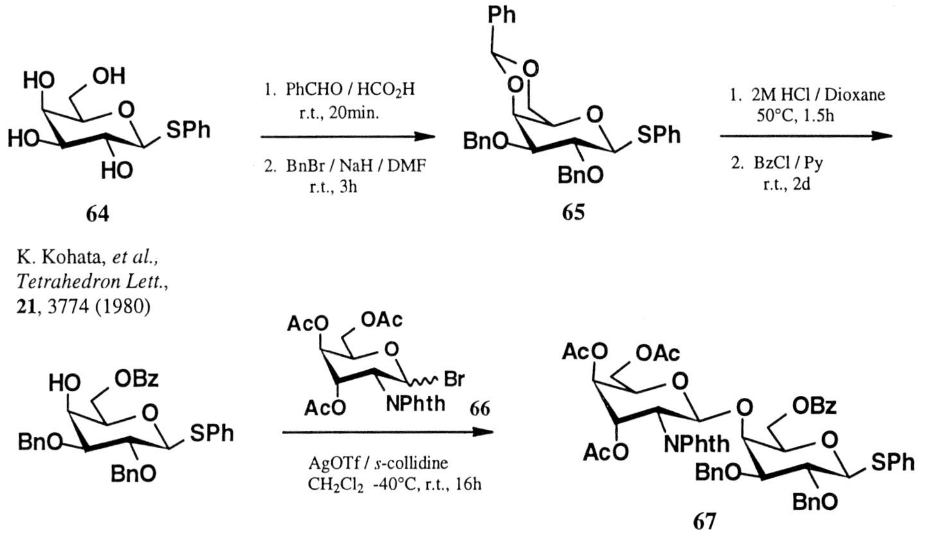

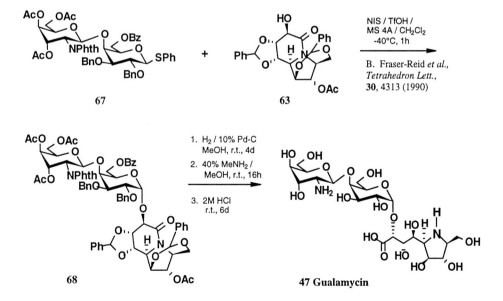

Scheme 15-19

Scheme 15-20

On the final stage, all attempts to open the Δ–lactam failed under alkaline conditions giving only a low yield of **47**. However, the convenience of using acids to catalyse the process was especially appealing to us. Hydrolysis of the lactam was successfully conducted in 2M HCl at room temperature for 6 days to give the dihydrochloride of gualamycin (**47**) in 86% yield without further hydrolysis, which was identical with the natural product in every respects including acaricidal activities.

When the glycosidase inhibiting activities were assayed, gualamycin (**47**) was found to inhibit maltase activity at IC_{50} 25 mg/ml. This inhibiting activity seems to be the origin of appearance of the acaricidal (anti-mite) activity, suggesting that a mite could get maltose as a source of life.

15.3 Conclusion

The glycosidase inhibitors, cyclophellitol, nagstatin and gualamycin, which are microbial metabolites, and their analogues were effectively synthesized from carbohydrates to clarify their structure - activity relationships.

As a result, new analogues having stronger activities than natural products were chemically designed and created. Their glycosidase inhibiting activities were quite substrate-specific, indicating that the α- and β-glycosidases recognize especially the C-1 positions and the residual portions as corresponding to those of α- and β-glycopyranosides, and consequently, the inhibitors serve as antagonists of the corresponding glycopyranosides. The anti-mite activity of gualamycin was suggested to be due to its maltase inhibiting activity.

Acknowledgements

The Mitsubishi Foundation, Shikoku Chemicals Co. and Yamanouchi Pharmaceutical Co., Ltd. are gratefully acknowledged for the generous support of this research.

15.4 References

1. K. Tatsuta, Y. Niwata, K. Umezawa, K. Toshima, M. Nakata, *Tetrahedron Lett.* **1990**, *31,* 1171-1172.
2. K. Tatsuta, Y. Niwata, K. Umezawa, K. Toshima, M. Nakata, *J. Antibiot.* **1991**, *44,* 456-458.
3. K. Tatsuta, Y. Niwata, K. Umezawa, K. Toshima, M. Nakata, *Carbohydr. Res.* **1991**, *222,* 189-203.
4. K. Tatsuta, Y. Niwata, K. Umezawa, K. Toshima, M. Nakata, *J. Antibiot.* **1991**, *44,* 912-914.

5. M. Nakata, C. Chong, Y. Niwata, K. Toshima, K. Tatsuta, *J. Antibiot.* **1993**, *46*, 1919-1922.

6. K. Tatsuta, S. Miura, S. Ohta, H. Gunji, *Tetrahedron Lett.* **1995**, *36*, 1085-1088.

7. K. Tatsuta, S. Miura, S. Ohta, H. Gunji, *J. Antibiot.* **1995**, *48*, 286-288.

8. K. Tatsuta, S. Miura, *Tetrahedron Lett.* **1995**, *36*, 6721-6724.

9. K. Tatsuta, Y. Ikeda, S. Miura, *J. Antibiot.*, **1996**, *49*, 836-838.

10. K. Tatsuta, M. Kitagawa, T. Horiuchi, K. Tsuchiya, N. Shimada, *J. Antibiot.* **1995**, *48*, 741-744.

11. K. Tatsuta, M. Kitagawa, *Tetrahedron Lett.* **1995**, *36*, 6717-6720.

12. S. Atsumi, K. Umezawa, H. Iinuma, H. Nakamura, Y. Iitaka, T. Takeuchi, *J. Antibiot.* **1990**, *43*, 49-53.

13. R. H. Schlessinger, C. P. Bergstrom, *J. Org. Chem.* **1995**, *60,* 16-17, and references cited therein.

14. K. Toshima, S. Mukaiyama, M. Kinoshita, K. Tatsuta, *Tetrahedron Lett.* **1989**, *30*, 6413-6416.

15. T. Aoyama, H. Naganawa, H. Suda, K. Uotani, T. Aoyagi, T. Takeuchi, *J. Antibiot.* **1992**, *45*, 1557-1561.

16. K. Tsuchiya, S. Kobayashi, T. Harada, T. Kurokawa, T. Nakagawa, N. Shimada, K. Kobayashi, *J. Antibiot.* **1995**, *48*, 626-629.

16 Synthesis of Enantiopure Azasugars from D-Mannitol: Carbohydrate and Peptide Mimics.

J. -C. Depezay

16.1 Introduction

Polyhydroxylated pyrrolidines, piperidines, indolizidines and also azepanes, the so-called azasugars, constitute an important family of glycosidase and glycosyltransferase inhibitors. These carbohydrate mimics have potential utility in the treatment of various diseases, such as diabetes, cancer and viral infections. Moreover these polyhydroxylated heterocycles are interesting nonpeptide scaffolds to build peptidomimics targeted towards specific receptors.

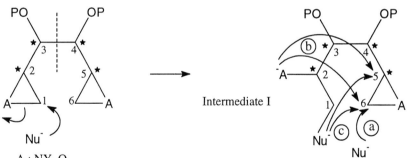

Scheme 16-1 Path a - Regiospecific opening of the small ring by a second nucleophile molecule produces an acyclic C_2-symmetric synthon.

Path b - The anion resulting from the first ring opening opens the second one through an intramolecular *6-endo-tet* or *5-exo-tet* process.

Path c - Intramolecular opening by the nucleophile previously introduced to form **I**. An enantiomerically pure six or seven membered heterocycle is built up.

Starting from D-mannitol we have developed a methodology to prepare enantiomerically pure polyhydroxylaed pyrrolidines, piperidines and azepanes through a double nucleophilic opening of C-2 symmetric bis-aziridines or bis-epoxides of D-*manno* and L-*ido* configuration.

Nucleophilic opening of one of the small ring leads to intermediate I (Scheme 16-1). Primary carbon atom C-1 or C-6 is generally attacked by the nucleophile ; these primary carbon atoms are equivalent according to the C_2-symmetry property of the starting material. Intermolecular (path a) or intramolecular (path b and c) nucleophilic opening of the second small ring yields acyclic or heterocyclic compounds (5, 6 or 7 membered ring).

16.2 Results

16.2.1 To Azasugars *via* Bis-Aziridines

16.2.1.1 Nucleophilic Opening of Bis-Aziridines

Nucleophilic opening of suitably N-activated D-mannitol derived bis-aziridines [1, 2, 3, 4, 5] can be oriented towards bis-opening or towards heterocyclisation (Scheme 16-2).

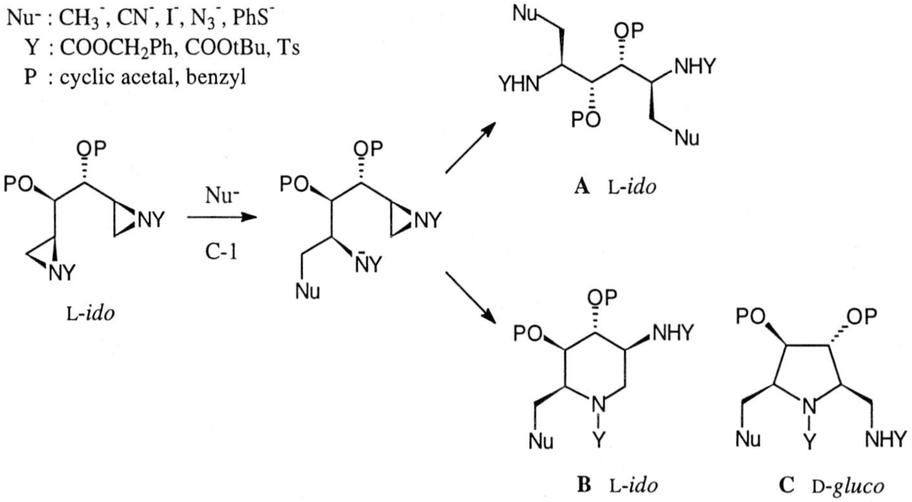

Scheme 16-2

Regioselective bis-opening leads to C_2-symmetric diaminodiols **A**, central core units of a class of non-hydrolyzable HIV-1-protease inhibitors. Ring opening followed by heterocyclisation provides access to polyhydroxylated piperidines and pyrrolidines. The orientation of the reaction either towards bis-opening or towards heterocyclisation depends on the substitution pattern at nitrogen, on the nature of the nucleophile and on the presence or abscence of Lewis acids or proton donors in the reaction medium. Moreover the regioselectivity of the intramolecular heterocyclisation step depends on the nature of the protecting group P of the C-3, C-4 diol. In case of restricted rotation (P = cyclic acetal) the cyclisation is forced to follow exclusively the *6-endo* pathway, providing enantiomerically pure tetrasubstituted piperidines **B** [3, 6], while conformationnaly flexible aziridines (P = benzyl) lead through the more favorable *5-exo-tet* process, to 2,5-disubstituted pyrrolidines **C** with high regioselectivity [4, 7].

Protection and activation of the ring nitrogen atom are essential to carry out the nucleophilic opening and can be achieved using benzyl, tosyl or alkoxycarbonyl groups.

To illustrate this methodology we describe the nucleophilic opening of conformationnally flexible N-benzyl and N-Boc bis-aziridines. The regioselectivity of the aziridine ring opening by cyanide ions depends on reaction conditions [7] (Scheme 16-3).

Starting from **1a** (L-*ido* configuration) it is possible to obtain either the C_2-symmetric β-aminonitrile **2**, the pyrrolidine **3** (azafuranose of D-*gluco* configuration) or the piperidine **4** (azapyranose of D-*manno* configuration). Reaction of **1a** with trimethylsilyl cyanide in the presence of a catalytic amount of ytterbium cyanide yielded mainly the symmetric β-amino-γ-hydroxynitrile **2**. Under these conditions the formation of a silylated amine is liable to minimize the intramolecular heterocyclisation towards pyrrolidine. Sodium cyanide opening of **1a** in aqueous dimethyl formamide provided the pyrrolidine **3** (70% yield).

Diethylaluminium cyanide, a currently used reagent for cyanhydrin synthesis, has never been reacted with aziridines. This reagent carries out the cyanide alkylation of **1a** at 0°C in 86% yield providing two nitrogen heterocycles as a 1/2 mixture of the pyrrolidine **3** (D-*gluco* configuration) and the piperidine **4** of D-*manno* configuration. This compound results from the nucleophilic attack of cyanide ion at the secondary carbon of the aziridine ring followed by an intramolecular substitution at C-5 following a *6-exo* mode cyclisation. This example constitutes the first case of nucleophilic attack occuring at a secondary carbon of bis-aziridines.

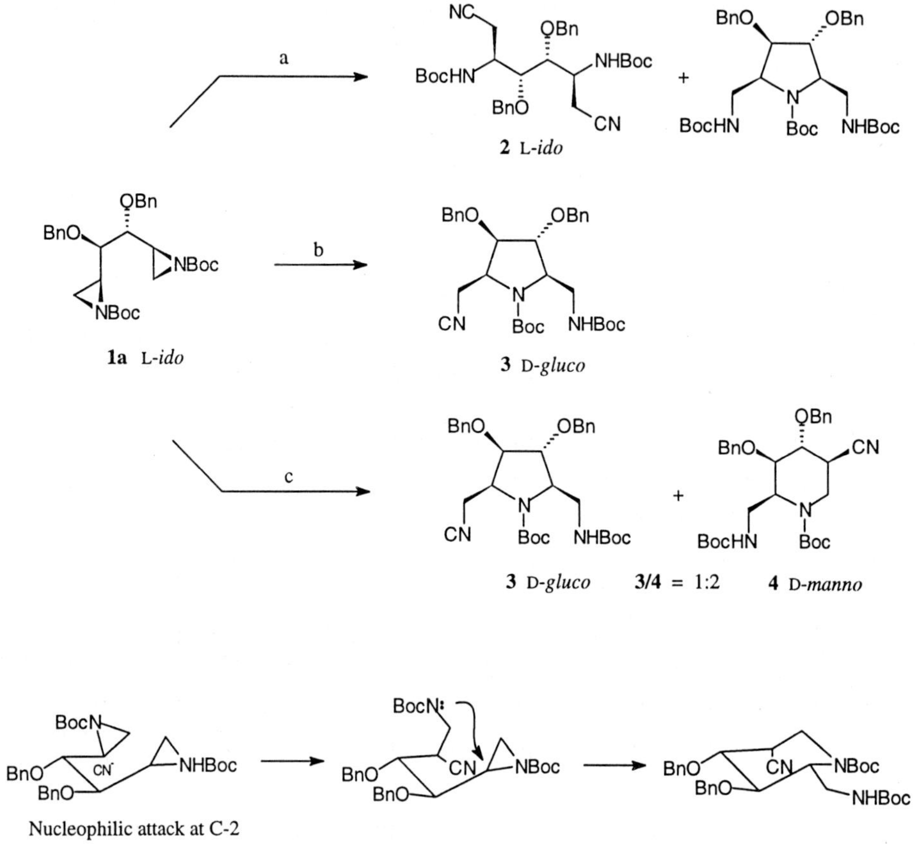

Scheme 16-3. Reagents: a) Me$_3$SiCN, Yb(CN)$_3$ cat, THF, 65°C, 24h, 67%; b) NaCN, DMF, H$_2$O 10%, 45°C, 8h, 70%; c) Et$_2$AlCN, toluene, 0°C, 3h, 86%.

16.2.1.2 Bis-Aziridines Precursors of Azadisaccharides

Polyhydroxylated pyrrolidines and piperidines often act as broad spectrum inhibitors of glycosidases. More selective inhibition of a particular glycosidase could be achieved by designing mimics of both portions of the natural substrate. Pseudodisaccharides in which a five-membered azasugar is linked to other sugars by a non hydrolysable link have potential of much specificity. It is therefore of interest to synthezise polyhydroxylated pyrrolidines substituted at C-2 by a functionnalized methyl group which might be used for incorporation of a second sugar.

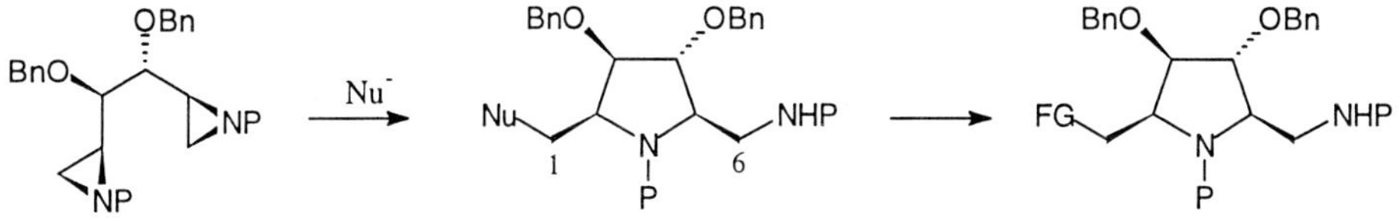

P = Boc **1a**	[4] →	Nu = N_3	**5**	a →	FG = NH_2 **6**
1a	b →	Nu = Br	**7**		
P = Bn **1b**	c →	Nu = SCN	**8**	d →	FG = SH **9**

Scheme 16-4. Reagents: a); H_2, Pd black, AcOEt, 20°C, 82% b) Li_2NiBr_4, THF, 0°C, 4h, 80%; c) KSCN, BF_3-Et_2O, CH_3CN, 60°C, 3h, 76%; d) $NaBH_4$, EtOH/H_2O/THF 4:1:6, 20°C, 3h, 75%.

New 2,5-imino hexitols of D-*gluco* configuration substituted at C-1 by an amino, a thio or a bromo group, potential candidates for coupling reactions, have been prepared, from D-mannitol derived bis-aziridines [8] (Scheme 16-4). The protected 1,6-diamino and aminothiol pyrrolidines **6** and **9** can be used as nucleophile for coupling.

The bromo iminoglucitol, electrophilic azasugar **7**, was obtained by Li_2NiBr_4 ring opening of the N-Boc bis-aziridine **1a**. Starting from N-benzyl bis-aziridine **1b**, intramolecular nucleophilic substitution at C-1 on the bromo substituted pyrrolidine formed, leads to a chiral bridged aminodiol **10** [8] (Scheme 16-5).

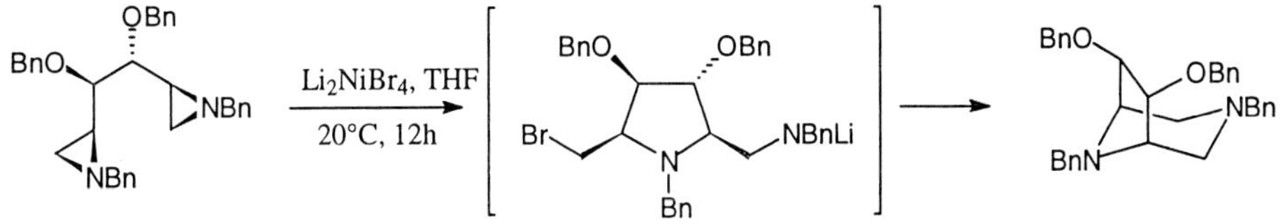

1b **10**

Scheme 16-5

The synthesis of azadisaccharides can be achieved by coupling activated hydroxypyrrolidines with glucose derivatives. For example coupling of the bromosubstituted pyrrolidine **7** with the sodium salt of 3-thioglucose **11** leads to thioether **12** [10]. Still more efficient is the preparation of the bisubstrate through nucleophilic opening followed by intramolecular aminocyclisation of bis-aziridines either by 3-deoxy-3-thio or 3-amino-D-glucose **11** or **13** (Scheme 16-6). The direct nucleophilic ring opening of the N-Boc bis-aziridine with thioglucose **11** already takes place at -20°C and is followed by slow intramolecular cyclisation into the thioglycosylsubstituted pyrrolidine **12** [10] (Scheme 16-7).

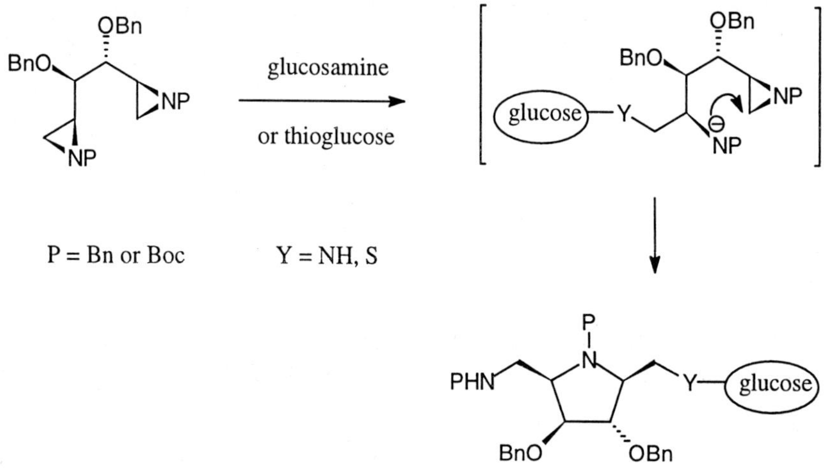

Scheme 16-6

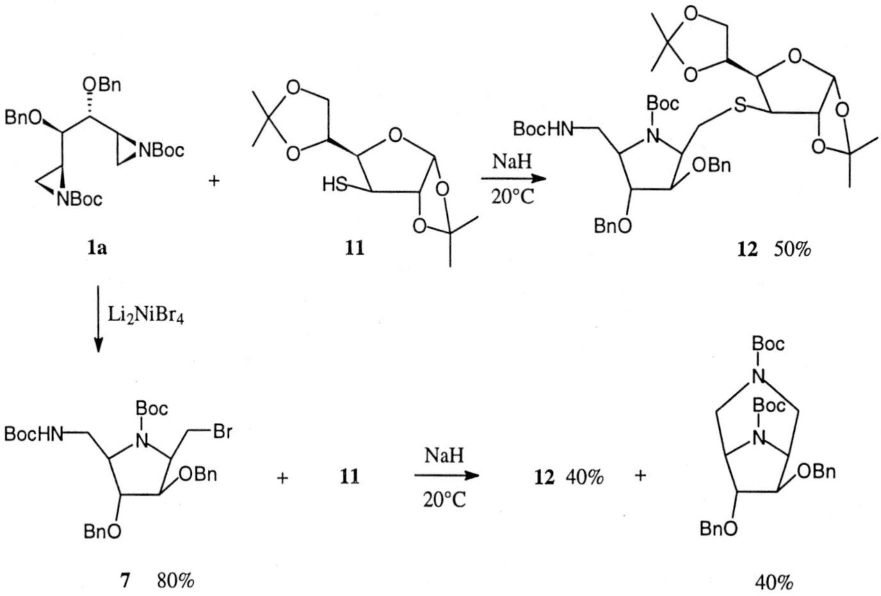

Scheme 16-7

The ring opening of activated aziridines by amines requiring elevated temperatures, the introduction of the amino group is generally better performed by nucleophilic ring opening with sodium azide [3, 4]. The ytterbium triflate catalyzed aminolysis of N-protected aziridines has been reported recently to proceed under smooth conditions with primary and secondary amines [9]. Both N-Boc and N-Bn bis-aziridines react with primary amines in the presence of ytterbium triflate [10]. Nucleophilic opening takes place regioselectively at one of the primary carbon atoms leading to a diamino intermediate apt to cyclize from both nitrogen atom and give either the pyrrolidine **A** (D-*gluco*), the piperidine **B** (L-*ido*) or the piperidine **C** (D-*gluco*) (Scheme 16-8).

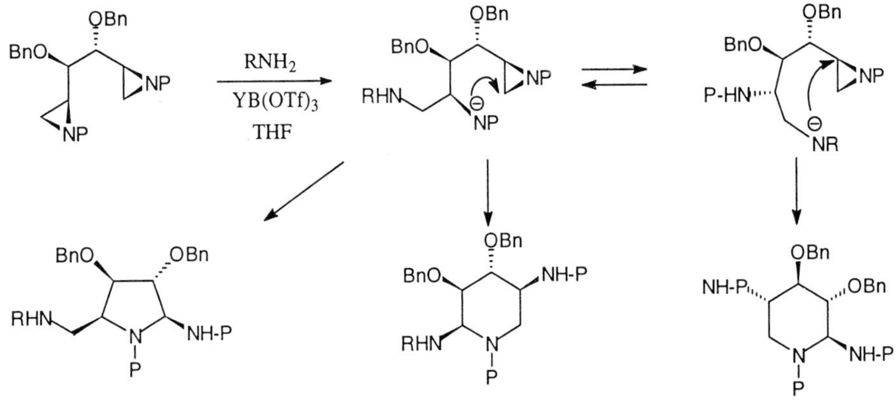

A D-*gluco* **B** L-*ido* **C** D-*gluco*

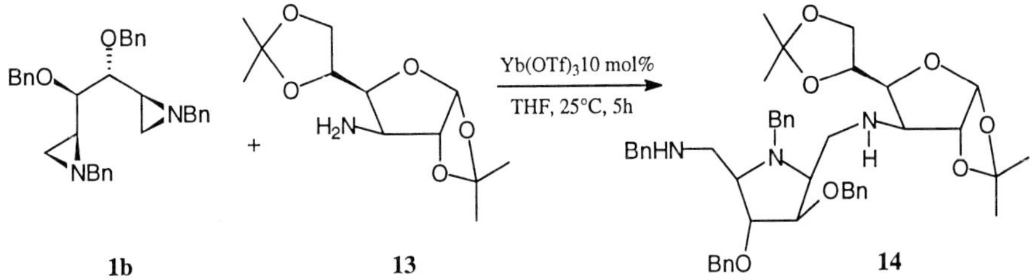

1b **13** **14**

Scheme 16-8

Nucleophilic opening of the N-Boc bis-aziridine **1b** by di-O-isopropylidene-2-deoxy-2-amino-α-D glucose **13** leads to a mixture (D-*gluco* pyrrolidine, L-*ido* and D-*gluco* piperidine). With the N-benzyl aziridine the non competitive formation of the N-glucosylpiperidines was observed. The azadisaccharide **14** is obtained as a single product in 65% yield (Scheme 16-8 bottom line).

16.2.2 To Azasugars *via* Bis-Epoxides

16.2.2.1 Nucleophilic Opening of Bis-Epoxides

A methodology to prepare azasugars has been developed simultaneously *via* bis-aziridines and *via* the opening of homochiral C_2-symmetric D-mannitol derived bis-epoxides by amines. This approach involves a regiospecific opening of one epoxy function by an amine followed by the expected aminocyclisation (Scheme 16-1, Path c) leading to polyhydroxy-piperidines and/or azepanes via a *6-exo-tet* or a *7-endo-tet* process respectively (Scheme 16-9).

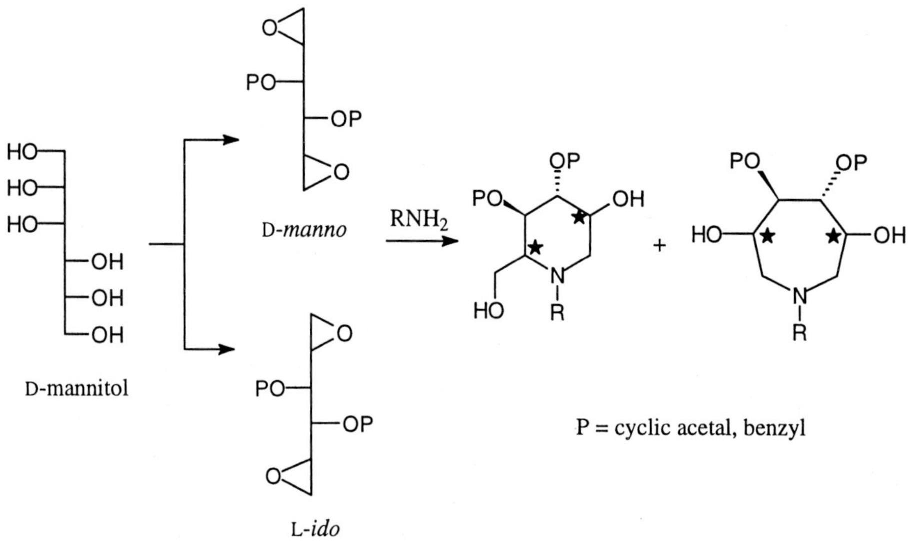

Scheme 16-9

Activation of the free hydroxyl groups of the resulting piperidines and azepanes should trigger isomerisation of the later into new interesting azasugars *via* a S_N process either by direct substitution or by neighbouring nitrogen participation.

Protected bis-epoxides, 1,2:5,6-dianhydro-3,4-*O*-isopropylidene-D-mannitol **15** and L-iditol **16**, 1,2:5,6-dianhydro-3,4-*O*-benzyl-D-mannitol **17** and L-iditol **18**, are easily obtained on a multigram scale from D-mannitol [2, 11, 12] (Scheme 16-10). Opening of isopropylidene bis-epoxide **16** by tryptamine or monoprotected 1,2-diaminoethane for example, gives only the corresponding azepanes in 65% yield [13, 14] (Scheme 16-10).

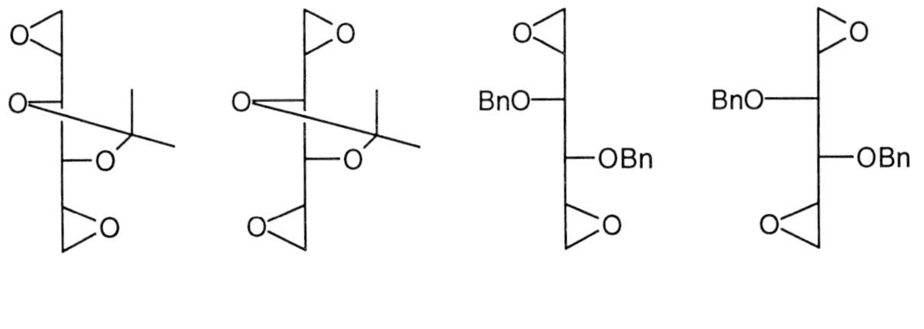

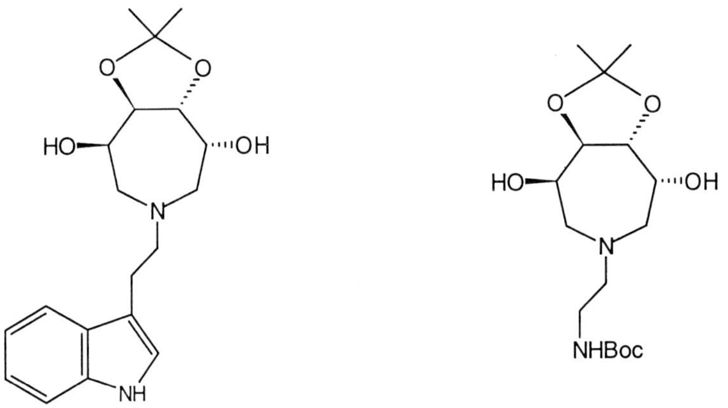

Scheme 16-10

Reaction of flexible bis-epoxides **17** or **18** with benzylamine in aprotic medium leads mainly to the *6-exo* adduct, whereas in protic medium or in aprotic medium in the presence of Lewis acid mainly the *7-endo* adduct is obtained [12] (Scheme 16-11).

Thus aminocyclisation can be directed towards piperidines **19** and **21** (45 and 50% yield of isolated compounds) or azepanes **20** and **22** (66 and 67% yield of isolated compounds) by selecting the appropriate conditions (Table 16-1). Hydrogenolytic removal of both N,O-benzyl protecting groups using Pd black in acetic acid, gave after purification the corresponding azasugars in quantitative yield (**23-26**). This aminocyclisation can be performed with other primary amines (Table 16-2) and particularly with glycine *tert*-butyl ester and with tryptamine [15].

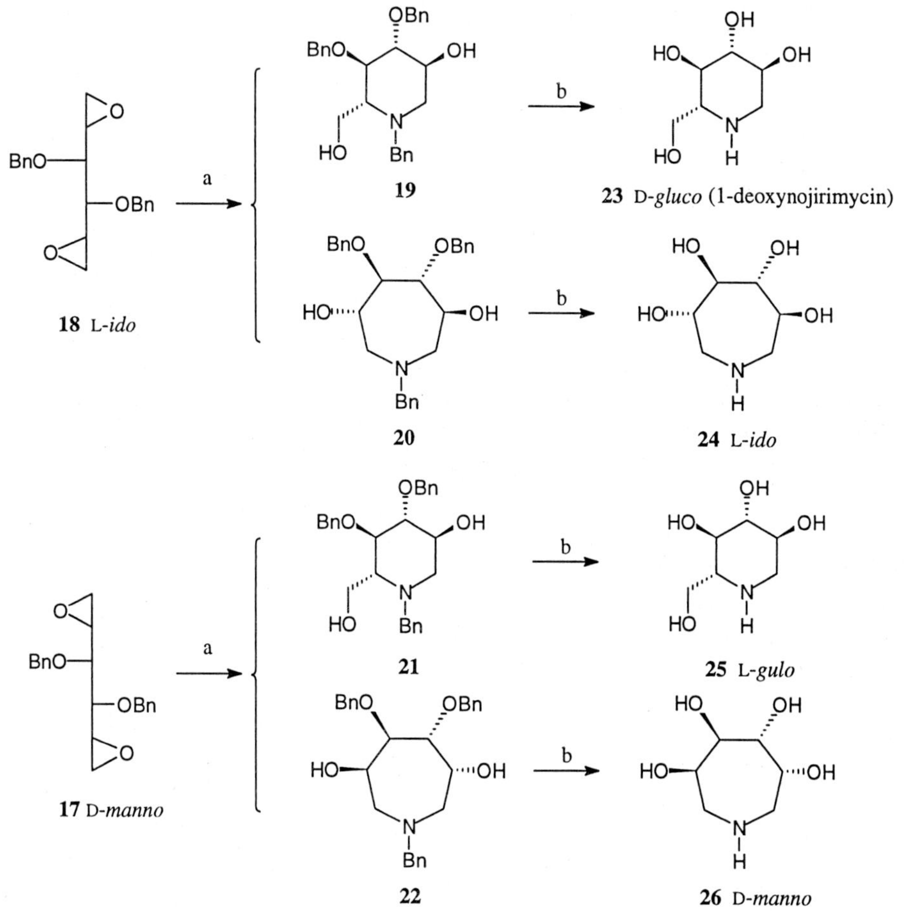

Scheme 16-11. Reagents: BnNH$_2$, see conditions in Table 16-1; b) H$_2$, Pd black, AcOH, 15h, 100%.

Table 16-1. Reaction of bis-epoxides **17** and **18** with benzylamine.

Entry	Epoxide	Reaction conditions	6-exo/7-endo ratio	Yield (%)[a]
1	18	BnNH$_2$ 5eq/CHCl$_3$/reflux/48h	55/45	**19** (45), **20** (33)
2	18	BnNH$_2$ 3eq/H$_2$O/25°C/48h	25/75	**19** (19), **20** (58)
3	18	BnNH$_2$ 10eq/H$_2$O/HClO$_4$ 5eq/25°C/4h	12/88	**19** (9), **20** (66)
4	17	BnNH$_2$ 5eq/CHCl$_3$/reflux/48h	55/45	**21** (50), **22** (45)
5	17	BnNH$_2$ 10eq/H$_2$O/HClO$_4$ 5eq/25°C/4h	30/70	**21** (28), **22** (67)

a: Yield of isolated compound after flash chromatography separation.

Table 16-2. Reaction of bis-epoxide **17** with RNH_2.

RNH_2	Conditions a,b or c	Yield[d] piperidine - azepane
$PhCH_2CH_2NH_2$	a	31% - 66%
$AcNH(CH_2)_4NH_2$	b	48% - 32%
$BnONH_2$, HCl	c	22% - 51%
$(Bn)_2NNH_2$	b	30% - 23%
$tBuO_2CCH_2NH_2$, HCl	b	40% - 33%
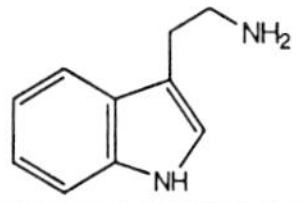 (indol-3-ylmethyl)-NH_2	a	26% - 64%
	b	50% - 45%

a) RNH_2 (10eq), $HClO_4$, H_2O, 25°C, 4h; b) RNH_2 (5eq), $CHCl_3$, Δ, 48h; c) RNH_2 (4eq), NEt_3 (5eq), 80°C, 15h; d) unoptimized yields of pure isolated compounds.

16.2.2.2 Isomerisation of Polyhydroxylated Piperidines

Skeletal rearrangements of pyrrolidines, piperidines or azepanes, via and aziridinium or azetidinium salt has already been reported but this method has found little applications in polysubstituted heterocycles synthesis [14,16].

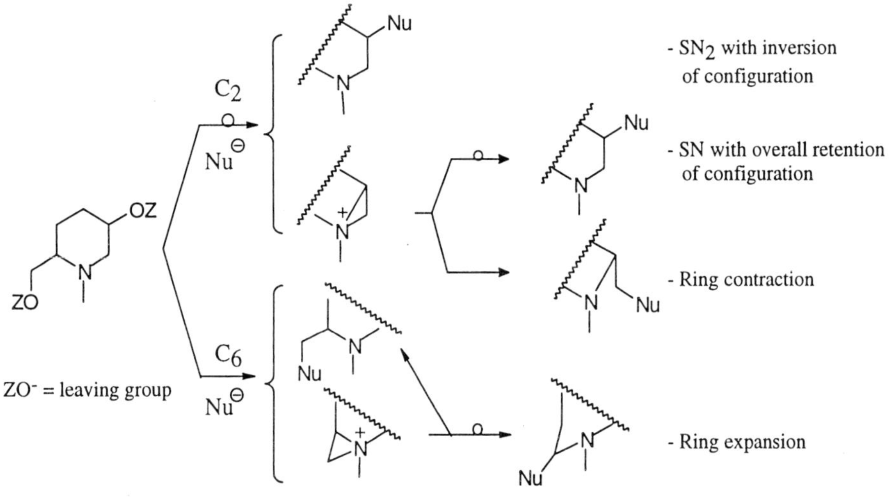

Scheme 16-12

Isomerisation of the 3,4-di-O-protected polyhydroxy-piperidines **19** and **21** which takes place after bis-hydroxyl activation at C-2 and C-6 can follow various pathways (Scheme 16-12) : S_N2, S_N with overall retention of configuration, ring contraction, ring expansion.

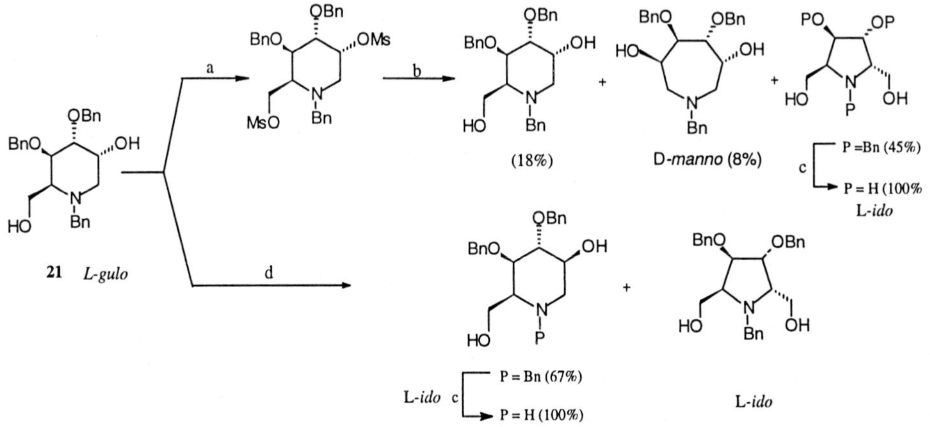

Scheme 16-13. Reagents: a) MsCl (2.3eq), Et$_3$N, CH$_2$Cl$_2$, 0°C, 100%; b) AcOCs, DMF, 40-50°C, then MeOH, K$_2$CO$_3$; c) H$_2$, Pd black then Dowex 100%; d) Ph$_3$P-DEAD-PhCOOH 4eq, THF, 0°C then MeOH, K$_2$CO$_3$.

Hydroxyl groups activation has been performed either by mesylation or formation of an alkoxyphosphonium salt under Mitsunobu conditions. From the L-*gulo*-piperidine, the L-*ido*-pyrrolidine is mainly obtained when mesylation is used for activation. Under Mitsunobu conditions the L-*ido*-piperidine is predominant (Scheme 16-13). S$_N$2 reaction competes with the aziridinium pathway (Scheme 16-14).

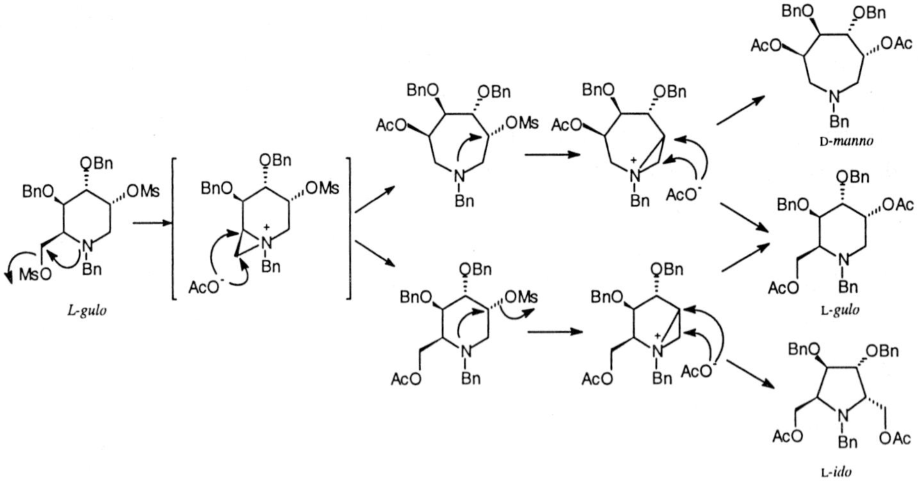

Scheme 16-14

Mesylates being poorer leaving groups than alkoxyphosphonium salts, substitution of the former requires more drastic conditions which allows aziridinium formation and subsequent ring contraction. In the case of Mitsunobu conditions (fast reaction at 0°C), the intermolecular S$_N$2 is easier than the intramolecular participation of nitrogen which requires and inversion of the chair conformation. Similar reactions were performed (Scheme 16-15) on the D-*gluco*-piperidine.

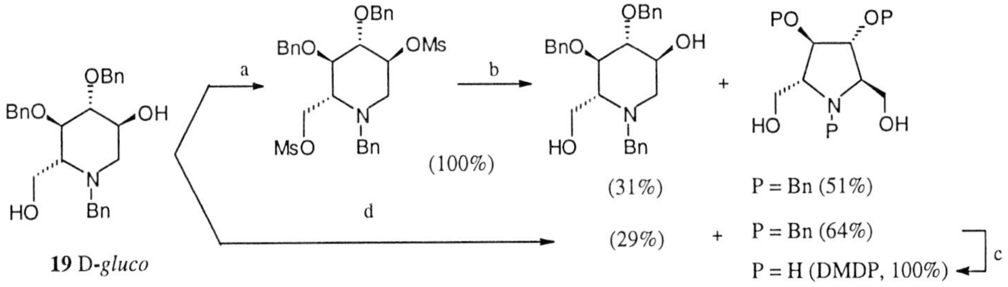

Scheme 16-15. Reagents: a) MsCl (2.3eq), Et$_3$N, CH$_2$Cl$_2$, 0°C, 100%; b) AcOCs, DMF, 40-50°C, then MeOH, K$_2$CO$_3$; c) H$_2$, Pd black then Dowex 100%; d) Ph$_3$P-DEAD-PhCOOH 4eq, THF, 0°C then MeOH, K$_2$CO$_3$.

In this case, the two procedures mainly gave 1) ring contraction to the D-*manno*-pyrrolidine, due to an easy neighbouring nitrogen participation since the leaving group is in equatorial position, and 2) subsequent ring opening of the aziridinium at the less substituted side. In order to avoid the neighbouring nitrogen participation N-benzyl-D-*gluco*-piperidine **19** has been converted into the N-carbonate **27**, which underwent substitution under Mitsunobu conditions to yield only **28** (91%) (Scheme 16-16). Deprotection by methanolysis and hydrogenolysis gives 1-deoxymannojirimycin (**29**).

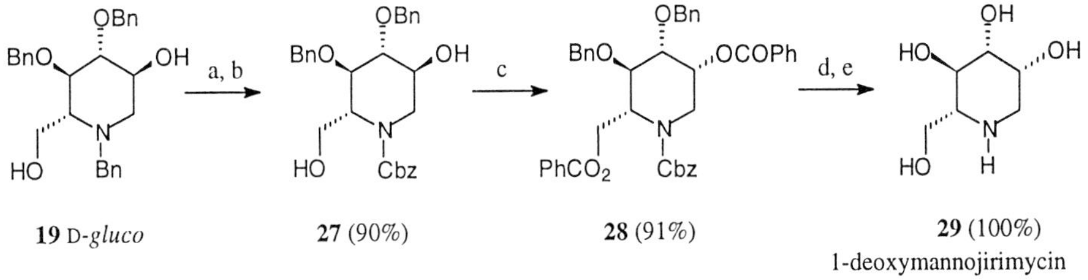

Scheme 16-16. Reagents: H$_2$, Pd(OH)$_2$/C, 20%, EtOH; b)BnOCOCl, K$_2$CO$_3$, DMF, 0°C; c) PPh$_3$-DEAD-PhCOOH, THF, 0°C; d) MeOH, K$_2$CO$_3$, in vacuo, 40°C; e) H$_2$, Pd black, AcOH.

16.2.3 Isomerisation of Polyhydroxylated Azepanes [14,17]

As previously described for the polyhydroxylated piperidines, isomerisation of D-*manno* and L-*ido* polyhydroxylated azepanes have been performed (Scheme 16-17 and Scheme 16-18).

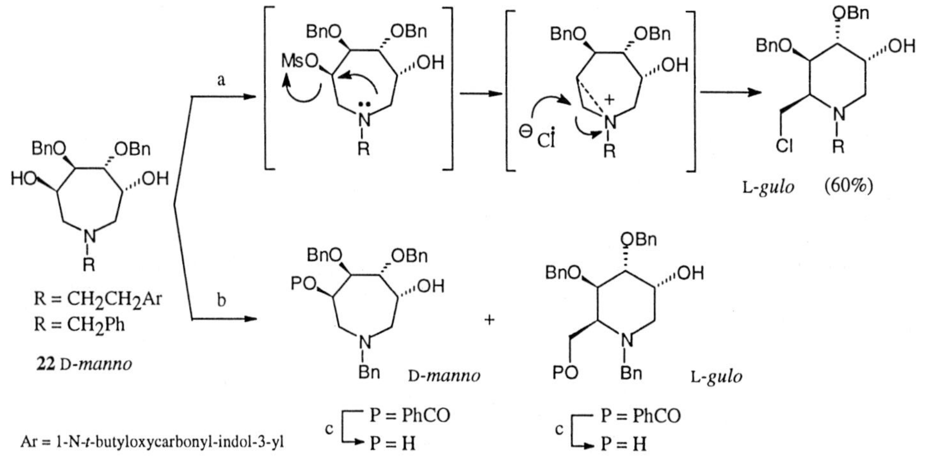

Scheme 16-17. Reagents: a) 1.4 eq MsCl, Et$_3$N, CH$_2$Cl$_2$, 0°C, 60%; b) 1.2 eq PPh$_3$-DEAD-PhCOOH, THF, 0°C, 64%; c) MeOH, K$_2$CO$_3$.

Ring contraction occurs in the presence of mesyl chloride to give chloromethylpiperidines which can be used directly in a chain extension reaction by nucleophilic substitution. Further uses of this methodology in the synthesis of non-peptide mimetics of somatostatin are presented below.

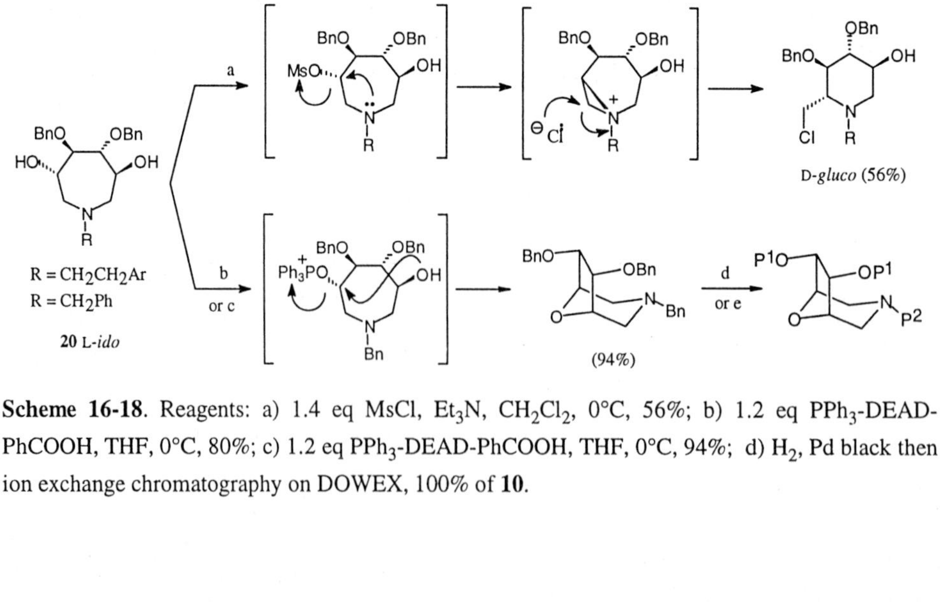

Scheme 16-18. Reagents: a) 1.4 eq MsCl, Et$_3$N, CH$_2$Cl$_2$, 0°C, 56%; b) 1.2 eq PPh$_3$-DEAD-PhCOOH, THF, 0°C, 80%; c) 1.2 eq PPh$_3$-DEAD-PhCOOH, THF, 0°C, 94%; d) H$_2$, Pd black then ion exchange chromatography on DOWEX, 100% of **10**.

Table 16-3. Inhibition of some glycosidases by azasugars

Compound	Yield (%)	α-Glu	β-Glu	α-Man	α-Fuc
22	50	55 / 70 (c)	1 / -	1 / -	94 / 28
	37	30 / -	13 / -	15 / -	94 / 22 (c)
21	17	55 / -	42 / -	40 / -	84 / 170 (c)
	25	74 / 65 (c)	11 / -	17 / -	50 / 260 (c)
	36	97 / 4.8 (c)	86 / 17 (c)	25 / -	4 / -
20	34	6 / -	0 / -	3 / -	29 / -
DNJ	24	100 / 0.44 (c)	64 / 1700	44 / 1000	27 / 500
DMDP	15	97 / 0.03 (c)	91 / 160	38 / -	35 / -
DMJ	19	100 / 18 (c)	39 / 1400 (c)	49 / 600 (c)	100 / 0.13 (c)
	14	88 / 53	4 / -	98 / 20 (c)	63 / 190 (c)
	15	66 / 83 (c)	8 / -	31 / -	2 / -

The formation of a bridged compound was observed upon treatment of **20** under Mitsunobu conditions (Scheme 16-18). This can be interpreted as an intramolecular displacement of the alkoxyphosphonium intermediate by the free hydroxyl group of the azepane, concurrently to the evolution towards the aziridinium. This type of chiral bridged morpholine is an interesting structure, N-substituted derivatives of 8-oxo-3-azabicyclo [3.2.1]-octane possessing various biological activities.

16.2.4 Inhibition Studies

Enantiomerically pure polyhydroxylated azepanes, piperidines, pyrrolidines and bridged morpholine obtained from D-mannitol derived bis-epoxides through nucleophilic opening, heterocyclisation and isomerisation were evaluated as inhibitors of different glycosidases (α-D-glucosidase, β-D-glucosidase, α-D-mannosidase and α-L-fucosidase) (Tableau 16-3) [13, 14]. Results obtained for DMDP, DMJ and DNJ are comparable with those reported. It is interesting to note that the C_2-symmetrical polyhydroxylated azepanes are inhibitors of glycosidases. The K_i values, in the low micromolar range, show that the greater flexibility of the seven membered ring makes these compounds capable of mimicking the putative oxonium ion transition state that is generated during oligosaccharide hydrolysis by glycosidases. By comparison, the rigid bridged morpholine showed no inhibition.

16.2.5 Synthesis of Non-Peptide Mimics of Somatostatin

Somatostatin-14 [Scheme 16-19) is a cyclic tetradecapeptide which serves not only to regulate the release of growth hormone or other pituitary hormones, but also plays a role in neuronal transmission.

Somatostatin-14: Ala[1]-Gly[2]-Cys[3]-Lys[4]-Asn[5]-Phe[6]-Phe[7]-Trp[8]-Lys[9]-Thr[10]-Phe[11]-Thr[12]-Ser[13]-Cys[14]

Sandostatin : H-D-Phe[1]-Cys[2]-Phe[3]-D-Trp[4]-Lys[5]-Thr[6]-Cys[7]-Thr[8]-OH

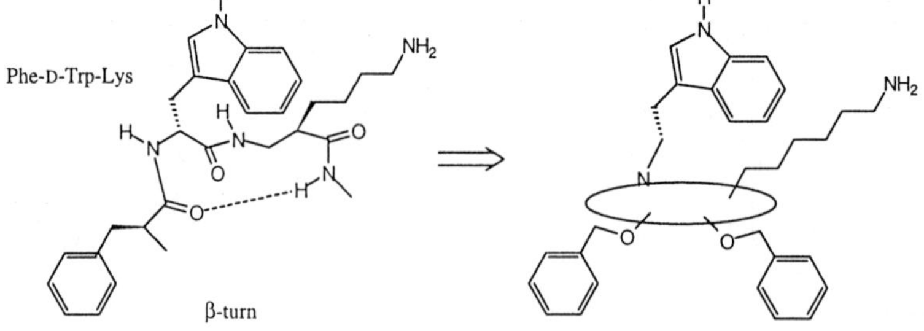

Scheme 16-19

The possible therapeutic applications of somatostatin-14 are limited by its rapid proteolytic degradation and its low selectivity. Several analogues or peptidomimetic compounds have been prepared. One of them, sandostatin® is used routinely in the treatment of acromegaly, carcinomas and gastroenteropancreatic tumors. Extensive investigation on the structure-activity relationships on somatostatin-14 showed that the tetrapeptide sequence Phe^7-D-Trp^8-Lys^9-Thr^{10} is the most relevant for biological activity. Nmr studies revealed that this tetrapeptide sequence adopts a β-turn structure.

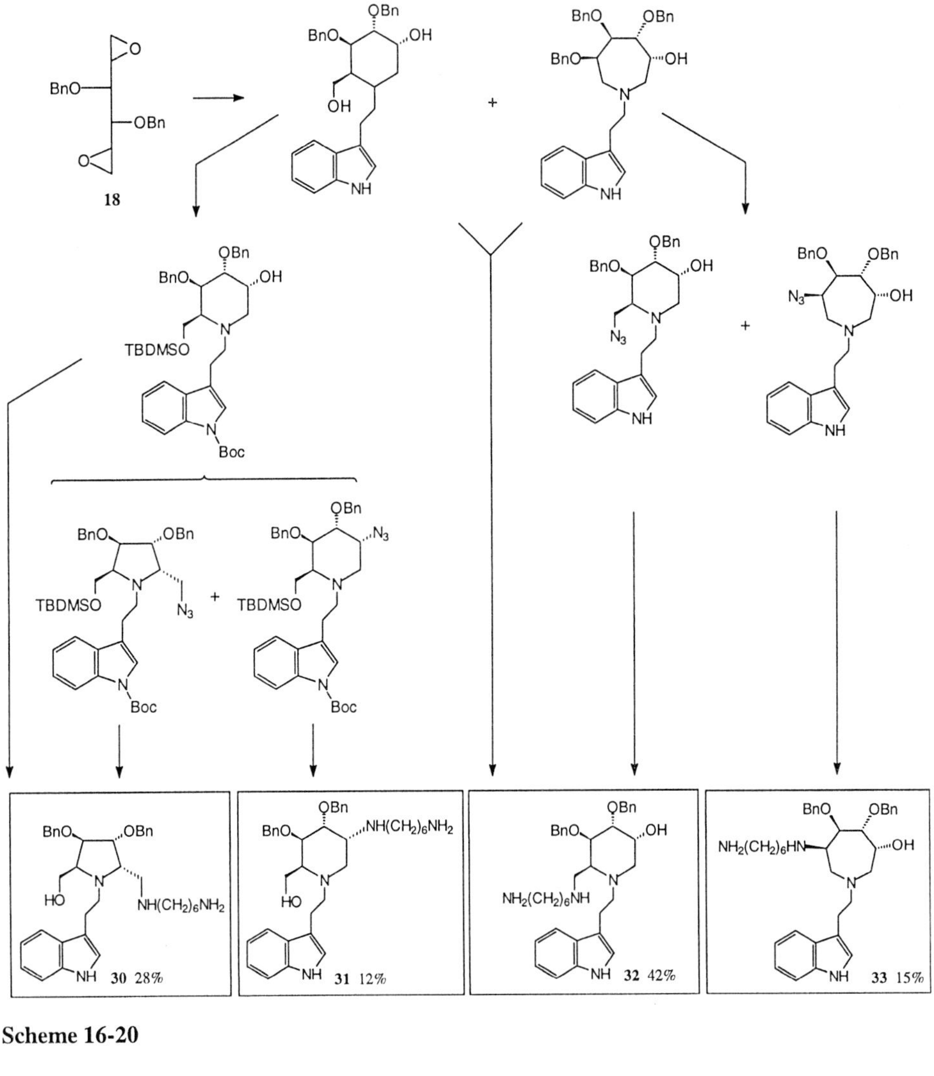

Scheme 16-20

We assumed that the above described azasugars may be used to mimic this β-turn, thus we tried to construct non-peptide mimics of somatostatin/ sandostatin® [13, 15, 18]. Access to four monocyclic mimics **30**, **31**, **32**, **33** is mentioned on Scheme 16-20.

For example, the piperidine **32** can be obtained in a few steps in 42% yield starting from D-*manno* bis-epoxide **18**. In a continuing effort to design structurally divergent classes of compounds acting on the somatostatin receptors we envisaged a new class of peptide mimics in which the scaffold unit is substituted by a rigid bicyclic system such as 2,5-diazobicyclo [2.2.2] octane backbone. The bicyclic derivatives **34**, **35** were prepared starting from the C_2-symmetrical azepane **20** (Scheme 16-21).

Binding assays have shown a moderate affinity of compounds **30-35** for somatostatin receptors ($IC_{50} \approx 12$ µM) regardless of size (five to seven membered rings) or the rigidity of the nitrogen heterocycle-based scaffolds used.

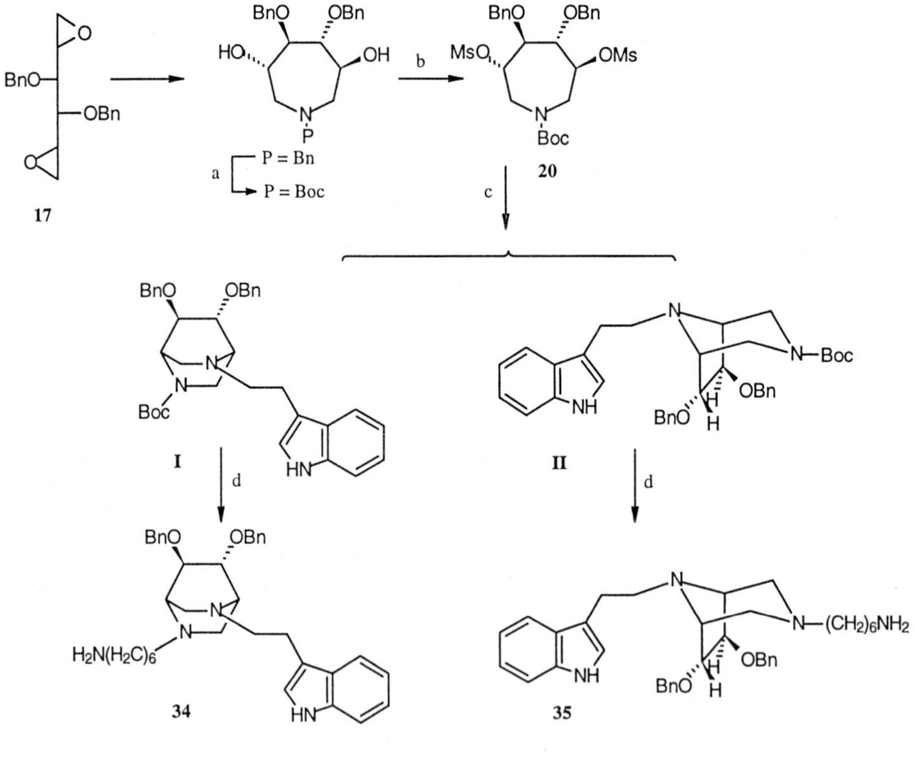

Scheme 16-21

16.3 Conclusion

The present work outlined efficient synthetic pathways to construct various enantiomerically pure azasugars with a pyrrolidine, piperidine or azepane framework starting from the one D-mannitol. These azasugars, carbohydrate mimics, are potent inhibitors of glycosidases and we has shown that they can be used as non-peptide scaffolds in the synthesis of peptide mimics.

Acknowledgements

It is a great pleasure to express my grateful thanks to Dr. A. Duréault and Dr. Y. Le Merrer who have supervised this research and to acknowledge the very significant contributions of graduate students and collaborators whose names appear in the references. Financial support of this project by Rhône-Poulenc (Bio-Avenir programme) with the participation of the French Ministry of Research and the French Ministry of Industry and by Roussel-Uclaf is gratefully acknowledged.

16.4 References

1. A. Dureault, C. Greck, J.-C. Depezay, *Tetrahedron Lett.* **1986**, *27*, 4157-4160.
2. Y. Le Merrer, A. Duréault, C. Greck, D. Micas-Languin, C. Gravier, J.-C. Depezay, *Heterocycles* **1987**, *25*, 541-548.
3. A. Duréault, I. Tranchepain, J.-C. Depezay, *J. Org. Chem.* **1989**, *54*, 5324-5330.
4. J. Fitremann, A. Duréault, J.-C. Depezay, *Tetrahedron Lett.* **1994**, *35*, 1201-1204.
5. J. Fitremann, A. Duréault, J.-C. Depezay, *Tetrahedron* **1995**, *51*, 9581-9594.
6. A. Duréault, I. Tranchepain, C. Greck, J.-C. Depezay, *Tetrahedron Lett.* **1987**, *28*, 3341-3344.
7. J. Fitremann, A. Duréault, J.-C. Depezay, *Synlett* **1995**, 235-237.
8. L. Campanini, A. Duréault, J.-C. Depezay, *Tetrahedron Lett.* **1995**, *36*, 8015-8018.
9. M. Meguro, N. Asao, Y. Yamamoto, *Tetrahedron Lett.* **1994**, *35*, 7395-7398.
10. L. Campanini, A. Duréault, J.-C. Depezay, *Tetrahedron Lett.* **1996**, *37*, 5095-5098.
11. Y. Le Merrer, A. Duréault, C. Gravier, D. Languin, J.-C. Depezay, *Tetrahedron Lett.* **1985**, *26*, 319-322.
12. a) L. Poitout, Y. Le Merrer, J.-C. Depezay, *Tetrahedron Lett.* **1994**, *35*, 3293-3296. b) J.-C. Depezay, Y. Le Merrer, L. Poitout, *French Patent Application* **94**, 01921.

13. L. Poitout, Thesis, Université Pierre et Marie Curie, Paris, **1995**.
14. Y. Le Merrer, L. Poitout, J. -C. Depezay, I. Dosbaa, S. Geoffroy, M-J. Foglietti, *Bioorg. Med. Chem. Lett* in press.
15. D. Damour, J. -C. Depezay, Y. Le Merrer, S. Mignani, G. Pantel, L. Poitout, *French patent Application* **1995**, 05510.
16. L. Poitout, Y. Le Merrer, J. -C. Depezay, *Tetrahedron Lett.* **1996**, *37*, 1609-1612.
17. L. Poitout, Y. Le Merrer, J. -C. Depezay, *Tetrahedron Lett.* **1996**, *37*, 1613-1616.
18. D. Damour, M. Barreau, J. -C. Blanchard, M-C. Burgevin, A. Doble, F. Herman, G. Pantel, E. James-Surcouf, M. Vuilhorgne, S. Mignani, L. Poitout, Y. Le Merrer, J. -C. Depezay, *Bioorg. Med. Chem. Lett.* **1996**, *6*, 1667-1672.

17 Furan-, Pyrrole-, and Thiophene-Based Siloxydienes *en route* to Carbohydrate Mimics and Alkaloids

G. Casiraghi, G. Rassu and F. Zanardi

17.1 Introduction

Monosaccharide compounds and the myriad of structurally related congeners and mimics, as well as highly hydroxylated alkaloidal derivatives constitute an ensemble of fascinating molecular entities, whose characteristics and functions are well documented [1].

The wide biological relevance and charming nature of these molecules combined with their constitutional and stereochemical diversity have stimulated considerable work directed towards the design and execution of novel, largely viable procedures to access them by synthesis [1]. Figure 17-1 displays some naturally occurring and "artificial" compounds which have been the subject of recent investigations in both the synthetic and the biomedical domains.

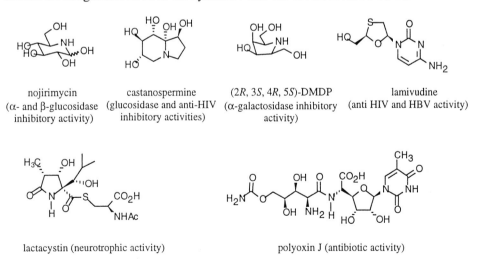

Figure 17-1. Relevant naturally occurring and synthetic structures and functions.

Owing to a great interest in these biofunctional substances and closely related both natural and unnatural structural variants, we embarked in a programme aimed at developing a chemically uniform synthetic strategy for the preparation of diverse classes of highly functionalized molecules, ranging from simple carbohydrate mimetics and amino acids to densely oxygenated monocyclic and bicyclic alkaloids and nucleosides. This account discusses our own synthetic plan and its application in synthesis, covering a selection of the most significant achievements by our research group in the period from 1990 to fall 1996 [1, 2]. After a short outline of the overall synthetic strategy (17.2), the emphasis will be placed upon the application of this chemistry to access simple amino sugar compounds (17.3.1), hydroxylated α-amino acids (17.3.2), bicyclic hydroxylated alkaloids (17.3.3), and nucleosides and nucleoside mimetics (17.3.4).

17.2 Outline of the Strategy

Flexibility and ergonomics are, among others, pivotal concerns in organic synthesis, especially where preparation of functional molecules is involved. Ideally, for the effects of shape and stereochemistry upon the functional profile to be evaluated, synthetic approaches to a given class of active compounds should be amenable to creation of maximal stereochemical variations and patterns of substitution in the series. In addition, to minimize the synthetic effort, truly viable schemes should exploit chemically unified protocols of a limited number of optimized transformations and employ readily available precursors and chemicals.

In searching for a synthetically efficient approach to densely functionalized compounds meeting the above mentioned criteria, our attention was attracted by a remarkable nucleophile, namely 2-(trimethylsiloxy)furan (TMSOF), first exploited by Ricci [3], Jefford [4], and others [5] during the 1980s. TMSOF, along with the close relatives 2-(*tert*-butyldimethylsiloxy)furan (TBSOF), *N*-*tert*-butoxycarbonyl-2-(*tert*-butyldimethylsiloxy)pyrrole (TBSOP), and 2-(*tert*-butyldimethylsiloxy)thiophene (TBSOT), subsequently developed by our research group, (Scheme 17.1) represent a remarkable ensemble of nucleophiles to be used as homologative reagents towards a wide variety of electrophilic substrates, ranging from various carbonyl precursors and imines to alkyl halides.

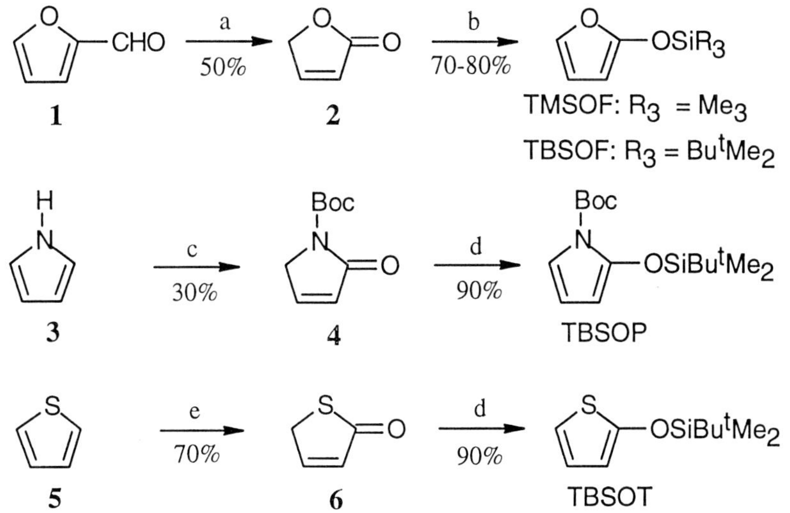

Scheme 17-1. Reagents: a) HCO$_2$Na, 30% H$_2$O$_2$; b) for TMSOF, Me$_3$SiCl, Et$_3$N; for TBSOF, TBSOTf, Et$_3$N, CH$_2$Cl$_2$; c) 30% H$_2$O$_2$; then Boc$_2$O, DMAP, CH$_2$Cl$_2$; d) TBSOTf, 2,6-lutidine, CH$_2$Cl$_2$; e) n-BuLi, methyl borate, 30% H$_2$O$_2$.

Indeed, these oxygen-, nitrogen-, and sulfur-containing heterocyclic siloxydienes constitute an ideal triad of chemically homogeneous elongation reagents. According to a parallel approach, these compounds are quickly obtainable in high isolated yields from inexpensive materials such as furfural **1**, pyrrole **3**, and thiophene **5**, respectively. The reagents display remarkable chemical stability and, in particular, TBS-derivatives can be easily subjected to chromatographic purification on silica gel without appreciable loss of their constitutional integrity. In addition, the materials are conveniently storable in a refrigerator for a long period of time.

Very importantly, as shown in Scheme 17.2, when reacted with suitable electrophilic precursors of type **B** in the presence of appropriate Lewis acid promoters, the siloxydiene reagents **A** regioselectively furnish γ-homologated adducts **C** whose heterocyclic component is manipulable on a large extent at all the carbon atoms, while the heteroatom may vary from oxygen to nitrogen, to sulphur. In addition, when a chiral information (S*) is embodied into the electrophilic reaction partner, chirality transmittal to the two newly created stereocenters takes place.

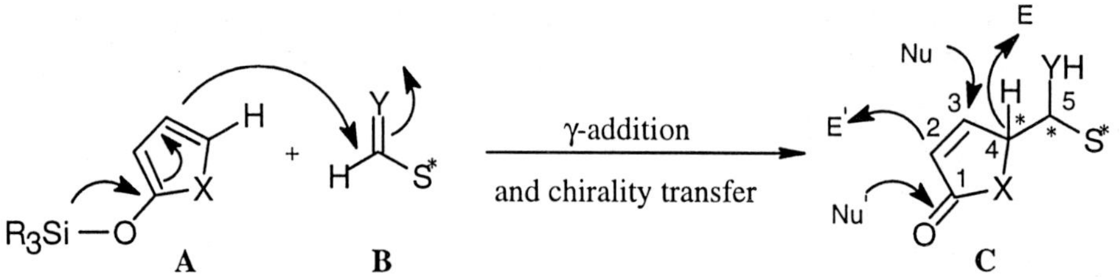

Scheme 17-2. The siloxydiene route to densely functionalised compounds.

The synthetic versatility of the maneuver is ensured by the wide choice of the reactants and promoters, as well as by the diversity of the chemical functions which can be incorporated into adducts **C**. As a result, large molecular diversity can emerge, comprising complex sugar units, nitrogen and sulphur sugar mimics, hydroxylated α-amino acids and alkaloids, modified nucleosides.

17.3 Synthetic Applications

17.3.1 Aminosugar Derivatives

Carbohydrate compounds in which one or more hydroxyl groups are replaced by amino functions represent a remarkable class of compounds, whose chemistry and biology has recently received intense scrutiny. The siloxydiene-based strategy was adopted to enter this class of compounds starting with TBSOP and suitable three-carbon enantiopure substrates.

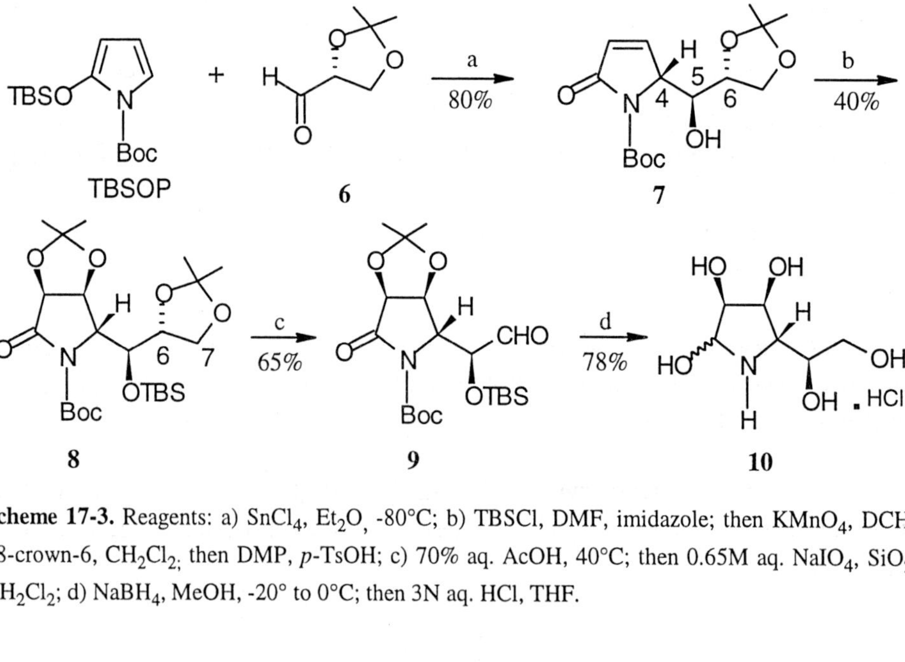

Scheme 17-3. Reagents: a) SnCl$_4$, Et$_2$O, -80°C; b) TBSCl, DMF, imidazole; then KMnO$_4$, DCH-18-crown-6, CH$_2$Cl$_2$; then DMP, p-TsOH; c) 70% aq. AcOH, 40°C; then 0.65M aq. NaIO$_4$, SiO$_2$, CH$_2$Cl$_2$; d) NaBH$_4$, MeOH, -20° to 0°C; then 3N aq. HCl, THF.

To access 4-amino-4-deoxy-D-talose **10** [6], the opening reaction was the diastereoselective SnCl$_4$-promoted coupling between isopropylidene protected D-glyceraldehyde **6** and TBSOP, furnishing crystalline 4,5-*threo*-5,6-*erythro*-configured α,β-unsaturated lactam **7** in high yield and diastereoselectivity (de > 90%) (Scheme 17-3).

Subsequent protection of the free C(5) hydroxy group as TBS-ether, followed by stereoselective *cis*, *anti*-dihydroxylation of the double bond and protection of the formed diol, then produced heptonolactam **8** in 40% yield for the three steps. One-carbon shortening at the right side of the molecule to aldehyde **9** was performed by first removing the terminal isopropylidene moiety of **8** (70% aq. AcOH) and then by cleaving the C(6)-C(7) bond using solid-liquid phase transfer conditions (NaIO$_4$-impregnated wet silica gel). Finally both the formyl group and the lactam carbonyl were reduced by NaBH$_4$ in methanol. Protected 4-amino-4-deoxy-D-talose was fully deprotected to the hydrochloride **10** by HCl treatment.

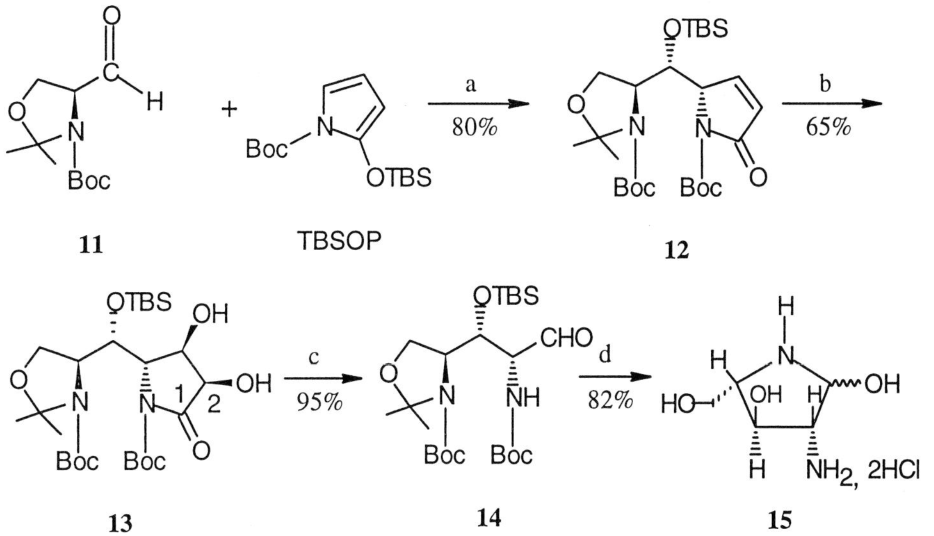

Scheme 17-4. Reagents: a) BF$_3$ etherate, CH$_2$Cl$_2$, -80°C; then TBSCl, DMF, imidazole; b) KMnO$_4$, DCH-18-crown-6, CH$_2$Cl$_2$; c) LiOH, THF, 0°C; then NaIO$_4$, SiO$_2$, CH$_2$Cl$_2$; d) citric acid, MeOH, 60°C; then 37% aq. HCl, THF.

A similar sacrificial synthetic plan led us to prepare the quite rare diaminoarabinose **15**, the sugar component of the naturally occurring antifungal antibiotic prumycin [7]. According to Scheme 17-4, the starting maneuver was the homologation of protected L-serinal **11** with TBSOP.

Remarkably, theBF$_3$ etherate-assisted Mukaiyama-aldol reaction proved regioselective, furnishing, after aqueous NaHCO$_3$ quenching, two 5-*O*-silylated adducts in 80% yield, wherein the 4,5-*threo*-lactam **12** predominated (75% de). At first, the formyl function of the five-carbon sugar **14** was installed by oxidative sacrifice of the C(1) and C(2) atoms of the seven-carbon precursor **12**. Thus, selective dihydroxylation of the carbon-carbon double bond of **12** yielded the diol **13** which was sequentially subjected to hydrolytic ring opening and diol fission. The protected diaminopentose **14** was obtained in high yield, and was used as such in the subsequent steps of the synthesis. The crucial annulation reaction was cleanly performed through selective deacetonidation by the mild reagent system citric acid-methanol, which also ensured concomitant pyrrolidinose formation. The final unmasking of the *N*-Boc and *O*-TBS protective groups was effected by HCl in THF, affording the bis-hydrochloride salt **15** which was isolated as an inseparable mixture of α- and β-anomers.

17.3.2 Hydroxylated α-Amino Acids

Hybrid structures with carbon-carbon linked amino acids and more or less hydroxylated moieties are often encountered in nature as individual molecules or as components of complex molecular assemblies. Pyrrole- and furan-based siloxydienes admirably served as precursors to form interesting acyclic and cyclic representatives of this important family of compounds. Successful implementation of our strategy to chiral syntheses of polyhydroxy-α-amino acids was achieved starting from enantiopure aldehydo sugars.

As an example, the total synthesis of 4-*epi*-polyoxamic acid **19** is illustrated in Scheme 17-5 [8,9]. According to an optimal protocol, protected lactam **16**, easily obtainable from TBSOP and aldehyde **6**, was first subjected to double bond dihydroxylation with KMnO$_4$ under solid-liquid phase transfer conditions. This gave lactam **17** as the sole isomer. Hydrolytic ring opening and subsequent oxidative diol fission at the C(2)-C(3) linkage (NaIO$_4$) afforded fully protected 2-amino-2-deoxy-D-arabinose **18**. Exposure of **18** to NaIO$_4$ in the presence of catalytic RuO$_2$ then provided a protected amino acid which was liberated to polyoxamic acid **19** by acidic treatment. The synthetic scope of this simple chemistry was then enlarged by successful synthesis of a wide number of six- and seven-carbon α-amino acid analogues, simply utilizing appropriate sugar precursors.

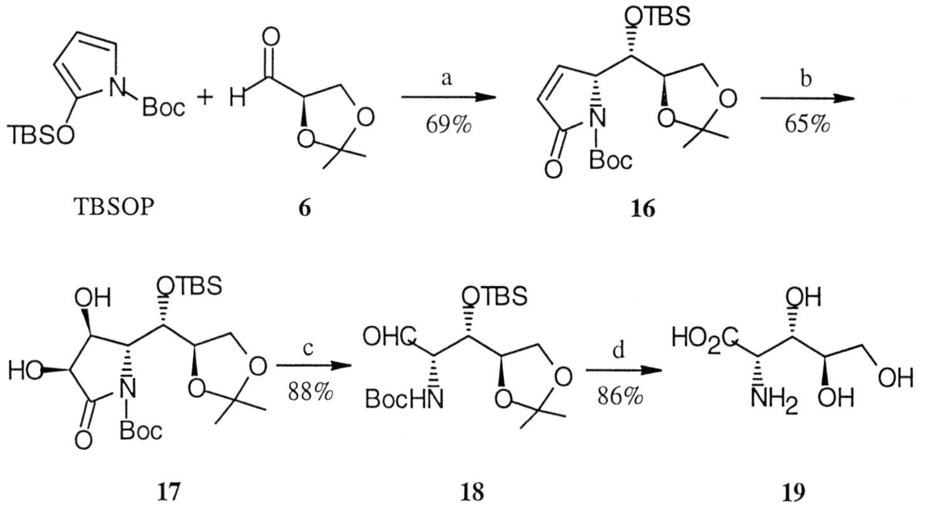

Scheme 17-5. Reagents: a) $SnCl_4$, Et_2O, -80°C; then TBSCl, DMF, imidazole; b) $KMnO_4$, DCH-18-crown-6, CH_2Cl_2; c) LiOH, THF, 0°C; then $NaIO_4$, SiO_2, CH_2Cl_2; d) $NaIO_4$, RuO_2; then TFA, MeOH; then SiO_2, AcOEt, MeOH, aq. NH_4OH.

A remarkable adaptation of this chemistry to the preparation of β-hydroxylated α-substituted α-amino acids is shown in Scheme 17-6 [10]. The plan includes a sequential γ,γ-double functionalization protocol of a common pyrrolin-2-one precursor **4**.

The first step was the creation of the γ-substituted pyrrolinone **20** via regioselective alkylation of TBSOPbeing the latter easily obtainable from **4**, as previously described. To access racemic α-methylthreonine **25**, pyrrolidone **20** was transformed into γ-methylsiloxydiene **21** according to the usual protocol, and this material was then reacted with acetaldehyde **22** by exposure to $SnCl_4$ in diethyl ether. After protection as TBS-ether, the γ,γ-disubstituted unsaturated lactam **23** was obtained with high diastereoselection (> 95% de). A five-step oxidative protocol, similar to that previously discussed for amino acid **19**, allowed the preparation of α-methylthreonine hydrochloride **25** to be prepared, via intermediate **24**, in 46% overall yield. According to these synthetic maneuvers, TBSOP and γ-substituted-TBSOP (*e.g.* **21**) can be regarded as glycyl anion and α-amino acyl anion equivalents, respectively.

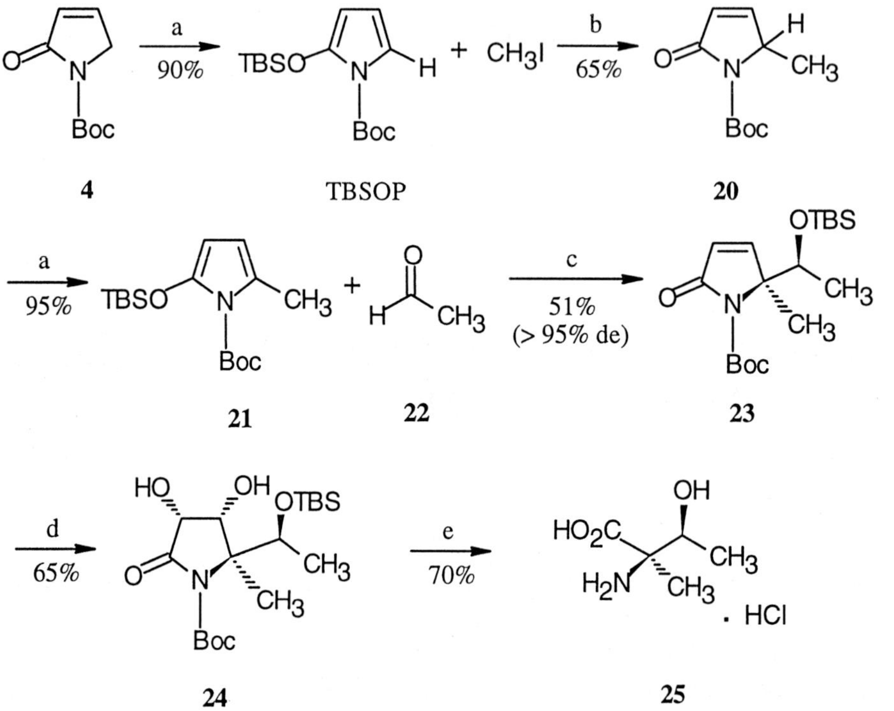

Scheme 17-6. Reagents: a) TBSOTf, 2,6-lutidine, CH$_2$Cl$_2$; b) CF$_3$CO$_2$Ag, CH$_2$Cl$_2$, 0°C; c) SnCl$_4$, Et$_2$O, -80°C; then TBSCl, DMF, imidazole; d) KMnO$_4$, DCH-18-crown-6, CH$_2$Cl$_2$; e) LiOH, THF, 0°C; then NaIO$_4$, SiO$_2$, CH$_2$Cl$_2$; then NaIO$_4$, RuO$_2$; then aq. 3N HCl.

A quite different approach was exploited to prepare terminal C-glycopyranosyl glycine derivatives, a scantly represented progeny of naturally occurring α-amino acids [11,12]. Scheme 17-7 depicts the total synthesis of α-D-talopyranosylglycine **30** by starting with TMSOF and unnatural D-serine derived aldehyde **26**. The first step was the synthesis of the butenolide intermediate **27** through BF$_3$ etherate-assisted coupling between **26** and TMSOF.

The lactone fragment of **27** was first elaborated according to a highly stereoselective three-step sequence consisting in protection of the free OH at C(5) as TMS-ether, *anti*, *cis*-dihydroxylation of the butenolide double bond using KMnO$_4$, and persilylation. This afforded heptonolactone **28** in high yield. The ring expansion to talopyranose **29** required three further operations. DIBALH reduction generated a γ-lactol intermediate, which, by citric acid/methanol treatment and subsequent peracetylation, was converted to **29** in 80% yield.

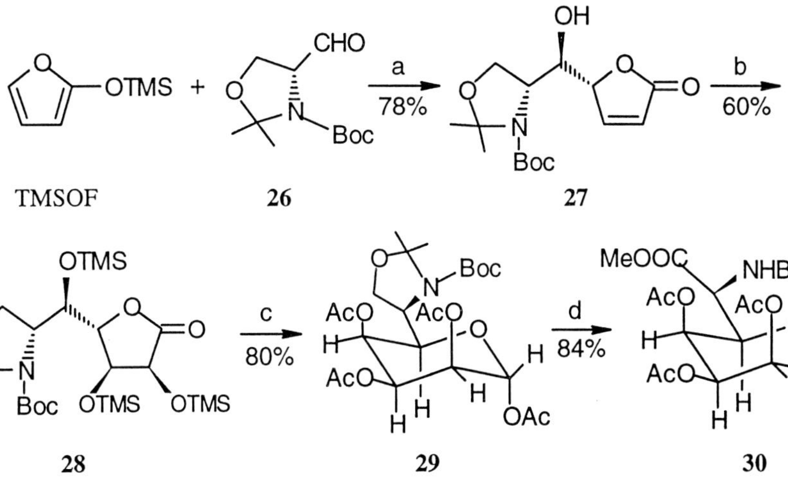

Scheme 17-7. Reagents: a) BF$_3$ etherate, CH$_2$Cl$_2$, -80°C; b) TMSCl, pyridine; then KMnO$_4$, DCH-18-crown-6; then TMSCl, pyridine; c) DIBALH, CH$_2$Cl$_2$, -80°C; then citric acid, MeOH; then Ac$_2$O, pyridine, DMAP; d) 70% aq. AcOH, 60°C; then NaIO$_4$, RuO$_2$; then CH$_2$N$_2$, Et$_2$O.

Treatment of **29** with 70% aqueous acetic acid resulted in selective removal of the terminal acetonide protection. The resulting primary alcohol was subjected to oxidation (NaIO$_4$, RuO$_2$), giving the expected carboxylic acid, which was finally transformed into the corresponding methyl ester **30** by treatment with diazomethane in diethyl ether. The potential of this approach was also exploited as an entry to structurally diverse congeners of both L- and D-sugar series.

In an effort to exploit the synthetic capability of TBSOP en route to natural and unnatural hydroxylated α-amino acids, our attention was attracted by *trans*-2,3-*cis*-3,4-dihydroxy-L-proline **36**, an unusual constituent of the adhesion protein of mussels (*Mytilus edulis*) [13]. The synthesis commenced with preparation of the adduct **31** by reaction of TBSOP with protected D-glyceraldehyde **6** under BF$_3$ etherate catalysis. As depicted in Scheme 17-8, treatment of D-*ribo*-configured lactam **31** with KMnO$_4$, as usual, followed by exposure of the diol so formed to dimethoxypropane in the presence of *p*-toluenesulfonic acid, directly afforded lactam **32** in 60% yield. The transformation of **32** into formyl pyrrolidinone **33** called for selective deprotection of the terminal acetonide (citric acid in methanol) and subsequent oxidative excision of the C(6) and C(7) carbon atoms (aq. NaIO$_4$).

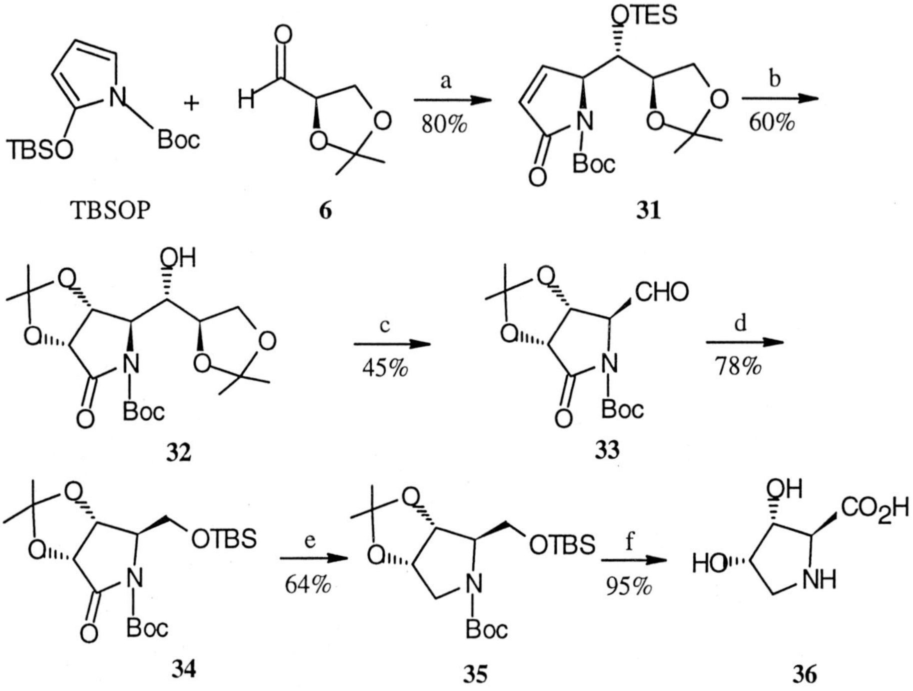

Scheme 17-8. Reagents: a) BF$_3$ etherate, CH$_2$Cl$_2$, -80°C; then TESOTf, 2,6-lutidine, CH$_2$Cl$_2$; b) KMnO$_4$, DCH-18-crown-6; then DMP, p-TsOH; c) citric acid, methanol, 65°C; then NaIO$_4$, SiO$_2$, CH$_2$Cl$_2$; d) NaBH$_4$, THF, H$_2$O, -30°C; then TBSCl, imidazole, DMF; e) LiEt$_3$BH, THF, -80°C; then Et$_3$SiH, BF$_3$ etherate, -80°C; f) TBAF, THF; then NaIO$_4$, RuO$_2$; then 3N aq. HCl; then DOWEX OH$^-$.

The formyl function of **33** was selectively reduced by NaBH$_4$ to a carbinol, which was protected as TBS-ether **34**. For optimal conversion, the reduction of the lactam carbonyl to methylene was performed according to a two-step protocol, consisting of partial reduction to an hemiaminal (LiEt$_3$BH) and subsequent deoxygenation (Et$_3$SiH, BF$_3$ etherate). L-prolinol **35** was then cleanly elaborated into the target amino acid **36** by conventional chemistry.

17.3.3 Bicyclic Hydroxylated Alkaloids

The chemistry, biochemistry and biological implications of polyhydroxylated bicyclic alkaloids including pyrrolizidine and indolizidine derivatives, have attracted wide interest in recent years. Because of the potential value of these alkaloids as glycoprocessing enzymes inhibitors and therapeutic agents, the discovery of versatile and viable procedures to access this important family of compounds has been of extreme interest to both organic and medicinal chemists. In our hands, both TMSOF and TBSOP proved highly efficient to create the skeleton of these alkaloidal compounds.

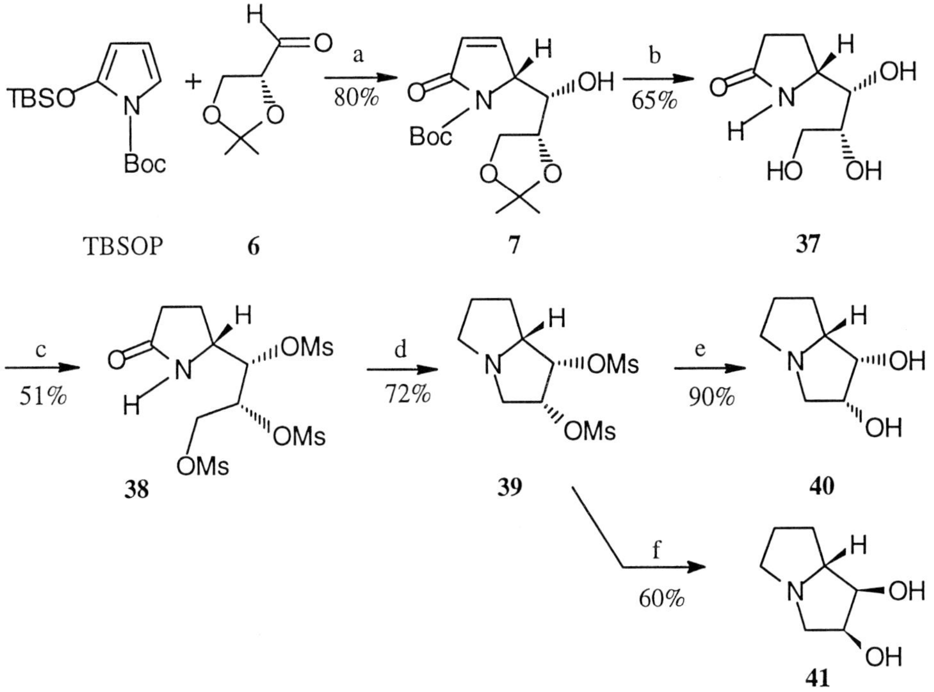

Scheme 17-9. Reagents: a) SnCl$_4$, Et$_2$O, -80°C; b) H$_2$, Pd/C, THF; then 6N aq. HCl, THF; c) MeSO$_2$Cl, pyridine; d) BH$_3$·DMS, THF; then DBU, 80°C; e) 6% sodium amalgam, isopropanol; f) tetrabutylammonium benzoate; then NaOMe.

As the first example, the divergent synthesis of diastereoisomeric *cis*-1,2-dihydroxypyrrolizidines **40** and **41** is detailed in Scheme 17-9 [14]. The common matrix, the unsaturated lactam **7**, was first prepared, as previously described, by starting with TBSOP and **6**. The double bond of **7** was then hydrogenated and the protective groups removed by acidic treatment. This afforded triol **37** in 65% isolated yield. Exposure of **37** to methanesulfonyl chloride in pyridine gave the trimesyl derivative **38**, which underwent annulation to **39** by carbonyl reduction (BH$_3$.DMS) . followed by treatment with DBU. For **40** to be prepared, a simple enantioconservative demesylation reaction was adopted (6% sodium amalgam in isopropanol) whereas for diastereomer **41**, a double nucleophilic displacement with benzoate anion was selected, resulting in inversion of configuration at the

C(1) and C(2) chiral centers. In a strictly analogous manner, the enantiomeric counterparts of **40** and **41** were accessible, by employing protected L-glyceraldehyde as a three-carbons chiral source.

An expeditious synthesis of (+)-1-deoxy-8-*epi*-castanospermine **48** involved the use of protected L-threose **42**, easily available from L-tartaric acid as a four-carbon chiral synthon and pyrrole-based siloxydiene TBSOP as a homologation reagent [15].

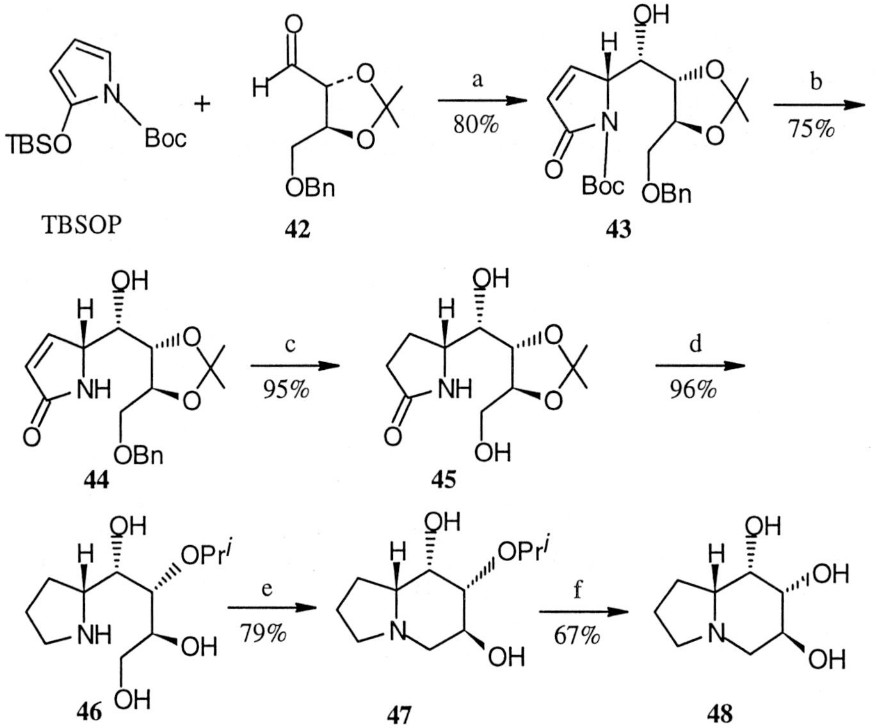

Scheme 17-10. Reagents: a) SnCl$_4$, Et$_2$O, -80°C; b) TMSOTf, PhSH; c) H$_2$, Pd/C, THF; d) BH$_3$·DMS, THF; then 3N aq. HCl, THF; then DOWEX OH$^-$; e) PPh$_3$, CCl$_4$, Et$_3$N, pyridine; f) BBr$_3$, CH$_2$Cl$_2$. .

The route (Scheme 17-10) envisaged double bond saturation in lactam **43**, followed by ring closure *via* intramolecular displacement of terminal C(8) hydroxyl by means of the nucleophilic C(4) nitrogen to create the indolizidine skeleton.

Optimally, the threose derivative **42** was treated with TBSOP in diethyl ether at -80°C in the presence of SnCl$_4$. The addition occurred regio- and stereoselectively at the C(5) carbon of TBSOP to form unsaturated lactam **43**, exclusively, whose absolute configuration was ascertained by X-ray analysis of a related saturated compound. Treatment of **43** with trimethylsilyl triflate in the presence of thiophenol cleanly afforded lactam **44**, which was hydrogenated and deprotected to compound **45**. Lactam **45** was directly exposed to an excess of BH$_3$·DMS. in THF and the amine-borane adduct thus formed was subjected to acidic treatment at room temperature. Quite surprisingly, this treatment afforded the isopropyl ether **46**, resulting from reduction of the lactam carbonyl with concomitant opening of the dioxolane ring and overreduction. Amino alcohol **46** was ready for the final annulation stage, and this was achieved by subjecting this compound to PPh$_3$-CCl$_4$-Et$_3$N in pyridine at room temperature. Indolizidine **47** was easily converted to castanospermine congener **48** by BBr$_3$ treatment.

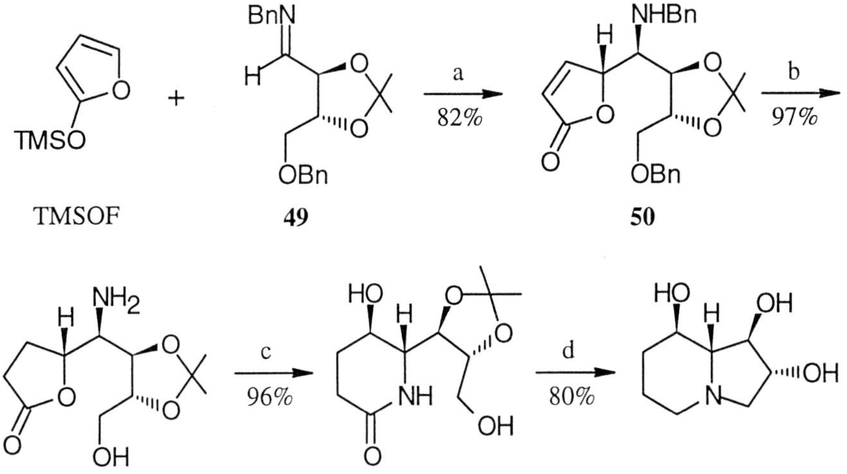

Scheme 17-11. Reagents: a) BF$_3$ etherate, CH$_2$Cl$_2$, -85°C; b) H$_2$, Pd/C, NaOAc, THF; c) DBU, C$_6$H$_6$, reflux; d) BH$_3$·DMS, THF; then 60% aq. TFA; then DOWEX OH$^-$; then PPh$_3$, CCl$_4$, Et$_3$N, DMF. .

A short route to **53** is outlined in Scheme 17-11. Protected D-threose imine **49** was the enantiomerically pure substrate which reacted with furan-based diene TMSOF, to afford the swainsonine derivative **53** [16]. The reaction of **49** with TMSOF in the presence of BF$_3$ etherate gave butenolide **50** with little if any stereoisomeric contamination. Double bond saturation with concomitant removal of the two benzyl protective groups to provide amine **51** was effected under conventional hydrogenation conditions. Next, upon treatment with DBU, amino lactone **51** underwent clean ring expansion to provide δ-lactam **52** in good yield.

Finally, treatment of this compound with an excess of $BH_3 \cdot DMS$ complex in THF effected the reduction of the lactam carbonyl to the corresponding amine-borane adduct. Deprotection of this adduct was accomplished by acidic treatment giving a densely oxygenated piperidine intermediate, which was directly cyclized to indolizidine **53** by exposure to PPh_3-CCl_4-Et_3N in DMF, as previously described for **48** (*vide supra*).

An attempt to extend this simple reaction protocol to the preparation of hydroxylated quinolizidines of type **59** resulted in a quite unexpected transformation (Scheme 17-12) [17]. Thus, preparation of the key homologue derivative **55** was accomplished without incidents by starting with TMSOF and D-*arabino* imine **54**, exactly adopting the reaction sequence previously illustrated for the synthesis of intermediate **52** (*vide supra*).

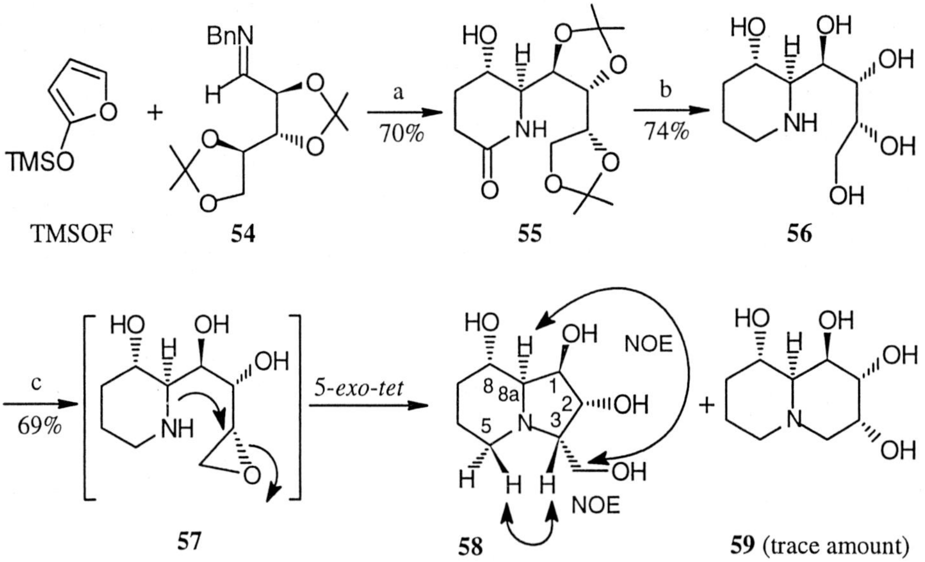

Scheme 17-12. Reagents: a) BF_3 etherate, CH_2Cl_2, -80°C; then H_2, Pd/C, NaOAc, THF; then DBU, benzene, reflux; b) $BH_3 \cdot DMS$, THF; then 60% aq. TFA; then DOWEX OH⁻; c) PPh_3, CCl_4, Et_3N, DMF.

The relative, and hence absolute configuration of **55** was established by a X-ray analysis, confirming the chirality of the five stereocenters. The subsequent transformation to piperidine **56** was achieved by reduction of the lactam carbonyl ($BH_3 \cdot DMS$) and subsequent acidic deprotection. Surprisingly enough, when **56** was subjected to the Mitsunobu-like annulation protocol using PPh$_3$-CCl$_4$-Et$_3$N mixture, the indolizidine **58** was mainly obtained, with only trace amounts of the expected quinolizidine **59**. This behaviour can be rationalized by assuming preliminary formation of the intermediate epoxide **57**, which undergoes preferential 5-*exo*-tetragonal ring closure to **58**, an expanded alexine derivative. The configurational assignment of **58** was mainly based on 2D-NOESY ^{1}H nmr experiments, showing diagnostic nOe contacts between H-C(3) and H-C(5) in the α position, and between H-C(8a) and both the protons of the C(3) hydroxymethyl substituent.

17.3.4 Nucleosides and Nucleoside Mimetics

Planning and synthetic exploitation of chemically uniform modular procedures for the construction of structurally related small organic compounds with specific functions are crucial issues of both life and material sciences. Parallel and repetitive executions of common synthetic protocols employing sets of equally reacting reagents or precursors ensure the production of individual basic structures for discovery of lead candidates. The siloxydiene triad in our hands fulfils these pivotal requirements, paving the way for the design of viable approaches to structurally and stereochemically diverse classes of biomolecules and mimics thereof. As an emblematic example, a unified synthesis of 2',3'-dideoxynucleosides and their sulfur and nitrogen mimics was planned, by exploiting the reagent triad TBSOF, TBSOT, and TBSOP [18-20]. A retrosynthetic analysis of this approach is highlighted in Scheme 17-13.

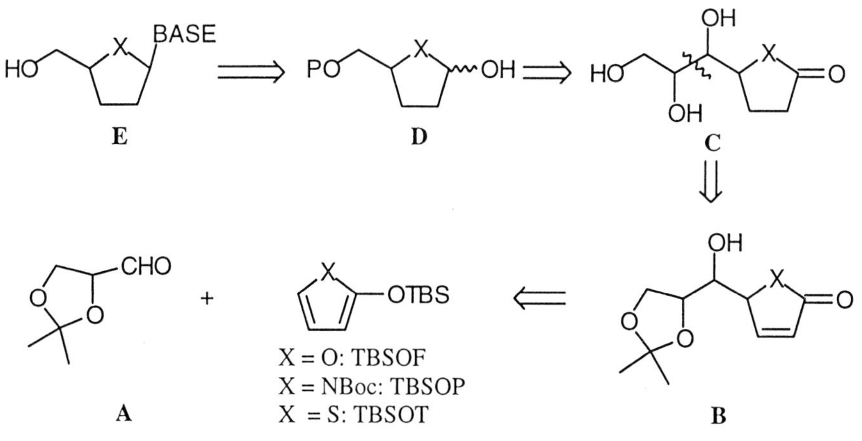

Scheme 17-13. Unified synthesis of nucleosides and nucleoside mimetics.

Accordingly, a given nucleoside of general formula **E** (X=O, S, NBoc) can be envisioned as being formed by three subunits, namely the core five-membered heterocycle, the nucleobase, and the hydroxymethyl arm. While the furanose unit directly correlates to the starting siloxydiene precursor, the hydroxymethyl moiety in **E** is provided by the glyceraldehyde precursor **A** via excision of two carbon atoms, as shown in **C**. Remarkably, the chirality present in **A** is transmitted into **D** and **E**, prior to loss of the native chiral information.

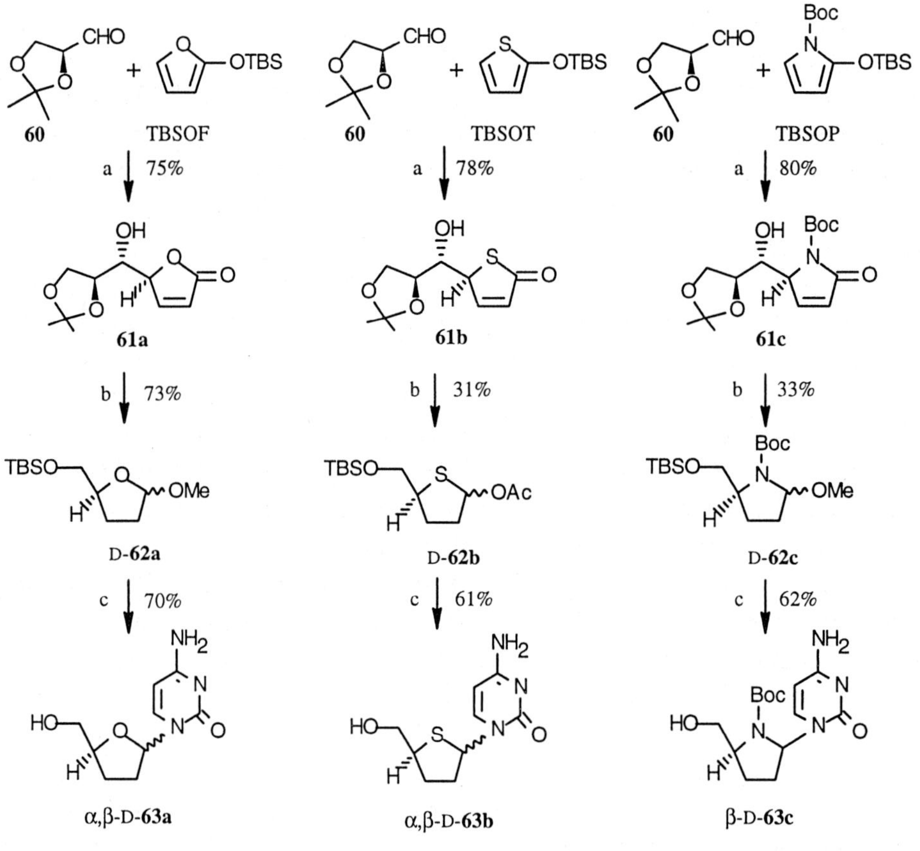

Scheme 17-14. Reagents: a) for **61a,b**, BF_3 etherate, CH_2Cl_2, -90°C; for **61c**, $SnCl_4$, Et_2O, -90°C; b) for **62a**, H_2, Pd/C, THF; then 80% aq. AcOH, 50°C; then $NaIO_4$, SiO_2, CH_2Cl_2; then $NaBH_4$, MeOH; then TBSCl, imidazole, CH_2Cl_2; then DIBALH, CH_2Cl_2, -90°C; then $CH(OMe)_3$, BF_3 etherate, Et_2O; for **62b**, H_2, Pd/C, THF; then 80% aq. AcOH, 50°C; then $NaIO_4$, SiO_2, CH_2Cl_2; then $NaBH_4$, MeOH; then TBSCl, imidazole, CH_2Cl_2; then Ac_2O, pyridine, DMAP; for **62c**, H_2, Pd/C, THF; then 80% aq. AcOH, 50°C; then $NaIO_4$, SiO_2, CH_2Cl_2; then $NaBH_4$, MeOH; then TBSCl, imidazole, CH_2Cl_2; then $LiEt_3BH$, THF, -78°C; then $CH(OMe)_3$, BF_3 etherate, Et_2O; (c)

silylated 4-*N*-acetylcytosine, SnCl$_4$/TMSOTf (1:1), 1,2-dichloroethane; then TBAF, THF; then K$_2$CO$_3$, MeOH.

According to this plan, D- and L-glyceraldehyde precursors can be regarded as "enantiomeric" hydroxymethylene cation equivalents, furnishing L-nucleosides from D-glyceraldehyde and D-nucleosides from L-glyceraldehyde (*vide infra*).

The parallel preparation of a 30-components collection of individual, pure pyrimidine nucleosides of both the D- and L-series was achieved according to an in-solution, unified chemical procedure. Scheme 17-14 illustrates, as an example, the preparation of five cytidine-related derivatives of the D-series, **63a-c**, from the three siloxydiene precursors TBSOF, TBSOT, and TBSOP and isopropylidene protected L-glyceraldehyde **60**. The overall synthetic maneuver comprises three parallel executions, adopting a rather uniform set of reactions.

The first step was the preparation of the unsaturated templates **61a-c** *via* stereoselective coupling between **60** and the individual siloxydienes. In the next stage of the sequence, the heterocyclic core unit within **61a-c** was hydrogenated, while the γ-triol appendage was shortened by two carbon atoms by the usual oxidative cleavage technique, followed by NaBH$_4$ reduction and protection. Suitable manipulation of the carbonyl function allowed the preparation of the key sugar units D-**62a-c**, ready for the final nucleobase coupling. A modified Vorbrüggen procedure was adopted to couple the silylated 4-*N*-acetylcytosine to the corresponding sugar components. To ensure kinetic convergence, an excess of nucleobase and a mixed Lewis acid catalytic system was employed, consisting in a 1:1 mixture of SnCl$_4$ and TMSOTf in 1,2-dichloroethane. This coupling furnished, after appropriate deprotection, the expected cytidine nucleosides D-**63a-c** in good yields. It should be noted that, while in the oxa and thia-series the nucleoside is formed as 1:1 α/β anomeric mixtures, the same transformation (D-**62c** → β-D-**63c**) is highly stereoselective, affording predominantly the β-anomer.

One can speculate that nitrogen pyramidalization within the heterocyclic iminium intermediate, with preferential *trans*-disposition of the *N*-Boc moiety, may play a role in hindering the α-face of the pyrrolidine nucleus, favoring the attack of the nucleobase on the less hindered β-face. Along these lines, successful preparation of five cytidine nucleosides of the L-series (L-**63a-c**), as well as 10 uridines (D- and L-**64a-c**) and 10 thymidines (D- and L-**65a-c**) was straightforwardly ensured. A summary of all the prepared nucleosides is depicted in Figure 17-2.

As an exciting extension of this strategy, a much more rapid preparation of a small library of the same 30 nucleosides was planned and carried out [20]. The first step of our plan was the preliminary synthesis of the appropriate sugar precursors D,L-**62a**, D,L-**62b**, and D,L-**62c** in a racemic form. Next, as shown in Figure 17-3, a "combine-split" in-solution approach was adopted, using the three oxygen, sulphur, and nitrogen sugars D,L-**62a-c** and the three pyrimidine bases, uracil (Ur), thymine (Th), and cytosine (Cy).

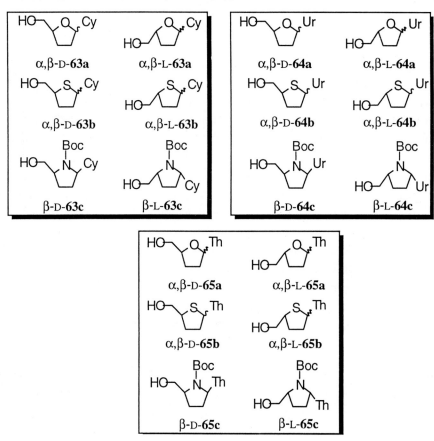

Figure 17-2. Cytidine, uridine, and thymidine compounds synthesized.

Thus, the carbohydrate precursors were mixed in equimolar quantity (0.2 mmol each) by dissolving them in anhydrous 1,2-dichloroethane and the resulting solution was divided into three equal portions. Each portion was independently reacted with one individual activated base, Ur, Th, Cy (4.0 equiv.) under SnCl$_4$/TMSOTf catalysis to give, after complete deprotection and separation of the nucleoside fraction, three diverse sublibraries (D,L-**SL$_{Ur}$**, D,L-**SL$_{Th}$**, and D,L-**SL$_{Cy}$**) which were finally combined to furnish a full library of nucleosides

(D,L-**L**), comprising the expected fifteen racemic compounds in 64% combined yield based on the average molecular weight of the nucleoside mixture.

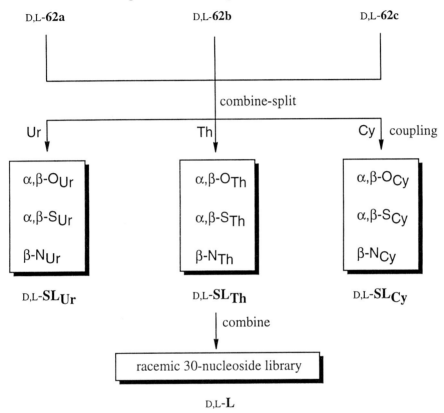

Figure 17-3. Schematic description of the construction of a 30-component full library, D,L-**L**, and three related sublibraries, D,L-**SL**$_{Ur}$, D,L-**SL**$_{Th}$, and D,L-**SL**$_{Cy}$ of racemic pyrimidine nucleosides. key: O$_{Ur}$ = D,L-**64a**, S$_{Th}$ = D,L-**65b**, N$_{Cy}$ = D,L-**63c**, etc.

In principle a truly random synthesis would furnish an ensemble of 18 racemates (36 different nucleosides), i.e. $3^2 \times 2^i$, where i = 2 is number of stereocenters in the resulting nucleosides. In practice, however, due to the stereoselective character of the azafuranose nucleobase coupling reaction, the total number of individuals in the full library was reduced to 15 major racemates out of 18. The three nucleoside mixtures, D,L-**SL**$_{Ur}$, D,L-**SL**$_{Th}$, and D,L-**SL**$_{Cy}$ were analyzed by reverse-phase HPLC. Having in hands all the pure components, strongly facilitated the analysis of the mixtures, by simply comparing the experimental HPLC traces with standard profiles of artificial mixtures of pure nucleosides. Remarkably, the HPLC analyses indicated the presence of the expected nucleosides with no detectable by-products. In addition, the analyses indicated that the nucleoside mixture in our hands was roughly equimolar.

17.4 Conclusion

Throughout this account, selected syntheses of densely functionalized molecules related to carbohydrates and carbohydrate mimics have been discussed, highlighting the invaluable merits of a triad of heterocyclic oxygen, sulfur, and nitrogen siloxydienes. The central issue in this work was the exploitation of a practical strategy based upon a rather uniform chemistry, employing few and simple synthetic transformations, and "popular" substrates and reagents. The rare flexibility of this technique led us to prepare a wide display of structurally diverse compounds of biological interest, ranging from monocyclic carbohydrates and nucleosides to polyhydroxylated amino acids and alkaloids. In spite of the numerous synthetic achievements, the full scope of this procedure is far from being completely explored, and future work should be addressed to further expand the arsenal of the synthetic maneuvers as well as the spectrum of the precursors, chemicals, and objectives.

Acknowledgements

We would like to express our sincere thanks to the effective members of our staff, Dr. Lucia Battistini, Dr. Mara Cornia, Dr. Pietro Spanu and Dr. Luigi Pinna for their skill co-operation throughout the work. We also like to thank the many colleagues, fellows, and students whose names are in the papers. Research in the authors' laboratories has been generously supported by the Consiglio Nazionale delle Ricerche (CNR) and Ministero dell'Università e della Ricerca Scientifica e Tecnologica (MURST).

17.5 References

1. G. Casiraghi, F. Zanardi, G. Rassu, P. Spanu, *Chem. Rev.* **1995**, *95*, 1677-1716.
2. G. Casiraghi, G. Rassu, *Synthesis* **1995**, 607-626.
3. M. Fiorenza, A. Ricci, M. N. Romanelli, M. Taddei, P. Dembech, G. Seconi, *Heterocycles* **1982**, *19*, 2327-2329.
4. C. W. Jefford, D. Jaggi, A. W. Sledeski, J. Boukouvalas, in *Studies in Natural Products Chemistry*; Atta-ur-Rahaman, Ed.; Elsevier B. V.: Amsterdam, 1989; Vol. 3, pp. 157-171.
5. D. W. Brown, M. M. Campbell, A. P. Taylor, X. Zang, *Tetrahedron Lett.* **1987**, *28*, 985-988.

6. P. Spanu, G. Rassu, F. Ulgheri, F. Zanardi, L. Battistini, G. Casiraghi, *Tetrahedron* **1996**, *52*, 4829-4838.

7. P. Soro, G. Rassu, P. Spanu, L. Pinna, F. Zanardi, G. Casiraghi, *J. Org. Chem.* **1996**, *61*, 5172-5174.

8. G. Casiraghi, G. Rassu, P. Spanu, L. Pinna, *Tetrahedron Lett.* **1994**, *35*, 2423-2426.

9. G. Rassu, F. Zanardi, M. Cornia, G. Casiraghi, *J. Chem. Soc., Perkin Trans. 1* **1994**, 2431-2437.

10. F. Zanardi, L. Battistini, G. Rassu, M. Cornia, G. Casiraghi, *J. Chem. Soc., Perkin Trans. 1* **1995**, 2471-2475.

11. G. Casiraghi, L. Colombo, G. Rassu, P. Spanu, *J. Chem. Soc., Chem. Commun.* **1991**, 603-604.

12. G. Casiraghi, L. Colombo, G. Rassu, P. Spanu, *J. Org. Chem.* **1991**, *56*, 6523-6527.

13. F. Zanardi, L. Battistini, M. Nespi, G. Rassu, P. Spanu,, M. Cornia, G. Casiraghi, *Tetrahedron Asymmetry* **1996**, *7*, 1167-1180.

14. G. Casiraghi, P. Spanu, G. Rassu, L. Pinna, F. Ulgheri, *J. Org. Chem.* **1994**, *59*, 2906-2909.

15. G. Casiraghi, F. Ulgheri, P. Spanu, G. Rassu, L. Pinna, G. Gasparri Fava, M. Ferrari Belicchi, G. Pelosi, *J. Chem. Soc., Perkin Trans. 1* **1993**, 2991-2997.

16. G. Casiraghi, G. Rassu, P. Spanu, L. Pinna, F. Ulgheri, *J. Org. Chem.* **1993**, *58*, 3397-3400.

17. G. Casiraghi, G. Rassu, F. Zanardi, unpublished results.

18. G. Rassu, L. Pinna, P. Spanu, F. Ulgheri, G. Casiraghi, *Tetrahedron Lett.* **1994**, *35*, 4019-4022.

19. G. Rassu, P. Spanu, L. Pinna, F. Zanardi, G. Casiraghi, *Tetrahedron Lett.* **1995**, *36*, 1941-1944.

20. G. Rassu, F. Zanardi, L. Battistini, E. Gaetani, G. Casiraghi, *J. Med. Chem.* in press.

18 Travelling through the Potential Energy Surface of Sialyl Lewis^x

A. Imberty and S. Perez

18.1 Introduction

The generation of new glycomimetic molecules may be achieved using different approaches, depending on the level of structural information which is available. In the absence of any detailed three-dimensional information about the protein receptor, at least three experimental ways may be considered. i) The rational design which requires the stepwise modifications of the physiologically relevant biomolecule. ii) The pharmacophore design, which requires the identification of key structural features from target molecule, and the identification of these features in any mimic molecule. iii) The library design which requires the synthesis of a large number of variant molecule around a structural theme. Obviously, the design of a « true » glycomimetic requires a reliable three-dimensional model for the carbohydrate pharmacophore to design a ligand, which binds to the 'true' receptor site.

When the three-dimensional structure of the true receptor site is known, the design of a molecule capable of mimicry can be attempted providing that the bioactive conformation is known. Determination of the receptor-bound conformation of a carbohydrate ligand for a receptor of known structure can be made directly through the elucidation of the three-dimensional structure of protein-carbohydrate crystalline complex. An alternative way is to obtain a view of the probable bound conformation by NMR spectroscopy via the measurement of transferred nOe effects on the complex in solution.

In general, carbohydrate molecules are known to exhibit significant conformational flexibility [1]. It has been demonstrated that it is not always the lowest energy minimum of the oligosaccharide which is the one selected by the protein receptor upon binding [2]. It is therefore desirable to determine all the possible conformations of the biologically active oligosaccharide before undergoing any mimetic design.

Since the discovery of its importance as ligand in leukocytes adhesion to endothelial cells, (see review [3]), the tetrasaccharide sialyl Le^x (αNeuNAc(2-3)βGal(1-4)[αFuc(1-3)]βGlcNAc) has been the subject of much interest. A chemo-enzymatic method has been developed for synthesising this oligosaccharide and related compound in large quantities [4]. The challenge is now in finding mimetics with simpler structure, higher affinity for the receptor and better stability against hydrolases (see [5] for examples). The known starting points for such studies are now 1/ the functional groups of importance [6] 2/ the crystal structure of the protein receptor, the E-selectin, in the un-liganded state [7] and 3/ an NMR study of the oligosaccharide conformation when bounded to the protein [8].

The present work addresses the completion of a library of low energy conformers of the tetrasaccharide sialyl Le^x, through an effective exploration of its conformational space, followed by the characterisation of molecular properties of each conformer identified.

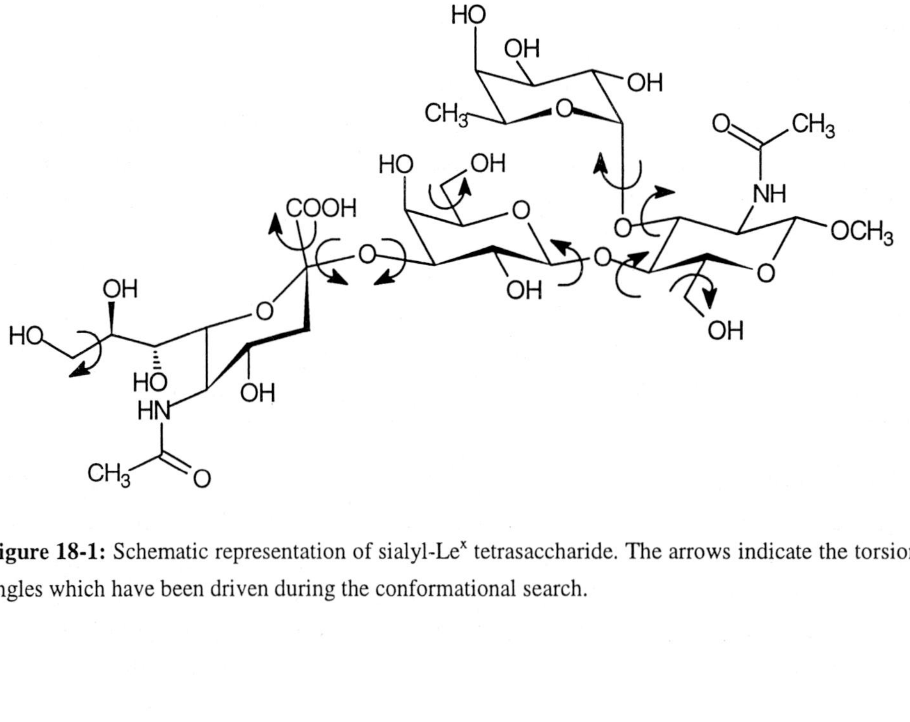

Figure 18-1: Schematic representation of sialyl-Le^x tetrasaccharide. The arrows indicate the torsion angles which have been driven during the conformational search.

The sialyl Le^x tetrasaccharide has been the subject of several NMR and conformational investigations [4, 9, 10, 11]. The difficulty in such conformational analysis resides in the number of degrees of freedom. The previously cited studies used either molecular dynamics or Monte Carlo search. An alternative approach is the use of an heuristic conformational search algorithm such as CICADA (Channels In Conformational space Analyzed by Driver Approach) method [12] interfaced with the MM3 program [13]. The CICADA approach explores the potential energy hyperspace in an intelligent fashion and low CPU consuming way. It has been recently shown to be very efficient for modelling oligosaccharides [14]. It allows for accurate determination not only of the global energy minimum and all secondary minima, but also of the low-energy conversion pathways. The results from CICADA also permit estimation of some flexibility indexes for molecules and for individual group rotations [15].

18.2 Computational Methods

18.2.1 Nomenclature

The torsion angles describing the glycosidic linkages are defined as Φ (O-5—C-1—O-1—C'-X) and Ψ (C-1—O-1—C'-X—C'-X+1). The signs of the torsion angles are defined according to the IUPAC rules [16]

18.2.2 Relaxed Maps of Disaccharides

The conformational space available to the three disaccharides constituting sialyl Le^x has been characterised by a systematic grid search method. Starting from a refined geometry, the procedure drives the two torsion angles of the glycosidic linkage in step of 20° over the whole angular range while the MM3 program provides full geometry relaxation except for the two driven torsion angles. Several maps were computed for each disaccharide in order to take into account the three staggered orientations of each monosaccharide hydroxymethyl group and the two orientations of the NeuNAc acidic group. Also the two possible networks of hydroxyl groups around each ring are considered.

18.2.3 Calculation of Potential Energy Surface Using CICADA

The potential energy hypersurface of the tetrasaccharide was explored by the CICADA program using the MM3 force-field. Several starting structures were used as input for the CICADA analysis with different orientations around the glycosidic torsion angles, based on the low-energy conformations found in the relaxed maps of the corresponding disaccharides. During the CICADA calculations, all glycosidic torsion angles were driven as well as those related to the orientation of the COOH group and of the primary hydroxyl groups. The remaining torsion angles were only monitored. A search with a 20° step was applied for all driven torsion angles. When CICADA detects a minimum, the conformation is fully optimised. The resulting structure is compared with the previously stored ones and stored if not yet detected. Structures corresponding to energy maxima, i.e. the transition states, are also stored.

For the tetrasaccharide, the dimensionality of the conformational hyperspace explored in the CICADA run was therefore 10. In addition, all the hydroxyl groups were monitored, which means that their orientations were taken in account for detecting new energy minima. The run was considered as complete when no new minima were detected in an energy window of 4 kcal/mol above the absolute minimum

18.2.4 Determination of Family of Low Energy Conformers

A home made program for family analysis in N dimensions [17] has been applied on the population of conformations. In this approach, inside a given energy window, a conformer is considered to belong to a family if all its torsion angles differ by no more than a given value to the torsion angles of at least one of the members of the family. In the present study, the value considered for determining the difference has been set to 15°.

18.2.5 Flexibility Indexes

The flexibility values can be expressed as the product of thermodynamic, kinetic and geometrical properties [15].

18.2.6 Calculating Molecular Properties

Calculations of several molecular properties : electrostatic potential, lipophilicity potential and hydrogen bonding sites capacities can be calculated for a given conformation. The lipophilicity potential is based on a modified Fermi type of function, with Crippen atomic partial values [18]. The projection of the molecular properties on the Connolly surface of the molecules has been performed using the MOLCAD option of the SYBYL program [19].

18.3 Results and Discussion

The CICADA run used to explore this 10 dimensions conformational space took 3 weeks computation on a Silicon Graphics R4000 processor. 2619 energy minima were stored, together with 4343 transition states. 34,280 MM3 energy minimisation's were required. This represents a significant computational task but had a grid search method been used over the 6 glycosidic torsion angles of the tetrasaccharide with a 20° increment, more than 34,000,000 energy minimisation's would have been required.

18.3.1 Analysis of the Linkages Conformation

Because of the multidimensional nature of the conformational space presented here, there is no simple fashion to describe the results. For example, the energy values of the conformers (either energy minima or transition states) can be plotted as a function of one torsion angles.

It is also possible to present the projection of the tetrasaccharide conformational hyperspace on the energy maps of the three constituting disaccharides. The orientations of the αNeuNAc(2-3)βGal linkage in the tetrasaccharide conformers, i.e. the energy minima, determined by the CICADA approach correspond closely to the lowest energy regions of the disaccharide map. This indicates that the conformational behaviour of the terminal moiety of the tetrasaccharide is almost not influenced by the presence of the Fuc and GlcNAc residues. NMR and modelling studies of the αNeuNAc(2-3)βGal disaccharide have demonstrated its high flexibility [20]. The flexibility occurs mainly around the Φ torsion angle as shown on Figures 18-2a and 18-3a.

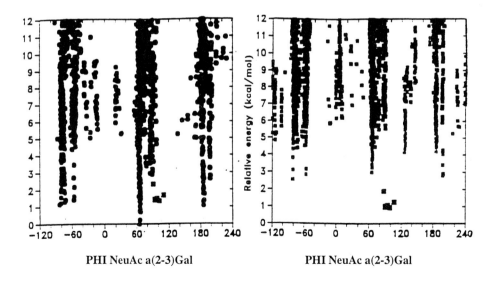

Figure 18-2. Energy of the conformers (a) and the transition states (b) of sialyl Le^x represented as a function of the Φ angle of the αNeuNAc(2-3)βGal linkage.

Analysis of the transition states (Figures 18-2b and 18-3b) gives an estimation of the height of the energy barriers. The projections of the tetrasaccharide transition states (Figure 18-3b) do not match exactly the paths between the minima as indicated by the disaccharide iso-energy contours. This is because the CICADA method analyses more degrees of freedom than the grid search. For example, the N-acetyl group orientation was not varied in the grid search method (since a *trans* orientation of the H-C-N-H segment is known to be the more stable one in all energy minima). However, the CICADA exploration demonstrates that a *cis* orientation of this segment allows to lower the energy of the transition states.

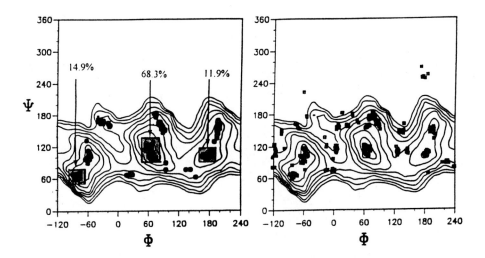

Figure 18-3 : Energy map of the disaccharide αNeuNAc(2-3)βGal with projection of the sialyl Le^x energy minima (a) and transition states (b). The isocontours, with 1 kcal.mol^{-1} separation, represent the energy levels of the disaccharide calculated using the grid search method. The dots and squares represent the energy minima and transition states respectively of the sialyl Le^x tetrasaccharide, within a 8 kcal.mol^{-1} energy window, as detected by the CICADA method. The regions with significant population (more than 3 %) have been framed and labelled.

The energy map of the αFuc(1-3) βGlcNAc disaccharide displays two main low energy regions centred around (Φ=-80, Ψ=80) and (Φ=-80, Ψ=150) respectively. The first one is slightly lower in energy. In the tetrasaccharide, this linkage displays quite a different behaviour. When looking at the CICADA conformers in a 8 kcal.mol^{-1}, all the different regions on the disaccharide maps are sampled. However, only the region centred around (Φ=-80, Ψ=150) is significantly populated. The other minimum, which was the favoured one for the disaccharide, is more than 4 kcal.mol^{-1} above the global minimum when looking at the tetrasaccharide. Such a reduction in flexibility is due to the interaction between the Fuc and the Gal residue. A 'stacking' interaction exists between these two non-bonded residues which stabilises this given conformation. This feature has been already proposed by the many NMR [4, 21] and theoretical studies [22] on the Lewis X trisaccharide and has been recently confirmed by the crystal structure elucidation of this oligosaccharide [23].

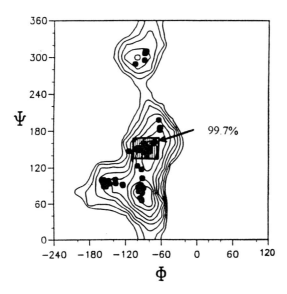

Figure 18-4 : Energy map of the disaccharide αFuc(1-3)βGlcNAc with projection of the sialyl Le^x conformers. See detailed explanations in the legend of Figure 18-3.

The lactosamine linkage of the sialyl Le^x tetrasaccharide displays the same reduction in flexibility as the one observed in the αFuc(1-3)GlcNAc linkage. In this case, only one edge of the main energy well is populated when looking at the tetrasaccharide population. Here also, this conformation results from the non-bonded interactions between Fuc and Gal. Such a feature has also been confirmed by the crystal structure of the Le^x trisaccharide [23].

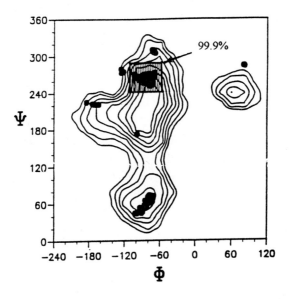

Figure 18-5 : Energy map of the disaccharide βGal(1-4)βGlcNAc with projection of the sialyl Le^x conformers. See detailed explanations in the legend of Figure 18-3.

18.3.2 Families of Conformers and Energy Minima

In all the low energy families, the conformational changes are restricted to the αNeuNAc(2-3)βGal linkage. Variations can exist for the two other linkages (see confG and confH) but they have a small percentage of occurrence (Table 18-1). Nevertheless, they could be more populated depending upon environmental changes. The three main conformational families detected by the analysis are displayed in Figure 18-6.

When looking at the three main conformational families, the one containing the lowest energy conformation (ConfA) is more populated than the two other ones which have energy levels about 1 kcal/mol above it. These three conformations correspond more or less closely to some which have been previously described in the literature [4, 9, 10, 11].

Table 18-1. Description of the main conformational families of the sialyl Le^x tetrasaccharide.

	Ener.	αNeuNAc2-3Gal Φ	αNeuNAc2-3Gal Ψ	αFuc1-3GlcNAc Φ	αFuc1-3GlcNAc Ψ	βGal1-4GlcNAc Φ	βGal1-4GlcNAc Ψ
ConfA : 68 %							
Best conformer	0.00	65.2	113.9	-80.7	150.5	-76.3	-104.1
Limit values		60.6	109.7	-88.5	146.2	-76.9	-105.7
(in 5 kcal.mol-1)		67.5	126.1	-79.2	151.0	-73.0	-98.0
ConfB : 12 %							
Best conformer	1.07	-174.8	108.4	-80.1	149.9	-76.5	-103.4
Limit values		-178.3	108.4	-83.4	147.9	-76.5	-105.7
(in 5 kcal.mol-1)		-172.8	113.1	-79.5	151.1	-72.3	-98.9
ConfC: 15 %							
Best conformer	1.15	-83.7	63.00	-80.6	149.9	-72.60	-105.3
Limit values		-83.7	61.0	-81.9	147.3	-75.1	-107.8
(in 5 kcal.mol-1)		-72.4	68.0	-79.4	150.9	-71.5	-102.4
ConfD: 3 %							
Best conformer	1.42	-56.6	115.0	-79.5	149.4	-77.0	-103.3
Limit values		-59.2	100.8	-81.0	149.3	-77.4	-103.9
(in 5 kcal.mol-1)		-53.8	115.0	-78.7	151.6	-75.6	-101.7
ConfE: 1 %							
Best conformer	2.64	-161.9	164.2	-78.6	147.8	-73.1	-100.7
Limit values		-163.4	147.7	-80.9	147.5	-74.6	-106.3
(in 5 kcal.mol-1)		-161.5	170.3	-78.6	151.3	-70.1	-99.7
ConfF: 1 %							
Best conformer	2.90	88.7	161.1	-79.5	149.5	-73.0	-102.4
Limit values		77.3	156.1	-80.4	147.2	-81.1	-106.2
(in 5 kcal.mol-1)		91.1	184.1	-77.2	152.0	-71.4	-99.5
ConfG: <0.1 %							
Best conformer	3.69	66.0	107.9	-145.2	98.8	-78.7	-97.2
Limit values		62.9	107.9	-154.0	96.6	-82.7	-101.6
(in 5 kcal.mol-1)		66.4	113.7	-145.2	100.1	-78.2	-97.2
ConfH: <0.1 %							
Best conformer	4.01	62.8	118.4	-94.4	85.6	-73.2	71.0
Limit values		60.8	118.4	-94.4	84.1	-80.0	61.0
(in 5 kcal.mol-1)		62.8	122.5	-92.2	92.1	-73.1	72.5

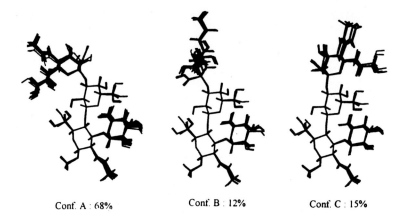

Conf. A : 68%   Conf. B : 12%   Conf. C : 15%

Figure 18-6 : For each of the three lowest energy families, the lowest energy conformers have been represented together with some other conformers of the family (within a energy window of 5/kcal/mol) in order to give an idea of the local flexibility.

Table 18-2. Comparison of literature data on sialyl Le^x conformation. The torsion angles have been all referred to the same convention by adding +/-120°.

	Ener	αNeuNAc2-3Gal		αFuc1-3GlcNAc		βGal1-4GlcNAc	
		Φ	Ψ	Φ	Ψ	Φ	Ψ
Lewis X crystal [23]				-72.5	139.2	-70.5	-104.6
(2 independent mol.)				-76.7	139.2	-80.0	-107.7
ConfA							
This work	0.0	65	114	-81	150	-76	-104
Ichikawa et al.[4]	0.7	41	127	-72	145	-65	-113
Mukhopadhyay [9]	2.6	52	126	-85	140	-60	-113
Hirayama et al. [11]	0.0	55	118	-63	142	-65	-116
Rutherford et al. [10]	0.6	44	115	-74	138	-70	-98
Scheffler et al. [8b]	nmr	44	126	-82	146	-81	-108
ConfB							
Best conformer	1.07	-175	108	-80	150	-76	-103
Ichikawa et al[4]	2.6	-172	100	-72	144	-66	-111
Hirayama et al. [11]	4.2	-178	108	-62	143	-63	-119
ConfC							
This work	1.15	-84	63	-81	150	-73	-105
Ichikawa et al[4]	4.8	-77	63	-72	145	-66	-112
ConfD							
This work	1.42	-57	115	-79	149	-77	-103
Ichikawa et al. [4]	0.0	-50	116	-72	144	-66	-111
Hirayama et al. [11]	0.5	-44	100	-63	142	-64	-117

Table 18-2 presents the comparison between the conformations predicted in this work, and the ones proposed previously by use of NMR and/or molecular modelling studies. Conformer A is now accepted to be probably the 'bioactive' conformation since it corresponds to the transferred nOe experiments observed for the complex with the E-selectin receptor [8]. It is generally predicted to occur in solution, whereas, not all methods predict it to be the most populated conformer.

If it now well established that other conformations are available, specially due to the flexibility of the αNeuNAc(2-3)βGal linkage, it is not easy to determine their respective populations.

18.3.3 Flexibility

The conformational flexibility is clearly mainly located around the glycosidic linkage of the NeuAc a(2-3) βGal moiety. Since the CICADA approach allows to describe the potential hypersurface, including not only the minima but also the transition states, it is possible to calculated flexibility indexes for the torsion angles which have been explored. The following values have been obtained for the sialyl Le^x.

Table 18-3. Fragmental flexibility indexes calculated for the sialyl Le^x tetrasaccharide.

Linkage or residue	torsion	flexibility index
αNeuNAc(2-3)βGal	Φ	41.7
	Ψ	44.8
αFuc(1-3)βGlcNAc	Φ	2.7
	Ψ	2.4
βGal(1-4)βGlcNAc	Φ	3.8
	Ψ	3.7
Gal	CH$_2$OH	23.5
NeuNAc	COOH	5.2
	CH$_2$OH	116.7

From these calculations, the αNeuNAc(2-3)βGal glycosidic linkage appears to be ten fold more flexible than the other ones. As for the substituent, the acidic group of NeuNAc is rather rigid, whereas the hydroxymethyl groups are flexible. The most flexible group is the hydroxymethyl group of NeuNAc. Further experimental studies have to be undertaken in order to validate the model proposed here

18.3.4 Molecular Properties

Lipophilicity is an important molecular property used in rational drug design. Because it expresses steric and polar intermolecular interactions between a solute and biphasic liquid system, it is more likely to mimic the interaction between the physiologically relevant biomolecule and its surrounding protein target. Recent advances in the field [24] have allowed the direct calculation of partitioned coefficient from molecular structures through the use of the Molecular Lipophilicity Potential (MLP). It offers a quantitative 3D computation of lipophilicity thereby taking into account the influence of all molecular fragments on the surrounding space.

The three lowest energy conformations of the most populated families, ConfA, ConfB and ConfC are represented on Plate 1, together with their molecular properties. Because of the conserved conformation of the Le^x core, all these conformers share some features such as a flat face. On such a face, a pattern of three hydroxyl groups capable of participating into hydrogen bonding is found. The conformational variations undergone by the sialic moiety generate not only variations in shapes for the tetrasaccharide but also very important modifications of the molecular properties in this region. The flat character of the molecule is enhanced by the particular orientation of the sialyl moiety in ConfA, from which an extra hydrogen bonding (donating/accepting) centre appears at the surface. Beside, the orientation of the primary hydroxyl group makes for a cavity having the proper geometry for complexing cations such as calcium. In contrast ConfB and ConfC display tortuous and convoluted shapes. The disk like shape of ConfA is very much complementary to the flat surface that E-selectin displays in its putative binding site [7].

Plate 1 (next page). Projection of molecular properties on the Connolly surface of the energy minima of the three lowest energy families, ConfA, ConfB and ConfC (from left to right). a) Electrostatic potential : coloured from blue (negative values) to red (positive values) , b) lipophilicity potential : coloured from red/brown (hydrophobic) to blue (hydrophilic) and c) hydrogen bonding capacity : blue (acceptor) and red (donor)

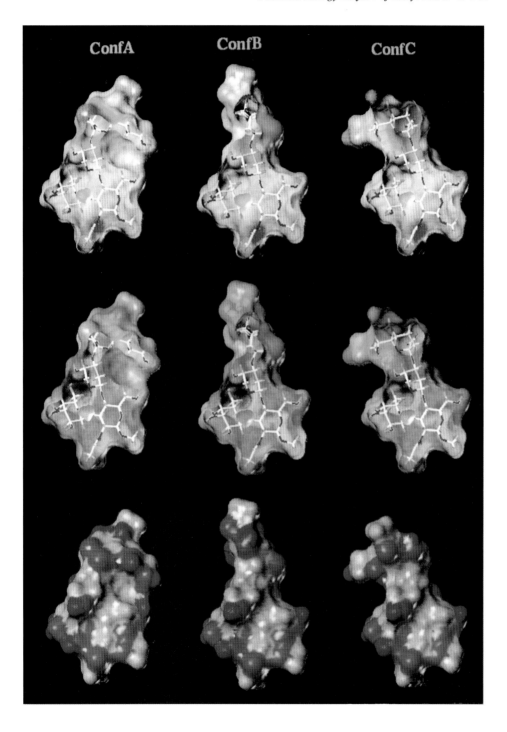

18.4 Conclusion

In summary, we have examined through molecular mechanics calculations the conformational features of the tetrasaccharide sialyl Le^x . The relaxed energy maps generated by MM3 of each of the constituent disaccharides have provided insights into the conformational possibilities and flexibilities of the different glycosidic linkages in the tetrasaccharide. The CICADA analysis of sialyl Le^x demonstrated a total of three conformational families, resulting primarily from the flexibility of the sialyl residues on the Le^x scaffold. The conformation of one of these low energy minima, derived from the modelling, corresponds closely to the conformation that sialyl Le^x takes upon binding to the E-selectin in solution. The analysis of the molecular shape and properties of what could be the « bioactive » conformation reveals some key structural features, such as, a disk-like shape of the molecule on which 5 to 6 anchoring points are located. We suggest that such a «bioactive surface » along with its key binding characteristics should be considered as a template for the generation of new glycomimetic compounds that would regulate the accumulation of leukocytes at sites of inflammation.

18.5 References

1. S. Pérez, *Curr. Opinion Struct. Biol.*, **1993**, *3*, 675-680.; S. Pérez, A. Imberty, J. P. Carver, *Adv. Computat. Biol.*, **1994**, *1*, 147-202.
2. A. Imberty, Y. Bourne, C. Cambillau, P. Rougé, S. Pérez, *Adv. Biophys. Chem.*, **1993**, *3*, 71-118.
3. L. A. Lasky, *Science* **1992**, *258*, 964-969.
4. Y. Ichikawa, Y.-C. Lin, D. P. Dumas, G.-J. Shen, E. Garcia-Junceda, M. A. Williams, R. Bayer, C. Ketcham, L. E. Walker, J. C. Paulson, C.-H. Wong, *J. Am. Chem. Soc.* **1992**, *114*, 9283-9298.
5. B. N. N. Rao, M. B. Anderson, J. H. Musser, J. H. Gilbert, M. E. Schaefer, C. Foxall, B. K. Brandley, *J. Biol. Chem.*, **1994**, *269*, 19663-19666; J. A. Ragan, K. Cooper, *Bioorg. Med. Chem. Lett.*, **1994**, *4*, 2563-2566; T. Uchiyama, V. P. Vassilev, T. Kajimoto, W. Wong, H. Huang, C.-C. Lin, C.-H. Wong, *J. Am. Chem. Soc.*, **1995**, *117*, 5395-5396.
6. D. Tyrrel, P. James, N. Rao, C. Foxall, S. Abbas, F. Dasgupta, M. Nashed, A. Hasegawa, M. Diso, D. Asa, J. Kidd, B. K. Brandley, *Proc. Natl. Acad. Sci. USA*, **1991**, *88*, 10372-10376; J. Y. Ramphal, Z.-L., Zeng, C. Perez, L. E. Walker, S.A. DeFrees, F. C. A. Gaeta, *J. Med. Chem*, **1994**, *37*, 3459-3463.
7. B. J. Graves, R. L. Crowther, C. Chandran, J. M. Rumberger, S. Li, K.-S. Huang, D. H. Presky, P. C. Familletti, B. A. Wolitzky, D. K. Burns, *Nature*, **1994**, *367*, 532-538.

8. R. M. Cooke, R. S. Hale, S. G. Lister, G. Shah, M. P. Weir, *Biochemistry*, **1994**, *33*, 10592-10596; K. Scheffler, B. Ernst, A. Katopodis, J. L. Magnani, W. T. Wong, R. Weisemann, T. Peters, *Angew. Chem. Int. Ed. Engl.*, **1995**, *34*, 1841-1844.
9. C. Mukhopadhyay, K. E. Miller, C. A. Bush, *Biopolymers*, **1994**, *34*, 21-29.
10. T. J. Rutherford, D. G. Spackman, P. J. Simpson, S. W. Homans, *Glycobiology*, **1994**, *4*, 59-68.
11. N. Hirayama, N. Yoda, T. Nishi, *Chem. Lett.*, **1994**, 1479-1482.
12. J. Koca, *J. Mol. Struct. (Theochem)*, **1994**, *308*, 13-24.
13. N. L. Allinger, Y. H. Yuh, J. H. Lii, *J. Am. Chem. Soc.*, **1989**, *11*, 8551-8566.
14. J. Koca, S. Pérez, A. Imberty, *J. Comp. Chem.*, **1995**, *16*, 296-310; A. Imberty, E. Mikros, J. Koca, R. Mollicone, R. Oriol, S. Pérez, *Glycoconj. J.*, **1995**, *12*, 331-349; S. B. Engelsen, S. Pérez, I. Braccini, C. Hervé du Penhoat, J. Koca, *Carbohydr. Res.*, **1995**, *276*, 1-29.
15. J. Koca, *J. Mol. Struct.*, **1993**, *91*, 255-269.
16. IUPAC-IUB Commission on Biochemical Nomenclature, *Arch. Biochem. Biophys.*, **1971**, *145*, 405-421.
17. A. Imberty, S. Pérez, *Glycobiology*, **1994**, *4*, 351-366.
18. V. N. Viswanadhan, A. K. Ghose, G. R. Revankar, R. K. Robins, *J. Chem. Inf. Comput. Sci.*, **1989**, *29*, 163-172; W. Heiden, G. Moeckel, J. Brickmann, *J. Comp.-Aided Mol. Design*, **1993**, *7*, 503-514.
19. SYBYL V6.2, Tripos Associates, 1699 S. Hanley Road, Suite 303, St Louis MO 63144, USA.
20. J. Breg, L. M. J. Kroon-Batenburg, G. Strecker, J. Montreuil, J. F. G. Vliegenthart, *Eur. J. Biochem.*, **1989**, *178*, 727-739; C. Mukhopadhyay, C. A. Bush, *Biopolymers*, **1994**, *34*, 11-20.
21. H. Thogersen, R. U. Lemieux, K. Bock, B. Meyer, *Can. J. Chem.*, **1982**, *60*, 44-57; K. E. Miller, C. Mukhopadhyay, P. Cagas, C. A. Bush, *Biochemistry*, **1992**, *31*, 6703-6709.
22. A. Imberty, E. Mikros, J. Koca, R. Mollicone, R. Oriol, S. Pérez, *Glycoconj. J.*, **1995**, *12*, 331-349.
23. S. Pérez, N. Mouhous-Riou, N. E. Nifant'ev, Y. E. Tsvetkov, B. Bachet, A. Imberty, *Glycobiology* **1996**, *6*, 537-542.
24. P.-A. Carrupt, P. Gaillard, F. Billois, P. Weber, B. Testa, C. Meyer, S. Pérez, in *Methods and Principles in Medicinal Chemistry*, Vol 5, (Eds. : V. Pliska, B. Testa, H. van de Waterbeened) VCH, 195-217.

19 Syntheses of Sulfated Derivatives as Sialyl Lewis^a and Sialyl Lewis^x Analogues

C. Augé, F. Dagron, R. Lemoine, C. Le Narvor and
A. Lubineau[*]

19.1 Introduction

Leukocyte migration into lymphatic tissues or inflammatory sites depends upon the expression of adhesion molecules. Among these molecules the selectin family consists of three members : E, L and P selectins, which share common structural features : an amino terminal lectin domain, an epidermal growth factor-like domain, a transmembrane sequence and a short cytoplasmic tail. E-selectin is transiently expressed on the surface of endothelial cells four hours after stimulation with cytokines, and recognizes carbohydrate ligands on leukocytes. L-selectin is constitutively expressed on most leukocytes and participates in the binding of these cells to the high endothelial venules of peripheral lymph nodes in the normal process of recirculation of lymphocytes, and in the recruitment of leukocytes at the site of inflammation as well. P-selectin stored in the membrane of secretory granules in platelets and endothelial cells, is immediately expressed on the cell surface upon activation by thrombogenic agents, and binds to carbohydrate ligands constitutively expressed on neutrophiles and monocytes. Since the discovery of the selectins, much work has been devoted to the characterization of the carbohydrate ligands involved in the adhesion processes either *in vivo* or *in vitro*. It is hoped that defining the minimum carbohydrate structures with the highest affinity will help to develop a new class of carbohydrate-based drugs as inhibitors of the interaction with selectins, in case of indesirable inflammatory responses.

Early studies have first shown that E-selectin binds to 3'-sialyl Lewis^x (**1**) and 3'-sialyl Lewis^a (**2**) (Figure 19-1) as well [1-5]. Furthermore the three-dimensional structures of sialyl Lewis^x hexasaccharide (**3**) and sialyl Lewis^a hexasaccharide (**4**) have been investigated and it has been shown that the positionning of the NeuAc, Gal and Fuc residues remains constant in both isomeric structures [4,6].

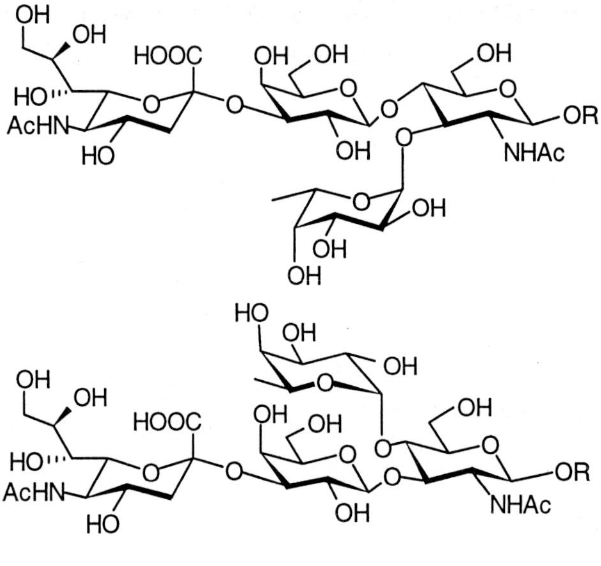

1 R = H
3 R = lactose

2 R = H
4 R = lactose

Figure 19-1. Sialyl Lewis^{a} and sialyl Lewis^{X} tetra- and hexasaccharides.

More recently an equimolar mixture of Lewis^{a} and Lewis^{X} tetrasaccharides **5** and **6**, sulfated at position 3 of the outer galactose and isolated from an ovarian cystadenoma glycoprotein (Figure 19-2) has been clearly established as a ligand for E-selectin with an intensity of binding at least equal to that observed with 3'-sialyl Lewis^{X} [7]. It is worth noting that this preparation of sulfated tetrasaccharides turned out to bind to L-selectin too [8].

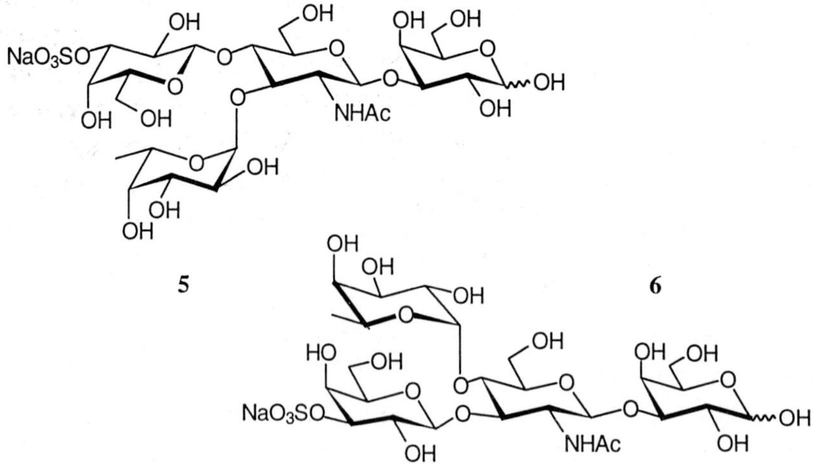

5

6

Figure 19-2. 3'-Sulfated Lewis^{a} and Lewis^{X} tetrasaccharides.

In order to ascertain which of the two sulfated oligosaccharides was the most potent ligand for selectins, we first synthesized the sulfated Lewis[a] tri- and pentasaccharides **7** and **8** [9,10] (Figure 19-3).

19.2 Synthesis of 3'-Sulfated Lewis[a] Trisaccharide and Pentasaccharide

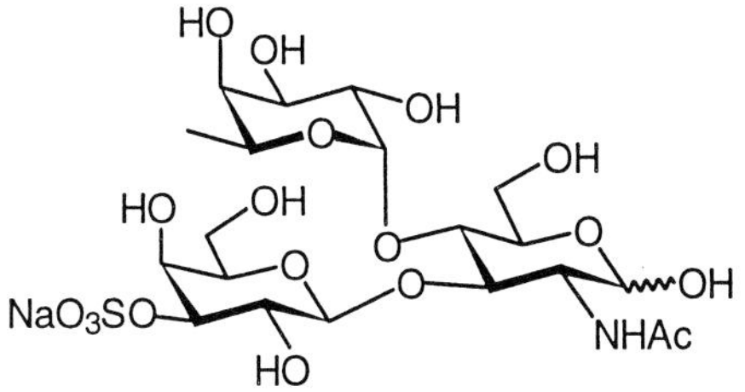

Figure 19-3. 3'-Sulfated Lewis[a] tri- and pentasaccharides.

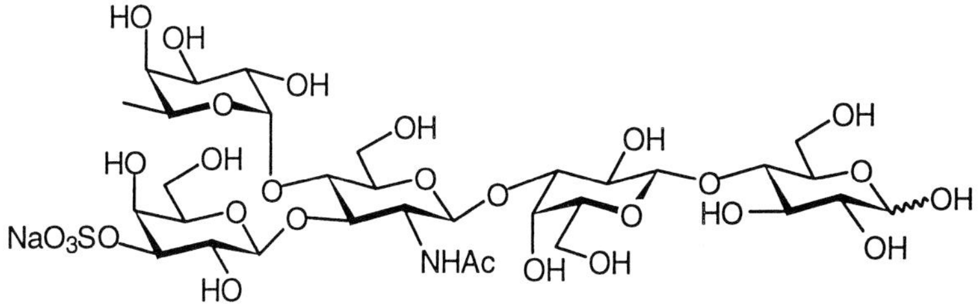

19.2.1 3'-Sulfated Lewis[a] Trisaccharide

The synthesis of the trisaccharide **7** is described in Scheme 19-1. The fully protected trisaccharide **11** was obtained in 89% yield by condensation of perbenzyl fucosyl bromide **10** with the disaccharide **9** under *in situ* anomerization conditions [11]. Two step deallylation (1 : RhCl(PPh$_3$)$_3$, 2 : HgCl$_2$-HgO, acetone-water) afforded compound **12** in 67% yield (90% based on starting material recovery). The sulfate group was then introduced using classical conditions (SO$_3$-NMe$_3$ complex in anhydrous DMF) giving **13** in 82% yield. Complete deprotection (10% Pd/C, H$_2$) afforded the sulfated Lewis[a] trisaccharide **7** in 80% yield after purification by silica gel chromatography, followed by cation exchange using resin (AG50W-X8, Na[+]) and freeze-drying of the aqueous solution.

Molecular dynamics simulations and NMR spectroscopy studies have indicated that Lewis[a], sialylated Lewis[a] and sulfated Lewis[a] trisaccharides adopt the same conformation which places the charged carboxylate group of the neuraminic acid of sialylated Lewis[a] and the charged sulfate group of sulfated Lewis[a] in approximately the same conformational orientation [12].

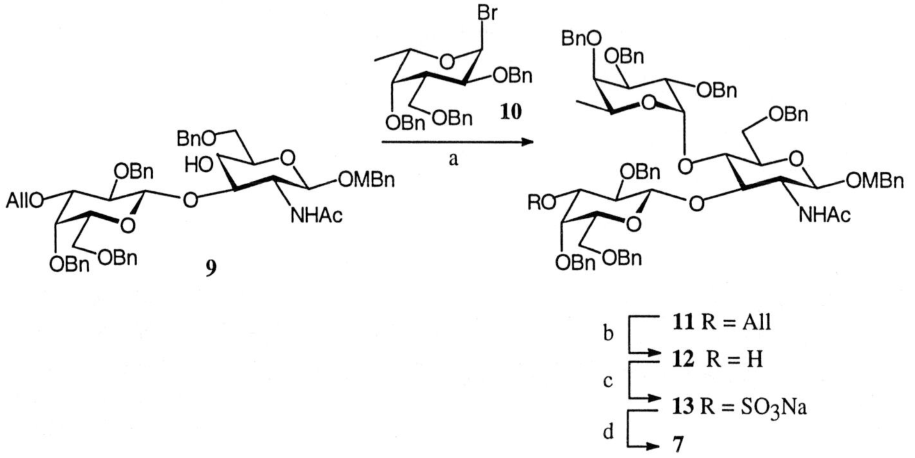

Scheme 19-1. Reagents: a) **10** (3eq), DIPEA (2.7eq), DMF, CH$_2$Cl$_2$, 1:4, 72h, 89%; b) Rh(Ph$_3$P)$_3$Cl, (0.5eq), EtOH-toluene-H$_2$O, 7:3:1, 70°C, 17h then HgO (2eq), HgCl$_2$ (2eq), acetone-H$_2$O, 9:1, rt, 1h, 67%; c) SO$_3$-NMe$_3$ (4eq), DMF, 55°C, 12h, 82%; d) Pd/C, H$_2$, 1atm, EtOH-H$_2$O, 9:1, rt, 60h, 80%.

19.2.2 3'-Sulfated Lewis[a] Pentasaccharide

The synthesis of the pentasaccharide **8** is delineated in Scheme 19-2. Selective removal of the anomeric paramethoxybenzyl group on trisaccharide **11**, performed by treatment with cerium ammonium nitrate, gave **14** (91%) which was then converted to the α-trichloroacetimidate **15** in 75% yield [13].

The crucial coupling of **15** with the lactose derivative **16** in the presence of BF$_3$-Et$_2$O gave the pentasaccharide **17** in 25% isolated yield. After acetylation of the free hydroxyl group giving **18,** removal of the allyl protecting group (67%) gave the hydroxy derivative **19** which was sulfated using classical conditions (SO$_3$-NMe$_3$ complex in anhydrous DMF) giving **20** in 92% yield. Finally deacetylation to **21** followed by hydrogenolysis gave the 3'-sulfated pentasaccharide **8** in 66% overall yield. The pentasaccharide **8** tested by Feizi and coworkers, turned out to be the most potent E-selectin ligand so far [14].

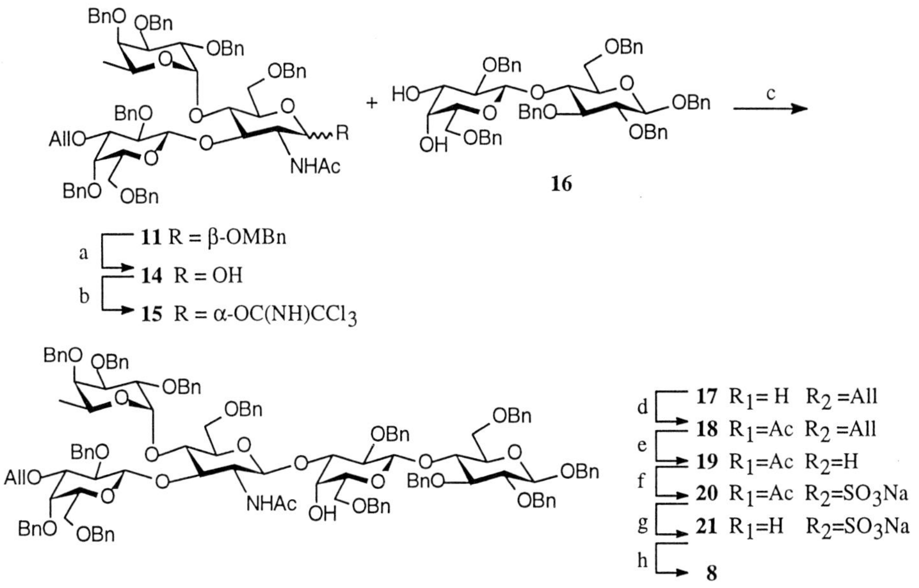

Scheme 19-2. Reagents: a) CAN (5eq), CH_3CN-H_2O, 9:1, -10°C, 10min, 91%; b) NaH, cat, CCl_3CN, (5eq), CH_2Cl_2, 0°C, 1.5h, 75%; c) **16** (4eq), BF_3-Et_2O (0.5eq), CH_2Cl_2, -40°C, rt, 25%; d) Ac_2O, pyr, 1:1; rt, 54h, e) 1. $Rh(Ph_3P)_3Cl$ (1eq), EtOH,-toluene-H_2O, 7:3:1, 700°C, 17.5h, 2. HgO (5eq), $HgCl_2$ (5eq), acetone-H_2O 9:1, rt, 1.5h, 67%; f) SO_3-NMe_3 (7eq), DMF, 55°C, 4.5h, 92%; g) 2M MeONa, MeOH, 60°C, 22h, h) Pd/C (1atm), rt, 9h, 66% (g and h steps).

19.3 Regioselective Sulfation

19.3.1 Stannylene Methodology

As previously shown, the use of sulfur trioxide complexes with pyridine or tertiary amine in solvents such as pyridine or N, N dimethyl-formamide gave good yields of sulfation, provided that only the hydroxyl groups to be sulfated are unprotected ; indeed regioselectivity is usually poor, except in favor of a primary alcohol [15]. Regioselectivity in sugar chemistry is often solved using the well-known stannylene methodology [16]. We therefore turned to this methodology in order to achieve regioselective sulfation. We first started from paramethoxybenzyl 6-O-*tert*-butyldimethylsilyl β-D-galactopyranoside **22** having three secondary unprotected hydroxyl groups at C-2, C-3 and C-4 ; reaction of the intermediate 3,4 di-O-butylstannylene (prepared *in situ* by treatment with dibutyltin oxide in toluene) with SO_3-NMe_3 afforded as the only product the 3-O-sulfate in 92% yield isolated in its peracetylated form **23** after silica gel chromatography (Scheme 19-3).

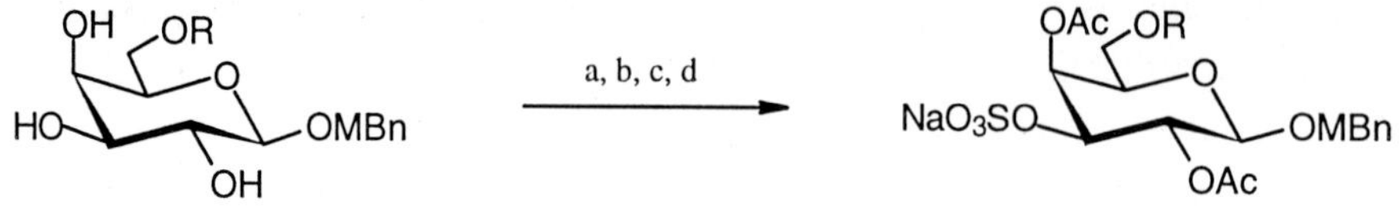

22 R = Si^{t}BuMe$_2$
24 R = H

23 R = Si^{t}BuMe$_2$
25 R = Ac

Scheme 19-3. Reagents: a) Bu$_2$SnO (1.1eq), toluene, 16h, reflux; b) SO$_3$-NMe$_3$ (1.2eq), DMF, 5h, rt; c) Ac$_2$O, pyr. 1-1, 16h, rt; d) AG5OW-X8 Na+, 92% from **22** (a, b, c, d steps), 81% from **24** (a, b, c, d steps).

The same reaction performed on galactoside **24** having four unprotected hydroxyl groups including a primary alcohol led also in a highly regioselective reaction to the 3-O-sulfate (81%) isolated in its peracetylated form **25** [17]. A similar regioselectivity in sulfation of a lactoside, based on the same methodology has been reported independantly [18]. This method was then successfully applied to the synthesis of 3'-sulfated Lewis^a trisaccharide **7**, described in Scheme 19-4, allowing a straightforward access to this compound, much quicker than the previously reported route.

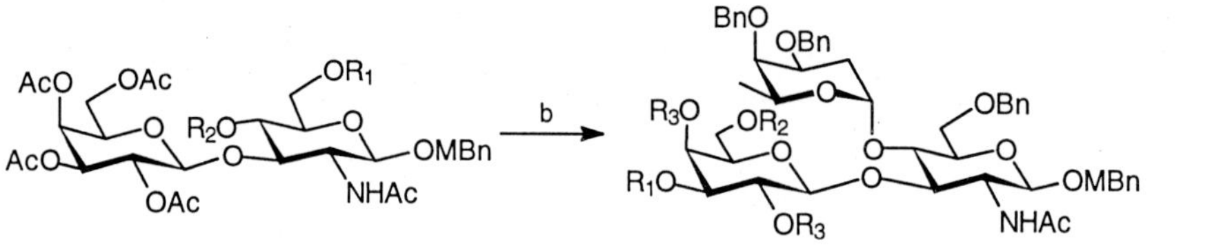

a
26 R$_1$, R$_2$ = CHPh
27 R$_1$ = Bn, R$_2$ = H

28 R$_1$ = R$_2$ = R$_3$ = Ac
29 R$_1$ = R$_2$ = R$_3$ = H
30 R$_1$ = R$_3$ = H, R$_2$ = Si^{t}BuMe$_2$
32 R$_1$ = NaSO$_3$, R$_2$ = Si^{t}BuMe$_2$, R$_3$ = H
7
31 R$_1$ = NaSO$_3$, R$_2$ = R$_3$ = H
7

Scheme 19-4. Reagents: a) NaBH$_3$CN, HCl g, THF, 2h, 0°C, 83%; b) **10**, Bu$_4$NBr, DIPEA, CH$_2$Cl$_2$-DMF 4:1), 18h, rt, 89%; c) NEt$_3$-MeOH-H$_2$O, 20h, rt, 100%; d) 1.Bu$_2$SnO, (1.1eq), tol, 16h, reflux, 2. SO$_3$-NMe$_3$ (1.2eq), DMF, 5h, rt, 69% from **29**, 94% from **30**;.e) tBuMe$_2$SiCl (1.1eq), pyr. 14h, rt, 85%; f) H$_2$ (1atm), 10% Pd/C, 16h, rt, 91%.

The trisaccharide **29** obtained in three high yielding steps from the disaccharide **26** afforded directly the sulfate **31** in 69% yield (80% based on starting material recovery). Alternatively, protection of the primary alcohol by a *tert*-butyldimethylsilyl protecting group gave **30** (85%) that led to the sulfate **32** in 94% isolated yield. Both compounds **31** and **32** were separately transformed into **7** by hydrogenation over 10% Pd/C (91%). It is worth noting that both the *tert*-butyldimethylsilyl and benzyl groups were removed during the reduction step.

19.3.2 Cyclic Sulfates

In the course of our investigation on regioselective sulfation, we were also interested in introduction of sulfate at the 4-position of a partially protected galactoside through cyclic sulfates. According to Sharpless methodology [19] formation of cyclic sulfates **34** and **36** from diols **33** and **35** involved two steps. In the first step each diol was converted into two optically active cyclic sulfites by treatment with thionyl chloride in the presence of a large excess of triethylamine to scavenge the hydrogen chloride generated in the reaction ; in the second step oxidation of the intermediate cyclic sulfites using ruthenium trichloride and sodium periodate led to cyclic sulfates **34** and **36** respectively (Scheme 19-5).

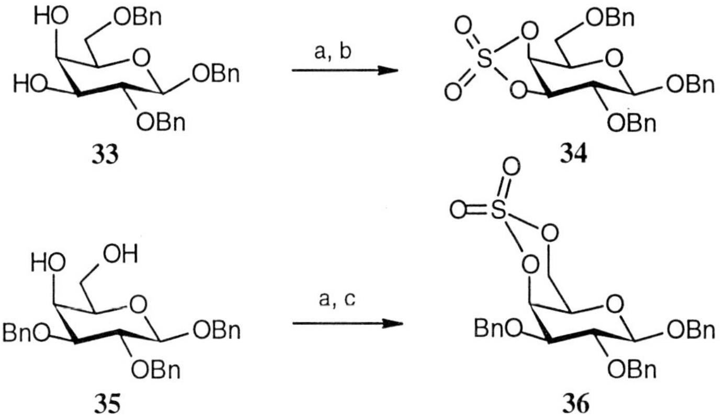

Scheme 19-5. Reagents: a) SOCl$_2$ (1.5eq), NEt$_3$ (4eq), CH$_2$Cl$_2$, 30mn, 0°C; b) RuCl$_3$, xH$_2$O (0.5%), NaIO$_4$ (2eq), CCl$_4$-CH$_3$CN-H$_2$O (2:2:3), 1.5h, 0°C, 91% (a and b steps); c) RuCl$_3$, xH$_2$O (2%), NaIO$_4$ (2eq), CCl$_4$-CH$_3$CN-H$_2$O (2:2:3), 4h, 0°C, 70% (a and c steps).

Both steps were achieved in one pot in good to excellent yields (**34** 91%, **36** 70%). Whereas nucleophilic ring opening of cyclic sulfates occurs at carbon atom with inversion of configuration [20] , monohydrolysis resulting from treatment in alcaline medium was attempted. Thus the 5-membering cyclic sulfate **34**, heated with aqueous sodium hydroxyde in THF afforded in 83%

yield a mixture of both monosulfated galactose derivatives **37** and **38** in a 68 : 32 ratio (see Scheme 19-6). On the other hand such a ring-opening in alkaline medium was not observed with the 6-membering cyclic sulfate **36** : under the above conditions, a mixture of two elimination products **39** and **40** (in a 35 : 65 ratio) was obtained in 93% yield. However the nucleophilic ring-opening of cyclic sulfate **36** was successfully achieved by treatment with tetrabutyl-ammonium benzoate in DMF at room temperature : attack on the primary carbon afforded quantitatively compound **41**, which was debenzoylated to give the monosulfate **42** (Scheme 19-6).

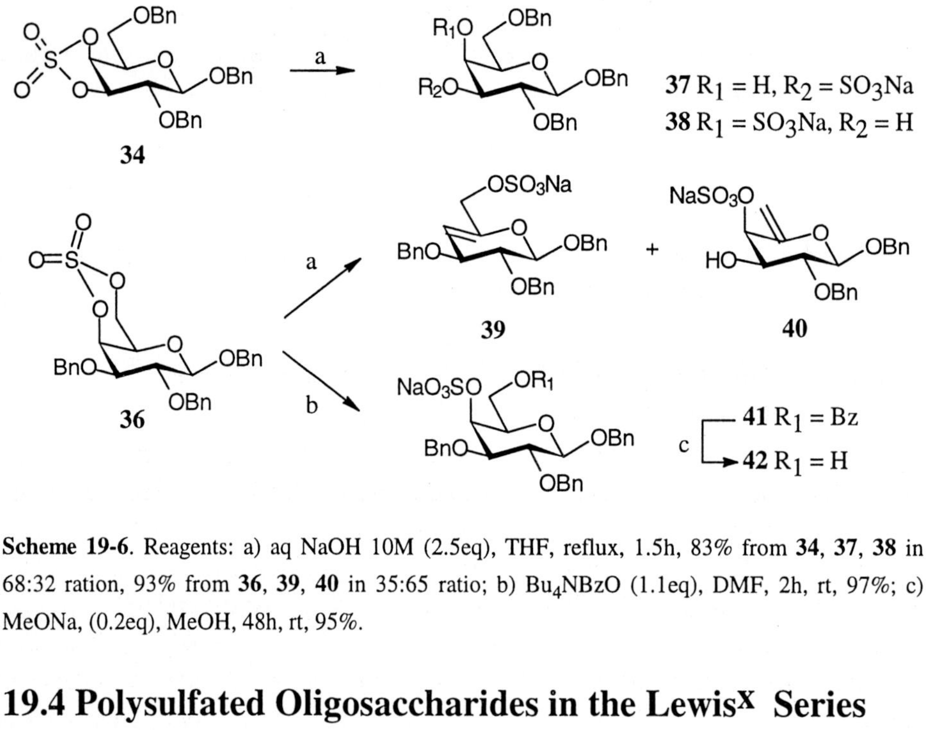

Scheme 19-6. Reagents: a) aq NaOH 10M (2.5eq), THF, reflux, 1.5h, 83% from **34**, **37**, **38** in 68:32 ration, 93% from **36**, **39**, **40** in 35:65 ratio; b) Bu$_4$NBzO (1.1eq), DMF, 2h, rt, 97%; c) MeONa, (0.2eq), MeOH, 48h, rt, 95%.

19.4 Polysulfated Oligosaccharides in the Lewis^X Series

19.4.1 Introduction

Recent studies concerning the *in vivo* specificity of L-selectin have led to partial characterization of the oligosaccharide structures of its biological ligand, the GlyCAM-1 glycoprotein, occuring in mouse lymph nodes. The 6'-sulfated sialyl Lewis^X (**43**) and 6-sulfated sialyl Lewis^X (**44**) sequences have been evidenced as the major capping groups of three O-linked chains of GlyCAM-1 [21,22] (Figure 19-4). However the 6-sulfated sialyl Lewis^X pentasaccharide (**44**, R$_3$ =

galactose), enzymatically prepared, has been reported as a weak inhibitor (IC_{50} 0.8 mM) although superior to sialyl Lewis^x tetrasaccharide (**1**, R = H), in the adhesion of L-selectin [23]. Likewise the chemically synthesized 6-sulfated sialyl Lewis^x tetrasaccharide (**43**, R_3 = H), turned out not to show greatly better inhibition than **1** [24,25]. Based on these findings we were interested in the synthesis of mono- and di-sulfated Lewis^x pentasaccharides as potential ligands for L-selectin.

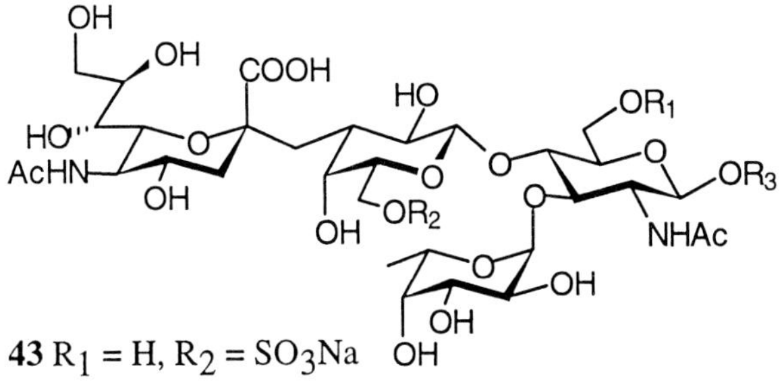

43 R_1 = H, R_2 = SO_3Na
44 R_1 = SO_3Na, R_2 = H

Figure 19-4. 6 and 6'-Sulfated sialyl Lewis^x sequences evidenced as the major capping groups of O-linked chains of the glyCAM-1 glycoprotein.

19.4.2 6'-Sulfated, 3'-Sulfated, 3',6'-Disulfated Lewis^x Pentasaccharides

The activated trisaccharide **45** was condensed onto the lactose derivative **16** according to a block synthesis, similar to that previously described in the Lewis^a serie. The synthesis of compound **45** is depicted in Scheme 19-7. Starting from the disaccharide **46** [26], or alternatively from the N-phtalimido derivative **47**, regioselective ring opening with sodium cyanoborohydride and HCl in THF afforded disaccharides **48** and **49** respectively in 81% and 74% yield. The fully protected trisaccharide **51** was obtained by condensation of peracetyl galactosyl trichloroacetimidate **50** with the disaccharide **49** in the presence of trimethylsilyl triflate at low temperature, whereas trisaccharide **52** was obtained by condensation of compound **50** with **48** in the presence of boron trifluoride etherate at room temperature, both compounds **51** and **52** in 77% yield.

Selective removal of the anomeric paramethoxybenzyl group as previously described, gave **53** (92%) and **54** (85%) which were then converted into the β-trichloroacetimidate **45** (92%) and the α-trichloroacetimidate **55** (69%) [13]. Coupling of **55** with **16** in the presence of BF$_3$-Et$_2$O afforded after deacetylation the pentasaccharide **56** in a low yield (24%) (Scheme 19-8). On the other hand, coupling of the phtalimido trisaccharide **45** with **16** under the same conditions afforded as expected the pentasaccharide **57** in a much better yield (70%) [27]. Conversion of the phtalimido group into the acetamido group according to standard procedure, followed by O-deacetylation led to the above pentasaccharide **56** in 77% yield.

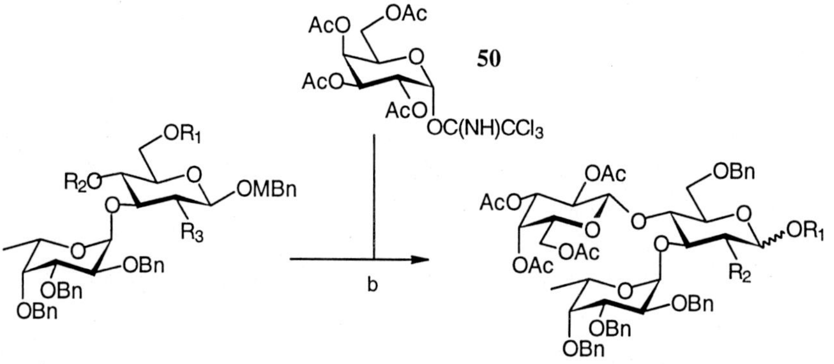

46 R_1, R_2 = CHPh
a
48 R_1 = Bn, R_2 = H, R_3 = NHAc

47 R_1, R_2 = CHPh, R_3 = NPht
a
49 R_1 = Bn, R_2 = H, R_3 = NPht

51 R_1 = β-MBn, R_2 = NPht
c
53 R_1 = H, R_2 = NPht

52 R_1 = β-MBn, R_2 = NHAc
c
54 R_1 = H, R_2 = NHAc
d
55 R_1 = α-C(NH)CCl$_3$, R_2 = NHAc

45 R_1 = β-C(NH)CCl$_3$, R_2 = NPht

Scheme 19-7. Reagents: a) NaBH$_3$CN, HClg, THF, 2h, 0°C, 81% form **46**, 74% from **47**; b) **50** (2eq), BF$_3$-Et$_2$O (0.5eq) CH$_2$Cl$_2$, 12h, rt, 77% for **52** or **50** (2eq) TMSOTf (0.1eq), CH$_2$Cl$_2$, 3h, -70°C to -30°C, 76% for **51**; c) CAN (10eq), CH$_3$CN/H$_2$O (9:1), 5min, -5°C, 85% for **54**, 92% for **53**; d) NaH (cat), CCl$_3$CN (5eq), CH$_2$Cl$_2$, 2h, 0°C, 92% for **45**, 69% for **55**.

Regioselective sulfation on the primary alcohol was then achieved by treatment of **56** with SO$_3$-NMe$_3$ complex in DMF-pyridine mixture affording, in 73% yield, the pentasaccharide **58**, which was further deprotected by hydrogenolysis to give the 6'-sulfated Lewis^x pentasaccharide **59** in quantitative yield (Scheme 19-9). The 3'-sulfated Lewis^x pentasaccharide **61** was synthesized according to the stannylene methodology.

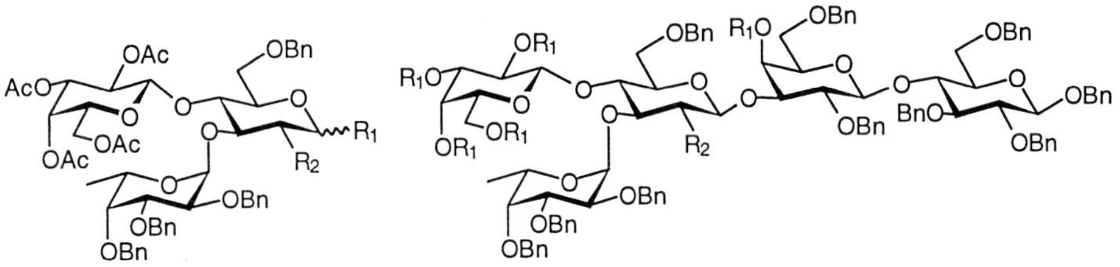

45 R$_2$ = NPht, R$_1$ = β-O-C(NH)CCl$_3$ $\xrightarrow{\text{b}}$ 57 R$_1$ = Ac, R$_2$ = NPht ⎤ c
55 R$_2$ = NHAc, R$_1$ = α-O-C(NH)CCl$_3$ $\xrightarrow{\text{a}}$ 56 R$_1$ = H, R$_2$ = NHAc ⎦

Scheme 19-8. Reagents: a) 1. **16** (4eq), BF$_3$-Et$_2$O (0.5eq), CH$_2$Cl$_2$, 0.5h, -50°C, 2. NEt$_3$-MeOH-H$_2$O, 12h rt, 24%; b) **16** (2eq), BF$_3$-Et$_2$O (0.5eq), CH$_2$Cl$_2$, 0.5h, -50°C, 70%; c)1. NH$_2$NH$_2$, EtOH, reflux, 2. Ac$_2$O, pyr., 3. MeONa, MeOH, 77%.

Treatment of pentasaccharide **56** with dibutyltin oxide in toluene afforded the intermediate 3',4'-di-O-butyl stannylene which was then allowed to react with SO$_3$-NMe$_3$ complex in DMF, giving after silica gel chromatography the pentasaccharide **60** in 73% yield. Conventional deprotection led to 3'-sulfated Lewis^x pentasaccharide **61** in quantitative yield.

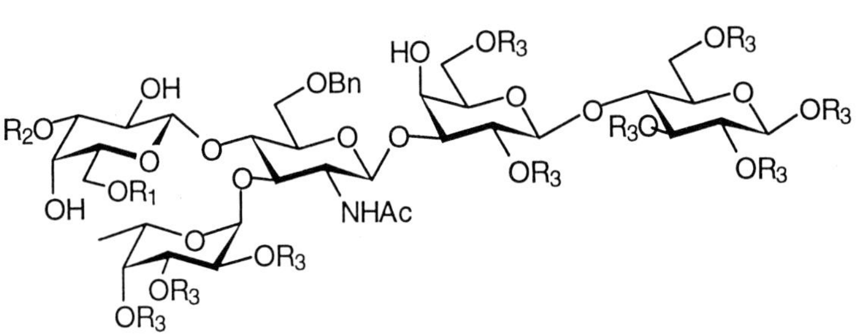

56 R$_1$ = H, R$_2$ = H, R$_3$ = Bn
58 R$_1$ = SO$_3$Na, R$_2$ = H, R$_3$ = Bn
59 R$_1$ = SO$_3$Na, R$_2$ = H, R$_3$ = H
60 R$_1$ = H, R$_2$ = SO$_3$Na, R$_3$ = Bn
61 R$_1$ = H, R$_2$ = SO$_3$Na, R$_3$ = H
62 R$_1$ = SO$_3$Na, R$_2$ = SO$_3$Na, R$_3$ = Bn
63 R$_1$ = SO$_3$Na, R$_2$ = SO$_3$Na, R$_3$ = H

Scheme 19-9. Reagents: .a) 1. SO$_3$-NMe$_3$ (3eq), DMF-pyr (1.5:1) 3h, 40°C, 2. AG50W-X8 Na$^+$, 73%; b) 1. Bu$_2$SnO (1.1eq), toluene, 15h, reflux, 2. SO$_3$NMe$_3$ (1.1eq), 3.AG50W-X8, Na$^+$; c) 1. Bu$_2$SnO (1.1eq), toluene, 15h, reflux, 2. SO$_3$-NMe$_3$ (6eq) DMF, 4h, 40°C, 3. AG50W-X8, Na$^+$, 78%; d) 1. H$_2$ (1atm), 10% Pd/C, 60h, rt, 2. AG50W-X8, Na$^+$, 95%.

A related strategy allowed the straightforward synthesis of 3',6'-disulfated pentasaccharide **62** (Scheme 19-9) ; the stannylene methodology was first applied to the pentasaccharide **56** affording under the above described conditions the intermediate monosulfated derivative **60** which after further addition of SO3-NMe3 complex was completely converted into the disulfate **62** ; both steps were achieved in one pot affording after silica gel chromatography **62** in 78% yield, which was quantitatively deprotected yielding the 3',6'-disulfated Lewis^x pentasaccharide **63**. NMR spectroscopic data showing characteristic downfield shifts for the carbon atoms bearing sulfate group and the adjacent protons, confirmed the structures.

19.4.3 3',6-Disulfated Lewis^x Pentasaccharide

The synthesis of 3',6-disulfated Lewis^x pentasaccharide was achieved according to a completely different approach. In a first experiment the 3'-sulfated Lewis^x trisaccharide glycoside **64** was successfully synthesized in two steps from β benzyl lactosamine **65** as shown in Scheme 19-10. First the free disaccharide **65** was regioselectively monosulfated at the 3'-position through the stannylene methodology, affording in 83% yield compound **66** that was enzymatically fucosylated in the second step, using crude soluble human milk α (1-3/4) fucosyltransferase and stoichiometric amount of GDP-Fucose chemically synthesized [28]. The trisaccharide **64** was thus obtained in 80% isolated yield.

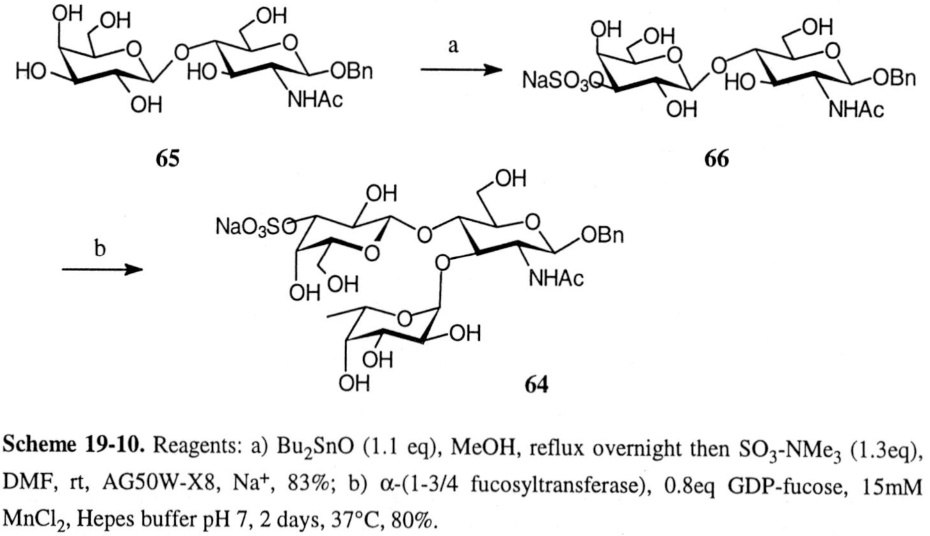

Scheme 19-10. Reagents: a) Bu$_2$SnO (1.1 eq), MeOH, reflux overnight then SO$_3$-NMe$_3$ (1.3eq), DMF, rt, AG50W-X8, Na$^+$, 83%; b) α-(1-3/4 fucosyltransferase), 0.8eq GDP-fucose, 15mM MnCl$_2$, Hepes buffer pH 7, 2 days, 37°C, 80%.

This result prompted us to consider the same chemo-enzymatic approach undoubtedly shorter than the chemical one for the synthesis of the 3',6-

disulfated Lewis^x pentasaccharide **67**. Condensation of the trichloroacetimidate **70** with the disaccharide **69** (obtained from the diol **68** via the intermediate 3,4 orthoester [29] in the presence of trimethylsilyl triflate gave the trisaccharide **71** (77%) (Scheme 19-11). Conversion of phtalimido group into acetamido under standard conditions afforded the trisaccharide **72** which was treated with SO$_3$-NMe$_3$ complex in DMF yielding the 6-monosulfated trisaccharide **73** (75%). Selective removal of benzyl groups in the presence of paramethoxy benzyl protecting group could be achieved by treatment with palladium hydroxyde and cyclohexene in methanol under reflux for 60 h giving the trisaccharide glycoside **74** in 70% yield. This trisaccharide was enzymatically galactosylated using β (1-4) galactosytransferase, UDP-galactose epimerase, alkaline phosphatase and stoichiometric amount of UDP-glucose, in cacodylate buffer (pH 7.5) in the presence of MnCl$_2$ (Scheme 19-12). The tetrasaccharide **75** was isolated in 84% yield.

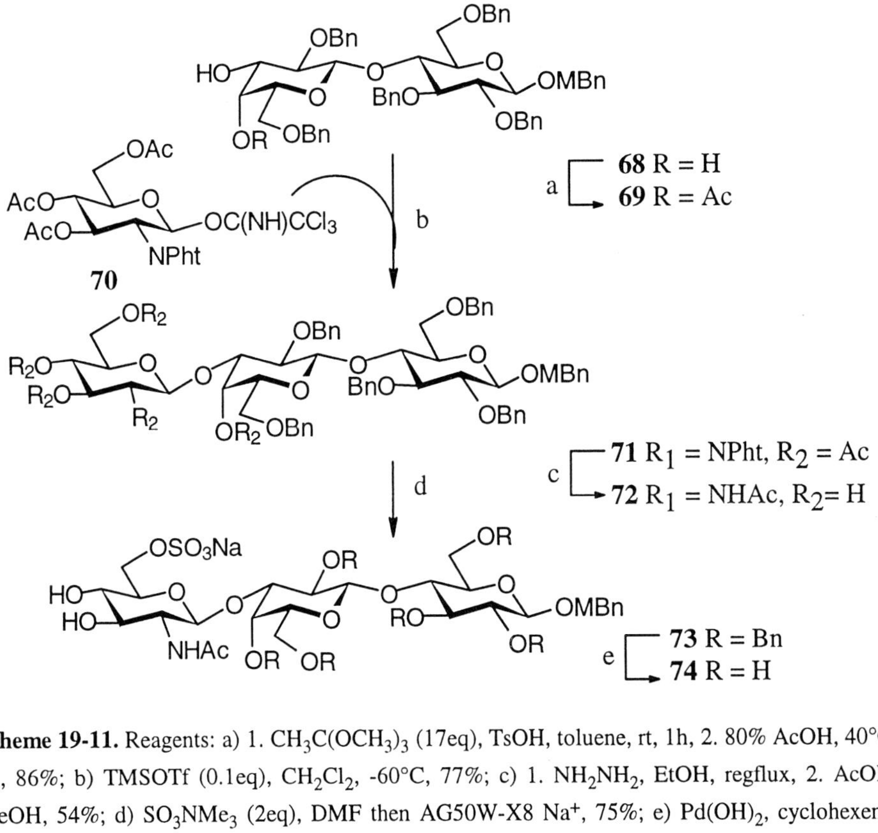

Scheme 19-11. Reagents: a) 1. CH$_3$C(OCH$_3$)$_3$ (17eq), TsOH, toluene, rt, 1h, 2. 80% AcOH, 40°C, 1h, 86%; b) TMSOTf (0.1eq), CH$_2$Cl$_2$, -60°C, 77%; c) 1. NH$_2$NH$_2$, EtOH, regflux, 2. AcOH, MeOH, 54%; d) SO$_3$NMe$_3$ (2eq), DMF then AG50W-X8 Na$^+$, 75%; e) Pd(OH)$_2$, cyclohexene, MeOH, reflux then AG50W-X8 Na$^+$, 70%.

The second sulfate group was then successfully introduced into **75** having 11 free hydroxyl groups, through the stannylene procedure as described for disaccharide **65**. The 3'-6 disulfated tetrasaccharide **76** was thus obtained in 60% yield.

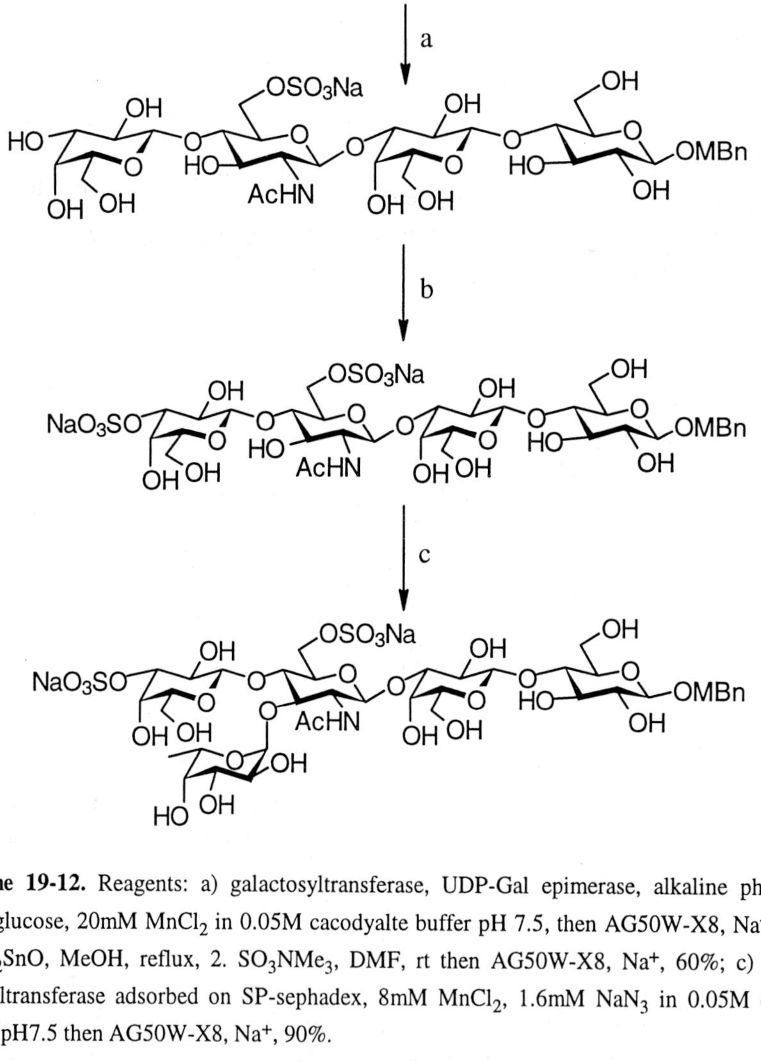

Scheme 19-12. Reagents: a) galactosyltransferase, UDP-Gal epimerase, alkaline phosphatase, UDP-glucose, 20mM MnCl$_2$ in 0.05M cacodyalte buffer pH 7.5, then AG50W-X8, Na$^+$, 84%; b) 1. Bu$_2$SnO, MeOH, reflux, 2. SO$_3$NMe$_3$, DMF, rt then AG50W-X8, Na$^+$, 60%; c) α-(1-3/4)-fucosyltransferase adsorbed on SP-sephadex, 8mM MnCl$_2$, 1.6mM NaN$_3$ in 0.05M cacodylate buffer pH7.5 then AG50W-X8, Na$^+$, 90%.

Finally **76** could be converted into the pentasaccharide **77** (90%) using human milk α (1-3/4) fucosyltransferase adsorbed on SP-Sephadex, in the presence of alkaline phosphatase, MnCl$_2$ and GDP-Fucose added portionwise.

Catalytic hydrogenation afforded the disulfated pentasaccharide target **67** in quantitative yield (Scheme 19-12).

19.5 Immunological studies

The 3' sulfated Lewis^a and Lewis^x oligosaccharides have been tested by Pr Ten Feizi *et al*, towards E-and L-selectins.

19.5.1 Specificity of the E-Selectin

The sulfated oligosacharides were investigated in two ways [14]. First, when converted into neoglycolipids immobilized on plastic microwells, they were evaluated for their ability to be bound by the CHO cells transfected to express high level of the E-selectin. In these binding assays the 3'-sulfated Lewis^a pentasaccharide **8** turned out a much better ligand than sialyl Lewis^a and sialyl Lewis^x analogues as shown in Figure 19-5.

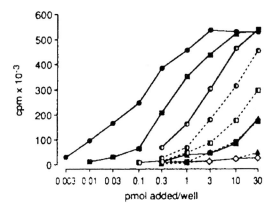

Figure 19-5. Quantitative binding assays using E-selectin expressing SC2 cells and lipid-linked oligosaccharides immobilized on plastic microcells. Symbols for Le^a -related sequences are joined by solid lines and those for Le^x -related sequences are joined by dotted lines. ● sulfated Le^a pentasaccharide; ■ sulfated Le^a (or Le^x) tetrasaccharide; ▲ sulfated Le^a (or Le^x) trisaccharide; ○ sialyl-Le^a (or Le^x) pentasaccharide; ▫ sialyl-Le^x tetrasaccharide; ✧ lacto-*N*-tetraose; reprinted from *J. Biol. Chem.*, **1994**, *269*, 1595-1598, by permission of the Journal Editorial Office.

Second, in an unusual sensitive assay, the free sulfated oligosaccharides were tested as inhibitors in the adhesion of the expressing E-selectin cells to sialyl Lewis^a pentasaccharide immobilized on plastic microwells. In these inhibition assays the sulfated Lewis^a pentasaccharide **8** and tetrasaccharide **6** were the most potent inhibitors (Figure 19-6) with IC$_{50}$ values (5.6 x 10^{-8} and 7.0 x 10^{-8} respectively) 5 and 15 times greater than those of the sialyl Lewis^a

and sialyl Lewis^X pentasaccharides and 100 times greater than that of the sulfated Lewis^X tetrasaccharide.

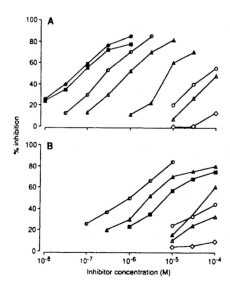

Figure 19-6. Inhibition assays using free oligosaccharides (Le^a-related in panel A and Le^X-related in panel B) as inhibitors of the binding of E-selectin expressing SC2 cells to sialyl-Lewis^a pentasaccharide immobilized in plastic microwells; for symbols see Figure 19-5; reprinted from *J. Biol. Chem.*, **1994**, *269*, 1595-1598, by permission of the Journal Editorial Office.

Inhibition assays were also performed using the sialyl Lewis^X pentasaccharide and the sulfated Lewis^a pentasaccharide as coats. The comparison of the inhibitory activities of the sulfated Lewis^a pentasaccharide **8** and the sialyl Lewis^X trisaccharide **1** is reported in Table 1-1.

Table 1-1. Comparisons of the inhibitory activities of sulfated Lewis^a pentasaccharide and sialyl Lewis^X trisaccharide.

Lipid-linked oligosaccharide coat	Inhibitor	
	Sulfated-Le^a pentasaccharide	Sialyl-Le^X trisaccharide
	Concentration (M) giving 50% inhibition	
Sialyl-Le^a pentasaccharide	5.6×10^{-8}	2.6×10^{-6}
Sialyl-Le^X pentasaccharide	4.8×10^{-8}	7.5×10^{-7}
Sulfated Le^a pentasaccharide	5.4×10^{-7}	1.9×10^{-5}

Again in all cases the concentrations of the oligosaccharides required to give 50% inhibition of binding were the lowest with the sulfated Lewis^a pentasaccharide.

19.5.2 Specificity of the L-Selectin

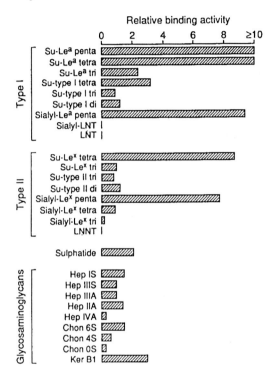

Figure 19-7. Relative L-selectin binding activities of the lipid-linked oligosaccharides investigated in the quantitative microwell binding assay. Abbreviations for oligosaccharide sequences are those used in ref. [30]; reprinted from *Glycobiology*, **1995**, *5*, 29-38, by permission of Oxford University Press.

The binding specificity of the recombinant soluble form of the rat L-selectin towards the 3'-sulfated Lewis^a and Lewis^x oligosaccharides has been also investigated [30]. These oligosaccharides were tested as neoglycolipids in microwell binding assays. Their activities were compared with those of sialyl Lewis^a and sialyl Lewis^x analogues, non fucosylated analogues, and also glycosaminoglycan derived sulfate oligosaccharides. The relative binding activities of all of the compounds tested are shown in Figure 19-7. The best activities were observed for both sulfated Lewis^a pentasaccharide **8** and tetrasaccharide **6**, slightly superior to the ones observed for the sialyl Lewis^a pentasaccharide, the sulfate Lewis^x tetrasaccharide and the sialyl Lewis^x pentasaccharide. It is worth noting that the sialylated derivatives do not tolerate the absence of fucose for binding, whereas binding still occurs with sulfated derivatives lacking fucose. Moreover the fact that glycosaminoglycans could act as ligands could be of biological significance in inflammatory lesions of glycosaminoglycan-rich tissues such as cartilage or cornea [30].

Acknowledgments

All biological evaluations of the compounds here described have been performed by Pr Ten Feizi and collaborators in the Glycosciences Laboratory, Imperial College of Medecine, Northwick Park Hospital, Watford Road, Harrow, Middlesex, HA1 3UJ U.K. They are fully acknowledged for these studies and fruitful discussions.

19.6 References

1. J. B. Lowe, L. M. Stoolman, R. P. Nair, R. D. Larsen, T. L. Berhend, R. M. Marks, *Cell* **1990**, *63*, 475-484.

2. M. L. Phillips, E. Nudelman, F. C. A. Gaeta, M. Perez, A. K. Singhal, S.-I. Hakomori, J. C. Paulson, *Science* **1990**, *250*, 1130-1132.

3. G. Walz, A. Aaruffo, W. Kolanus, M. Bevilacqua, B. Seed, *Science* **1990**, *250*, 1132-1135

4. E. L. Berg, M. K. Robinson, O. Mansson, E. C. Butcher, J. L. Magnani, *J. Biol. Chem.* **1991**, *266*, 14869-14872.

5. M. Larkin , T. J. Ahern, M. S. Stoll, M. Shaffer, D. Sako, J. O'Brien, C-T. Yuen, A. M. Lawson, R. A. Childs, K. M. Barone, P. R. Langer-Safer, A. Hasegawa, M. Kiso, G. R. Larsen, T. Feizi, *J. Biol. Chem.* **1992**, *267*, 13661-13668.

6. J. L. Magnani, *Glycobiology* **1991**, *1*, 318-320.

7. C. -T. Yuen, A. M. Lawson, W. Chai, M. Larkin , M. S. Stoll, A. C. Stuart, F. X. Sullivan, T. J. Ahern, T. Feizi, *Biochemistry* **1992**, *31*, 9126-9131.

8. P. J. Green, T. Tamatani, T. Watanabe, M. Miyasaka, A; Hasegawa, M. Kiso, C.-T. Yuen, M. S. Stoll, T. Feizi, *Biochem. Biophys. Res. Commun.* **1992**, *188*, 244-251.

9. A. Lubineau, J. Le Gallic, R. Lemoine, *J. Chem. Soc., Chem. Commun* **1993**, 1419-1420.

10. A. Lubineau, J. Le Gallic, R. Lemoine, *Biorg. Med. Chem.* **1994**, *2 ,* 1143-1151.

11. R. U. Lemieux, H. Driguez, *J. Am. Chem. Soc.* **1975**, *97*, 4063-4069.

12. H. Kogelberg, T. J. Rutherford, *Glycobiology* **1994**, *4*, 49-57.

13. R. R. Schmidt, *Angew. Chem. Int. Ed. Engl.* **1986**, *25*, 212-235.

14. C. -T. Yuen, K. Bezouska, J. O'Brien, M. S. Stoll, R. Lemoine, A. Lubineau, M. Kiso, A. Hasegawa, N. J. Bockovich, K. C. Nicolaou, T. Feizi, *J. Biol. Chem.* **1994**, *269*, 1595-1598.

15. J. R. Turvey, *Adv. Carbohydr. Chem.* **1965**, *20*, 183-218.

16. S. David, S. Hanessian, *Tetrahedron* **1985**, *41*, 643-663.

17. A. Lubineau, R. Lemoine, *Tetrahedron Lett.* **1994**, *35*, 8795-8796.

18. B. Guilbert, N. J. Davis, S. L. Flitsch, *Tetrahedron Lett.* **1994**, *35*, 6563-6566.

19. B. M. Kim, K. B. Sharpless, *Tetrahedron Lett.* **1989**, *30*, 655-658.

20. B. B. Lohray, *Synthesis* **1992**, 1035-1051.

21. S. Hemmerich, S. D. Rosen, *Biochemistry* **1994**, *33*, 4830-4835.

22. S. Hemmerich, H. Leffler, S. D. Rosen, *J. Biol. Chem.* **1995**, *270*, 12035-12047.

23. P. R. Scudder, K. Shailubhai, K. L. Duffin, P. R. Streeter, G. S. Jacob, *Glycobiology* **1994**, *4*, 929-933.

24. R. K. Jain, R. Vig, R. Rampal, E. V. Chandrasekaran, K. L. Matta, *J. Am. Chem. Soc.* **1994**, *116*, 12123-12124.

25. P. Crottet, Y. J. Kim, A. Varki, *Glycobiology* **1996**, *6*, 191-208.

26. G. Balavoine, S. Berteina, A. Gref, J-C. Fischer, A. Lubineau, *J. Carbohydr. Chem.*. **1995**, *14*, 1237-1249.

27. S. Sato, Y. Ito, T. Nukada, Y. Nakahara, T. Ogawa, *Carbohydr. Res.* **1987**, *167*, 197-210.

28. B. M. Heskamp, H. J. G. Broxterman, G.A. van der Marel, J.H. van Boom, *J. Carbohydr. Chem.*. **1996**, *15*, 611-622.

29. G. V. Reddy, R. K. Jain, B. S. Bhatti, K. L. Matta, *Carbohydr. Res.* **1994**, *263*, 67-77.

30. P. J. Green, C. T. Yuen, R. A. Childs, W. Chai, M. Miyasaka, R. Lemoine, A. Lubineau, B. Smith, H. Ueno, K. C. Nicolaou and T. Feizi, *Glycobiology* **1994**, *5*, 29-38.

20 Architectonic Neoglycoconjugates: Effects of Shapes and Valencies in Multiple Carbohydrate-Protein Interactions

D. Zanini and R. Roy

20.1 Introduction

The search for ligands of high affinity and specificity is at the forefront of activity in a large number of glycobiological investigations [1]. Traditional methods of structure-activity relationship (SAR) have led to the development of effective inhibitors for various carbohydrate-protein interactions. These strategies, while they constitute valuable tools with which to explore lead compounds, neglects another class of interactions. These interactions generally occur when extracellular proteins bind to carbohydrates.

Carbohydrate-protein interactions have critical roles in numerous biological functions such as cancer cell metastasis, fertilization, pathogenic infections and inflammation [2]. Despite their importance, individual carbohydrate-protein interactions are often of low affinity with dissociation constants ($K_{D}s$) in the mMolar range [3]. It is now recognized that multiple carbohydrate-protein interactions may cooperate in some recognition events to amplify affinity, thereby explaining the roles that these interactions play.

These clusters or cooperative effects rely on the arrangement of multiple receptors in such a way as to bind to multiple carbohydrates having well organized architectures. It is known that numerous receptors have clustered carbohydrate recognition domains [4-6]. Consequently, neoglycoproteins [7-9] and glycopolymers [8, 10] have been used successfully to demonstrate that multivalency does indeed amplify carbohydrate-protein binding interactions by factors as high as thousands. Figure 20-1 illustrates possible ways in which this could happen [4-6].

Still, neoglycoproteins and glycopolymers, by their very nature, have ill defined chemical structures. They are heterogeneous in size and carbohydrate content. Thus, while they can demonstrate the role of multivalency in recognition processes, they fail to allow precise biophysical analyses of the cluster effect.

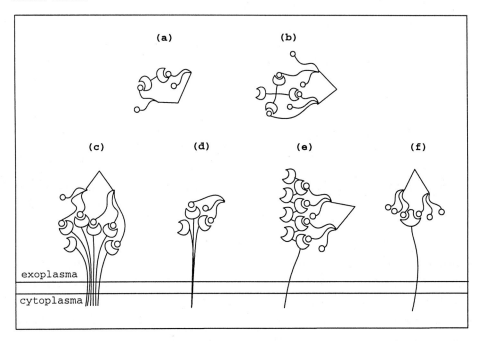

Figure 20-1. Amplification of specific carbohydrate-protein recognition processes via multivalent interactions: (a) and (b) soluble di- and tetra-valent receptors, *e.g.* galectins and phytohemagglutinins, (c) clustered monovalent receptors at the cell surface, *e.g.* hepatic asialoglycoprotein receptor, (d) and (e) oligomeric receptors, *e.g.* macrophage mannose receptor, and (f) receptors that bind more than one carbohydrate simultaneously.

Chemically well defined glycoconjugates with systematically varied shapes and carbohydrate densities would constitute useful tools for a complete understanding of these important recognition processes. In addition, the use of chemically well defined glycoconjugates would offer better insights into the geometrical organization of the receptors under investigation.

Recently, much work has been devoted to the design and synthesis of such neoglycoconjugates. Compounds spanning from glycoclusters, with as little as two or three conjugated carbohydrate haptens, to spherical glycodendrimers, with numerous surface carbohydrate residues, have been reported. These novel glycoconjugates, shown schematically in Figure 20-2, are being employed to probe these multiple interactions and the roles they play. This chapter will focus on current syntheses of neoglycoconjugates with varying shapes and carbohydrate valencies.

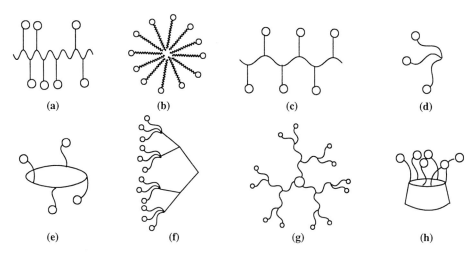

Figure 20-2. Multivalent carbohydrate ligands: possible shapes and valencies. (a) glycopolymers, (b) liposomes, (c) glycotelomers and glycopeptoids, (d) glycoclusters, (e) cyclic glycopeptides, (f) glycodendrimers (dendrons), (g) spherical glycodendrimers, and (h) cyclodextrins and glycocalixarenes.

20.2 Neoglycoproteins and Glycopolymers

Although it is not the intention of this chapter to focus on the synthesis of neoglycoconjugates with variable compositions, any manuscript describing the cluster effect would be remiss in not ascribing importance to neoglycoproteins, neoglycolipids, and glycopolymers that have given credence to the role of multivalency. Many reviews currently exist on the subject [7-16].

Neoglycoproteins, neoglycolipids, and glycopolymers represent classical multivalent carbohydrate clusters. Increased binding avidity of these synthetic glycoconjugates have led to their use as models in a number of carbohydrate-protein interaction studies [17, 18], vaccines [19], inhibitors of cell adhesions by viruses, bacteria, mycoplasma, and toxins [20], ligands in affinity chromatography [21, 22], and as drug carriers [23].

Many strategies exist for the preparation of glycopolymers [10-16] among which polysialosides have been extensively studied as Influenza virus hemagglutination inhibitors. Polymers with O- [24-28], S- [29] and C- [30, 31] α-sialosides have been made using copolymerization or grafting techniques. Other strategies too have been employed to generate polymers containing more complex sialyloligosaccharides including sialyl Lewis^x [32], sialyl Lewis^a [33], 3'-sulfo-Le^X-(Glc) [34, 35], and even polymerized liposomes [36].

Because neoglycoproteins and neoglycolipids are still widely used and also because glycopolymers are gaining recognition, it is strategically appealing to synthesize the necessary glycan precursors in forms suitable for the preparation of all of these kinds of neoglycoconjugates [10-12, 37, 38]. This strategy has been followed in our laboratory. In this common precursor methodology (Scheme 20-1), N-acryloylated derivatives such as **3**, prepared by conventional glycosylation chemistry, were used. These were coupled to proteins such as bovine serum albumin (BSA) or tetanus toxoid via Michael-type additions of ε-amino groups of the proteins onto the N-acryloylated monomer to give a T-antigen-protein conjugate (*e.g.* T-antigen-BSA **4**). T-Antigen derivative **3** was also copolymerized with acrylamide to give T-antigen-based copolymer **5**. Glycoconjugates **4** and **5** have been instrumental in the preparation of human monoclonal antibodies against a breast cancer marker composed of T-antigen (Galβ-(1,3)-GalNAcα) [39].

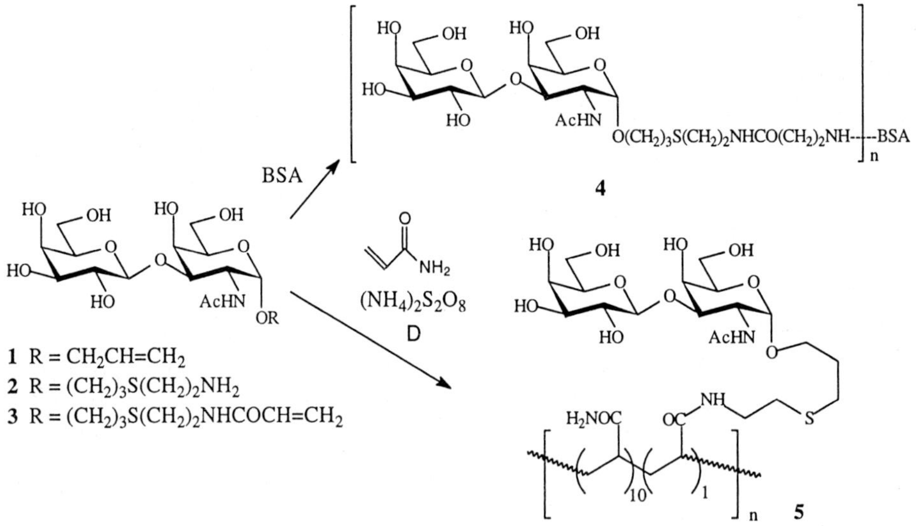

Scheme 20-1. Common precursor strategy for the synthesis of neoglycoproteins and copolyacrylamides containing T antigen.

The above methodology was used to synthesize potent multivalent sialoside inhibitors (**6** and **7**) of the hemagglutination of human erythrocytes by influenza A and C viruses [29, 40] and other more complex glycopolymers containing lactose (**8**) [41], N-acetyllactosamine (**9**), lacto-N-tetraose (**10**) [42], GM$_3$ saccharide (**12**) [43], and a 3'-sulfo-Lewis^X-(Glc) analog (**11**) [34] (Scheme 20-2). Water soluble copolyacrylamide **11** combined glycomimetic and multivalent strategies to inhibit binding of both L- and E-selectins in the µMolar range. Glycopolymer **12** is the most potent L- and E-selectin inhibitor presently available and demonstrates that SAR and multivalent strategies may work cooperatively to give ligands with greater binding affinities.

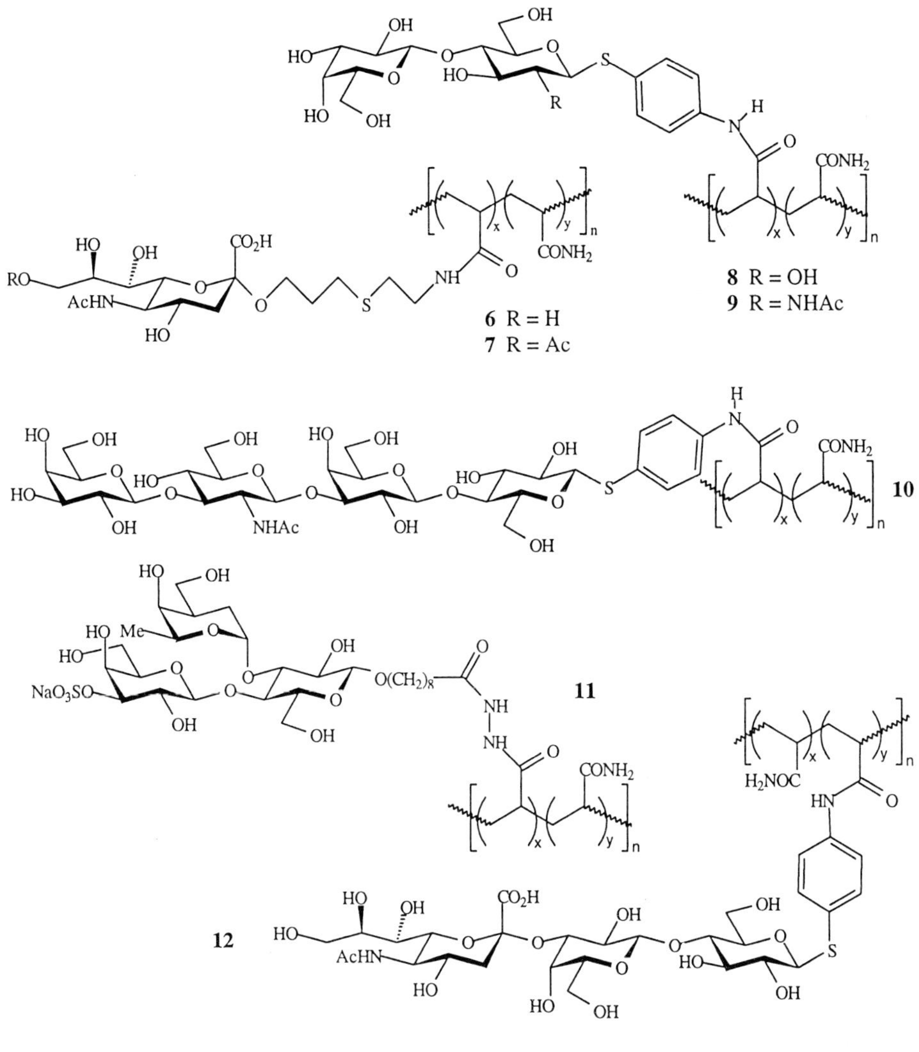

Scheme 20-2. Examples of copolyacrylamide glycopolymers.

20.3 Glycotelomers

Telomers (or oligomers) are short homopolymers of less than ≈10-12 residues [44]. Glycotelomers represent linear glycosidic clusters of discrete carbohydrate densities. They resemble glycopolymers in that they are linearly arranged, hypervalent glycoconjugates, but they differ in that carbohydrate densities are now controllable. As stated above, a thorough understanding of the cluster effect requires chemically well defined molecules with known relative positioning and carbohydrate content.

Very few examples of glycotelomers exist so far and this is surprising given that they are readily available in single step reactions via the quenching of polymerization reactions by radical scavengers (or telogens) such as thiols. The number of repeating units is easily controllable by altering the concentration of the thiol in the initial reaction mixtures.

Using lactose as a model, two series of telomers incorporating short and long spacer arms between the telomer backbones have been synthesized (Scheme 20-3) [45, 46].

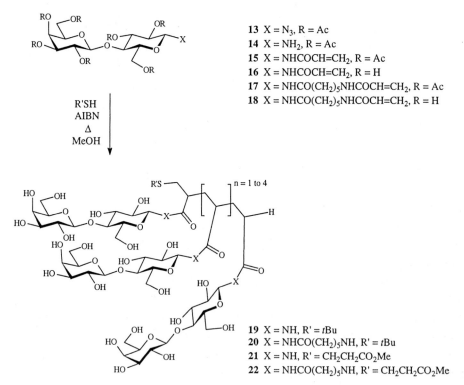

Scheme 20-3. Synthesis of lactose-containing glycotelomers.

Reduction of lactosyl azide **13**, obtained under phase transfer catalysis [47, 48], afforded anomerically unstable lactosylamine **14** which was rapidly treated with either acryloyl chloride or 6-acrylamidohexanoic acid to give monomer precursors **15** and **17**. After de-O-acetylation, monomers **16** and **18** were telomerized in the presence of either *tert*-butyl mercaptan or methyl mercaptoprorionate to provide families of discrete telomers **19-22** separable by size exclusion chromatography [45, 46]. These telomers were only slightly better ligands than lactose in the solid-phase inhibition assays between peanut lectin and a polymeric lactoside used as model coating antigen [45].

20.4 Glycopeptoids

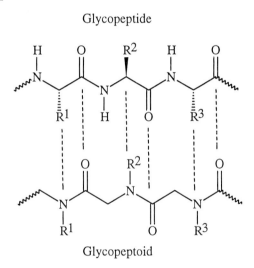

Figure 20-3. Structural similarities between glycopeptides and glycopeptoids.

In keeping with the design of discrete, linear multivalent glyco-conjugates, attention has been given to the synthesis of glycopeptoids. Peptidomimetics are known to be metabolically stable, as compared to peptides. Peptoids are N-substituted oligoglycines (NGs) whose carbonyl and side chain residues are superimposable to those of natural peptides (Figure 20.3). Oligomeric units, readily prepared by a convergent blockwise approach using orthogonally protected derivatives [49-53], yielded novel multivalent structures. In addition, secondary amide linkages at every branching point generated libraries of conformational rotamers (E/Z) which may be used to screen a wide area of multivalent receptor topographies (induced fit).

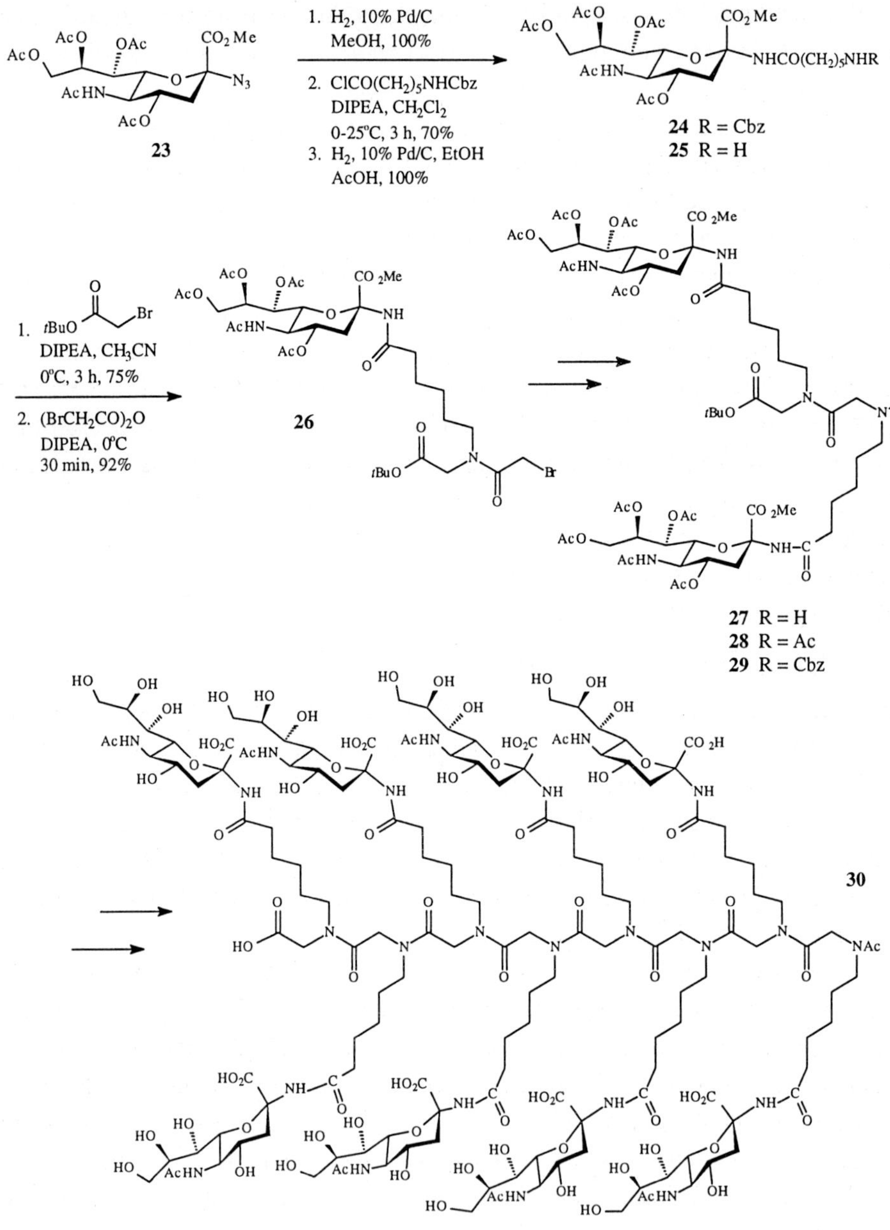

Scheme 20-4. Synthesis of multivalent α-sialopeptoids.

The synthesis of sialic acid-based peptoids has been used as an example (Scheme 20-4). In the first model, each N-linked sialoside residue was interspaced by one glycine unit and 6-aminocaproic acid was used as spacer in order to distance the sialic acids away from the backbone [54].

α-Sialosyl azide **23** [55] was derivatized into protected amine derivative **24** which provided free amine **25** after hydrogenolysis (Scheme 20-4). Amine **25**, by treatment with *tert*-butyl bromoacetate followed by N-bromoacetylation (bromoacetic anhydride) was transformed into N-substituted key building block **26**. Alkylation of **26** with **25** gave orthogonally protected dimer **27**. The secondary amine of **27** was then N-acetylated using acetyl chloride to provide end group dimer **28** or alternatively transformed into N-benzyloxycarbonyl protected derivative **29** (CbzCl). Sequential deprotection of *tert*-butyl ester and Cbz group followed by amide coupling of the amine and acid fragments using dicyclohexylcarbodiimide (DCC) afforded trimer, tetramer, hexamer, and octamer **30** after protecting group removal under standard basic conditions (Scheme 20-4). These sialoside-containing glycopeptoids are currently undergoing biological investigation.

20.5 Glycoclusters

While the optimum valency requirement for efficient binding may be established by linear neoglycoconjugates of known carbohydrate densities, such as those described above, other factors too must be considered. The requirements for conformations and geometries, such as bond angles and intramolecular glycosidic distances, must be considered in the study of carbohydrate-protein interactions. This understanding is possible via the use of glycosidic clusters of different sizes and shapes.

Much work on the design and synthesis of galactosyl, sialosyl, and mannosyl clusters has been done (reviewed in [37, 56]). Investigations by Lee and Lee [57-59] on rat and rabbit hepatocytes have resulted in the generation of potent trivalent GalNAc inhibitors. Continued studies by Kichler and Schuber [60] on trivalent galactosyl clusters further confirmed that trivalent oligosaccharides containing Gal/GalNAc moieties situated at the apexes of a triangle with sides 1.5, 2.2, and 2.5 nm constituted the optimum geometry for rat and rabbit hepatocyte receptors [61].

By adjusting the intra-sialosyl distance in a number of divalent clusters, Glick and Knowles [62, 63] have obtained a sialosyl conjugate in which the two sialic acids were 5.7 nm apart. The best dimer was 100 fold more potent than methyl α-sialoside in influenza virus X-31 inhibitions.

Other sialyl Lewis^X clusters have been synthesized, either enzymatically or chemically. 1,4-Butanediol [64], 1,5-pentanediol [64], galactoside [64], and nitromethane-trispropionic acid [65] have all been used successfully as scaffolding elements to demonstrate that multivalent interactions are involved in selectin binding processes. Unverzagt *et al.* [66] have branched sialyl-α-(2,6)-β-LacNAc dimers at different positions of synthetic peptides, including compact glycine-rich and helical proline-rich peptides, to give clusters which were eight and four fold more potent over the corresponding monovalent trisaccharide, respectively.

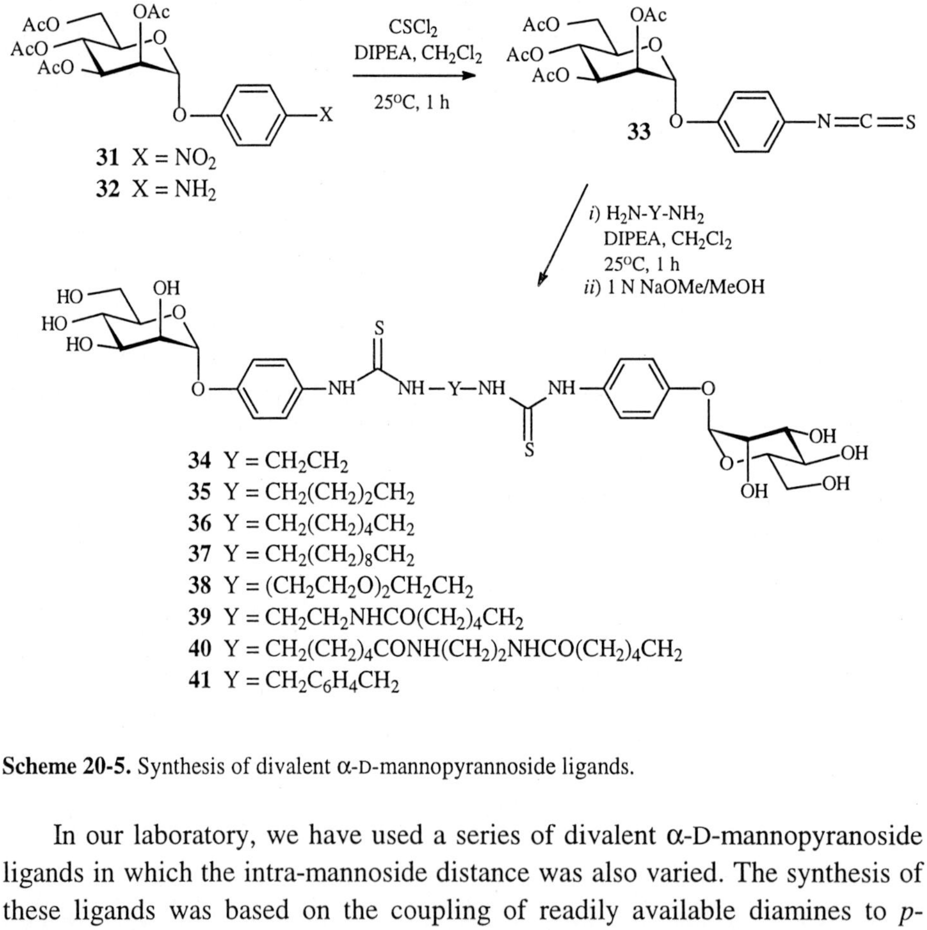
AcO
AcO
AcO
OAc
O
O
X
31 X = NO2
32 X = NH2
CSCl2
DIPEA, CH2Cl2
25°C, 1 h
AcO
AcO
AcO
OAc
O
33
O
N=C=S
i) H2N-Y-NH2
DIPEA, CH2Cl2
25°C, 1 h
ii) 1 N NaOMe/MeOH
HO
HO
HO
OH
O
S
O
NH
NH—Y—NH
NH
O
S
OH
OH
OH
OH
34 Y = CH2CH2
35 Y = CH2(CH2)2CH2
36 Y = CH2(CH2)4CH2
37 Y = CH2(CH2)8CH2
38 Y = (CH2CH2O)2CH2CH2
39 Y = CH2CH2NHCO(CH2)4CH2
40 Y = CH2(CH2)4CONH(CH2)2NHCO(CH2)4CH2
41 Y = CH2C6H4CH2

Scheme 20-5. Synthesis of divalent α-D-mannopyrannoside ligands.

In our laboratory, we have used a series of divalent α-D-mannopyranoside ligands in which the intra-mannoside distance was also varied. The synthesis of these ligands was based on the coupling of readily available diamines to *p*-isothiocyanatophenyl 2,3,4,6-tetra-*O*-acetyl-α-D-mannopyranoside **33** (Scheme 20-5) [67]. These dimers were tested for their relative binding potency by solid-phase enzyme linked lectin assays (ELLA, described in Section 20.10) in the interaction between yeast mannan and Concanavalin A (Con A) lectin, using methyl α-D-mannopyranoside as a standard. The results showed dimers **34** to **41** to be ≈10 to 90 fold more potent than methyl α-D-mannopyranoside (Table 20-1). Mannosylated cluster **36**, having a spacer length of six carbon units, was found to be the most effective inhibitor. Dimers with shorter or longer spacers than that of **36** were less efficient [67].

These data indicate that cooperative binding interactions depend not only on the overall number of carbohydrate ligands, but also on their relative positioning with respect to each other.

Table 20-1. Inhibition of binding of yeast mannan to Concanavalin A by divalent α-D-mannopyranoside ligands [63].

Compound	IC$_{50}$ (μM)	Relative Potency^a to Me α-D-Man	Relative Potency^a to pNO$_2$-Ph α-D-Man
D-Mannose (α,β)	>2500	-	-
Me α-D-Man	924	1.0	0.1
pNO$_2$-Ph α-D-Man	106	8.8	1.0
34	38 (76)	24.3 (12.2)	2.8 (1.4)
35	28 (56)	33.2 (16.6)	3.8 (1.9)
36	10 (20)	88.9 (44.5)	10.6 (5.3)
37^b	28 (56)	33.6 (16.8)	3.8 (1.9)
38	47 (94)	19.7 (9.9)	2.3 (1.2)
39	41 (82)	22.5 (11.3)	2.6 (1.3)
40	45 (90)	20.7 (10.4)	2.4 (1.2)
41	49 (98)	18.8 (9.4)	2.2 (1.2)

^{a} Values in parentheses are expressed relative to monomeric content.
^{b} Compound dissolved in 2% DMSO solution.

20.6 Cyclic Glycopeptides

With methods of discovering necessary valencies and geometries comes the task of combining these requirements. Conjugates that optimize valency and intra-glycosidic distance are presently being synthesized.

One approach is the design of cyclic glycopeptides. Sprengard et al. [68] have synthesized novel glycoconjugates containing three sialyl Lewis^x (sLe^x) residues. The peptidic sequence was Ala$_{(1)}$-Asp$_{(3)}$-Ser-Asp$_{(2)}$-D-Ala$_{(2)}$-Asp$_{(1)}$-Gly-OH and the three aspartic acid residues were coupled with an aminated sLe^x precursor. The number of aspartic acid residues controlled the valency, while the aminated sLe^x precursor could be modified to adjust the spacer length, i. e. the intra-glycosidic distance. These conjugates exhibited IC$_{50}$s in the 0.35-

0.6 mM range when tested as inhibitors against E-selectin-Ig fusion proteins. This represents a 2 to 3 fold increase as compared to the sLe^x monomer.

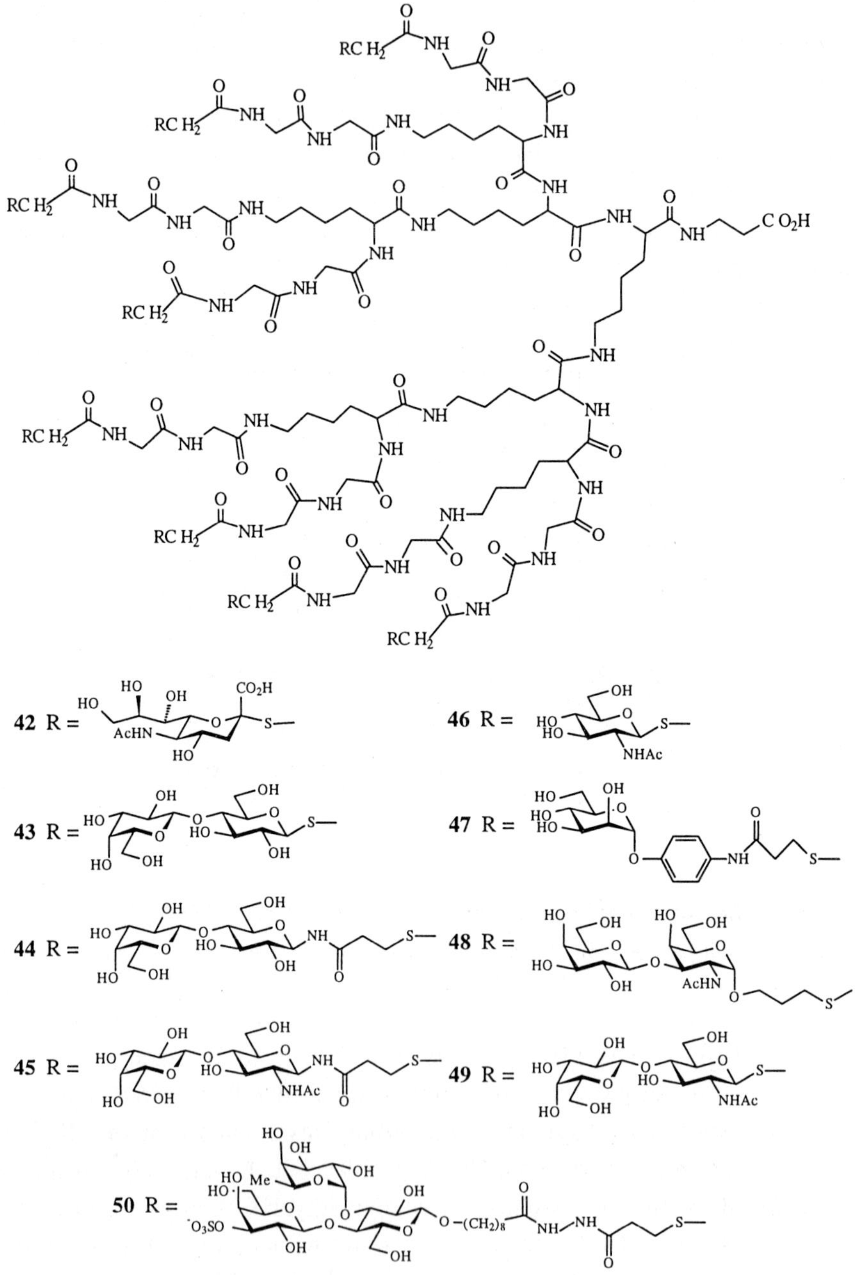

Scheme 20-6. Typical structures of the first glycodendrimers based on an L-lysine core.

20.7 Glycodendrimers

Another approach for combining valency, conformation, and geometric requirements stems from the preparation of glycodendrimers. Bi-directional glycodendrimers or dendrons, are multibranched structures with covalently attached carbohydrate residues. They represent novel, monodisperse biopolymers with known glycoside densities. They bridge the gap between small clusters and large polymers, since they are generally synthesized with valencies between these two extreme classes of neoglycoconjugates. The lengths of the branches, as well as the choice of the carbohydrate precursors, allows for control of intra-glycosidic bond distances [69].

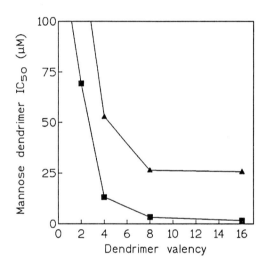

Figure 20-4. Effect of dendrimer valency on the inhibition of binding of Con A to yeast mannan for mannose dendrimers **47** (■). IC$_{50}$s expressed relative to monomeric α-D-mannopyranosdie content (▲).

The first glycodendrimer synthesis in which up to 16 sialic acid residues were attached at the periphery of a multi-branched L-lysine core appeared in 1993 [70]. Solid-phase synthesis on Wang resin, along with HOBt/DIC chemistry, was used to generate the hyperbranched L-lysine core which was functionalized with an N-chloroacetylglycylglycine spacer. Peracetylated thioglycosides were added to the solid phase for nucleophilic substitution of the N-chloroacetyl groups. In this manner, glycodendrimers containing α-thiosialosides **42** [70], β-D-lactosides **43** and **44** [71], N-acetyllactosaminides **45**

[71], N-acetylglucosaminides **46** [71], α-D-mannosides **47** [72], T-antigen **48** [73], and 3'-sulfo-Lewis^X-(Glc) analog **50** [74] were prepared (Scheme 20-6).

In a chemoenzymatic strategy, GlcNAc-based dendron **46** was treated with UDP-glucose, GlcNAc β-1,4-galactosyl transferase, and UDP-glucose 4'-epimerase to give N-acetyl-lactosaminyl dendrons **49** [75]. This shows that glycodendrimers are amenable to enzymatic techniques and opens the door to further enzymatic manipulations in order to prepare glycodendrimers containing even more complex oligosaccharide sequences.

When tested in enzyme linked lectin assays (ELLA, section 20.10), all glycodendrimers showed enhanced binding affinity as compared with their monomeric precursors. Figures 20.4 and 20.5 demonstrate the relationships generally observed between IC$_{50}$ and valency.

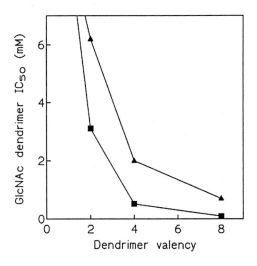

Figure 20-5. Effect of dendrimer valency on the inhibition of binding of WGA to porcine stomach mucin type III for GlcNAc dendrimers **46** (■). IC$_{50}$s expressed relative to monomeric N-acetylglucosamine content (▲).

With an increase in valency comes a corresponding increase in inhibitory potential, *i.e.* a corresponding decrease in IC$_{50}$. In addition, α-thiosialosyl dendrons such as **42** were shown to be as potent as their analogous polymers in the inhibition of hemagglutination of Influenza virus [70] and dendrons containing 3'-sulfo-Lewis^X-(Glc) (**50**) are presently the most effective E- and L-selectin antagonists known (Table 20-2) [74].

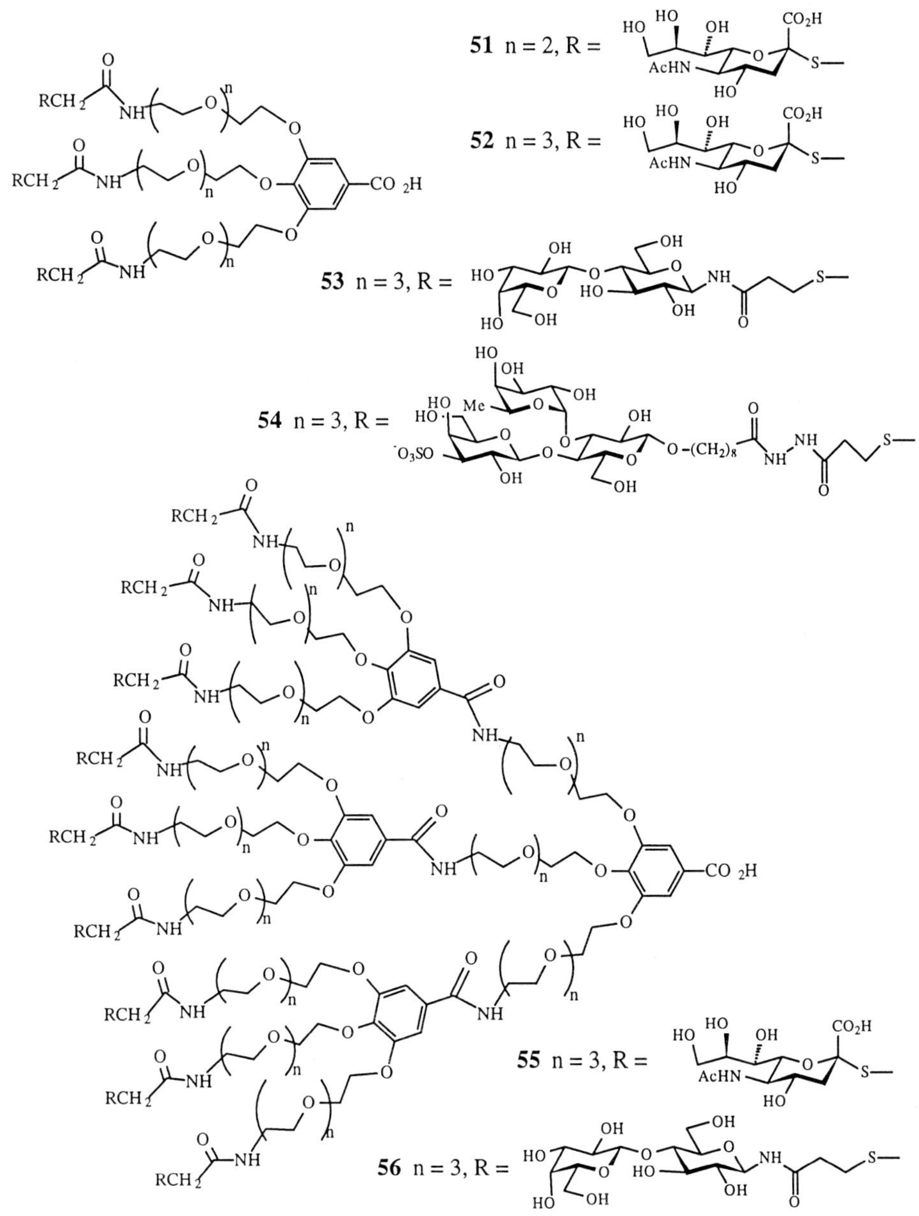

Scheme 20-7. Representative structures of gallic acid-based glycodendrimer.

Once again, the combined glycomimetic and multivalent strategies have established the potential for the synthesis of powerful carbohydrate ligands.

Table 20-2. Inhibitory Potency of 3'-Sulfo-Lewis^X-(Glc) Dendrimers (50) Using Chimeric Ig Fusion Proteins and a Reference Hexasaccharide Glycolipid.

Compound	IC$_{50}$ (µM) L-Selectin[a]	IC$_{50}$ (µM) E-Selectin[a]
monomeric 3'-sulfo'Le^X-(Glc)	>5000	2300
dimeric 3'-sulfo'Le^X-(Glc)	10 (20)	300 (600)
tetrameric 3'-sulfo'Le^X-(Glc)	2 (8)	70 (280)
octameric 3'-sulfo'Le^X-(Glc) **50**	1 (8)	20 (160)

[a] Values in parentheses are expressed relative to monomeric content.

The above design afforded glycodendrimers with carbohydrate densities corresponding to 2^n, where n represents the n'th generation of the L-lysine scaffold. The success of this approach in the generation of viable inhibitors, has given our group the impetus to concentrate on the synthesis of other glycodendrimers having various shapes, spacer lengths, and carbohydrate densities with values other than 2^n.

Glycodendrimers based on a gallic acid core bearing α-thiosialosides **51**, **52**, and **55** [76], β-D-lactosides **53** and **56** [77], and 3'-sulfo-Lewis^X-(Glc) **55** [74] (Scheme 20-7) were prepared. Bi-directional glycodendrimers based on a phosphotriester backbone containing N-acetylgalactosaminides were also synthesized [69, 78, 79]. All of these glycoconjugates demonstrated again that an increase in valency was effective in enhancing the binding affinity of carbohydrate-protein interactions, using ELLA inhibition experiments.

Another synthesis of a glycodendrimer used a dendritic backbone based on a 3,3'-iminobis(propylamine) core [80]. A divalent 3,3'-iminobis(propylamine) derivative (**57**) bearing orthogonally protected amines (Cbz) and a *tert*-butyl protected acid was selectively deprotected to give amine **58** and acid **59** as key precursors. These two fragments were coupled using HOBt/DIC strategy to provide dendrons with Cbz functionalities of up to 16 in valency. Cbz-protected precursors were transformed into N-chloroacetylated dendrons and treated with an α-thiosialoside which generated sialodendrimers such as **64** (Scheme 20-8).

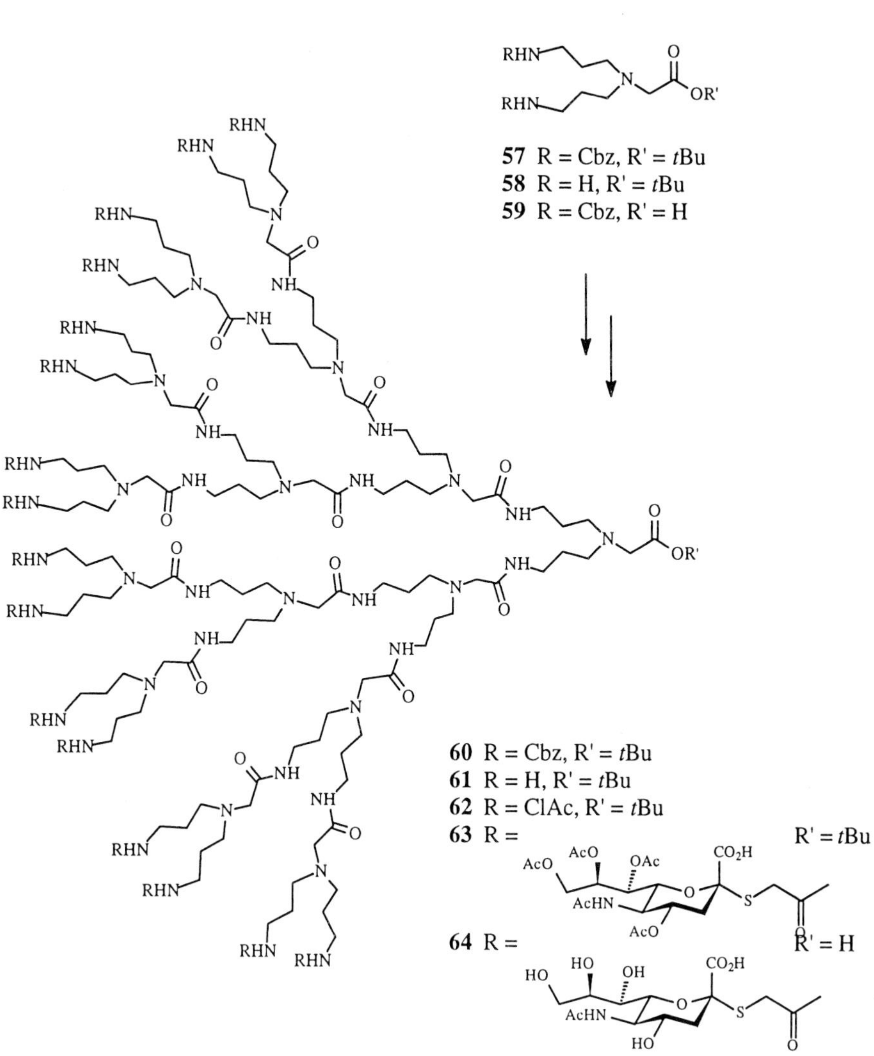

Architectonic Neoglycoconjugates... 401

57 R = Cbz, R' = *t*Bu
58 R = H, R' = *t*Bu
59 R = Cbz, R' = H

60 R = Cbz, R' = *t*Bu
61 R = H, R' = *t*Bu
62 R = ClAc, R' = *t*Bu
63 R = [AcO, AcHN, OAc, CO₂H sialic acid-S-propanoyl group] R' = *t*Bu
64 R = [HO, AcHN, OH, CO₂H sialic acid-S-propanoyl group] R' = H

Scheme 20-8. Synthesis of α-sialodendrimer based on a 3,3'-iminobis(propylamine) core.

In solid phase inhibition assays using neoglycoconjugates **64** as inhibitors of the binding of human α_1-acid glycoprotein (orosomucoid) to the lectin *Limax flavus*, glycodendrimers such as **64** were found to be 3 to 130 times better than monomeric 5-acetamido-5-deoxy-D-*glycero*-α-D-*galacto*-2-nonulopyranosyl azide (NeuAcαN$_3$) used as a standard (Table 20-3) [81].

Table 20-3. Inhibition of Binding of Human α_1-Acid Glycoprotein (Orosomucoid) to *Limax flavus* by Sialodendrimers.

Compound	IC$_{50}$ (nM)[a]	Relative Potency[a]
α-NeuAcN$_3$	1500	1
dimeric sialodendrimer	176 (352)	8.5 (4.2)
tetrameric sialodendrimer	11.8 (47.2)	127 (32)
octameric sialodendrimer	206 (1650)	7.3 (0.91)
hexadecameric sialodendrimer **64**	425 (6800)	3.5 (0.22)

[a]Values in parentheses are expressed relative to monomeric content.

20.8 Spherical Glycodendrimers

Spherical glycodendrimers also manage to effectively combine valency and geometry requirements. They do this is the same manner as for glycodendrimers, but they surpass glycodendrimers in that their potential for higher valencies is increased.

Spherical Starburst™ PAMAM dendrimers [82] made of polyamidoamine are commercially available and they have already been used to attach various carbohydrate derivatives. Disaccharide lactones of lactose and maltose have been directly conjugated to the amine terminated dendrimers and these spherical dendrimers (with 12, 24, and 48 glycan residues, **65** and **66**, Scheme 20-9) were shown to bind strongly and reversibly to Concanavalin A and pea lectins [83]. Toyokuni and Singhal [84] have attached a dimeric T$_N$-antigen (α-D-GalNAc-O-Ser) peptide to fifth generation (48 amines) PAMAM dendrimers to give spherical dendrimers **67** which were found to be non-immunogenic.

We [85] and others [86] have used an isothiocyanate coupling reaction to attach various glycosides to spherical dendrimers. α-D-Mannopyranoside containing *p*-thiophenyl urea functionality [67, 72, 85] and peracetylated glycosyl isothiocyanates of β-D-glucose, α-D-mannose, β-D-galactose, β-cellobiose, and β-lactose [86, 87] have been conjugated to PAMAM dendrimers in this fashion (**69** to **73**, Scheme 20-9). Mannose-based dendrimers **68** were found to have enhanced binding affinity for Concanavalin A lectin [85] and α-D-mannopyrannoside dendrimers **69** showed improved inhibitory potencies in the agglutination of a type 1-fimbriated *Escherichia coli* clone with yeast cells [88].

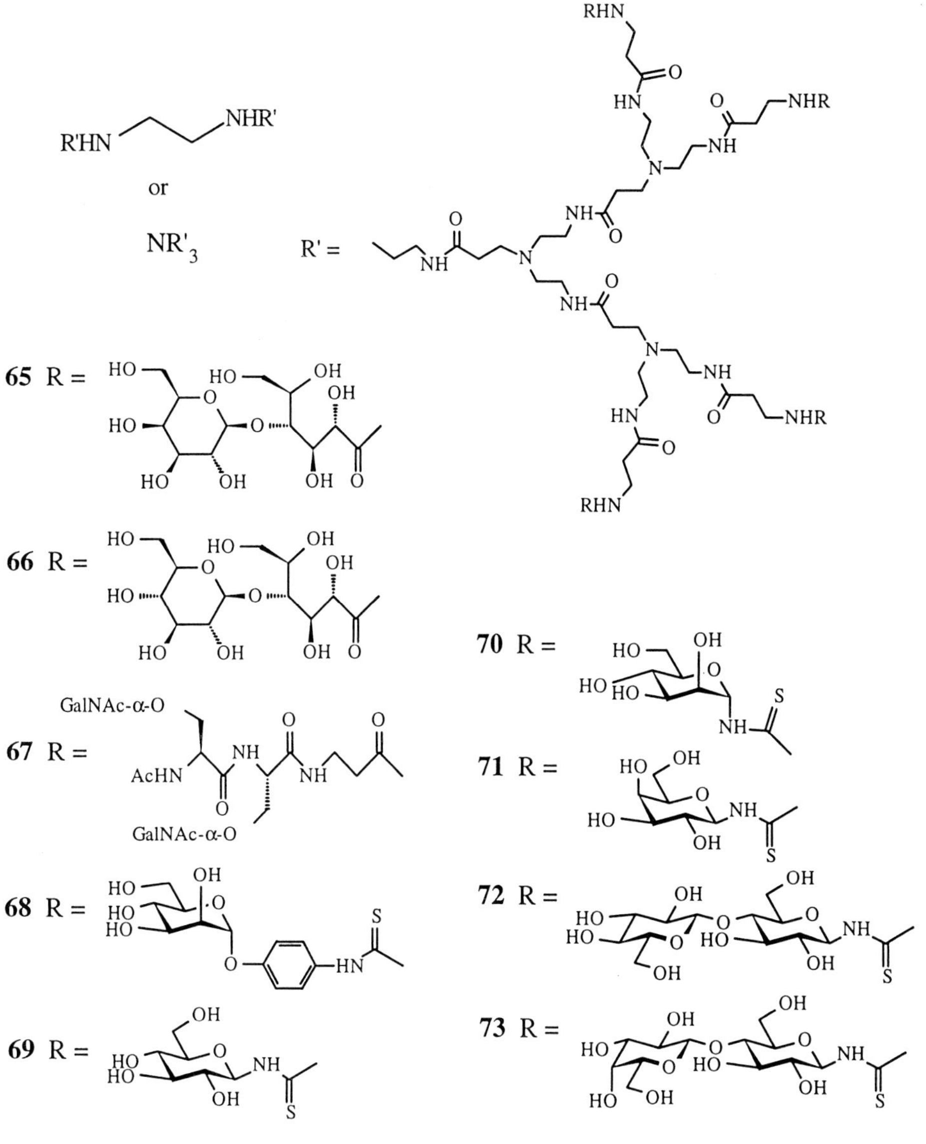

Scheme 20-9. Representative structures of PAMAM based glycodendrimers.

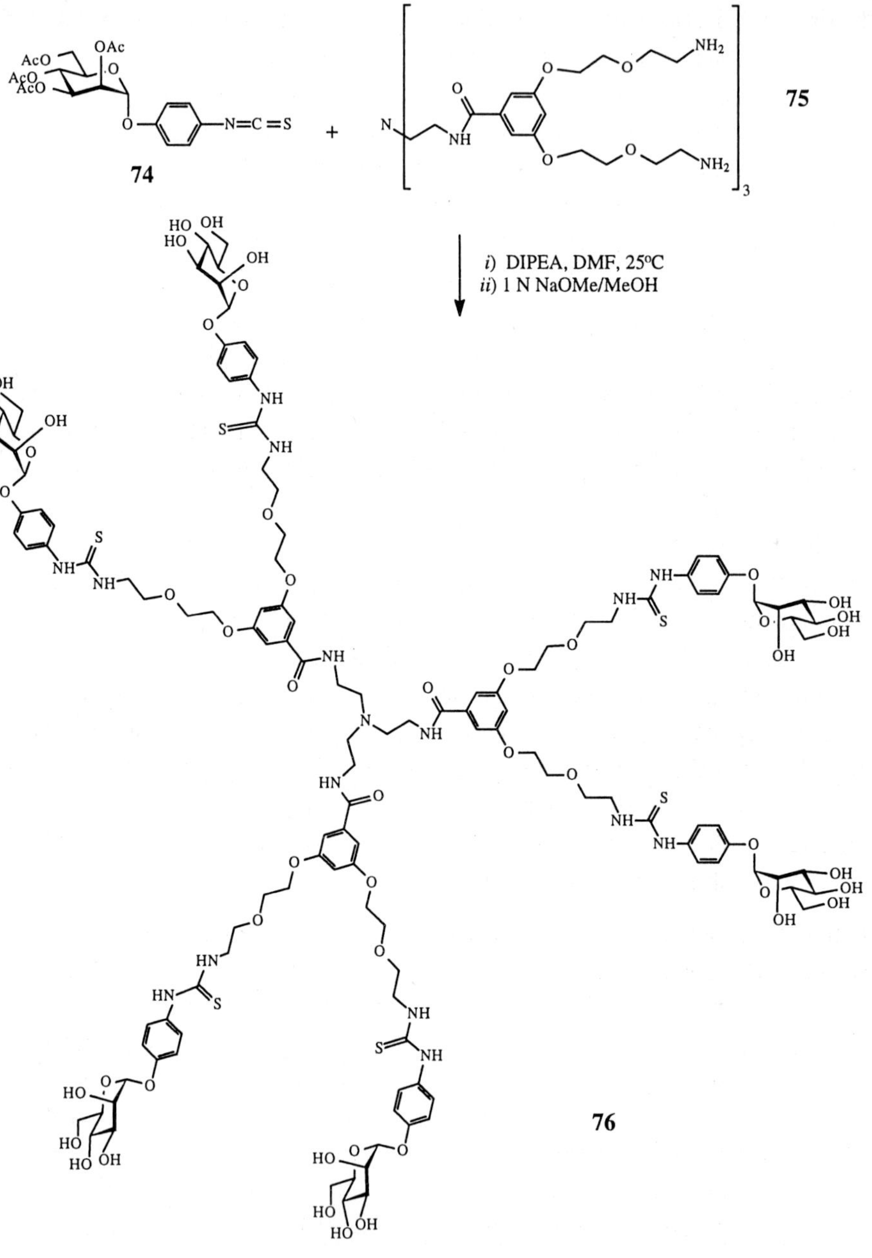

Scheme 20-10. Synthesis of hexavalent α-D-mannopyranoside dendrimer.

Using this same isothiocyanate strategy, α-D-mannopyranoside derivative **74** was attached to hexavalent amine core **75** to give spherical mannosylated dendrimer **76** (Scheme 20-10) [89]. In solid phase inhibition assays of the binding of yeast mannan to both Con A and *Pisum Sativum* (pea) lectins, slight improvements over *p*-nitrophenyl α-D-mannopyranoside used as a standard were observed for glycoconjugate **76** (1.7 fold for Con A and <3 fold for pea lectin).

Other groups have recently looked to the generation of spherical dendrimers based on core structures other than PAMAM dendrimers. Ashton *et al.* [90] have condensed tri- and hexa-glucosylated derivatives onto a trifunctional 1,3,5-benzene tricarboxylic acid core via amide bond formation to generate spherical dendrimers with valencies of nine and eighteen.

In order to avoid base catalyzed retro-Michael degradations to which PAMAM dendrimers are susceptible, we have used a convergent approach to tether a dimer and tetramer such as **77** to both hexamethylenediamine and tris-(2-aminoethyl)amine to give glycodendrimers of up to twelve in valency (**83**, Scheme 20-11) [81].

In ELLA testing similar to that described for glycodendrons **64**, the spherical glycodendrimers exhibited increased inhibitory potentials with an increase in valency (Table 20-4). Sialodendrimer **83** was shown to be 180 times more potent than monomeric α-NeuAcN$_3$ standard. Comparing directly glycodendrimers and spherical dendrimers (Tables 20.3 and 20.4), it is evident that geometrical requirements such as intra-glycosyl distance and conformation must be considered in the design of various glycoconjugates. Hexadecavalent sialodendrimer **64** exhibited a relative inhibitory potency of only 3.5, whereas spherical sialodendrimer **83**, with a valency of twelve, showed a 182 fold increase in inhibitory capacity [81].

Table 20-4. Inhibition of Binding of Human α$_1$-Acid Glycoprotein to *Limax Flavus* by Spherical Sialodendrimers

Compound	IC$_{50}$ (nM)[a]	Relative Potency[a]
NeuAcαN$_3$	1500	1
tetrameric sialodendrimer	58.7 (235)	26 (6.4)
hexameric sialodendrimer	16.9 (101)	89 (15)
octameric sialodendrimer	17.5 (140)	86 (11)
dodecameric sialodendrimer **83**	8.22 (98.6)	182 (15)

[a]Values in parentheses are expressed relative to monomeric content.

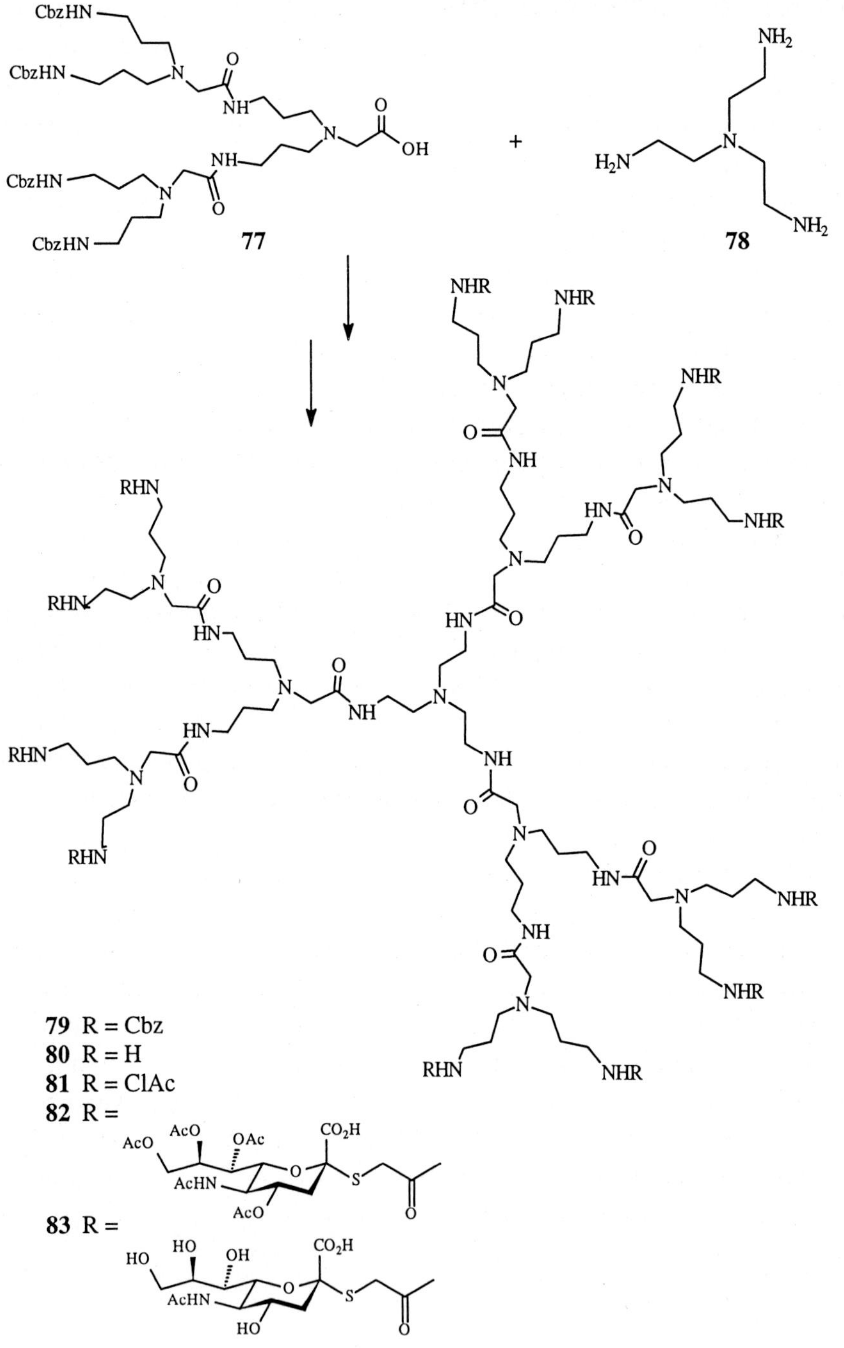

406 D. Zanini and R. Roy

79 R = Cbz
80 R = H
81 R = ClAc
82 R = [tetra-O-acetyl-N-acetyl sialic acid thioether]
83 R = [N-acetyl sialic acid thioether]

Scheme 20-11. Synthesis of spherical α-sialodendrimers based on 3,3'-iminobis(propylamine).

20.9 Cyclodextrins and Glycocalixarenes

In order to provide multivalent glycoconjugates with more rigid frameworks cyclodextrins substituted with carbohydrate residues and glycocalix[n]arenes were prepared. In some ways, cyclodextrins and cone-shaped calixarenes can be considered structurally similar. They both have symmetrical shapes defined by hydrophobic cavities. Cyclodextrins have this cavity enclosed between two hydrophilic rims whereas calixarenes may be conferred with hydrophilicity by conjugation to carbohydrates. Both glycoconjugates may be considered good models to study and understand weak intermolecular interactions and recognition mechanisms. Their ability to form inclusion complexes makes them attractive as drug targeting molecules. In addition they offer glycosidic clusters with still different valencies and conformations.

De Robertis *et al.* [91] have generated cyclodextrins fully substituted with glycosidic residues. β-Cyclodextrin incorporating up to seven galactose moities were prepared from a (6-bromo-6-deoxy)cyclomaltoheptaose derivative and either the sodium salts of 1-thio-(α and β)-D-galactopyranose or 1,2-ethanedithio-α-D-galactopyranose. These novel glycoconjugates were found to inhibit the flocculation of yeast cell wall lectin KbWCl and had better recognition capacities to KbWCl over mono-substituted cyclodextrin derivatives.

Cyclodextrins substituted with α-D- [92] and β-D-glucose [92, 93], α-D-mannose [94], β-D-galactose [94], and lactosamine [95] have been similarly described.

The first carbohydrate linked calix[4]arene was reported by Marra *et al.* [96, 97]. These glycoconjugates however were not water soluble and did not consider spacer arm requirements for efficient carbohydrate-protein interactions. We have prepared *p*-tert-butylcalix[4]arene substituted with four α-thiosialoside groups [98]. *p*-Tert-butylcalix[4]arene **84** was initially transformed into the known tetraacid chloride **85**, which was treated with mono-Boc protected 1,4-butanediamine to give derivative **86**. After Boc-protecting group removal, tetraamine **87** was converted to tetra-N-chloroacetylated derivative **88**. Via nucleophilic substitution with 2-thio-α-sialic acid derivative **89**, fully protected glycocalix[4]arene **90** was obtained (Scheme 20-12).

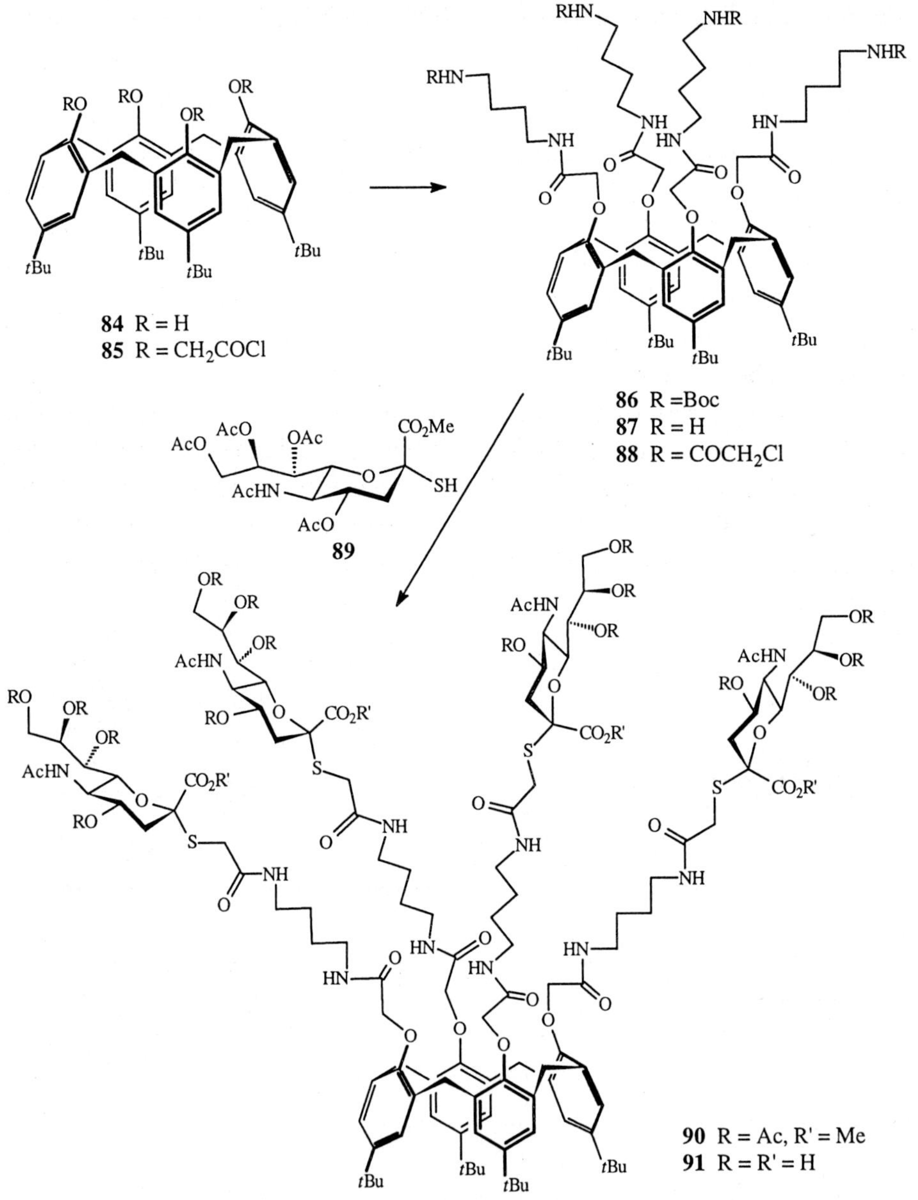

84 R = H
85 R = CH2COCl
86 R =Boc
87 R = H
88 R = COCH2Cl
89
90 R = Ac, R' = Me
91 R = R' = H
tBu
OR
RO
NHR
RHN
AcHN
AcO
OAc
CO2Me
SH
CO2R'
NH
HN

Scheme 20-12. Synthesis of α-sialo-*p*-tertbutyl-calix[4]arene.

Turbidimetric analysis showed that sialocalix[4]arene **91** bound strongly and reversibly to wheat germ agglutinin (WGA). These novel glycoconjugates constitute the first water soluble and biologically active glycocalix[4]arenes [98].

20.10 Biological Testing

In order to assess the specific role of multivalency and its biological implications in carbohydrate-protein interactions, these synthetic neoglycoconjugates must be evaluated both *in vitro* and *in vivo*. Commonly employed techniques in our laboratories include double immunodiffusion, turbidimetric analysis, and competitive enzyme linked lectin assays (ELLA) with either plant lectins or antibodies. As ELLA was used extensively in the above studies, it will be discussed briefly here.

Lectins are carbohydrate recognizing proteins of non-immune origin. They are often arranged in clusters (Figure 20-1) and, as such, they may be effectively used as models for other glycoside binding receptors where multivalent factors are important. Competitive inhibition assays make it possible to obtain an estimate (IC_{50}) of a particular hapten concentration necessary to inhibit a given interaction.

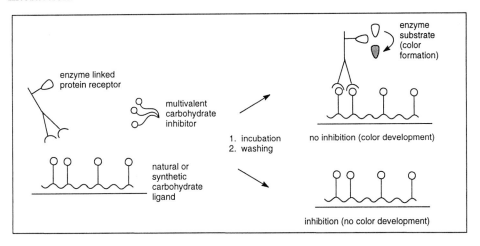

Figure 20-6. Solid phase competitive enzyme linked lectin assays (ELLA)

In competitive solid phase ELLA, a microtitre plate is coated with natural (*e.g.* glycoproteins, polysaccharides) or synthetic (*e.g.* glycopolymers) carbohydrate ligands. An enzyme linked receptor (*e.g.* peroxidase or phosphatase linked lectin) is added to the plate together with the glycoconjugate inhibitor. If the glycoconjugate is an effective inhibitor, it will prevent the protein receptors from binding to the carbohydrate ligand coated to the plate. Hence, the solid phase coated ligand competes with the free ligand in the binding to the enzyme-linked receptor (Figure 20-6). A multivalent neoglycoconjugate with improved inhibitory potential binds effectively to the receptor, and thus reduces receptor-plated ligand interactions. Once plotted as per cent inhibition versus inhibitor concentration, these data provide an estimation of the concentration of multivalent neoglycoconjugate necessary for 50% inhibition (IC_{50}).

20.11 Conclusion

A better understanding of the role of multivalent carbohydrate-protein interactions in biological systems is necessary. Presently novel neoglycoconjugates with varying degrees of valencies, conformations, and geometries are being synthesized. Evaluation of the cluster effect via ELLA, turbidimetric analysis, and double immunodiffusion are the current modes of biological testing and many such data are very encouraging. They reveal that with a proper multivalent design, binding affinity may be greatly enhanced depending on the biological systems under investigation. The novel glycoconjugates being currently synthesized may be fine-tuned in their development as potent ligands from which critical information about carbohydrate-protein interactions and receptor topography may be obtained. Understanding of the cluster or multivalent effect will bring the possibility of combining this binding enhancement methodology with that of structure-activity relationship strategies. With this possibility comes the potential for the development of truly effective therapeutic agents with applications in the areas of inhibitors of viral and bacterial infections, anti-inflammatory agents, and in cell specific drug targeting.

20.12 References

1. B. E. Eaton, L. Gold, D. A. Zichi, *Chem. Biol.* **1995**, *2*, 633-638.
2. A. Varki, *Glycobiology* **1993**, *3*, 97-130.
3. E. J. Toone, *Curr. Opin. Struct. Biol.* **1994**, *4*, 719-728.
4. K. Drickamer, M. E. Taylor, *Annu. Rev. Cell Biol.* **1993**, *9*, 237-264.
5. S. H. Barondes, D. N. W. Cooper, M. A. Gitt, H. Leffler, *J. Biol. Chem.* **1994**, *269*, 20807-20810.
6. L. L. Kiessling, N. L. Pohl, *Chem. Biol.* **1996**, *3*, 71-77.
7. C. P. Stowell, Y. C. Lee, *Adv. Carbohydr. Chem. Biochem.* **1980**, *37*, 225-281.
8. Y. C. Lee, R. T. Lee, *Neoglycoconjugates: Preparations and Applications*, Academic Press, San Diego, **1994**.

9. Y. C. Lee, R. T. Lee, *Methods Enzymol.* **1994**, *242* and *247*.

10. R. Roy in *Carbohydrate Chemistry* (Ed.: G. J. Boons), Chapman and Hall, Glasgow, **1997**, in press.

11. R. Roy, *Trends Glycosci. Glycotechnol.* **1996**, *8*, 79-99.

12. R. Roy in *Modern Methods in Carbohydrate Synthesis* (Eds.: S.-H. Khan, R. O'Neil), Harwood Academic, Amsterdam, **1996**, p. 378.

13. N. V. Bovin, E. Yu Korchagina, T. V. Zemlyanukhina, N. E. Byramova, A. E. Ivanov, U. P. Zubov, L. V. Mochalova, *Glycoconjugate J.* **1993**, *10*, 142-151.

14. N. V. Bovin, H.-J. Gabius, *Chem. Soc. Rev.* **1995**, *24*, 413-421.

15. G. Magnusson, A. Ya Chernyak, J. Kihlberg, L. O. Kononov in *Neoglycoconjugates: Preparations and Applications* (Eds.: Y. C. Lee, R. T. Lee), Academic Press, San Diego, **1994**, p. 53.

16. A. Ya Chernyak, *ACS Symp. Ser.* **1994**, *560*, 133-156.

17. I. J. Goldstein, R. O. Poretz in *The Lectins, Properties, Functions, and Applications in Biology and Medicine* (Eds.: I. E. Liener, N. Sharon, I. J. Goldstein), Academic Press, Orlando, **1986**, p. 35.

18. R. U. Lemieux, *Acc. Chem. Res.* **1996**, *29*, 373-380.

19. W. E. Dick, M. Beurret in *Contribution to Microbiology and Immunology* (Eds.: J. M. Cruse, R. E. Lewis, Jr.), Karger, Basel, **1989**, *10*, p. 48.

20. J. C. Paulson in *The Receptors* (Ed.: M. Conn), Academic Press, New York, **1985**, *2*, p. 131.

21. J. H. Pazur, *Adv. Carbohydr. Chem. Biochem.* **1981**, *39*, 405-447.

22. R. L. Schnaar, *Anal. Biochem.* **1984**, *143*, 1-13.

23. a) L. W. Seymour, *Adv. Drug Deliv. Rev.* **1994**, *14*, 89-111; b) G. Molema, D. K. F. Meijer, *Adv. Drug Deliv. Rev.* **1994**, *14*, 25-50; c) M. Monsigny, A.-C. Roche, P. Midoux, R. Mayer, *Adv. Drug Deliv. Rev.* **1994**, *14*, 1-24.

24. a) R. Roy, C. A. Laferrière, *Can. J. Chem.* **1990**, *68*, 2045-2054; b) R. Roy, C. A. Laferrière, A. Gamian, H. J. Jennings, *J. Carbohydr. Chem.* **1987**, *6*, 161-165.

25. R. Roy, C. A. Laferrière, *Carbohydr. Res.* **1988**, *177*, C1-C4.

26. R. Roy, C. A. Laferrière, *J. Chem. Soc., Chem Commun.* **1990**, 1709-1711.

27. N. E. Byramova, L. V. Mochalova, I. M. Belyanchikov, M. N. Matrosovich, N. V. Bovin, *J. Carbohydr. Chem.* **1991**, *10*, 691-700.

28. W. J. Lees, A. Spaltenstein, J. E. Kingery-Wood, G. M. Whitesides, *J. Med. Chem.* **1994**, *37*, 3419-3433.

29. R. Roy, F. O. Andersson, G. Harms, S. Kelm, R. Schauer, *Angew. Chem., Int. Ed. Engl.* **1992**, *31*, 1478-1481.

30. a) G. B. Sigal, M. Mammen, G. Dahmann, G. M. Whitesides, *J. Am. Chem. Soc.* **1996**, *118*, 3789-3800; b) M. A. Sparks, K. W. Williams, G. M. Whitesides, *J. Med. Chem.* **1995**, *38*, 4179-4190.

31. a) D. H. Charych, J. O. Nagy, W. Spevak, M. D. Bednarski, *Science* **1993**, *261*, 585-588; b) W. Spevak, J. O. Nagy, D. H. Charych, M. E. Schaeffer, J. H. Gilbert, M. D. Bednarski, *J. Am. Chem. Soc.* **1993**, *115*, 1146-1147; c) J. O. Nagy, P. Wang, J. H. Gilbert, M. E. Schaeffer, T. G. Hill, M. R. Callstrom, M. D. Bednarski, *J. Med. Chem.* **1992**, *35*, 4501-4502.

32. N. E. Nifant'ev, Yu E. Tsvetkov, A. S. Shashkov, A. B. Tuzikov, I. V. Meslennikov, I. S. Popova, N. V. Bovin, *Bioorg. Khim.* **1994**, *20*, 552-555.

33. N. E. Nifant'ev, A. S. Shashkov, Yu E. Tsvetkov, A. B. Tuzikov, I. V. Abramenko, D. F. Gluzman, N. V. Bovin, *ACS Symp. Ser.* **1994**, *560*, 267-275.

34. R. Roy, W. K. C. Park, O. P. Srivastava, C. Foxall, *Bioorg. Med. Chem. Lett.* **1996**, *6*, 1399-1402.

35. T. V. Zemlyanukhina, N. E. Nifant'ev, A. S. Shashkov, Y. E. Tsetkov, N. V. Bovin, *Carbohydr. Lett.* **1995**, *1*, 277-284.

36. W. Spevak, C. Foxall, D. H. Charych, F. Dasgupta, J. O. Nagy, *J. Med. Chem.* **1996**, *39*, 1018-1020.

37. R. Roy in *Topics Curr. Chem.* (Eds.: J. Thiem, H. Driguez), Springer-Verlag, Heidelberg, **1997**, *187*, 241-274.

38. R. Roy, *Curr. Opin. Struct. Biol.* **1996**, *6*, 692-702.

39. R. Roy, M.-G. Baek, L. Filion, S. Ogunnaike, unpublished data.

40. A. Gamian, M. Chomik, C. A. Laferrière, R. Roy, *Can. J. Microbiol.* **1991**, *37*, 233-237.

41. R. Roy, F. D. Tropper, A. Romanowska, *Bioconjugate Chem.* **1992**, *3*, 256-261.

42. R. Roy, F. D. Tropper, A. Romanowska, R. K. Jain, C. F. Piskorz, K. L. Matta, *Bioorg. Med. Chem. Lett.* **1992**, *2*, 911-914.

43. S. Cao, R. Roy, *Tetrahedron Lett.* **1996**, *37*, 3421-3424.

44. C. M. Starks in *Free Radical Telomerizations*, Academic Press, New York, **1974**.

45. W. K. C. Park, S. Aravind, A. Romanowska, J. Renaud, R. Roy, *Methods Enzymol.* **1994**, *242*, 294-304.

46. S. Aravind, W. K. C. Park, S. Brochu, R. Roy, *Tetrahedron Lett.* **1994**, *35*, 7739-7742.

47. R. Roy, F. D. Tropper, S. Cao, J.-M. Kim, *ACS Symp. Ser.* **1997**, in press.

48. R. Roy in *Handbook of Phase Transfer Catalysis* (Eds.: Y. Sasson, R. Neumann), Chapman and Hall, Glasgow, **1997**, in press.

49. J.-M. Kim, R. Roy, *Carbohydr. Res.* **1997**, in press.

50. R. Roy, U. K. Saha, *Chem. Commun.* **1996**, 210-212.

51. J. -M. Kim, R. Roy, *Carbohydr. Lett.* **1996**, *1*, 465-468.

52. U. K. Saha, R. Roy, *Tetrahedron Lett.* **1995**, *36*, 3635-3638.

53. U. K. Saha, R. Roy, *J. Chem Soc., Chem. Commun.* **1995**, 2571-2573.

54. U. K. Saha, R. Roy, unpublished data.

55. F. D. Tropper, F. O. Andersson, S. Braun, R. Roy, *Synthesis* **1992**, 618-620.

56. R. Roy in *Carbohydrates: Targets for Rational Drug Design* (Ed.: Z.Witczak), Marcel Dekker Inc., New York, **1997**, p. 84.

57. R. T. Lee, Y. C. Lee in *Neoglycoconjugates: Preparations and Applications* (Eds.: Y. C. Lee, R. T. Lee), Academic Press, San Diego, **1994**, p.23.

58. Y. C. Lee in *Carbohydrate Recognition in Cellular Function, CIBA Foundation Symposium 145* (Eds.: G. Bock, S. Harnette), Wiley, Chichester, **1989**, p. 80.

59. R. T. Lee, Y. C. Lee, *Glycoconjugate J.* **1987**, *4*, 317-328.

60. A. Kichler, F. Schuber, *Glycoconjugate J.* **1995**, *12*, 275-281.

61. R. T. Lee, P. Lin, Y. C. Lee, *Biochemistry* **1984**, *23*, 4255-4261.

62. G. D. Glick, J. R. Knowles, *J. Am. Chem. Soc.* **1991**, *113*, 4701-4703.

63. G. D. Glick, P. L. Toogood, D. C. Wiley, J. J. Skehel, J. R. Knowles, *J. Biol. Chem.* **1991**, *266*, 23660-23669.

64. S. A. DeFrees, W. Kosch, W. Way, J. C. Paulson, S. Sabesan, R. L. Halcomb, D.-H. Huang, Y. Ichikawa, C.-H. Wong, *J. Am. Chem. Soc.* **1995**, *117*, 66-79.

65. G. Kretzschmar, U. Sprengard, H. Kunz, E. Bartnik, W. Schmidt, A. Toepfer, B. Hörsch, M. Krause, D. Seiffge, *Tetrahedron* **1995**, *51*, 13015-13030.

66. C. Unverzagt, S. Kelm, J. C. Paulson, *Carbohydr. Res.* **1994**, *251*, 285-301.

67. D. Pagé, R. Roy, *Glycoconjugate J.* **1997**, in press.

68. U. Sprengard, M. Schudok, W. Schmidt, G. Kretzschmar, H. Kunz, *Angew. Chem., Int. Ed. Engl.* **1996**, *35*, 321-324.

69. R. Roy, *Polymer News* **1996**, *21*, 226-232.

70. a) R. Roy, D. Zanini, S. J. Meunier, A. Romanowska, *ACS Symp. Ser.* **1994**, *560*, 104-119; b) R. Roy, D. Zanini, S. J. Meunier, A. Romanowska, *J. Chem. Soc. Chem.Commun.* **1993**, 1869-1872.

71. D. Zanini, W.K. C. Park, R. Roy, *Tetrahedron Lett.* **1995**, *36*, 7383-7386.

72. D. Pagé, D. Zanini, R. Roy, *Bioorg. Med Chem.* **1996**, *4*, 1949-1961.

73. R. Roy, D. Zanini, M. -G. Baek, unpublished data.

74. D. Zanini, R. Roy, W. K. C. Park, C. Foxall, O.P. Srivastava, *Carbohydr. Lett.* **1997**, in press.

75. D. Zanini, R. Roy, *Bioconjugate Chem.* **1997**, in press.

76. S.-N. Wang, S. J. Meunier, Q.-Q. Wu, R. Roy, *Can. J. Chem.* **1997**, in press.

77. R. Roy, W. K. C. Park, Q.-Q. Wu, S.-N. Wang, *Tetrahedron Lett.* **1995**, *36*, 4377-4380.

78. D. Zanini, W. K. C. Park, S. J. Meunier, Q.-Q. Wu, S. Aravind, B. Kratzer, R. Roy, *Polym. Mater. Sci. Eng.* **1995**, *73*, 82-83.

79. W. K. C. Park, B. Kratzer, D. Zanini, Q.-Q. Wu, S. J. Meunier, R. Roy, *Glycoconjugate J.* **1995**, *12*, 456-457.

80. D. Zanini, R. Roy, *J. Org. Chem.* **1996**, *61*, 7348-7354.

81. D. Zanini, R. Roy, *J. Am. Chem. Soc.* **1997**, in press.

82. a) P. R. Dvornic, D. A. Tomalia, *Curr. Opin. Coll. Inter. Sci.* **1996**, *1*, 221-235; b) D. A. Tomalia, H. D. Durst, *Topics Curr. Chem.* **1993**, *165*, 193-313; c) D. A. Tomalia, A. M. Naylor, W. A. Goddard III, *Angew. Chem., Int. Ed. Engl.* **1990**, *29*, 138-175.

83. K. Aoi, K. Itoh, M. Okada, *Macromolecules* **1995**, *28*, 5391-5393.

84. T. Toyokuni, A. K. Singhal, *Chem. Soc. Rev.* **1995**, 231-242.

85. D. Pagé, R. Roy, submitted.

86. T. K. Lindhorst, C. Kieburg, *Angew. Chem., Int. Ed. Engl.* **1996**, *35*, 1953-1956.

87. T. K. Lindhorst, C. Kieburg, *Synthesis* **1995**, 1228-1230.

88. U. Krallmann, T. K. Lindhorst, *Proceedings of the XVIIIth Int. Carbohydr. Symp.* **1996**, Milan, July 21-26, p. 587.

89. D. Pagé, S. Aravind, R. Roy, *Chem. Commun.* **1996**, 1913-1914.

90. P. R. Ashton, S. E. Boyd, C. L. Brown, N. Jayaraman, S. A. Nepogodiev, J. F. Stoddart, *Chem. Eur. J.* **1996**, *2*, 1115-1128.

91. L. De Robertis, C. Lancelon-Pin, H. Driguez, F. Attioui, R. Bonaly, A. Marsura, *Bioorg. Med. Chem. Lett.* **1994**, *4*, 1127-1130.

92. S. Cottaz, H. Driguez, *Synthesis* **1989**, 755-758.

93. H. Parrot-Lopez, H. Galons, A. W. Coleman, J. Maheteau, M. Miocque, *Tetrahedron Lett.* **1992**, *33*, 209-212.

94. C. Lancelon-Pin, H. Driguez, *Tetrahedron Lett.* **1992**, *33*, 3125-3128.

95. E. Leray, H. Parrot-Lopez, C. Augé, A. W. Coleman, C. Finance, R. Bonaly, *J. Chem. Soc., Chem. Commun.* **1995**, 1019-1020.

96. A. Marra, M.-C. Schermann, A. Dondoni, A. Casnati, P. Minari, R. Ungaro, *Angew. Chem., Int. Ed. Engl.* **1994**, *33*, 2479-2481.

97. A. Marra, A. Dondoni, F. Sansone, *J. Org. Chem.* **1996**, *61*, 5155-5158.

98. S. J. Meunier, R. Roy, *Tetrahedron Lett.* **1996**, *37*, 5469-5472.

21 From Glycosaminoglycans to Heparinoid Mimetics with Antiproliferative Activity

H. P. Wessel, A. Chucholowski, J. Fingerle, N. Iberg, H. P. Märki, R. Müller, M. Pech, M. Pfister-Downar, M. Rouge, G. Schmid and T. Tschopp

21.1 Introduction

Glycosaminoglycans are a family of natural polysaccharides [1, 2] composed of disaccharide repeating motives which consist of - as the name indicates - a glycosamine and another glycose derivative, which is in most cases a uronic acid (cf. Scheme 21-1). Especially the heparinoid glycosaminoglycans with glucosamine and uronic acid repeating motives, heparan sulfate and heparin, are complex polysaccharides with molecular weights up to 30'000 D and dispersion of molecular weights. The repeating units display microheterogeneity in that the uronic acid may be D-glucuronic acid or the epimeric L-iduronic acid, and the repeating unit may be sulfated at various positions (at O-2 of iduronic acid, O-3 of glucosamine, N-2 of glucosamine) or acetylated at N-2 of glucosamine. Heparinoid glycosaminoglycans are able to bind to a high number of proteins, the so-called heparin-binding proteins, and are thus capable of influencing many biological functions [3, 4].

We were interested in the inhibition of migration and proliferation of smooth muscle cells (SMC), important processes in the formation of arteriosclerotic lesions and restenosis [5, 6]. Heparan sulfates isolated from SMC [7] or endothelial cells [8] were shown to potently inhibit SMC growth, and are believed to endogenously regulate this process. Also the related heparin inhibits SMC growth [9-12] and migration [13] *in vitro* and neointimal thickening after vascular injury *in vivo* [14].

21.2 Sulfated Carbohydrates

The goal of the our investigations was the preparation of heparinoid mimetics, less complex than heparin and which should selectively inhibit SMC growth and migration. The starting point was the readily available polysaccharide heparin which is in clinical use as an anticoagulant since decades. Bleeding complications, however, associated with the anticoagulant effect of heparin represent the main side-effect and limit the use of heparin at higher concentrations. Heparin's anticoagulant activity is mainly mediated by antithrombin III (AT_{III}); we, therefore, routinely monitored the AT_{III}-mediated anticoagulant effect of new compounds for anti-thrombin (anti-factor II_a) and anti-factor X_a activity in chromogenic assays.

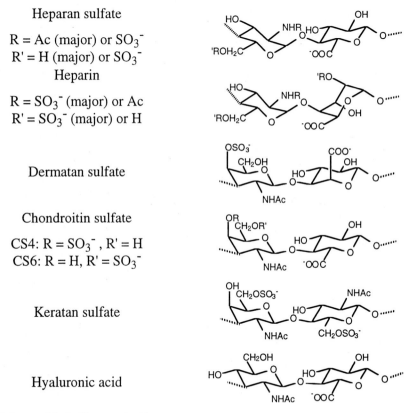

Scheme 21-1. Glycosaminoglycans.

Carboxyl-reduction of heparin (**1** in Scheme 21-2 represents the regular repeating unit) followed by sulfation of the newly formed primary hydroxyl groups resulted in a heparin derivative **2** (CRS-heparin, carboxyl-reduced sulfated heparin) [15] with antiproliferative activity similar to that of heparin in an *in vitro* SMC proliferation assay (Table 21-1).

1

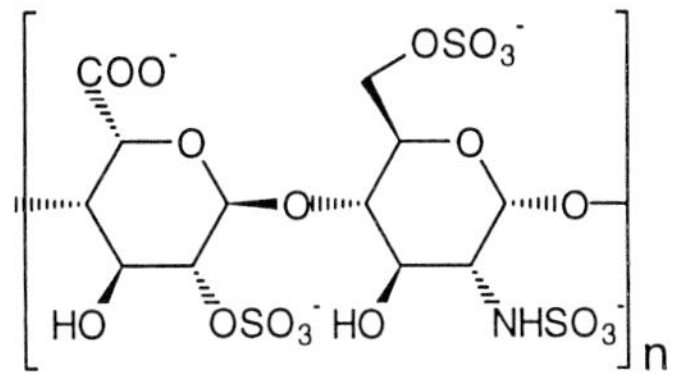

2

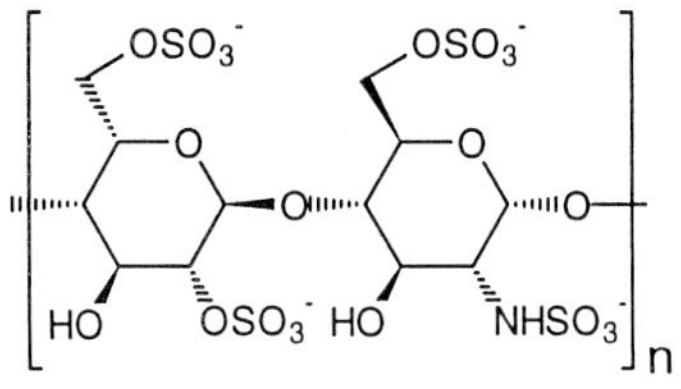

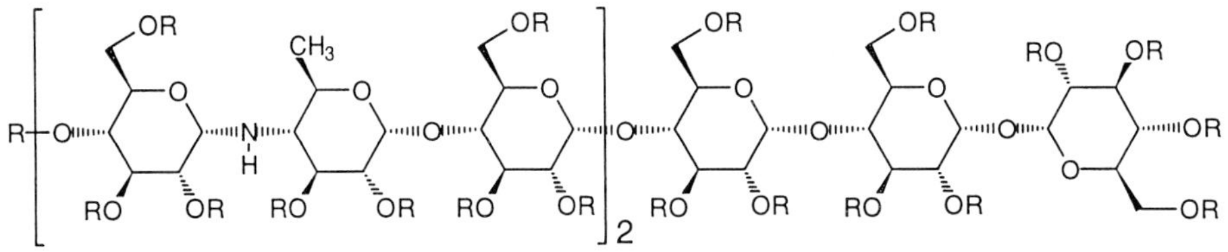

3 DS ≈ 2.4

4 n = 1 DS ≈ 3.0
5 n = 2 DS ≈ 2.9
6 n = 3 DS ≈ 2.4

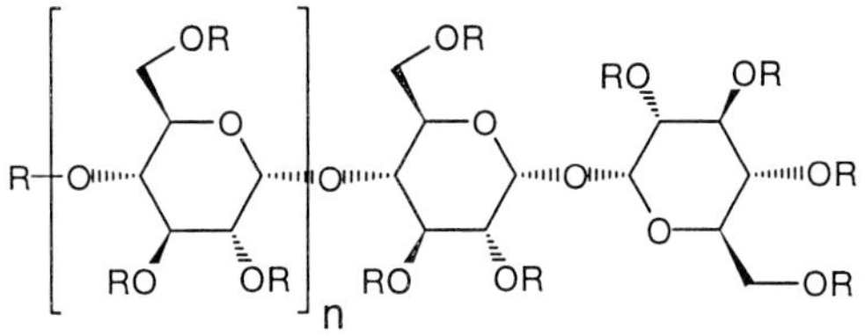

7 DS ≈ 3.0

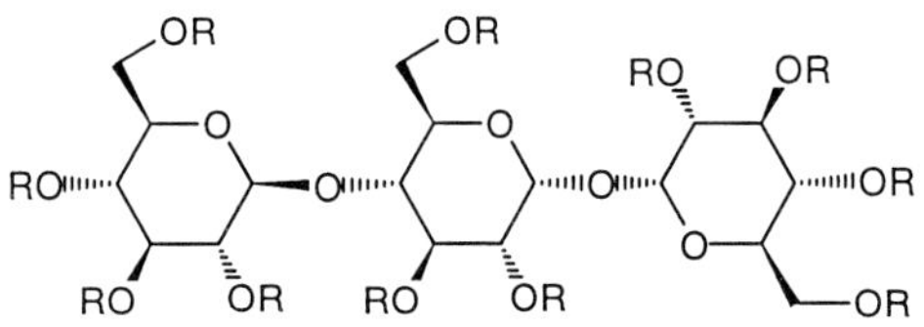

8 n = 1 DS ≈ 2.8
9 n = 2 DS ≈ 2.6

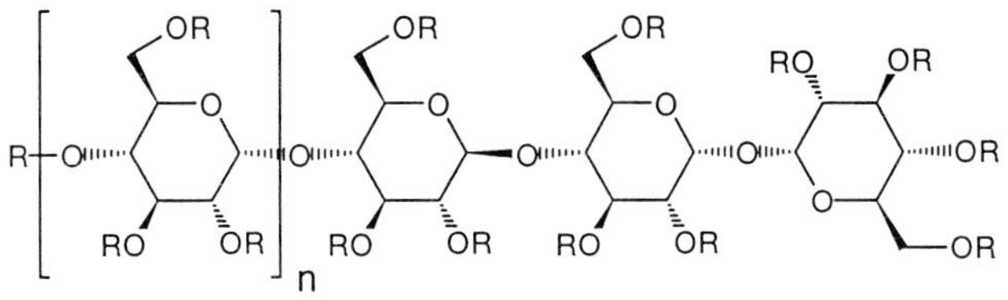

Scheme 21-2. Sulfated saccharides, DS (degree of sulfation) denotes the mean number of sulfates per monosaccharide unit.

Table 21-1. Activities of sulfated carbohydrates.

	Compound	SMC Growth inhibition [% at 100mg/ml] [a]	IC$_{50}$ [mg/ml] [b]
1	Heparin	50	2.2 / 2.7
2	CRS-Heparin	75	170 / 680
3	Trestatin A sulfate	81	> 1000 / > 1000
4	α-Linked trisaccharide	inactive	> 1000 / > 1000
5	α-Linked tetrasaccharide	< 12	> 1000 / > 1000
6	α-Linked pentasaccharide	57	> 1000 / 900
7	β-Linked trisaccharide	38	> 1000 / > 1000
8	β-Linked tetrasaccharide	59	> 1000 / > 1000
9	β-Linked pentasaccharide	64	> 1000 / > 1000

a. Relative antiproliferative activity
b. Anticoagulant activity anti-IIa/anti-Xa

The AT$_{III}$-mediated anticoagulant effect was reduced by two orders of magnitude confirming [16] that the antiproliferative effect does not depend on AT$_{III}$-mediated anticoagulant properties. Thus the carboxyl groups of heparin are not required for its antiproliferative activity, and a number of sulfated non-uronic oligosaccharides were screened. Trestatin A sulfate (**3**) was found to be highly active in the inhibition of SMC proliferation [15] (Table 21-1). Since Trestatin A had been developed as an amylase inhibitor [17, 18] and, as such, probably mimics the helical conformation of amylose, it was obvious to study amylose substructures. Sulfated maltooligosaccharides exhibited, however, no appreciable antiproliferative activity. On the other hand, molecular modelling experiments had suggested that the trehalose end of Trestatin A bends out from the helical conformation, so that we investigated substructures of this moiety of Trestatin A sulfate to simplify the active structure [19].

Table 21-2. Biological activities of sulfated tetrasaccharides of the type (disaccharide)-(1→4)-α,α–trehalose

Compound (DS)[a]	Disaccharide structure	Disaccharide configuration	r_i [b]
8 (DS ≈ 2.8)		αDGlc(1→4)βDGlc (Mal)	0.9
10 (DS≈3.0)		αDGlc(1→6)βDGlc (Iso-Mal)	0.9
11 (DS ≈ 3.0)		αDGal(1→6)βDGlc (Mel)	0.8
12 (DS ≈ 3.0)	CH₃	αLRha(1→6)βDGlc (Rut)	0.7
13 (DS ≈ 2.7)		βDGal(1→3)αDAra	0.7
14 (DS ≈ 3.3)		βDGlc(1→6)βDGlc (Gen)	0.6
15 (DS ≈ 2.9)		βDGal(1→4)βDGlc (Lac)	0.5
16 (DS ≈ 2.8)		βDGlc(1→2)βDGlc (Soph)	0.5
17 (DS ≈ 2.7)		βDGlc(1→4)βDGlc (Cel)	0.3

a. Determined as reported in ref.15
b. Anti-proliferative activity r_i = relative inhibitory activity compared to heparin, at 100 µg/ml

For the preparation of these all-α-D-linked oligosaccharides we chose a block synthesis approach with a suitably blocked trehalose derivative as glycosyl acceptor [20, 21]. Effective and stereoselective glycosylation reactions were made possible by the employment of triflic anhydride as a new promoter [22-24]. Out of the α-D-linked oligosaccharides **4**, **5**, and **6** (Scheme 21-2) only the pentasaccharide **6** showed a heparin-like antiproliferative effect (Table 21-1). An investigation of related compounds revealed that the analogous β-D-linked oligosaccharides **7** - **9** (Scheme 21-2) prepared in block syntheses [25] using standard methodology are even more active: already the trisaccharide **7** was active in the growth inhibition assay, and the tetrasaccharide **8** had a heparin-like effect, which was only slightly increased with pentasaccharide **9** (Table 21-1) [19].

So far, the sulfated maltosyl trehalose tetrasaccharide **8** is the smallest oligosaccharide derivative with an SMC antiproliferative effect comparable to heparin. Modification on either the maltose or the trehalose moiety did not increase the activity. Examples of tetrasaccharides with modifications in the β-maltosyl moiety [26] are summarized in Table 21-2, antiproliferative activities are expressed as relative inhibitory activity compared to heparin (r_i) [26]. The antiproliferative activity can nearly be completely abolished by relatively small structural variation such as a formal anomerization of the terminal saccharide - comparing sulfated maltosyl trehalose **8** with sulfated cellobiosyl trehalose **17**. These findings together with the earlier observations on the activity of β-D-linked *vs* α-D-linked trehalose oligosaccharides indicated that the activity is not just mediated by an unspecific charge-charge interaction but that sulfates at defined positions in space are necessary for the interaction with the target protein(s).

The oligosaccharides discussed so far could be highly, but not completely sulfated so that the sulfated carbohydrates are mixtures of compounds with a defined backbone. With the goal to further simplify the structures and to finally arrive at defined compounds we studied the importance of sulfates at the various positions of the sulfated tetrasaccharide **8**. To this end, selectively deoxygenated tetrasaccharide analogues were prepared; the main synthetic strategy was to introduce the respective deoxygenation site on the disaccharide level and to build up the tetrasaccharides in a [2+2] - block synthesis (a synthetic example is depicted in Scheme 21-3) [27]. The investigation of SMC antiproliferative activities of the deoxygenated β-maltosyl trehalose tetrasaccharides gave antiproliferative effects between r_i = 1.0 and r_i = 0.5, values of r_i < 0.7 (Scheme 21-4) were regarded as significantly below the reference tetrasaccharide **8**. Although it is not possible to deduce precise locations of essential sulfates from these results, it seems that sulfates at the inner glucose ring of the molecule are not essential and, thus, might be replaced by non-charged moieties.

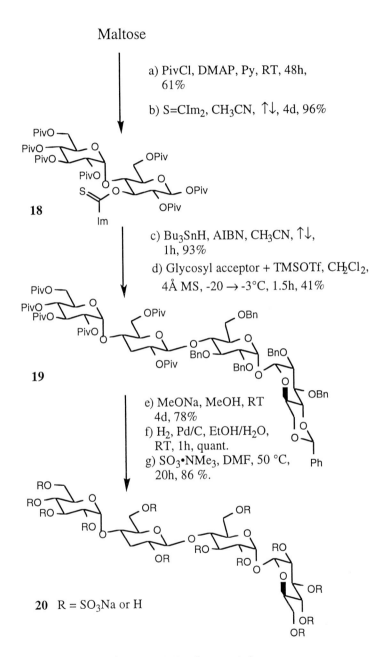

Scheme 21-3. Synthesis of 3"H-β–maltosyl-(1→4)-α,α-trehalose.

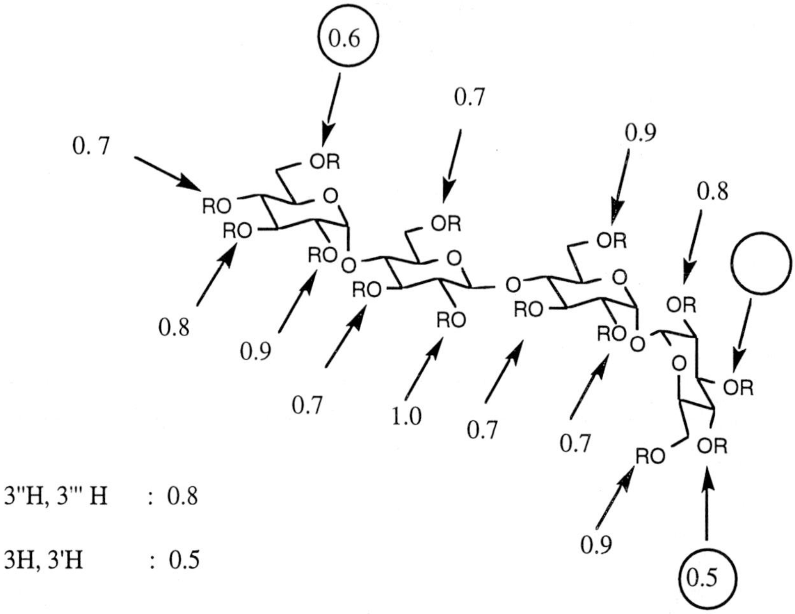

3"H, 3'''H : 0.8

3H, 3'H : 0.5

Scheme 21-4. r_i-Values of deoxygenated β–maltosyl-(1→4)-α,α-trehalose derivatives, essential sulfate positions are circled. Arrows indicate single deoxygenated positions. The missing r_i value for the tetrasaccharide deoxygenated in position 3 can be approximated from a dideoxygenated tetrasaccharide, since r_i values of the deoxygenated tetrasaccharides are sufficiently 'additive'.

Addressing the question of random sulfation, also smaller saccharides were investigated. While trisaccharides such as **7** could be sulfated only incompletely, it was possible to entirely sulfate disaccharides such as trehalose or methyl maltoside [28].

21.3 Spaced Sugars

Based on these results we prepared 'spaced saccharides' (Scheme 21-5) in which disaccharides are separated by a spacer unit, preferentially of aromatic nature. Consistent with our earlier experience, the spaced saccharides could be completely sulfated and rendered chemically defined entities. Examples are described in Table 21-3, remarkably, derivatives with SMC antiproliferative activity significantly higher than heparin's were obtained, which points at a contribution of the spacer to the overall binding. Similar compounds had been investigated with regard to complement modulating activity by others [29].

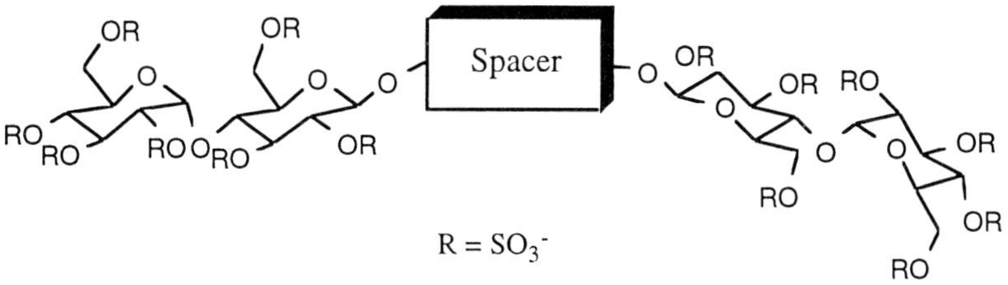

Scheme 21-5. Spaced saccharides, example showing maltosyl residues.

Table 21-3. Activities of selected persulfated spaced maltosides.

	Compound	r_i [a]	IC_{50} [µg/ml] [b]
1	Heparin	1.0	2.2 / 2.7
21	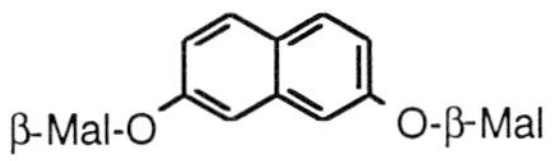 β-Mal-O—(naphthalene)—O-β-Mal	1.0	> 1000 / > 1000
22	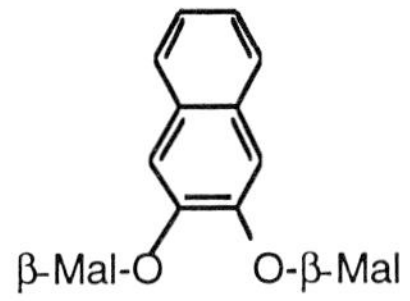 β-Mal-O (naphthalene) O-β-Mal	1.2	> 1000 / > 1000

a. r_i Relative antiproliferative activity
b. Anticoagulant activity anti-IIa/anti-Xa

C6 open chain sugars have approximately the same spatial extension as disaccharides. Thus, a corresponding replacement leads to sulfated 'spaced open chain sugars' (Scheme 21-6), an extremely versatile class of new compounds, since the open chain sugar moiety can be attached by various linkers to the spacer. While sugars of different chain length and configuration can be employed, glucitol and especially glucamine (1-amino1-deoxy-glucitol) are particularly readily accessible starting materials. The syntheses are straightforward, for example, the amide linked open chain sugars can be prepared by reaction of an activated dicarboxylic acid with glucamine followed by sulfation. Also the open chain sugars can be completely sulfated leading to chemically defined end-products. As shown in Table 21-4, the *in vitro* antiproliferative effect was further enhanced in this class of compounds yielding activities up to $r_i = 1.6$.

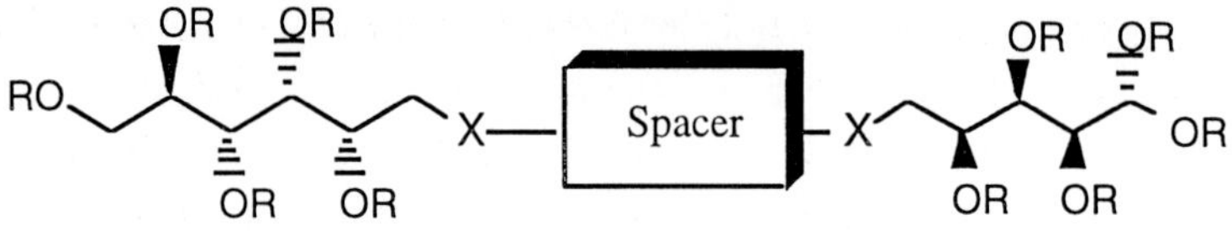

Scheme 21-6. Spaced open chain sugars. Example showing glucityl residues.

R = SO$_3^-$, X = -O-, -NH-, -NH-CO-, -CO-NH-, -NH-CO-NH-,
-NH-CS-NH-, -NH-SO$_2^-$,..

Table 21-4. Activities of selected persulfated spaced open chain sugars with different linkers.

	Compound	r$_i$ [a]	IC$_{50}$ [mg/ml][b]
1	Heparin	1.0	2.2 / 2.7
23	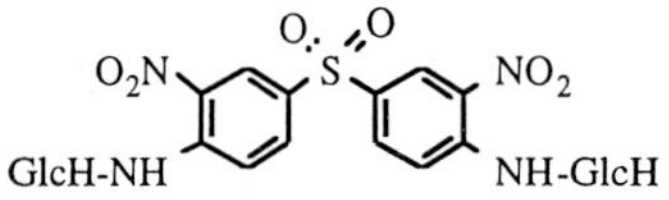 O$_2$N–⌬–SO$_2$–⌬–NO$_2$ GlcH-NH / NH-GlcH	1.0	>1000 / >1000
24	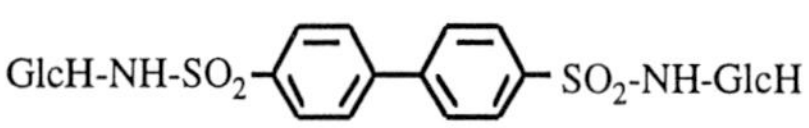 GlcH-NH-SO$_2$–⌬–⌬–SO$_2$-NH-GlcH	1.0	>1000 / >1000
25	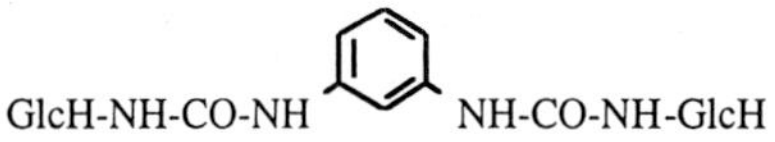 GlcH-NH-CO-NH–⌬–NH-CO-NH-GlcH	1.5	>1000 / >1000
26	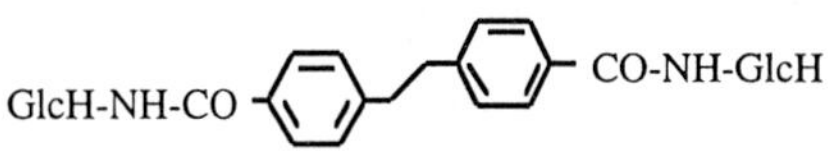 GlcH-NH-CO–⌬–⌬–CO-NH-GlcH	1.6	>1000 / >1000

GlcH = (glucityl with OR groups), R = SO$_3$Na

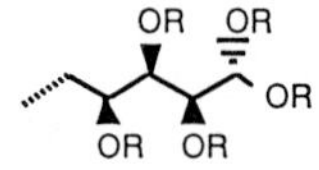

a. r$_i$ Relative antipro-liferative activity

b. Anticoagulant activity anti-IIa/anti-Xa

For selectivity we measured inhibition of AT$_{III}$-mediated anticoagulative effects, which are mainly responsible for the anticoagulant heparin action. The clotting process is, however, additionally influenced by heparin cofactor II as well as by other factors of the coagulation cascade. These multifactorial conditions are better reflected in a clotting assay for the 'activated partial thromboplastin time' (aPTT). In this assay, the selectivity of CRS-heparin or the active sulfated carbohydrates amounts to a factor of 5-10, which could be increased by another factor of ten with the spaced saccharides (Table 21-5). With the spaced open chain sugars, high selectivities of more than three orders of magnitude could be achieved.

Table 21-5. aPTT Selectivities of selected compounds.

	Compound	r_i [a]	IC_{50} [µg/ml] [b,c]
1	Heparin	1.0	1.3
2	CRS-Heparin	1.2	11
3	Trestatin A sulfate	1.2	7
8	Sulfated β-Mal-Tre tetrasaccharide	0.9	12
21	Sulfated 2,7-naphthalene spaced maltose	1.0	77
26	Sulfated 4,4'-diphenylethyl spaced glucamine	1.5	> 1000

a. r_i Relative antipro-liferative activity

b. Anticlotting activity (aPTT)

c. IC_{50} = compound concentration leading to a clotting time of twice the control

In conclusion, we could show that potent inhibitors of SMC proliferation with selectivity against AT_{III}-mediated anticoagulant effects could be obtained by modification of heparin. A consequent size reduction of the polysaccharide lead compound led to sulfated tetrasaccharides with antiproliferative effects similar to heparin. The problem of random sulfation was successfully addressed by the introduction of spaced oligosaccharides. Finally, high antiproliferative activities and considerable selectivites against clotting parameters were achieved with the spaced open chain sugars. This new class of compounds represents non-glycosidic heparinoid mimetics of low molecular weight.

21.4 A New Approach to Oligosaccharide Mimetics

Recent approaches to carbohydrate mimetics mainly focused on the preparation of mono- or disaccharide analogues. Strategies for the construction of oligosaccharide mimetics include the replacement of the interglycosidic oxygen by a methylene or substituted methylene group or also more extended units to give the so-called C-disaccharides [30] and analogues. The build-up of such mimetics is synthetically comparatively demanding, and oligomers up to C,C-trisaccharides have been prepared [31]

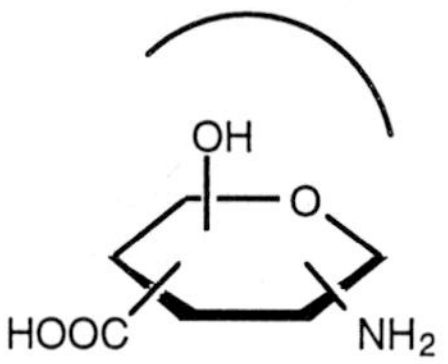

Scheme 21-7. Carbohydrate amino acid exposing a carbohydrate epitope.

We have now assembled saccharide units *via* peptide linkages using carbohydrate amino acids of type **26** (Scheme 21-7) as building blocks. The advantage of this approach is that numerous carbohydrate amino acids are readily available, and oligomerization is based on established peptide chemistry and can be performed in solution or on solid phase. The feasibility of the concept was demonstrated by a [2+2] block synthesis of normuramic acid tetramer **27** (Scheme 21-8) in solution [32] using 2-chloro-4,6-dimethoxy-1,3,5-triazine [33] as a coupling reagent; the glucosaminuronic acid tetramer **28** was obtained by solid phase homologation [34] employing the fluorenylmethoxycarbonyl (Fmoc) [35, 36] protection scheme and TATU activation [37]. With this methodology, libraries of compounds with carbohydrate epitopes can be constructed.

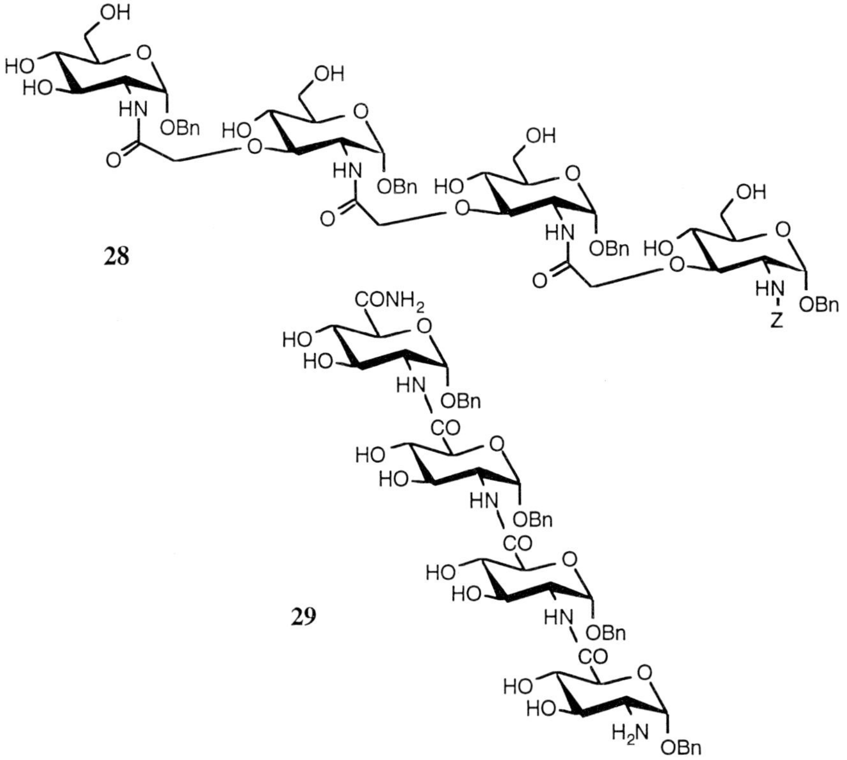

Scheme 21-8. Amide-linked tetrasaccharide mimetics prepared by solution chemistry (**27**) or solid phase synthesis (**28**).

21.5 References

1. W. D. Comper in *Heparin (and Related Polysaccharides)* (Ed.:M. B. Huglin), Gordon and Breach Science Publishers, New York, **1981**.

2. U. Lindahl, M. Höök, *Ann. Rev. Biochem.* **1978**, *47*, 385-417.

3. R. L. Jackson, S. J. Busch, A. D. Cardin, *Physiological Reviews* **1991**, *71*, 481-538.

4. F. Zhou, T. Höök, J. A. Thompson in *Heparin and Related Polysaccharides* (Ed.: D. A. Lane et al.), Plenum Press, New York, 1992.

5. R. Ross, *Nature* **1993**, *362*, 801-809.

6. R. Ross, *New Engl. J. Med.* **1986**, *314*, 488-500.

7. L. M. S. Fritze, C. F. Reilly, R. D. Rosenberg, *J. Cell Biol.* **1985**, *100*, 1041.

8. W. E. Benitz, R. T. Kelley, C. M. Anderson, D. E. Lorant, M. Bernfield, *Am. J. Respir. Cell Mol. Biol.* **1990**, *2*, 13-24.

9. R. L. Hoover, R. Rosenberg, W. Haering, M. J. Karnovsky, *Circ. Res.* **1980**, *47*, 578-583.

10. J. J. Castellot, jr., M. L. Addonizio, R. Rosenberg, M. J. Karnovsky, *J. Cell Biol.* **1981**, *90*, 372-379.

11. J. J. Castellot, jr., L. V. Favreau, M. J. Karnovsky, R. D. Rosenberg, *J. Biol. Chem.* **1982**, *257*, 11256-11260.

12. J. P. T. Au, R. D. Kenagy, M. M. Clowes, A. W. Clowes, *Haemostasis* **1993**, *23(suppl.1)*, 177-182.

13. R. A. Majack, A. W. Clowes, *J. Cell. Physiol.* **1984**, *118*, 253-256.

14. A. W. Clowes, M. J. Karnovsky, *Nature* **1977**, *265*, 625-626.

15. H. P. Wessel, M. Hosang, T. B. Tschopp, B.-J. Weimann, *Carbohydr. Res.* **1990**, *204*, 131-139.

16. J. R. Guyton, R. D. Rosenberg, A. W. Clowes, M. J. Karnovsky, *Circ. Res.* **1980**, *46*, 625-634.

17. K. Yokose, K. Ogawa, T. Sano, K. Watanabe, H. B. Maruyama, Y. Suhara, *J. Antibiot., Ser. A* **1983**, *36*, 1157-1165.

18. K. Yokose, K. Ogawa, Y. Suzuki, I. Umeda, Y. Suhara, S. A. J. Antibiot., 36, 1166, *J. Antibiot., Ser. A* **1983**, *36*, 1166-1175.

19. H. P. Wessel, T. B. Tschopp, M. Hosang, N. Iberg, *BioMed. Chem. Lett.* **1994**, *4*, 1419-1422.

20. H. P. Wessel, G. Englert, P. Stangier, *Helv. Chim. Acta* **1991**, *74*, 682-696.

21. H. P. Wessel, B. Mayer, G. Englert, *Carbohydr. Res.* **1993**, *242*, 141-151.

22. H. P. Wessel, *Tetrahedron Lett.* **1990**, *31*, 6863-6866.

23. H. P. Wessel, N. Ruiz, *J. Carbohydr. Chem.* **1991**, *10*, 901-910.

24. A. Dobarro-Rodriguez, M. Trumtel, H. P. Wessel, *J. Carbohydr. Chem.* **1992**, *11*, 255-263.

25. H. P. Wessel, R. Minder, G. Englert, *J. Carbohydr. Chem.* **1995**, *14*, 1101-1106; ibid, **1996**, *15*, 201-216.

26. H. P. Wessel, E. Vieira, M. Trumtel, T. B. Tschopp, N. Iberg, *BioMed. Chem. Lett.* **1995**, *5*, 437-442.

27. H. P. Wessel, N. Iberg, M. Trumtel, M.-C. Viaud, **1995**, *BioMed. Chem. Lett* **1996**, *6*, 27-36.

28. H. P. Wessel, S. Bartsch, *Carbohydr. Res.* **1995**, *274*, 1-9.

29. T. G. Miner, S. Bernstein, J. P. Joseph, EPA 082 927 to American Cyanamid, **1983**.
30. D. Rouzaud, P. Sinaÿ, *J. Chem. Soc., Chem. Commun.* **1983**, 1353-1354.
31. T. Haneda, P. G. Goekjian, S. H. Kim, Y. Kishi, *J. Org. Chem.* **1992**, *57*, 490-498.
32. H. P. Wessel, C. M. Mitchell, C. M. Lobato, G. Schmid, *Angew. Chem. Int. Ed. Engl.* **1996**, *34*, 2712-2713.
33. Z. J. Kaminsky, *Synthesis* **1987**, 917-920.
34. C. Müller, E. Kitas, H. P. Wessel, *J. Chem. Soc., Chem. Commun.* **1995**, 2425-2426.
35. L. A. Carpino, G. Y. Han, *J. Org. Chem.* **1972**, *37*, 3404-3409.
36. E. Atherton, H. Fox, D. Harkiss, C. J. Logan, R. C. Sheppard, B. J. Williams, *J. Chem. Soc., Chem. Commun.* **1978**, 537-539.
37. L. A. Carpino, *J. Am. Chem. Soc.* **1993**, *115*, 4397-4398.

22 Conduritols and Analogues as Insulin Modulators

D.C. Billington [1], F. Perron-Sierra, I. Picard, S. Beaubras, J. Duhault, J. Espinal and S. Challal

22.1 Introduction

Diabetes is recognised as probably the third largest killer disease in the world and today is also the leading cause of adult blindness. Diabetes sufferers have increased risk of heart disease, cerebro vascular disease, and suffer long term complications including renal impairment. Leaves of the Indian shrub, *Gymnena Sylvestre* have been used as a herbal remedy for diabetes for at least 2500 years [2,3]. The leaves contain numerous complex natural products and clinical trials with powdered leaf extract in 1930 showed significant glucose lowering effects at a dose of 4g per patient per day. Recent reports of the hypoglycaemic activity of the natural product, conduritol A, and speculation that conduritol A is the active principal of *Gymnena Sylvestre* [4] prompted us to examine the chemistry and pharmacology of the conduritols.

Conduritols A and F are naturally occurring 5-cyclohexene-1,2,3,4-tetrols [5]. Optically active conduritol F is found in small quantities in almost all green plants as the **L** isomer whereas the occurrence of the *meso* conduritol A is restricted to specific sub-families of tropical plants [6]. The other four isomeric conduritols are designated B, C, D, and E and exist as one *meso* compound (conduritol D) and three **D/L** pairs (conduritols B, C and E) and do not occur naturally. These compounds and related structures may provide new leads for the development of modulators of insulin action [7]. The structures of the conduritols A-E are presented in Scheme 22-1.

The synthesis of the conduritols and related compounds was comprehensively reviewed in 1990 [5] and other synthetic approaches have appeared more recently concentrating mainly on enantiospecific routes [8-11].

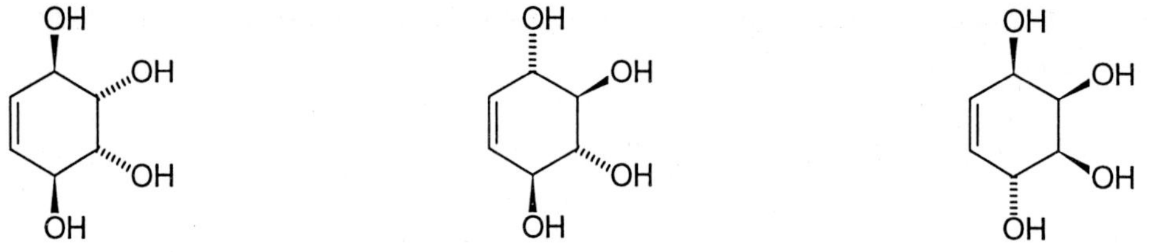

Conduritol A

Conduritol B

Conduritol C

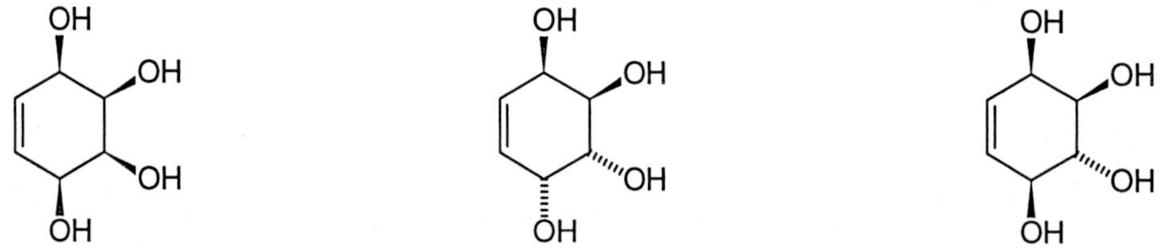

Conduritol D

Conduritol E

Conduritol F

Scheme 22-1

22.2 Results

We chose to synthesise the six conduritols for preliminary testing as *meso* compounds or racemic pairs by modification of published routes as shown in Schemes 22-2 to 22-6. Treatment of the commercially available *cis*-diol **1**[12] with one equivalent of mCPBA in water at room temperature over 16h gave a mixture of racemic conduritols C (**2**) and F (**3**), Scheme 22-2.

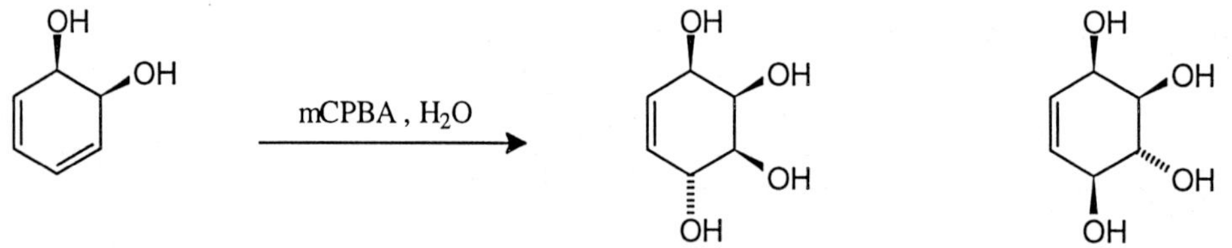

1   mCPBA, H$_2$O   Conduritol C (**2**)   Conduritol F (**3**)

Scheme 22-2

These two compounds were easily obtained in pure form by direct crystallisation followed by washing with ethanol. Conduritol C crystallises first, from the reaction mixture followed by conduritol F.

Protection of the known [13] dibromo diol **4** as its bis-TBS ether, followed by base catalysed elimination of HBr gave the protected diene **5**, Scheme 22-3.

Epoxidation of **5** using mCPBA in dichloromethane gave a 1:1 mixture of two separable epoxides **6** and **7** which were recovered in good yield following chromatography Scheme 22-3. Removal of the TBS protecting group from **6** followed by epoxide opening in water gave a 1:1 mixture of conduritols B (**8**) and F (**3**) which were separated by chromatography.

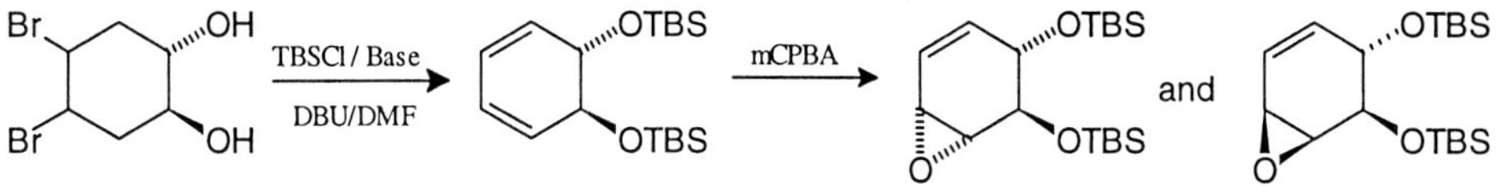

4 **5** **6** **7**

Scheme 22-3

A similar ring opening of epoxide **7** in water gave pure conduritol A **9** [14].

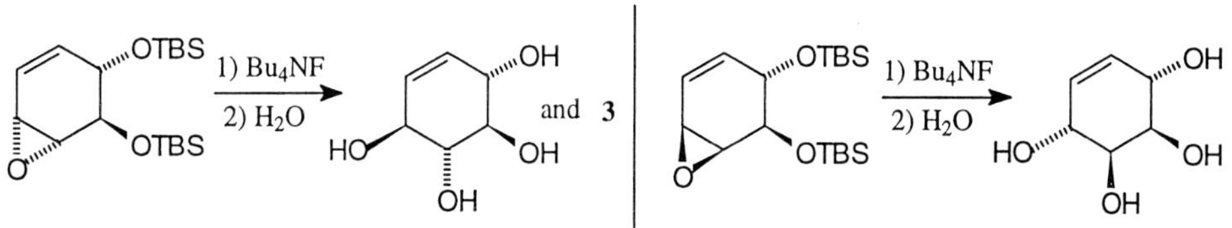

6 **Conduritol B (8)** **7** **Conduritol A (9)**

Scheme 22-4

Fully protected conduritol D was obtained from the high pressure Diels-Alder reaction between vinylene carbonate **10** and diene **11**. Subsequent treatment of the Diels-Alder adduct **12** with a basic ion exchange resin then gave the pure conduritol D (**13**) [15].

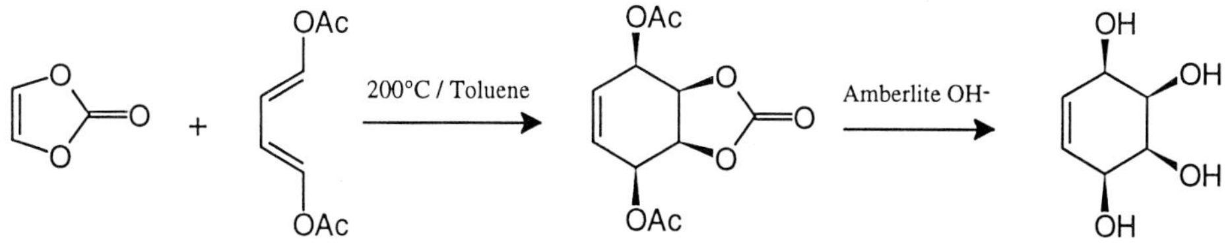

10 **11** **12** **Conduritol D (13)**

Scheme 22-5

The final isomer in this series, conduritol E was also prepared from the *cis*-diol **1** by protection as the isopropylidene acetal, followed by oxidation of one of the double bonds using catalytic osmium tetroxide to give the protected tetrol **14**.

Removal of the acetal protecting group then gave pure conduritol E (**15**) Scheme 22-6 [10].

Having all of the isomeric conduritols in hand, we then examined their ability to modulate the release of insulin from isolated pancreatic islets at two different glucose concentrations and the results of these studies are shown in Table 22-1.

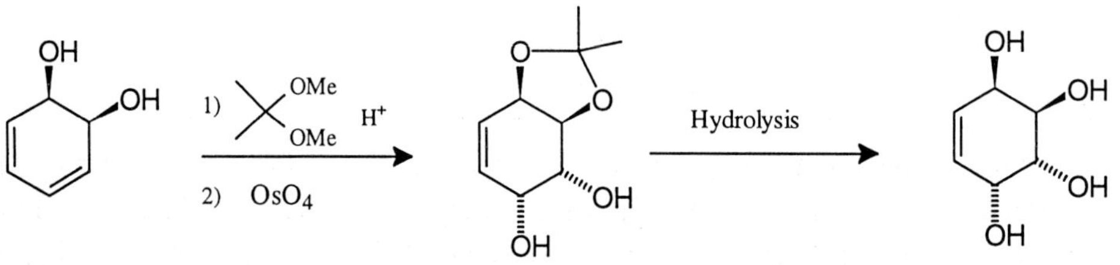

Scheme 22-6

Briefly, islets of Langerhans were obtained from male Sprague-Dawley rats by the method of Lacy [16]. These islets were then incubated at 37°C in the presence of either 2.8 mM or 16.7 mM glucose for 4h in the presence or absence of the test drug and insulin secretion from the islets was measured [17]. Data are presented in the table as the percentage increase or decrease in insulin secretion compared to the mean secretion obtained at each glucose concentration in the absence of drug. All compounds were tested at 10^{-4}M with eight replicates per incubation. It can be seen from Table 22-1 that conduritol A stimulates the production of insulin at 2.8 mM glucose but inhibits the production of insulin at 16.7 mM glucose. Conversely, conduritol B stimulates the production of insulin at both glucose concentrations. Conduritol C shows a profile similar to conduritol A whereas conduritol D and F are considered to be inactive in this assay. Conduritol E shows a weak simulation of insulin secretion at both glucose levels. A saturated version of conduritol A in which the double bond has been removed by hydrogenation shows weak inhibition of insulin secretion at both glucose concentrations.

Table 22-1. Modulation of insulin secretion by the conduritols.

Conduritol Isomer	% change at 2.8 mM glucose	% change at 16.7 mM glucose
Conduritol A	+45	- 30
Conduritol B	+41	+50
Conduritol C	+27	- 20
Conduritol D	0	0
Conduritol E	+12	+15
Conduritol F	+ 9	- 3
Saturated A	- 16	- 13

Given these interesting effects of conduritols on insulin secretion, we embarked on the synthesis of a series of analogues of conduritol A in which fused rings took the place of the conduritol double bond. These analogues were based on a working hypothesis that the four hydroxyl groups of conduritol A in the α,β,β,α stereo chemistry were largely responsible for the biological activity, while the double bond of conduritol A probably served as a conformational restraint. This was backed up by the observation that the saturated analogue of conduritol A is significantly less active than the parent natural product.

We adopted a Diels-Alder based strategy [18] for the synthesis of analogues of conduritol A followed by sequential stereoselective reductions (NaBH$_4$/CeCl$_3$) and/or oxidations (OsO$_4$ or mCPBA or Vanadyl acac) to give the required pattern of hydroxyl substituents.

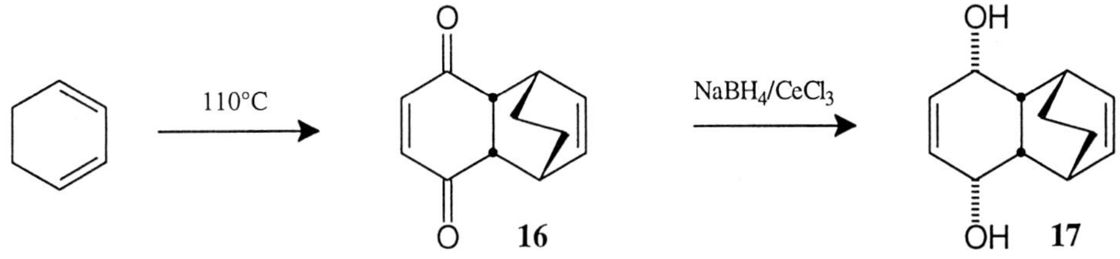

Scheme 22-7

Cycloaddition reaction between 1,3-cyclohexadiene and benzoquinone at 110°C in toluene gave the adduct **16**. Stereoselective reduction of **16** using NaBH$_4$/CeCl$_3$ in 50:50 dichloromethane/methanol gave the *cis* diol **17**. This compound served as a key intermediate for subsequent syntheses. Oxidation of **17** with catalytic osmiumtetroxide at elevated temperature resulted in the *cis* hydroxylation of both double bonds giving the hexol **18**.

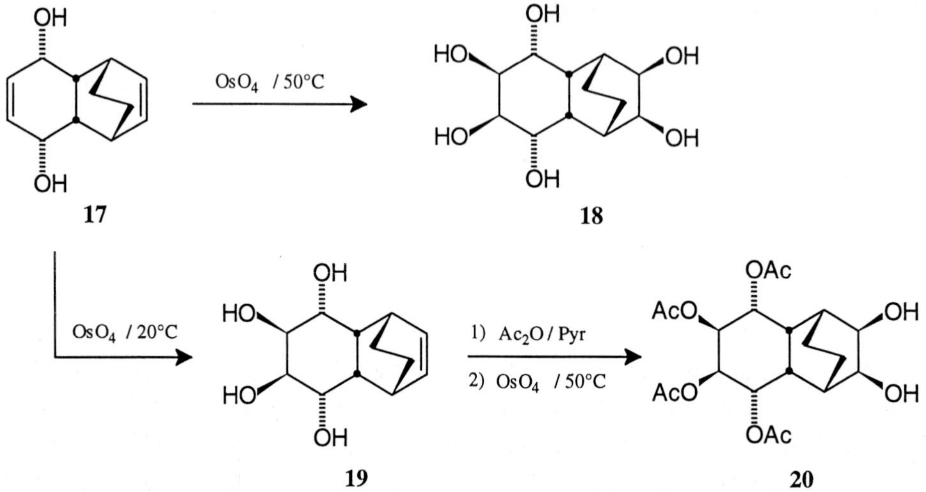

Scheme 22-8

Alternatively, treatment of **17** with catalytic osmiumtetroxide at 20°C and careful control of the reaction time led selectively to the tetrol **19** in which only the more reactive double bond had been hydroxylated.

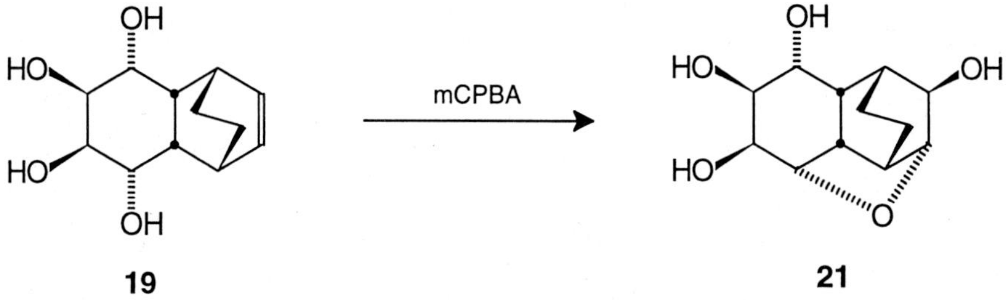

Scheme 22-9

Acetylation of **19** with acetic anhydride in pyridine followed by a second treatment with catalytic osmiumtetroxide, this time at 50°C, gave the tetraacetylated hexol **20** having two free hydroxyl groups. Attempted formation of a *trans* diol system *via* mCPBA treatment of **19** led to *in-situ* intramolecular opening of the generated epoxide by the proximal OH group and isolation of compound **21**.

Table 22-2. Modulation of insulin secretion.

Compound number	% change at 2.8 mM glucose	% change at 16.7 mM glucose
18	- 18	+44
20	- 60	- 23
21	0	- 70
22	- 60	0
23	- 45	- 35
24	- 30	- 30
25	- 60	- 45

Compounds **24** and **25** were synthesised by routes identical to those described for compounds **18** and **21** but starting from the Diels-Alder adduct obtained from benzoquinone and γ-terpinene as starting material. The biological activity of these compounds is presented together with their chemical structures in Table 22-2 and Scheme 22-10.

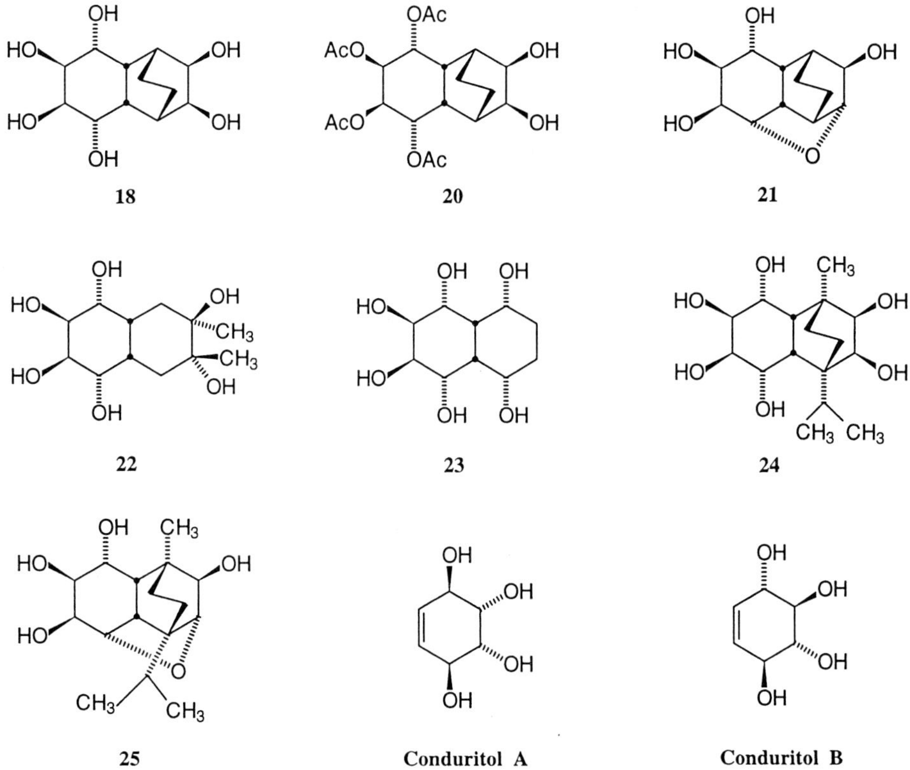

Scheme 22-10

The compounds were tested under identical conditions to those used for the biological evaluation of conduritols A to E, and once again the results are presented as a percentage change in insulin secretion at two different glucose concentrations in the presence of 10^{-4}M drug in Table 22-2. Several interesting points of comparison arise from the data presented in Tables 22-1 and 22-2. In contrast to the conduritols which show a range of both stimulatory and inhibitory effects, the bicyclic analogues presented are almost all inhibitors of insulin secretion at all glucose concentrations. Compounds **24** and **25** for example show marked suppression of insulin secretion at both high glucose and low glucose concentration. In contrast, compound **22** shows marked inhibition at low glucose but appears to have no effect at high glucose, whereas compound **21** has no effect at low glucose but shows marked suppression of insulin secretion at high glucose concentrations.

The biological activity reported here show that the conduritol isomers and our synthetic analogues provide useful pharmacological tools for studies of the modulation of insulin secretion *in vitro* and may provide new leads for the development of modulations of insulin secretion of therapeutic potential.

22.3 References and Notes

1. *Present address* : Department of Pharmaceutical and Biological Sciences, Aston University, Aston Triangle, Birmingham, B4 7ET, UK.

2. K. R. Shanmugasundaram, C. Panneerselvam, P. Samudram, E. R. B. Shanmugasundaram, *J. Ethnopharmacology* **1983**, *7*, 205.

3. K. R. Shanmugasundaram, C. Panneerselvam, P. Samudram, E. R. B. Shanmugasundaram, *Pharm. Res. Com.* **1981**, *13*, 475.; C. Day, *New Antidiabetic Drugs*; (Eds C. J. Bailey, P. R. Flatt), Smith-Gordon : London, **1990** ; pp. 267.

4. *European Patent* **EP-474-358-A** to DAI-Nippon Sugar Co, **1992**; *Derwent* **92-081887/11.**

5. M. Balci, Y. Sutbeyaz, H. Secen, *Tetrahedron* **1990**, *46*, 3715.

6. H. Kindl, O. Hoffmann-Ostenhof, *Fortschrit. Chem. Organ. Naturst.* **1966**, *24*, 149.

7. In the general context of cyclitols and insulin action, derivatives of *chiro*-inositol derived from the GPI anchor molecule have been shown to exert weak insulin like activity, and these compounds may represent mediators of insulin action . See for example : S. V. Ley, L. L. Yeung, *Synlett.* **1992**, 997-998. K. K. Reddy, J.R. Falck, J. Capdevila, *Tetrahedron Lett.* **1993**, *34*, 7869-7872.

8. See for example : S. Takano, T. Yoshimitsu, K. Ogasawara, *J. Org. Chem.* **1994**, *59*, 54-57. H. A. J. Carless, K. Busia, Y. Dove, S. S. Malik, *J. Chem. Soc., Perkin Trans. I* **1993**, 2505. T. Akiyama, H. Shima, M. Ohnari, T. Okazaki, S. Ozaki, *Bull. Chem. Soc. Jpn* **1993**, *66*, 3760. C. R. Johnson, P. A. Ple, J. P., Adams, *J. Chem. Soc., Chem. Commun.* **1991**, 1006-1007. T. Hudlicky, H. Luna, H. F. Olivo, C. Andersen, T. Nugent, J. D. Price, *J. Chem. Soc., Perkin Trans. I* **1991**, 2907.

9. H. A. J. Carless, *J. Chem. Soc., Chem. Commun.* **1992**, 234-235.

10. T. Hudlicky, J.D. Price, H.F. Olivo, *Synlett.* **1991**, 645-647.

11. See for example : H.A.J. Carless, *Tetrahedron Asymmetry* **1992**, *3*, 795. T. Hudlicky, J. Rouden, H. Luna, S. Allen, *J. Am. Chem. Soc.* **1994**, *116*, 5099-5107. T. Hudlicky, H. F. Olivo, B. P. McKibben, *J. Am. Chem. Soc.* **1994**, *116*, 5108-5115.

12. ICI (Zeneca) France , 1 Avenue Newton, 92140, Clamart ; *cis* - 1,2-dihydrocatechol Cat No **31064.**

13. K. L. Platt, F. Oesch, *Synthesis* **1977**, *7*, 499.

14. R. A. Aleksejczyk, G. A. Berchtold, *J. Am. Chem. Soc.* **1985**, *107*, 2554-2555.

15. R. Criegee, P. Becher, *Chem. Ber.* **1957**, *90*, 2516.

16. P. E. Lacy, M. Kostianowski, *Diabetes* , **1967**, 35.

17. **Method :** Islets of Langerhans were obtained from male Sprague-Dawley rats by collagenase digestion, according to the method of Lacy [15]. Routinely 150 islets were obtained from each rat. Following isolation, islets were incubated at 37°C in Kreb's Ringer bicarbonate buffer at pH7.4, containing 0.5% bovine serum albumin and either 2.8 or 16.7 mM glucose. These concentrations were chosen to achieve basal (2.8 mM) or maximal (16.7 mM) stimulation of insulin secretion from the islets.After incubation for 4 h two 50 ml aliquots from each incubate were sampled and used for measurement of insulin (Pasedeph kit, Pharmacia, France). Data are presented as the percentage increase or decrease in insulin secretion compared to the mean secretion obtained at each glucose concentration, in the absence of drug. Under these conditions, the mean basal and maximally stimulated rates of insulin secretion were 14.5 +/- 2.6 (n=8) and 27.5 +/- 3.2 (n=8). All synthetic compounds were tested at 10^{-4} M with 8 replicates per incubation.

18. For a related Diels-Alder strategy which appeared after this work was completed, see : T. Hudlicky, B. P. McKibben, *J. Chem. Soc., Perkin Trans. 1* **1994**, 485.

23 Strategies for the Synthesis of Inositol Phosphoglycan Second Messengers

N. Khiar and M. Martin-Lomas

23.1 Introduction

The role of membrane lipids in transmembrane signalling is now well established. The action of extracellular agonists results in the activation of several lipases which give rise to the generation of lipid-derived second messengers. These second messengers comprise a growing list of compounds such as inositol 1,4,5-triphosphate-diacyl-glycerol, ceramide, lysophosphatidic acid, eicosanoids, etc [1,2]. A new pathway for receptor-mediated signal transduction seems to be related to the production of inositolphosphoglycans (IPGs) after phospholipase-mediated hydrolysis of glycosyl-phosphatidyl inositols (GPIs) [3-5]. (Scheme 23-1)

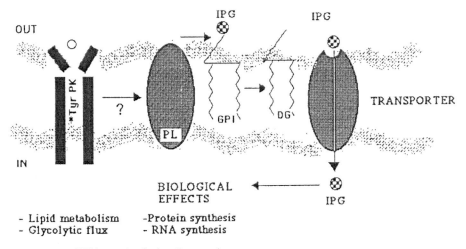

Scheme 23-1

A number of glycosyl-phosphatidylinositols serve to attach proteins or polysaccharides to the outer face of cellular membranes through a covalent linkage [6-9]. A second group of mammalian GPIs, the free GPIs, serves as precursors to biologicaly active IPG whose phospholipase-mediated production is modulated by insulin and by a variety of other hormones, cytokines, and neurotrophic and growth factors [3-5].

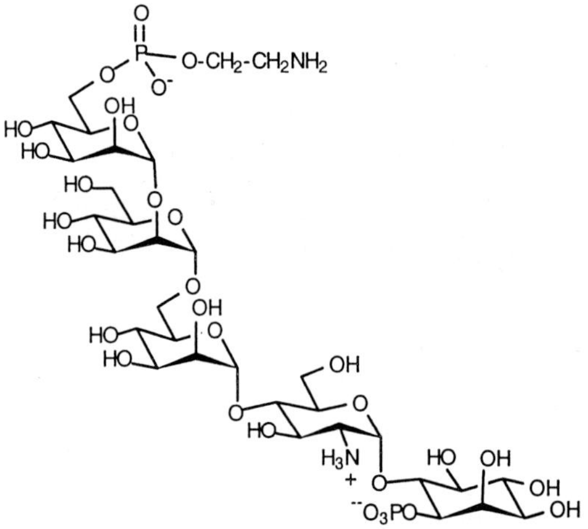

Figure 23-1

The detailed structures of several GPI protein anchors have been determined in recent years. These include those of *Trypanosoma brucei* variant surface glycoprotein (VSG), rat brain Thy-1 antigen [11], human erythrocyte acetylcholinesterase [12] *Leishmania major* promastigote surface protease (PSP) [13] and *Trypanosoma cruzi* IG7 antigen [14]. On the basis of these structures a consensus GPI-anchor structure can be postulated (Figure 23-1) with conserved structural features. As examples of these anchoring structures those corresponding to the VSG of *Trypanosoma brucei* and to rat brain Thy-1 antigen are shown in Figures 23-2 and 23-3.

The precise chemical structure of free GPI, however, remains to be elucidated. Present evidence indicates that there exist at least two distinct families of molecules. The first family gives rise to the generation of IPGs that inhibit c-AMP-dependent protein kinase and are composed of *myo*-inositol, non-acetylated glucosamine, galactose and phosphate [15,16].

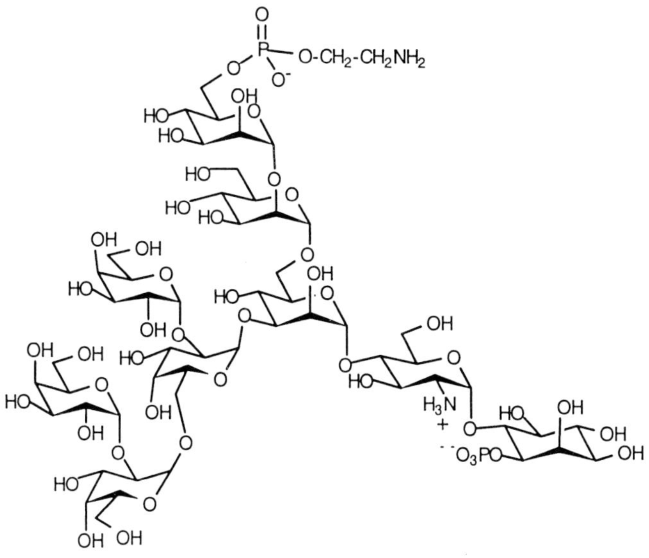

Figure 23-2

The second family produces IPGs that activate pyruvate dihydrogenase phosphatase and are composed of D-*chiro*-inositol, non-acetylated galactosamine, mannose and phosphate [17,18].

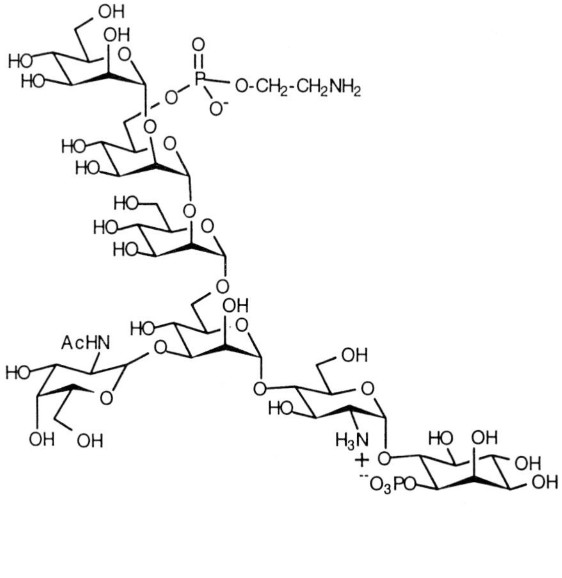

Figure 23-3

23.2 The Structure of the Inositol Phosphoglycan which Mediates the Intracellular Post-Receptor Action of Insulin

The complete structural determination of these IPG-mediators has been difficult because of the scarcity of purified material that can be obtained from mammalian tissues. In an attempt to circumvent this problem it was possible to obtain sufficient glycolipid from bovine liver to permit a partial structural characterization (Scheme 23-2).

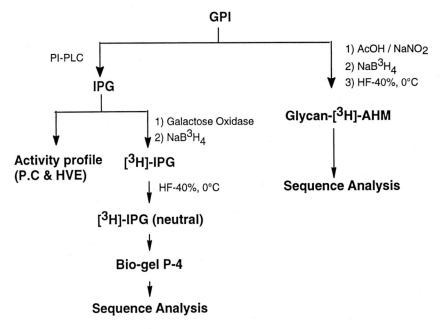

Scheme 23-2

Sequential exoglycosidase digestion of the fragments containing labelled glycans after derivatization confirmed that the label was present at the reducing end of several oligosaccharides that differed in the number of α-Gal residues (2-4 molecules) and phosphate groups. All the oligosaccharides also contained a N-acetylhexosamine moiety and an undetermined hexose residue [19]. These results can be accounted for by the partial structures indicated in Figure 23-4.

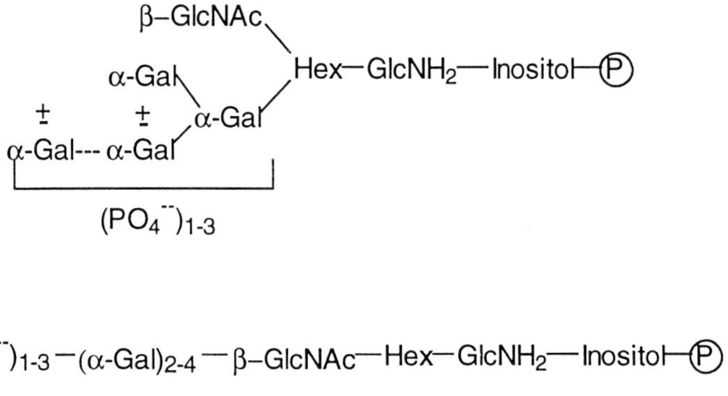

```
         β-GlcNAc
              \
     α-Gal         Hex—GlcNH₂—Inositol—Ⓟ
   ±    ±    α-Gal
 α-Gal---α-Gal
 |_____|
     (PO₄⁻⁻)₁₋₃
```

$$(PO_4^{--})_{1-3} - (\alpha\text{-Gal})_{2-4} - \beta\text{-GlcNAc} - \text{Hex} - \text{GlcNH}_2 - \text{Inositol} - \text{Ⓟ}$$

Figure 23-4

23.3 The Synthesis of Building Blocks for the Preparation of Inositolphosphoglycan Second Messengers

The elucidation of the inositolphosphoglycan pathway of signal transduction requires the development of effective strategies for the synthesis of building blocks which are needed for structural and biological investigations. The main challenges for an effective synthetic approach to the inositolphosphoglycans are:

1) Synthesis of conveniently functionalised enantiomerically pure D-*myo* or D-*chiro* inositol derivatives.

2) Stereoselective synthesis of the oligosaccharide moiety to be linked to the inositol residue.

3) Stereoselective glycosylation of the inositol moiety.

23.3.1 Synthesis of Conveniently Functionalised Enantiomerically Pure D-*myo* or D-*chiro* Inositol Derivatives

With regard to the synthesis of conveniently functionalised D-*myo* or D-*chiro* inositol, recent interest in inositol phosphate chemistry, and particularly in the synthesis of potential inhibitors of the phosphatidyl inositol cycle, has led to an impressive development in the chemistry of *myo*-inositol [20]. However very few useful regioselective reactions have been reported and most of the work which has appeared involves quite a number of protection and deprotection steps. The main problems when developing effective regioselective reactions with *myo*-inositol are the slight differences of reactivity among the hydroxyl groups, and, of practical importance, the very low solubility of *myo*-inositol in most organic solvents.

Following the reported work by Köster and Dahloff [21] on the quantitative O-stannylation of polyhydroxy compounds *via* per-*O*-diethylboryl derivatives, we have developed an efficient route to conveniently functionalised enantiomerically pure *myo*-inositol derivatives (Scheme 23-3). Starting from *myo*-inositol (**1**), reaction with "activated" triethyl borane at room temperature gave the hexane-soluble hexa(diethylboryl)derivative **2** which was submitted to partial transmetallation using the tributylstannyl enolate of acetyl acetone. The intermediate partially stannylated-borylated species reacted in toluene with different alkylating and acylating agents to give, in good yield, a regioselectively O-alkylated or O-acylated *myo*-inositol derivative **3** [22]. When the acylating agent was an optically active compound, this method permitted the resolution of *myo*-inositol. This has been achieved with success by using menthyl chloroformate as the chiral source [23]. Better yield and selectivity were obtained when using dibutyltin-bis-acetyl-acetonate as stannylating agent. Thus 1-*O*-menthyloxy and 3-*O*-menthyloxycarbonyl *myo*-inositol (**4**) and (**5**) could be obtained in 40 and 30% yield respectively, and fractional recrystallisation afforded these derivatives in optically pure form. Acetalation of the desired 1-*O*-menthyloxycarbonyl derivative **4** led to the building blocks **6** and **7** ready to be glycosylated either at position 6 or 4 and phosphorylated at position 1.

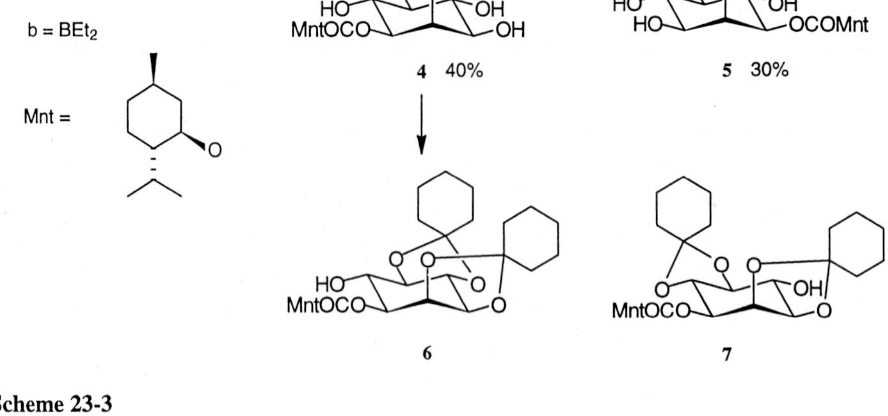

Scheme 23-3

We have also developed an alternative route to conveniently functionalised *myo-* or D-*chiro*-inositol derivatives based on a Ferrier carbocyclisation strategy (Scheme 23-4).[24]

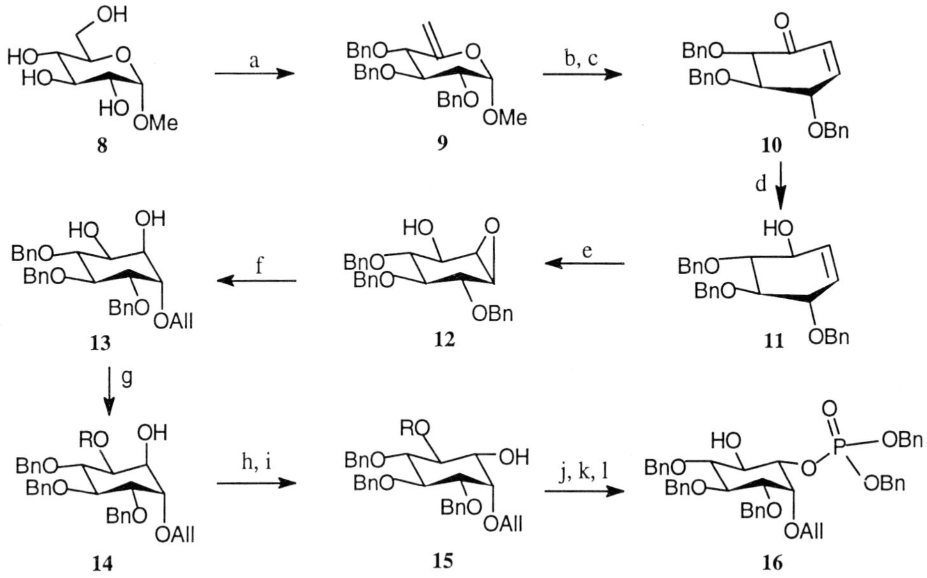

Scheme 23-4: Reagents: a) 5 steps, 52%; b) HgCl$_2$, Me$_2$CO/H$_2$O, 100°C; c) MsCl, DMAP, Pyridine, 63%; d) CeCl$_2$.7H$_2$O, NaBH$_4$, -50°C, 97%; e) mCPBA, CH$_2$Cl$_2$, rt, 93%; f) AllOH, BF$_3$.Et$_2$O, CH$_2$Cl$_2$, rt, 75%; g) Bu$_2$SnO, Bu$_4$NBr, RBr, 3Å MS, MeCN, 80°C, 85%; h) PCC, CH$_2$Cl$_2$, rt; i) (R)Alpine hydride, THF, -40°C, 58%; j) i-Pr$_2$NP(OBn)$_2$, tetrazole, MeCN, CH$_2$Cl$_2$; k) RuCl$_3$.3H$_2$O (cat), NaIO$_4$, CH$_2$Cl$_2$, MeCN, H$_2$O; l) CF$_3$COOH, CH$_2$Cl$_2$, 70%.

Thus starting from methyl α-D-glucopyranoside **8**, enone **10** was obtained in seven steps on a multigram scale. Stereoselective reduction of the carbonyl group gave the allylic alcohol **11** in 97% yield. Epoxidation of **11** with m-CPBA afforded the key epoxide **12** [25], an important intermediate for the synthesis of the glucosidase inhibitor conduritol B epoxide. Acid-catalysed trans-diaxial opening of the epoxide ring of **12** afforded the D-*chiro*-inositol derivative **13**, which was transformed into **14**, a key intermediate in the preparation of different glycosylated and phosphorylated D-*chiro*-inositol derivatives [26]. Compound **14** was transformed into the enantiomerically pure *myo*-inositol derivative **15** by inversion at C-1 through a two step oxidation-reduction reaction. Phosphorylation and subsequent deprotection afforded the *myo*-inositol derivative **16** ready for glycosylation at position 6 [27].

23.3.2 Synthesis of Conveniently Functionalised Glucosamine Unit.

2-Azido-2-deoxy derivatives of mono and disaccharides are frequently used intermediates in the synthesis of 2-amino-2-deoxy-oligosaccharides. We have developed a new and general strategy to access these compounds from the corresponding amine based on a diazo transfer from trifluoromethanesulfonyl azide (TfN$_3$) (Scheme 23-5). Thus treatment of D-glucosamine hydrochloride **17** with sodium methoxide in methanol followed by triflyl azide, and subsequent acetylation led to peracetylated 2-azido-2-deoxy-D-glucopyranose **18** in 91% yield. In the same way the 2-azido-2-deoxy derivatives of mannose and galactose were prepared in 65 and 70% yield respectively [28].

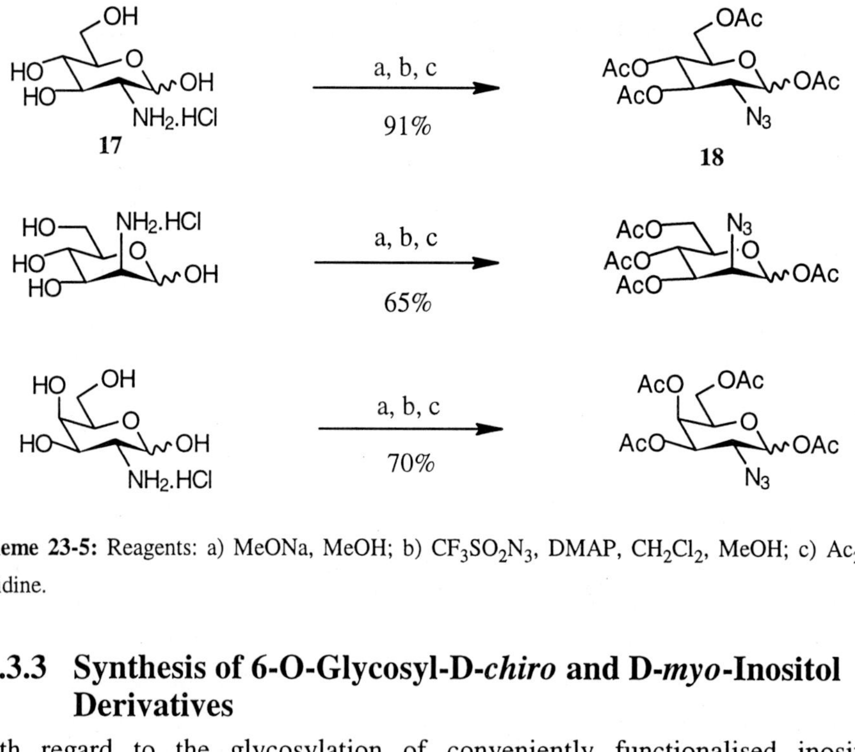

Scheme 23-5: Reagents: a) MeONa, MeOH; b) CF$_3$SO$_2$N$_3$, DMAP, CH$_2$Cl$_2$, MeOH; c) Ac$_2$O, pyridine.

23.3.3 Synthesis of 6-O-Glycosyl-D-*chiro* and D-*myo*-Inositol Derivatives

With regard to the glycosylation of conveniently functionalised inositol derivatives, after extensive investigation it was found that the trichloroacetimidate method [29] gave the best yields and selectivity. Thus using the trichloroacetimidate derivative **19** as glycosyl donor and trimethylsilyl triflate as promoter, the *myo*- inositol acceptor **16** was glycosylated in 65% yield. The disaccharide **20** was then fully deprotected to give 6-O-glycosyl-D-*myo*-insitol-1-phosphate **I** in 40% yield [24].

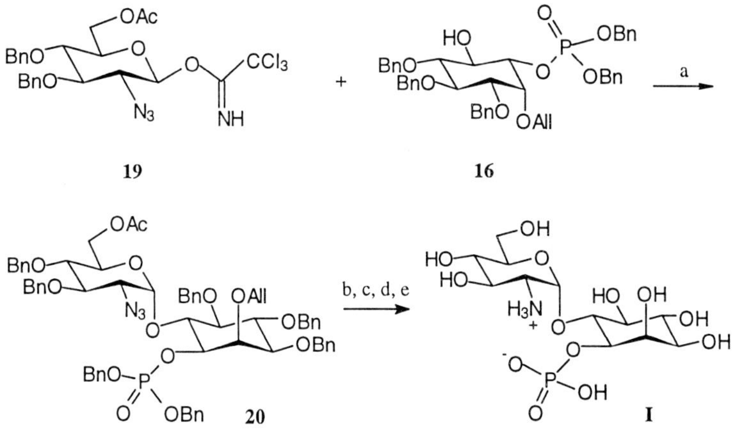

Scheme 23-6: Reagents: a) TMSOTf, CH$_2$Cl$_2$, -20°C; b) NH$_3$, MeOH; c) [Ir(COD) (Ph$_2$PMe)$_2$] PF$_6$, H$_2$, THF; d) I$_2$, THF, H$_2$O; e) H$_2$, Pd/C, EtOH, H$_2$O, 28%.

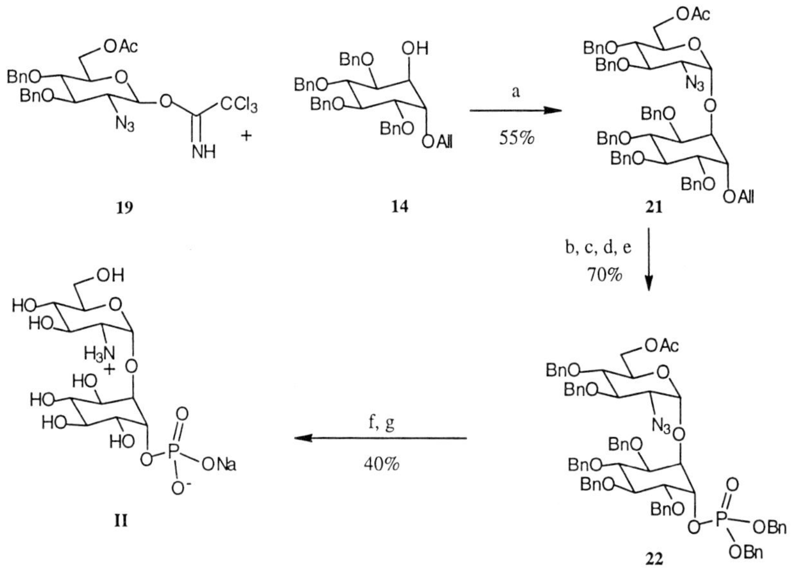

Scheme 23-7: Reagents: a) TMSOTf, CH$_2$Cl$_2$, -20°C, 65%; b) [Ir(COD)(Ph$_2$PMe)$_2$]PF$_6$, H$_2$, THF; c) I$_2$, THF, H$_2$O; d) i-Pr$_2$NP(OBn)$_2$, tetrazole, MeCN, CH$_2$Cl$_2$; e) RuCl$_3$, 3H$_2$O, NaIO$_4$, CH$_2$Cl$_2$/MeCN/H$_2$O; f) NH$_3$, MeOH, THF g) H$_2$, Pd/C, MeOH, NaOAc/AcOH.

Glycosylation of the D-*chiro*-inositol derivative **14** using the same conditions (Scheme 23-7) gave the disaccharide **21** in 55% yield. Deallylation and subsequent phosphorylation using the phosphoramidite procedure, gave the fully protected intermediate **22** in 70% yield. Deacylation and hydrogenation afforded 6-*O*-glycosyl-D-*chiro*-inositol-1-phosphate **II** [27].

23.4 Evaluation of the Insulin-Like Activity of 4-*O*-Glucosamine-*myo*-Inositol-1-Phosphate, 6-*O*-Glucosamine-*myo*-Inositol-1-Phosphate, 6-*O*-Glucosamine-*myo*-Inositol-1,2-Cyclic Phosphate

By using the above synthetic strategies several glycosyl inositols were obtained (Figure 23-5). Disaccharides **I**, **II**, and **III** were then evaluated for their insulin mimetic activities by investigating their capacity to induce proliferation of the otic vesicle of chick embryo and the associated cochleovestibular ganglion.

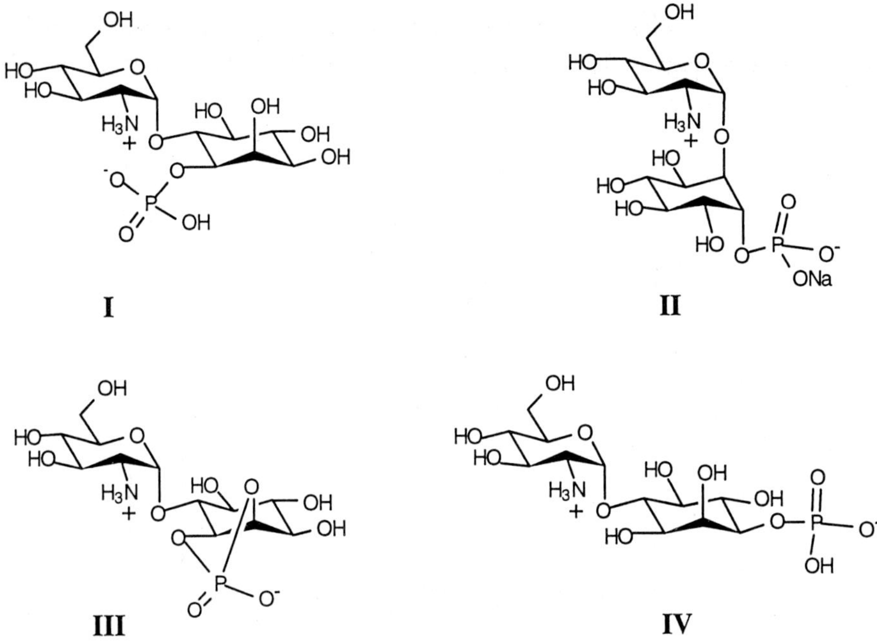

Figure 23-5

Figure 23-6 shows the vesicular growth in the presence or absence of foetal calf serum, the proliferation induced by insulin like growth factor-I and by cyclic phosphate **III**. Cyclic phosphate **III** was able to stimulate cell proliferation 2.2-fold by itself.

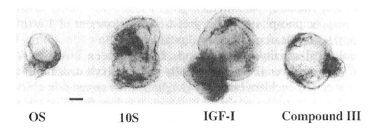

OS   10S   IGF-I   Compound III

Figure 23-6

Figure 23-7 shows the parallel measurement of the incorporation of [^{3}H] thymidine. Compound **III** stimulated 2.2 fold the incorporation of [^{3}H] thymidine into the otic vesicle [30].

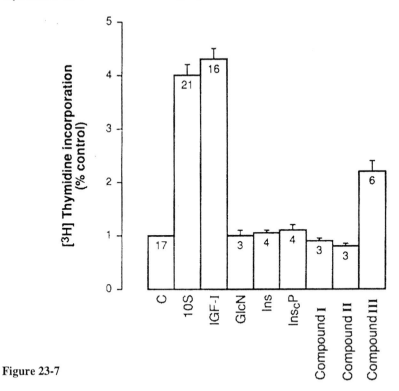

Figure 23-7

23.5 Toward the Total Synthesis of the Pseudo octa-saccharide V, Putative Second Messenger of the Hormone Insulin.

The next step was directed towards the total synthesis of model oligosaccharide **V**, one of the possible structures which can be drawn from the branched general structure in Figure 23-4

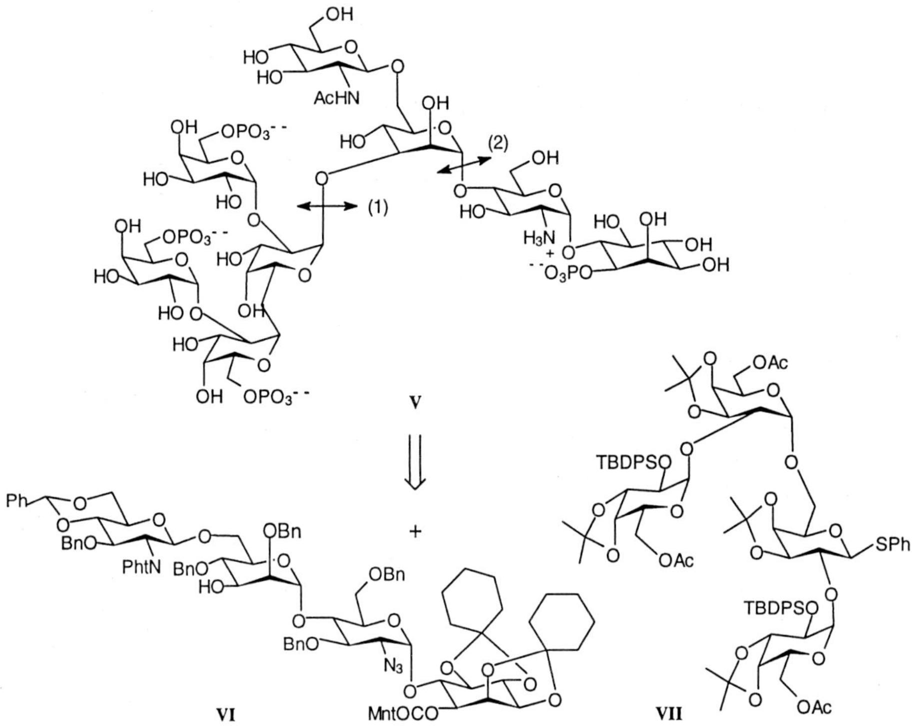

Scheme 23-8

Scheme 23-8 shows the retrosynthetic analysis of compound **V**. The first disconnection (1) generates two tetrasaccharide blocks **VI** and **VII**. With regard to the phosphorylated sites of the inositol ring and in the tetragalactose moiety, the ideal strategy would be to use protective groups that could be selectively removed in any order once the entire molecule has been assembled.

23.5.1 Synthesis of the Tetrasaccharide VII [31]

The second disconnection (2) shows that tetrasaccharide **VI** can be made by a 2+2 coupling between disaccharide **VIII** and **IX** (Scheme 23-9).

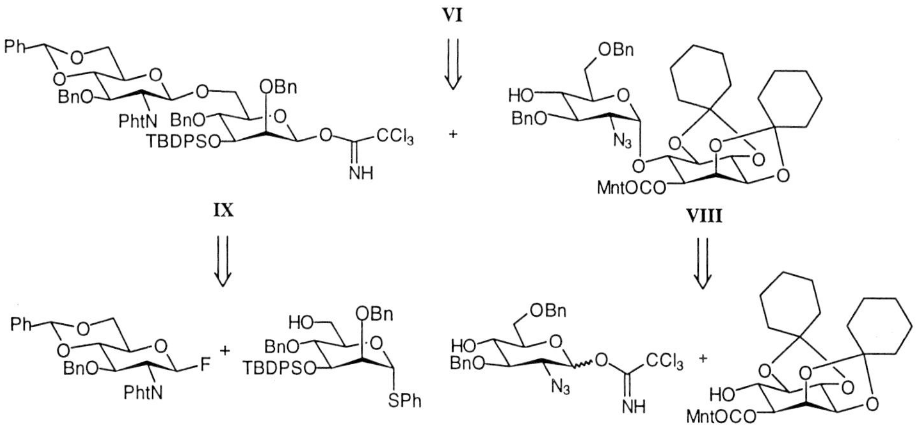

Scheme 23-9

With regard to the synthesis of disaccharide **VIII**, as we have seen before, the trichloroacetimidate glycosylation method demonstrated once again its efficiency (Scheme 23-10).

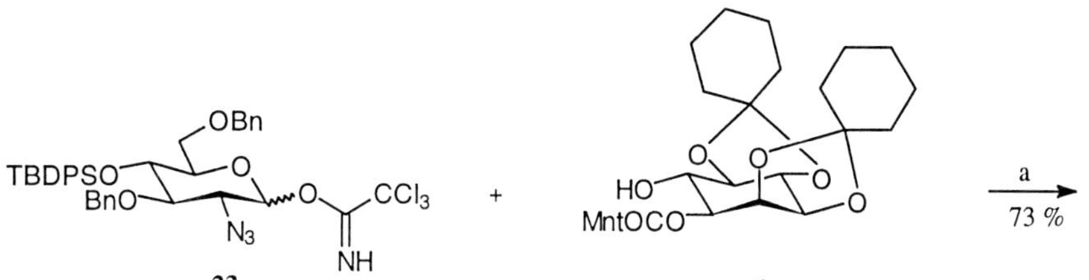

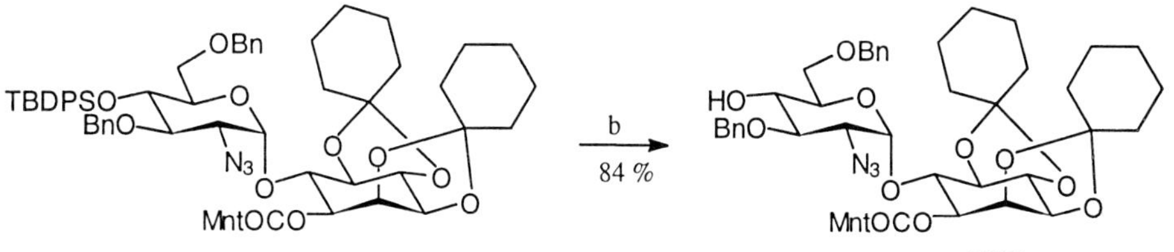

Scheme 23-10: Reagents: a) TMSOTf, Et$_2$O; b) TBAF, THF.

Using the conveniently functionalised trichloroacetimidate donor **23**, the inositol acceptor **6** was glycosylated in 73% isolated yield (95% based on the recovered alcohol). Removal of the silyl group using tetrabutylammonium fluoride proceeded in 84% yield giving the desired disaccharide acceptor **VIII** ready for glycosylation at the 4 position.

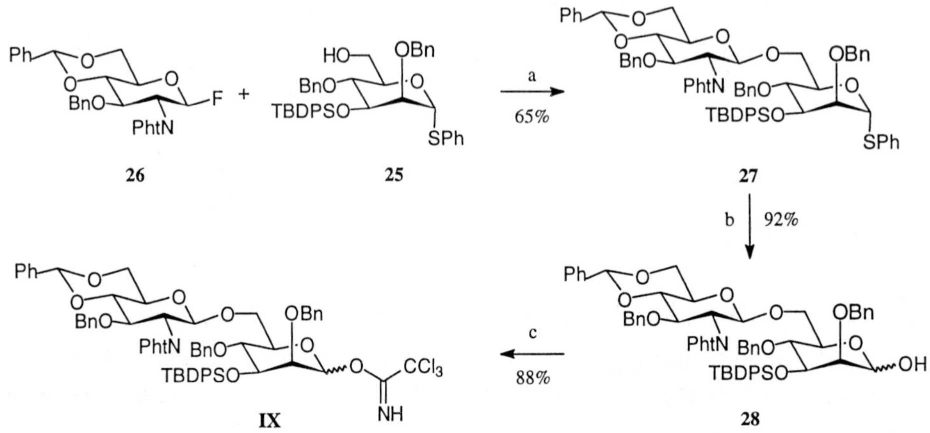

Scheme 23-11: Reagents: a) Cp_2ZrCl_2, AgOTf, CH_2Cl_2,; b) NBS, Me_2CO/H_2O; c) CCl_3CN, K_2CO_3, CH_2Cl_2.

For the synthesis of the second disaccharide **IX** a different glycosylation reaction, using a glycosyl fluoride as the glycosyl donor, was prefered [32] (Scheme 23-11). Thus using Suzuki´s conditions ($ZrCl_2$ and AgOTf) [33] the primary alcohol of the mannose derivative **25** was glycosylated in 63% yield with the fluoride donor **26**. It has been shown that the low yield of the glycosylation reaction is due to the presence of the diphenylsilyl group at position 3. When the protective group at that position was chloroacetate, the yield of the glycosylation reaction was practically quantitative. Unfortunately the chloroacetate group did not suit the experimental conditions required in the following steps. Hydrolysis of the thioglycoside using Nicolaou´s conditions (NBS in acetone and water) led to **28** in 92 %. Reaction of **28** with trichloacetonitrile and potassium carbonate gave the trichloroacetimidate derivative **IX** in 88% yield ready for the next step. The 2+2 coupling of disaccharide acceptor **VIII** and disaccharide donor **IX** took place with complete stereoselection and gave tetrasaccharide **VI** in 65% isolated yield (Scheme 23-12)

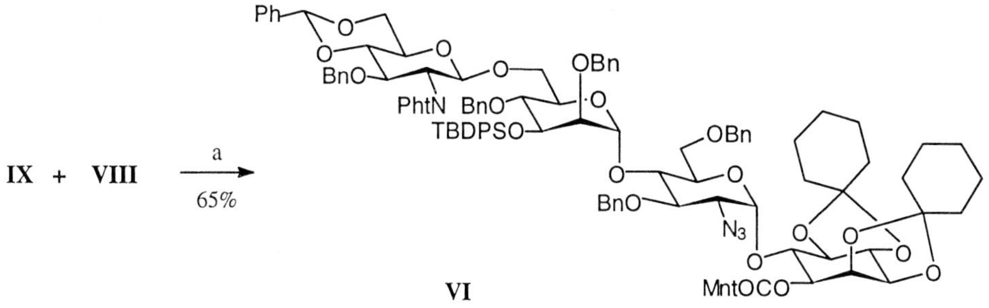

IX + VIII $\xrightarrow[65\%]{a}$ VI

Scheme 23-12: Reagents: a) TMSOTf, Et$_2$O, 22°C.

23.5.2 Synthesis of the Tetrasaccharide VII

In seeking a short and highly convergent route to tetrasaccharide **VII**, a new approach based on the sulfoxide glycosylation reaction [34] was developed. A retrosynthetic analysis shows that the target molecule **VII** could be obtained from diol **X** via a double glycosylation reaction (Scheme 23-13).

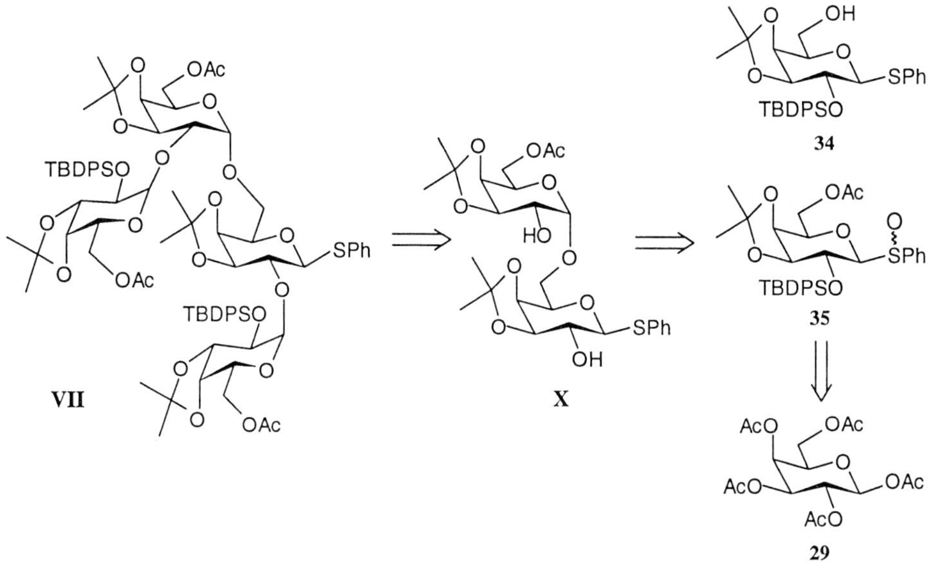

Scheme 23-13

Diol **X** was obtained from the galactosyl acceptor **34** and the galactosyl donor **35**(**S, R**), both accessible from β-D-galactose pentaacetate **29**. The three glycosidic bonds were formed stereoselectively using the sulfoxide methodology in only two glycosylation steps using a single galactosyl donor **35**.

Additionally an advanced intermediate **33** (Scheme 23-14) obtained on a multigram scale, was used for the synthesis of both galactosyl acceptor **34** and the sole galactosyl donor **35** by a single transformation in each case, making this approach highly convergent. Using triflic anhydride as promoter both sulfoxides **35R** and **35S** reacted similarly, so the glycosylation reaction could be performed using either an epimerically pure sulfinyl donor or the mixture of the two diastereomeric sulfoxides. Thus when sulfoxides **35**(*R, S*) were treated with triflic anhydride followed by the acceptor **34** and 2,6-di-tert-butyl-4-methylpyridine, disaccharide **36** was obtained in 78% yield and in 86:14 α:β ratio. Treatment of the desired disaccharide **36α** with large excess of TBAF in dry THF gave quantitatively the desired diol **X** suitable for the next reaction.

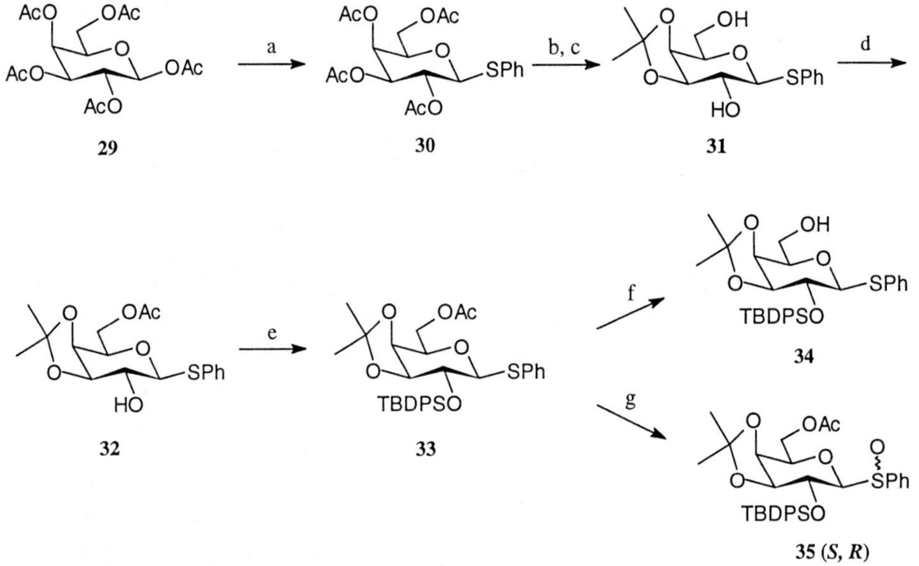

Scheme 23-14: Reagents: a) PhSH, BF$_3$.Et$_2$O, CH$_2$Cl$_2$, rt; b) MeONa, MeOH, rt; c) DMP, Acetone, H$^+$, rt, 75% three steps; d) AcCl, collidine, CH$_2$Cl$_2$, -78°C, 88%; e) TBDPSCl, Imidazole, DMAP, DMF, 85%; f) MeONa, MeOH, 86%; g) mCPBA, CH$_2$Cl$_2$, -78°C, 95%.

Double glycosylation of the acceptor **X** with the same glycosyl donor **35**(*S,R*) using triflic anhydride as promoter in the presence of 2,6-di-tert-butyl-4-methylpyridine and 4Å molecular sieves at -78°C, gave the target tetrasaccharide **VII** in 40% isolated yield (Scheme 23-15) [35].

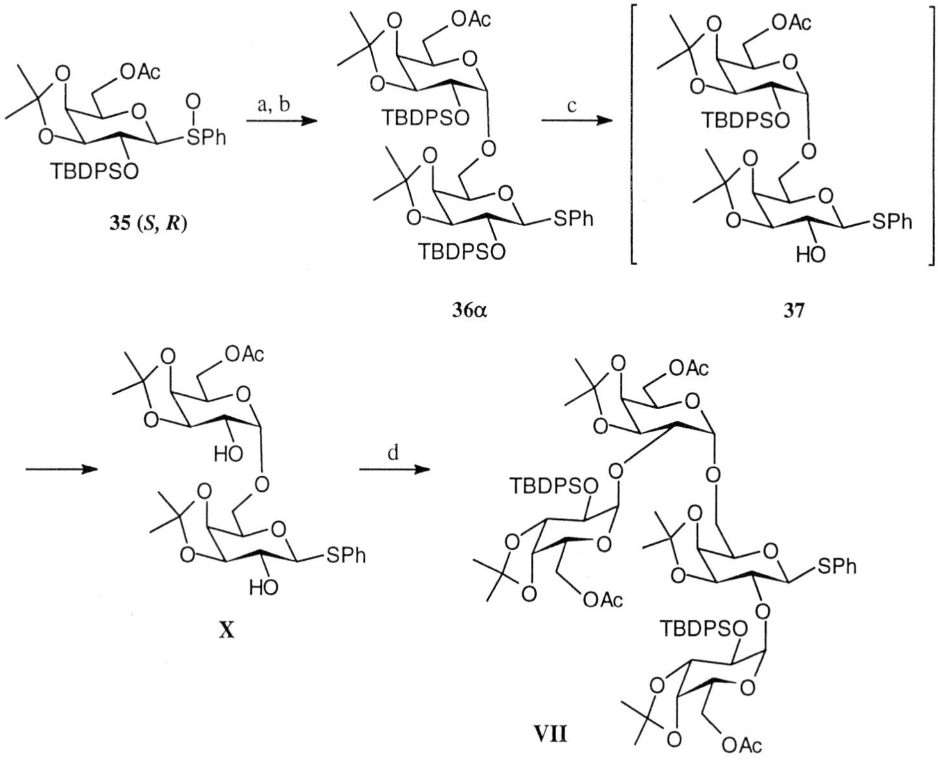

Scheme 23-15: Reagents: a) Tf$_2$O, 2,6-di-*tert*-butyl-4-methylpyridine, **34**, Et$_2$O:CH$_2$Cl$_2$, 3:1, -78°C, 78%, α:β 84:16; b) cc separation; c) TBAF, THF, quant; d) **35(S, R)**, 4 eq, Tf$_2$O, 2,6-di-*tert*-butyl-4-methylpyridine Et$_2$O-CH$_2$Cl$_2$, 3:1, -78°C to -40°C, 40%.

It is worth noting the useful and interesting stereochemical outcome of the double glycosylation step: only tetrasaccharide **VIII** with the desired α linkages in the two newly created glycosidic bonds has been obtained and no other isomer has been observed. Finally tetrasaccharide **VII** was oxidized to the corresponding sulfoxide in 80% yield and is now ready to be linked to the rest of the IPG.

23.6 Conclusion

Inositolphosphoglycans are involved as second messengers in a new pathway of receptor-mediated intracellular signal transduction. These second messengers are generated by the phospholipase-mediated hydrolysis of glycosyl-phosphatidylinositols. The turnover of these molecules is modulated by insulin, cytokines and neurotrophic and growth factors. The elucidation of this new mechanism of signal transduction requires the development of effective and versatile strategies for the synthesis of inositolphosphoglycan building blocks, particulary the regioselective manipulation of optically active *myo*- and *chiro*-inositol derivatives and highly convergent methodologies for oligosaccharide synthesis.

Suitably protected *myo*-inositol derivatives have been prepared making use of a transmetallation reaction of perborylated *myo*-inositol with dibutyl-tin-bis-acetylacetonate and subsequent treatment with an optically active acylating agent. Thus, regioselective substitution and optical resolution have been achieved simultaneously in one pot. D-*Chiro* inositol derivatives were prepared using a different approach from methyl α-D-glucopyranoside through a Ferrier carbocyclization reaction.

These D-*chiro* and D-*myo*-inositol compounds have been glycosylated with 2-azido-2-deoxy-D-glucose-derived glycosyl donors to obtain the 2-amino-2-deoxy-α-D-(*myo*- or *chiro*-)- inositols **I**, **II**, and **III**. The glycosyl donors have been prepared from D-glucosamine hydrochloride using a diazo transfer reaction with triflyl azide, which has now been extended and investigated in the carbohydrate field. Some of the glycosyl inositols thus prepared induced proliferation in the inner ear of chick embryo mimicking an insulin-like activity in this system.

Using these glycosyl inositol derivatives as building blocks, the synthesis of a model octasaccharide (**V**) whose structure, according to structural data, may represent one member of the families of inositolphosphoglycan second messengers, was attempted. Trichloroacetimidate, and fluoride, based glycosylation reactions afforded tetrasaccharide **VI** in good yield. The sulfoxide glycosylation method allowed the preparation of tetrasaccharide **VII** in only two glycosylation steps and from a common precursor.

Summarizing, a number of effective and versatile strategies have been developed that permit the preparation of building blocks for inositolphosphoglycan synthesis. These building blocks allow the preparation of a variety of compounds that are needed for structural and biological investigation within the framework of a wide research program devoted to the elucidation of the inositolphosphoglycan-mediated signalling pathway.

23.7 References

1. J. M. Mato, *Phospholipid Metabolism in Cellular Signalling*, CRC Press, Boca Raton **1990**

2. N. Divecha, R. F. Irvine, *Cell*, **1995**, *80*, 269-278.

3. E. Kilgour, *Cell Signal*. **1993**, *5*, 97-105.

4. G. N. Gaulton, J. C. Pratt, *Seminar in Immunol.* **1994**, *6*, 97-104.

5. T. W. Rademacher, H. N. Caro, S. Kunjara, D. Y. Wang, A. L. Greenbaum, P. Mc Lean, *Braz. J. Med. Biol. Res.* **1994**, *27*, 327-341.

6. M. A. J. Fergusson, A. F. Williams, *Ann. Rev. Biochem.* **1988**, *57*, 285-320.

7. J. R. Thomas, R. A. Dwek, T. W. Rademacher, *Biochemistry.* **1990**, *29*, 5413-5422.

8. M. A. J. Fergusson, J. S. Brimacombe, S. Cottaz, R. A. Field, L. S.Guther, S. W. Homans, M. J. Mc Conville, A. Mehlert, K. G. Milne, J. E. Ratton, Y. A. Roy, P. Schneider, N. Zitzmann, *Parasitol,* **1994**, *108,* S 45-54.

9. M. A. J. Fergusson, *Glycobiology: A Practical Approach*, (Ed. : M. Fakuda, A. Kobata), IRL Press, **1993**, 349-383.

10. M. A. J. Fergusson, S. W. Homans, R. A Dwek, T. W. Rademacher, *Science*, **1988**, *239*, 753-759.

11. S. W. Homans, M. A. J. Fergusson, T. W. Rademacher, R. Anad, A. F. Williams, *Nature*, **1988**, *333*, 269-272.

12. W. L. Roberts, S. Santikarn, V. N. Reenhold, T. L. Rosenbary, *J. Biol. Chem.* **1988**, *263*, 18776-18784.

13. P. Schneider, M. A. J. Fergusson, M. J. Mc Conville, A. Melhert, S. W. Homans, C. Bordier, *J. Biol. Chem.* **1990**, 265, 16955-16964.

14. M. L. S. Guther, M. L. Cardoso de Almeida, N. Yoshida, M. A. J. Fergusson, *J. Biol. Chem.* **1992**, *267,* 6820-6828.

15. J. M. Mato, K.L. Kelly, A. Alber, L. Jarret, B. E. Corkey, J. A. Cashel, D. Zopf, *Biochem. Biophys. Res. Commun.* **1987**, *146*, 764-770.

16. J. M. Mato, K.L. Kelly, A. Alber, L. Jarret, *J. Biol. Chem.* **1987**, *262,* 2131-2137.

17. J. Larner, L. C. Huang, C. F. W. Scwartz, A. S. Oswald, T. Y. Shen, M. Kinter, G. Tang, K. Zeller, *Biochem. Biophys. Res. Commun.* **1988**, *151*, 1416-1426.

18. V. Pale, J. Larner, *Biochem. Biophys. Res. Commun.* **1992**, *184*, 1042-1047.

19. H. N. Caro, R. A. Dwek, T. V. Rademacher, J. M. Mato, M. Martin-Lomas , Unpublished results.

20. D. C. Billington, *Chem. Soc. Rev.* **1989**, *18*, 83-122; B. V. L. Potter, D. Lampe *Angew. Chem. Int. Ed. Engl.* **1995**, *34*, 1933-1972 and references cited therein.

21. K. M. Taba, R. K. Köster, W. V. Dahloff, *Synthesis*, **1984**, 399-401.

22. A. Zapatta, R. Fernandez de la Pradilla, M. Martin-Lomas, S. Penadés, *J. Org. Chem.* **1991**, *56*, 445-447.

23. A. Aguilo, M. Martin-Lomas, S. Penadés, *Tetrahedron Lett.* **1992**, *33*, 401-404

24. R. Ferrier, *J. Chem. Soc., Perkin Trans 1*, **1979**, 1455-1458; R. J. Ferrier, S. Middleton, *Chem. Rev.* **1993**, *93*, 2779-2831.

25. C. Jaramillo, R. Fernandez de La Pradilla, M. Martin-Lomas, *Carbohydr. Res.* **1991**, *209*, 296-298.

26. C. Jaramillo, M. Martin-Lomas, *Tetrahedron Lett.* **1991**, *32*, 2501-2504.

27. C. Jaramillo, J-L. Chiara, M. Martin-Lomas, *J. Org. Chem.* **1994**, *59*, 3135-3141.

28. A. Vasella, C. Witzig, J-L Chiara, M. Martin-Lomas, *Helv. Chim. Acta*, **1991**, *74*, 2073-2077.

29. R.R. Schmidt, *Angew. Chem. Int. Ed. Engl.* **1986**, *25*, 212-235.

30. A. Zapata, Y. León, J. M. Mato, I. Varela-Nieto, S. Penadés, M. Martin-Lomas *Carbohydr. Res.* **1994**, *264*, 21-23.

31. M. Flores, J-L Chiara, M. Martin-Lomas, unpublished Results.

32. (a) T. Mukaiyama, Y. Murai, S. Shoda, *Chem. Lett.* **1981**, 431-435; (b) K. C. Nicolaou, R. E. Dolle, D. P. Papahatjis, J. L. Randall , *J. Am. Chem. Soc.* **1984**, *106*, 4189-4192; (c) K. C. Nicolaou, J. L. Randall, G. T. Furst, *J. Am. Chem. Soc.* **1985**, *107*, 5556-5558.

33. K. Suzuki, H. Maeta, T. Suzuki, T. Matsumoto, *Tetrahedron Lett.* **1989**, *30*, 6879-6882.

34. D. Kahne, S. Walker, Y. Cheng, D. Van Engen, *J. Am. Chem. Soc.* **1989**, *111*, 6881-6882.

35. N. Khiar, M. Martin-Lomas, *J. Org. Chem.* **1995**, *60*, 7017-7021.

24 The Catalytic Efficiency of Glycoside Hydrolases and Models of the Transition State Based on Substrate Related Inhibitors

G. Legler

24.1 Introduction

Enzymes are distinguished from simple, man-made catalysts by their great *specificity* with respect to reaction type and substrate structure and by their *efficiency* which causes selected reactions to proceed in seconds that would otherwise take years or centuries. Both aspects are linked by the transition state, a fleeting, energy rich species which has to be passed on the route from substrate(s) to product(s). According to current theory, the transition state is stabilized over the ground state by a multitude of weak interactions with the enzyme, thereby accelerating the reaction by lowering the free energy of activation.

The combination of a large number of weak interactions rather than a few strong ones ensures a greater specifity because more structural features of the substrate contribute to the binding energy and, more importantly, to the small conformational alterations of the enzyme summarized as "induced fit' which are required for efficient catalysis. With glycosidases, this is nicely demonstrated by the specific replacement of a hydroxyl group by hydrogen which does not introduce the steric repulsions shown by its acetylation or methylation. 2-Deoxy-ß-glucosides, for example, are hydrolyzed by ß-glucosidases from almonds [1] and *Asp. wentii* [2] 10^3- and $> 10^5$-fold slower than the corresponding glucosides. Binding constants, on the other hand, are not or much less affected. This effect on hydrolysis rates is all the more remarkable as 2-deoxyglucosides are hydrolyzed by dilute acids about three orders of magnitude *faster* than glucosides.

Table 24-1. Rate constants and activation parameters for the hydrolysis of ß-glucosides[3]

a) Without catalyst at pH 5.0

Aglycon	k (100 °C), s^{-1}	k (25 °C), s^{-1}	Δ H*, kJ/mol	Δ S*, kJ/mol
4-Nitrophenol	0.64 · 10^{-6}	52 · 10^{-12}	118	- 32
4-Methylumbelliferone	0.28 · 10^{-6}	7.7 · 10^{-12}	135	- 2.9
D-Glucose (Cellobiose)*	0.22 · 10^{-6}	13 · 10^{-12}	123	- 11

b) Enzyme catalyzed hydrolysis
  ß-Glucosidase A₃ from *Aspergillus wentii* at pH 5.0 and 25 °C

	k$_{cat}$, s^{-1}	k$_m$, mol/l	Δ H*, kJ/mol	Δ S*, kJ/mol
4-Nitrophenol	210	0.62 · 10^{-3}	70	+ 146
4-Methylumbelliferone	350	0.57 · 10^{-3}	68.5	+ 142
D-Glucose (Cellobiose)*	330	0.16 · 10^{-3}	not determined	
ß-Glucosidase B from almonds at pH 6.0 and 25 °C				
4-Nitrophenol	750	1.7 · 10^{-3}	28	+ 35
4-Methylumbelliferone	26	1.3 · 10^{-3}	46	+ 59

* The uncatalyzed cleavage of cellobiose is probably not hydrolytic but due to an elimination reaction. No hydrolysis could be observed with simple alkyl ß-glucosides and phenyl ß-glucoside during 24 h at up to 110 °C

24.2 Catalytic Efficiency and the Transition State

An estimate of the catalytic efficiency of an enzyme requires the comparison of rate constants for the enzymic reaction with those for the reaction without or with a simple inorganic catalyst. The uncatalyzed hydrolysis of glycosides can be observed only at high temperatures but from its temperature dependence the rate can be extrapolated to conditions where enzymes are stable, using the Arrhenius equation. With this approach we get acceleration factors exceeding 10^{12} from a comparison of k$_{cat}$ (cleavage rate of a substrate molecule bound at the active site) with the extrapolated hydrolysis rate in water (Table 24-1). In order to grasp the meaning of 10^{12} we should realize that this is the number of seconds in about 30,000 years. It should be noted that the uncatalyzed reactions can be observed only with glycosides having a good leaving group propensity like p-nitrophenol or 4-methylumbelliferone.

One may object to this mode of comparison that it exaggregates the enzymic efficiency because it is based on substrate molecules bound at the active site but at any moment only a small fraction is in this environment. Another mode of estimate would be a comparison of the enzymic reaction with the acid catalyzed one (Figure 24-1) under conditions where both follow the same rate law. Because of the extremely low basicity of glycosides ($pK_a \sim -6$) only a tiny fraction is present in its protonated form which may be likened to the enzyme substrate complex. The rate of hydrolysis is of first order with respect to the substrate *and* hydronium ion concentrations (eq. 1). For enzymes this rate law holds at substrate concentrations $[S] \ll K_m$ (eq. 2)

$$(1) \quad -\frac{d[S]}{dt} = k_{H_3O^+}[H_3O^+][S]$$

$$(2) \quad -\frac{d[S]}{dt} = k_{cat}/K_m [E][S]$$

Using published data [4] for the activation energy to extrapolate the second order rate constant $k_{H_3O^+}$ to ambient temperature we get values from 10^{-8} to 10^{-7} $M^{-1} s^{-1}$ for alkyl and aryl glucosides. Values for k_{cat}/K_m (Table 24-1), on the other hand, range from 2×10^4 to 2×10^5 $M^{-1} s^{-1}$ which shows enzymes to have a more than 10^{13}-fold higher catalytic efficiency than strong acids even though their only functional groups are the carboxyl groups of aspartic and glutamic acid or, possibly, the protonated imidazole residue of histidine. Even when the comparison is based on a weight basis rather than on molar concentrations glycosidases are still seen to be up to 10^{10}-fold better catalysts than acids.

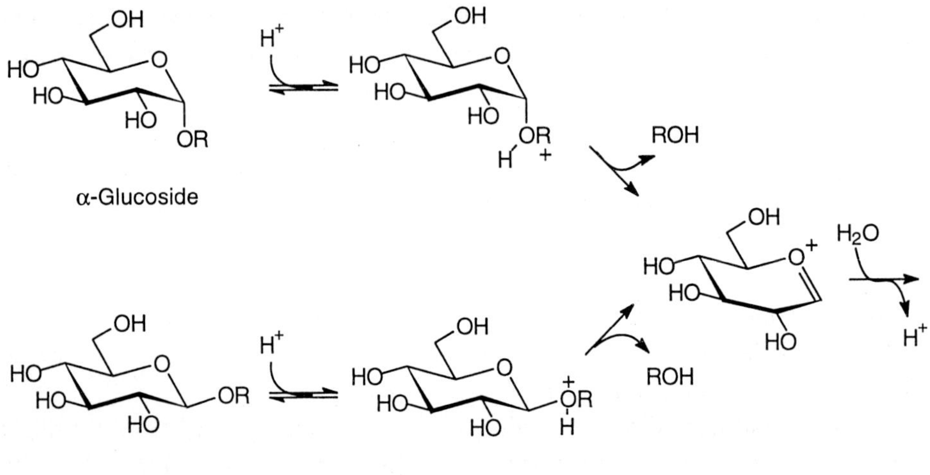

Figure 24-1. Acid catalyzed hydrolysis of α- and ß-D-glucopyranosides. All steps are reversible and the method can be used to prepare glycosides from sugars and alcohols.

According to the Eyring theory [5], the transition state is treated as an energy rich molecular species which decomposes with the frequency of molecular vibrations into products or back to the ground state. This species is considered to be in equilibrium with the ground state so that kinetics can be linked with thermodynamics. The number of molecules N* in the transition state is then related to their total number N_t by equ. 3, where ΔG^* is the Gibbs free energy of activation (equ. 4) with ΔH^* as the enthalpy (heat) of activation and ΔS^* as entropy of activation (entropy difference between transition and ground states).

(3) $N^*/N_t = \exp(-\Delta G^*/Rt)$ (4) $\Delta G^* = \Delta H^* - T \Delta S^*$

With reactions in solution, both terms of equ. 4 include the interactions of the ground and transition states with the solvent which cause reaction rates to be strongly solvent dependent. This is of great importance for enzymic reactions where the solvent shell is replaced by the interactions with the active site.

As reaction rates are proportional to N* the observed rate enhancement reflects the difference $\Delta\Delta G^* = \Delta G^*(enz) - \Delta G^*(uncat)$ between the enzymic and the uncatalyzed reactions. At 25°C (298 K) a 10^{12}-fold acceleration requires $\Delta\Delta G^*$ 16.3 kcal/mol (68.1 kJ/mol). A comparison of the activation parameters for the enzymic and uncatalyzed hydrolysis of glycosides shows (Table 24-1) for the enzymic and uncatalyzed hydrolysis of glycosides shows (Table 24-1) that glycosidases greatly reduce ΔH^* but, in addition, shift ΔS^* from negative to positive values. When the comparison is made with acid catalysis we find a shift from small to larger positive values. A shift to a larger positive ΔS^* can easily be rationalized for k_{cat}/K_m because a large contribution to ΔS^* comes from the entropy contribution of water molecules released from the hydration shell when the enzyme substrate complex is formed. A possible explanation for a positive ΔS^* when k_{cat} is compared with k_{uncat} is the negative entropy of the bound substrate relative to the freely dissolved state. The reaction within the ES-complex then involves a transition from a greatly restricted ground state to the more flexible transition state shortly before aglycon departure.

It was first pointed out by Pauling [6] that the active site of enzymes might have evolved towards optimal complementarity to the transition state rather than to the substrate in its ground state. Interactions with such an active site would stabilize the transition state, thereby lowering ΔG^* and enhancing the reaction rate. Based on the Eyring postulate of an equilibrium between the ground and transition states (equilibrium constant K^*_s) and the direct relation of transition state population with reaction rates Wolfendenhas derived an equation which gives the *virtual* dissociation constant of the transition state relative to the dissociation constant of the substrate in the ground state K_s (equ. 5, Scheme 24-1): [7]

$$
\begin{array}{ccccc}
 & K^*_s & & k_{uncat} & \\
 S & \rightleftharpoons & S & \longrightarrow & P \\
 + & & + & & \\
 E & & E & & \\
 K_s \updownarrow & K^*_{ES} \updownarrow & K_{TS} \updownarrow & & \\
 ES & \rightleftharpoons & ES & \xrightarrow{k_{cat}} & P + E \\
\end{array}
\qquad \frac{K_{TS}}{K_s} = \frac{k_{cat}}{k_{uncat}} \qquad (5)
$$

Scheme 24-1

Thus, K_{TS} should be smaller than K_s by the same factor as the reaction rate is enhanced by the enzyme. This means that a hypothetical compound having exactly the same structure as the transition state but lacking its reactivity should inhibit the enzyme with a K_i that is smaller than K_s (or K_m) by the same factor. Of course, it is not possible to synthesize such a compound but one can judge the resemblance of a proposed transition state with the "real' counterpart from the extent by which K_i of an analogue approaches K_{TS}. As will be shown later, we still have not even made half the way (expressed in powers of ten) towards a "perfect" analogue of the transition state of glycosidases.

As early as 1926 H. v. Euler [8] concluded from the pH-dependence of invertase activity that an acidic group with pK_a 6.6 might be involved in the hydrolysis of saccharose. In 1960 Fischer and Stein [9] extended the catalytic functionalities of glycosidases by proposing a concerted action (bifunctional catalysis) of an acidic (electrophilic) and a basic (nucleophilic) group. These more or less hypothetical ideas were put on a firm basis by Phillips and his group [10] by their now paradigmal model of lysozyme action derived from the crystal structure of an enzyme inhibitor complex (Figure 24-2). The essential features of this model are: i) a carboxyl group (Glu-35) to protonate the oxygen of the glycosidic bond when it is broken (general acid catalysis), ii) a carboxylate (Asp-52) to stabilize the positive charge remaining on the anomeric carbon on aglycon departure (nucleophilic catalysis), iii) facilitation of glycosyl cation formation by distortion of the ground state chair conformation towards a half-chair with sp^2-character of C-1. The glycosyl cation remaining after aglycon departure might thus be stabilized long enough to react stereospecifically with a water molecule to give the ß-configurated product.

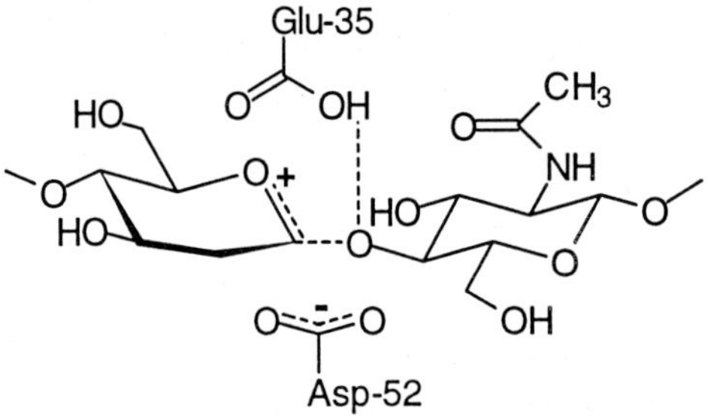

Figure 24-2. Model of the transition state for the cleavage of a $(GlcNAc)_n$ chito-oligosaccharide chain by lysozyme as proposed by Phillips [10]. The N-acetyl group of the "left" GlcNAc-residue has been omitted for clarity

24.2 Evidence for a Glycosyl Cation-Like Transition State

Extension of this model to other glycosidases received experimental support with respect to point iii) from inhibition studies with aldono-1,5-lactones, mainly by Levvy and coworkers[11] in the late sixties. The lactones were found to inhibit ß-specific enzymes up to 50,000-fold better than the corresponding sugars; α-glycosidases were only moderately inhibited.

Lysozyme, too, was inhibited only 120-fold better by chitotetraono-1,5-lactone than by chitotetraose. It was argued by Leaback[12] that the geometry of the lactones resembled that of a glycosyl oxocarbonium ion intermediate or cationic transition state and that the energy required to crowd the tetrahedral sugar (or substrate) into the binding site would be available as additional binding energy with the lactones. Considerations of protein flexibility, however, and later crystallographic studies with other lysozyme inhibitors [13, 14] made it unlikely that substrate distortion provides a significant contribution to the catalytic efficiency of glycosidases. The strong inhibition by aldonolactones could also result from polar interactions with their C=O dipole [15] or even from a reversible covalent reaction with one of the catalytic groups [16].

Regardless of the molecular details of the strong inhibition by lactones there is, however, another line of evidence for a transition state with a planar sp^2 geometry at C-1. Reactions on a saturated carbon atom are slower by as much as 20% if a hydrogen which does not itself take part in the reaction has been replaced by deuterium, provided a change in the hybridisation from sp^3 to sp^2 occurs on the reaction pathway. This secondary α-deuterium kinetic isotope effect" results from a lower bending force constant of the C-D relactive to the C-H bond;[17] it is usually characterized by the ratio of rate constants k_H/k_D. Examples are solvolytic reactions of secondary tosylate esters and benzylic halides under S_N1 conditions for which k_H/k_D = 1.15 to 1.20 has been found [18].

The results obtained from isotope effect studies on the enzymic and non-enzymic hydrolysis of some glycosides are given in Table 24-2. Note that product analysis has shown that these enzymes act, like most other glycosidases, with retention of the anomeric configuration, i.e. with at least two distinct chemical steps (Scheme 24-2):

$$E + Glc\text{-}OR \underset{k_{-1}}{\overset{k_1}{\rightleftharpoons}} E\text{-}Glc\text{-}OR \xrightarrow[ROH]{k_2} E\text{-}Glc \xrightarrow[H_2O]{k_3} E + Glc\text{-}OH$$

Scheme 24-2

Table 24-2. Secondary α-deuterium isotope effects during hydrolysis of glycosides

Catalyst	Substrate	k_H/k_D	Reference
2 N HCl, 50 °C	β Glc → Phe	1.13	[19]
3 M NaOMe, 70 °C	β Glc → Phe	1.03	[19]
ß-Glucosidase (almonds) pH 5.0, 25 °C	β Glc → Phe	1.01	[19]
Lysozyme (hen's egg) pH 5.5, 40 °C	1,4-β GlcNAc → β Glc → Phe	1.11	[19]
ß-Galactosidase (E. coli) pH 7.0, 25 °C	β Gal → 4 Br-Phe	1.00	[20]
ß-Galactosidase (E. coli) pH 7.0, 25 °C	β Gal → 2,4 (NO$_2$)$_2$Phe	1.25	[20]
ß-Galactosidase (E. coli) pH 7.0, 25 °C	β Gal → Pyridinium ion	1.19	[20]
none, water, 100 °C	β Gal → Pyridinium ion	1.13	[20]
ß-Glucosidase A$_3$ pH 4.0 (Asp. wentii)	β Glc → 4-MU	1.14	[21]
ß-Glucosidase A$_3$ pH 4.0 (Asp. wentii)	β Glc → 4 Br-Isoquinolin.	1.11	[21]

The data given in Table 24-2 can be discussed as follows: lysozyme and ß-glucosidase A$_3$ from *Asp. wentii*, where aglycon bond cleavage (k$_2$) is rate limiting, act *via* a transition state which resembles that of the acid catalyzed reaction in its hybridisation change from sp^3 to sp^2, i.e. a glycosyl cationlike species. Such ions have an estimated life-time of 10^{-10} s in water;[22] on the enzyme they are likely to form a covalent intermediate with a closely positioned carboxylate (Asp-52 in lysozyme) rather than an ion pair as originally proposed in the Phillips mechanism. Formation of the intermediate by nucleophilic displacement is unlikely on chemical grounds and is ruled out by the isotope effect which should be k$_H$/k$_D$ = 1.00 as for the base catalyzed hydrolysis of phenyl ß-glucoside. ß-Glucosidase from almonds could act by such a mechanism but then k$_{cat}$ should vary in a straightforward manner with the leaving group propensity of the aglycon, which is not the case [23]. A possible explanation would be a rate determining conformation change as has been shown to occur on complex formation with D-glucono-1,5-lactone and the active site directed inhibitor conduritol B-epoxide [24].

The isotope effects for O-galactoside hydrolysis by the enzyme from *E. coli* are limiting cases for slow and fast substrates, respectively. Whereas the rate limiting step with k_H/k_D 1.00 is represented with the former by aglycon release (k_2) it is k_3 (hydrolysis of the galactosyl enzyme) for the latter. The explanation for $k_H/k_D = 1.00$ for k_2 [19] was similar to that for almond ß-glucosidase. With the faster substrates the rate of bond breaking was supposed to be similar to that of the conformation change to give intermediate values for k_H/k_D. With ß-galactosides of acidic phenols k_{cat} was considered to be represented by k_3. The isotope effect $k_H/k_D = 1.25$ was explained by a galactosyl enzyme consisting of an unreactive α-acylal in equilibrium with a small proportion of an ion pair of the galactosyl cation and the carboxylate of the acylal. Formation of ß-D-galactose would only result from the stereospecific reaction of a water molecule with the galactosyl cation. The observed isotope effect would then be the result of the cation formation.

ß-Galactosyl pyridinium and isoquinolinium salts are hydrolyzed by ß-galactosidase considerably slower than aryl ß-galactosides but they still experience acceleration factors of 10^{10} to 10^{12} over the uncatalyzed reaction. Their low rate of hydrolysis shows that k_2 is rate limiting. However, in contrast with the slow aryl O-ß-galactosides, k_H/k_D was 1.19, i.e. a galactosyl cation-like transition state can be formed without acid catalysis which is, of course, not possible with the pyridinium salts. ß-Galactosidase from *E. coli* requires Mg^{2+} for an efficient hydrolysis of O-galactoside but not for the cleavage of ß-galactosyl pyridinium salts. From this and a substantial isotope effect of the (slow) hydrolysis of O-galactosides by the Mg^{2+}-free enzyme Sinnott et al. [25] concluded that Mg^{2+} is required for a conformation change which permits proton transfer to the glycosidic oxygen and which is rate limiting with the slow aryl galactosides.

Secondary α-deuterium kinetic isotope effects ranging from k_H/k_D 1.08 to 1.20 have been found with yeast α-glucosidase [26] and intestinal sucrase/isomaltase [27]. Despite the complications seen with ß-galactosidase from *E. coli* there is thus ample evidence for a transition state of sp^2, glycosyl cation-like character. Aldono-1,5-lactones can, therefore, be regarded as transition state mimics as proposed by Leaback [12] even though they bind, with few exceptions, only a few hundred- to several thousandfold better than the parent aldoses. The relevance of a slow approach to the inhibition equilibrium observed in a few cases [28, 29] will be discussed later.

The lability of the aldono-1,5-lactones with respect to isomerisation to the 1,4-isomer and to hydrolysis has prompted the synthesis of other glycon analogues having a flat, sp²-like geometry at C-1. Examples are glycono-1,5-lactams (**1**) [30, 31], hydroxyimino lactones (**2**) [32] and lactams (**3**) [33], and nojiritetrazoles (**4**) [34].

In general, α-glycosidases are much less susceptible to inhibition by this group of inhibitors than ß-specific enzymes which may be due to the different orientation of the -C=O and -C=N dipoles and of the anomeric oxygen of α- and ß-glycosides, respectively. Published data [15, 30-34] also show that these compounds have a similar or an as much as 30-fold lesser inhibitory potency than the corresponding lactones. This was unexpected for the lactams because they best resemble the lactones in structure and their dipole moments are of similar magnitude (µ(valerolactone) 4.22 D; µ(valerolactam) 3.83 D). The good inhibition by the nojiritetrazoles demonstrates that enzyme-inhibitor (or -substrate) interactions cannot be very close with the ring oxygen or -NH- of the lactams as hardly any steric effects are introduced by the tetrazole ring. A good linear correlation with a slope of 1.1 of $\log(k_{cat}/K_m)$ with $\log(1/K_i)$ of D-*gluco*- and D-*manno*-tetrazoles was seen with six α-glycosidases studied with the respective p-nitrophenyl glucoside and mannoside [34]. This correlation, being in contrast with that of substrate affinity ($\log(1/K_m)$ with $\log(1/K_i)$), was taken as evidence for the classification of nojiritetrazoles as true transition state mimics.

24.3 Evidence for Active Site Carboxylate Groups

A crucial feature of the transition state which is only marginally addressed by the lactone type of inhibitors is the positive charge shared between C-1 and the ring oxygen. Glycon analogues bearing a positive charge in this region should have an enhanced inhibitory potency, provided other glycosidases also have the two catalytic carboxyl groups of the lysozyme model.

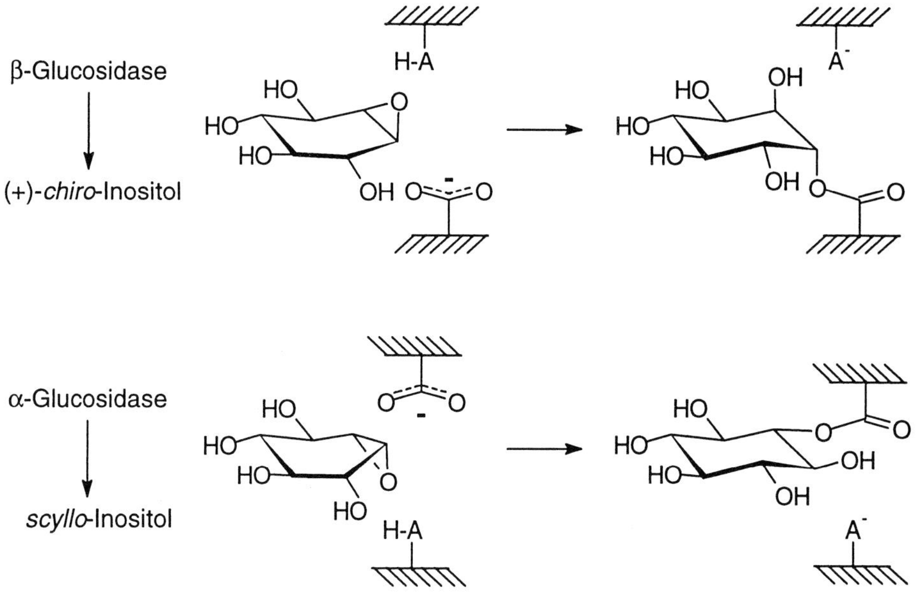

Figure 24-4. Active site directed inactivation of ß- and α-glucosidases by 1-D-1,2-anhydro-*myo*-inositol (Conduritol B epoxide)

That this is indeed the case was shown by the active site directed inactivation of numerous glycosidases by 1,2-anhydro-inositols (conduritol epoxides) with an appropriate hydroxylation pattern (Figure 24-4). [15] The inactivation requires a suitably positioned acid to protonate the rather unreactive oxirane which, in the case of ß-specific enzymes, forms an ester with the catalytic carboxylate by trans-diaxial ring opening. With α-glycosidases the ester formation occurs by trans-diequatorial ring opening. The inverse reaction of the catalytic carboxylate with α-glycosidases nicely demonstrates the inverse orientation of the carboxylate with respect to the anomeric carbon of the bound substrate.

The other line of evidence for an active site carboxylate and a glycosyl cation-like transition state comes from studies with basic (cationic) glycon derivatives which inhibit up to several ten thousandfold better than the parent sugars, i.e. with inhibition constants well within or even below the micromolar range [15]. This strong inhibition is thought to be due to the formation of an intimate ion pair consisting of the protonated inhibitor and a carboxylate group. As ionic interactions are weak in aqueous solutions we have to conclude that access of solvent water is greatly restricted to this part of the active site. This was confirmed by inhibition studies in the presence of buffer concentrations from 5 to 300 µM where K_i was constant or varied in the same way as K_m for the neutral substrate.

Important representatives of this group of inhibitors are glycosylamines (**5**) [35], hydroxylated piperidines like nojirimycin (**6**) [36], 1,5-dideoxy-1,5-iminohexitols (**7**) [30], and hydroxylated pyrrolidines (**8**) [37]. The hydroxylated piperidines and pyrrolidines were first isolated from molds [36], bacteria,[38], and plants [37], but the interest raised by their successful application in studies of glycoprotein biosynthesis [39] has greatly stimulated the development of syntheses for these and related inhibitors. Basic (cationic) analogues are now known for all major sugars [15].

The following generalisations can be made about the inhibition characteristics structural features required for strong inhibition:

i) Position of the basic (cationic) *centre:* the -NH$_2$ or -NH- groups must be linked directly to the anomeric carbon atom or its equivalent. 2-Amino-2-deoxy-hexoses (**9**) inhibit no [21, 40] or at most 20-fold better [41, 42] than the corresponding sugars. Glycosylmethylamines (**10**) are less efficient inhibitors than glycosylamines (Table 24-3). For some enzymes like ß-galactosidase from E. coli [43] or ß-glucosidase from almonds [41] this could be due to the high basicity of the inhibitors (pK$_a$ 9.1 for ß-D-galactosylmethylamine [43] *vs.* pK$_a$ 5.6 for D-glucosylamine [38]) because these enzymes require the free base rather than the protonated inhibitor for strong inhibition. Calculation for K_i for the inhibition of ß-galactosidase by *galacto*-**9** based on the concentration of the free base [43] gave pH-independent values which were almost identical to K_i for D-galactosylamine (7.8 µM) [44].

For enzymes which are strongly inhibited by cationic glycon analogues but weakly by the type **10** inhibitors like ß-glucosidase from *Asp. wentii* [21] or from bovine spleen [45] we assume that the flexibility of the active site carboxylate is greatly restricted, thus preventing a close approach to the cationic centre.

A comparison [15] of inhibitors of type **5, 6** and **7** shows that ß-glycosidases are inhibited from 10- to 800-fold better by nojirimycins (**6**) than by their 1-deoxy analogues (**7**), regardless of their susceptibility to the basic or cationic form of the inhibitor. It may well be that a hydrogen bond from the active site acid to the hydroxyl group on C-1 of the nojirimycin makes a significant contribution to the binding energy. The inhibitory potency of the glycosylamines (**5**) is similar to that of the 1,5-iminohexitols (**7**) or up to 80-fold lower. No such generalisations can be made for α-glycosidases except that many of them are less susceptible to these basic glycon analogues.

An interesting aspect of how the position of the basic centre effects the inhibitory potence of these inhibitors was addressed by Jespersen et al. [52] who synthesized isofagomine (**11**) where the anomeric carbon rather than the ring oxygen was replaced by the -NH- group. Whereas **11** inhibits α-glucosidase from yeast about threefold less than 1-deoxynojirimycin (**7**), it is an about 50-fold better inhibitor than **7** for ß-glucosidase from almonds (K_i 0.1 µM vs. 5.0 µM at pH 6.8), even though **11** lacks the OH-group on C-2.

ii) Basicity: Strong inhibition by ion-pair formation with glycon analogues requires that these are of sufficient basicity to accept a proton from solvent water (ß-glucosidase from *Asp. wentii* [21], α-glucosidase I from the endoplasmatic reticulum[53]) or from an acidic group of the enzyme (ß-glucosidase from almonds [41], ß-galactosidase from *E. coli* [44], α-glucosidase II from the endoplasmatic reticulum) [53]. In fact, the low inhibitory potency of N-aryl (pK_a 1.5 [41]) and N-acyl derivatives was taken as evidence that ion-pair formation makes an important contribution to the binding energy [15]. This was confirmed with ß-glucosidase from almonds by the 500-fold increase in K_i for castanospermine (**12**) by its conversion to the N-oxide.

11   12   13

Effects of similar magnitude were seen on N-acylation of 1-deoxynojirimycin with lysosomal ß-glucosidase [55] and N-arylation of 1,4-dideoxy-1,4-

imino-D-lyxitol (13) with lysosomal α-mannosidase [56]. Similar results were obtained with N-substituted ß-galactosylmethylamines with ß-galactosidase from *E. coli* where only derivatives with $pK_a \geq 8.0$ inhibited much better (K_i based on the free base, see above) than their non-basic analogues, N-substituted C-(ß-D-galactosyl) formamides [43].

Table 24-3. Relative inhibitory potency of basic sugar analogues expressed as K_i(Hexose)/K_i(Analogue). Inhibitor numbers refer to structures in the text

Enzyme	Type of inhibitor			
	5	9	10	7
ß-Glucosidases				
Sweet almonds [41, 46]	260	13	15	1,700
Asp. wentii A₃ [21]	1,750	0.15	20	1,040
Bovine, lysosomal [45]	3,400	1.2	10	1,200
Bovine, cytosolic [27]	230	---	3.4	260
α-Galactosidases				
E. coli [48]	625	---	---	210,000
Coffee beans [48]	24	---	---	212,000
ß-Galactosidases				1,680
E. coli [42-44, 48]	2,850	16	42	
ß-N-Acetylglucosaminidases				
Bovine kidney [48]	930	---	---	2,600
Asp. niger [49]	3.6	---	---	3.0
α-L-Fucosidase				
Bovine kidney [51]	42	---	---	12,100

As the presumed catalytic mechanism involves proton transfer from an acidic group and facilitation of glycosyl cation formation by a carboxylate the pH-dependence of k_{cat}/K_m and k_{cat} is governed by pK_a-values of these two groups in the free enzyme and the enzyme transition state complex, respectively. The pH-dependence of $1/K_i$ of a transition state mimic should thus resemble that of k_{cat}/K_m. Where such studies have been made they generally showed an increase of $1/K_i$ with increasing pH, often with a limiting slope of +1 in plots of $\log(1/K_i)$ vs. pH. With ß-glucosidase A₃ from *Asp. wentii* we found [21] that the pH-dependence calculated with protonated glycosylamine and nojirimycin was identical with that for ß-glucosylpyridinium ion, i.e. a curve reflecting the deprotonation of an acidic group with pK_a 5.6. The plot of log (k_{cat}/K_m), on the other hand, had a slope of - 0.9 and leveled off at low pH, indicating the requirement of an *acid* with pK_a 5.8. From the inverse pH-dependence of $1/K_i$ and k_{cat}/K_m we have to conclude that, in this case, the inhibitors of type **5** and **6** are not transition state mimics but product analogues which derive their enhanced affinity over glucose from ion-pair formation at the active site.

The pH-dependence of the response of ß-glucosidase from almonds to D-glucono-1,5-lactam (**3**) and 1-deoxynojirimycin (**7**) was studied by Dale et al [46]. who found that the inhibition by **3** showed a sigmoidal decrease with increasing pH which could be described by the dissociation of an acid with pK_a 6.8. The inhibition by **7**, on the other hand, increased with increasing pH because only the free base (pK_a 6.7) is inhibitory and then decreased as found for **3** due to dissociation of an acid with pK_a 6.8. Thus, inhibition by both the neutral **3** and the basic glycon analogue **7** shows the same dependence on pH which did not, however, resemble the pH-dependence of V_{max}/K_m (bell-shaped with an basic group of pK_a 4.4 and an acidic one with pK_a 6.7). For this reason, the authors ruled out a transition state resemblance for the two inhibitors. This judgement might have to be revised, however, because they did not take account of the fact that bond breaking is probably not rate limiting with this enzyme (see α-deuterium kinetic isotope effects).

14   15

An enzyme inhibitor system which fulfilled the criterion of pH-dependence for transition state resemblance has been described by Axamawaty et al [57]. who studied the inhibition of α-L-arabinofuranosidase from *Monilinia fructigena* by the hydroxylated pyrrolidines **14** and **15**. Both inhibitors showed a bell-shaped pH-dependence of $1/K_i$ which could be described by the dissociation of an acid with pK_a 5.9 (acid catalytic group of the enzyme) and the dissociation of the protonated inhibitor (pK_a 7.8 for **14** and pK_a 7.6 for **15**).

When K_i is based on the free base the pH-dependence of $1/K_i$ bears a close resemblance to that of k_{cat} and k_{cat}/K_m. The unprotonated forms of **14** and **15** can thus be considered as true transition state mimics by the pH-criterion. It should be noted, however, that because of the unfavourable relation of pK_a (inhibitor) to pK_a (catalytic acid) the maximal inhibition (K_i(**14**) 1600 µM and K_i(**15**) 1,3 µM) is only about 1% of the values to be expected if both the acid and the inhibitor were fully in the correct state of ionisation. For more effective inhibition pK_a(acid) should be larger than pK_a(inhibitor).

Table 24-4. Inhibition of ß-glucosidases by D-glucose and its neutral, weakly basic and strongly basic analogues at pH 5.0 and 25°; K_i-values in µM (G. Legler, unpublished data; **17** was kindly provided by B. Ganem, (Ithaca, NY)

	Inhibitors			
Enzyme source	α/ß-D-Glucose	D-Glucono-1,5-lactone	Nojirimycin (**6**)	D-Gluconamidrazone (**17**)
Almonds	30,000	5.0	1.3	4.7
Asp. wentii	2,800	6.5	0.070	0.031
Bovine kidney (lysosomal)	220.000	5.0	0.9	3.3

The detrimental effect of high basicity on the inhibitory potency has already been discussed for the glycosylmethylamines [43]. These arguments are also valid for a group of inhibitors which were introduced by Ganem and coworkers [33] and which have a planar sp²-geometry at C-1 and a positive charge combined in the same molecule. Glucamidine (**16**, pK_a 10.6) and glucamidrazone (**17**, pK_a 8.7) are, in spite of their better resemblance to a glycosyl cation-like transition state, less potent inhibitors for a number of glucosidases than nojirimycin (**6**) (Table 24-4). This may well be due to their inability to form a hydrogen bond with the catalytic acid as donor and the exo- or endocyclic nitrogen as acceptor. Depending on the extent of proton transfer in the transition state (general acid catalysis), hydrogen bond formation could make a larger contribution to the binding energy with a good transition state mimic than ion-pair formation.

iii) Hydroxyl groups, ring structure, and specificity: The response of glycosidases to glycon derived inhibitors corresponds, in general, to their sugar specifity, i.e. enzymes having a broad specificity like ß-glucosidase from almonds or the cytosolic ß-glucosidase/ß-galactosidase from mammals are strongly inhibited by compounds with both D-*gluco*- and D-*galacto*-configuration. However, in order to qualify as true transition state mimics, their values for $1/K_i$ should vary in proportion to the corresponding values for k_{cat}/K_m which express catalytic efficiency rather than to $1/K_m$ expressing substrate affinity where bond breaking with the aglycon is rate limiting.

Table 24-5. Comparison of hydroxylated pyrrolidine and piperidine (**7**) inhibitors. K_i-values are in µM, type **7** inhibitors have the glycon configuration corresponding to the enzyme specificity

Enzymes	Inhibitors		
	(pyrrolidine)	**7**	(piperidine)
α-Glucosidase, yeast [37]	0.18	12.6	0.73
ß-Glucosidase, almonds [37, 58]	200	47	1.7
ß-Glucosidase, Asp. wentii [58]	---	2.7	57
ß-Glucosidase, bovine, lysos. [58]	---	240	44
	(pyrrolidine)	**7**	(piperidine)
α-Mannosidase, jack beans [37, 59]	0.5	68	0.5
α-Galactosidase, coffee beans [37]	0.2	0.0016	500

For enzymes with rate limiting deglycosylation a low K_m will fake a high substrate affinity and where a conformational change is rate limiting k_{cat} will be slower than the bond breaking step. Hydroxylated pyrrolidines of appropriate configuration have often been found better inhibitors than the corresponding type **7** piperidines (Table 24-5). This may seem surprising because there is one hydroxyl group less in the 1,4-dideoxy-1,4-iminopentitols and in the 2,5-dideoxy-2,5 iminohexitols there is a hydroxymethyl group in a position which appears not compatible with the structure of aldohexoses. According to Sinnott [59] the better inhibition may well reflect a greater similarity of hydroxy group orientation with the transition state in the five-membered inhibitors. As can be seen from the data in Table 24-5 this is, however, not a general phenomenon. Even larger variations in the response to a certain type of inhibitor which also may reflect differences in transition state conformation have been reported for the hydroxylated indolizine alkaloids castanospermine (**12**) and swainsonine (**18**). The susceptibilities range from $K_i > 20$ nM (no inhibition with 5 mM **12**) for yeast α-glucosidase [61] to K_i 0.55 nM for intestinal sucrase and an apparently irreversible but non-covalent inhibition of isomaltase [62]. Whereas mammalian α-mannosidase I A/B which is involved in the biosynthesis of N-linked glycoproteins is not measurably inhibited by **16** ($K_i > 1,000$ μM), α-mannosidase II of the same pathway and also located in the Golgi compartment is inhibited with K_i 0.1 μM [63]. Note that strong inhibition by **12** and **18** is approached slowly, i.e. on the time-scale of minutes. That glycosidases show a much wider spread in their susceptibilities to the indolizine type inhibitors than the monocyclic ones is probably caused by a more rigid structure of the former which is expected to enhance differences in active site geometry.

Whereas the cyclic, glycon based inhibitors have a spatial arrangement of their hydroxyl groups which resemble the transition state only to a moderate extent it should be possible for open chain analogues to adopt a conformation with an optimal correspondence to the glycon binding site. That this is, indeed possible has recently been shown by Fowler et al [64]. who studied the inhibition of α-glucosidase from yeast by a number of open chain and, in part, truncated derivatives of nojirimycin and its 1-deoxy derivative. Compounds **19**, **20** and **21** were similar or up to 4-fold better inhibitors than **6** and **7**. In order to appreciate these results we should take account of the large entropy loss resulting from freezing out four or five rotational degrees of freedom when these flexible inhibitors bind to the enzyme which puts them at a disadvantage over their cyclic analogues. According to Page and Jencks [65] the entropy loss per degree of freedom amounts to about 5 cal/K, corresponding to an increase in the binding energy of 6 to 7.5 kcal/mole. Thus, if these open chain aminoalcohols could be fixed in the conformation they adopt in the enzyme inhibitor complex they would inhibit with K_i-values in the sub-nanomolar range.

24.4 Interactions with the Aglycon Site

The catalytic activity of glycosidases results from more or less specific interactions with the sugar moiety of the substrate, from interactions with the aglycon part which are generally of much lower specificity, and from the bond to be broken where specificity appears to be absolute with respect to the anomeric configuration and the atomic species (oxygen, nitrogen, sulfur) linking sugar and aglycon. Both glycon and aglycon interactions have larger effects on k_{cat} than on K_m, indicating that it is not substrate binding *per se* which is important for maximal stabilization of the transition state but the subtle conformational alterations induced by the substrate.

Table 24-6. Inhibition enhancement by ion-pair formation and interactions with the aglycon site. K_i-values in µM; last column gives -fold enhancement of affinity expressed by $1/K_i$ over the neutral reference compound.

Enzyme	Inhibitors		-fold enhancement.
β-Glucosidase	Glc ~ OH	Glc ~ NH_2	
sweet almonds B[41]	80,000	310	270
	Glc $\underline{\beta}$ S - CH_2Ph	Glc ~ NH - CH_2Ph	
	1,400	0.32	4,400
β-Glucosidase	Glc ~ OH	Glc ~ NH_2	
bovine, lysosomal [41]	170,000	65	2,600
	Glc $\underline{\beta}$ O - $C_{12}H_{25}$	Glc $\underline{\beta}$ NH - $C_{12}H_{25}$	
	15	0.0005	30,000
β-Galactosidase	Gal ~ OH	Gal ~ NH_2	
Escherichia coli [44]	21,000	7.0	3,000
	Gal $\underline{\beta}$ S - C_8H_{17}	Gal $\underline{\beta}$ NH - C_7H_{15}	
	28	0.00057	48,000

As aglycon departure is not yet complete in the transition state of the first bond breaking step, good transition state analogues should include the aglycon, even though the correct bond lengths and bond angles will be difficult to realize in a stable molecule. In contrast with the large number of glycon based inhibitors described in the literature there are much fewer studies which address the aglycon as well. As mimics of most natural substrates require rather involved synthetic procedures we have first used simple alkyl derivatives of glycosylamines[41, 44] and 1-deoxynojirimycins[55, 47] to provide interactions with the aglycon site. Our results, some of which are given in Table 24-6, bring about the following aspects:

i) The affinity increase over the non-basic reference compound brought about by ion-pair formation is enhanced up to 30-fold by simultaneous interactions with the aglycon site. Perhaps this is due to entropic factors resulting from a reduced flexibility of the bound inhibitor by the aglycon.

ii) Simple hydrophobic interactions are quite strong. This is not unexpected with the lysosomal ß-glucosidase which has ß-glucosyl ceramide (**22**) as the natural substrate where the hydrocarbon chain of the sphingosine moiety starts four C-C-bonds from the glycosidic bond (the acyl chain makes only a small contribution to the binding energy [66]). The natural substrate for almond ß-glucosidase, amygdalin (**23**), is hydrophilic as well as hydrophobic. Two glycosidic bonds have to be hydrolyzed to release mandelic acid nitrile on cell damage which is then cleaved by oxynitrilase to benzaldehyde and hydrocyanic acid. The natural substrate of ß-galactosidase from *E. coli*, on the other hand, is the disaccharide lactose. This, however, does not undergo simple hydrolysis but is rearranged to a large extent to *allo*-lactose [67], the natural inducer for ß-galactosidase in this bacterium. Perhaps glucose is flexibly held in the aglycon site by hydrophobic interactions with the CH-groups of the pyranose ring faces to give sufficient time for the transglycosylation reaction.

iii) The effect of the N-alkyl substituent on K_i was less pronounced with the 1-deoxynojirimycin derivatives than with the glycosylamines. This is understandable because the substituent is in an "unnatural" position and has to "fold back" in order to interact with the aglycon site.

A combination of the trigonal planar geometry at C-1 with an aglycon directed substituent was realized by Vasella's group [32, 68] who added the N-phenylcarbamoyl group to the hydroximinolactones **2** corresponding to D-glucose and N-acetylglucosamine. The inhibition by the N-phenylcarbamoyl derivatives **24** was enhanced 27-fold over the parent compound with ß-glucosidase from almonds, 500-fold with the N-acetylglucosaminidase from the fungus *M. rouxii*, but only 3- to 4-fold with N-acetylglucosaminidase from jack beans and bovine kidney.

24.5 N¹-Alkyl Gluconaminidines as "Perfect" Transition State Mimics?

In order to test the combined contributions of a planar geometry and a strongly basic centre at C-1 and interaction with the aglycon site we have synthesized two N¹-alkyl gluconamidines by the route given in Figure 24-5:[69]

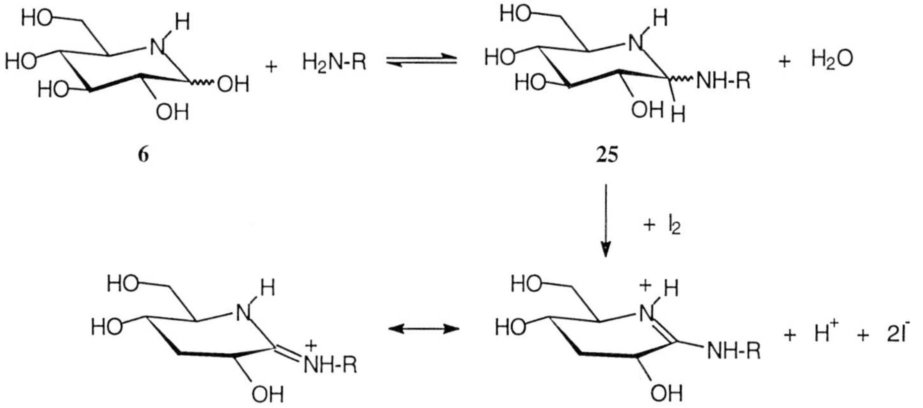

Figure 24-5. Synthesis of N-alkyl gluconamidines from nojirimycin by oxidative alkyl amination.

This route was based on the idea that the anomeric hydroxyl group of nojirimycin could be replaced by an amine with similar ease as by hydrogen sulfite[30] or hydrocyanic acid [70]. We reasoned that the initially formed substitution product **25** would be very labile but that it might possibly be stabilized by dehydrogenation to the N¹-substituted gluconamidine **26**. As the lability of **25** would preclude its detection by t. l. c. and as chemical test for amidines are only moderately sensitive we used the increase in the inhibitory potency against bovine lysosomal ß-glucosidase to monitor the reaction and aid the isolation of the desired product. This enzyme is inhibited by **6** with K_i 0.9 µM and we expected an at least 1000-fold better inhibition by **26** when dodecylamine is employed in the reaction.

By employing the inhibition of bovine lysosomal or cytosolic ß-glucosidase to monitor the reactions of Figure 24-5 we could prepare N¹-dodecyl D-gluconamidine (C_{12}-**26**) in ~ 30% yield. Concomitant oxidation of **6** gave ~ 50% of D-glucono-1,5-lactam. Tests with reaction mixtures prior to the addition of iodine appeared to indicate that less than 1% of the intermediate **25** had been formed. We ascribe this to an unfavourable equilibrium in the first reaction and/or a rapid hydrolysis of **25** during the dilution steps required for the inhibition measurements.

Purified C_{12}-**26** and the N^1-butyl derivative (C_4-**26**) were characterized by ^{1}H-NMR-spectroscopy and by the reaction with ninhydrin before and after hydrolysis with dilute NaOH. The latter was standardized with the unsubstituted amidine, 2-amino-1,2-dehydropiperidine. Preliminary inhibition data are given in Table 24-7.

Table 24-7. Inhibition of ß-glucosidases by the N^1-alkyl gluconamidines C_4-**26** and C_{12}-**26** and by the corresponding N-alkyl glucosylamines and ß-glucosides (K_i-values in µM at pH 5.0 except where indicated).

Enzyme source	C_4 - **26**	C_4 - Glucosylamine	C_4 - ß - Glucoside	K_s/K_i
Sweet almonds	4.8	12	20,000	4,100
Asp. wentii (pH 6.0)	0.0043	0.065	12,000	$2.8\ 10^6$
Bovine, cytosolic (pH 7.0)	0.13	0.74	210	1,600
	C_{12} - **26**	C_{12} -Glucosylamine	C_{12}-ß-Glucoside	
Bovine, lysosomal	0.0007	0.0005	15	22,000
Bovine, cytosolic (pH 7.0)	0.0002	0.0015	0.1	500

Our results (Table 24-7) show that the strongly basic N^1-alkyl gluconamidines ($pK_a \sim 10.5$) have a similar or only moderately higher inhibitory potency than the corresponding weakly basic N-alkyl glucosylamines (pK_a 6.5). This was even seen with the enzyme from *Asp. wentii* and the bovine lysosomal ß-glucosidase which are inhibited by cationic glycon derivatives to the same extent or even better than by their neutral counterparts [21, 45]. We conclude that a hydrogen bond from the catalytic carboxyl group (donor) to the exocyclic nitrogen of the inhibitor (acceptor) makes a large contribution to the binding energy. This bond cannot be formed with the fully protonated amidines; their advantage over the glucosylamines by their sp^2-structure of C-1 is thus largely abolished. This hypothesis could be tested with glyconamidines prepared with weakly basic amines, e.g. perfluoralkylamines or nitrobenzylamines.

A noteworthy point is the slow approach to the inhibition shown by the gluconamidines with all four ß-glucosidases. With the N-alkyl glucosylamines this phenomenon was observed only in the inhibition of the lysosomal ß-glucosidase by the C_{12}-derivative. The slow inhibition by many transition state mimics has been discussed by Wolfenden and Fink [71] in terms of an active site adapted to the ground state where the transition state mimic would experience a large number of non-productive encounters before it could bind tightly to the weakly populated active conformation of the enzyme. The substrate in the ground state would cause a change to the catalytically active conformation by a rapid induced fit mechanism except with enzymes where a conformational change is assumed to be rate-limiting. Another explanation for slow inhibition has been advanced by Schloss [72] who assumed a requirement for tightly binding reactive (metastable) intermediates. As the affinity of the enzyme is higher for the transition state than for these intermediates their dissociation constant will be larger. In order to prevent their rapid dissociation the rate constants for both the dissociation and the association must be small [15]. A difficulty in relating the rate constants for the approach to the inhibition equilibrium with transition state (or reactive intermediate) resemblance comes from a lack of correlation with their binding constants.

As discussed in the first part, Wolfenden [7] has shown that the resemblance of an inhibitor with the transition state may be judged from the relation of its inhibition constant to the binding constant of the substrate K_s/K_i, which, for a perfect mimic of the transition state, should approach the acceleration factor k_{cat}/k_{uncat}. This criterion is best fulfilled by the enzymes from *Asp. wentii* and bovine lysosomes. Interestingly, both re strongly inhibited by glycon related cations. Nevertheless, their values for K_s/K_i are many orders of magnitude lower than the observed acceleration factors.

24.6 Conclusion

Studies with non-covalent, glycon derived inhibitors have, in principle, confirmed for most glycosidases a model based on the Phillips mechanism for lysozyme (Figure 24-2). Essential features to be incorporated into effective transition state mimics are:

i) A *(partial) positive charge on the anomeric carbon* to interact with a closely positioned carboxylate of the active site, e.g. glycon or substrate analogue having a **nitrogen group of intermediate basicity** (pK_a 4 to 8, see below) at or in the position of the anomeric carbon atom.

ii) In the transition state the ground state $^4C_1$ chair conformation is deformed towards a **planar (sp^2-like) arrangement of C-2, C-1, O-5, and C-5**. Mimics are exemplified by sugar lactones and related compounds.

iii) The aglycon is still part of the first transition state, i.e. inhibitors which permit **interactions with the aglycon binding site**, even if they are non-specific, are much more potent than glycon analogues without this feature.

iv) A group on the anomeric carbon atom or its equivalent with **hydrogen bond acceptor capability**. General acid catalysis (i.e. proton transfer in the transition state) by an acidic group in close proximity to the glycosidic oxygen makes a substantial contribution to catalysis. In the enzyme inhibitor complex this arrangement can be replaced by a hydrogen bond linking the catalytic acid and the basic nitrogen on C-1. With inhibitors having $pK_a > 8$, however, this interaction cannot occur because they are fully protonated under the assay conditions. Possible exceptions are enzymes like ß-glucosidase A_3 from *Asp. wentii* [21], the bovine lysosomal ß-glucosidase [45], and the mammalian α-glucosidase I from the endoplasmatic reticulum [73] which are strongly inhibited by permanently cationic glycon derivatives by ion-pair formation.

Weakly basic ($pK_a < 3$) and non-basic derivatives of otherwise potent inhibitors bind poorly because the lone electron pair on the nitrogen is delocalized and not available for efficient hydrogen bonding.

If we apply Wolfenden's [7] criterion for the resemblance of inhibitor structure with the transition state to the many hundreds of enzyme inhibitor combinations published for glycosidases we notice that only very few of the values for K_s/K_i exceed 10^6. With the exception of C_4-**26** with ß-glucosidase from *Asp. wentii* (Table 24-7) and a disaccharide mimic derived from 1-deoxynojirimycin with intestinal sucrase [74] only castanospermine with the same enzyme [62] and swainsonine with α-mannosidase from jack beans [63] reach this inhibitory potency. As non-enzymic hydrolysis rates of different glycosides are similar [4] and k_{cat}-values of most glycosidases differ by not more than two orders of magnitude the observed rate enhancement factor exceed 10^{11}. For what reasons, then, are the K_i-values of our inhibitors still five or more orders of magnitude from that of a "perfect" transition state mimic?

Even though the model of the first transition state given in Figure 24-2 seems to be valid for the vast majority of glycosidases it can only be considered a first approximation because details like bond lengths, bond angles, and orientation of hydroxyl groups are not precisely specified. This can be seen from a spread of more than 10^4 in the K_i-values reported for a specific type of inhibitor with individual enzymes [15]. Glycosidases are probably unique with respect to details of the transition state as they are with respect to structure. Differences are most pronounced with inhibitors having a restricted conformational flexibility as with aldonolactones, hydroxylated pyrrolidines, and, in particular, indolizine alkaloids. This, and the unexpectedly large inhibitory potency of the open chain analogues of 1-deoxynojirimycin [64] which are free to adopt an optimal orientation of their hydroxyl groups, point to variations in transition state conformation among individual enzymes. As transition state stabilization requires the cooperation of many weak interaction with complementary groups of the active site only a few of the inhibitors studied so far provide a satisfactory match. Because each enzyme appears to be unique with respect to the details of transition state structure good mimics which fulfill all requirements will thus only be obtained with good luck or by the intuition of the designer.

Acknowledgment

The financial support of his research work by the *Fonds der chemischen Industrie* and the *Deutsche Forschungsgemeinschaft* is gratefully acknowledged by the author.

24.7 References

1. G. Legler, *Acta Microbiol. Acad. Sci. Hung.* **1975**, *22*, 403-409.
2. K. -R. Roeser, G. Legler, *Biochem. Biophys. Acta* **1981**, *657*, 321-333.
3. G. Legler, *Forschungsber. Nordrhein-Westfalen*, Nr. 2846, Westdeutscher Verlag Opladen (FRG), **1979**.
4. J. N. BeMiller, *Adv. Carbohydr. Chem.*, **1967**, *22*, 25-108.
5. H. Eyring, *J. Chem. Phys.* 1935, *3*, 107-110; M.G. Evans, M. Polanyi, *Transact. Farad. Soc.*, **1935**, *31*, 875-880.
6. L. Pauling, *Chem. Eng. News* **1946**, *24*, 1375-1378; *Nature (Lond.)* **1948**, *161*, 707-710.
7. R. Wolfenden, *Acc. Chem. Res.* **1972**, *5*, 10-18.
8. H. v. Euler, *Chemie der Enzyme*, 3rd ed., J.F. Bergmann, München, **1927**, p. 306.
9. E. H. Fischer, F. A. Stein, *The Enzymes*, 2nd. ed., Vol. 4, p. 303, Acad. Press, New York, **1960**.
10. D. C. Phillips, *Sci. Am.* **1966**, *215*, (5) 78-90; C.C.F. Blake, G.A. Mair, A.T.C. North, D. C. Phillips, V. R. Sarma, *Proc. Roy. Acad. Sci (London)* Ser. B, **1967**, *167*, 378-387.
11. G. A. Levvy, S. M. Snaith, *Adv. Enzymol.* **1972**, *36*, 158-181.
12. D. H. Leaback, Biochem. *Biophys. Res. Commun.* **1968**, *32*, 1025-1030.
13. J. A. Kelly, A. R. Sielecki, B. D. Sykes, M. N. G. Sykes, D. C. Phillips, *Nature (London)* **1979**, *282*, 875-879.
14. M. Schindler, Y. Assaf, N. Sharon, D. M. Chipman, *Biochemistry* **1977**, *16*, 423-428.
15. G. Legler, *Adv. Carbohydr. Chem. Biochem.* **1990**, *48*, 319-348.

16. M. L. Sinnott, I. J. L. Souchard, *Biochem. J.* **1973**, *133*, 89-98.

17. E. A. Halevi, *Progr. Phys. Org. Chem.* **1963**, *1*, 1-28.

18. V. J. Shiner, jr., W. E. Buddenbaum, B. L. Murr, G.Lamaty, *J. Am Chem. Soc.* **1968**, *90*, 418-426.

19. F. W. Dahlquist, T. Rand-Meir, M. A. Raftery, *Biochemistry*, **1969**, *8*, 4214-4221.

20. M. L. Sinnott, O. Viratelle, *Biochem. J.* **1973**, *100*, 81-86.

21. G. Legler, M. L. Sinnott, S. G. Withers, *J. Chem. Soc., Perkin Trans. 2,* **1980**, 1376-1383.

22. D. Locker, L. E. Jukes, M. L. Sinnott, *J. Chem. Soc., Perkin Trans. 2,* **1973**, 190-195.

23. R. L. Nath, H. N. Rydon, *Biochem. J.* **1954**, *57*, 1-10.

24. G. Legler, F. Witassek, *Hoppe-Seyler's Z. Physiol. Chem.* **1974**, *355*, 617-625.

25. M. L. Sinnott, S. G. Withers, O. M. Viratelle, *Biochem. J.* **1978**, *175*, 539-546.

26. L. Hosie, M. L. Sinnott, *Biochem. J.* **1985**, *226*, 437-441.

27. A. Cogoli, G. Semenza, *J. Biol. Chem.* **1975**, *250*, 7802-7809.

28. A. Tanaka, M. Ito, K. Hiromi, *J. Biochem. (Tokyo)* **1986**, *100*, 1379-1385.

29. G. Legler, E. Lüllau, E. Kappes, F. Kastenholz, *Biochem. Biophys. Acta* **1991**, *1080*, 89-95.

30. S. Inouye, T. Tsuruoka, T. Ito, T. Niida, *Tetrahedron* **1968**, *23*, 2125-2144.

31. F. Kastenholz, *Thesis (Diplom, Chemistry)*, Univ. Cologne, **1990**.

32. D. Beer, A. Vasella, *Helv. Chim. Acta* **1986**, *69*, 267-270; M. Horsch, L. Hoesch, A. Vasella, *Eur. J. Biochem.* **1991**, *197*, 815-818.

33. B. Ganem, G. Papandreou, *J. Am. Chem. Soc.* **1991**, *113*, 8984-8985.

34. Ph. Ermert, A. Vasella, *Helv. Chim. Acta* **1991**, *74*, 2043-2053.

35. H. Y. Lai, B. Axelrod, *Biochem. Biophys. Res. Commun.* **1973**, *54*, 463-468.

36. T. Niwa, S. Inouye, t. Tsuruoka, Y. Koaze, T. Niida, *Agric. Biol. Chem.* **1970**, *34*, 966-972.

37. G. W. J. Fleet, S. J. Nicholas, P. W. Smith, S. V. Evans, L.B. Fellows, *Tetrahedron Lett.* **1985**, 3127-3130.

38. D.D. Schmidt, W. Frommer, L. Müller, E. Truscheidt, *Naturwiss.* **1979**, *66*, 584-583.

39. A. D. Elbein, Ann. Rev. Biochem. **1987**, *56*, 497-534; B. Winchester, G. W. J. Fleet, *Glycobiology* **1992**, *2*, 199-206.

40. K. Osiecki-Newman, G. Legler, M. Grace, T. Dinur, S. Gatt, R. J. Desnick, G. A. Grabowski, *Enzyme* **1988**, *40*, 1376-1383.

41. G. Legler, *Biochem. Biophys. Acta* **1978**, *529*, 94-101; W. Lai, O. R. Martin, *Carbohydr. Res.* **1993**, *250*, 185-193.

42. R. E. Huber, M. T. Gaunt, *Canad. J. Biochem.* **1982**, *60*, 608-612.

43. J. N. BeMiller, R. J. Gilson, R. W. Myers, M. M. Santoro, M. P. Yadav, *Carbohydr. Res.* **1993**, *250*, 93-100.

44. G. Legler, M. Herrchen, *Carbohydr. Res.* **1983**, *116*, 95-103.

45. H. Liedtke, *Thesis,* University of Cologne, **1987**; G. Legler in `Lipid Storage Disorders', NATO ASI Ser. A.R. Salvayre, L. Douste-Blazy, S. Gatt, ed., Plenum Press, New York, **1987**, p. 63-72.

46. M. P. Dale, H. E. Ensley, K. Kern, K. A. R. Sastry, L. D. Byers, *Biochemistry* **1985**, *24*, 3530-3539.

47. G. Legler, E. Bieberich, *Arch. Biochem. Biophys.* **1988**, *260*, 437-442.

48. G. Legler, S. Pohl, *Carbohydr. Res.* **1986**, *155*, 119-129.

49. E. Kappes, G. Legler, *J. Carbohydr. Chem.* **1989**, *8*, 371-388.

50. U. Petzold, *Thesis*, University of Cologne, **1984**.

51. G. Legler, A. E. Stütz, H. Immich, *Carbohydr. Res.* **1995**, *272*, 17-30.

52. T. M. Jespersen, W. Dong, M. R. Sierks, T. Skydstrup, I. Lundt, M. Bols, *Angew. Chem.* **1994**, *106*, 1858-1860.

53. J. Schweden, C. Borgmann, G. Legler, E. Bause, *Arch. Biochem. Biophys.* **1986**, *248*, 335-340.

54. R. Saul, K. J. Molyneux, A.D. Elbein, *Arch. Biochem. Biophys.* **1984**, *230*, 668-675.

55. G. Legler, H. Liedtke, *Biol. Chem. Hoppe-Seyler, Biol. Chem.* **1985**, *366*, 1113-1122.

56. P. DeShong, D. R. Sidler, D. A. Kell, N. N. Aronson, Jr., *Tetrahedron Lett.* **1985**, 3747-3748.

57. M. T. H. Axamawaty, G. W. J. Fleet, K. A. Hannah, S. G. Namgoog, M.L. Sinnott, *Biochem. J.* **1990**, *266*, 245-249.

58. G. Legler, A. Korth, A. Berger, Ch. Ekhart, G. Gradnig, A.E. Stütz, *Carbohydr. Res.* **1993**, *250*, 67-77.

59. M. J. Eis, C. J. Rule, B. A. Wurzburg, B. Ganem, *Tetrahedron Lett.* **1985**, *26*, 5392-5398.

60. M. L. Sinnott in *Enzyme Mechanisms*, M. I. Page, A. Williams, ed., Royal Soc. **1987**, p. 255-297.

61. L. Hosie, P. J. Marshall, M. L. Sinnott, *J. Chem. Soc., Perkin Trans. 1* **1984**, 1121-1124.

62. C. Danzin, A. Erhard, *Arch. Biochem. Biophys.* **1987**, *257*, 472-475.

63. D. R. P. Tulsiani, T. M. Harris, O. Touster, *J. Biol. Chem.* **1982**, *257*, 7936-7939 and *Arch. Biochem. Biophys.* **1985**, *236*, 427-434.

64. P. A. Fowler, A. H. Haines, R. J. K. Taylor, E. J. W. Chrystal, M. B. Gravestock, *J. Chem. Soc., Perkin Trans. I*, **1994**, 2229-2235.

65. M. I. Page, W. P. Jencks, *Proc. Natl. Acad. Sci. US*, **1971**, *68*, 1678-1683.

66. K. Osiecki-Newman, D. Fabbro, G. Legler, R. J. Desnick, G. A. Grabowski, *Biochem. Biophys. Acta* **1987**, *915*, 87-100.

67. R. E. Huber, G. Gurz, K. Wallenfels, *Biochemistry* **1976**, *15*, 1994-2001.

68. M. Horsch, L. Hoesch, A. Vasella, D. M. Rast, *Eur. J. Biochem.* **1991**, *197*, 815-818.

69. G. Legler, M. -Th. Krauthoff, S. Felsch, *Carbohydr. Res.* **1995**, *292*, 91-101.

70. H. Böshagen, W. Geiger, B. Junge, *Angew. Chem.* **1981**, *93*, 800-801.

71. R. Wolfenden, L. Finck in `Enzyme Mechanisms', M. I. Page, A. Williams, ed., Royal Soc. **1987**, p. 97-122.

72. J. V. Schloss, *Acc. Chem. Res.* **1988**, *21*, 348-353.

73. H. Hettkamp, G. Legler, E. Bause, *Eur. J. Biochem.* **1984**, *142*, 85-90.

74. C. Danzin, J. B. Ducep, A. Ehrhard, P. Zimmermann, **1995**, *Proc. First Euro-Conf. on Carbohydr. Mimics*, C32.

25 Design and Synthesis of Potential Fucosyl Transferase Inhibitors

G. A. van der Marel, B. M. Heskamp, G. H. Veeneman,
C. A. A. van Boeckel and J. H. van Boom*

25.1 Introduction

25.1.1 Role of Selectins in Leucocyte Extravasation

A few years ago it was found that the selectins, a family of three cell adhesion glycoproteins (*i.e.* E-, L-, and P-Selectin) mediate local inflammatory responses by recognition of carbohydrate determinants [1-6]. The first selectin discovered, L-Selectin [7-9], is present on circulating leucocytes and is not only involved in lymphocyte recirculation between blood and lymph but also in neutrophil and monocyte migration to sites of inflammation. Preformed P-Selectin [10-12] is stored in intracellular secretory granules of both endothelial cells and platelets.

P-Selectin is mobilized to the plasma membrane of the endothelial cells or platelets within a few minutes after thrombin or histamine activation and binds to neutrophils and monocytes, thus functioning in the process of blood coagulation.

E-Selectin [13,14], expressed on the vascular endothelial cells two to four hours after cytokine induction, plays a key role in the early stage of leucocyte extravasation to inflamed tissue. Carbohydrate determinants on circulating leucocytes are *inter alia* recognized by E-Selectin and become adhered to the vascular endothelium (see Figure 1) [15-19]. As a result of this initial low affinity attachment (**I**), the velocity of the leucocytes is decreased and they start rolling along the wall of the blood vessel (**II**). The rolling adhesion precedes the stationary adhesion (**III**), which involves the non-carbohydrate interaction of chemoattractant activated leucocyte integrins with members (*e.g.* intercellular adhesion molecule-1, ICAM-1) of the immunoglobulin superfamily (IgSF). In the final stage, the leukocytes migrate through the vascular wall (**IV**) to the site of inflammation.

492 G. A. van der Marel et al

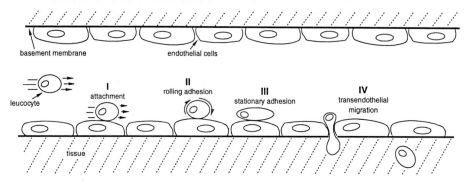

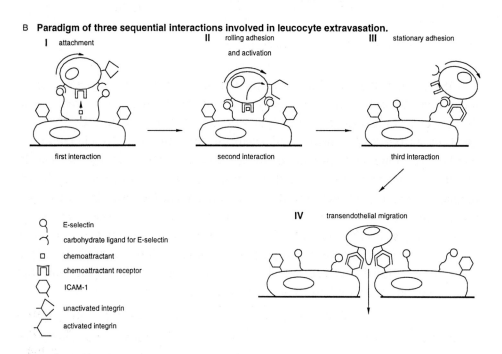

Figure 25-1

Recently, several naturally occurring carbohydrate determinants have been identified as ligands for the selectins [20-34]. Thus, it has been demonstrated that sialyl Lewis X (SLe^x: NeuAcα2→3Galβ1→4[Fucα1→3]GlcNAc, **1** in Figure 2) [28, 29, 32-34] and related structures (see Table 1) [20] are recognized by the selectins. SLe^x is the terminal structure of the glycan of various glycosphingolipids and glycoproteins, which are *inter alia* located on the surface of certain leucocytes (*i.e.* granulocytes and monocytes) [5,28,32,33]. The E-Selectin-SLe^x interaction appears to be crucial for the leucocyte adhesion to the vascular endothelium [33].

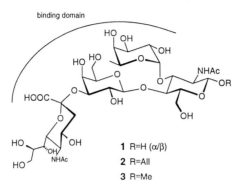

Figure 25-2. Structure and binding domain of SLe^x and derivatives.

It is evident that the above described mode of leucocyte migration from the blood stream to areas of injury or inflammation is essential for the host defence against pathogenic microorganisms as well as tissue repair. On the other hand, excessive and/or misdirected leucocyte recruitment is associated with a number of acute and chronic diseases [35], including cardiogenic shock, reperfusion injury, rheumatoid arthritis, psoriasis and ARDS (adult respiratory distress syndrome). In addition, the SLe^x determinant has been identified as a tumor-associated antigen [36,37]. Moreover, the increased expression of sialyl-dimeric-Le^x (see Table 1, entry 4) on the surface of carcinoma cells has been correlated to the metastatic potential of these cells [38,39]. On the basis of the latter information it may be postulated that adhesion inhibitors are potential anti-inflammatory as well as anti-metastatic agents.

Adhesion inhibitors can be divided in three categories. The first category comprises receptor-based adhesion inhibitors, such as monoclonal antibodies (MAbs) raised against the receptors involved in the adhesion process. The second category of adhesion inhibitors are the so-called ligand-based inhibitors. Ligand-based adhesion inhibition involves treatment with ligands or analogues thereof, which competitively block the endothelium-leucocyte interaction by occupying the selectin binding sites. The last category of adhesion inhibitors consists of inhibitors of the biosynthesis of the carbohydrate ligands. Such inhibitors may prevent the formation of ligands on leucocytes and thus preclude the leucocyte adhesion to the activated endothelium.

Table 25-1. Structure of some carbohydrate ligands and their recognition by selectins

Entry	Name	Structure	Recognized[a] by L-Selectin	E-Selectin	P-Selectin
1	3'-sialyl-LN	NeuAcα2→3Galβ1→4GlcNAcβ1→R	-	-	-
2	sialyl Le^x	NeuAcα2→3Galβ1→4(Fucα1→3)GlcNAcβ1→R	+	++	++
3	VIM-2 (CD65)	NeuAcα2→3Galβ1→4GlcNAcβ1→3Galβ1→4(Fucα1→3)GlcNAcβ1→R	+	+/-	+
4	sialyl-dimeric-Le^x	NeuAcα2→3Galβ1→4(Fucα1→3)GlcNAcβ1→3Galβ1→4(Fucα1→3)GlcNAcβ1→R	+	++	++
5	sialyl Le^a	NeuAcα2→3Galβ1→3(Fucα1→4)GlcNAcβ1→3	+	++	++
6	3'-sulfo-Le^x	SO$_4$-3Galβ1→4(Fucα1→3)GlcNAcβ1→R	++	+	++
7	3'-sulfo-Le^a	SO$_4$-3Galβ1→3(Fucα1→4)GlcNAcβ1→R	++	+	++

[a] Recognition is indicated on a relative scale of ++ (best) to - (none)

25.2 Receptor-based Adhesion Inhibition

The earliest described leucocyte adhesion inhibitors are MAbs directed to integrins, e.g. lymphocyte function associated antigen-1 (LFA-1), or members of the IgSF, e.g. ICAM-1 [35, 40-42]. Furthermore, several examples of selectin specific MAbs which function as therapeutic agents have been reported [43-47]. For instance, treatment of rats with MAbs, raised against human E-Selectin, markedly reduced the immunoglobulin G (IgG) immune complex induced neutrophil extravasation in the lungs [45].

25.2.1 Ligand-based Adhesion Inhibition

In principle, ligand-based adhesion inhibition can be effected by infusion with ligands or analogues thereof, which will block the interaction between selectins and leucocytes. For example, Mulligan et al [48]. reported that pretreatment of rats with SLe^x diminished the IgG immune complex induced lung injury as well as the neutrophil accumulation in the lung tissue. The protective effect of SLe^x has been postulated to originate from blocking the binding of leucocytes to E-Selectin. In contrast, non-fucosylated SLe^x (entry 1, Table 1, R=OH) proved to be inactive, indicating that the fucose moiety is essential for the recognition of SLe^x by E-Selectin.

At present, research is mainly focused on the design and synthesis of modified SLe^x derivatives [49-62]. Wong and coworkers described the partially enzymatic preparation [59, 61] of compounds **4** and **5**. In a cell adhesion assay the affinity of **4** for E-Selectin was found to be similar to that of SLe^x, whereas bivalent derivative **5** showed a fivefold better inhibitory activity [61]. The latter finding supports the proposed multivalent interaction of SLe^x with E-Selectin. In order to obtain simplified SLe^x analogues several groups [54-56] prepared tethered mimics (*e.g.* **6-8**). Unfortunately, the inhibitory effect on the E-Selectin-SLe^x binding of these derivatives proved to be 10-40 fold lower than that of SLe^x. The latter finding may be explained by the increased conformational flexibility of the tethered derivatives.

Meanwhile, the domain of interaction of SLe^x with E-Selectin was determined with the aid of both conformational and structure-activity analyses. NMR and molecular modelling studies revealed that SLe^x **1** [29,63] as well as

SLe^x-All **2** [61,64] and SLe^x-Me **3** [63] adopt predominately the conformation (in aqueous solution) portrayed in Figure 25-2. Furthermore, structure-activity experiments [27,29,34,51,59,65-67] revealed that crucial recognition sites of SLe^x include the carboxyl group of the sialic acid moiety, the hydroxyl functions of the fucose residue and the HO-4 and HO-6 groups of the galactose moiety. On the basis of these findings it may be concluded that interaction with E-Selectin occurs at the "top-face" of SLe^x, as shown in Figure 25-2 [68].

In our laboratory [69] we prepared SLe^x mimic **9**, in which the galactose moiety of naturally occurring SLe^x is substituted by a six atom bridge *i.e.* a propoxymethoxy residue. The latter spacer may adopt a conformation similar to that of the galactose moiety.

The synthetic route to mimic **9** is depicted in Scheme 25-1. The crucial step consists of the introduction of a methylene acetal linkage between the hydroxypropyl spacer of methyl 4,7,8,9-tetra-*O*-acetyl-*N*-acetyl-2-*O*-(3-hydroxypropyl)-α-D-neuraminate (**17**) and the HO-4 of hexyl 2-acetamido-6-*O*-benzyl-3-*O*-(tribenzyl-α-L-fucopyranosyl)-2-deoxy-β-D-glucopyranoside (**16**). The required sialic acid derivative **17** was readily available following a recently reported procedure [70]. The synthesis of disaccharide **15** could be realized *via* fucosylation of the secondary hydroxyl group in hexyl 2-acetamido-4,6-*O*-benzylidene-2-deoxy-β-D-glucopyranoside (**12**). The latter was obtained from known oxazoline **10** in three steps. Thus, glycosidation of **10** with hexanol under the agency of catalytic trifluoromethanesulfonic acid (TfOH) gave the fully acetylated derivative **11**. Subsequent Zemplén deacetylation of **11** and ensuing acetalisation with benzaldehyde dimethylacetal furnished the required glucosamine derivative **12**.

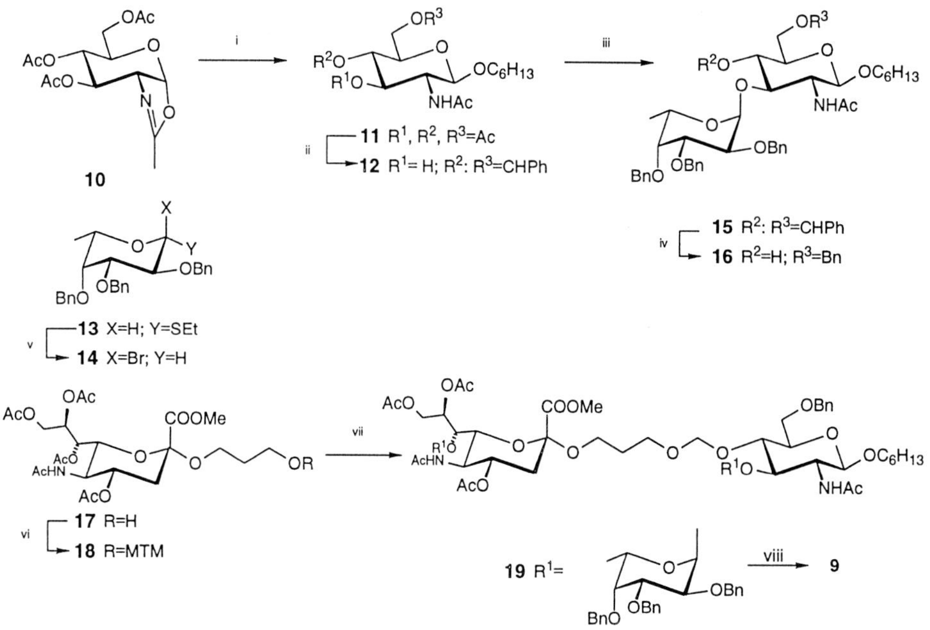

Design and Synthesis of Fucosyl Ttransferase... 497

Scheme 25-1 Reagents see [a].

[a]**Key:**
(i) $C_6H_{13}OH$, TfOH, CH_2Cl_2 (74%); (ii) (a) NaOMe, MeOH; (b) $PhCH(OMe)_2$, p-TsOH, DMF, (85%, 2 steps); (iii) **13**, IDCP, DCE/Et_2O, 5/1 (61%), α/β=1/1 or **14**, Bu_4NBr (82%, α); (iv) Me_3NBH_3, $AlCl_3$, THF (59%); (v) Br_2, CH_2Cl_2; (vi) Me_2S, $(BzO)_2$, CH_3CN (84%); (vii) **16**, NIS/TfOH, DCE (59%); (viii) (a) KOt-Bu, MeOH; (b) 20% $Pd(OH)_2/C$, 0.4 MPa H_2, i-$PrOH/H_2O/HOAc$, 5/4/1; (c) 4 N NaOH/MeOH/dioxane, 2/9/29 (60%, 3 steps).

In the first instance perbenzylated ethyl 1-thio-β-L-fucopyranoside **13** was employed as fucosyl donor. However, condensation of donor **13** with glucosamine acceptor **12** under the promotion of iodonium di-*sym*-collidine perchlorate (IDCP) gave disaccharide **15** as an inseparable mixture of anomers (α/β=1/1). On the other hand tetrabutylammonium bromide promoted glycosylation of **12** with fucosyl bromide **14**, attained by treatment of thioethyl donor **13** with bromine, afforded exclusively α-linked dimer **15**. Regioselective ring opening of the benzylidene acetal with borane trimethylamine and aluminium trichloride in THF gave the requisite dimer **16**.

The introduction of the methylene acetal linkage between building blocks **16** and **17** was accomplished as follows. Reaction of the primary hydroxyl group in the sialic acid derivative **17** with dimethylsulfide and benzoyl peroxide afforded the corresponding methylthiomethylether **18**. Condensation of the latter with dimer acceptor **16** under the agency of NIS and catalytic TfOH furnished the fully protected target molecule **19**. Removal of the protective groups in **19** was effected by a three-step procedure. Zemplén type deacylation, hydrogenolysis of the benzyl ethers and saponification of the methyl ester was followed by purification of the crude compound by gel filtration to afford the homogeneous target compound **9**, the identity of which was firmly corroborated by ^{1}H and ^{13}C NMR spectroscopy.

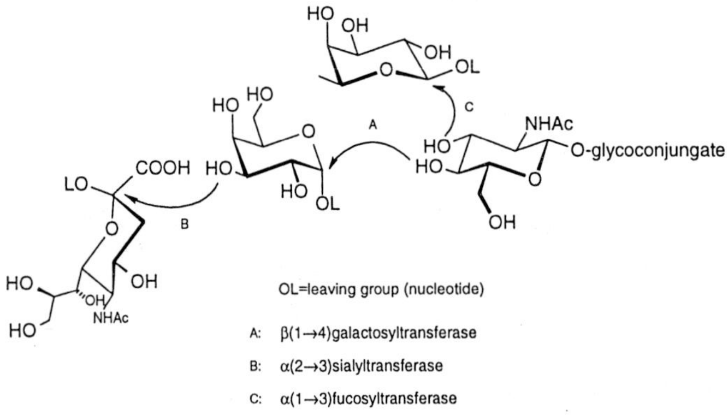

Figure 25-3. Biosynthesis of SLe^x.

25.2.2 Fucosyltransferase inhibitors

The third category of adhesion inhibition involves inhibition of the biosynthesis of selectin ligands, in particular the SLe^x determinant. As depicted in Figure 25-3, the biosynthesis of the SLe^x tetrasaccharide is mediated by three glycosyltransferases. In the first step (A) β(1→4) galactosyltransferase adds a galactose moiety to the terminal glucosamine residue of a glycoconjugate. Subsequently a sialic acid unit is transferred (step B) to the galactose moiety by α(2→3) sialyltransferase. The final enzymatic step C comprises the attachment of a fucose residue to the glucosamine by α(1→3) [34] [α(1→3)FT]. As mentioned earlier, the presence of a fucose moiety is crucial for the recognition of the SLe^x determinant by E-Selectin [27,34,48,65]. Therefore, inhibitors of α(1→3)FT may suppress the leucocyte adhesion process.

Up to now five different human α(1→3)FTs designated FT-III [72], FT-IV [73-75], FT-V [77], FT-VI [78], and FT-VII [79] have been cloned and characterized. The latter transferase is assumed to participate in the biosynthesis of SLe^x [79]. The donor substrate for all FTs is guanosine 5'-(β-L-fucopyranosyl)-diphosphate (GDP-Fuc), whereas the carbohydrate acceptor varies with the enzyme [80]. One of the products of the enzymatic fucosyltransferprocess is GDP (see Figure 25-4), which presents the first reported FT inhibitor [81].

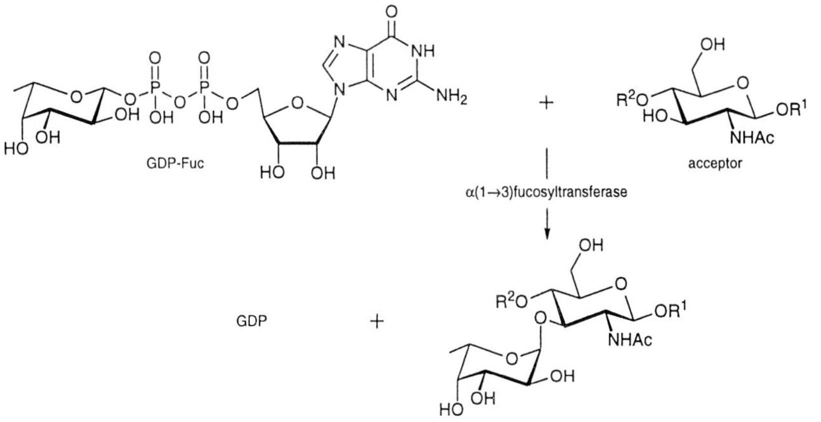

Figure 25-4. Enzymatic fucosyltransfer.

An important strategy towards potential FT inhibitors is the design and synthesis of substrate analogues. For instance, several metabolically stable GDP-Fuc analogues have been prepared. Toyokuni and collaborators [82] described the synthesis of carbocyclic GDP-Fuc analogue **20**, in which the fucosyl moiety is replaced by a more stable 5a-carbafucopyranosyl (pseudo fucose) residue. This compound exhibited an inhibitory effect on α(1→3/4)FT (FT-III) about twice as potent as the endogenous inhibitor GDP. Luengo and Gleason [83] reported the preparation of C-fucosidic analogues **21-24**. Up to now information on the biological activities of these potential inhibitors is not available.

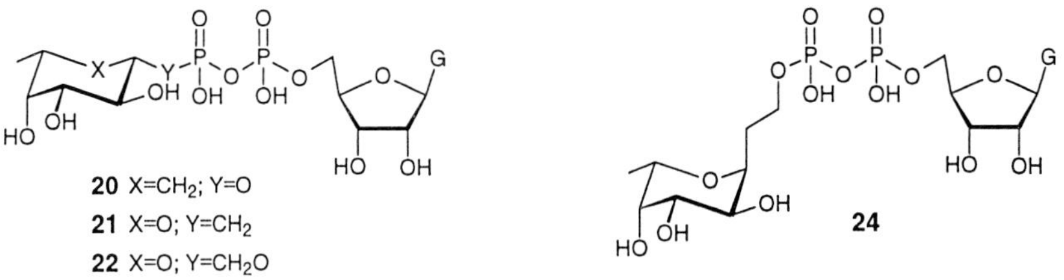

20 X=CH$_2$; Y=O
21 X=O; Y=CH$_2$
22 X=O; Y=CH$_2$O
23 X=O; Y=CH$_2$CH$_2$O

Another class of substrate analogues consists of acceptor analogues, *e.g.* Galβ(1→4)deoxynojirimycin (**25**). The latter compound was tested for FT-V inhibition by Wong and collaborators [84,85] and proved to be a poor inhibitor compared to GDP (IC$_{50}$= 8-40 mM vs. 0.05 mM) [86].

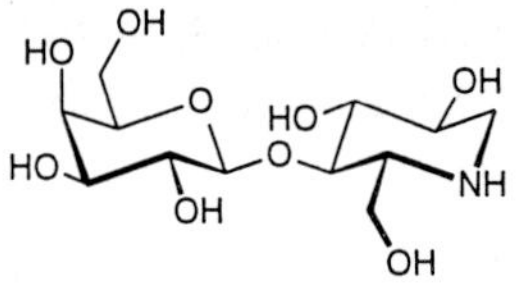

25

Another well-known approach to procure effective enzyme inhibitors is the construction of transition state (TS) analogues of a specific enzymatic reaction [87-89]. With respect to fucosyltransferases the latter strategy is hampered by the limited knowledge of the TS of this enzymatic reaction. It has been proposed that the fucosyltransfer proceeds *via* an ion-pair mechanism [90,91] as shown in **Figure 25-5**, which implies that in the TS donor and acceptor residues are simultaneously bound to the transferase. In addition the TS is characterized by a positively charged fucose moiety adopting a flattened conformation, *i.e.* presumably a half-chair conformation. On the basis of this proposed TS various potential inhibitors of α(1→3)FT have been developed.

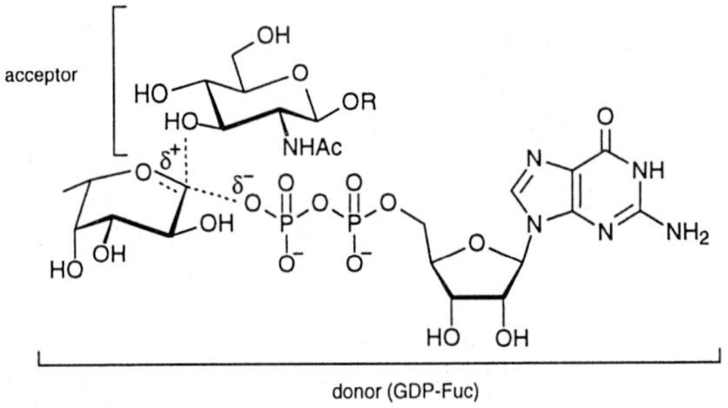

Figure 25-5. Postulated transition state of α(1→3)FT.

Several fucose and GDP-Fuc analogues having a flattened ring conformation have been reported. For example, Toyokuni and collaborators [82] described the synthesis and biological activity of compound **26**. It was shown that inhibitor **26** is approximately twice as active against FT-III as substrate analogue **20** and four times more active than GDP. The increased inhibitory activity of **26** has been attributed to the half-chair conformation of the cyclohexene ring. The preparation L-fucal-1-ylmethylphosphonate (**27**) was reported by Frische and Schmidt [92]. Wong and collaborators[93] proposed an alternative manner to mimic the half-chair conformation. Furanose aza-sugar containing compounds **28** and **29** proved to be moderate inhibitors of FT-V (IC_{50}=80 and 34 mM, respectively) [94]. On the other hand, an increased inhibition was observed when **28** and **29** were tested in the presence of GDP, *e.g.* enzyme activity was reduced to 10% at concentrations of **29** and GDP as low as 34 mM and 0.05 mM, respectively. This interesting synergistic effect was explained by the possible formation of a complex (**A**) of GDP and the aza-sugar in the active site of the enzyme.

Hindsgaul and collaborators [90] described a second type of inhibitor based on the TS, a so called bisubstrate analogue, which contains structural features of both donor (GDP-Fuc) and acceptor. In compound **30**, a bisubstrate analogue for α(1→2)FT, the terminal GDP-phosphorous is attached via an ethylene spacer to the HO-2 of phenyl β-D-galactopyranoside, i.e. the acceptor hydroxyl function to which the enzyme normally transfers a fucose residue.

Kinetic studies revealed that analogue **30** is a competitive inhibitor of α(1→2)FT, with respect to the donor as well as the acceptor, with an inhibitory activity comparable to that of GDP.

We reasoned that a compound in which three important recognition sites for the fucosyltransferase are embedded, i.e. a guanosine, a fucose and an acceptor (glucosamine) moiety, may be an active and selective inhibitor.

In order to validate this concept, trisubstrate analogue **31**, in which a L-fucopyranosyl residue is α-linked to the HO-3 of cyclohexyl 2-acetamido-2-deoxy-β-D-glucopyranoside while a guanosine moiety is anchored to the HO-2 of the fucose *via* a phosphate-ethylene bridge, was designed and prepared [95]. The flexibility of the spacer may enable the guanosine and fucose residues to adopt the correct relative orientation for optimal recognition by the enzyme. The synthetic route to target compound **31** is outlined in Scheme 25-2 and comprises four stages: (i) preparation of the hydroxyethyl spacer containing fucosyl donor **35**, (ii) coupling of the donor **35** to the glucosamine acceptor **36**, introduction of the guanosine 5'-phosphate residue, and (iv) cleavage of the protective groups.

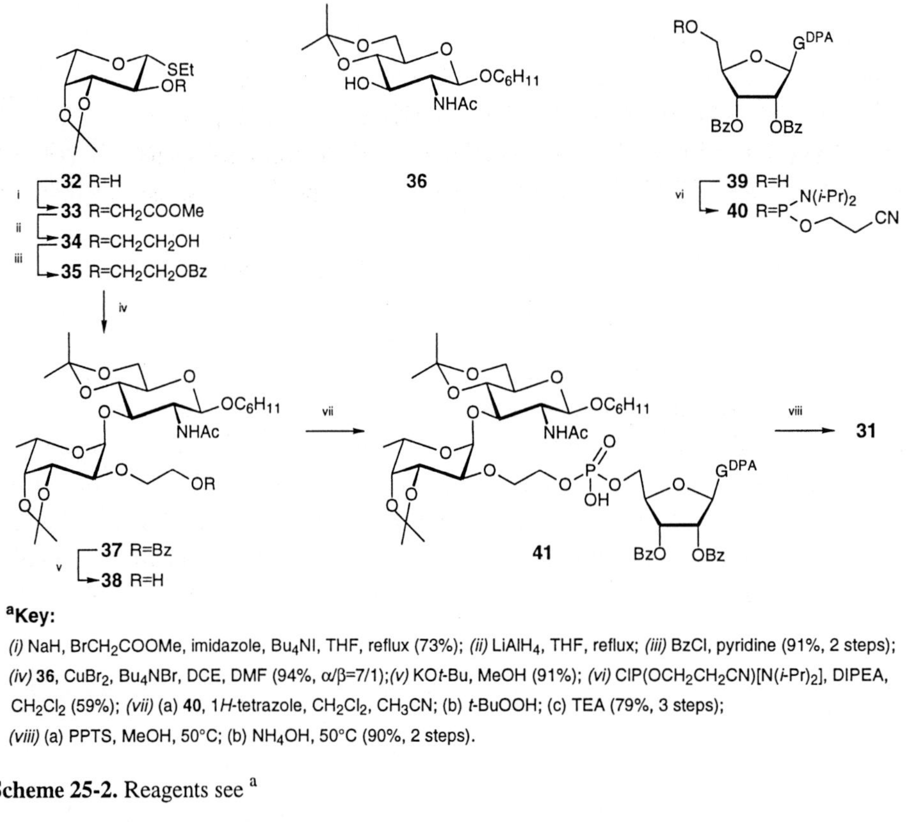

^a**Key:**
(i) NaH, BrCH$_2$COOMe, imidazole, Bu$_4$NI, THF, reflux (73%); (ii) LiAlH$_4$, THF, reflux; (iii) BzCl, pyridine (91%, 2 steps); (iv) **36**, CuBr$_2$, Bu$_4$NBr, DCE, DMF (94%, α/β=7/1); (v) KO*t*-Bu, MeOH (91%); (vi) ClP(OCH$_2$CH$_2$CN)[N(*i*-Pr)$_2$], DIPEA, CH$_2$Cl$_2$ (59%); (vii) (a) **40**, 1*H*-tetrazole, CH$_2$Cl$_2$, CH$_3$CN; (b) *t*-BuOOH; (c) TEA (79%, 3 steps); (viii) (a) PPTS, MeOH, 50°C; (b) NH$_4$OH, 50°C (90%, 2 steps).

Scheme 25-2. Reagents see ^a

Fucosyl donor **35** was prepared from readily accessible ethyl 3,4-*O*-isopropylidene-1-thio-β-L-fucopyranoside (**32**). Treatment of **32** with excess methylbromoacetate and sodium hydride gave **33**, the ester function of which was reduced with lithium aluminium hydride to give **34**. Subsequent benzoylation of the hydroxyl function in **34** afforded fucosyl donor **35**. Condensation of donor **35** with glucosamine acceptor **36** under agency of cupric bromide and tetrabutylammonium bromide gave disaccharide **37** as a mixture of anomers (α/β=7/1) in a high yield. Separation of the individual anomers by silica gel chromatography and subsequent removal of the benzoyl group in **37** afforded disaccharide **38**. The next stage in the assembly of **31** entailed the introduction of the 5'-phosphate guanosine moiety in disaccharide **38**. To this end, the primary hydroxyl group of guanosine derivative **39** was phosphitylated with chloro 2-cyanoethyl-*N*,*N*-diisopropyl-phosphoramidite in the presence of *N*,*N*-diisopropylethylamine to give **40**. Phosphodiester **41** was obtained *via* a one-pot three-step procedure. Condensation of dimer **38** with **40** under the agency of 1*H*-tetrazole was followed by oxidation with *tert*-butyl hydroperoxide. Subsequent elimination of the cyanoethyl group from the intermediate phosphotriester gave phosphodiester **41**. Deacetonisation of **41** was effected by treatment with pyridinium *p*-toluenesulfonate (PPTS) in methanol. Ammonolysis of the ester functions and subsequent purification by gel filtration furnished target compound **31**, the structure of which was firmly established by NMR spectroscopy.

Preliminary biological experiments revealed that **31** did not display an inhibitory effect on α(1→3)FT-VI. The latter disappointing outcome urged us to design a trisubstrate analogue, the structure of which resembles more closely the postulated transition state (Figure 25-5). The newly designed analogue **42** [96] is characterized by an anomerically branched fucose residue the anomeric centre of which is α-L-linked to a glucose acceptor, and the equatorial aminomethyl group is connected to guanosine through a malondiamido bridge. The replacement of the pyrophosphate linkage by the neutral malondiamido bridge may facilitate membrane transport and may chelate with Mn^{2+}, which is required by the enzyme.

Trisubstrate analogue **42** is in principle accessible by coupling of a protected anomerically branched L-fucopyranosyl donor with an appropriate acceptor and subsequent introduction of 5'-amino-5'-deoxyguanosine *via* a malonic spacer at the equatorial branch of the fucose residue (Scheme 25-3).

Donor phenyl 1-azido-3,4,5-tri-*O*-benzyl-1-deoxy-2-seleno-L-fuco-hept- ulo pyranoside (**45**) was prepared in two steps from tri-*O*-benzyl-L-fucono-1,5-lactone (**43**). Treatment of **43** with dimethyltitanocene gave **44**. Addition of *N*-

phenylselenophtalimide and azidotrimethyl silane to **44** afforded donor **45** in 30% yield. A better yield (68%) of the *anti*-Markovnikov addition product **45** was obtained by treatment of **44** with sodium azide and diphenyldiselenide in the presence of (diacetoxyiodo) benzene.

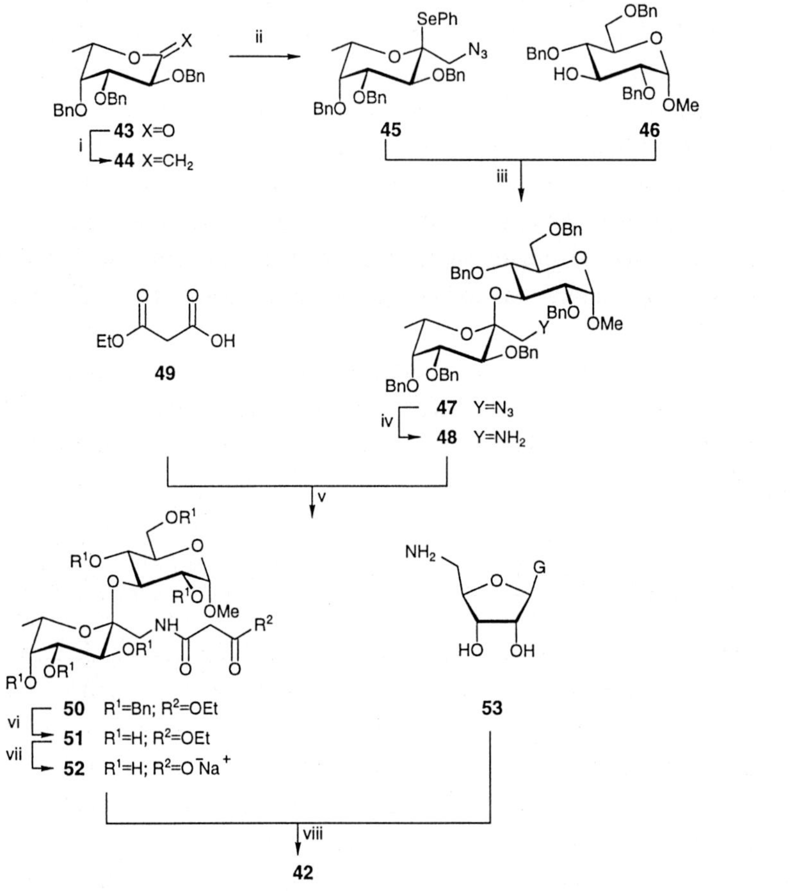

^aKey:

(*i*) Cp$_2$TiMe$_2$, toluene (74%); (*ii*) NaN$_3$, (PhSe)$_2$, PhI(OAc)$_2$, CH$_2$Cl$_2$ (68%); (*iii*) NIS, DCE, Et$_2$O (73%); (*iv*) LiAlH$_4$, Et$_2$O (100%); (*v*) DIPC, NMM, H$_2$O/DMF, 2/98 (67%); (*vi*) 20% Pd(OH$_2$)/C, H$_2$, TEAB, *i*-PrOH/H$_2$O, 1/1; (*vii*) 1 N NaOH (88%, two steps); (*viii*) DIPC, NMM, H$_2$O/DMF 2/98 (27%).

Scheme 25-3. Reagents see ^{a}

Having the requisite donor in hand, the coupling of **45** with acceptor methyl 2,4,6-tri-*O*-benzyl-α-D-glucopyranoside (**46**) was undertaken. Conden-sation of **45** with **46** under the influence of NIS afforded exclusively the α-linked dimer **47**, the anomeric configuration of which was ascertained by NMR spectroscopy. Final assembly of trisubstrate analogue **42** comprises the introduction of the malonic acid bridge between the *fuco*-ketodisaccharide and 5'-amino-5'-deoxyguanosine. First the azido function in **47** was reduced with lithium aluminium hydride to yield the corresponding amine **48**. Subsequent coupling of **48** with partially protected malonic acid **49** in the presence of *N,N*-diisopropylcarbodiimide (DIPC) and *N*-methylmorpholine (NMM) afforded compound **50**. Debenzylation of **50** by hydrogenolysis under buffered conditions (triethylammonium bicarbonate, TEAB) gave compound **51**. In this respect it is of interest to note that hydrogenolysis of **50** in the absence of TEAB buffer led to cleavage of the glycosidic bond. Saponification of the ethyl ester in **51** furnished intermediate **52**, which was purified by gel-filtration and condensed with 5'-amino-5'-deoxy guanosine (**53**) under influence of DIPC and NMM. Purification by gel filtration and silica gel chromatography afforded target compound **42**, the ^{13}C NMR, ^{1}H NMR and mass spectral data of which are in full accordance with the proposed structure. Preliminary biological experiments reveal that compound **42** did not show an inhibitory effect on $\alpha(1\rightarrow3)$FT-VI. The inhibitory effect of **42** on other fucosyltransferases is currently under investigation.

25.3 Conclusion

In summary, in this chapter the successful syntheses of several potential adhesion inhibitors are reported. First an efficient route to the new SLe^x mimic **9**, characterized by the presence of a propoxymethoxy moiety instead of the naturally occurring galactose residue, has been presented. In addition, attention was focussed on the design of TS analogues for $\alpha(1\rightarrow3)$FT, a new type of trisubstrate analogues. Two trisubstrate analogues *i.e.* compound **31** and **42** were prepared. Unfortunately, in preliminary biological experiments neither **31** nor **42** displayed any inhibitory effect on $\alpha(1\rightarrow3)$FT-VI.

The lack of inhibitory effect on analogue **31** may be due to the inaccessibility of the HO-2 of the fucose residue for recognition by the enzyme. Since not much is known about the structure of the active site of the transferase enzyme, it is difficult to explain the disappointing inhibitory activity of analogue **42**. A more effective inhibitor may be designed by taking into consideration the following aspects: (i) modification of the fucose residue, (ii) incorporation of different acceptor moieties, (iii) application of spacers, having different length and molecular properties, for tethering the fucose, guanosine, and acceptor moieties. The flexible synthetic route to compound **42** facilitates future design and synthesis of trisubstrate analogues of fucosyltransferases.

25.4 References

1. D. E. Levy, P. C. Tang, J. H. Musser, *Ann. Rep. Med. Chem.*, **1994**, 215.
2. N. Sharon H. Lis, *Sci. Am* **1993**, *269*, 74.
3. L. A. Lasky, *Science* **1992**, *258*, 964.
4. M. Bevilacqua, E. Butcher, E. Furie, B. Furie, M. Gallatin, M. Gimbrone, J. Harlan, K. Kishimoto, L. Lasky, R. McEver, J. Paulson, S. Rosen, B. Seed, M. Siegelman, T. Springer, L. Stoolman, T. Tedder, A. Varki, D. Wagner, I. Weissman, G. Zimmerman, G. , *Cell* **1991**, *67*, 233.
5. T. A. Springer, L. A.; Lasky, *Nature* **1991**, *349*, 196.
6. L. Osborn, *Cell* **1990**, *62*, 3.
7. L. M. Stoolman, *Cell* **1989**, *56*, 907.
8. L. A. Lasky, M. S. Singer, T. A. Yednock, D. Dowbenko, C. Fennie, H. Rodrigues, T. Nguyen, S. Stachel, R. D. Rosen, *Cell* **1989**, *56*, 1045.
9. M. H. Siegelman, M. van de Rijn, I. L. Weismann, *Science* **1989**, *243*, 1165.
10. G. I. Johnston, R. G. Cook, R. P. McEver, *Cell* **1989**, *56*, 1033.
11. E. Larsen, A. Celi, G. E. Gilbert, B. C. Furie, J. K. Erban, R. Bonfanti, D. D. Wagner, B. Furie, *Cell* **1989**, *59*, 305.
12. J. -G. Geng, M. P. Bevilacqua, K. L. Moore, T. M. McIntyre, S. M. Prescott, J. M. Kim, G. A. Bliss, G. A. Zimmerman, R. P. McEver, *Nature* **1990**, *343*, 757.
13. M. P. Bevilacqua, J. S. Pober, D. L. Mendrick, R. S. Cotran, M. A.Jr. Gimbrone, *Proc. Natl. Acad. Sci. U.S.A.* **1987**, *84*, 9238.
14. M. P. Bevilacqua, S. Stengeling, M. A. Jr. Gimbrone, B. Seed, *Science* **1989**, *243*, 1160.
15. T. A. Springer, *Cell* **1994**, *76*, 301.
16. R. O. Hynes, A. D. Lander, *Cell* **1992**, *68*, 303.
17. E. C. Butcher, *Cell* **1991**, *67*, 1033.
18. M. B. Lawrence, T. A. Springer, *Cell* **1991**, *65*, 859.
19. T. A. Springer, *Nature* **1990**, *346*, 425.
20. A. Varki, *Proc. Natl. Acad. Sci. U.S.A.* **1994**, *91*, 7390.
21. T. P. Patel, S. E. Goelz, R. R. Lobb, R. B. Parekh, *Biochemistry* **1994**, *33*, 14815.
22. K. L. Moore, S. F. Eaton, D. E. Lyons, H. S. Lichtenstein, R. D. Cummings, R. P. McEver, *J. Biol. Chem.* **1994**, *269*, 23318.
23. C. -T. Yuen, K. Bezouska, J. O'Brien, M. Stoll, R. Lemoine, A. Lubineau, M. Kiso, A. Hasegawa, N. J. Bokovich, K. C. Nicolaou, T. Feizi, *J. Biol. Chem.* **1994**, *269*, 1595.
24. Y. Shimizu, S. Shaw, *Nature* **1993**, *366*, 630.

25. C. -T. Yuen, A. M. Lawson, W. Chai, M. Larkin, M. S. Stoll, A. C. Stuart, F. X. Sullivan, T. J. Ahern, T. Feizi, *Biochemistry* **1992**, *31*, 9126.

26. E. L. Berg, M. K. Robinson, O Mansson, E. C. Butcher, J. L.; Magnani, *J. Biol. Chem.* **1991**, *266*, 14869.

27. M. Tiemeyer, S. J. Swiedler, M. Ishihara M. Moreland, H. Schweingruber, P. Hirtzer, B. K. Brandley, *Proc. Natl. Acad. Sci. U.S.A.* **1991**, *88*, 1138.

28. M. J. Polley, M. L. Philips, E. Wayner, E. Nudelman, A. K. Singhal, S. -I. Hakomori, J. C. Paulson, *ibid.*, 6224.

29. D. Tyrrell, P. James, N. Rao, C. Foxall, S. Abbas, F. Dasgupta, M. Nashed, A. Hasegawa, M. Kiso, D. Asa, J. Kidd, B.K. Brandley, *ibid.*, 10372.

30. E. Larsen, T. Palabrica, S. Sajer, G.E. Gilbert, D.D. Wagner, B.C. Furie, B. Furie, *Cell* **1990**, *63*, 467.

31. B. K. Brandley, S. J. Swiedler, P. W. Robbins, *ibid.*, 861.

32. G. Walz, A. Aruffo, W. Kolanus, M. Bevilacqua, B. Seed, *Science* **1990**, *250*, 1132.

33. M. L. Philips, E. Nudelman, F. C. A. Gaeta, M. Perez, A. K. Singhal, S.-I. Hakomori, K. C. Paulson, *ibid*, 1130.

34. J. B. Lowe, L. M. Stoolman, R. P. Nair, R. D. Larsen, T. L. Behrend, R. M. Marks, *Cell* **1990**, *63*, 475.

35. J. M. Harlan, D. Y. Liu in Adhesion: It's role in inflammatory disease (Ed.: W.H. Freeman and Company), New York, **1992**.

36. R. Kannagi, Y. Fukushi, T. Tachikawa, A. Noda, S. Shin, K. Shigeta, N. Hiraiwa, Y. Fukuda, T. Inamato, S. -I. Hakomori, H. Imura, *Cancer Res.* **1986**, *46*, 2619.

37. K. Fukushima, M. Hirota, P. I. Terasaki, A. Wakisaka, H. Togashi, D. Dhia, N. Suyama, Y. Fukushi, E. Nudelman, S. -I. Hakomori, *Cancer Res.* **1984**, *44*, 5279.

38. Y. Matsushita, S. Nakomori, E. A. Seftor, M. J. C. Hendrix, T. Irimmura, *Exp. Cell. Res.* **1991**, *196*, 20.

39. T. Matsusako, H. Muramatsu, T. Shirahama, T. Muramatsu, Y. Ohi, *Biochem. Biophys. Res. Commun.* **1991**, *181*, 1218.

40. P. Hutchings, H. Rosen, L. O'Reilly, E. Simpson, S. Gordon, A. Cook, *Nature* **1990**, *348*, 639.

41. N. B. Vedder, R. K. Winn, C. L. Rice, E. Y. Chi, K. -E. Arfors, J. M. Harlan, *Proc. Natl. Acad. Sci. U.S.A.* **1990**, *87*, 2643.

42. C. D. Wegner, R. H. Gundel, P. Reilly, N. Haynes, L. G. Letts, R. Rothlein, *Science* **1990**, *247*, 456.

43. R. K. Winn, D. Liggit, N. B. Vedder, J. C. Paulson, J. M. Harlan, *J. Clin. Invest.* **1993**, *92*, 2042.

44. M. S. Mulligan, M. J. Polley, R. J. Bayer, M. F. Nunn, J. C. Paulson, P. A. Ward, *J. Clin. Invest.* **1992**, *90*, 1600.
45. M. S. Mulligan, J. Varani, M. K. Dame, C. L. Lane, C. W. Smith, D. C. Anderson, P. A. Ward, *J. Clin. Invest.* **1991**, *88*, 1396.
46. R. H. Gundel, C. D. Wegner, C. A. Torcellini, C. C. Clarke, N. Hayens, R. Rothlein, C. W. Smith, L. G. Letts, *ibid.*, 1407.
47. S. R. Watson, C. Fennie, L. A. Lasky, *Nature* **1991**, *349*, 164.
48. (a) M. S. Mulligan, J. B. Lowe, R. D. Larsen, J. Paulson, Z. -I. Zheng, S. DeFrees, K. Maemura, M. Fukuda, P. A. Ward, *J. Exp. Med.*, **1993**, *178*, 623; (b) For similar experiments with SLe^x directed against P-Selectin; Mulligan *et al.*, *Nature* **1993**, *364*, 149.
49. S. A. DeFrees, W. Kosch, W. Way, J. C. Paulson, S. Sabesan, R. L. Halcomb, D.-H. Huang, C. -H. Wong, *J. Am. Chem. Soc.* **1995**, *117*, 66.
50. R. L. Halcomb, D. -H. Huang, C. -H. Wong, *J. Am. Chem. Soc.* **1994**, *116*, 11315.
51. W. Stahl, U. Sprengard, G. Kretzschmar, H. Kunz, *Angew. Chem. Int. Ed. Engl.* **1994**, *33*, 2096.
52. A. Giannis, *Angew. Chem. Int. Ed. Engl.*, **1994**, *33*, 178.
53. Y. Ichikawa, R. L. Halcomb, C.-H. Wong, *Chem. Brit.* **1994**, 117.
54. S. Hanessian, H. Prabhanjan, *Synlett* **1994**, 868.
55. N. M. Allanson, A. H. Davidson, C. D. Floyd, F. M. Martin, *Tetrahedron: Assym.* **1994**, *5*, 2061.
56. J. A. Ragan, K. Cooper, *Bioorg. Med. Chem. Lett.* **1994**, *4*, 2563.
57. P. V. Nikrad, M. A. Kashem, K. B. Walsichuk, G. Alton, A. P. Venot, *Carbohydr. Res.*, **1994**, *250*, 145.
58. K. Singh, A. Fernandez-Mayoralas, M. Martin-Lomas, *J. Chem. Soc., Chem. Commun.* **1994**, 775.
59. S. A. DeFrees, F. C. A. Gaeta, Y. -C. Lin, Y. Ichikawa, C.-H. Wong, *J. Am. Chem. Soc.* **1993**, *115*, 7459.
60. K. C. Nicolaou, N. J. Bockovich, D. R. Carcanague, *ibid.*, 8843.
61. Y. Ichikawa, Y. -C. Lin, D. P. Dumas, G. -J. Shen, E. Garcia-Junceda, M. A. Williams, R. Bayer, C. Ketcham, L. E. Walker, J. C. Paulson, C. -H. Wong, *J. Am. Chem. Soc.* **1992**, *114*, 9283.
62. K. C. Nicolaou, C. W. Hummel, U. Iwabuchi, *ibid.*, 3126.
63. Y. -C. Lin, C. W. Hummel, D. -C. Huang, Y. Ichikawa, K. C. Nicolaou, C. -H. Wong, *ibid.*, 5452.
64. G. E. Ball, R. A. O'Neill, J. E. Schultz, J. B. Lowe, B. W. Weston, J. O. Nagy, E. G. Brown, C.J. Hobbs, M.D. Bednarski, *ibid.*, 5449.

65. J. Y. Ramphal, Z. -L. Zheng, C. Perez, L. E. Walker, S. A. DeFrees, F. C. A. Gaeta, *J. Med. Chem.* **1994**, *37*, 3459.

66. J. H. Musser, N. Rao, M. Nashed, D. Dasgupta, S. Abbas, A. Nematalla, V. Date, C. Foxall, D. Asa, P. James, D. Tyrrell, B. K. Brandley, *Trends Receptor Res.* **1993**, *20*, 33.

67. B. K. Brandley, M. Kiso, S. Abbas, P. Nikrad, O. Sivasatava, C. Foxall, Y. Oda, A. Hasegawa, *Glycobiology* **1993**, *3*, 633.

68. A recent study revealed that the conformation of SLe^x, when bound to E-Selectin differs from the commonly perceived solution conformation. R. M. Cooke, R. S. Hale, S. G. Lister, G. Shah, M. P. Wier, *Biochemistry* **1994**, *33*, 10591.

69. B. H. Heskamp, G. H. Veeneman, G. A. van der Marel, C. A. A. van Boeckel, J. H. van Boom, *Recl. Trav. Chim. Pays-Bas* **1995**, *114*, 398.

70. G. H. Veeneman, R. G. A. van der Hulst, C. A. A. van Boeckel, R. L. A. Philipsen, G. S. F. Ruigt, J. A. D. M. Tonnaer, T. M. L. van Delft, P. N. M. Konings, *Bioorg. Med. Chem. Lett.* **1995**, *5*, 9.

71. B. N. N. Rao, M. B. Anderson, J. H. Musser, J. H. Gilbert, M. E. Schaefer, C. Foxall, B. K. Brandley, *J. Biol. Chem.* **1994**, *269*, 19963.

72. J. F. Kukowska,-Latallo, R. D. Larsen, R. P. Nair, J. B. Lowe, *Genes & Dev.* **1990**, *4*, 1288.

73. S. Goelz, R. Kumar, B. Potvin, S. Sundaram, M. Brickelmaier, P. Stanley, *J. Biol. Chem.* **1994**, *269*, 1033.

74. R. Kumar, B. Potvin, W. A. Muller, P. Stanley, *J. Biol. Chem.* **1991**, *266*, 21777.

75. J. B. Lowe, J. F. Kukowska-Latallo, R. P. Nair, R. D. Larsen, R. M. Marks, B. A. Macher, R. J. Kelly, L. K. Ernst, *ibid.* 17467.

76. S. E. Goelz, C. Hession, D. Goff, B. Griffiths, R. Tizard, B. Newman, G. Chi-Rosso, R. Lobb, *Cell* **1990**, *63*, 1349.

77. B. W. Weston, R. P. Nair, R. D. Larsen, J. B. Lowe, *J. Biol. Chem.* **1992**, *267*, 4152.

78. B. W. Weston, P. L. Smith, R. L. Kelly, J.B. Lowe, *ibid.*, 24575.

79. K. Sasaki, K. Kurata, K. Funayama, M. Nagata, E. Watanabe, S. Ohta, N. Hanai, T. Nishi, *J. Biol. Chem.* **1994**, *269*, 14730.

80. B. A. Macher, E. H. Holmes, S. J. Swiedler, C. L. M. Stults, C. A. Srnka, C. A., *Glycobiology* **1991**, *1*, 577.

81. A. J. R. Bella, Y. S. Kim, *Biochem. J.* **1971**, *125*, 1157.

82. S. Cai, M. R. Stroud, S. Hakomori, T. Toyokuni, *J. Org. Chem.* **1992**, *57*, 6693.

83. J. I. Luengo, J. G. Gleason, *Tetrahedron Lett.* **1992**, *33*, 6911.

84. C. Gautheron-Le Narvor, C. -H. Wong, *J. Chem. Soc., Chem. Commun.* **1991**, 1130.

85. (a) C. -H. Wong, D. P. Dumas, Y. Ichikawa, K. Koseki, S.J. Danishefsky, B.W. Weston, J. B. Lowe, *J. Am. Chem. Soc.* **1992**, *114*, 7321; (b) D. P. Dumas, Y. Ichikawa, C. -H. Wong, J. B. Lowe, R. P. Nair, *Bioorg. Med. Chem. Lett.* **1991**, *1*, 425

86. It is difficult to compare the inhibitory activities of for instance compounds **20** and **25** in a quantitative manner, due to the different assay conditions and $\alpha(1\rightarrow 3)$FTs which are used. This is illustrated by the fact that the IC50 for GDP (0.05 mM) reported by Wong (ref. 85) is much lower than the corresponding IC50 (0.55 mM0 that can be deduced from the biological activities described by Toyokuni (ref. 82).

87. P. A. Bartlett, C. K. Marlowe, *Biochemistry* **1983**, *22*, 4618.

88. G. E. Lienhard, *Science* **1973**, *180*, 149.

89. R. Wolfenden, *Acc. Chem. Res* **1972**, *5*, 10.

90. M. M. Palcic, L. D. Heerze, O. P. Srivastava, O. Hinsgaul, *J. Biol. Chem.* **1989**, *264*, 17174.

91. It has been postulated that the reaction catalyzed by glucoronyltransferase proceeds *via* a S_N2 mechanism, resulting in a similar TS. D. Noort, M.W.H. Coughtrie, B. Burchell, G.A. van der Marel, J.H. van Boom, *Eur. J. Biochem.* **1990**, *188*, 309.

92. (a) K. Frische, R. R. Schmidt, *Liebigs Ann. Chem.*, **1991**, 297; (b) K. Frische, R. R. Schmidt, *Bioorg. Med. Chem. Lett.* **1993**, *3*, 1747.

93. Y. -F. Wang, D. P. Dumas, C. -H. Wong, *Tetrahedron Lett.* **1993**, *34*, 403.

94. The same enzyme as used previously for acceptor analogue **25** was employed to determine the biological activity [ref. 85(a)].

95. B. M. Heskamp, G. A. van der Marel, J. H. van Boom, *J. Carbohydr. Chem.* **1995** *14*, 1265.

96. B. M. Heskamp, G. H. Veeneman, G. A. van der Marel, C. A. A. van Boeckel, J. H. van Boom, *Tetrahedron* **1995**, *51*, 8397.

26 Enzymatic Synthesis of Lactose Analogues Using Glycosidases

A. Fernández-Mayoralas

26.1 Introduction

Orientals, blacks, and certain other races, in quite high numbers, show an inability to digest lactose as readily beyond infancy. Intestinal lactase is an enzyme responsible for splitting lactose into its monosaccharide components. If the individual is deficient in intestinal lactase, a significant amount of lactose remains intact within the intestine. Some of it may be fermented by ever-present bacteria, then, organic acids and gases are formed. By osmosis, its presence draws water from the surrounding intestinal tissues. The result is bloating, pains, and, with all the water, diarrhea. Therefore, the evaluation of the intestinal lactase is important in gastroenterology as well as in pediatrics.

We have previously developed [1] a non-invasive method to evaluate the activity of intestinal lactase based on oral administration of 4-O-β-D-galactopyranosyl-D-xylose (**1**, Figure 26-1), a disaccharide very similar to lactose which lacks the hydroxymethyl group at position 5. This compound was found to be a substrate of the enzyme, yielding upon its action D-galactose, which is metabolized, and D-xylose. The latter is passively absorbed from the small intestine, not phosphorylated, and excreted in the urine from where it can be determined by a simple colorimetric procedure. The amount of excreted xylose is then correlated with the activity of intestinal lactase.

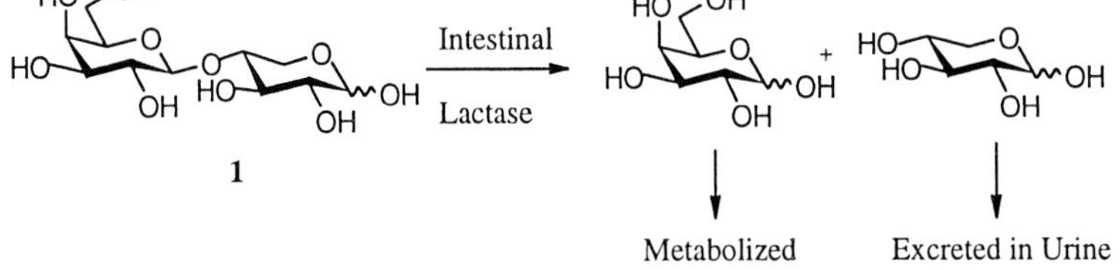

Figure 26-1: Hydrolysis *in vivo* of **1** by intestinal lactase.

Compound **1** can be obtained from 2,3,4-tri-*O*-acetyl-α-D-xylopyranosyl bromide in a synthetic sequence that involves seven reaction steps with about 9% overall yield [2]. This synthesis is not practical for an application of the diagnostic method and a simpler preparation method was sought. In this context, enzymes offer the opportunity of one-step preparations under mild conditions in regio- and stereoselective manner. Glycosidases are commercially available enzymes which have been frequently used for the synthesis of disaccharides [3]. Under physiological conditions they catalyze the hydrolysis of glycosidic bonds [4]. In Figure 26-2 the reaction pathway for the hydrolysis and the formation of disaccharides catalyzed by a β-galactosidase is schematically depicted. The enzyme recognises terminal units of β-galactosides. It is supposed that an intermediate galactosyl-enzyme is formed after the release of the aglycon ROH. Water molecules react with the galactosyl residue to give galactose, the hydrolysis product, which is the main reaction under physiological conditions.

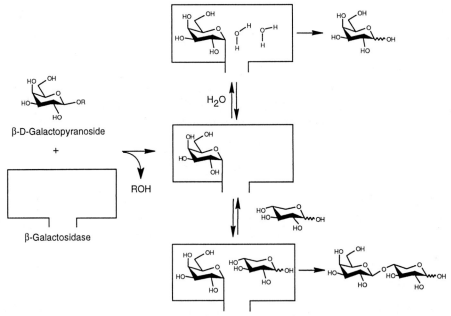

Figure 26-2: Reaction pathways for the hydrolysis and transglycosylation of a β-D-galactopyranoside catalyzed by β-galactosidase.

However, when a monosaccharide is present in the medium in high concentration, it could bind in place of the aglycon [5]. The galactose unit could be transferred to this monosaccharide giving, in this case, disaccharides as the reaction products. Since the disaccharides formed are also substrates of the enzyme, their synthesis using this approach must be carried out as a kinetically controlled process. A systematic study on the β-galactosidation of xylose derivatives (**3**, Xyl(β)-R in Figure 26-3) with its anomeric position protected, using β-galactosidase enzymes and o-nitrophenyl β-D-galactopyranoside (**2**, Gal(β)-ONP) as galactosyl donor was carried out (Figure 26-3). We wanted to know the influence of different factors on the yield and the regioselectivity of the possible disaccharides formed, such as the changes in the reaction medium, the effect of using immobilized enzymes, the origin of the enzyme, and the substituents in the xylose acceptor. The optimized reaction could provide a direct access to galactosyl-xylose disaccharides, which may be tested as substrates of the intestinal lactase. Only the effect of the origin of the β-galactosidase and, in more detail, the effect of substituents in the xylose acceptor will be evaluated in this chapter. Changes in the reaction conditions and the immobilization of the enzyme did not alter the selectivity of the reaction [6].

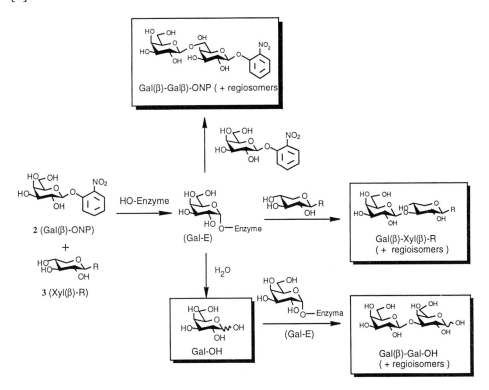

Figure 26-3: All possible products that are formed from the reaction of **2** and **3** in the presence of a β-galactosidase.

26.2 Results and Discussion

All possible products of the aforementioned reaction are shown in Figure 26-3. Once the galactosyl-enzyme intermediate is formed, several nucleophiles can act as acceptors. The desired one, D-xylose, to give Gal(β)-Xyl(β)-R disaccharides as mixture of regioisomers, the donor, Gal(β)-ONP, to give Gal(β)-Gal(β)-ONP disaccharides; and water, to give galactose (GalOH). Even galactose arising from hydrolysis may, after reaching a certain concentration, attack to the galactosyl-enzyme to give Gal(β)-Gal-OH. Gas chromatography and HPLC was used to follow the formation of all these products.

We first examined the relative formation of disaccharides with three β-galactosidases of different sources in the presence of benzyl β-D-xylopyranoside (Xyl(β)-OBn) (table 1) [6]. With the β-galactosidase from *Aspergillus oryzae*, the non-desired Gal(β)-Gal(β)-ONP were formed rapidly, reaching the maximum of ~36%, and then decreased being hydrolysed. The desired products, Gal(β)-Xyl(β)-OBn, had lower rate of formation, maximum yield being lower than 8%. Even galactose coming from the hydrolysis of the donor gave galactosyl-galactose dimers more rapidly. With the β-galactosidase from Jack bean hydrolysis of **2** was observed as main pathway. However, β-galactosidase from *E. coli* showed a marked preference to bind β-D-xylopyranoside as acceptor. Gal(β)-Xyl(β)-OBn disaccharides were most rapidly formed, reaching a maximum of 30%. The other disaccharides were also formed but at lower rates.

Table 26-1: Maximum percentage of disacharides formed using different β-galactosidases

Enzyme Source	Gal(β)-Xyl(β)-OBn	Gal(β)-Gal(β)-OBn	Gal-GalOH
A. oryzae	7.3	35.7	23.7
E. coli	30.0	8.8	18.0
Jack bean	3.0	-	4.0

Having found the best enzyme we next turned our attention to the effect of substituents at the anomeric carbon of D-xylose. We prepared [7, 8] a variety of xylopyranosides (**3a-i**, Table 26-2), including alkyl xylopyranosides with short and long chains and with aromatic rings, aryl xylopyranosides having electron donating or electron withdrawing groups, and *S*- and *C*-xylopyranosides. The results obtained using these acceptors are summarised in Table 26-2 [7, 8]. Two β-D-galactopyranosyl-β-D-xylopyranosides were formed in all cases, the 1,3 and the 1,4 linked glycosides (compounds **4** and **5**, respectively); no appreciable amount of the corresponding 1,2-glycoside was observed.

Table 26-2: Yield and regioselectivity of the disaccharides **4** and **5** from the β-galactosidase-catalyzed reaction of **2** (100 mM) and **3** (50 mM). regioselectivity

Compound	R in 3	Yield (%) 4 + 5	Regioselectivity 4:5
3a	CH$_3$O	6	0:1.0
3b	CH$_3$(CH$_2$)$_7$O	8	-
3c	C$_6$H$_5$-CH$_2$O	30	1:0.4
3d	C$_6$H$_5$-O	23	1:1.4
3e	C$_6$H$_5$-(CH$_2$)$_2$O	25	1:1.2
3f	C$_6$H$_5$-S	19	1:2.9
3g	C$_6$H$_5$-CH$_2$	15	1:5.0
3h	NO$_2$-C$_6$H$_4$-O	18	1:5.5
3i	MeO-C$_6$H$_4$-O	20	1:1.6

As a general feature, the yield was higher when the aglycon in **3** had a phenyl group. For instance from the methyloxy **3a** to the benzyloxy derivative **3c** there was a fivefold increase in yield. A long hydrophobic chain, such as the octyloxy group (compound **3b**) did not cause any important improvement as compared to the methyloxy.

Among the substrates containing a phenyl group, there are some small differences in the yield. More interesting was the effect of the substituent on the regioselectivity regioselectivity, which was significantly altered by small variations in the aglycon. Thus, from the methyloxy **3a** to the benzyloxy derivative **3c** the regioselectivity of the reaction reverses: while with **3a** a 1,4 disaccharide was the only product, with **3c** the 1,3-disaccharide was the main product. The absence of the methylene group in the phenyloxy derivative **3d** changed the regioselectivity observed with the benzyloxy **3c**, however the presence of two methylenes in the phenylethyloxy **3e** leads to a regioselectivity similar to that obtained with **3d**. More remarkable, the replacement of the oxygen atom of the phenyloxy aglycon by a sulfur atom (compound **3f**) increased two times the formation of the 1,4 with respect to the 1,3 linked disaccharide. In the same way, the replacement by a carbon atom **3g** or the *para*-substitution of the phenyloxy by a nitro group **3h** resulted in a fourfold increase of the relative formation of the 1,4 disaccharide. However, the *para*-substitution with an electron-donating methyloxy group **3i** gave similar regioselectivity as the non-substituted phenyloxy **3d**.

All these results show a clear dependence of the yield and regioselectivity on small variations in the aglycon of the xylose acceptor. However, all these values are taken at the maximum formation of galactosyl-xylose disaccharides. Therefore, we do not know whether the disaccharide regioisomers are hydrolysed as they are formed at different rates, so that the values could be different at the initial moments of the reaction. In order to get a more exact information about the selectivity of the enzymatic reaction, we decided to measure initial rates of formation with some of the xylopyranosides [8].

The values of the initial rates of formation of the disaccharides **4** and **5** (V_4 and V_5, respectively) allowed us to calculate the relative specificity [8] of the enzyme for each of the xylosides and the regioselectivity as the ratio between the rates of formation of 1,3- and 1,4-galactosyl-xylosides (Table 26-3).

The specificity is higher when the aglycon of **3** has a phenyl group as compared to the methyl derivative. For instance, from the methyloxy (**3a**) to the benzyloxy derivative (**3c**) there is a fivefold increase in specificity. This is in agreement with the higher yields obtained at the maximum formation of disaccharides. Therefore, we can conclude that the increase in yield is a consequence of the higher specificity of the enzyme for the xylosides containing a phenyl ring. Among the acceptors with only one atom between the aromatic ring and the anomeric carbon (**3d**, **3f**, and **3g** in Table 26-3), there are some differences in specificity.

Table 26-3: Initial rate, specificity, and regioselectivity for the reaction of Gal(β)-ONP (**2**, 50 mM) and some Xyl(β)-R (**3**, 50 mM) in the presence of β-galactosidase from *E. coli* at pH 7.0.

Compound	R in 3	Initial rate $V_4 + V_5$	Relative specificity	Regioselectivity $V_4:V_5$
3a	CH₃O	10	0.05	0:10.0
3c	Ph–CH₂O	10	0.25	8.4:1.6
3d	Ph–O	16	0.20	8.7:7.3
3e	Ph–(CH₂)₂O	18	0.20	7.5:10.5
3f	Ph–S	10	0.11	2.7:7.3
3g	Ph–CH₂	9	0.08	1.7:7.3

For a better understanding of these differences we have to look at the regioselectivity. The rate of formation of 1,3-disaccharides decreases in the same direction in which the electronegativity of the atom linked to the anomeric carbon decreases, whilst the rate of formation of the 1,4-disaccharide remains constant. Considering these results, we postulate that two different complexes between the xyloside and the galactosyl-enzyme produce, respectively, the 1,3- and the 1,4-galactosyl-xyloses (Figure 26-4). We suggest that the pyranoid ring undergoes 180 degrees flip from one to the other complex; in this way, the position of the reactive hydroxyl group (HO-3 or HO-4) is about the same in each complex. In the 1,3-complex there might be an important stabilizing interaction between the lone pair electrons of the ring oxygen in the xyloside and an electrophilic centre in the enzyme that would be affected by the substitution at the anomeric carbon, that is the X atom in Figure 26-4. In such a way, the higher the electron density in the X atom (oxygen, sulfur or carbon), the higher is the interaction between ring oxygen and electrophilic centre. This interaction would also explain the regioselectivity obtained with the *p*-nitrophenyl derivative **3h** (Table 26-2). On the other hand, in the 1,4-complex, HO-2 would occupy the same position as the ring oxygen in the 1,3-complex. The nucleophilicity of the O-2 would not be affected by the substitution at the anomeric carbon.

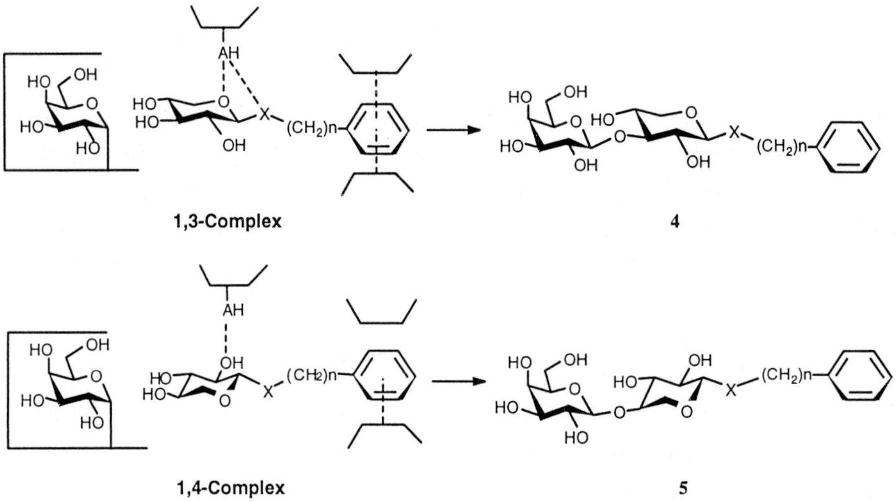

Figure 26-4: Postulated complexes for the formation of disaccharides **4** and **5**.

In these hypothetical complexes there is an additional stabilizing interaction of the phenyl ring of the acceptor with the enzyme, which must be more important in the 1,3-complex. From the data in Table 26-3 and comparing the regioselectivities of the benzyl, phenyl and phenylethyl derivatives (**3c**, **3d**, and **3e** respectively) with that of the methyl derivative (**3a**), from which the 1,3-disaccharide was not obtained, this interaction in the case of the benzyl derivative seems to be optimum in the 1,3-complex and minimum in the 1,4-complex. However, the differences in each complex for the phenyl and phenylethyl derivatives are less important, and both display similar regioselectivity. We calculated the minimized more stable conformers of the benzyl, phenyl and phenylethyl derivatives [8]. The conformations of the phenyl and phenylethyl glycosides show a similar orientation of the aromatic ring and the reacting hydroxyls, being in contrast to that of the benzyl derivative, in which the aromatic ring is in a different orientation in the space. Hence, the regioselectivity could be also influenced by the relative orientation of the aromatic ring and the reacting hydroxyls of the acceptor.

In comparing the regioselectivity of the galactosyl-xylosides formed at the initial moments of the reactions (Table 26-3) and at their maximum formation (Table 26-2), one can see that in some cases this varies appreciably. For instance, for the benzyl β-D-xylopyranoside (**3c**) the ratio of 1,3- and 1,4-galactosyl-xyloses changes from 1:0.2 at the initial moments to 1:0.4 at the time of maximum formation. Two mechanisms can account for this effect: either one of the regioisomers being formed is subsequently more rapidly hydrolysed or there is disaccharide intramolecular isomerization catalysed by the β-galactosidase in which the xylose portion does not become free of the enzyme during the isomerization. A similar mechanism has been proposed for the lactose allolactose isomerization in the bacterium *E. coli*; in this case the allolactose formed keeps the lac operon induced [9]. To test this second possibility we carried out an experiment in which a solution of benzyl 1,3-galactosyl-xyloside (**4c**), the disaccharide whose formation experiments the highest regioselectivity variation, was submitted to the action of β-galactosidase. Only the hydrolysis products, galactose and benzyl xylopyranoside, were formed, the 1,4-galactosyl-xyloside regioisomer was not observed at any moment of the reaction. This result proves the nonexistence of intramolecular isomerization and is consistent with our proposed hypothesis which postulated the existence of two complexes for the formation of 1,3 and 1,4-galactosyl-xylosides: for the intramolecular isomerization the xylopyranoside has to turn 180 degrees, a process which must have important steric barriers, thereby driving in this case diffusion into free solution.

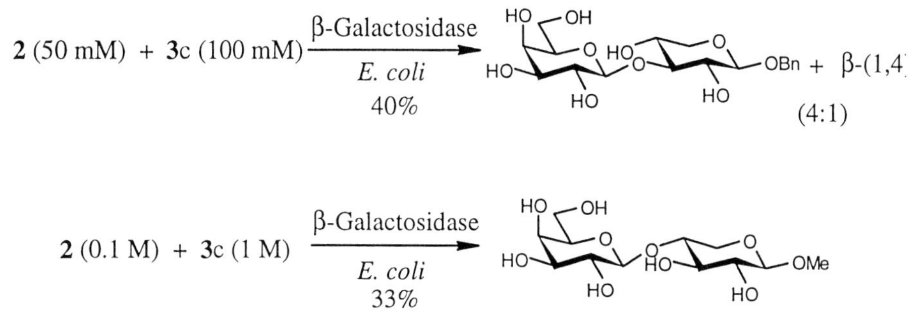

Figure 26-5: Regioselective synthesis of Gal(β)-Xyl(β)R disaccharides.

At this moment we have collected enough information to synthesize disaccharides with a control of the regioselectivity by choosing the convenient enzyme and the substituent at the anomeric position of the acceptor. Given the results obtained with the β-galactosidase from *E.coli*, benzyl β-D-xylopyranoside (**3c**) was used to prepare selectively a 1,3-galactosyl-xylose disaccharide, and methyl β-D-xylopyranoside (**3a**) for the preparation of the corresponding 1,4-regioisomer (Figure 26-5). Using **3c** at 100 mM concentration we obtained in 40% isolated yield the 1,3-disaccharide as a main product, and from **3a** at 1M concentration the 1,4-disaccharide was the only galactosyl-xylose containing product, obtained in 33% isolated yield.

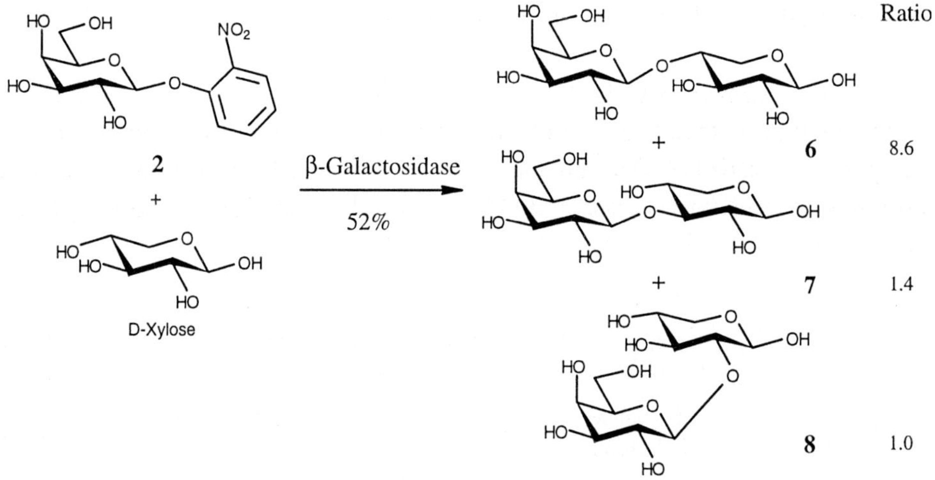

Figure 26-6: β-Galactosidase-catalyzed synthesis of β-D-galactopyranosyl-D-xyloses from D-xylose.

We finally faced the galactosidation of unprotected xylose [10], which may directly afford free Gal(β)-Xyl-OH disaccharides to be tested as substrates of intestinal lactase. After varying the xylose concentration, the best result was obtained when the reaction was carried out at 500 mM xylose concentration, to give a mixture of 1,4-, 1,3- and 1,2-galactosyl-xyloses (compounds **6, 7,** and **8**, respectively, in Figure 26-6) in 8.6:1.4:1.0 ratio in 52% isolated yield. Under these conditions we could prepare several grams of these products, which were isolated by carbon column chromatography using water as eluent. Although the mixture of regioisomers could not be efficiently separated, experiments carried out with pure disaccharides showed that all of them were substrates of the intestinal lactase and gave xylose after hydrolysis; therefore, the mixture can be used to evaluate the lactase activity. Experiments with rats using the mixture of galactosyl-xyloses showed a correlation between the percentage of D-xylose excreted in urine and the levels of intestinal lactase [10].

26.4 Conclusion

In summary, the work presented here shows the influence of substituents in the acceptor substrate on the selectivity of the enzymatic galactosidation of monosaccharides, allowing the controlled synthesis of disaccharides. Glycosidases, in spite of being non-specific, can be useful for the synthesis of bulk disaccharides. In our case we could prepare galactosyl-xylose disaccharides in large amounts in one reaction step, under mild conditions and using commercially available catalysts and substrates.

Acknowledgement

I would like to thank the people who has participated in this work: Drs. R. Lopez, J. Cañada, M. Martin-Lomas, J. Aragón and D. Villanueva. Financial support by DGICYT (Grant PB93-0127-C02-01) and Comunidad de Madrid (Grant AE00049/95) are gratefully acknowledged.

26.5 References

1. J. J. Aragón, A. Fernández-Mayoralas, J. Jiménez-Barbero, M. Martín-Lomas, A. Rivera-Sagredo, D. Villanueva, *Clin. Chim. Acta* **1992**, *210*, 221-226.

2. A. Rivera, A. Fernández-Mayoralas, J. Jiménez-Barbero, M. Martín-Lomas, D. Villanueva, J. J. Aragón, *Carbohydr. Res.* **1992**, *228*, 129-135.

3. C.-H. Wong, G. M. Whitesides, *Enzymes in Synthetic Organic Chemistry*, 1st ed., Elsevier Science Ltd, Oxford, **1994**, p. 283; K. G. I. Nilsson, *Trends Biotechnol.* **1988**, *6*, 256-264.

4. K. Wallenfels, R. Weil, *The Enzymes*, 3rd ed. **1972**, vol. 7, 617.

5. R. E. Huber, M. T. Gaunt, K. L. Hurlburt, *Arch. Biochem. Biophys.* **1984**, *234*, 151-160.

6. R. López, A. Fernández-Mayoralas, M. Martín-Lomas, J. M. Guisán, *Biotechnol. Letters* **1991**, *13*, 705-710.

7. R. López, A. Fernández-Mayoralas, *Tetrahedron Lett.* **1992**, *33*, 5449-5452.

8. R. López, A. Fernández-Mayoralas, *J. Org. Chem.* **1994**, *59*, 737-745.

9. R. E. Huber, G. Kurz, K. Wallenfels, *Biochemistry* **1976**, *15*, 1994-2001; R. W. Franck, *Bioorg. Chem.* **1992**, *20*, 77-88.

10. J. J. Aragón, F. J. Cañada, A. Fernández-Mayoralas, R. López, M. Martín-Lomas, D. Villanueva, *Carbohydr. Res.* **1996**, *290*, 209-216.

27 Synthesis of Nitrogen Containing Linkers for Antisense Oligonucleotides

Yogesh S. Sanghvi

27.1 Introduction

In the last decade, modification of the sugar-phosphate backbone in nucleic acids has become one of the major areas of research in developing novel constructs for oligonucleotide-based therapeutics [1]. These molecules are short (~ 20-mer), synthetically prepared single-stranded mimics of natural nucleic acids with potential for their use as sequence specific inhibitors of the gene expression. Such an interference is caused by hybridization (*via* Watson-Crick hydrogen bonding) between the chemically modified nucleic acid piece (antisense oligonucleotide) and the targeted cellular mRNA (sense), and thereby blocking the translation to the corresponding protein [2]. In principle, this antisense oligonucleotide and target mRNA binding process provides many orders of magnitude higher affinity and specificity than the traditional drug-design approach.

Successful drug development based on antisense technology requires the synthesis and use of chemically modified oligonucleotides that render stability to the cellular nucleases and maintain high fidelity in terms of affinity and base-pair specificity toward the target mRNA. These requirements have resulted in significant efforts toward chemical modifications of the natural phosphodiester linkage **1** of the DNA. One of the simplest and widely studied first generation modifications is replacement of one of the non-bridging oxygen atoms with sulfur, providing a phosphorothioate linkage **3**. Currently, more than a dozen antisense phosphorothioates are undergoing human clinical trials as antitumor, antiviral and anti-inflammatory agents [3]. In addition to the backbone modification, 2′-OMe RNA **4** has been utilized extensively in antisense constructs, due to their enhanced nucleolytic stability and improved affinity for the mRNA target, compared to the PS oligonucleotides **3**.

We believe that useful second and third generation antisense drugs may emerge through replacement of the natural PO linkage **1** with achiral and neutral linkages. In this arena, our research has been focused on the nitrogen-containing isostere. This chapter will describe the synthesis and properties of nucleosidic building blocks and their coupling to furnish novel nucleosidic dimers containing a wide range of backbones (Figure 27-1, entries **5-13**).

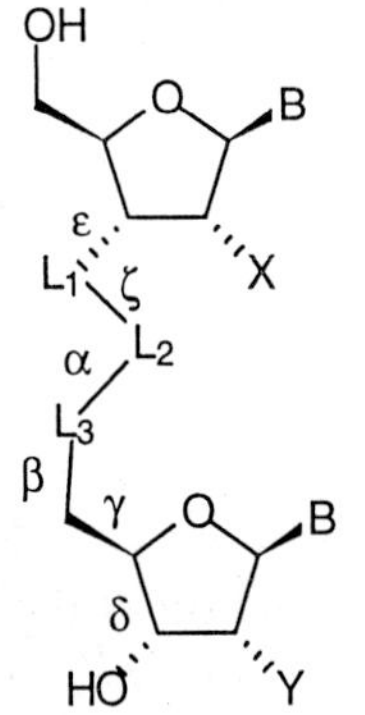

B = T, C, 5-MeC, A, G, U, 5-MeU

Torsion angle notation (IUPAC) for backbone linkages of nucleic acids

	L_1	L_2	L_3	X	Y	Name [Ref] [a]
1	O	PO_2	O	H	H	DNA [1]
2	O	PO_2	O	OH	OH	RNA [2]
3	O	P(S)O	O	H	H	PS Oligo [3]
4	O	PO_2	O	OCH_3	OCH_3	2'-OMe RNA [1]
5	CH	N	O	H	H	Oxime [4]
6	CH_2	NH	O	H	H	MI [4]
7	CH_2	NCH_3	O	H	H	MMI [14]
8	CH_2	NCH_3	NCH_3	H	H	MDH [12]
9	CH_2	O	NCH_3	H	H	MOMI [13]
10	O	NCH_3	CH_2	H	H	HMIM [15]
11	CH_2	NCH_3	CH_2	H	H	Amine 1 [17]
12	CH_2	CH_2	NCH_3	H	H	Amine 2 [17]
13	CH_2	NCH_3	O	OCH_3	OCH_3	2'-OMe MMI [11]

[a] MI: methyleneimino, MMI methylene (methylimino); MDH methylene (dimethylhydrazo); MOMI methyleneoxy (methylimino); HMIM hydroxy (methyiminomethylene)

Figure 27-1. Structure of natural and synthetic backbone linkages.

27.2 Results

27.2.1 3'-C-N-O Linked Backbones

Replacement of the central phosphorus atom of a phosphodiester linkage with a nitrogen atom has several distinct advantages in terms of the antisense properties.

First, it provides a very high degree of nuclease stability, and the modified hydroxylamino linkage will obviously not be a substrate for natural nucleases. Second, it is neutral and achiral. This may help reduce the net negative charge of the oligo and alter the pharmacokinetics in a favorable manner. Furthermore, enhanced nucleophilicity of an hydroxylamine compared to an amine allows convenient post-oligo conjugation capacity for the introduction of a variety of functionalities. Considerations of the synthetic accessibility and hydrolytic stability led us to believe that this may be a linkage of choice for advanced studies in antisense constructs.

27.2.1.1 Synthesis of Oxime Dimer 5

Scheme 27-1 outlines the preparation of conformationally restricted (via E and Z isomers) oxime linked **5**, for which the key reaction is the dehydrative coupling of 3´-deoxy-3´-C-formyl-5´-O-tritylthymidine **14** and 5´-O-amino-3´-O-(tert-butyldiphenylsilyl)thymidine **15** under acidic conditions [4]. We have reported an efficient and stereoselective synthesis of **14** using an intermolecular radical C-C bond formation reaction [5]. Interestingly, this method is widely applicable towards the preparation of C-branched nucleosides and various carbohydrate precursors [6].

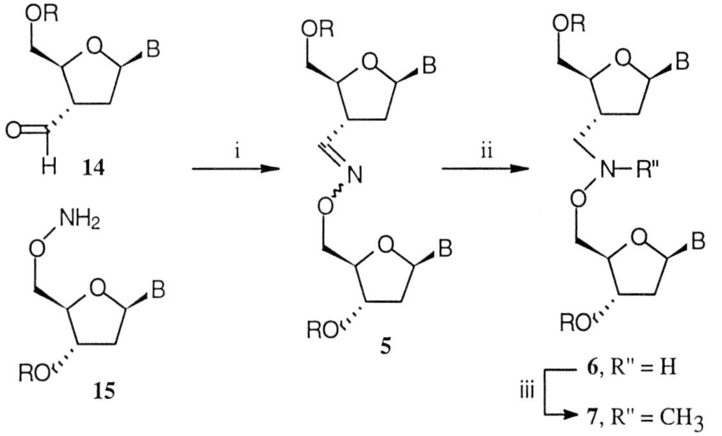

Scheme 27-1. Synthesis of dimeric nucleoside analogs **5-7**. Reagents: i) 1% AcOH in CH_2Cl_2, rt, 4-6 hr; ii) $NaBH_3CN/AcOH$, rt, 2 hr; in all structures of schemes 27-1 to 27-9: R, R' = Protecting groups, e.g. 4,4´-Dimethoxytriphenylmethyl or tert-Butyldiphenylsilyl ; B = T, C', 5-MeC', A', G', 5-MeU, ' = appropriate base protecting group; iii) aq. $HCHO/NaBH_3CN/AcOH$, rt, 2hr.

The 5´-O-N-bond formation in **15** was accomplished via a regioselective Mitsunobu reaction in excellent yield [7]. The coupling of **14** with **15** furnished **5** in quantitative yield as a mixture of E/Z isomers in 1:1 ratio. The two isomers can be separated by HPLC and have been characterized by ^{1}H NMR studies [4]. The two isomers were not separated for incorporation into oligos due to their rapid equilibration under basic conditions utilized for the deprotection of oligos. The hybridization data is summarized in Table 27-1, indicating a large destabilization effect, possibly due to loss of flexibility (around torsion angle) and isomeric mixture of oxime linkages.

Table 27-1. Hybridization data of backbone modified oligomers[a].

Backbone	Sequence 5´ → 3´	Tm (ΔTm)[b]
5 (oxime)	CTCGTACCT*TTCCGGTCC	60.2 (-3.1)
6 (MI)	CTCGTACCT*TTCCGGTCC	55.0 (-0.5)
7 (MMI)	GCGT*TT*TT*TT*TT*TGCG	50.8 (+0.1)
8 (MDH)	GCGT*TT*TT*TT*TT*TGCG	51.3 (+0.2)
9 (MOMI)	GCGT*TT*TT*TT*TT*TGCG	40.2 (-2.1)
10 (HMIM)	GCGT*TT*TT*TT*TT*TGCG	43.5 (-0.9)
11 (amine 1)	GCGT*TT*TT*TT*TT*TGCG	37.1 (-2.6)
12 (amine 2)	GCGT*TT*TT*TT*TT*TGCG	25.9 (-4.4)
13 (2'-OMe MMI)	GCGT*TT*TT*TT*TT*TGCG	67.0 (+3.7)

[a] Oligonucleotides **5-13** were hybridized with complement RNA; [b] Absorbance vs temperature profiles were measured at 4 mM of each strand in 100 mM Na$^+$, 10 mM phosphate, 0.1 mM EDTA, pH 7.0, ΔTm represents the difference in the Tm per modification in the given sequence.

27.2.1.2   Synthesis of MI Dimer 6

In order to obtain a more flexible linkage, reduction of **5** with NaBH$_3$CN/AcOH provided **6** in 80% yield. Additionally, synthesis of **6** was accomplished in a single step *via* a free radical coupling of an iodo nucleoside **16** and oxime nucleoside **17** in a stereoselective manner [8]. This pinacolate mediated radical reaction as depicted in the Scheme 27-2 has recently been optimized to obtain **6** in over 80% yield with a small amount of **18** [9]. In general, we believe that this reaction is not limited to the synthesis of dimers, but also applicable to the assembly of higher homologues of nucleic acids and carbohydrates [10]. Incorporation of **6** into oligos was accomplished *via* a phenoxyacetyl protection of the imino group which comes off on NH4OH treatment at the end of oligo synthesis. The Tm of the modified oligo (Table 27-1) was indeed better than the oxime linked **5**, indicating that increased flexibility was helpful towards binding to RNA target. However, it was not as good as the natural PO linked oligo. We reasoned that there may be an inversion-rotation occurring around the N-O bond and free NH group may be involved in some H-bonding with free water.

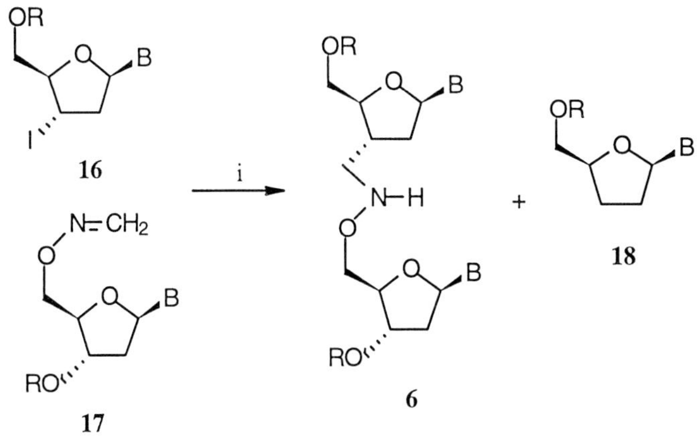

Scheme 27-2. Synthesis of MI linked dimer **6**; Reagent i) bis(trimethylstannyl)benzopinacolate, 0.2 M benzene or chlorobenzene, 80 °C, 4-6 hr; iii) aq. HCHO/NaBH$_3$CN/AcOH, rt, 2hr.

27.2.1.3 Synthesis of MMI dimer 7

Therefore, an alkyl substituent, such as a small methyl group on the nitrogen atom may increase the hydrophobicity of the linkage and provide a small inversion-rotation barrier. Reductive methylation of dimer **6** with HCHO/NaBH3CN/AcOH furnished **7** in quantitative yield [4]. Additionally, **7** was also synthesized in one-pot reaction from **5** using a similar procedure. Hybridization studies (Table 27-1) indicated that incorporation of MMI linked dimer **7** had remarkably no destabilizing effect compared to the natural PO linked oligo. Moreover, there was a clear enhancement in the affinity when the nitrogen of **6** was substituted with a methyl group. This suggests that a perfect balance of flexibility, rigidity and hydrophobicity may provide backbone linkages that mimic the affinity and specificity of the natural PO linkage [11].

27.2.2 3´-C-N-N Linked Backbone

Next, we decided to test the limits of hydrophobicity in the backbone linkage by introducing two methyl substituents *via* a hydrazine backbone. We reasoned that tetra-substituted hydrazine with a p*k*a of 6.5 should remain neutral at physiological pH and allow introduction of a nonionic and hydrophobic functionality in antisense constructs.

27.2.2.1 Synthesis of MDH dimer 8

An acid catalyzed coupling of **14** with 5´-deoxy-5´-hydrazino-thymidine **19** gave the putative hydrazone dimer **20**. Attempted isolation of **20** resulted in extensive decomposition of the product. Therefore, *in situ* reduction followed by formylation provided a stable dimethyl hydrazino linked **8** (Scheme 27-3) in overall 60% yield [12]. The structure of dimer **8** was confirmed by ^{1}H, ^{13}C NMR, MS and elemental analysis. Utilizing standard phosphoramidite protocol **8** (R = DMT, R' = Amidite) was incorporated into oligos. The Tm studies indicated that five incorporations of MDH linkage in an oligo enhanced the affinity by 0.2 °C/modification compared to the unmodified DNA (Table 27-1). It is interesting to note that additional hydrophobicity was beneficial for the hybridization to the complement RNA. Hence, oligos containing the MDH linkage may have distinct advantages as antisense molecules.

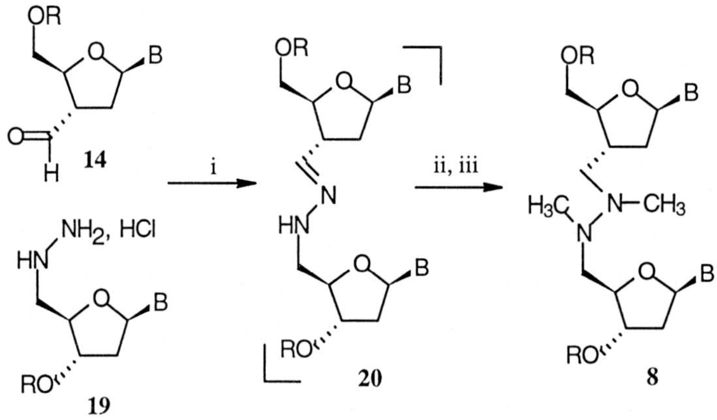

Scheme 27-3. Synthesis of MDH linked dimer **8**; Reagents: i) 1% AcOH in CH$_2$Cl$_2$, rt, 4-6 hr; (ii) NaBH$_3$CN/AcOH, rt, 2 hr.

27.2.3 3´-C-O-N Linked Backbone

In order to optimize and better understand the role on the *N*-methyl substituent within the four atom space of the backbone linkage, we synthesized MOMI linked dimer **9** as a positional isomer of MMI linkage **7**.

27.2.3.1   Synthesis of MOMI dimer 9

A NaBH$_4$ reduction of **14** furnished 3´-*C*-hydroxymethyl nucleoside analog in 83% yield which underwent a clean Mitsunobu reaction to provide an *N*-phthalimido derivative in 80% yield. The latter compound on hydrolysis furnished **21** (70%) after purification. Previously reported **22** was coupled with **21** under standard acidic conditions to give an oxime linked dimer **23** which was then reduced to **24** and subsequently methylated to furnish **9** in good yields (Scheme 27-4) [13]. Following the standard protocol **9** (R = DMT, R' = Amidite) was prepared and incorporated into oligos. The Tm analysis of the duplex indicated that MOMI linked oligo was substantially destabilizing compared to the MMI linked oligo (ΔTm/modification -2.1 °C, Table 1). Clearly, a subtle change in the positioning of the *N*-methyl functionality had a dramatic effect on the affinity towards complement RNA.

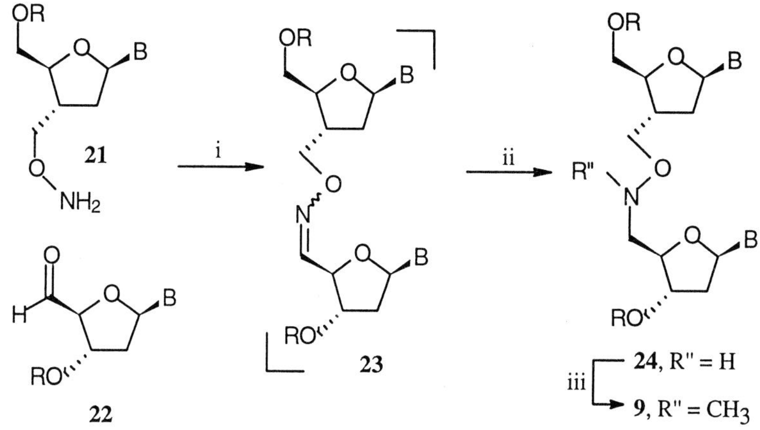

Scheme 27-4. Synthesis of MOMI linked dimer **9**; Reagents: i) 1% AcOH in CH_2Cl_2, rt, 4-6 hr; ii) $NaBH_3CN$/AcOH, rt, 2 hr; iii) aq. $HCHO/NaBH_3CN$/AcOH, rt, 2 hr.

27.2.4 3'-O-N-C Linked Backbone

We believe that the role of 3'-C-C bond in the MMI linkage **7** is of prime importance in providing uncompromised affinity for RNA target. We attribute this to a preorganization into a preferred A-geometry for duplex formation *via* 3'-endo sugar conformation of the 3'-C-C linked residue [14]. Additionally, substitution of 3'-O by a CH_2 group reduces the ring gauche effects and contributes to the enhanced conformational stability. Therefore, in order to clearly define the role of 3'-CH_2 in MMI linkage, we replaced it with a 3'-O substituent and studied the hybridization of resulting HMIM linked dimer **10** in oligos.

27.2.4.1   Synthesis of HMIM dimer 10

A straightforward synthesis of HMIM linked **10** was visualized by an extension (Scheme 27-5) of the pinacolate mediated intermolecular radical coupling reaction, developed for the synthesis of **6**. The two key components, radical acceptor **25** and radical precursor **26** were prepared in the following manner. Mitsunobu reaction of 5´-O-TBDPS protected xylothymidine resulted in the inversion of configuration at 3´-position giving a 3´-O-phthalimido thymidine. Hydrazinolysis of the latter compound followed by formylation with aq. HCHO of the product furnished **25** in excellent overall yields. The synthesis of **26** was accomplished by iodination of 3´-O-TBDPS protected thymidine in 75% yield. Radical coupling of **25** with **26** provided **27** in 55% yield after purification. A small amount (~10%) of 5´-deoxy nucleoside **28** was also obtained as a side product.

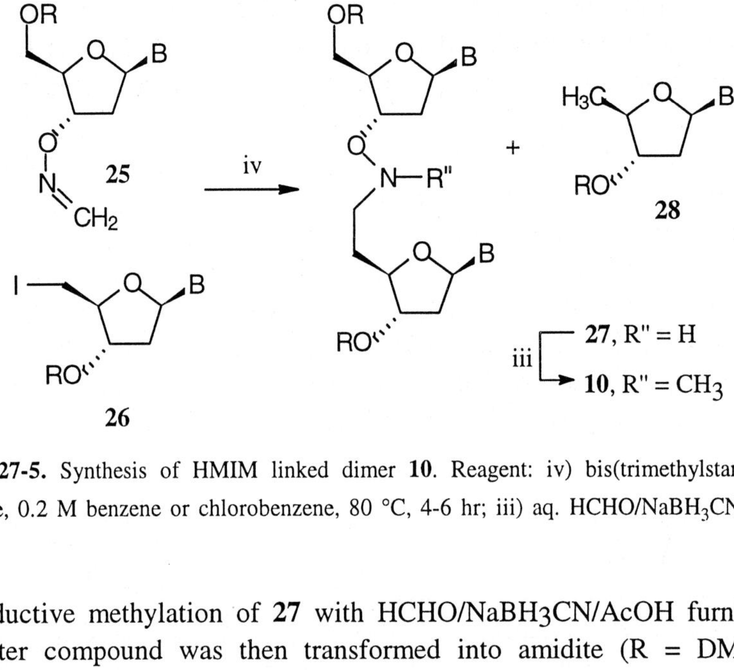

Scheme 27-5. Synthesis of HMIM linked dimer **10**. Reagent: iv) bis(trimethylstannyl)benzopinacolate, 0.2 M benzene or chlorobenzene, 80 °C, 4-6 hr; iii) aq. HCHO/NaBH$_3$CN/AcOH, rt, 2hr.

Reductive methylation of **27** with HCHO/NaBH3CN/AcOH furnished **10**. The latter compound was then transformed into amidite (R = DMT, R' = Amidite) in good yield following a standard protocol. Incorporation of the amidite into oligos was accomplished via automated DNA synthesizer. The Tm analysis of the oligo containing five HMIM linked dimers indicated reduced affinity compared to the MMI linked oligo (ΔTm/modification -1.0 °C) towards complement RNA (Table 27-1). In addition, ^{1}H NMR studies of the free dimer **10** (R = R' = H) indicated a preference towards 3´-exo conformation for the top sugar residue [15]. These results taken together demonstrate the function and importance of preorganization towards 3´-endo conformation in the sugar residue and its application in designing novel backbone linkages.

27.2.5 3′-C-N-C and C-C-N Linked Backbone

Our interest in utilizing amino-linked backbone arose from the ability to generate a cationic oligonucleotide analog. We believed that tertiary amines should be protonated at physiological pH and may interact with polyanionic backbone of RNA complement reducing the electrostatic repulsion between the two strands. This may eventually assist in improved affinity and cellular uptake. We have prepared two such linkages **11** and **12** and studied their antisense properties.

27.2.5.1 Synthesis of Amine 1 and 2 Dimers (11 and 12)

Aldehyde **14** and amine **28** are the precursors to the dimer **11**. Similarly, **30** and **31** are required for the synthesis of dimer **12**. Therefore, a reductive coupling [16] of **14** with **28** and **30** with **31** in the presence of NaBH(OAc)$_3$ furnished a dimeric amine **29** and **32**, respectively. Both of the latter compounds were methylated (HCHO/NaBH$_3$CH/AcOH) to furnish the desired dimer **11** and **12**, respectively in good yields (Schemes 27-6 and 7). These dimers were successfully incorporated into oligonucleotides and their Tms measured. Surprisingly, both of these modifications were equally destabilizing on duplex formation, when compared to the unmodified DNA. One plausible explanation could be the presence of a *C-C* bond within the four atom backbone which allows free rotation around the torsional angles, leading to perhaps unwanted flexibility. Interestingly, restriction of flexibility *via* substitution of a carbonyl group in place of a methylene group resulted in an enhancement of Tm [17].

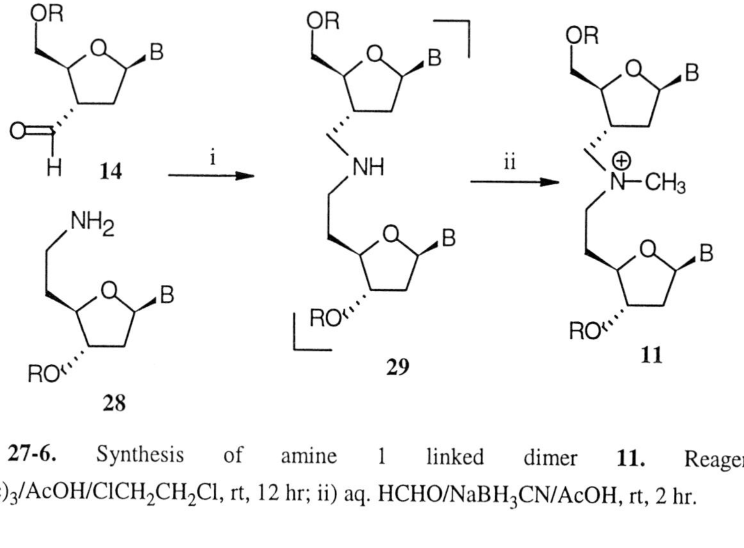

Scheme 27-6. Synthesis of amine 1 linked dimer **11**. Reagents: i) NaBH(OAc)$_3$/AcOH/ClCH$_2$CH$_2$Cl, rt, 12 hr; ii) aq. HCHO/NaBH$_3$CN/AcOH, rt, 2 hr.

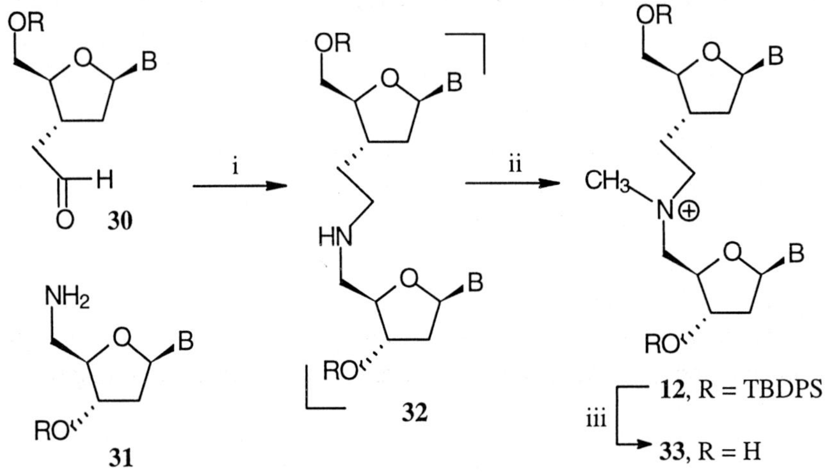

Scheme 27-7. Synthesis of amine 2 linked dimer **12**. Reagents: i) NaBH(OAc)$_3$/AcOH/ClCH$_2$CH$_2$Cl, rt, 12 hr; ii) aq. HCHO/NaBH$_3$CN/AcOH, rt, 2 hr; iii) TBAF/THF, rt, 1h.

27.2.6 2´-OMe 3´- C-N-O Linked Backbone

The SAR on various nitrogen containing backbone linkages **5-12** provided the 3´-5´ MMI Linkage **7** as a clear choice for further studies due to desirable antisense properties, such as nuclease stability and uncompromised affinity, and base pair specificity towards target RNA. However, we believe that it was necessary to further improve the affinity for complement RNA because increased Tm translated well into increased *in vitro* and *in vivo* activities [18]. It has been demonstrated that higher affinity oligomers can be constructed *via* incorporation of a 2´-electronegative substituent uniformly in an oligonucleotide. Much evidence has been provided to confirm that a 2´-OMe substitute remains in the minor groove, forces the sugar to adopt a 3´-endo pucker, distorts minor groove hydration at the local and global levels, resulting in higher Tms of hybrids formed between bis-2´-OMe oligomers and complement RNA [19].

Therefore, it was a rational choice for us to combine the best backbone modification (i.e. MMI) with the 2´-OMe sugar substituent to create novel oligomers that may have a synergistic effect on the thermodynamics of duplex formation between bis 2´OMe MMI oligo and complement RNA.

27.2.6.1   Synthesis of bis 2´-OMe MMI Linked dimer 13

The intermolecular radical coupling procedure [8, 9] described (Scheme 27-5) for the synthesis of 2´-deoxy MMI linked dimer was also applicable to the 2´-OMe series. Therefore, a convergent free radical coupling of iodo nucleotide **34** and oxime **35**, mediated by pinacolate has been tried to furnish a mixture of dimer **36** (R" = H) in modest (~ 30%) to excellent yield (~ 80%) depending on the nature of base substitutent, like before, a small amount of 3´-deoxy nucleoside **37** was also obtained as a side product. Interestingly, the radical reaction in 2´-OMe series was found to be less stereoselective compared to the 2´-deoxy series [9]. An average of 5 - 25% of the *xylo* or β-diastereomer of the MI dimer **36** (R = H) was formed with the desired 3´ - 5´ α-linked MI dimer. All attempts to separate the two stereoisomers failed in our hands. We attributed this result to the steric and electronic effects of 2´-OMe group adjacent to the 3´- radical which dictated the accessibility of the π–radical and subsequent addition of the acceptor. In brief, this method was not suitable for the scale-up of dimeric building blocks.

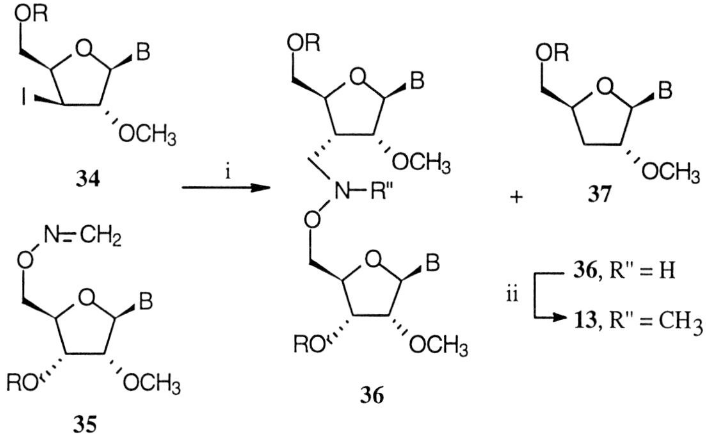

Scheme 27-8. Synthesis of 2'-OMe MMI linked dimer **13**. Reagents: i) bis(trimethylstannyl)benzo pinacolate, 0.2 M benzene or chlorobenzene, 80 °C, 4-6 hr; ii) aq. HCHO/NaBH$_3$CN/AcOH, rt, 2 hr.

In order to circumvent the stereochemical problems, we have now developed an alternative route [20] which utilizes the coupling of R-CHO nucleoside **38** with R'-O-NHCH$_3$ nucleoside **39** to furnish the desired dimer **13** *via* in situ reduction of **40** with BH$_3$. Pyridine/PPTS (Scheme 27-9). This procedure is now well established and allows the rapid synthesis of all sixteen bis-2´-OMe MMI linked dimers in high yield. We believe at the present time this is the best procedure for the preparation and scale-up of MMI linked dimers in high yield compared to the solution and solid-phase synthesis described earlier [14]. The dimer **13** was transformed to the corresponding amidite and incorporated into oligomer using DNA synthesizer. The results of the Tm studies with oligomer containing dimer **13** (B = T) indicated a substantial enhancement in the affinity (Δ Tm +3.6 °C/modifications) for the complement RNA compared to the oligomer containing MMI linkage **7** alone. The boost in the affinity was attributed to the additional hydrophobicity due to the presence of 2´-OMe substituents in the N-puckered sugar conformation. There two effects are believed to work in concert to provide an A-type of duplex formation with least entropic penalty.

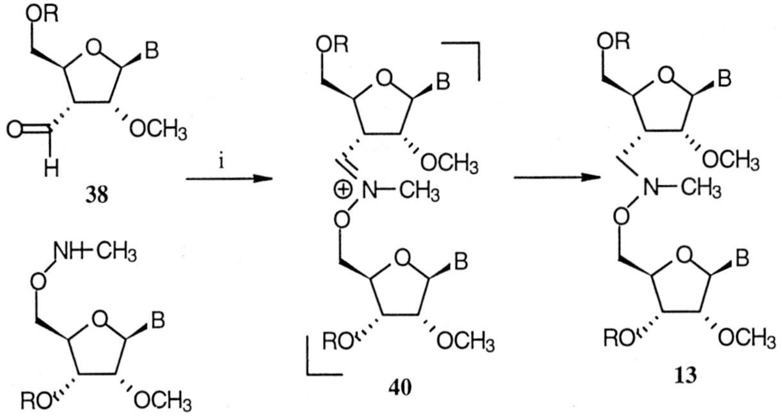

Scheme 27-9. Synthesis of 2'-OMe MMI linked dimer **13**. Reagents: i) Pyridine•BH$_3$/PPTS/ MeOH/ THF, rt, 4 hr.

27.3 Conclusion

Several nitrogen-containing backbone modifications of oligomers have been synthesized and their affinities for RNA complement studied. As a result, we have identified bis-2´-OMe MMI **13** as the most promising backbone surrogate for incorporations into antisense molecules. The syntheses of sixteen dimers (**13**, B = T, 5-MeC, A, G) have been accomplished on a multigram scale. The routes described herein should be amenable to kilo quantities of these materials making accessibility of modified antisense constructs easy. It appears that an alternating phosphodiester -bis-2´-OMe MMI (**13**) backbone motif in an antisense oligo may provide the optimal design features for the highest affinity and the nucleolytic stability [21]. Detailed biological evaluations of these novel second/third generation antisense oligos is in progress.

Another important aspect of this study is in our improved understanding of the backbone design elements and their effects on the binding affinity to target

RNA. Clearly, introduction of 2´ OMe substituent produces an approximate trans conformation of the C2´ - C3´ and O2´ - CH$_3$ bond, which stabilizes the ground state conformation and increases the barrier to interconversion among the N-S conformers. This reduction in unproductive motion may contribute to the exceptional RNA affinity observed with the incorporations of MMI Linkage **13**. Additional gain in the affinity may be derived from the local and global changes in the hydrodynamics of duplex formation. These results suggest that the affinity of backbone modified antisense oligonucleotides for RNA target can be tuned *via*: (i) Decreasing the entropic motion of the sugar; (ii) Choosing modifications with a higher degree of *N*-pucker; (iii) Preorganizing duplex structure to an A-type helix; (iv) Reducing the flexibility of the 3´-5´ linker; (v) Avoiding free rotation of atoms within the backbone; (vi) Identifying atom placement of hydrophobic groups into the major and minor groove. In summary, a fine tuning between the gain in entropy and the loss in enthalpy is the key to successful design of high affinity antisense oligomers.

Acknowledgment

I wish to thank my present and former colleagues whose name appear in the references below for their hard work and dedication in accomplishing our objectives.

27.4 References

1. (a) Y. S. Sanghvi, P. D. Cook in *Carbohydrate Modifications in Antisense Research* (Eds.: Y. S. Sanghvi, P. D. Cook), American Chemical Society Symposium Series #580, Washington, DC, USA, **1994**, p 1 - 22, and references cited therein. (b) A. DeMesmaeker, R. Haner, P. Martin, H. Moser, *Acc. Chem. Res.* **1995**, *28*, 366 - 374. (c) P. Herdewijn, *Liebigs Ann.* **1996**, 1337 - 1348. (d) R. S. Varma in *Molecular Biology and Biotechnology: A Comprehensive Desk Reference* (Ed.: R. A. Meyers), VCH Publishers, NY, USA, **1995**, p 617 - 621.

2. (a) S. T. Crooke, Therapeutic Applications of Oligonucleotides, R. G. Landes Co., Austin, TX, USA, 1995, p 1 - 134. (b) K.-H. Altmann, N. M. Dean, D. Fabbro, S. M. Freier, T. Geiger, R. Haner, D. Husken, P. Martin, B. P. Monia, M. Muller, F. Natt, P. Nicklin, J. Phillips, V. Pieles, H. Sasmor, H. E. Moser, *Chimia* **1996**, *50*, 168 - 176.

3. (a) S. T. Crooke, *Chem. Ind.* **1996**, 90 - 92. (b) M. D. Matteucci, R. D. Wagner, *Nature* **1996**, *384*, 20 - 22.

4. J. -J. Vasseur, F. Debart, Y. S. Sanghvi, P. D. Cook, *J. Am. Chem. Soc.* **1992**, *114*, 4006 -4007.

5. Y. S. Sanghvi, R. Bharadwaj, F. Debart, A. DeMesmaeker, *Synthesis* **1994**, 1163 - 1166.

6. Y. S. Sanghvi, B. Ross, R. Bharadwaj, J. -J. Vasseur, *Tetrahedron Lett.* **1994**, *35*, 4697 - 4700.

7. M. Perbost, T. Hoshiko, F. Morvan, E. Swazye, R. Griffey, Y. S. Sanghvi, *J. Org. Chem.* **1995**, *60*, 5150 - 5156.

8. F. Debart, J. -J. Vasseur, Y. S. Sanghvi, P. D. Cook, *Tetrahedron Lett.* **1992**, *33*, 2645 - 2648.

9. B. Bhat, E. E. Swayze, P. Wheeler, S. Dimock, M. Perbost, Y. S. Sanghvi, *J. Org. Chem.* **1996**, *61*, 8186 - 8199.

10. P. D. Cook, Y. S. Sanghvi, P.-P. Kung, PCT Int. Appl. WO 95/18623, July 13, **1995**.

11. Y. S. Sanghvi, E. E. Swayze, D. Peoc'h, B. Bhat, S. Dimock, *Nucleosides Nucleotides* **1997**, *16*, in press.

12. Y. S. Sanghvi, J. -J. Vasseur, F. Debart, P. D. Cook, *Collect. Czech. Chem. Commun. Symposium issue*, **1993**, *58*, 158 - 162.

13. F. Debart, J. -J. Vasseur, Y. S. Sanghvi, P. D. Cook, *Bioorg. Med. Chem. Lett.* **1992**, *2*, 1479 - 1482.

14. F. Morvan, Y. S. Sanghvi, M. Perbost, J. -J. Vasseur, L. Bellon, *J. Am. Chem. Soc.* **1996**, *118*, 255 -256.

15. Y. S. Sanghvi et al., unpublished results.

16. A. F. Abdel-Magid, K. G. Carson, B. D. Harris, C. A. Maryanoff, R. D. Shah, *J. Org. Chem.* **1996**, *61*, 3849 - 3862.

17. A. DeMesmaeker, A. Waldner, Y. S. Sanghvi, J. Lebreton, *Bioorg. Med. Chem. Lett.* **1994**, *4*, 395 - 398.

18. K.-H. Altmann, D. Fabbro, N. M. Dean, T. Geiger, B. P. Monia, M. Muller, P. Nicklin, *Biochem. Soc. Trans.* **1996**, *24*, 630 - 637.

19. M. Egli, *Angew. Chem. Int. Ed. Engl.* **1996**, *35*, 1894 - 1909.

20. E. E. Swayze, B. Bhat, D. Peoc'h, Y. S. Sanghvi, *Nucleosides Nucleotides*, **1997**, *16*, in press.

21. Y. S. Sanghvi in *Comprehensive NaturalProducts Chemistry* (Eds.-in-chief, D. H. R. Barton, K. Nakanishi), Vol. 7: *DNA Aspects of Molecular Biology* (Ed. E. T. Kool), Pergamon Press, 1997, in press.

28 Nucleosides and Nucleotides Containing 5-Alkyl Pyrimidines. Chemistry and Molecular Pharmacology

L. Ötvös

28.1 Chemistry

Pharmacologically significant 5-substituted-2'-deoxyuridine derivatives have been known since the early period of nucleoside chemistry. The iodo compound was the first clinically effective antiviral nucleoside analogue, while the 5-fluoro-deoxyuridine got later a role in cancer chemotherapy (Figure 28-1). The mechanism of action of these compounds is summarised in a very informative review published in 1992 [1].

The first alkyl substituted pyrimidine nucleoside - besides thymidine - was 5-ethyl-2'-deoxyuridine (EDU) synthesised by Gauri [2] as well as Swierkowski and Shugar [3]. The compound became an antiherpetic drug, since, despite its rather low antiviral selectivity, its incorporation into the DNA of human cells does not involve mutagenicity. Its successful application in antiviral chemotherapy stimulated the synthesis and investigation of series of 5-alkyl-, 5-alkenyl- and 5-alkynyl-2'-deoxyuridines as well as thymidine analogues containing branched- and cycloalkyl substituents in position 5 [4-7].

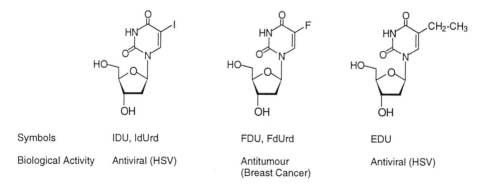

Figure 28-1. Some 5-substituted-2'-deoxyuridines in clinical use.

28.1.1 Synthetic Methods

Since the first synthesis of EDU, numerous methods were developed over 25 years for the preparation of 5-alkyl-2'-deoxyuridines, in general. The most important methods are divided into two main types. In the first the key step is to establish an N-glycosidic bond from two components (Method 1), while in the second type of methods the ready nucleosides were transformed to the alkyl derivatives. (Method 2; Figure 28-2)

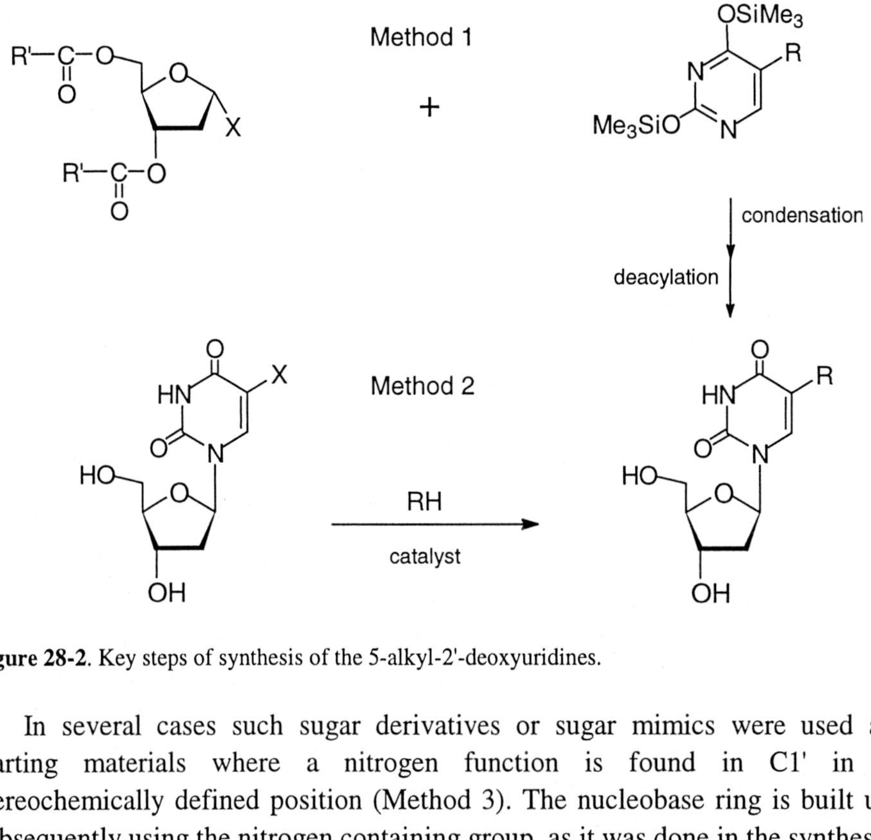

Figure 28-2. Key steps of synthesis of the 5-alkyl-2'-deoxyuridines.

In several cases such sugar derivatives or sugar mimics were used as starting materials where a nitrogen function is found in C1' in a stereochemically defined position (Method 3). The nucleobase ring is built up subsequently using the nitrogen containing group, as it was done in the synthesis of thymidine by Smejkal, Farkas and Sorm [8] and of optically active carbocyclic thymidine in our laboratory [9]. These methods are special, not generally applied (Figure 28-3). The classic approach (Method 1) is based on the reaction of a suitably protected halogeno- (usually chloro) sugar with a uracil, preferably the 2,4-bis-trimethylsilyl derivative. Over several decades a large variety of reaction conditions (catalyst, temperature, solvent, protecting group of the components, etc.) were tried. The role of these factors were explained by Hubbard, Jones and Walker in 1984 [10].

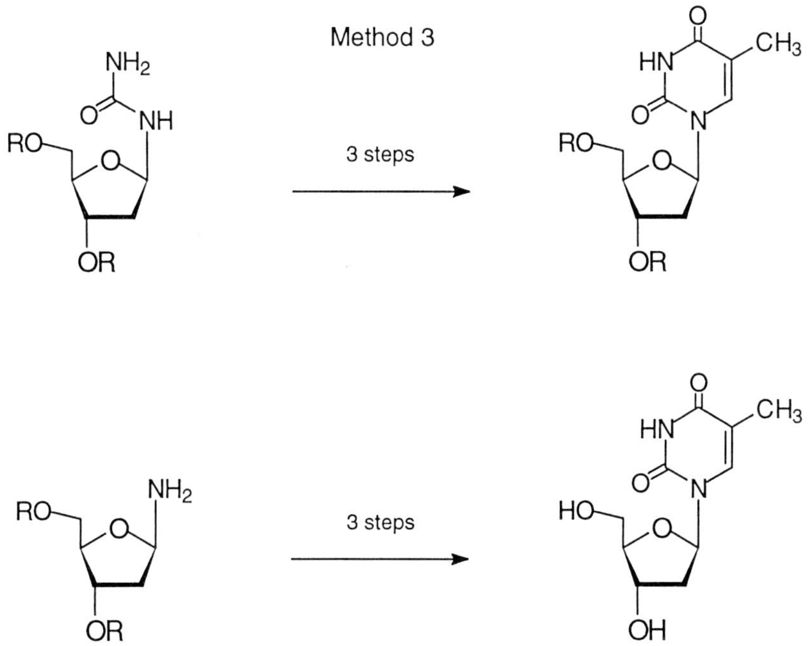

Figure 28-3. Synthesis of nucleosides from 1'-amino derivatives.

One of the last modifications was published by Basnak et al. [11] in 1994. As the biologically active nucleoside analogues are always ß-anomers, the main purpose was to obtain the nucleoside in this stereoisomeric form. The authors synthesised the 5-alkyl-2'-deoxyuridines in high yields. In the case of the pharmacologically most interesting 5-isopropyl- and 5-(1-adamantyl) derivatives, the products, besides 90 % along with ß anomer contained 10 % α-nucleoside. The isomers could be separated by chromatography.

The formation of the α-anomer from the α-chlorosugar can occur by S_N1 reaction, or by isomerisation of the halogenosugar into the ß-anomer before the S_N2 type attack by the N_1-atom of the uracil would take place. Elimination or the lowering of the possibility of these reactions results in an increased ß to α anomeric ratio, because the S_N2 type substitution of α-halogen will be dominant. Taking into consideration that the reaction is bimolecular, this condition can be ensured if the chlorosugar is present in a low concentration and the base is in excess.

Based on this idea, lately we developed a new method [12]. According to this, the α-halogeno sugar is mainly in undissolved form. Only a small amount of chlorosugar is in the solution containing the 5-substituted 2,4-O-trimethylsilylated uracil, thus the base is always in excess. The concentration of the halogenosugar in the solvent strongly depends on the structure of its acyl groups and on the nature of the solvent. In the case of 5-isopropyl-2'-deoxyuridine 96 % of the product is in the form of ß-anomer and it can be purified without chromatography. The patented method is now being used on industrial scale to produce the compound which is one of the most active antiherpetic drugs for topical application. The molecular pharmacology will be discussed later.

The transformation of an isolated or previously prepared nucleoside into 5-alkyl-, 5-alkenyl- and 5-alkynyl- 2'-deoxyuridines represents the second type of the synthetic methods. The great advantage of this procedure is that from one single nucleoside a series of compounds can be prepared. In our laboratory most frequently 5-iodo-2'-deoxyuridine has been used as the starting material. The compound was transformed into 5-(1-alkynyl)-deoxyuridines by method of Robins [13]. Partial reduction gave the 5-(1-alkenyl) derivatives. The ratio of the E and Z isomers depended on the catalyst and the R substituent. Total reduction of the triple bond using Pd catalyst and hydrogen gave 1-alkyl-dUs. The method allows the simple synthesis of tritium-containing radioactive nucleosides. The introduction of four ^{3}H atoms results in a product of high specific radioactivity (Figure 28-4).

Reactions of deoxyuridines containing saturated and unsaturated 5-alkyl substituents have been studied in our laboratory in several chemical transformations, e.g. synthesis of 5-substituted 2'-deoxycytidines, formation of 3',5'-cyclophosphate derivatives [14], and preparation of 5-mono- and triphosphates [15, 16]. An unusual reaction was found in the of 2,2'-anhydro-2'-deoxyuridines in the sugar moiety, discussed below.

28.1.2 Deacylation of Derivatives in the Sugar Moiety

The nucleosides selectively acylated in position 5' or 3' play an important part in various synthetic approaches. They differ from each other in their biological activity such as their pro-drug properties; they also differ from the diacyl compounds. The same applies to the 2,2'-anhydropyrimidine derivatives. For example, there is a great difference between the acylated and non-acylated compounds in the inhibitory effect of phosphorylase [17], or in antiviral activity.

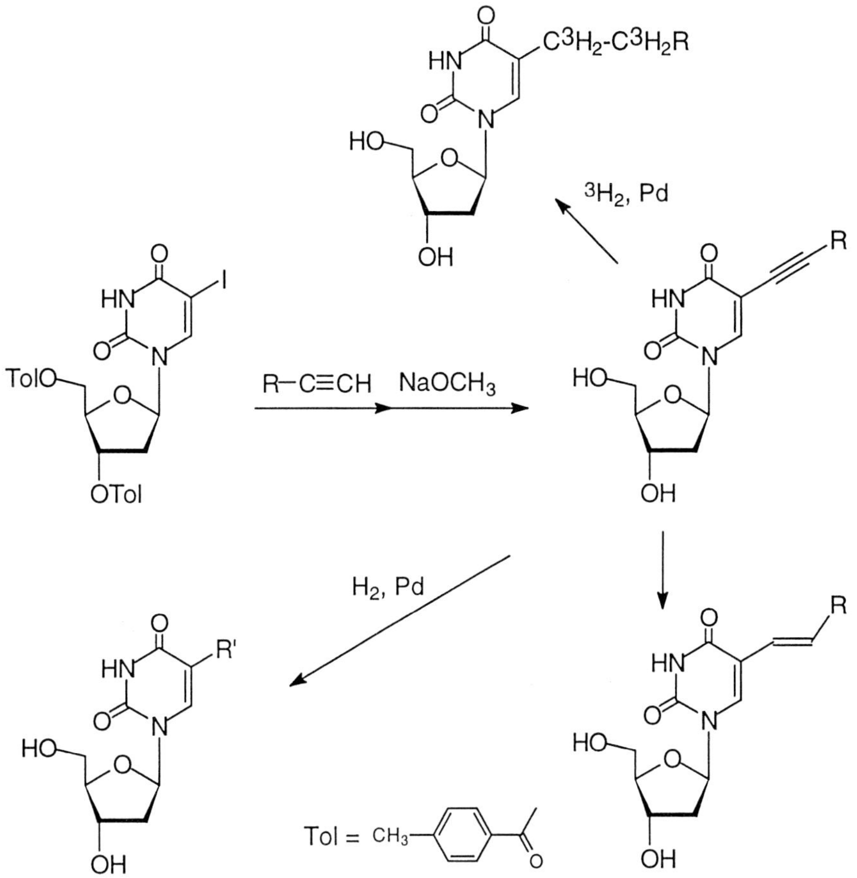

Figure 28-4. Synthesis of 5-alkyl-substituted nucleosides by transformation of IDU.

The ammonolysis of 3',5'-diacyl-5-alkyl-2'-deoxyuridines takes place at the 3'- and 5'-positions practically at the same rate. The reactivity of the 2,2'-anhydro derivatives is basically different from that of the former compounds. Our experiments have shown that the deacylation of these compounds starts primarily in the 3'-position. Thus, 3',5'-diacetyl-2,2'-anhydro-5-ethyl-2'-deoxyuridine gives the 5'-monoacetyl derivative about forty times more rapidly than yielding the 3'-monoacyl compound (Figure 28-5). This result is rather surprising, and the explanation is still a puzzle. The most likely supposition is that the sugar ring is strained in the anhydro compound, and this strain is decreased when the 3'-acyl group is eliminated. This selectivity in deacylation is so high that it can well be utilised for preparative purposes.

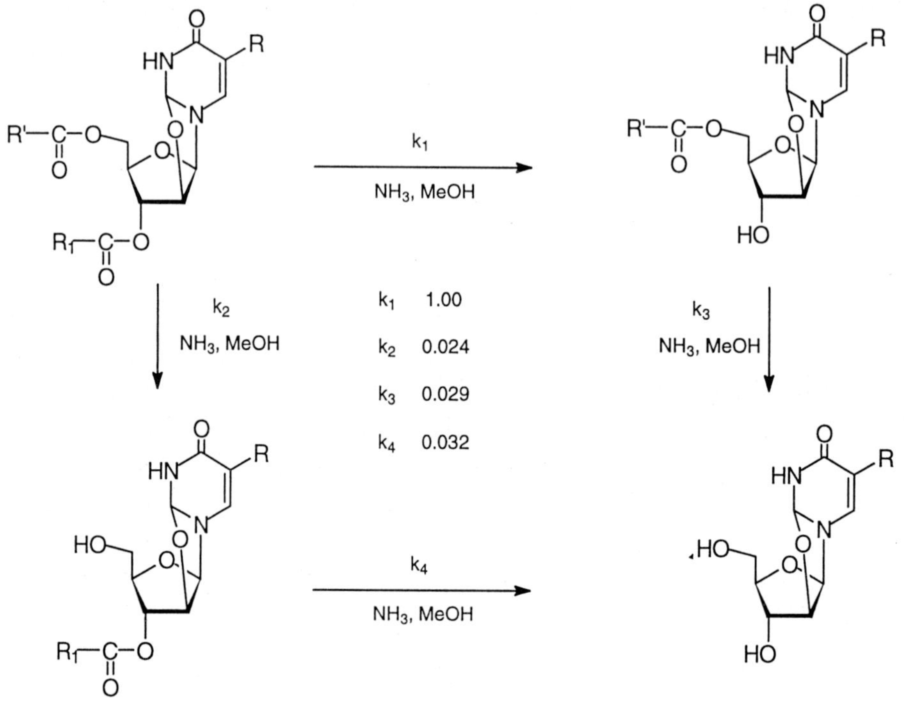

Figure 28-5. Relative deacylation rates of 5-ethyl-2,2'-anhydro-2'-deoxyuridine derivatives.

k_1	1.00
k_2	0.024
k_3	0.029
k_4	0.032

28.2 Bioorganic Chemistry

All synthetic studies discussed above served mainly one purpose, namely to provide compounds for bioorganic and molecular pharmacological investigations.

28.2.1 Substrate Specificity in Polymerase Catalysed Reactions

The antiviral and the anticarcinogenic actions of deoxyribonucleoside analogues are exerted by enzymatic incorporation into DNA. The reaction is catalysed by DNA polymerases. The substrates of the enzymes are nucleoside triphosphates. Substrate specificity of the enzyme-catalysed incorporation of 5-alkyl-, 5-alkenyl- and 5-alkynyl-2'-deoxyuridines was investigated in detail between 1977 and 1988 in our laboratory. A summarised explanation of substrate specificity was published in 1987 [18]. After 1990 the investigations were resumed because of our interest in the following points.

1. Confirmation of our explanation of substrate specificity in view of recently published knowledge about the structure of polymerases.
2. Computer modelling of polymerase substrate-complexes.

3. Substrate properties of dUTPs containing a long-chain substituent in position 5, apparently inconsistent with earlier results.
4. Increased importance of 5-alkyl-deoxyuridines (e.g. 5-hexyl-dU) in cancer chemotherapy.
5. Development of 5-isopropyl-2'-deoxyuridine (IPDU) into an antiviral drug (Epervudin).
6. Research of antisense oligonucleotides.

28.2.1.1 5-n-Alkyl-, 5-(n-1-alkenyl)- and 5-(n-1-alkynyl)dUTPs

Reinvestigation of the relative incorporation in the series of 5-n-alkyl-deoxyuridines into poly(dA-r^{5}dU) in *E. coli* DNA polymerase catalysed reaction gave very similar results we got earlier (Figure 28-6). Hundred per cent incorporation is assigned to the natural substrate thymidine, where the methyl group is in position 5 on the pyrimidine ring. 5-Ethyl-, 5-n-propyl- and 5-n-butyl-deoxyuridine triphosphates are only slightly less good substrates than dTTP. The reactivity significantly decreased when the substituent had five carbon atoms, and very low incorporation was observed in the case of 5-n-hexyl-dUTP.

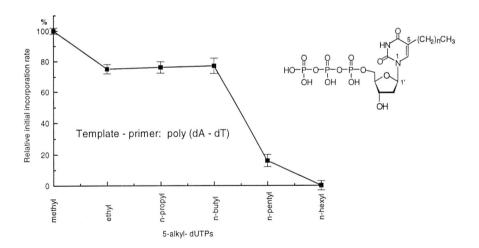

Figure 28-6. Relative incorporation rates of 5-alkyl-dUTPs.

A similar tendency was observed in the investigation of (E)-5-(1-alkenyl)-2'-deoxyuridine substrates, but here the reactivity drops at seven carbon atoms of the substituent. In the series of 1-alkynyl substituents the reactivity significantly decreases at the octynyl side chain attached to the pyrimidine ring. (Figure 28-7)

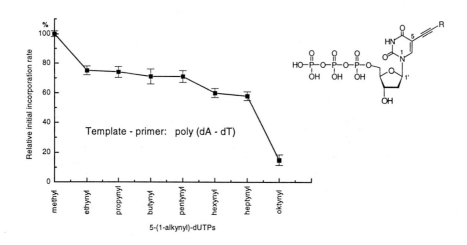

Figure 28-7. Relative incorporation rates of 5-(1-alkynyl) dUTPs.

The tendency in the substrate specificity was similar when instead of poly(dA-dT), we studied activated calf thymus, but the relative reactivities were different. In these reactions the n-hexyl-, 1-heptenyl- and 1-octynyl 5-dUTPs showed small but well defined incorporation.

The substrate specificity can be explained by the following mechanism. Figure 28-8 shows a very schematic representation of the active site of DNA polymerases. The binding of deoxyuridine triphosphate is ensured by the base pairing to the template on one hand and by the binding of the triphosphate group to the enzyme, on the other. In this way, the steric position of the substituent R is determined. The group R is oriented toward the 5'-triphosphate group of the substrate and the 3'-OH end group of the primer. If the substituent is small, like a methyl, an ethyl, a vinyl or an ethynyl group, there is no hindering of the nucleophilic attack of the 3'-OH group on the α-phosphorus atom of the triphosphate moiety. When the substituent is longer, at a certain chain length the substituent may hinder the attack of the hydroxyl group of the primer.

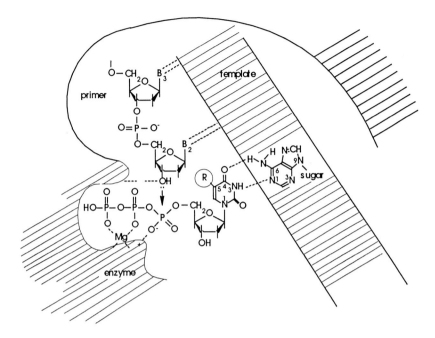

Figure 28-8. Position of the R substituent in the active site of DNA-polymerases.

The steric hindrance decisively depends on the conformation of the substituent R. It was mentioned that the 5-n-hexyl-dUTP is a very poor substrate of *E. coli* polymerase I, when employing poly(dA-dT) template primer. In the zig-zag conformation of 5-n-hexyl substituents there would be no steric hindrance as evidenced by a Dreiding model and computer modelling. The experimental data can only be explained in that case when the 5-n-alkyl substituents are in twisted conformation around the C_1-C_2 atoms. At this position the end methyl group of the n-hexyl substituent gives rise to a perfect steric hindrance. The effect of the end methyl group of the n-pentyl substituent is significant, but not complete. The n-propyl or n-butyl substituent are too short to sterically hinder the attack of the hydroxyl group even in twisted conformation.

The presence of a triple bond in 5-(1-alkynyl)-dUTPs will fix the position of the first three carbon atoms of the substituents, thus formation of the twisted conformation of the substrate must start by rotation around the bond between the third and fourth carbon atoms. In consequence, in this steric arrangement significant steric hindrance begins with the 1-octenyl as the shortest substituent, in good agreement with the experimental data.

The analysis of dUTPs containing 1-alkenyl groups is more complicated because of the possibility of E and Z stereoisomerism, however, the experimental data can be well explained also in these cases.

Table 28-1. K_M and/or K_I values of several dU derivatives.

Substrate or inhibitor	K_M	K_I
dTTP	4,1±0.50	
E-5-(1-heptenyl)dU	0.8±0.05	0.65±0.08
E-5-(1-octenyl)dU		0.35±0.08

To prove that the substrate specificity mainly depends on kinetics rather than on binding factors, we investigated the inhibitory properties of poor substrates. It was found that 5-(E)-(1-heptenyl)- and 5-(E)-(1-octenyl)-dUTPs are good inhibitors of the E. coli polymerase-catalysed poly(dA-dT) synthesis. The K_I constants are one order of magnitude smaller than the K_M value of dTTP. The experiments proved that the low reactivity of a given substrate is not due to the absence of their binding ability (Table 28-1).

These studies are interesting not only from the theoretical bioorganic, but also from the practical pharmacological points of view. For example, in the last three years 5-n-hexyl-2'-deoxyuridine became very important in the development of anticarcinogenic compounds. The compound inhibits proteoglycan biosynthesis and by this mechanism reduces the metastatic potency of the malignant tumour cells [19-20].

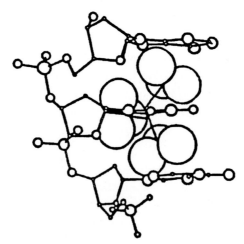

Figure 28-9. Steric compression between the nearest bases and the 5-*tert*-butyl-substituent of a dU unit incorporated in DNA.

28.2.1.2 Deoxyuridines Containing Branched Substituents in Position 5

The dUTPs containing α-branched-chain 5-substituents like isopropyl and tertiary butyl groups, do not inhibit the attack of the 3' OH group of the increasing DNA chain on the triphosphate α-phosphorus atom. In this case, however, another type of steric hindrance is set up. Figure 28-9 shows that in the course of incorporation there will be a steric compression arising between the methyl groups of the substituent and one or two rings of the nearest bases. This steric interaction is so high that theoretically it will completely block the incorporation of the 5-tert-butyl-dUTP. In spite of this prediction, a low incorporation was found in repeated experiments both in the cases of 5-isopropyl-dUTP and the 5-t.bu-dUTPs (Table 28-2).

Table 28-2. Relative incorporation of dUTPs containing branched-chain substituents in position-5 : Template: Activated calf thymus DNA.

Substrate r^5dUTP	Incorporation
dTTP	100
ip^{5}dUTP	3.5
t.bu^{5}dUTP	2.7

The experimental data can only be explained by some structural change in the isopropyl-deoxyuridine-containing DNA, formed in the reaction. UV absorption - temperature melting profiles (Figure 28-10) of this polynucleotide show that there is a big structural change in the DNA structure due to the IPDU incorporation.

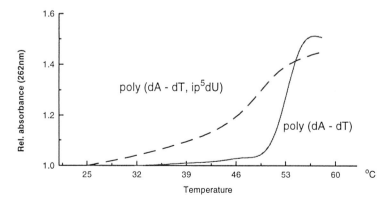

Figure 28-10. UV melting profiles of poly(dA-dT) and poly (dA-dT, ip^{5}dU).

The change in UV absorption is very sharp in the case of poly(dA-dT) and increased only very gradually in the case of the mixed polymer. Electron microscopic analysis [21] proved that the double helix structure is strongly perturbed. The main structural difference is that loops or hairpins are formed at the place of the incorporation of the 5-isopropyl-deoxyuridine.

28.2.2 Structural Properties and Stability of Polynucleotides Containing 5-Alkyl-2'-deoxyuridines

The consequence of the structural properties appears in several enzymatic transformations. The presence of loops or hairpins in the IPDU containing polymer is well compatible with its strongly increased sensitivity to the simple strand-specific nuclease A1 as compared to poly(dA-dT). More than 95 % of the poly(dA-dT) remained resistant after a two-hour reaction, independently from the enzyme concentration. Under the same conditions, the polymer containing 5-isopropyl-deoxyuridine disintegrated in up to 82 %.

The thermal stability of polynucleotides and the reactivity in enzyme-catalysed reactions also depend on the incorporated unnatural nucleotides in the series of 5-n-alkyl-dU-containing polymers: Table 28-3 shows the T_m values of the poly(dA-d5U)s which could be isolated. The decrease between the thymidine- and 5-ethyl-deoxyuridine-containing polymer is significant, but the other polynucleotides have similar thermal stabilities.

Table 28-3. T_m values of poly (dA-r^{5}dU) copolymers in physiological buffer.

r^5 = 5-n-alkyl group in substituted dU

r group	T_m	ΔT_m
me	58.1	0
et	47.1	-11.0
pr	44.0	-14.1
bu	43.3	-14.8
pe	42.4	-15.7

A significant difference can be observed in the enzymatic cleavage. E. coli DNA polymerase has not only polymerase activity but exonuclease activity, too. Table 28-4. demonstrates that the relative hydrolysis rates of 2'-deoxyuridine-containing polymers strongly depend on the structure of the polynucleotides. The polymer containing 5-n-pentyl-deoxyuridine undergoes hydrolysis about twenty times slower than poly(dA-dT) does.

Table 28-4. Relative enzymatic hydrolysis rate of poly (dA-r^{5}dU) nucleotides.

r in poly (dA-r^{5}dU)	Relative hydrolysis rate
me	100
et	57.5
pr	34.7
bu	20.8
pe	5.5

28.3 Molecular Pharmacology

28.3.1 Molecular Pharmacology of the Antiherpetic Activity of 5-Isopropyl-2'-deoxyuridine

Results of the synthetic and bioorganic studies discussed above have been used in drug development on molecular pharmacological basis. It is generally accepted that the selective mode of action of antiherpetic nucleoside derivatives (Acyclovir, BVDU, EDU) is based on the appearance of induced thymidine kinase in the infected cells [22]. The real antiviral effect, however, is manifested in the loss of the biochemical function of the viral DNA. The well known anti-HSV drug Acyclovir produces this effect by termination of the DNA strains. The molecular mechanism of 5-isopropyl-2'-deoxyuridine (Epervudine, IPDU) is different. The cause of the loss of viral DNA function is due to the easy cleavage of DNA by nucleases that had been discussed in connection with the bioorganic investigations (Table 28-4). The selectivity of the drug is promoted by the existence of viral-induced polymerase IPDU triphosphate which is a good substrate for this enzyme.

Table 28-5. Accumulation of 5-isopropyl-dU from Hevizos® ointment in the skin.

Time (min)	Total radioactivity resorbed (A)	Radioactivity in the skin (B)	B/A x 100
15	27.9	9.25	33.1
30	30.5	6.60	21.6
60	32.9	7.70	23.3
120	30.8	6.00	19.4
180	26.0	7.25	27.8
240	20.7	5.90	28.5
480	26.7	5.40	20.2

Hevizos® (Biogal, Hungary) is an ointment containing Epervudine (5-isopropyl-deoxyuridine). Besides the molecular pharmacological properties, the very good efficiency of this drug in the topical treatment of herpes simplex, herpes genitalis, and herpes zoster can be attributed to its unusually high accumulation in the skin (Table 28-5).

28.3.2 Antisense Drug Properties of Oligonucleotides Containing 5-Alkyl-2'-deoxyuridines

A good chance of the application of 5-alkyl-substituted-dUTPs in pharmacology is their use as building units in antisense oligonucleotides. Antisense oligonucleotides, in general, can be regarded as a great promise in chemotherapy.

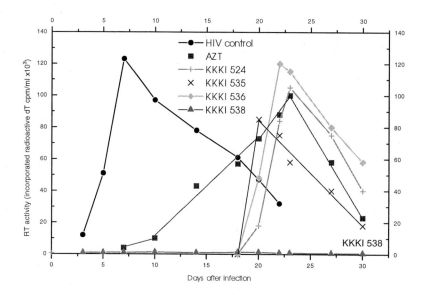

Figure 28-11. Anti-HIV activity of AZT and some oligonucleotides *in vitro*.

The most important trend in antisense technology is the blocking of translation by complex formation between the messenger RNA and the oligonucleotide. The main requirements of antisense oligomers are the following:
1. sequence specificity including good hybridisation with mRNA
2. nuclease resistance
3. appropriate transportation through the cell membranes.

The results of the thermostability of polynucleotides discussed above have encouraged us to investigate in detail the hybridisation ability of 5-alkyl- and 5-(1-alkynyl)dUs containing oligonucleotides [23].

The resistance of the compound against 3'-exonucleases has been found very high. Hydrolysis effected by other enzymes are under investigation. Based on the results mentioned, we studied the anti-HIV activity with more than two hundred oligonucleotides, *in vitro*. Highest inhibition has been found when alkynyl deoxycytidines were employed. If the oligomers were modified by alkynyl groups in the pyrimidine bases and using thiophosphates instead of phosphates, the inhibition increased further.

Figure 28-11 shows the antiviral effect, measured by reverse transcriptase activity, of several oligomers in comparison with AZT. The compounds were added to the cell culture before and after 4 hours of the HIV infection. The most active compound, code number KKKI-538 containing 5-(1-hexynyl)dUs inhibited virus replication throughout 40 days. It is in the last stage of toxicological studies and before clinical trial.

28.4 References

1. C. Perigand, G. Gosselin, *Nucleosides, Nucleotides* **1992**, *11*, 903-945.
2. Gauri K.K. Brit. Patent 1 170 565, **1969**, *Chem. Abstr.* **1970**, *72*, 448.
3. M. Swierkowski, D. Sugar, *J. Med. Chem.* **1969**, *12*, 533-534.
4. L. Ötvös, A. Szabolcs, J. Sági, A. Szemző, *Nucleic Acids Res., Spec. Publ.* **1975**. *1*, 49-52.
5. A. Szabolcs, J. Sági, L. Ötvös, *J. Carbohydr. Nucleosides, Nucleotides,* **1975**, *2*, 197-211.
6. M. Dranski, A. Zgit-Wroblewska, *Pol. J. Chem.* **1980**, *54*, 1085-1092.
7. I. Basnak, J. Farkas, J. Zajucek, Y. Hvlas, *Coll. Czech. Chem. Commun,* **1986**, *51*, 1764-1771.
8. J. Smejkal, J. Farkas, F. Sorm, *Coll. Czech. Chem. Commun.* **1966**, *31*, 291-297.
9. L. Ötvös, J. Béres, Gy. Sági, I. Tömösközi, L. Gruber, *Tetrahedron Lett.* **1987**, 6381-6384.
10. A.J. Hubbard, A.S. Jones, R.T. Walker, *Nucleic Acids Res.* **1984**, *12*, 6827-6837.
11. I. Basnak, A. Balkan, P.L. Coe, R.T. Walker, *Nucleosides, Nucleotides* **1994**, *13*, 177-196.
12. L. Ötvös, J. Rákóczi, Gy. Nagy, J. Nagy, É. Ruff, Hung. Patent 204 840, **1992**.
13. J. Robins, J. Barr, *J. Org. Chem.* **1983**, *48*, 1854-1862.

14. J. Béres, L. Ötvös, G.B. Wasley, J. Balzarini, E. DeClercq, *J. Med. Chem.* **1986,** *29,* 494-499.

15. A. Szemző, A. Szabolcs, J. Sági, L. Ötvös, *J. Carbohydr. Nucleosides, Nucleotides,* **1980,** *7,* 365-379.

16. T. Kovács, L. Ötvös, *Tetrahedron Lett.* **1988,** 6207-6216.

17. Zs. Veres, A. Szabolcs, I. Szinai, G. Dénes, M. Kajtár-Peredy, L. Ötvös, *Biochem. Pharmacol.* **1985,** *34,* 1737-1740.

18. J. Ötvös, J. Sági, T. Kovács, R.T. Walker, *Nucleic Acids Res.* **1987,** *15,* 1763-1777.

19. A. Jeney, J. Timár, G. Pogány, S. Paku, E. Moczár, M. Mareel, L. Ötvös, L. Kopper, K. Lapis, *Tokai J. Exp. Clin. Med.* **1990,** *15,* 167-177.

20. A. Jeney, L. Kopper, E. Hídvégi, K. Lapis, A. Szabolcs, L. Ötvös, *Int. J. Exp. Clin Chemoter.* **1991,** *4,* 32-39.

21. J. Sági, J. Stokrová, M. Vorlickova, A. Spanova, J. Kypr, É. Ruff, L. Ötvös, *Biochem. Biophys. Res. Commun.* **1992,** *185,* 96-102.

22. E. DeClercq in *Antiviral Drug Development* (Ed.: E. DeClercq, R.T. Walker) Plenum Pres, New York **1988,** 97.

23. J. Sági, A. Szemző, K. Ebinger, G. Sági, É. Ruff, L. Ötvös, *Tetrahedron Lett.* **1993,** *35,* 2191-2194.

29 Anhydrohexitols as Conformationally Constrained Furanose Mimics, Design of an RNA-Receptor.

P. Herdewijn

29.1 Introduction

Oligonucleotides might, theoretically, be used in therapy on three different levels (Figure 29-1). They can be targeted against proteins (aptamer approach), against double stranded DNA (antigene therapy) and against single stranded RNA (antisense oligomers). Targeting double stranded DNA with oligonucleotides is the most difficult task because this biological polymer is not easily accessible and also because it is not trivial to design molecules that are both tight-binders and selective binders. Proteins can be targeted by several other approaches, so RNA seems to be the most appropriate target for oligonucleotides. This can be done in two ways: one possibility is to design an oligonucleotide able to cleave the RNA either by introducing a catalytic core into the oligomer (ribozyme) or by inducing RNase H (an enzyme that cleaves the RNA part of an RNA-DNA duplex). This approach has the advantage that one oligonucleotide is able to destruct several targetted RNAs, making the inhibition process catalytic. The other possibility is to design an oligonucleotide which binds tightly to its target and sterically blocks the RNA so that it will not be able to exert its biological function.

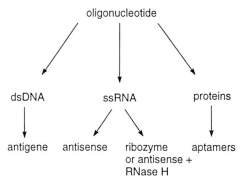

Figure 29-1. Common targets for oligonucleotides.

Natural oligonucleotides are composed of nucleotide units where the individual nucleosides are bonded to each other by a phosphodiester linkage (Figure 29-2). This bond is very susceptible to cleavage by phosphodiesterases which makes natural oligonucleotides unsuitable as such for therapeutic use. Just as we learned during the last decade to design peptidomimetics based on the structure of oligopeptides, we are now in the process of developing nucleotidomimetics using natural oligonucleotides as models.

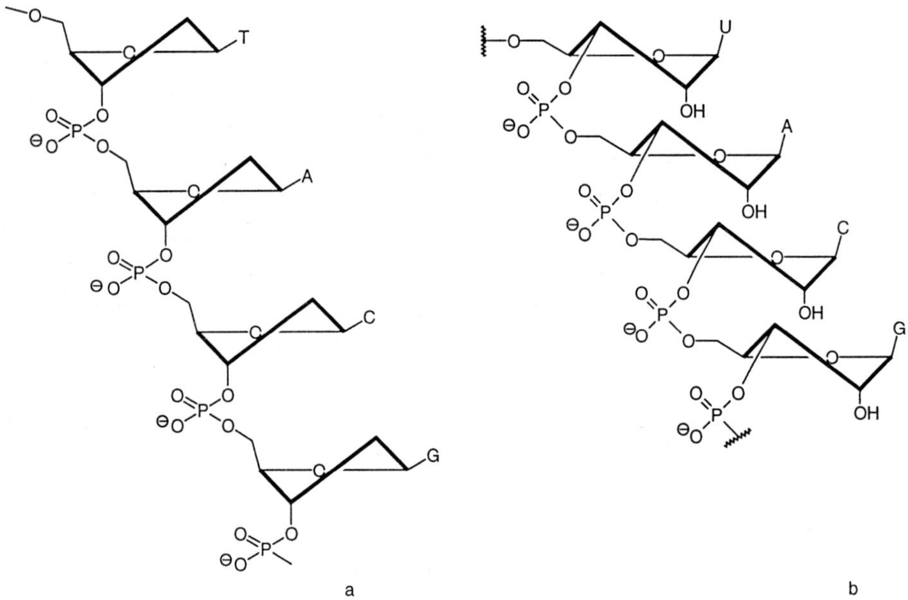

Figure 29-2. a) Structure of an oligodeoxynucleotide in a 2'-endo/3'-exo conformation of the 2-deoxy-D-*erythro*-pentofuranosyl moieties; **b)** Structure of an oligoribonucleotide in a 2'-exo/3'-endo conformation of the D-ribofuranosyl units.

The aim of this research is to generate molecules which are sterically and electronically complementary to natural single stranded RNA or double stranded DNA, so that they can function as a receptor for single stranded RNA or double stranded DNA. When these nucleotidomimetics can be introduced into a cell, they have the potential to capture the complementary RNA or double stranded DNA and down-regulate gene expression (Figure 29-3).

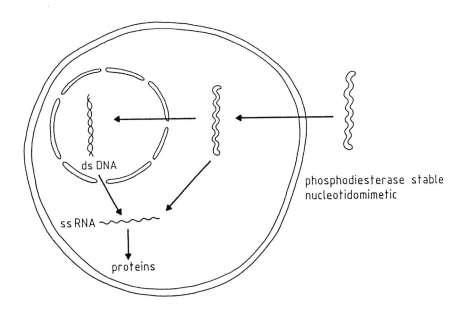

Figure 29-3. Targets of a nucleotidomimetic into a cell.

According to the size of the rings, the naturally occurring monosaccharides can be divided into two major groups: five membered furanoses and six-membered pyranoses. While pyranose carbohydrates are ubiquitous, furanose sugars are predominantly found in the nucleic acids content of cells, i.e. deoxynucleic acid (DNA) and ribonucleic acid (RNA). The backbone of DNA is made up of phosphorylated 2-deoxy-D-*erythro*-pentofuranosyl units while RNA is a linear phosphorylated D-ribofuranosyl polymer. Both nucleic acids have anomerically positioned purine and pyrimidine heterocycles in the β-configuration. When compared with pyranose saccharides, an important characteristic of a furanose is its conformational flexibility. Indeed, although preferred puckering modes are found in nucleosides, conformational changes in the five membered furanose ring occur with low energy barriers. This flexibility of the furanose ring is important for allowing conformational changes of DNA and RNA. Substantial conformational changes in pyranose sugars, on the other hand, cost more energy. Although the nature and the position of substituents highly influence conformational changes in ring structures, generally pyranose rings are more constrained than furanose rings.

This difference may be exploited to introduce oligomeric structures that can function as conformationally constrained RNA receptors. Indeed, to function as an antisense oligomer, able to sterically block the RNA target, the nucleotidomimetic should bind strongly to its target RNA. The stability of molecular association is determined by both enthalpic and entropic factors. Enthalpy stabilization can be achieved by diminishing interstrand phosphate repulsion, increasing stacking interactions, and, increasing the strength of the hydrogen bonding network. Interstrand phosphate repulsion can be reduced by synthesizing oligomers with neutral backbones. However, this may also interfere with the solubility of the oligomer. Increasing stacking and hydrogen bond interactions may result in less specific interactions. Therefore, entropic stabilization could be a useful alternative. Theoretically, this can be achieved by synthesizing a conformationally restricted oligomeric structure with a preorganized helical shape complimentary to that of a helical single stranded RNA. As six-membered rings are more rigid than five membered rings, phosphorylated pyranose oligomers seem to be the prime candidates for this purpose. The question arises as to which pyranose sugar can be considered an ideal mimic of a furanose ring in the conformation found in double stranded RNA. A second question is, will this furanose → pyranose replacement in the nucleotide unit of an oligonucleotide lead to a helical structure of the polymer with the correct pitch height, unit height and unit twist to allow hybridization with RNA. This increase in conformational purity will have consequences in the selectivity of its mode of action (RNA vs DNA, sequence selectivity) which will have to be investigated. The present research therefore looks for a link between furanoses and pyranoses on the nucleoside level.

29.2 Double Stranded RNA Structure

Before starting the design and synthesis process, it is important to understand the three-dimensional structure of the RNA target in its double stranded form. Knowledge about this structure is, however, incomplete and comes mainly from X-ray diffraction studies. It is beyond the scope of this chapter to give a complete description of the RNA structure, however, we will mention some important aspects necessary to understand the further development process. Right handed helical double-stranded nucleic acid structures are classified in different families of which the A and B families are the most important. The solution conformation of double stranded RNA helices is dependent on several factors, although it is generally accepted that double stranded RNA adopts an A-form conformation. The furanose sugar residues in A-type helices adopt a C3'-endo conformation and those in B-type helices a C2'-endo conformation (Figure 29-2a-b). Therefore, it would be advantageous if our pyranose nucleoside, used to build up a phosphodiester linked oligomer, could mimic a furanose sugar in the C3'-endo conformation. This alone is of course not sufficient to obtain a helical structure with about 11 base pairs per helix turn, a pitch height of 30 Å, an axial rise per nucleotide of about 2.75 Å, a base pair tilt angle of 16° to 19°, a rotation per nucleotide of about 32° and a dislocation of base pairs from helix axis of 4.4 to 4.9 Å.

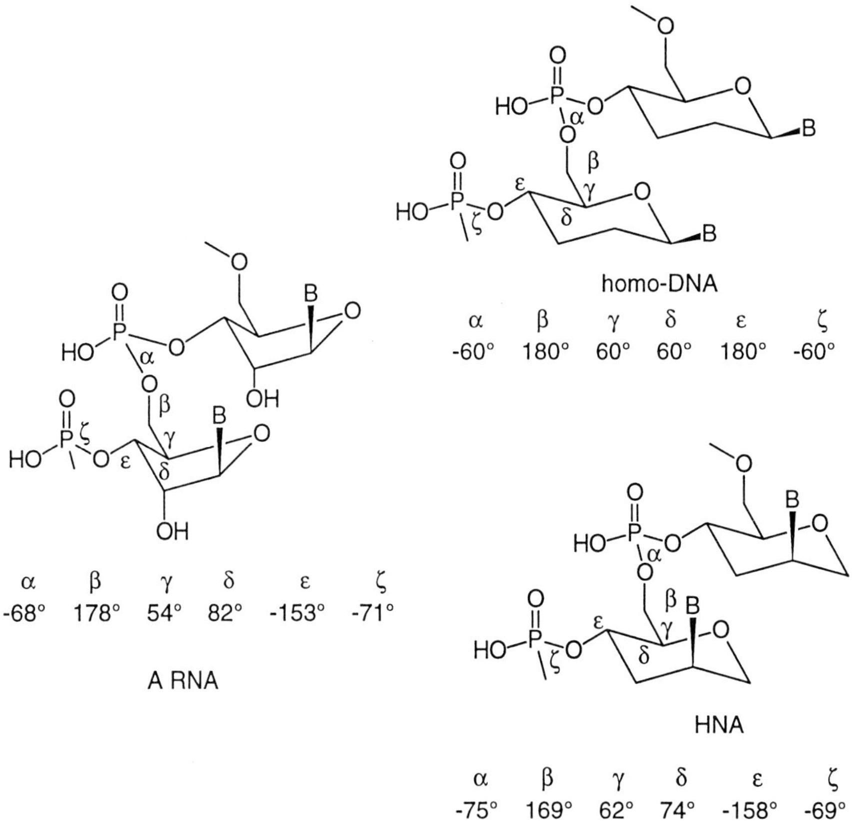

	α	β	γ	δ	ε	ζ
homo-DNA	-60°	180°	60°	60°	180°	-60°
A RNA	-68°	178°	54°	82°	-153°	-71°
HNA	-75°	169°	62°	74°	-158°	-69°

Figure 29-4. Structure and torsion angles of A-RNA, homo-DNA and HNA.

In order to be able to understand the structural changes of an oligonucleotide which take place when modified nucleosides are incorporated, we start by giving a short explanation of the structure of phosphorylated 2',3'-dideoxyglucopyranose oligomers (homo DNA) as described by Eschenmoser and Dobler [1] (Figure 29-4). The conformation of the sugar-phosphate backbone is defined by torsion angles α, β, γ, δ, ε, ζ. In homo-DNA these torsion angles have values around 60° or 180°. This quasi-ideal staggered conformation of the backbone structure results in a quasi-linearity of the homo-DNA double strand. Because of this linearity base pairs are too far from each other for efficient stacking (Figure 29-5).

In these pyranose nucleosides, with an anomerically positioned base moiety, the base is oriented equatorially, which means that it can hardly influence the conformation of the backbone structure. In order to obtain a helical structure, we will have to introduce substituents on the six membered carbohydrate ring which will force the geometry of the oligomer away from ideality.

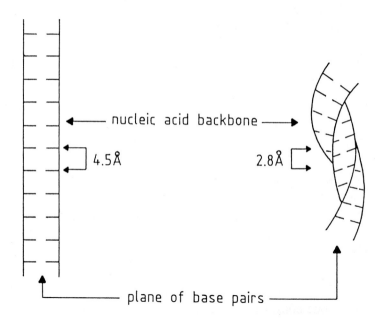

Figure 29-5. Change of base pair distance during winding of a double stranded nucleic acid.

29.3 Design of Oligonucleotides with a Six Membered Carbohydrate Moiety.

Our research on nucleoside analogues with a six membered carbohydrate moiety started with the preparation of several pyranosyl nucleosides [2, 3] i.e. 2,3-dideoxyglucopyranoses, 2,4-dideoxyallopyranoses, 3,4-dideoxyglucopyranoses and their incorporation in oligonucleotides (Figure 29-6). These molecules were synthesised in order to be able to study the general structure of a pyranose nucleoside and to assemble the necessary parameters needed to begin a conformational search experiment. These three nucleosides have a 1,5-relationship between the base moiety and the hydroxymethyl group. The heterocycle base is anomerically positioned. The secondary hydroxyl group is implanted on position 4, position 3 and position 2, respectively. This means that, when oligonucleotides are built up using these phosphorylated pyranose nucleosides, the backbone of the repeating unit contains six, seven and eight bonds, respectively. X-ray analysis of the monomeric pyranose nucleoside analogues led to the following conformational analysis: the pyranosyl ring of the nucleosides adopts a slightly flattened $^4C_1$ chair conformation for β-anomers so that the base moiety is always oriented equatorially; the $C_{1'}$-$O_{5'}$ bond is shorter than the $C_{5'}$-$O_{5'}$ distance; the glycosidic torsion angle (χ) in β-nucleosides is located in the region 180°-270° (-anticlinal).

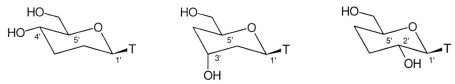

Figure 29-6. Initial series of pyranosyl nucleosides used for oligonucleotide synthesis.

Homo-oligomers containing these pyranosyl nucleosides are not able to hybridize with natural DNA due to their inability to form helix-like structures as found in the natural furanosyl-oligonucleotides. But the assembled data provided us with enough information to start a conformational search experiment [4] with the aim of finding a way to reorganize the three substituents (hydroxymethyl, hydroxyl, base moiety) on a six membered ring. The aim of this reorganization is to obtain a preorganized helical oligomer mimicking a natural nucleic acid structure. When one considers a pyranosyl ring with an anomerically positioned base moiety in the β configuration, there are four positions left on the six membered ring which can be substituted. To keep the system as simple as possible, we only considered positioning one hydroxymethyl group and one secondary hydroxyl group. But even in this case there are 56 possible combinations and this is too many to consider for synthesis. By employing a simple manual elimination procedure using Dreiding models, 15 pyranosyl residues with some configurational similarities to furanosyl nucleosides were retrieved (Figure 29-7). Some of them, however, are chemically unstable and may therefore be eliminated.

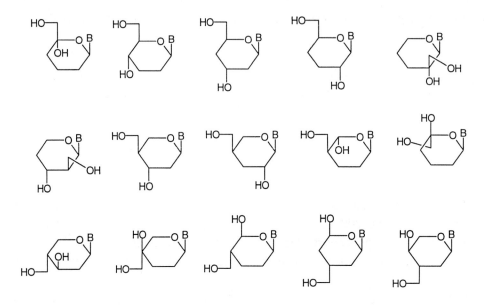

Figure 29-7. Pyranosyl nucleosides proposed for synthesis after a manual elimination procedure using Dreiding models.

In the next experiment, the ability of tetramers of the 15 pyranosyl nucleosides to form stable double helical complexes with unmodified deoxyadenylate tetramer strands was checked by exploring the conformational space of the isolated tetramers using high temperature molecular dynamics (MD) and energy minimization techniques [4]. This method retrieved two oligomers with promising hybridization capabilities. The reliability of the theoretical calculations could be checked due to the availability of experimental data on three of the proposed nucleosides (those represented in Figure 29-6 were used as negative controls because they did not hybridize very well). Peptide nucleic acids (PNA) were used as positive controls because it was known that PNA forms a stable duplex with natural nucleic acids. The two selected nucleosides both have a 1,3,4 relationship between the base moiety, the hydroxymethyl group and the secondary hydroxyl group (Figure 29-8).

Anhydrohexitols as Conformationaly Constrained... 561

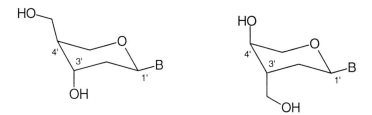

Figure 29-8. Pyranosyl nucleosides proposed for synthesis after a molecular dynamic studies.

Molecular dynamics in an aqueous environment on DNA complexes as hexamers proved that one duplex was unstable but that the other duplex remains very stable during 200 ps of MD simulations. This latter oligonucleotide was built up from phosphorylated 2',3'-dideoxy-3'-C-hydroxymethyl-α-L-*threo*-pentopyranosyl nucleosides. Despite the presence of two axial substituents and one equatorial substituent, this nucleoside might have the $^4C_1$ conformation, if one takes into account the anomerically positioned base moiety, the "supposed" greater bulkiness of a base moiety compared with a hydroxymethyl group and the gauche effect between the secondary hydroxyl group and the ring oxygen atom (Figure 29-9).

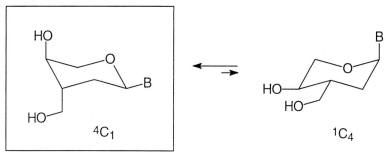

Figure 29-9. Structure of 2',3'-dideoxy-3'-C-hydroxymethyl-α-L-*threo*-pentopyranosyl nucleoside.

Until now, we only considered anomerically positioned nucleoside bases. By moving the ring oxygen, however, to its neighboring position and simultaneously changing the position of the hydroxymethyl group and hydroxyl group (otherwise a hemi-acetal group is obtained) a completely different kind of structure is obtained. The ring oxygen is now in a 1,3-relationship with the base moiety and because less steric hindrance is created around the base moiety than around the hydroxymethyl group, the conformation is now predominantly $^4C_1$ (Figure 29-10). The base moiety is oriented axially despite the loss of a gauche effect between the ring oxygen atom and the secondary hydroxyl group. Two of the three substituents are equatorially oriented.

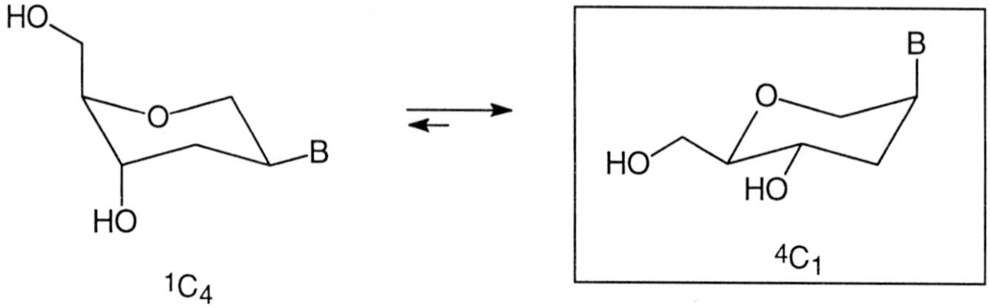

Figure 29-10. Structure of 1,5-anhydro-2,3-dideoxy-D-*arabino*-hexitol nucleoside..

We now have two molecules to synthesize with similar constitution but with a completely different organization of the functional groups.

29.4 Synthesis of 2',3'-Dideoxy-3'-C-Hydroxymethyl-α-L-*threo*-Pentopyranosyl Nucleoside and 1,5-Anhydro-2,3-Dideoxy-D-*arabino*-Hexitol Nucleoside.

The synthesis of a 3-C-branched-chain nucleoside may start from a Ferrier rearrangement followed by a Stork C-C bond formation reaction and a Mitsunobu-type inversion of configuration (Figure 29-11).

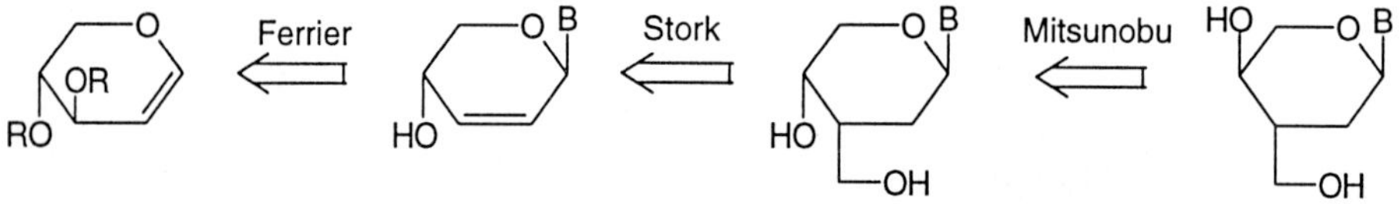

Figure 29-11. Retrosynthetic analysis of 3'-branched pentopyranosyl nucleoside.

Indeed, the synthetic plan has to take into consideration the fact that the hydroxymethyl group dictates the anomeric configuration during sugar-base condensation reaction of pyranoses [5]. Under equilibrium conditions, the nucleoside mainly formed has both an equatorially oriented hydroxymethyl group and base moiety. This means that in D-pyranose nucleosides with a 5-CH_2OH function, the β-anomer is favoured and in 3-C-hydroxymethyl-α-L-*threo*-pentopyranose the α-anomer is preferred (Figure 29-12).

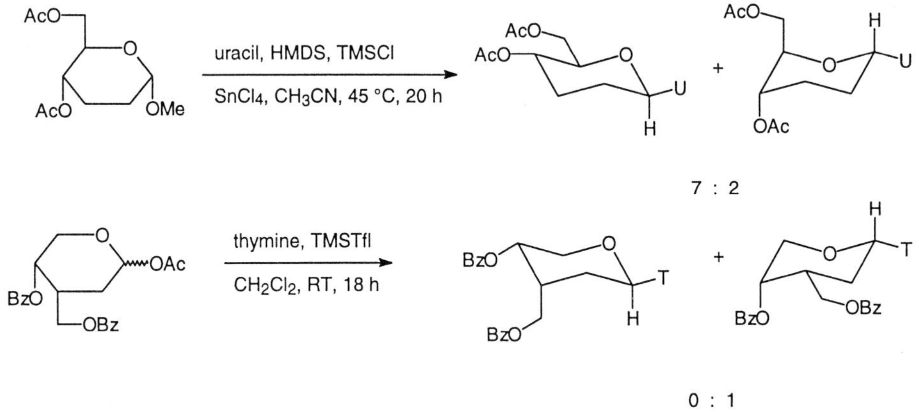

Figure 29-12. Sugar-based condensation reactions predominantly lead to pyranose nucleosides with an equatorial oriented hydroxymethyl group and base moiety.

However, we need the β-anomer for the latter compound. Therefore we have to start our synthetic strategy to the 3-branched nucleoside by first introducing the base moiety and in a second step, introducing 3-C-branching. A useful reaction for this strategy is the Ferrier rearrangement. An acid catalyzed Ferrier rearrangement with D-xylal using nucleoside base, however, gives a complex reaction mixture and the obtained results are very dependent on the conditions used. Therefore the reaction conditions to make this Ferrier rearrangement useful for preparative purposes had to be changed. Two modifications were introduced [6]. First, the hydroxyl groups of D-xylal were protected with a p-nitrobenzoyl group instead of more common acetyl groups. Because of the electron withdrawing character of the p-nitrobenzoyl group, the intermediate cation is less stabilized by anchimeric assistance and no 3-substituted side-compounds are formed. Secondly, no acid catalyst is added and the reaction is carried out under fusion conditions in boiling DMF. The liberated p-nitrobenzoic acid is too weak to force the anomerization reaction and the α/β mixture of the D-*glycero*-pent-2-enopyranosyl nucleosides is formed exclusively in multigrams quantities (Scheme 29-1).

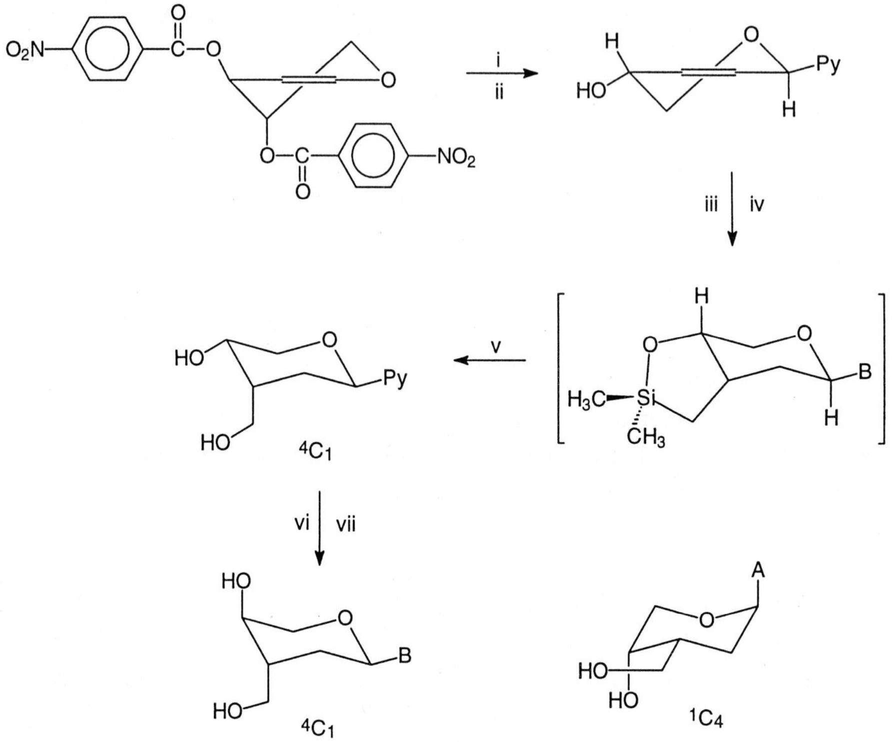

Scheme 29-1. Reagents: i) TMS(base), DMF, bp; ii) NaOMe, MeOH, dioxane; iii) ClSi(CH$_3$)$_2$CH$_2$Br, imidazole, DMF; iv) Bu$_3$SnH, AIBN, toluene; v) KF, KHCO$_3$, H$_2$O$_2$, DMF; vi) DEAD, Ph$_3$P, BzOH, dioxane, xylene; vii) NH$_3$, MeOH.

At this stage we carried out an initial study on the conformational behaviour of purine and pyrimidine nucleosides with a six membered carbohydrate moiety. As was derived from X-ray studies as well as from nmr studies, the purine base of the β-D-glycero-pent-2-enopyranosyl nucleoside preferentially adopts a pseudoaxial position while the pyrimidine base preferentially adopts a pseudoequatorial orientation (Figure 29-13). This means that the driving forces for the purine nucleosides are mainly stereoelectronic in origin (π-σ* overlap, gauche effect, anomeric effect) while steric effects (avoiding 1,3-pseudoaxial interactions) determine the preferred conformation of pyrimidine nucleosides.

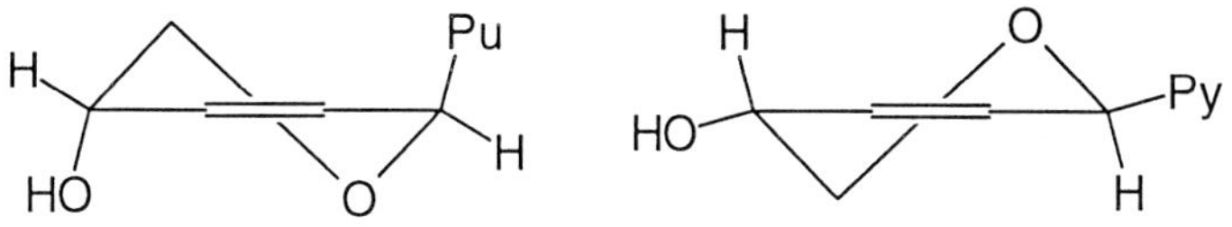

Figure 29-13. Preferential conformation of β-D-*glycero*-pent-2-enopyranosyl nucleosides.

The 3-C-branching was carried out using bromomethyldimethylsilyl chloride in a radical C-C-bond formation reaction followed by an oxidative carbon-silicon cleavage [7] (Scheme 29-1). Again a difference in conformation is observed between the obtained purine and pyrimidine nucleosides [7, 8]. The β-configured adenine analogue adopts a $^1C_4$ conformation with an axial oriented adenine [8] (Scheme 29-1). This conformation is additionally stabilized by a gauche effect and an anomeric effect, which do not exist in the $^4C_1$ conformation. The latter conformation is preferred with a pyrimidine (thymine) [7] base moiety (Scheme 29-1). The more bulky pyrimidine base is preferentially equatorially oriented despite unfavourable axial interactions between the 3'-hydroxymethyl group and the protons H1' and H5'. The above mentioned observation in regard to the driving forces for the conformational preferences of some pyranose nucleosides was confirmed here.

Inversion of configuration at C-4 using Mitsunobu type conditions gives us the title compound [7, 8] (Scheme 29-1). This inversion induces a conformational change of the adenine nucleosides, and, purine as well as pyrimidine nucleosides now adopt the same ($^4C_1$) conformation [7, 8]. This may be explained by the re-establishment of a gauche effect (O5'-C5'-C4'-OH) which disappears upon 4'-epimerization of the adenine nucleoside. The configuration and conformation of the final compound was confirmed by X-ray studies [7].

For the synthesis of the 1,5-anhydrohexitol nucleoside, D-glucose is used as starting material either in its protected pyranose form or in its protected furanose form. The first synthetic scheme (starting from the pyranose form [9]) has the advantage that no furanose → pyranose conversion is necessary during the synthesis, but it has the disadvantage of a regioselective protection procedure needed to remove the 3-hydroxyl group (Scheme 29-2). In the second scheme (starting from the furanose form [10]) the 3-hydroxyl group is removed selectively, but a furanose → pyranose conversion is needed to obtain the final compound (Scheme 29-3). The second scheme, however, is preferred because it can be carried out without intermediate chromatographic purifications in a final yield of 30-40%. This scheme is therefore more useful for large scale synthesis.

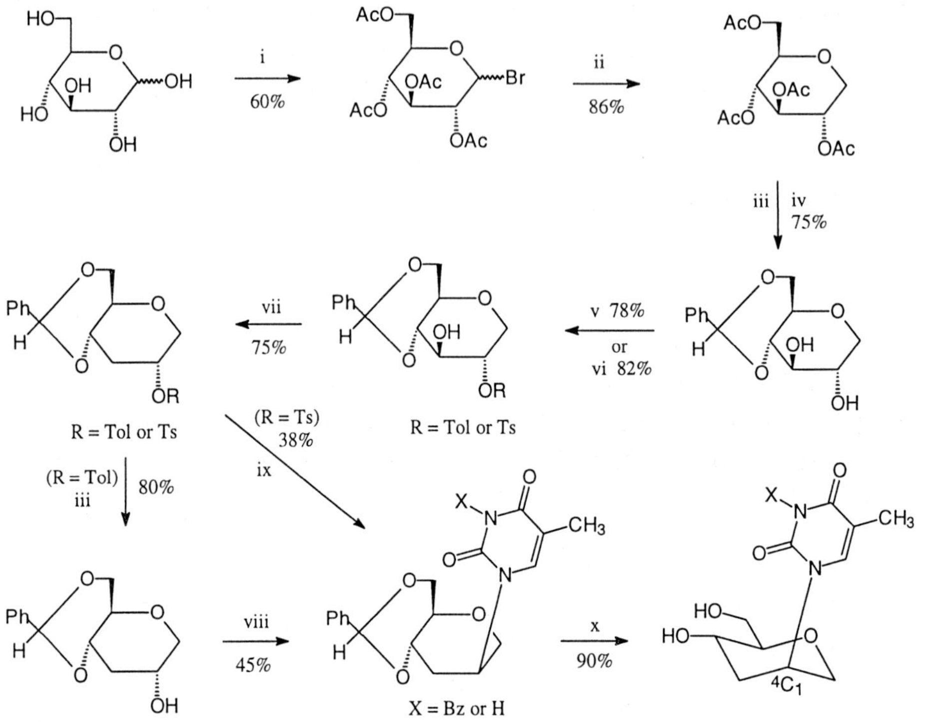

Scheme 29-2. Reagents: i) Ac$_2$O, HBr, rt; ii) Bu$_3$SnH, Et$_2$O, rt, KF, H$_2$O; iii) NaOMe, MeOH, rt; iv) PhCH=O, ZnCl$_2$, toluene, rt; v) Bu$_2$SnO, C$_6$H$_6$, dioxane, p-toluoyl chloride, rt; vi) Bu$_2$SnO, C$_6$H$_6$, dioxane, TsCl, 50 °C; vii) S=CCl$_2$, DMAP, CH$_2$Cl$_2$, RT; 2,4-Cl$_2$C$_6$H$_3$OH, rt; Bu$_3$SnH, AIBN, C$_7$H$_8$, 80 °C; viii) N^3-benzoylthymine, Ph$_3$P, DEAD, dioxane; ix) thymine, LiH, DMF, 120°; x) NH$_3$, MeOH, RT; 80% HOAc, 80 °C.

2,3,4,6-tetra-*O*-acetylglucopyranosyl bromide was reductively dehalogenated using tributyltin hydride (Scheme 29-2). The acetyl groups were removed and a benzylidene protecting group was introduced affording 1,5-anhydro-4,6-*O*-benzylidene-D-glucitol. Selective reaction of the hydroxyl function at position 2 was feasible following activation with dibutyltin oxide. Position 2 was either selectively protected as an ester with toluoyl chloride or was functionalized with a leaving group using tosyl chloride, affording 1,5-anhydro-1,6-*O*-benzylidene-3-deoxy-2-*O*-p-toluoyl-D-glucitol or 1,5-anhydro-4,6-*O*-benzylidene-3-deoxy-2-*O*-p-(tolylsulfonyl)-D-glucitol. The hydroxyl group in position 3 was removed by conversion to the 2,4-dichlorophenylthio carbonate derivatives followed by a

Barton-type reduction. The nucleoside base can be introduced either by a nucleophilic displacement of the tosylate or using Mitsunobu reaction conditions on the deprotected alcohol. Finally the benzylidene protecting group was removed with acetic acid. X-ray diffraction analysis and NMR analysis confirm the predicted conformation of the 1,5-anhydrohexitol nucleosides [9,11] as depicted in Scheme 29-2.

The second route (Scheme 29-3) starts from commercially available diacetone D-glucose [10]. Barton deoxygenation removed the 3-hydroxyl function. Acid deprotection of 3-deoxy-D-glucose and peracetylation gave a α/β anomeric mixture of the pyranoses and furanoses in a ratio 5/1. Treatment of this mixture with HBr/HOAc and reduction of the crude glycosyl bromides with Bu$_3$SnH gave 70% isolated yields of the deoxygenated pyranoses. The key intermediate was obtained by deacetylation followed by benzylidene formation.

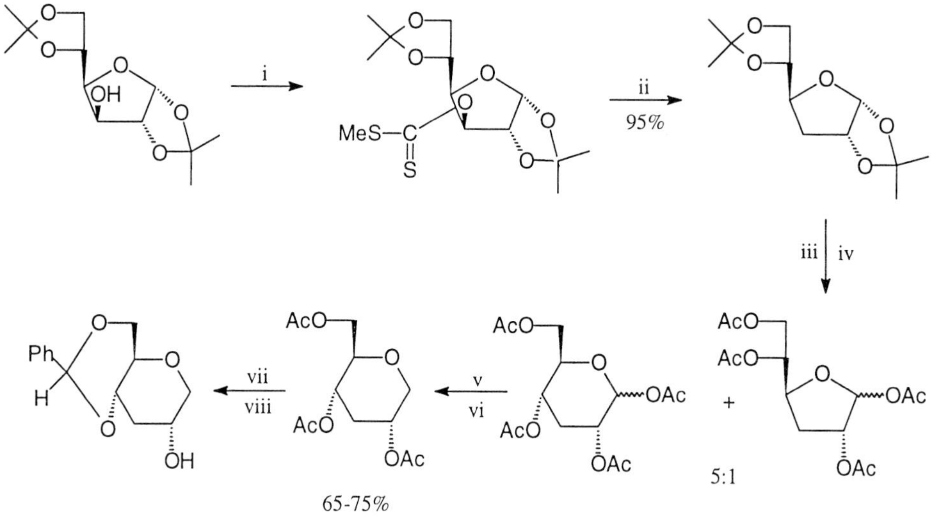

Scheme 29-3. Reagents: i) NaH, imidazole, CS$_2$, MeI, THF, distillation; ii) Bu$_3$SnH, toluene, Δ, distillation; iii) IRA-120(H$^+$), EtOH, H$_2$O; iv) Ac$_2$O, C$_5$H$_5$N, RT; v) HBr, HOAc; vi) Bu$_3$SnH, Et$_2$O, RT; vii) NaOMe, MeOH; viii) PhCH(OCH$_3$)$_2$, dioxane, TsOH, RT

Introduction of the different bases require different reaction conditions and different starting materials. To be able to introduce the nucleoside into an oligonucleotide, again the protecting group strategy, is dependent on the base moiety being considered. As the description of this research is beyond the scope of this chapter we will restrict ourselves by giving one selected example (that of the guanine analogue). The 1,5-anhydro-4,6-O-benzylidene-3-deoxy-D-glycitol is activated as a triflate (Scheme 29-4) and reacted with the tetrabutylammonium salt of 2-amino-6-iodopurine in dichloromethane giving regioselective N-9 alkylation of the purine base.

Treatment with 10% HCl hydrolysed the 6-iodo substituent as well as the benzylidene protecting group. Protection of the amino-group was carried out using a transient protection (silylation) procedure. Finally the primary hydroxyl group was protected with monomethoxytrityl chloride and the secondary hydroxyl group was phosphitylated leading to the phosphoramidite, ready for incorporation in an oligonucleotide.

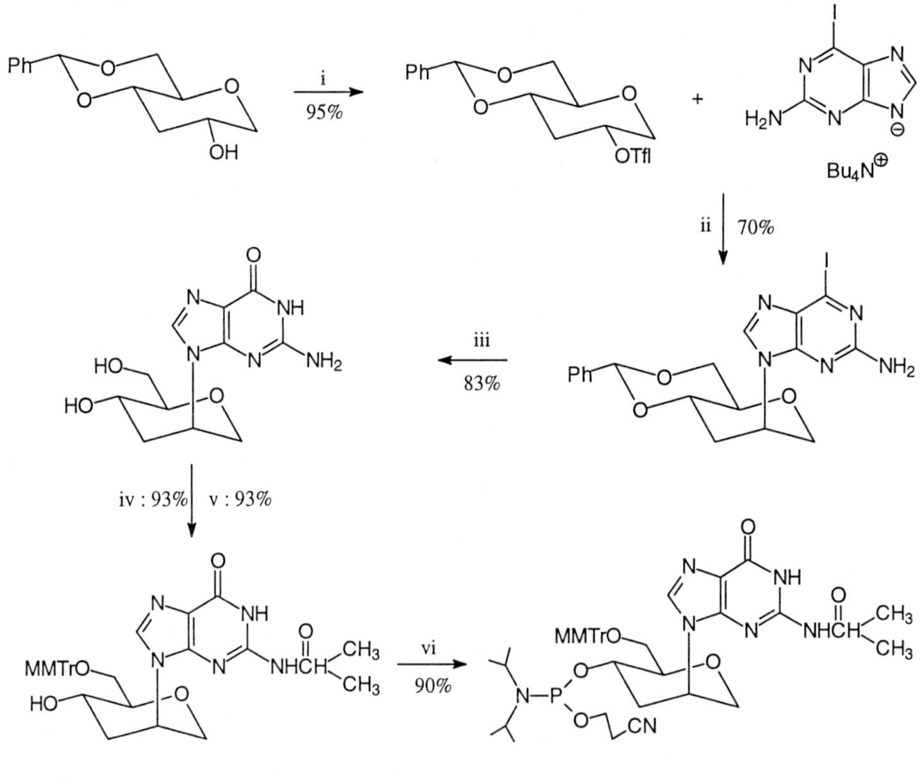

Scheme 29-4. Reagents: i) $(CF_3SO_2)_2O$, C_5H_5N, -5 °C; ii) CH_2Cl_2, 0→20 °C; iii) 10% aq. HCl, 100 °C; iv) BSA, C_5H_5N, 12°C; [$(CH_3)_2CHCO]_2O$, RT; NH_3, H_2O; v) MMTrCl, Py, DMF, RT; vi) Et_3N, CH_2Cl_2, $NCCH_2CH_2OP(Cl)N[CH(CH_3)_2]_2$, RT.

29.5 Oligonucleotide Built up from Nucleosides with a Six-membered Carbohydrate-like Moiety.

Using 2',3'-dideoxy-3'-C-hydroxymethyl-α-L-*threo*-pentopyranosyl nucleosides protected with a monomethoxytrityl group at the primary hydroxyl group and phosphitylated at the secondary hydroxyl group (exemplified by the guanine building block in Scheme 29-4) as building blocks, oligonucleotides were synthesized on an automated synthesizer, then subsequently deblocked and purified by HPLC [8]. These oligonucleotides are indeed able to form duplexes with natural DNA but at a lower stability than originally calculated, which means that the modelling calculations partially failed (Table 29-1). Several reasons can be proposed for this. First, many factors involved in the hybridization of oligomers are still not completely defined and the available parameters for carrying out the calculations are therefore inadequate. A second reason is that a conformational search of 200ps using high temperature MD is inadequate. The order of destabilization of the duplex may be reversed just by increasing the calculation time. It is therefore clear that a lot of research still needs to be done before ending up with an accurate model to predict the hybridizing capabilities of oligonucleotides.

Table 29-1. Melting temperatures at 0.1 M NaCl of oligonucleotides containing 3-deoxy-3-C-hydroxymethyl aldopentopyranosyl nucleosides(*) [8].

Duplex	Tm	ΔTm/modification
T13.dA13	33.2	-
TT*T9T*T.dA13	22.8	-5.2 °C
T$^*_{13}$.dA13	-	-
T13.dA6A*dA6	24.0	-9.2 °C
T13.A$^*_{13}$	11.7	-1.65 °C

The study of the anhydrohexitol-incorporated oligonucleotides was more successful, although initial results using oligodeoxynucleotides (DNA) were confusing [12]. Oligo(hexitol)thymidylates are self associating at high salt concentration (1M NaCl) (Tm between 45 °C and 50 °C for a thirteen mer). Oligo(hexitol)thymidylates also form very stable duplexes with oligo(hexitol)adenylates at high (1M NaCl) as well as at low (0.1M NaCl) salt concentrations (Tm of 80 °C, compared with 33 °C for a natural DNA duplex). Natural oligothymidylates form duplexes with oligo(hexitol)adenylates but at a decreased stability when compared with the natural oligoT.oligo dA duplexes (Tm of 21 °C compared to 33 °C for the natural system). At high salt concentration (1M NaCl), duplexes between natural oligodeoxyadenylates and oligo(hexitol)thymidylates are formed (Tm about 65 °C), but not in the presence of high $MgCl_2$ concentrations (100 mM) where the oligo(hexitol)thymidylate self-association dominates. At low salt concentration (0.1M NaCl), not only duplexes but also triplexes between natural oligoadenylate and oligo(hexitol)thymidylates are observed.

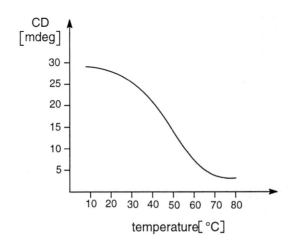

Figure 29-14. Temperature dependent CD measurement of oligoT*.oligoT* (at 270 nm).

The data obtained with these homopolymers is highly dependent on the conditions used and it must be made clear that the preorganization of the anhydrohexitol oligomers is not well suited to accommodate a natural single DNA strand. It should be noted that this results was confirmed by temperature dependent CD measurements (for example Figure 29-14) and by a plot of the Tm (melting temperature of the duplex) versus the logarithm of the concentration of the oligonucleotides (for example Figure 29-15). This last experiment may be used to distinguish between intramolecular and intermolecular interactions.

Anhydrohexitols as Conformationaly Constrained... 571

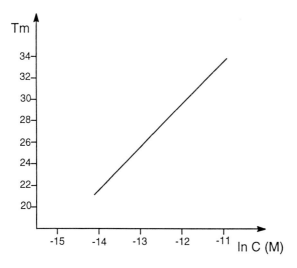

Figure 29-15. Tm (°C) versus ln C for the oligoA*.oligothymidylate duplex at 1 N NaCl.

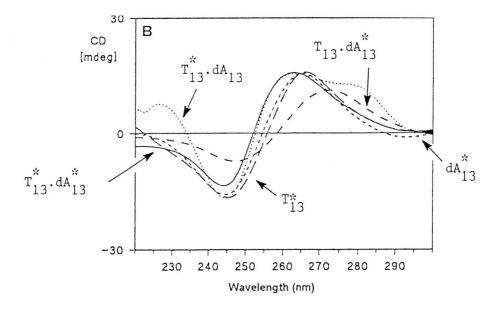

Figure 29-16. CD spectra recorded at 30 °C measured at 0.1 M NaCl, 10 mM cacodylate pH 7.0, 10 mM MgCl$_2$. .

572   P. Herdewijn

We can also demonstrate that the oligonucleotides containing hexitol monomers at their end (or completely modified oligomers) are completely resistant to degradation by phosphodiesterase which is, of course, a prerequisite for good antisense construction. Given the anomalous behaviour of the aforementioned oligoadenylate and oligothymidylate duplexes, we first undertook a structural analysis using CD spectroscopy (Figure 29-16). The CD spectra of the different duplexes, the single stranded oligo(hexitol)adenylates and of the oligo(hexitol)thymidylates were surprisingly similar and demonstrate similarity with the CD spectrum of double stranded A-RNA. Because oligodeoxyadenylate.oligothymidylate are conformationally "stiff" and prefer the B-form, we turned our attention to other sequences: in the first instance, mixed polypurine sequences followed by adding cytosine bases, and secondly, a mixed sequence containing all four nucleoside bases.

The T_m of fully modified oligonucleotides containing guanine and adenine base moieties demonstrate increased stability of the HNA-DNA and HNA-RNA duplex when compared to their natural counterparts (DNA-DNA and DNA-RNA). This was demonstrated for a six mer (sequence AGGAGA) (ΔT_m/modification: +3.5°C for HNA-DNA and +5.7°C for HNA-RNA) as well as for a twelve mer (sequence AGGGAGAGGAGA) (ΔT_m/modification: +1.3°C for HNA-DNA and +3.0°C for HNA-RNA). The stabilization pro nucleotide, however, decreases with chain length and is also sequence dependent (data not shown). The data also demonstrates that the hexitol oligomers form more stable duplexes with RNA than with DNA. This difference is even more striking when pyrimidine(cytosine) base moieties are introduced into the sequence (Table 29-2). When G/A/C mixed sequences (8 and 12 mer were tested) are considered, the HNA-DNA duplex is even less stable than the DNA-DNA duplex (ΔT_m/modification: -0.65°C for 8 mer and -0.85°C for 12 mer) and is of about equal stability as for RNA-DNA duplexes (ΔT_m/modification: +0.07°C for 8 mer and -0.3°C for 12 mer). The HNA-RNA duplexes are more stable than the DNA-RNA duplexes (ΔT_m/modification: +2.45°C for 8 mer and +1.40°C for 12 mer) and DNA-DNA duplexes (ΔT_m/modification: +1.72°C for 8 mer and +0.87°C for 12 mer). These results can be expected assuming that HNA functions as mimic of the natural RNA oligomer.

The presence of cytosine bases apparently reduces the ΔT_m/modification (compare termal stability of duplexes containing G/A/C sequences to duplexes containing only G/A sequences). In the next step, a modified 8 mer sequence containing all four bases (G/A/C/T) was investigated (sequence 13

GCGTAGCG). Although the sequence is somewhat different from the mixed 8 mer G/A/C sequence, the ratio between the pyrimidine and purine bases is the same. Also the number of G and A bases is the same in both sequences. Only one cytosine base is replaced by one thymine base. The melting temperature of this HNA-RNA sequence is still higher than that of the DNA-RNA (ΔTm/modification: +4.2°C) and RNA-RNA duplex (ΔTm/modification: +0.86°C), although the difference between HNA-RNA and RNA-RNA is less pronounced than expected.

Table 29-2 : Tm of fully modified hA/hG/hC oligonucleotides.

Sequence	DNA complement	RNA complement
h(CGACGGCG)	46.3 °C	65.3 °C
d(CGACGGCG)	51.5 °C	45.7 °C
h(CACCGACGGCGC)	59.2 °C	79.8 °C
d(CACCGACGGCGC)	69.4 °C	63.1 °C

The effect of hexitol nucleosides on hybridization seems to be very sequence specific and, after a first evaluation cycle, a purine base in the HNA sequence seems better suited than a pyrimidine base, which might be due to different stacking interactions or different hydration.

Strong hybridization and nuclease stability are two factors which may benefit the antisense effect of an oligonucleotide. However, these characteristics serve no purpose when the action is not sequence specific. Therefore, we investigated the influence of all 12 possible mismatches on the thermal behaviour of the aforementioned 8 mer oligonucleotide. In every case the stability of the duplex decreases from 13°C to 38°C making these HNA-RNA interactions very specific (Table 29-3) [13].

29.6 1,5-Anhydrohexitol Nucleoside as a Mimic of a Furanose Nucleoside in its 2'-endo/3'-exo Conformation.

Although the energy of interconversion is low, the carbohydrate moiety of nucleosides preferentially adopts a C3'-endo conformation (with phase angles 0°≤p≤36°) or a C2'-endo conformation (pseudorotational phase angles 144°≤p≤190°) (Figure 29-17). Right handed helical double stranded nucleic acids are normally classified in A and B families.

Table 29-3. Influence of mismatches for HNA-RNA interactions.: example given for hG and hC (at 0.1 M NaCl).

	hexitol sequence 6'G[CG]TAGCG3' RNA complement 3'C[GC]AUCGC5'	
	Tm	ΔTm
hG-C	54.5	-
hG-U	40.8	-13.7
hG-A	16.3	-38.2
hG-G	30.6	-23.9
hC-G	54.5	-
hC-C	36.2	-18.3
hC-U	37.4	-17.1
hC-A	41.7	-12.8

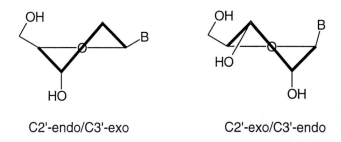

Figure 29-17. Preferred conformation of furanose nucleosides.

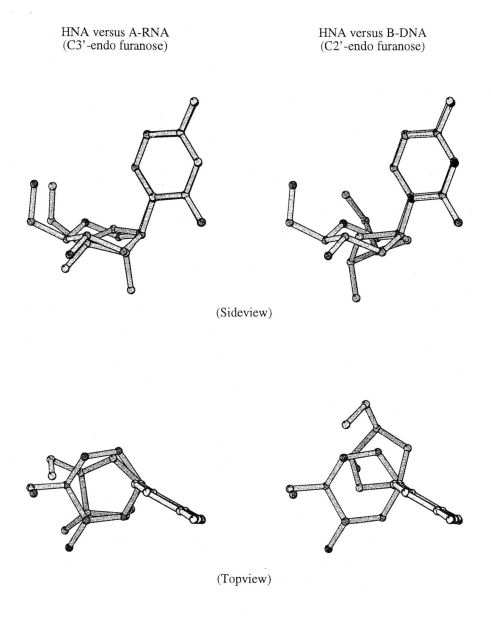

Figure 29-18. Overlap of anhydrohexitol nucleoside with a furanose nucleoside in the C3'-endo and C2'-endo conformations.

The formation of a helical structure restricts the conformational freedom of the furanose rings and pushes them into a C3'-endo conformation in A-type helices and in a C2'-endo conformation in B-type helices. It is generally accepted that in solution double stranded DNA adopts a conformation close to the B-type family of structures. Double stranded RNA sequences, on the other hand, preferentially adopt the A-conformation. It seems therefore, that when targeting RNA, it would be advantageous to use modified nucleotides, to build a phosphodiester-linked oligomer, which are able to mimic a furanose sugar in the C3'-endo conformation. Indeed, when trying to overlap the anhydrohexitol nucleoside with a furanose nucleoside in the 2'-endo and 3'-endo conformation, an acceptable fit is only found with the 3'-endo form (Figure 29-18). It therefore seems that the anhydrohexitol nucleoside can be considered as a conformationally constrained mimic of a ribonucleoside. However, in order to be considered as an RNA-receptor, the oligomer built from anhydrohexitol nucleotides should have an intramolecular order that prepares the molecule for duplex formation with natural RNA, which is clearly not the case with the 2,3-dideoxyglucopyranose oligomers.

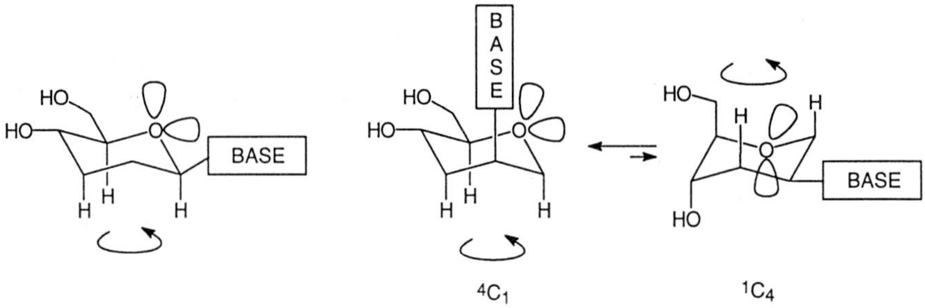

Figure 29-19. Preferred conformation of nucleosides with a six-membered carbohydrate moiety showing unfavourable 1,3-diaxial interactions.

In 2,3-dideoxyglycopyranose nucleosides all bulky substituents are placed equatorially and only hydrogen atoms are placed axially (Figure 29-19). When the base moiety is moved to the 2'-position, less steric hindrance is created in the $^4C_1$ conformation than in the $^1C_4$ conformation (Figure 29-19). The first conformation is therefore the favoured one. In the $^1C_4$ conformation 1,3-diaxial interactions between the 5'-hydroxymethyl moiety and the hydrogen atoms on C-1' and C-3' are present. So the position of the ring oxygen atom with respect to the base moiety and the hydroxymethyl group determines the preferred conformation.

The displacement of the base moiety from the 1'-position to the 2'-position, places the heterocycle in a 1,3-relationship with the ring oxygen atom and creates less steric hindrance around the base than around the hydroxymethyl group. Due to this positioning of the base moiety in the anhydrohexitol, the inclination of the base moiety with respect to the plane of the sugar ring comes into the range of the angles found in furanose nucleosides (Figure 29-4). This then induces helicity in the strand as a result of the influence of steric effects and gauche effects. This can be observed by a change in the backbone torsion angles which now comes near to those found in dsRNA (Figure 29-4). The single stranded anhydrohexitol oligomer shows a preorganized structure in a right handed rotation with stacking potentialities as found in dsRNA. Its electronic and Van der Waals surface is complimentary to that of a single stranded RNA which can be bound tightly to the hexitol oligomers. A molecule is created with a preorganized helical structure that may function as RNA receptor. This observation is further sustained by CD experiments [13]. The CD spectrum of a HNA-RNA duplex is similar to that of a RNA-RNA-duplex showing a positive cotton effect at 270 nm and a negative cotton effect at 245 nm (Figure 29-20). Also, modelling experiments show a A-like shape for the HNA-RNA duplex with an open cylinder along the helix axis (Figure 29-21) comparable with the structure of dsRNA.

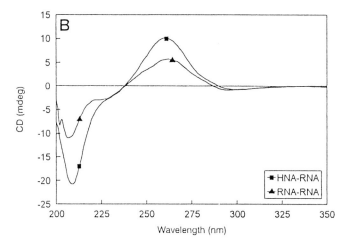

Figure 29-20. CD spectra of an HNA-RNA and RNA-RNA duplex at 0.1 M NaCl, 10 mM cacodylate pH 7.0, 10 mM $MgCl_2$ (HNA sequence : 6'GCGTAGCG3').

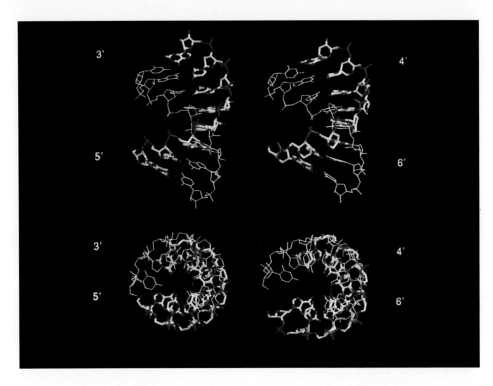

idealized ds A-RNA
5'-GCGUAGCG-3'

minimized HNA-RNA
6'-GCGTAGCG-4'

Figure 29-21

Finally we investigated the antiviral activity of the anhydrohexitol nucleosides. Normal furanose nucleosides are accepted by kinases of several herpes viruses. Due to this selective recognition, the nucleosides are phosphorylated in herpes virus infected cells and, hence, demonstrate antiviral activity. Likewise, the anhydrohexitol nucleosides with modified base moieties (for example 5-iodouracil-1-yl and 5-ethyluracil-1-yl base) show antiviral activity and are phosphorylated by viral kinases [9,11]. This not only further sustains the hypothesis that anhydrohexitol nucleosides are good mimics for furanose nucleosides, but also suggests that the 3'-endo/2'-exo conformation of a furanose nucleoside might be a conformational intermediary form playing a role during the metabolic activation of an antiviral nucleoside. It demonstrates that the study of constrained nucleosides may also be very useful for unravelling the conformational alterations that take place during metabolic activation of natural nucleosides within cells.

29.7 References

1. A. Eschenmoser, M. Dobler, *Helv. Chim. Acta* **1992**, *75*, 218-259.
2. K. Augustyns, A. Van Aerschot, C. Urbanke, P. Herdewijn, *Bull. Soc. Chim. Belg.* **1992**, *101*, 119-130.
3. K. Augustyns, F. Vandendriessche, A. Van Aerschot, R. Busson, C. Urbanke, P. Herdewijn, *Nucleic Acids Res.* **1992**, *20*, 4711-4716.
4. P. Herdewijn, B. Doboszewski, H. De Winter, C. De Ranter, I. Verheggen, K. Augustyns, C. Hendrix, A. Van Aerschot, *Am. Chem. Soc. Symp. Ser.* **1994**, *580*, 80-99.
5. K. Augustyns, J. Rozenski, A. Van Aerschot, P. Claes, P. Herdewijn, *Tetrahedron* **1994**, *50*, 1189-1198.
6. B. Doboszewski, N. Blaton, P. Herdewijn, *J. Org. Chem.* **1995**, *60*, 7909-7919.
7. B. Doboszewski, N. Blaton, J. Rozenski, A. De Bruyn, P. Herdewijn, *Tetrahedron* **1995**, *51*, 5381-5396.
8. B. Doboszewski, H. De Winter, A. Van Aerschot, P. Herdewijn, *Tetrahedron* **1995**, *51*, 12319-12336.
9. I. Verheggen, A. Van Aerschot, S. Toppet, R. Snoeck, G. Janssen, P. Claes, J. Balzarini, E. De Clercq, P. Herdewijn, *J. Med. Chem.* **1993**, *36*, 2033-2040.
10. M. Anderson, S. Daluge, L. Kerremans, P. Herdewijn, *Tetrahedron Lett.* **1996**, *37*, 8147-8150.
11. I. Verheggen, A. Van Aerschot, L. Van Meervelt, J. Rozenski, L. Wiebe, R. Snoeck, G. Andrei, J. Balzarini, P. Claes, E. De Clercq, P. Herdewijn, *J. Med. Chem.* **1995**, *38*, 826-835.
12. C. Hendrix, I. Verheggen, H. Rosemeyer, F. Seela, A. Van Aerschot, P. Herdewijn, *Chemistry Eur. J.* **1997**, *3*, 110-120.
13. C. Hendrix, H. Rosemeyer, B. De Bouvere, A. Van Aerschot, F. Seela, P. Herdewijn, *J. Am. Chem. Soc.* (accepted).

Index

(+)-5-norbornen-2-one 225

1,1′-bistetrahydropyran, synthesis of 7
1,3-diaxial interactions 120
1,3-diaxial strain 120
1,3-dithianes *see* dithiane
1,4-addition
– of radical on enone 27
1,6-*epi*-cyclophellitol
– synthesis of 286
1,8-diazabicyclo[5.4.0]-undec-7-ene *see* DBU
1-cyanovinyl esters
– in the Diels-Alder reaction 29
1-deoxy-1-nitromethylpyranose, in the Henry condensation 25
1-stannyl glycal 7

2-(*tert*-butyldimethylsiloxy)furan 328, 341, 343
2-(*tert*-butyldimethylsiloxy)thiophene 328, 341, 343
2-(trimethylsiloxy)furan 328-339
2,2′-anhydro-5-alkyl-2′-deoxyuridines
– 3′,5′-diacyl derivatives 541
– ammonolysis of acyl derivatives 542
– phosphorylase inhibitory effect 540
– reduction with tri-*n*-butyltin hydride 566
2,3,4,6-tetra-*O*-benzyl-α-D-gluconolactone 71
2,3,4,6-tetra-*O*-acetylglucopyranosyl bromide
– 2,3,4,6-tetra-*O*-benzyl-α-D-glucopyranosyl bromide 70
2,3,5-tri-*O*-benzyl-D-arabinose
– vinylation with divinylzinc 81
2,3,5-tri-*O*-benzyl-L-ribofuranose

– as starting material 291
2,3-dideoxy hex-2-enopyranosides
– in nitrile oxide cycloaddition 53
2,3-*O*-isopropylidene-5-*O*-triphenylmethyl-D-ribose 71
2,4,4,6-tetrabromo-2,5-cyclohexadien-1-one 296
2,5-anhydrohexonic acid
– as precursor of *C*-nucleosides 29
2,5-imino hexitols
– in azadisaccharides synthesis 311
2,6,7-trideoxy-2,6-imino-D-*glycero*-L-mannose
– cross-aldolisation of 40
2-acetamido-2-deoxy-3,4,6-tri-*O*-benzyl-D-glucopyranose 79
2-acetamido-2-deoxy-D-glucosyl cation
– mimics of 93
2-chloro-4,6-dimethoxy-1,3,5-triazine 428
2-deoxy-D-ribose
– as starting material 109
2-deoxy-D-ribose, as starting material 4
2′-deoxyuridine
– 5-ethyl 537
– 5-iodo 537
– 5-isopropyl 549
– 5-*n*-hexyl 546
2-isoxazoline *see* isoxazoline
2′-OMe RNA
– in antisense oligonucleotides 523

3,4,5,7-tetra-*O*-benzyl-D-glucoheptenitol 74
3,4,6-tri-*O*-benzyl-D-arabinohexenitol 74
3,4-di-*O*-butylstannylene
– in regioselective sulfation 369
3′,6′-disulfated Lewis^x pentasaccharide

- synthesis of 373
3',6-disulfated Lewis^x pentasaccharide
- synthesis of 376
3'-sialyl Lewis^a
- as ligand of E-selectin 365
3'-sialyl Lewis^x
- as ligand of E-selectin 365
3'-sulfated Lewis^a oligosaccharides
- binding to L-selectin 381
- biological activity 379
3'-sulfated Lewis^a pentasaccharide
- biological activity 368
- synthesis of 367, 368
3'-sulfated Lewis^a tetrasaccharide
- structure 366
3'-sulfated Lewis^a trisaccharide
- synthesis of 367, 370
3'-sulfated Lewis^x oligosaccharides
- binding to L-selectin 381
- biological activity 379
3'-sulfated Lewis^x pentasaccharide
- synthesis of 373
3'-sulfated Lewis^x tetrasaccharide
- structure 366

4,5-dihydroisoxazole *see* isoxazoline
$^4C_1$ conformation 561
4-*epi*-polyoxamic acid 332

5,6-dideoxy-D-*xylo*-hex-5-enofuranoside
- as dipolarophile 56
5a-carba D-talopyranose
- synthesis of 216
5a-carba-α-D-mannopyranose
- synthesis of 216
5a-carba-allopyranoses
- synthesis of 216
5a-carba-mannopyranoses
- synthesis of 216
5a-carba-sugars *see* carba-sugars
5-alkyl-2'-deoxyuridines 538
- synthesis 538
5-amino-1,2,3,4-cyclopentanetetraols
- as glycosidase inhibitors 102
5-deoxy-carba-D-*arabino*-furanose
- synthesis of 227

5-deoxy-carba-D-*lyxo*-furanose
- synthesis of 227
5-deoxy-carba-D-*ribo*-furanose
- synthesis of 227
5-deoxy-carba-D-*xylo*-furanose
- synthesis of 227
5-ethyl-2'-deoxyuridine
- 2,2'-anhydro 541
- antiviral activity 537
- synthesis 537
5-*exo-trig* cyclisation 125
5-iodo-2'-deoxyuridine
- antiviral activity 540
- in the synthesis of 5-alkyl-dUs 540
5-isopropyl-2'-deoxyuridine
- antiviral activity 549
- synthesis 540

6,6-dimethylfulvene
- as starting material 252
- photooxidation of 253
6'-sulfated Lewis^x pentasaccharide
- synthesis of 373
6'-sulfated sialyl Lewis^x
- in GlyCAM-1 372
6-amino-6-deoxy heptenitols 266
6-*exo-tet* ring closure 115
6-*exo-trig* ring closure 115
6-sulfated sialyl Lewis^x
- in GlyCAM-1 372

7-*endo-trig* ring closure 115
7-oxabicyclo[2.2.1]hept-5-en-2-yl
  derivative 29

A$^{(1,3)}$ strain 253
α$_1$-acid glycoprotein 401
α-amino acids
- hydroxylated 332, 335
- substituted 333
acarbose 88
- synthesis of 90
acaricide 304
acceptor
- in glycosylation 302

acetal 152
- as protecting group 210
- benzylidene 302
- benzylidene, methods for introducing 176
- oxidative cleavage 152
acetaldehyde 333
acetonitrile
- lithium anion 146
acrylamide 388, 389
activated partial thromboplastin time *see* aPTT
Adenophostins 191, 192, 193, 197, 199
α-D-glucosidase
- inhibition by azasugars 322
adhesion 350
adhesion inhibitors 493, 494, 505
adhesion molecules 365
affinity 385, 387, 398, 400, 410
α-galactosidase
- of *Aspergillue orizae* 40
- of *Aspergillus niger* 40
- of bovine liver 40
- of *Escherichia coli* 40
α-glucosidase
- inhibition by carba-sugars 95
- inhibition by cyclic isoureas 103
- inhibitor of 104, 270
α-homonojirimycin 266, 270
aldehyde
- sugar-derived, in intramolecular aldolisation 149
aldol
- intramolecular aldolisation 149
alexine
- ring expanded 341
alkaloids 327
- bicyclic 328
alkoxyphosphonium salt
- in the activation of hydroxyl groups 318
alkyne
- nickel-chrome mediated coupling 24
allosamidin 123, 124
allosamizoline 241

- synthesis of 249
α-mannosidase
- inhibition of 476
α-methyl acarviosin 89
- synthesis of 93
α-methylthreonine 333
amide
- reduction 151
amide linkage 425
amination 266, 271
- reductive 266, 267, 269, 270, 271, 274, 277
aminocyclitol
- allosamizoline 250
- reaction with phenylisocyanate 103
- synthesis of 124
aminocyclohexanes 240
aminocyclopentitols 133, 157, 159, 163, 240
aminocylisation 315
aminoglucoenitol 80
aminosugars 330
amylose 420
anhydrohexitol 553
anomeric carbanion
- by reductive cleavage 21
anomeric effect 71
anomeric lithio derivative 21
anomeric samarium derivative 21
antagonist
- of β-D-glycopyranosides 298
- of glycopyranoside 304
antibiotic 215
anticoagulant 417-420, 426-427
antifungal action 1
antigene therapy 553
anti-inflammatory 493
anti-mite *see* acaricide
antiperiplanar arrangement 35
antiproliferative 420-427
antisense 556
antisense oligomer 553
antisense oligonucleotide 550
- anti-HIV activity 550
- base modified 550

- nitrogen containing linkers 523
antithrombin III 418
antitumor 523
antiviral 215, 239, 523
antiviral activity
- of nucleoside analogues 578
aPTT 427
Arbuzov reaction 75, 76
aristeromycin
- structure 143
arteriosclerosis 417
Aspergillus oryzae
- galactosidase of 514
AT_{III} *see* antithrombin III
australine 270
aza-*C*-disaccharide 19, 40 262-268
- (1→6) 40
aza-*C*-glycosyl *see also* azasugars
aza-*C*-glycosyl compounds 266, 270
- aza-*C*-glycosides 261
- azaglycosyl cyanides 272
- β-linked 274
azadisaccharides
- from bis-aziridines 310
- synthesis of 311
azaglycoside 263
azaseptanose
- 1-deoxy 278
azaseptanoses
- 1-deoxy 261, 276, 278
azasugar
- 1-deoxynojirimycin 260
- aza-disaccharides 260
azasugars 40, 240, 259
- 1-deoxynojirimycin 274
- as β-turn mimics 324
- as inhibitors of glycosidases 321
- azaglycoside 262
- azaglycosyl cyanide 264
- by nucleophilic ring opening of bis-aziridines 308
- enantiopure, synthesis from D-mannitol 307
- piperidine 259, 261, 271
- pyrrolidine 259

- structures of 262
- synthesis from bis-epoxides 314
azepane 273, 277
- containing- pseudodisaccharides 274
azepanes
- as inhibitors of glycosidases 322
- as mimics of oxonium ion 322
- polyhydroxylated 307
  - isomerisation of 320
- synthesis from epoxides 314
azidosphingosine 99
aziridine 275, 283
- containing analogue of cyclophellitol 288
aziridines
- bis-, as precursors of azadisaccharides 310
- bis-, *N*-Boc and *N*-benzyl protected 309
- bis-, nucleophilic ring opening of 308, opening with aminosugars 313
- bis-, opening with cyanide 309
- bis-activation of nitrogen atoms 309
- bis-*N*-Boc, opening by Li_2NiBr_4
- ring opening of 99
aziridinium
- cation, as intermediate 268, 273
- intermediate, in the synthesis of piperidines 317

β (1-4) galactosytransferase
- in enzymatic galactosylation 377
bacterial cell wall 77
bacterial polysaccharides 79
Baeyer-Villiger
- epoxidation 228
Baldwin rules 115
Barbier reaction 268
Barton-McCombie deoxygenation 4
β-D-fructose-2,6-bisphosphate
- activator of phosphofructokinase 67
- inhibitor of phosphatase 67
β-D-galactosidase
- inhibition of 297
β-D-glucosidase

- inhibition of 286
β-D-glycosidase
- inhibition of 298
β-D-mannosidase
- inhibition of 297
benzotriazolyl
- as leaving group 264
benzoylation
- regioselective 269
β-galactosidase 514
- inhibition of 476
β-glucocerebrosidase
- inhibition of by PDMP analogues 100
β-glucosidase
- inhibition of 475
- of almond 40
- of *Caldocellum saccharolyticum* 40
β-hydroxyketone
- from isoxazoline 50, 61
biosynthesis 498
bisglycal systems, synthesis of 7
bis(trimethylstannyl)benzopinacolate 526, 529
bisubstrate analogue 501
block synthesis 421, 422, 428
BODIPY FL *See* fluorescent labelling
borane
- dimethylsulfide complex 337, 338, 339
boric acid
- in the cleavage of isoxazolines 58
boron tribromide
- dealkylation with 338
boron trifluoride 331-334, 339-343
branched-chain carbohydrates, through cross-aldolisation 38
branched-chain cyclohexenone
- synthesis 150
brevetoxin A 1
brevetoxin B 1
brevetoxin B, synthesis of 3, 4
brevetoxin B3 1
brevetoxins
- biological activity 1

- structures of 2
bromomethyl dimethyl silyl ether
- radical cyclisation of 216
bromonium ions
- PM3 calculations 248, 249
β-xylopyranosylnitromethane
- preparation of 51

C,C-trisaccharides 19
caesium fluoride
- in stannylene-mediated benzylation 179
calcitriol
- synthesis of 117
calcium channels 1
calixarenes 407
- sialic acid 407
C-arabinosyl bromide 74
carba-hexofuranose
- conversion into pentofuranose 228
carba-maltoses 94
- with ether link 94
- with imino link 94
- with sulfide link 94
carbamate cyclic 276
carba-nucleosides 157, 224
- nomenclature 224
carba-oligosaccharides
- synthesis of 94
carba-pentofuranose
- strategy of synthesis 226
carba-sugars 87, 143, 146, 283
- 1,2-anhydro-5a-carba-β-D-mannopyranose 95
- 1,6-anhydro
  - as inhibitors 93
- 4a-amino 102
- 5a-carba-α-D-galactopyranose 88
- amino 88
- as glycocerebrosidase inhibitors 99
- carba-maltose
  - inhibitors of α-glucosidase 95
- chemo-enzymatic total synthesis 209
- in glycolipids analogues 97
- in glycosphingolipids analogues 98

- in glycosylceramides analogues 98
- nomenclature 223, 224
- structure of 87, 88
- synthesis of 89, 214
- synthesis of ether-linked carba-disaccharides 96

carba-ulose 231
carba-uridine 229
carbocycle 143
- chiral, formation of 149
- synthesis 143, 162

carbocycles
- from carbohydrates 107

carbocyclic vicinal diol 124
carbocyclic thymidine
- synthesis 538

carbocyclisation 150
- using dithiane 108

carbohydrate amino acid 428
carbohydrate epitope
- non-hydrolysable 19

carbohydrate ligands
- of selectins 365
- three-dimensional structure 349

carbohydrates
- conversion into carbocycles 107
- metabolism 239

castanospermine
- 1-deoxy-8-*epi* 338

C-branched nucleosides 525
- 3'-C-hydroxymethyl-, synthesis of 528

carbohydrate determinants 491, 492
carbohydrate epitope 428
C-disaccharide 19, 428
- (1→5) 32
- 2,6:8,12-bis(anhydro)-7-deoxytredecitols 20
- 3-deoxy-3-(hexopyranosyl)methyl
- 4-deoxy-4-(hexopyranosyl)methyl
- 5-deoxy-5-(hexofuranosyl)methyl 20
- 8,12-anhydro-7-deoxytredecose 19, 20
- β-(1→6) 19
- of 1,5-imino-galactoside 41

C-disaccharides
- aldol route to 49
- aminomethylene-linked 56
- carbonyl-linked 52, 59
- Diels-Alder route to 50
- having functionalised bridge 50
- hydroxymethylene-linked 49, 57, 59
  - conformation of 60
- isoxazoline route to 49
- methylene-linked 49
- nitrile oxide route to 49
- nitroaldol route to 49
- radical route to 49
- Wittig route to 49
- X-ray crystal structure of 54

cell surface glycans
- mimics of 95

cell-mediated processes 19
cellobiose 275, 402, 422
- analogue of 273

cellular nucleases
- stability of nucleotides mimics to 523

C-galactopyranoside
- of D-mannose 32
- of D-mannoside 35
- of L-mannose 32, 33

C-glycoside
- of carbapyranoses 19
- of carbasugars 35

C-glycosyl aldehyde 73
C-glycosyl halide 73
C-glycosylation 69
C-glycosyl-mercurioderivative 74
chair conformation
- of pyranosyl nucleoside analogues 559, 561

Channels In Conformational space Analyzed by Driver Approach *see also CICADA*
chitin 250
chitinase
- inhibitors 241
  - as fongicides 250

cholesterol 6
- thiobenzoate of 6

chondroitin sulfate 418
chromium-nickel coupling 4
CICADA 352
Ciguatoxin 2, 15
circular dichroïsm
- of nucleotides analogues 570, 571
cis-dihydroxylation
- using osmium tetroxide 228
C-kojibiose 28
C-lactose 25
clotting 426, 427
cluster of
- galactose 407
- mannose 397, 402
- sialic acid 393
- sialyl Lewis^X 388, 394, 395
cluster effect 386-395, 410
- calixarenes 407
- cyclodextrins 387, 407
- glycoclusters 386-387
- glycodendrimers 386-387, 396-405
- glycopeptides 387, 391
- glycopeptoids 387, 391, 393
- glycopolymers 385-390, 409
- glycotelomers 387, 390
- liposomes 387, 388
- neoglycoproteins 385-388
- receptors 385-386, 393, 409
combine-split 344
complex
- enzyme-substrat 518
concanavalin A 395, 402
conduritol A
- synthesis of 211
conduritol B
- epoxide of 449
conduritol C
- synthesis of 214
conduritol epoxides
- as glycosidases inhibitor 473
conduritol F
- synthesis of 214
conduritol F-(-) 130
conduritols 433
- A-F, structures of 434
- chemo-enzymatic total synthesis 209

- naturally occurring 213
- synthesis of 213, 433
conformation 420
- optimisation of, by computational methods 352
conformational changes
- of DNA, RNA 555
- of sugars 555
conformational flexibility 349
Connolly surface 352
copper(I) 7, 10
C-ribosyl phosphonate 71
C-ristobiose 28
CRS-(carboxyl-reduced sulfated) heparin 427
C-sucrose 24
cuprate addition 4
cyclic enol ether 11
- from olefinic ester 11
- preparation via metathesis 11
- preparation via Stille coupling 7, 8
cyclic ether 1
- synthesis of 3
   - from thionolactone 4
   - via olefin ester metathesis 11
   - via Stille coupling 7
cyclic sulfates
- 5-membered, monohydrolysis of 371
- in regioselective sulfation 371
- nucleophilic opening 371
cyclic sulfites
- oxidation in sulfates 371
cyclisation 262-270
- *5-exo-tet* 308
- *6-endo-tet* 308
- mercury(II)-mediated 271
- NIS mediated- 267
cyclitol
- diamino 218
cycloaddition 437
- [3+2] 283
- [4+2] 242
- 1,3-dipolar 26
- intramolecular 284
- of nitrile oxides 50
cycloalkene

- functionalisation of 241
cyclobutane 109, 111
- synthesis of 113, 114
cyclodextrin
- galactose 407
- glucose 407
- mannose 407
cyclodextrins 387, 407
cycloheptane
- synthesis of 114, 119
cyclohexane 214
- 1,2,3-triol, synthesis of 209
- branched-
  - synthesis of 144
- synthesis of 114
cyclohexene
- branched- 144
cyclohexenone 150
- branched-chain, as carbasugar precursor 144
- synthesis 145
cyclopentadiene
- as starting material 250
cyclopentane 109
- formation of 137
- pentaol- 252
- synthesis of 109
cyclopentanes
- from D-glucose derivatives 173, 184-189
cyclopentenes 241
- electrophilic additions to 247
cyclopentenone 152, 153
- synthesis 152
cyclopentitol
- synthesis of
  - by pinacol coupling 130
cyclophellitol 283-286
- inhibition of glycosidases by 288, 289
- synthesis of 286
- synthesis of analogues 286
cyclopropane 110
- synthesis of 110, 112

D-2-deoxy-*myo*-inositol 1,3,4,5-tetrakisphosphate 173
D-arabinose 133
- imine 340
DBU
- catalyst for intramolecular aldolisation 150
- ring expansion with 337, 340
deallylation 451
degree of sulfation 419
DEIPS *see* diethylisopropylsilyl
dendrimers
- 3,3'-iminobis(propylamine) 400
- 3'-sulfo Lewisx 398
- bi-directional dendrimers 400
- cellobiose 402
- dendrons 405
- galactose 402
- gallic acid 400
- lactose 402
- L-lysine 396
- maltose 402
- *N*-acetylglucosamine 398
- *N*-acetyllactosamine 397
- PAMAM 402
- phosphotriester 400
- sialic acid 397
- spherical 405
- T-antigen 398
- T_N-antigen 402
deoxygenation 422
deoxynucleic acid *see* DNA
deoxyribonucleic acid *see* DNA
dermatan sulfate 418
Dess-Martin reagent 14
desulfurization 6
- cholesterol, preparation of 6
- mechanism 6
- of dithiane 113, 116
- tin hydride mediated 6
D-gluconamidrazone
- as glycosidases inhibitor 478
D-glucono-1,5-lactam
- as glycosidases inhibitor 477
D-glucose 132
D-glyceraldehyde

- 2,3-O-isopropylidene 331, 335, 343
D-gulonolactone
- 2,3:5,6-di-O-isopropylidene 146
diabete 433, 441
- complications 433
dialdose
- 7-deoxytrideca-
  - via cycloaddition 61
dialdoses 60 see also higher-carbon dialdoses
- 6-deoxydodeca-
  - via cycloaddition 63
- trideca- 61
diazo transfer 450, 460
diborane
- reduction of oximes with 83
dibutyltin-bis-acetyl-acetonate 448
dibutyltin oxide 109, 369
dicarboxylic acid 425
Diels-Alder 435, 439
- between furan and acrylic acid 89
- endo adduct 89
diethylaluminium cyanide
- opening of bis-aziridines 309
diethylisopropylsilyl
- as protecting group 285
dihydroxylation 211, 331, 332
- of cyclopentene derivatives 244
- of olefins 243
- with $KMnO_4$ 332
- syn 214
- with OsO_4 245
diisobutylaluminium hydride
- reduction with 334
di-O-isopropylidene-2-deoxy-2-amino-α-D glucose
- opening of bis-aziridines with 313
diol
- formation from olefin 214
diphenyl phosphite 73
disaccharide
- initial rate of formation 516
- staggered orientation 351
- synthesis of 512
disaccharide mimics 35, 40
dithiane

- epoxy 117
- in the conversion of carbohydrates into carbocycles 107
- lithio derivative 109, 118
dithioacetal
- of 3,4-di-O-benzyl-5-O-trityl-D-arabinose, lithium salt 20
- synthesis of 109
dithionolactone, bridging reaction of 3
divinylzinc 75, 81, 269
DL-6-deoxy-6-hydroxymethyl-scyllo-inositol-1,2,4-trisphosphate
- as mimic of $Ins(1,4,5)P_3$ and adenophostin A 198
δ–lactam 303
D-mannitol
- as starting material 126, 308
D-mannose 132
- as starting material 4
D-mannosyl cation
- mimics of 93
DMDP 327
D-myo-inositol 1,3,4,5-tetrakisphosphate see $Ins(1,3,4,5)P_4$
D-myo-inositol 1,4,5-trisphosphate see $Ins(1,4,5)P_3$
DNA 523
- conformational changes 555
- structure 555
dolichyl diphosphate 68
dolichyl-diphosphate-oligosaccharide 77
double glycosylation 457, 458
double stranded DNA 553, 554
D-ribofuranose
- as starting material 295
D-ribose 133
DS see degree of sulfation
D-serinal 334
D-sorbitol
- as starting material 128
D-threo-1-phenyl-2-decanoyl-amino-3-morpholino-1-propanol see PDMP
D-xylose
- as starting material 51

electrostatic potential 352

ELLA *see* enzyme linked lectin assay
endoperoxide 251
- reduction of 251
endothelial cells 350, 365
energy maps 353
- of αFuc(1-3)βGlcNAc 355
- of αNeuNAc(2-3)βGal 354
- of βGal(1-4)βGlcNAc 356
energy minimization 560
enol ether olefin 11
enol triflate 8
enolate
- lithium, reaction with sugar lactones 146
enone
- 3-methylidene-7-oxabicyclo[2.2.1]heptan-2-one 30
enzyme 394, 398, 409
- specificity 516
enzyme linked lectin assay 394, 398, 409
epimerisation 56
epoxidation 167, 214, 449
- by peracids 212, 251
- *syn*- 218
epoxide
- 1,2:5,6-dianhydro-3,4-*O*-benzyl-D-iditol 314
- 1,2:5,6-dianhydro-3,4-*O*-benzyl-D-mannitol 314
- 1,2:5,6-dianhydro-3,4-*O*-isopropylidene-D-mannitol 314
- 1,2:5,6-dianhydro-3,4-*O*-isopropylidene-L-iditol 314
- bis-, nucleophilic ring opening of
- BIS, opening with 1,2-diaminoethane 314
- bis-, opening with amines 317
- bis-, opening with benzylamine 316
- bis-, opening with tryptamine 314
- bis-, synthesis of 314
- formation of 119
- hydrolytic opening 210
- opening of 218
- opening with azide ion 251
- ring opening of 90, 91
- ring opening with alcohols 90
- ring opening with amines 97
- synthesis of 110
Escherichia coli 402
- β-galactosidase of 514
E-selectin 350, 365
- binding site 360
- crystal structure 350
- ligand, conformation in solution 350
- ligands of 365
- sulfated Lewis[a] oligosaccharides, affinity for 379
- sulfated Lewis[x] oligosaccharides, affinity for 380
ester
- enzymatic formation of acetate 211
- enzymatic hydrolysis 210
esterphosphonate-ketone condensation 3
ethyl diphenyl phosphite
- Arbuzov reaction with 75
exoglycosidase 446

Ferrier carbocyclisation 144, 449, 460
Ferrier rearrangement 53
- in preparation of chiral inositol polyphosphates 173, 174, 175, 176, 180, 181, 183
- influence of protecting groups on stereochemical outcome of 181
- of D-xylal 563
flexibility values 352
fluorescent labelling
- of Ins(1,4,5)P$_3$ mimic 199
Fmoc 428
formaldehyde 527
fortamine-(+) 219
- chemo-enzymatic total synthesis 209
- synthesis of 218
fortimycin B 219
fortimycin A
- structure of 218
free radical cyclisation 157, 168
fructose 1,6-bisphosphatase 67
fuco-ketodisaccharide 505
fucosidase
- inhibition by azasugars 322

fucosyl bromide 367
fucosylation 496
fucosyltransferase 498
- human milk α-(1-3/4) 376, 379
- human milk α-(1-3/4), in glycosylation 376
- inhibitors 491, 498
furan 327
furanose
- conformational flexibility 555
furanose mimics 553
furfural 329

gabosine
- structure 143
galactose 402, 407
galactosyl-enzyme intermediate 512
galactosylation
- enzymatic 377
galactosyltransferase 76
gallic acid 399, 400
gambieric acid 2
γ-aminoalcohol
- from isoxazoline 50
GDP-fucose 498
- in enzymatic glycosylation 376
- analogue 499, 500
gentamycin 240
Giese's glycosylation *see* radical, glycosylation
glucamine 425, 427
gluconamidine
- N-alkyl, inhibition of glycosidases by 484
- N-alkyl, synthesis of 483
gluconaminidine
- N1-alkyl, as inhibitor of glycosidases 483
glucosamine 417
glucose 402, 407
glucosidase
- inactivator of 275
glucosylceramide synthase
- inhibitor of 100
glucuronic acid 417

glycal
- 1-C-lithio derivative 25
glycal-1-ylmethyl phosphonate 76
GlyCAM-1 372
glycocerebrosidase
- inhibitors of 99
glycocerebroside synthase
- inhibitors of 99
glycocluster 386, 387
glycodendrimer 397
glycohydrolase
- inhibition of 40
glycol cleavage 149
glycolipids
- marine-, structure of 252
glycolipids analogues
- as immunomodulators 97
glycomimetics 239
- as carbohydrate metabolism inhibitors 239
glyconolactone 70
glycoproteins 491-492
glycosaminoglycans 417, 418
glycosidases 239, 512
- active site 472
- endoglycosidases 239
- exoglycosidases 239
- inactivator of 264
- inhibition of 260, 276, 277, 284, 297, 322
    - catalytic mechanism 476
    - transition state 466
glycosidase inhibitors 87, 123, 157, 165, 283, 284, 463
- azasugars 307, 321
- basic sugar analogues 476
- catalytic efficency 464
- effect of basic centre 474
- immunoregulatory effects of 242
- substrate related 463
glycosidic linkage
- conformational flexibility around 359
- torsion angles 351
glycosphingolipids 98, 492
- analogues of 97

glycosyl acceptor 422
glycosyl bromide 21
glycosyl cation
– mimics of 93
– transition state analogue 468
glycosyl donor 302
glycosyl fluoride 460
glycosyl phosphates
– as glycosyl donors 67
– as metabolic regulators 67
glycosyl phosphatidyl inositol 123, 443, 459
– of rat brain Thy-1 antigen 444
– of *Trypanosoma brucei* 444
– of VSG 444
– structure of 444
glycosyl samarium derivative 29
– addition to *C*-formyl sugars 29
glycosyl transferases 239
glycosylamides 98
– as analogues of glycolipids 97
glycosylamines 264
glycosylation 422, 447, 449-451, 455, 456, 496
– by enzymes 68
– effect of substituent on the regioselectivity of 516
– regioselectivity of 513-517
– with sulfoxides 457
– with trichloracetimidates 450
glycosylceramide analogues
– as inhibitors of β-galactocerebrosidase 99
– as inhibitors of β-glucocerebrosidase 99
glycosylceramides 98
glycosylphosphatidylinositol 123
glycosyltransferase 498
– inhibition by azasugars 307
GM$_3$ 389
GPI *see* glycosylphosphatidylinositol
Grignard reagent 70
gualamycin 283
– anti-mite activity 284

guanosine 5'-(β-L-fucopyranosyl) diphosphate *see* GDP
Gymnena sylvestre 433

halocyclization 74
halogenoses
– carba-analogues 229
halopyranoside
– ring contraction 125
hemibrevetoxin B 2
Henry's condensation 25
heparan sulfate 417-418
heparin 417-427
heparin cofactor II 426
heparin-binding proteins 417
heparinoid 417
hepatocytes 393
hetero-coupling 8, 9
Hevizos 549
– enzymatic cleavage 549
– poly(dA-d5U)s 548
– synthesis by polymerase I 545
– thermal stability 548
hexamethylphosphoramide 131
hexitol, 1-deoxy-1-nitro
– synthesis of *51*
hexopyranosides
– 6-deoxy-6-iodo
  – reaction with samarium diiodide 136
– ring contraction of 135
higher-carbon dialdoses
– synthesis of 60
– via cycloaddition 61
HMPA see hexamethylphosphoramide
homoazaglycosides 270
homoazasugars 261, 264, 270-276
– β-homonojirimycin 271-274
– homoazadisaccharide 272, 269-273
– homoazaglycosides 269-273
– homogalactostatin 274-276
homo-coupling 7
Horner-Emmons olefination 71
– intramolecular 145
HPLC

- analysis of nucleosides 345
hybridization 527
- of antisense oligonucleotides 523
hydroboration 14
- of cyclohexene derivatives 216
hydrogen bonding 352
hydrogen bond interactions
- in RNA 556
hydrogenation 452
hydrogenolysis
- of benzyl ethers 219
- of isoxazoline 54
hydrolysis
- of ester by enzymes 209
hydrolysis of β-glucosides
- activation parameters 464
- isotope effect 470
- rate constant 464
hydroxy dithioketal cyclization 3
hydroxy epoxide cyclization 3
hydroxy ketone, reductive cyclization 3
hydroxyl group
- activation of 314, 318
hydroxylamine 124, 132
- reduction of 134

iduronic acid 417
imidazole 284
indolizidine 336
indolizidines
- polyhydroxylated 307
Influenza virus 387, 398
inhibitor
- of α-galactosidase 259
- of β-glucosidase 259
- of α-mannosidase 259
- of β-galactosidase 259
- of glycosidase 259, 272, 276
inositol
- D-*chiro*- 126
- D-*chiro*, 6-*O*-glycoside of 450, 451
- D-*chiro*, enantiomerically pure 447
- D-*chiro*, synthesis of 447
- D-*myo*, enantiomerically pure 447
- L-*chiro*- 128

- *myo* 125, 444
- *myo*, 1-O-menthyloxy derivative of 448
- *myo*, 3-O-menthyloxy derivative of 448
- *myo*, 4-O-glucosaminyl-1-phosphate derivative of 452
- *myo*, 6-O-glucosaminyl-1,2-cyclic phosphate derivative of 452
- *myo*, 6-O-glucosaminyl-1-phosphate derivative of 452
- *myo*, 6-O-glycosyl derivative of 449
- *myo*, acylation of 448
- *myo*, alkylation of 448
- *myo*, diethyl boryl derivative of 448
- *myo*, glycosylation of 448, 449
- *myo*, regioselective protection 447
- *myo*, resolution of 448
- *myo*, stannylation of 448
- *myo*, transmetallation of 448
- *scyllo*- 126
inositolphosphoglycans 443, 446, 459
- inhibition of c-AMP by 444
- inhibition of pyruvate deshydrogenase by 445
- mediators of insulin action 446
- synthetic approach to 447
- sequence analysis 446
- structural characterisation 446
Inositol polyphosphates 171
Ins(1,3,4,5)P$_4$ 171-175, 184
Ins(1,4,5)P$_3$ 171-176, 184-92, 198, 199
- ring-contracted mimics of 184
inside alkoxy effect
- in cycloaddition 58
insulin 433-460
- putative second messengers of 123
- secretion 436-441
intercellular communication 19
interstrand phosphate repulsion 556
intestinal lactase 511
- deficiency 511
- evaluation of activity 511
intramolecular aldol 143
isourea 101
- *N*-alkyl cyclic derivatives 103

- as glycosidase inhibitors 103
isoxazole, 4,5-dihydro 49
isoxazoline 49, 284
- hydrolytic cleavage 58, 61
- in the synthesis of C-disaccharides
- fiflir analysis of 54
- N-O bond cleavage 50
- reductive cleavage 54
- X-ray crystallography 61

Jack bean
- galactosidase of 514

keratan sulfate 418
keruffarides
- synthetic studies 252
ketals
- temporary connection using 29
ketones
- reduction 59
- epoxidation of 233-235
ketyl radical
- reactions of 123
Knoevenagel condensation
- intramolecular 154

labelled glycans 446
lac operon 519
lactacystin 327
lactase activity 521
lactol
- oxime from sugar- 124
lactone
- 1,4-anhydrourono-, Bayer-Villiger oxidation of 29
- 2,3:6,7-di-O-isopropylidene-D-glycero-D-gulo-heptono-1,4- 40
- 2,3-O-isopropylidene-D-ribono-1,4-, reaction with enolates 147
- 2-deoxy-2-methylidene-D-erythro-pyrano- 27
- carbafuranose from 153
- formation of 3
- hydroxyimino, as inhibitor of glycosidases 472

- sugar-derived, reaction with enolates 148
lactone, conversion to thionolactone 7
lacto-N-tetraose 389
lactose 389, 390, 391, 402, 511
lamivudine 327
L-arabinose
- 2,4-diamino-2,4-dideoxy 331
lectin 391, 394, 398, 401-409
- Concanavalin A 395, 402
- *Limax flavus* 401, 402
- *Pisum Sativum (pea):* 405
- Wheat germ agglutinin 408
lectin domain 365
Leloir pathway 68
leukocyte 491-494
- adhesion to endothelial cells 350
- in inflammation 362
- extravasation 491
- migration of, in inflammation 365
Lewis acid
- catalysis by 329
Lewis[a] tetrasaccharide
- 3-sulfate 366
Lewis[a] trisaccharide
- conformation by nmr 367
Lewis[x] tetrasaccharide
- 3-sulfate 366
Lewis[x] trisaccharide
- crystal structure 356
L-glucose 284
L-glyceraldehyde
- 2,3-O-isopropylidene 338, 343
L-iditol
- from D-mannitol 126
Limax flavus 401, 402
linker 425, 426
lipophilicity potential 352
- Fermi type function 352
liposomes 387, 388
lithium
- chelation control 291
lithium anion
- condensation with lactones 153
lithium azide
- in the opening of cyclic sulfates 218

lithiumtriethylborohydride 336
long-chain carbohydrates 19-20
long-chain sugar *see* long-chain carbohydrates
low energy conformers 352
L-proline
- *trans*-2,3-*cis*-3,4-dihydroxy 335
L-selectin 365, 372
L-serinal 332
L-threose
- protected derivative 338
L-xylofuranose
- as starting material 295
lymphatic tissues
- in inflammation 365

Maitotoxin 1, 7, 8, 10, 15
- biological activity of 1
- NO-PQ ring system, synthesis of 8
- structure of 2
maltase
- inhibition of 284, 304
maltose 402, 422, 427
- analogue of 273
- homoaza analog of 278
mandelic acid
- as starting material 243
mannopyranosyl cation 242
mannose 386, 397, 402, 407
mannosidase
- inhibition by azasugars 322
mannostatin
- synthesis of 241
mannostatin A 123, 124, 143, 241
- as inhibitor of α-mannosidases 241
- as inhibitor of mannosidase II 241
- structure 241
- total syntheses 242
mannostatin B 241
marine toxins 1 *see* brevetoxins
- ciguatoxin 2
- gambieric acid 2
- maitotoxin 2
- red tide 2
mCPBA (m-chloroperbenzoic acid) 434-438

membrane anchors 123
mercuric trifluoroacetate 80
mercurio-cyclization 73, 81
mercury (II) salts
- in conversion of hex-5-enopyranosides to inososes 180
mercury (II) trifluoroacetate
- in cyclisation 276
messenger RNA *see* mRNA
metathesis 11
- molybdenum-mediated 13
- norbornene skeleton 13
- olefin-ester 11
- one pot procedure 13
- ring closing metathesis (RCM) 11
- Tebbe reagent 11
- titanacyclobutane 12
- titanium methylidene 13
methyl 4,6-*O*-benzylidene-α-D-glucopyranoside
- improved preparation of 176
- selective alkylation and esterification of 178
methyl 4,6-*O*-benzylidene-α-D-mannopyranoside
- improved preparation of 176
methyl acarviosin
- chemical modifications 93
methyl α-D-glucopyranoside 176
methylenetriphenylphosphorane 73
methyloxime 82
Michael addition 3
Michael cyclization 71
mimetic 417, 428, 429
mimetic design 350
minor groove
- of hybrids with RNA 532
Mitsunobu reaction 181, 276, 525-528
MM3 351
modelling studies
- of αNeuNAc(2-3)βGal 353
molecular dynamics 560-561
- simulations 367
molecular mechanics
- calculations 362

mRNA 523
Mukaiyama esterification 13
Mukaiyama aldolisation 332
multivalency *see* cluster effect
multivalent effect *see* cluster effect
mureine 77
Mytilus edulis 335

N,N-diethylacetamide
- lithium anion 146
N-acetyl-α-D-glucosamine 1-phosphate 77
N-acetyl-α-D-mannosamine 1-phosphate 77
N-acetyl-β-D-glucosaminidase
- inhibition of 297, 298
N-acetyl-D-galactosamine 293
N-acetylgalactosamine 400
N-acetylglucosamine 398
N-acetyllactosamine 389
N-acetylneuraminic acid 78
nagstatin 283
- as inhibitor of N-acetyl-β-D-glucosaminidase 290
- de-branched analogue 293
- D-*gluco* analogue 295
- D-*manno* analogue 295
- D-*talo* analogue 295
- inhibition of glycosidases by 289
- L-*galacto* analogue 295
- N-acetyl-D-glucosamine analogue 295
- N-acetyl-L-galactosamine analogue 295
- structure of 290
naked sugar 29
naked sugars 19
N-benzyl-(2,3,5-tri-O-benzyl-D-arabinofuranosyl)amine
- reaction with vinyl Grignard 80
N-bromosuccinimide 180
N-butyl deoxynojirimycin 103
n-butyl lithium 109
neoglycolipid 387
neoglycolipids 381
neoglycoprotein 385, 387

neplanocin-A 143
neurotoxicity 1
N-iodosuccinimide (NIS)
- as glycosylation promoter 302
- in cyclization 267
nitrile oxide
- cycloaddition 26, 49, 50, 57
  - regioselectivity 53
  - stereoselectivity 53
- cycloaddition with olefin 284
- derived from D-arabinose 52
- derived from D-galactose 52
- derived from D-mannose 52
- derived from D-xylose 51, 53, 56
- of sugars, preparation of 51
- pyranose-1-carbo- 52
- pyranose-1-carbo-, cycloaddition of 53
- TDI-mediated preparation of 57
nitro sugar
- 2,3:5,6-di-O-isopropylidene-1-deoxy-1-nitro-D-manno-furanose 20
nitrogen heterocycles
- 5, 6, or 7 membered-ring, synthesis of 308
- as scaffolds 324
nitromethane
- condensation of, on D-xylose 51
N-linked glycoprotein 77
N-methyl, N-methoxy acetamide
- lithium anion 153
NMR studies
- of αNeuNAc(2-3)βGal 353
nojirimycin 327
- 1-deoxy
  - as glycosidases inhibitor 475
- as glycosidases inhibitor 475
normuramic acid 428
N-tert-butoxycarbonyl-2-(tert-butyldimethylsiloxy)pyrrole 328-343
N-tritylimidazole
- lithio derivative 291
nucleoside
- cytidine derivatives 343
- furanose ring conformation 574
- library of 343

- mimics 328
- nitrogen-containing mimics 341
- polyhydroxylated 328
- pyrimidine derivatives 343
- sulfur-containing mimics 341
- thymidine derivatives 343
- uridine derivatives 343

nucleoside analogues
- 1,5-anhydrohexitol 562
- 1,5-anhydrohexitol, antiviral activity 578
- 1,5-anhydrohexitol, synthesis of 565
- 1,5-anhydrohexitols, conformation of 567
- 3'-C-branched-chain analogues 562
- 3'-branched pentopyranosyl analogues 562
- hexitol-containing 573
- pyranosyl 558, 559, 560, 564
- pyranosyl, conformation of 565
- pyranosyl, orientation of the base 564

nucleoside diphospho sugar 68

nucleotide analogues
- 1,5-anhydrohexitol, structure of 574

nucleotide diphosphate 68
nucleotide mimic 554, 556

olefin-olefin metathesis 11
oligomerization 428
oligonucleotides 553
- cleavage of, by phosphodiesterases 554
- with six-membered carbohydrate rings 558

oligonucleotides analogues
- 2'OMe 3'-O-N-C linkage 532
- 2'OMe methylene(methylimino) linkage 532
- 3'-C-C-N linkage 531
- 3'-C-N-C linkage 531
- 3'-C-N-N linkage 527
- 3'-C-N-O linkage 525
- 3'-C-O-N linkage 528
- 3'-O-N-C linkage 529
- automated synthesis 569
- building blocks 569
- duplex with DNA, stability of 569
- hybridization of 526
- hydroxy(methyliminomethylene) linkage 529
- linkage, structure 524
- methylene(dimethylhydrazo) linkage 527
- methylene(methylimino) linkage 527
- methyleneoxy(methylimino) linkage 528
- methylenimino linkage 526
- oxime linkage 525

oligosaccharides
- biologically active conformation 350
- lowest energy minimum 349

o-nitrophenyl β-D-galactopyranoside 513
open chain sugar 425, 426, 427
optical purity
- determination of 210

osmium tetraoxide
- in conversion of allyl glycoside to 2-hydroxyethyl glycoside 197

osmylation 166
ovarian cystadenoma glycoprotein 366
oxidation
- Bayer-Villiger 29
- with mCPBA 38
- of sulfites with $RuCl_3$-$NaIO_4$ 371
- with osmium tetroxide 437, 438
- Swern oxidation 216, 301
- with Fetizon's reagent 296

oxidative cleavage
- of acetal-protected diols using periodic acid 152
- of diols 149
- with RuO_4-$NaIO_4$ 39
- with sodium periodate 331

oxime 525
- catalytic reduction of 82
- ether-
  - as precursor of carbocycles 124, 131, 138
  - cross-coupling 131
  - cyclisation of 157
- reduction of 254

- reductive cross-coupling 124
oxocene 4
oxonium ion
- mimics of 322
oxygenated cyclohexane 284
ozone
- double bond cleavage with 253

palladium (0) 7
palladium black 315
palladium tetrakis-triphenylphosphine 9
PAMAM 402, 403, 405
pancreatic islets 436
paramethoxybenzyl
- anomeric protecting group 368, 373
PDMP
- analogues 100
- as inhibitor of glucosylceramide synthase 100
peptides, cyclic glyco- 387, 395
- sialyl Lewis^x 395
peptoids 387, 393
- sialic 393
periodic acid
- oxidative cleavage 152
π-facial discrimination
- in cycloaddition of nitrile oxides 58
phosphitylation
- in inositol polyphosphate preparation 183, 188, 190
phosphodiester linkage
- analogues 523
- in nucleotides 523
phosphodiesterase 554
- degradation of oligonucleotide analogues by 572
phosphofructokinase 67
phospholipase 443, 459
phosphonate 268
phosphono analogue 68
- angles 69
- bond lenght 69
- isosteric 69
- non-isosteric 69
- of α-D-glucose 1-phosphate 70, 76

- of α-D-mannosamine 1-phosphate 81
- of α-D-ribose 1-phosphate 76
- of β-D-arabinose 1-phosphate 76
- of β-D-fructose 2,6-bisphosphate 73, 75
- of β-D-mannose 1-phosphate 76
- of glucose 6-phosphate 69
- of N-acetyl-α-D-glucosamine 1-phosphate 82
- of N-acetyl-α-D-mannosamine 1-phosphate 82
- of ribose 5-phosphate 69
phosphorylation 448, 449
pinacol coupling 132 see reductive coupling of C=O/C=O
- intramolecular 138
- SmI$_2$ promoted 124
piperidine 262, 264, 267, 274, 275
- α-alkoxy 263
- C-formyl derivative 265
- polyhydroxy- 264
piperidines
- polyhydroxylated 307
- polyhydroxylated, isomerisation of 317, 318
- ring contraction of 319
Pisum Sativum (pea) 405
platelets 365
p-methoxybenzylidene acetal
- selective cleavage of 186
polyacrylamide 389
polyamidoamine dendrimers see PAMAM
polyether
- synthesis 1
polyether natural products 1
polymerase
- substrate specificity 542
polymers 397, 398
- 3'-sulfo-Lewis^x 388
- GM$_3$ 389
- lacto-N-tetraose 389
- lactose 389
- N-acetyllactosamine 389
- sialyl Lewis^A 388

- sialyl Lewis^X 388
- T-antigen 388

polyoxin J 327
potassium carbonate 9
promoter 422
protective groups 497, 502
protein receptor 349
prumycin 332
P-selectin 365
pseudo-oligosaccharide
- as inhibitors 88

pseudo-sugars 87
pseudo-trisaccharide 88, 250
pyranose
- as mimics of furanose 556
- conformational changes in 555
- ring contraction of 125

pyranosylnitromethanes 51
pyrogallol 210
- catalytic hydrogenation of 210

pyrrole 327
pyrrolidine 264, 271, 284
- 2,5-disubstituted 309
- aglycon 300
- by intramolecular heterocyclisation 309
- by ring contraction of polyhydroxylated piperidines 319
- C-vinyl derivative of 270
- hydroxylated, as inhibitor of glycosidase 479
- polyhydroxylated 307
  - substituted at C-2 310
- thioglycosyl substituted 311

pyrrolizidine 336
- cis-1,2-dihydroxy 337

radical
- glucopyranosyl 27
- glycosidation using 27, 29
- pyranosyl 28

radical coupling 529, 532
- of iodo nucleosides 526
- of oxime nucleosides 526

radical cyclisation 123
- of bromoacetals 216

radical deoxygenation 118, 566
radical desulfurization 6
- of thionoester 6
- of thionolactone 7

radical glycosidation 34
Raney nickel
- desulfurization of dithianes 113
- reductive cleavage of isoxazoline 54
- reduction of oximes 82

rational design 349
RCM see ring closing metathesis
red tide 2
reduction 437
- of amide 151
- Luche's conditions 151
- of lactone to lactol 132
- of substituted cyclopentanone 255
- sodiumborohydride 343
- stereoselective, with sodium borohydride 437

reductive
- carbocyclisation 124, 125
  - of carbohydrates 123
- cleavage 134
  - of N-O bonds 138
- coupling of C=O/C=NOR 125
- coupling of C=O/C=O 125
- dealkoxyhalogenation of 6-halopyranosides 135
- dealkoxyhalogenation using samarium diiodide 135
- elimination 125

reductive amination 266 see also amination
- of cyclopentanones 252
- of ketone 56

reductive cyclization 3
reductive methylation 530
- of amine 527

regioselective
- sulfation 369, 374
- glycosylation 516

restenosis 417
r_i value 424, 425, 426
ribofuranosyl phosphonates 72
ribonucleic acid see RNA

ribozyme 553
ribonolactone D and L
- as starting material 242
ring closing metathesis 11
RNA
- conformational changes 555
- double stranded,
  - A and B families 556
  - helical conformation of 556
  - three dimensional structure of 556
- helical single stranded 556
- receptor, conformationally constrained 556
- structure 555
RNA-DNA duplex 553
RNA-receptor
- design of 553
RNase H 553
rolling adhesion 491
ruthenium dioxide 332, 335

salbostatin 89
- synthesis of 91, 92
samarium diiodide 123, 165, 170
- in the formation of cyclitols 138
- in the reduction of hydroxylamines 134
- mediated ring contraction 189, 190
samarium(III) alkoxide intermediate 137
second messenger 443-454, 459
- of cytokin 443
- of growth factors 443
- of insulin 443
selectin 389, 394-398, 491-494
selective bromination 295
selective tosylation
- using tin derivatives 109
selectivity 426-427
semiempirical calculations
- on bromonium ion 248
sialic acid 78, 393, 397, 496-498
sialyl Lewis^x 349-350, 492
- bioactive conformation 359
- conformation by nmr 351

- conformation, comparison of data 358
- conformational families 357
- conformational features 362
- conformational investigation, by molecular dynamics 351
- conformational space studies 350-361
- conformers 356
- conformers, energy of 353
- flexibility indexes 359
- lactosamine linkage 355
- low energy conformers 350
- mimics 496, 505
- non-bonded interactions in 356
- potential energy calculations 351
- refined geometry 351
sialylated Lewis^a trisaccharide
- conformation by nmr 367
sialyloligosaccharides 388
sigmatropic shift [2,3] 216
signal transduction mechanism 123
silkworm trehalase
- inhibition of , by trehazolin analogues 101
siloxydienes 327
- γ-substituted 333
- strategy 330
silver triflate
- as glycosation promoter 302
single stranded RNA 553-554
single-stranded nucleic acids mimics 523
singlet oxygen 251
SmI$_2$ see samarium diiodide:in Barbier reaction see samarium diiodide
SMC see smooth muscle cell
smooth muscle cell (SMC) 417-424
S$_N$2 displacement
- by dithiane 108, 110
SO$_3$-NMe$_3$ complex
- sulfation with 367-377
sodium amalgam 337
sodium channels 1
sodium hydroxide
- in the opening of cyclic sulfates 371
solid phase synthesis 428-429
somatostatin 320, 323

- non-peptide mimics 320
spaced saccharide 424, 427
stabilized ylide 71
stacking interactions
- in RNA 556
stannyl enol ether 8, 9
stannylene
- in regioselective sulfation 374-378
stannylene acetal
- for selective alkylation/esterification 178-179, 195-196
stereoselective reduction 449
Stille coupling 7, 9
- 1,1'-bistetrahydropyran system, synthesis of 7
- 1-stannyl glycal 7
- bisglycal system, preparation of 7
- copper(I) facilitated 10
- copper(I) chloride 10
- enol triflate 8
- hetero-coupled product 9
- palladium tetrakis-triphenylphosphine 9
- palladium(0) 7
- potassium carbonate 9
- stannyl enol ether 9
streptomycin 240
sugar aldehyde
   - 1,2:3,4-di-O-isopropylidene-α-D-galacto-hexodialdo-1,5-pyranose 20
sugar lactone 146
sulfate 417-420, 424-427
- cyclic, opening of 218
- essential sulfates 422
sulfated Lewis[a] trisaccharide
- conformation by nmr 367
sulfation 427
sulfite
- formation of 119
- nucleophilic opening 119
sulfur ylide
- epoxidation with 231
swainsonine
- 1-*epi* 339
sweetener 215

Swern oxidation 126, 128
syn-bromonium ions 247
synthesis
- combinatorial 345
- parallel 343
synthesis of
- 1,1'-bistetrahydropyran, 7
- 1,5-anhydrohexitol nucleoside 565
- 1,6-*epi*-cyclophellitol. 286
- 3',6'-disulfated Lewis[x] pentasaccharide 373
- 3',6-disulfated Lewis[x] pentasaccharide 376
- 3'-*C*-hydroxymethyl-*C*-branched nucleosides 528
- 3'-sulfated Lewis[a] pentasaccharide 368
- 3'-sulfated Lewis[a] trisaccharide 367, 370
- 3'-sulfated Lewis[x] pentasaccharide 373
- 5a-carba D-talopyranose 216
- 5a-carba-α-D-mannopyranose 216
- 5a-carba-allopyranoses 216
- 5a-carba-mannopyranoses 216
- 5-deoxy-carba-D-*arabino*-furanose 227
- 5-deoxy-carba-D-*lyxo*-furanose 227
- 5-deoxy-carba-D-*ribo*-furanose 227
- 5-deoxy-carba-D-*xylo*-furanose 227
- 5-isopropyl-2'-deoxyuridine 540
- 6'-sulfated Lewis[x] pentasaccharide 373
- acarbose 90
- allosamizoline 249
- α-methyl acarviosin 93
- allosamizoline 250
- azadisaccharides from bis-aziridines 310
- bisglycal systems, 7
- brevetoxin B, 3, 4
- calcitriol 117
- carba-oligosaccharides 94
- carba-maltose 89, 214
- conduritol A 211

- conduritol B epoxide 449
- conduritol C 214
- conduritol F 214
- cyclic ether 3
- cyclobutane 109, 111
- cycloheptane 114, 119
- cyclohexane 214
    - 1,2,3-triol, 209
    - branched- 114, 144
- cyclopentane 109
    - pentaol- 252
- cyclopentenone 152
- cyclopentitol 130
- cyclophellitol 286
- cyclopropane 110
- disaccharide 512
- dithioacetal 109
- bis-epoxide 110
- fortamine-(+) 219
- N-alkyl-gluconamidine 483
- hemibrevetoxin B 1
- hexitol, 1-deoxy-1-nitro 51
- higher-carbon dialdoses 60
- inositol D-*chiro* 447
- maitotoxin, NO-PQ ring sytem of 8
- mannostatin 241
- nitrogen heterocycles 308
- polyether framework 13
- salbostatin 89-92
- trehazolin 252
- truncated brevetoxin B 1
- validamycin 89
- validamycin A 90
- validatol 115

talose
- 4-amino-4-deoxy 331
T-antigen 388
TATU (O-(7-azabenzotriazol-1-yl)-1,1,3,3-tetramethyluronium hexafluorophosphate 428
TBSOF see 2-(*tert*-butyldimethylsiloxy)furan
TBSOP see N-*tert*-butoxycarbonyl-2-(*tert*-butyldimethylsilyloxy)pyrrole

TBSOT see 2-(*tert*-butyldimethylsiloxy)thiophene
Tebbe reagent 11
telogen 390
telomers 390, 391
tetrahydropyranyl ether 151
tetrasaccharide 420-424, 427, 429
tetrazole
- analogue of nojirimycin 472
thallous cyclopentadienide 251
theicoic acid 77
thiirane 283
- analogues of cyclophellitol 288
thiocarbonyl group 4
thioglycoside 456
thionoester 6
thionolactone 7
thiooxazolidinone 251
thiophene 327
thymidine 453
- 3'-deoxy-3'-C-formyl-5'-O-trityl 525
- 3'-O-phthalimido- 530
- 5'-deoxy-5'-hydrazino- 528
- 5'-O-amino-3'-O-(*tert*-butyldiphenylsilyl) 525
tin tetrachloride
- catalysis with 331, 343
titanacyclobutane 12
titanium methylidene 12
titanium-mediated metathesis 11
Tm 528-531
- of hexitol-containing oligonucleotides 573
- of oligonucleotide analogues 571
T_N-antigen 402
tobramycin 240
tolylene diisocyanate (TDI)
- in the preparation of nitrile oxides 52
torsion angles
- in oligosaccharides 352
- in pyranosyl nucleoside analogues 559
- in sugar phosphate backbone 558
transacetalization 112

transferred nOe effects 349
transition state 352
- analogue 500
- energy barriers 354
- model
  - for glycosidase inhibitor 468
  - for free radical cyclisation 164
- of nitrile oxides cyloaddition 58
transition state analogue *see* transition state
transition state mimic
- of glycosidases 483
translation
- in protein synthesis 523
trehalase
- inhibition by trehazolin 241
- inhibitors of 92
trehalose 420-424
trehalostatin 157, 160, 165
trehazolin 123, 124, 133, 157, 160, 165, 241
- 5a'-carba analogue 101
- 5a'-carba analogue as inhibitor of trehalase 101
- as trehalase inhibitor 100
- chemical modification of 100
- structure of 101
- synthesis of 252
- unsaturated 5a'-carba analogue 101
trestatin A 420, 427
triazole 284
- analogue of nagstatin 291
tributyltin hydride *see* tri-*n*-butyltin hydride
trichloroacetimidate 368, 373, 377
triethylphosphite 80
triethylsilane
- deoxygenation with 336
triflic anhydride 422
trifluoromethanesulfonyl azide 450
trimethylsilyl cyanide
- opening of bis-aziridines 309
trimethylsilyl triflate
- as glycosylation promoter 373
- as promoter 377
tri-*n*-butyltin hydride 6

- reduction with 566
- desulfurization with 6
tri-*O*-acetyl-D-glucal
- as starting material 9
tri-*O*-benzyl-L-xylofuranose
- as starting material 291
triphenylphosphine
- annulation with 340
triphenyltin hydride 6, 161, 167
tris(trimethylsilyl)phosphite
- Arbuzov reaction with 75
trisubstrate analogue 501-505
truncated brevetoxin B 1
tumor-associated oligosaccharides 78
tunicamine 61
tunicamycin 61
turbidimetric analysis 409, 410

UDP-galactose epimerase
- in enzymatic galactosylation 377
UDP-GlcNAc 77
UDP-glucose
- as glycosyl donor 377
uronic acid 417

validamine 88
- as inhibitor of glycosidases 92
- hydroxy 88
validamycin
- synthesis of 89
validamycin A 88
- structure of 115
- synthesis of 90
validatol
- synthesis of 115
validoxylamine
- as inhibitor of glycosidases 92
validoxylamine A 89
valienamine 88, 99, 100
- as inhibitor of glycosidases 92
- structure 143
valiolamine 88
vinylation 75
virus 387, 398
vitamin D_3 117
voglibose 89

603

wheat germ agglutinin 408
Wittig olefination 4, 24, 284
- intramolecular 144

X-ray analysis
- of pyranosyl nucleosides 559
xylopyranoside
- β-D-galactopyranosyl 1,3 515
- benzyl- 514
xylose 511
- 4-*O*-β-D-galactopyranosyl 511

yeast 395-407
- cell 402
- mannan 394, 395
ytterbium cyanide
- opening of bis-aziridines 309

Zimmerman-Traxler model 38
zirconocene-mediated ring contraction 184, 187